T0166258

METEOROLOGICAL MONOGRAPHS

VOLUME 32 | NOVEMBER 2004 | NUMBER 54

NORTHEAST SNOWSTORMS

VOLUME II: THE CASES

Paul J. Kocin
Louis W. Uccellini

American Meteorological Society
45 Beacon Street, Boston, Massachusetts 02108

Credits for photographs on the dust jacket

Front cover
Hempstead, Long Island, 6 Feb 1978 (photo reprinted courtesy of *Newsday*).

Back cover
Top row: (left) Skiers at Radio City Music Hall, Feb 2003 (courtesy of Rob Gardiner, www.nyclondon.com); (middle) Hamilton St, 30 Jan 1966 (*Blizzards and Snowstrorm of Washington, D.C.,* Historical Enterprises, 1993, p. 63); (right) Snarled traffic on the Northern State Pkwy, 12 Feb 1983 (*Great Blizzards of New York City,* Historical Enterprises, 1994, p. 73).
Upper-middle row: (left) Cars buried in snow near the Capitol, 16 Feb 1958 (*Washington Weather: The Weather Sourcebook for the D.C. Area,* Historical Enterprises, 2002); (middle) New York City, 6 Apr 1982 (photo reprinted courtesy of the *New York Times*); (right) Snowplows on the Long Island Expressway, 11 Feb 1983 (photo reprinted courtesy of *Newsday*).
Lower-middle row: (left) Student walks home from school, 13 Jan 1964 (courtesy of *Washington Weather: The Weather Sourcebook for the D.C. Area,* Historical Enterprises, 2002); (middle) Walking across the Memorial Bridge, 7 Feb 1967 (*Washington Weather: The Weather Sourcebook for the D.C. Area,* Historical Enterprises, 2002); (right) After the snowstorm of 15 Feb 1958 (*Washington Weather: The Weather Sourcebook for the D.C. Area,* Historical Enterprises, 2002).
Bottom row: (left) Times Square, 7 Feb 1967 (reprinted courtesy of the *New York Times*); (middle) Knickerbocker Snowstorm, 28 Jan 1922 (*Washington Weather: The Weather Sourcebook for the D.C. Area,* Historical Enterprises, 2002); (right) A hotel at Rockaway Beach, 1920 (*Great Blizzards of New York City,* Historical Enterprises, 1994, p. 3).

ISBN 1-878220-64-0
ISSN 0065-9401

Support for this monograph has been provided by the National Oceanic and Atmospheric Administration's National Weather Service and the National Centers for Environmental Prediction.

Published by the American Meteorological Society
45 Beacon St., Boston, MA 02108

For a catalog and ordering information for AMS Books, which include meteorological and historical monographs, see www.ametsoc.org/pubs/books.

Printed in the United States of America
by Allen Press, Inc., Lawrence, KS

TABLE OF CONTENTS

INTRODUCTION

This volume contains descriptions and analyses of over 100 snowstorms that provide the supporting foundation for volume I. The main goal of this volume is to provide the analyses to show how these important weather systems organize and evolve prior to and during the development of snowfall, or snowfall changing to rain, freezing rain, or ice pellets along the northeast coast of the United States. This comprehensive assortment allows the reader to assess and compare many cases, to observe any unique aspects a case may present, as well as identify similarities to other cases. Detailed examinations of one or two cases, although they can provide many important insights, have limited applicability in other situations with only superficial similarity. As noted throughout volume I, the evolutions of the surface and tropospheric conditions prior to and during these storms are difficult to generalize because of their case-to-case variability. Palmén and Newton (1969, p. 273) emphasize this problem by stating, "Cyclonic disturbances appear in such diverse forms that it is impossible to give a description that is uniformly applicable to all cases." Problems with compositing antecedent conditions for East Coast storms have also been cited by Brandes and Spar (1971).

Four chapters are included in this volume. In chapter 9, a historical overview of heavy snowstorms in the Northeast from the Colonial days through 1950, with detailed descriptions of several outstanding cases, is presented. In chapter 10, 30 heavy snowfall cases summarized in chapters 4 and 5 of volume I are examined from organized sequences of weather charts, which include snowfall totals, surface analyses, 850- and 500-hPa analyses, and 250-hPa wind analyses. Two additional and more recent cases, the "Presidents' Day II" snowstorm of 15–18 February 2003 and the snowstorm of 5–7 December 2003, are also included. In chapter 11, 37 cases described as "near misses" in chapter 5 of volume I, are described, also with snowfall, surface, and upper-level analyses. The 37 cases represent three groups of near-miss cases. The first 15 cases are representative of "interior snowstorms," storms in which the heavy snow bands form inland and snow changes to rain, freezing rain, or ice pellets in the Northeast urban corridor. The second group is composed of 15 additional cases termed moderate snowstorms, in which heavy snowfall amounts are relatively isolated while more moderate amounts of 4–10 inches (10–25 cm) predominate. The third group of seven cases represents cases in which freezing rain or ice pellets are the most significant aspect of the storms in the urban corridor. The monograph concludes with chapter 12, presenting several analyses of early and late season heavy snow events that occur during autumn or spring and illustrate cases that define the margins of the snow season.

All upper-level analyses for cases, starting with the December 1947 snowstorm, are produced throughout this volume with a reanalysis technique first described by Kalnay et al. (1996) for the 50-year period 1948–98. The reanalyses are generated for a 5-day period for each event using the T254, 64-level version of the National Centers for Environmental Prediction's (NCEP) Global Forecast System. As such, the details in the structure of upper- and lower-level jet streaks, the evolution of short-wave features and associated vorticity patterns, and the lower-level thermal fields, can be well represented. The gridded data that provide the basis for these reanalyzed fields are available on a DVD provided with this monograph, which includes temperature, winds, moisture, and geopotential height at 40-km resolution and over 44 levels. This dataset allows researchers, students, and weather enthusiasts the opportunity to investigate any aspect of more than 70 storms in more detail than can be produced in this monograph. It is hoped that this dataset will form the basis of research studies for years to come.

As a reminder of what draws many of us to the wonder, beauty, danger, and widespread impact of these impressive snowstorms, we offer the following piece that was written by Bennett Noble for the *Christian Science Monitor* in the 1970s (L. Bosart 2003, personal communication).

Meteorological Ambition

"The sun that brief December day
Rose cheerless over hills of gray. . . ."

It is a strangely beautiful trait of most of us that, when called upon to do so, we can conjure up the lines of a rusty old poem. We can remember stuttering through "Snowbound" in a breathless monotone back in the fourth grade—hands clutched at the seams of corduroy knickers, cowlick trembling, eyes riveted to the portrait of Francis P. Hurd which hung at the back of the room, and inwardly cursing the genius of John Greenleaf Whittier.

Of course, we are not perfect enough to remember all

of "Snowbound," although we are perfect enough to soften our opinion of Mr. Whittier; and we forget who Francis P. Hurd was—and the cowlick, like the corduroy knickers, is long gone.

But the singular joy of anticipating a good snowstorm remains, and for that we are very grateful.

There is nothing quite like the promise and anticipation of a good snowstorm. It makes Christmas pale, birthdays insignificant, and is rivaled only by Election Day for the joy of anxious hope.

We take our grasp of this situation from our great intellect; we take our license from the words of Mr. Mark Twain, who promised us weather is a literary specialty and that no untrained hand can turn out a good article on it.

Plainly, we have "Meteorological Ambitions," as James Russell Lowell put it; that is, we are of that breed of people who like to be "hotter and colder, more deeply snowed up, to have more trees and larger blown down than their neighbors." Well, he puts it as well as we've seen it put, but it goes deeper than that.

Further, our grasp of the subject includes a magnificent mental file of the Great Storms. The grandest, for instance, was the Great Valentine's Day Blizzard of '39. Now, there are those who will tell you it doesn't snow anymore the way it did that marvelous February 14, but this is not so. It is that we were shorter then, and the snow seemed deeper. It is that there were fewer plows and fewer roads then, and fewer houses to break the wind. It may also be, but we doubt it. "Things look fairer when we look back at them." But we digress.

There was a time when a man merely looked at the sky, knew it was a weather breeder, sniffed at and hefted the air, made his forecast without shouting, and set about preparing for the storm. Now we are blessed with meteorologists, climatologists and satellites to tell us of weather. Even an amateur can play now.

We thrill at it. We long to hear of those great rivers of arctic air sweeping down from the Dakotas, through Nebraska and Texas to mate with the warm circulations over the Gulf of Mexico. We become ecstatic when the two forces meet and commence dancing together. We get a nervous twitch when the mass moves northeastward, breaks across the Appalachians, and tumbles across the coastal plain to the Atlantic. There in that unholy breeding ground between the Jersey shore and the Virginia capes, new energy spawns a secondary storm. It lurches northward! Great God, it's coming!

We are close to fainting now, as we scan the southern sky for cirrus twists and the thickening middle clouds that follow. And the swing of the wind from calm and variable to early and peckery. Quickly now! Has that high drifted out to Bermuda so it can pump our storm north? Is that cold front lying just far enough west to channel our storm properly–not to pull it too far west, but to guide it just enough east so that the storm center will skip and bump along the coast as it bends to the northeast.

Is it moving slow enough so that it gathers all the available energy from the Atlantic? Is that Bermuda high strong enough to keep pumping? Or will it dissipate and let that storm out to sea south of us?

And what time is it? You never saw a good snowstorm start in the early morning. Late forenoon at the earliest and better after lunch. Watch out for those first few spits of snow. Let it act like that for an hour or so, but no longer. It should start and get about seriously falling without dawdling. Watch the wind. Don't let it swing too quickly. Let it pick up from the southeast while the snow spits, then it can climb (slowly, slowly) to the northeast. Now, all is right! Sit back and wait twelve hours, then you can go shovel.

Ah it is a mysterious bit of chemistry . . . . a magnificent brew!

If we had our way we would reduce New England seasons to two; the Hurricane Season and the Snowstorm Season. That would leave the rest of you with the months of April through June to worry about, the blandest most uninteresting time of the year.

It serves you right if you don't think as we do and end up getting stuck with April through June.

Chapter 9

HISTORICAL OVERVIEW

Through the centuries, severe winter storms have become part of the historical folklore of many regions. Comprehensive accounts of major winter storms in the northeast United States over the past three centuries have been assembled by weather historian David Ludlum (1966, 1968, 1976, 1982, 1983).[1] Ludlum's contribution to the study of meteorology in the United States has been to document the great weather events of American history, writing about numerous phenomena such as hurricane and tornadoes, and providing historical accounts of the distribution, depth, and duration of great snowfalls, along with their wind velocity, degree of cold, and effects on the general population. Nearly 300 years of such information provides compelling historical and climatological insight into the levels of severity these storms are capable of attaining. A sampling of the region's legendary snowstorms from the 18th, 19th, and first half of the 20th centuries follows in this chapter.

1. A review of major snowstorms from the Colonial period to 1900

In the year 2000, it is likely that there was no human alive who remembered and experienced the March 1888 "Blizzard of '88." However, the stories of those who lived through that blizzard have enabled the storm to become the "benchmark" for Northeast snowstorms that persists to this day. However, historical accounts have also documented earlier storms that have been at least as severe, and, in one or two cases, may have been more extreme. For example, the "great snow" of 1717 occurred nearly 60 years before the Declaration of Independence was signed but was remembered and cited by generations of people. This "great snow" and other cases that have been recorded by historians are reviewed in this section, followed by descriptions of several historic snowstorms that have affected the northeastern United States, beginning with the "Blizzard of '88."

[1] Historical discussions of storms also appear in selected issues of *Weatherwise,* 1948 to the present.

a. Selected snowstorms: 1700–1899

1) 27 February–7 March 1717: The "Great Snow"

The great snow of 1717 was a series of four snowstorms, two relatively minor and two major, which left depths in excess of 3–4 ft (90–120 cm) of snow across most of southern New England, with drifts exceeding 25 ft (750 cm). The effects of these storms were so impressive that local historical accounts still singled out this storm period as the "great snow" more than 100 years following the event.

2) 24 March 1765

This storm affected the area from Pennsylvania to Massachusetts. From Philadelphia came this report: "On Sunday night last there came on here a very severe snowstorm, the wind blowing very high, which continued all the next day, when it is believed there fell the greatest quantity of snow that has been known for many years past; it being generally held to be two feet [(60 cm)], or two feet and a half [(75 cm)], on the level, and in some places deeper." (Ludlum 1966, p. 62.)

3) 27–28 January 1772: "The Washington and Jefferson Snowstorm"

George Washington, living in Mount Vernon, and Thomas Jefferson, living in Monticello, were both marooned by this storm. Snowfall was estimated at 3 ft (90 cm) on a level across Virginia and Maryland. Washington wrote, "the deepest snow which I suppose the oldest living ever remembers to have seen in this country." (Ludlum 1966, p. 145)

4) 26 December 1778: "The Hessian Storm"

The storm of 26 December 1778 was a severe blizzard accompanied by heavy snows, high winds, and bitter cold extending from Pennsylvania to New England, with drifts reported to 15 ft (500 cm) in Rhode Island. The storm was named for troops occupying Rhode Island during the Revolutionary War who were stranded by the deep snows.

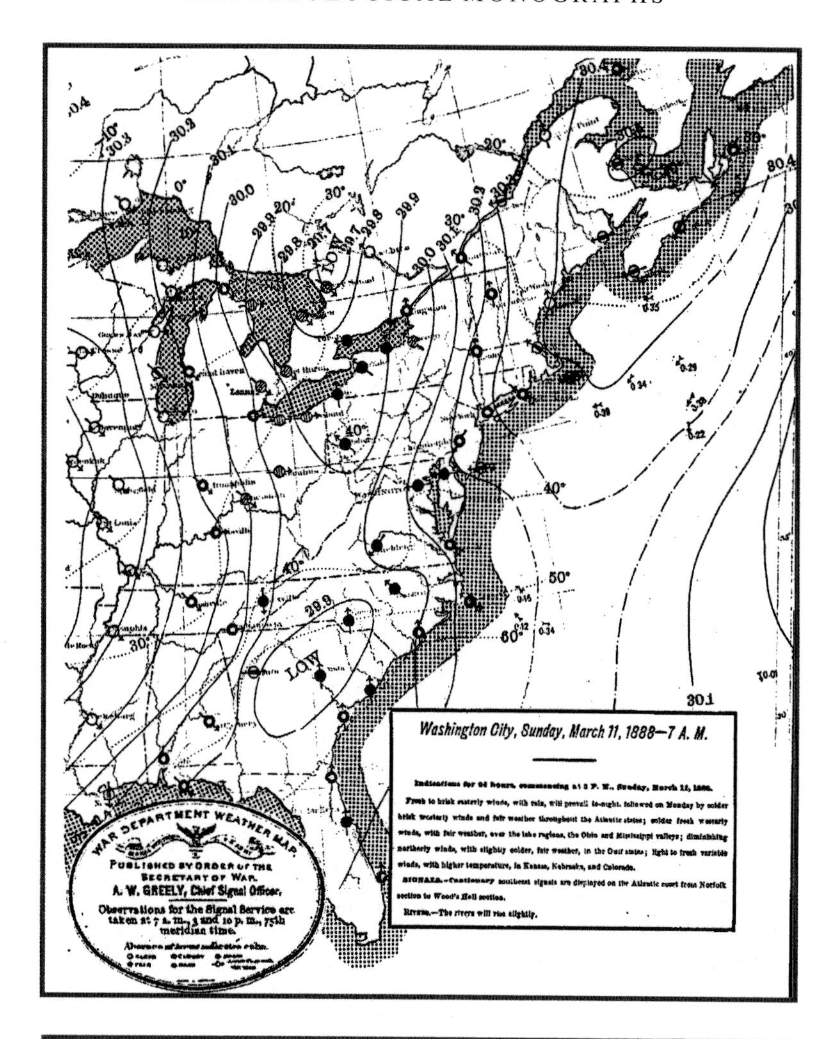

Washington City, Sunday, March 11, 1888—7 A. M.

Indications for 24 hours, commencing at 3 P. M., Sunday, March 11, 1888.

Fresh to brisk easterly winds, with rain, will prevail to-night, followed on Monday by colder brisk westerly winds and fair weather throughout the Atlantic states; colder fresh westerly winds, with fair weather, over the lake regions, the Ohio and Mississippi valleys; diminishing northerly winds, with slightly colder, fair weather, in the Gulf states; light to fresh variable winds, with higher temperature, in Kansas, Nebraska, and Colorado.

SIGNALS.—Cautionary southeast signals are displayed on the Atlantic coast from Norfolk section to Wood's Holl section.

RIVERS.—The rivers will rise slightly.

FIG. 9-1. (top) Daily Weather Map for 11 Mar 1888. (bottom) Forecast or "indications" made by the U.S. Signal Service Corps on 11 Mar 1888.

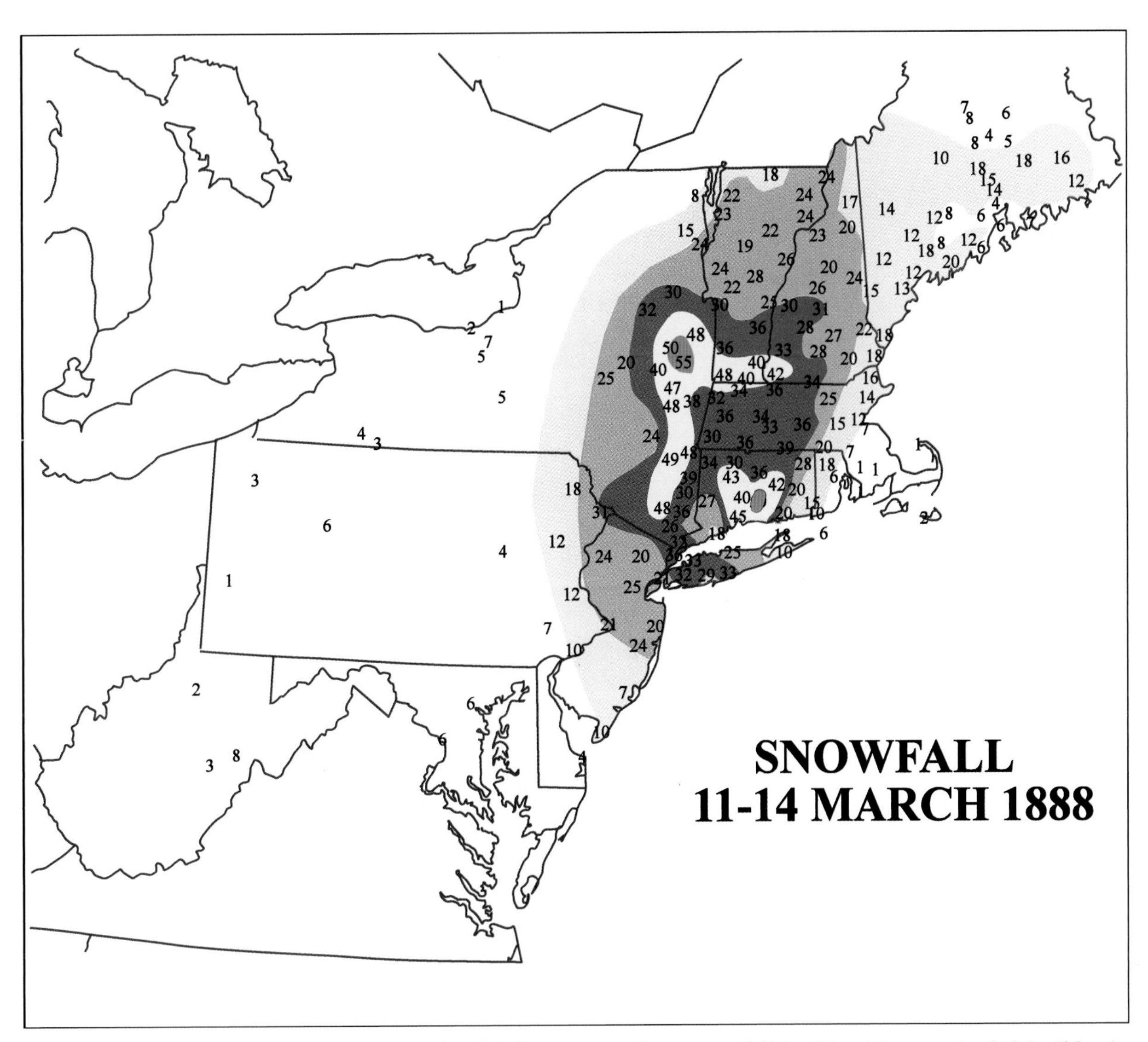

FIG. 9-2. Total snowfall (in.) 11–14 Mar 1888. Colored shading represents increments of 10 in. (25 cm) for amounts of 10 in. (25 cm) and greater.

5) 28 DECEMBER 1779–7 JANUARY 1780

Three storms during one of the coldest winters of the past three centuries produced deep snows in much of New England. The first storm produced rain from New York City southward, but snow occurred in New England with 18 in. (45 cm) at New Haven, Connecticut. The second system was a violent snowstorm with extremely high tides from the Carolinas northward. The third storm was confined primarily to eastern New England. Snow depths in the wake of the three storms ranged between 2 and 4 ft (60 and 120 cm) from Pennsylvania to New England.

6) 4–10 DECEMBER 1786

This period featured another succession of three crippling snowstorms. Estimates in the local press placed snow depths from Pennsylvania to New England at 2–4 ft (60–120 cm). In describing the third storm, a Boston newspaper notes, "a snowstorm equally severe and violent with that we experienced on Monday and Tuesday preceding. The quantity of snow is supposed to be greater, now, than has been seen in this country at any time since that which fell seventy years ago, commonly termed "The Great Snow" (Ludlum 1966, p. 71).

7) 19–21 NOVEMBER 1798: "THE LONG STORM"

The storm described by Ludlum as the heaviest November snowstorm in the history of the coastal Northeast from Maryland to Maine occurred during 19–21 November 1798. Eighteen in. (45 cm) of snow reportedly fell in New York City.

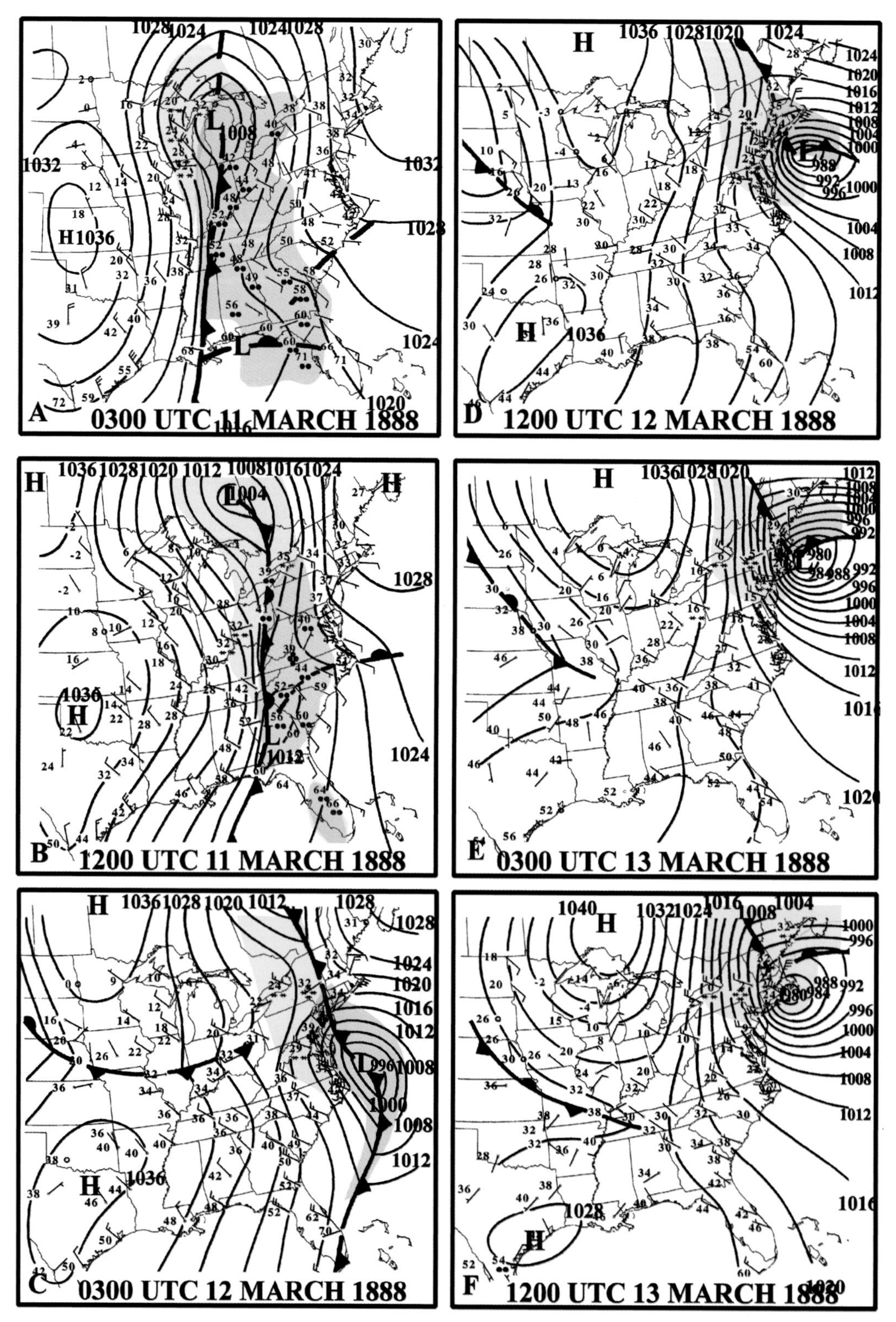

FIG. 9-3. Sequence of surface weather maps between 0300 UTC 11 Mar and 1200 UTC 13 Mar 1888, including fronts, isobars (hPa at 4-hPa intervals); and station symbols (temperatures °F, winds, current weather).

PLATE. 9-1a. Wall Street, New York City, 12 or 13 Mar 1888 (New York Historical Society; from Caplovich 1987, p. 54).

8) 26–28 JANUARY 1805

This cyclone brought a very heavy snowstorm to New York City and New England. Snow fell continuously for 48 h in New York City, where 2 ft (60 cm) accumulated.

9) 23–24 DECEMBER 1811

Temperatures fell from well above freezing on 23 December to near 0°F (−18°C) on 24 December, while a storm intensified explosively off Long Island, New York. Snowfall in New York City, Long Island, and southern New England averaged 12 in. (30 cm), as severe blizzard conditions prevailed. Strong winds and high tides caused extensive damage to shipping.

10) 5–7 JANUARY 1821

An extensive snowstorm spread from Virginia to southern New England, leaving 12 in. (30 cm) at Washington, D.C.; 14 in. (35 cm) at Baltimore, Maryland; 18 in. (45 cm) at Philadelphia, Pennsylvania; and 14 in. (35 cm) in New York City.

11) 14–16 JANUARY 1831: "THE GREAT SNOWSTORM"

This storm produced the heaviest snowfall over the largest area of any storm studied by Ludlum. Accumulations exceeded 10 in. (25 cm) from the Ohio Valley across much of the Atlantic coast north of Georgia. Washington reported 13 in. (33 cm), with 18 in. (45 cm) at Baltimore, 18–36 in. (45–90 cm) near Philadelphia, 15–20 in. (37–50 cm) at New York City, and 20–30 in. (50–75 cm) over southern New England. The snowfall distribution of this storm was rivaled, and likely exceeded, by the 12–14 March 1993 "Superstorm" (see chapter 10.24).

12) 8–10 JANUARY 1836

This event became known as the "The Big Snow" for interior New York, northern Pennsylvania, and western New England, where 30–40 in. (75–100 cm) fell. The storm also buried the coastal plain, with 15 in. (37 cm) at Philadelphia, 15–18 in. (37–45 cm) at New York City, and 2 ft (60 cm) across southern New Jersey.

PLATE 9-1b. Church Street, New Haven, CT, Mar 1888 (New Haven Colony Historical Society; from Caplovich 1987, p. 204).

13) 18–19 JANUARY 1857: "THE COLD STORM"

The cold storm combined snowfall in excess of 10 in. (25 cm) with temperatures near or below 0°F (−18°C) and high winds to produce severe blizzard conditions from North Carolina to Maine. Snow totals ranged from 15 in. (37 cm) near Norfolk, Virginia, to 18–24 in. (45–60 cm) in Washington, and 24 in. (60 cm) at Baltimore, with 12 in. (30 cm) at New York City, and 14 in. (35 cm) at Boston, Massachusetts.

b. 11–14 March 1888: "THE BLIZZARD OF '88"

The most legendary of all storms to affect the northeast United States occurred with an unprecedented combination of heavy snows, high winds, and bitterly cold temperatures. The effects of the storm were not as widespread as many other notable snowstorms, but its impact was immense over eastern New York, western New England, New York City, and northern New Jersey. More than 400 people were estimated to have perished on land and at sea as the storm raged along the coastline for 3 days. The storm has been the subject of numerous studies and reports (Hayden 1888; Upton 1888; Werstein 1960; Hughes 1976, 1981; Ludlum, 1976, 1983; Kocin 1983; Caplovich 1988) describing the significant effects of the storm, its impact on how forecasts were made and transmitted in the late 19th century, and its influence in eliminating aboveground electrical wiring, which was devastated during the course of the storm. Much of the following discussion is derived from Kocin (1983).

The storm began innocuously. On 11 March 1888, the "indications" or weather forecasts from the Washington office of the United States Signal Service (Fig. 9-1), the predecessor of today's National Weather Service, indicated little forewarning of a major storm: "Fresh to brisk easterly winds, with rain, will prevail tonight followed on Monday by colder, brisk westerly winds and fair weather throughout the Atlantic states." A trough of low pressure neared the East Coast, accompanied by two separate low pressure centers: one moving north into Canada, and another new center developing over the southeast United States. A moderate rainstorm enveloped the East Coast on 11 March but conditions changed dramatically overnight as the system reached the Atlantic Ocean.

Cold air began to pour eastward toward the coastline. As the southeast low pressure center passed off the Carolina coastline during the evening of 11 March, the storm began to intensify explosively as it moved slowly northeastward toward southern New England. At the same time, cold air continued moving eastward and reached Washington by evening and New York City by midnight. Overnight, rain changed to snow and by the morning of 12 March, residents of New York City awoke to a blinding blizzard marked by heavy snowfall,

PLATE 9-1c. Main Street, Northampton, MA, Mar 1888 (New York Historical Society; from Caplovich 1987, p. 221).

downed overhead wires, and temperatures spiraling downward toward 0°F (−18°C). Heavy snow, high winds, and bitterly cold temperatures would continue throughout the day and the following night. By the time the snow ended on 14 March 1888, 2–4 ft of snow (50–100 cm) had fallen across much of eastern New York, northern New Jersey, and western and central New England (Fig. 9-2). The severity of the storm was such that it achieved more notoriety than all the other great winter storms that have affected the northeast United States.

The local record of New York City's weather office included the following remarks about the storm (U.S. Weather Bureau 1888):

> The light rain turned to snow at 12:10 a.m. and continued throughout the day. High northwesterly winds, maximum velocity 48 MPH. The high winds began during the early morning, snapping off anemometer wire at 3:00 a.m.; during the driving snow and low temperature it was impossible to get instrument in working order until 10:00 a.m. and then only with the greatest efforts. It is supposed that the highest velocity is properly recorded by self register. The storm is the most severe ever felt in this vicinity. The high wind and fine cutting snow made it almost impossible for travel. Travel by street railroads was entirely suspended by 7: a.m. and by 8: a.m. all business suspended, railway trains and ferries ceased to run, all vehicles snow bound in the streets and abandoned. Many accidents and one fatal collision occurred on the elevated

PLATE 9-1d. Clark Street, Brooklyn, NY, Mar 1888 (Brooklyn Historical Society; from Caplovich 1987, p. 156).

> railway. At some points the snow had drifted to an elevation of 15 to 20 ft. Snowploughs that started to clean the railway tracks and abandoned were buried. All telegraph and telephone wires down and at night the city was in darkness. Trees, signposts and awnings were blown down. Several persons found in snow banks that had evidently fallen exhausted from the severity of the storm. The streets of the city were deserted and only a few of the hundreds of thousands that are seen daily on the streets made their appearance.
>
> The assistants on duty at the office are deserving the greatest praise for their promptness and most willing efforts under such trying circumstances to keep the instruments in proper working order and all routine duties both in and out of the office, reflects credit upon their courage.

For many areas of the Northeast, this was (and still is) the heaviest snowstorm on record with 20 to more than 50 in. (50–125 cm) common from northern New Jersey into eastern New York and much of western New England. Near-record late winter cold temperatures and high winds combined with the snow to create prolonged blizzard conditions over a section of the country populated by newly arrived immigrants and "old-timers" who had never experienced such a storm and assumed that blizzards only occurred in the plains. The storm resulted in tremendous personal suffering and caused severe damage to transportation, shipping, and communications.

Snowfall did not extend much farther south than northern Virginia, but the Washington area was paralyzed by wind-blown snow and ice that accumulated up to 6 in. (15 cm) while the Philadelphia area received 10 in. (25 cm). Snow totals and duration increased dramatically farther north in northeastern Pennsylvania, northern New Jersey, eastern New York, Long Island, Vermont, New Hampshire, western and central Massachusetts, most of Connecticut, and parts of Rhode Island and Maine (Fig. 9-2), where 1.5 to nearly 5 ft (38–150 cm) of snow fell. New York City officially received 21 in. (53 cm) while communities across other portions of New York City and Long Island measured anywhere from 18 to 36 in. (45–90 cm). North of New York City, 30–50 in. (75–127 cm) were common throughout New York's Hudson Valley, with Albany and Troy, New York, measuring 46.7 and 55 in. (119 and 138 cm), respectively, with the greatest totals reported near Saratoga Springs, where up to 58 in. (145 cm) fell. Over New England, snow totals were equally impres-

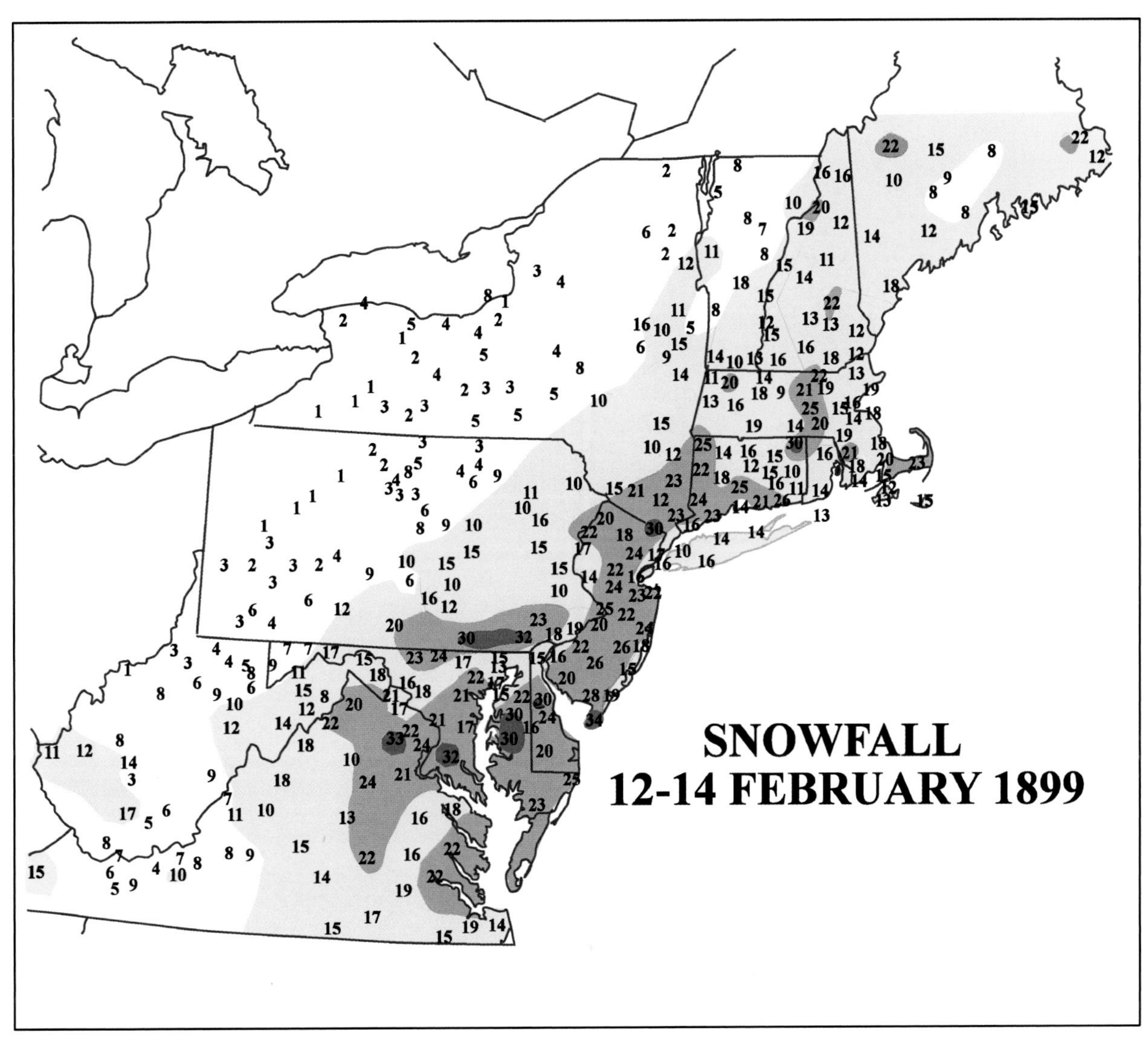

FIG. 9-4. Total snowfall (in.) 11–14 Feb 1899 for the northeastern United States. Colored shading represents increments of 10 in. (25 cm) for amounts of 10 in. (25 cm) and greater.

sive, with widespread 20–40-in. (51–102 cm) accumulations, including 44.7 in. (114 cm) in New Haven, Connecticut; 32 in. (80 cm) at Worcester, Massachusetts; and 27 in. (68 cm) in Concord, New Hampshire. Much of this region saw the snow accompanied by gale force winds and temperatures that fell to near 0°F (−18°C).

A series of surface weather analyses documents the evolution of the storm (Fig. 9-3). High pressure and weather conditions typical of mid-March characterized the weather across the East Coast late on Saturday, 10 March, as a cold front marched across the Midwest and Gulf coast states accompanied by a widespread region of mostly rain from Alabama to Michigan. The dominant surface cyclone was located over northern Michigan while a new low pressure center had developed near Mobile, Alabama. By Sunday afternoon (Fig. 9-3b), the cold front continued moving to the east as the northern low pressure center moved northward into Canada and as the southern low moved northeastward and deepened to 1004 hPa along the North Carolina coast. Still, rain is primarily observed with this system from North Carolina northward into New England while westerly winds and colder temperatures were found behind the front.

Following this time, weather conditions deteriorated rapidly from Virginia northward as the cyclone continued to intensify off the East Coast, cold air filtered eastward, and the heavy rains changed over to freezing rain and then to heavy snow. By Monday morning, 12 March (Fig. 9-3c), the blizzard was in full swing from New Jersey northward. A reinforcing cell of cold high pressure north of the Great Lakes resulted in colder temperatures feeding southward toward the storm center, and by evening (Fig. 9-3d) the storm had reached its maximum intensity near southeastern Massachusetts with a central sea level pressure of 978 hPa while tem-

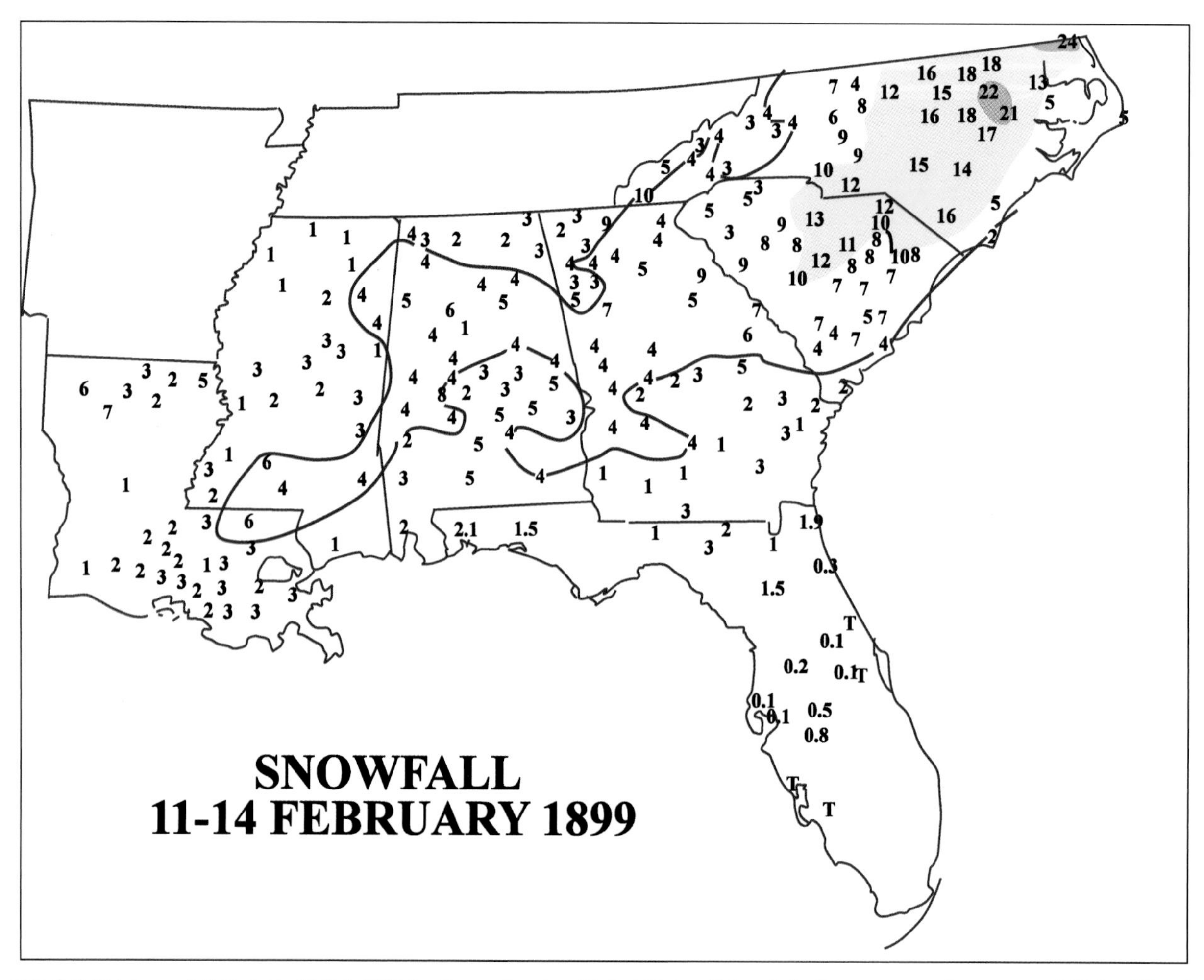

FIG. 9-5. Total snowfall (in.) 11–14 Feb 1899 for the southeastern United States. Colored shading represents increments of 10 in. (25 cm) for amounts of 10 in. (25 cm) and greater. Solid line represents 4 in. (10 cm) contour of snowfall.

peratures plunged across much of the region. Overnight, the storm continued to rage and temperatures fell to the single digits to near 0°F (−18°C) over a large area. By the morning of 13 March, the storm showed the first signs of weakening. By midafternoon, the storm was clearly weakening (Fig. 9-3e) but snows continued over much of the region. By the following morning (Fig. 9-3f), the center of low pressure had continued to weaken considerably and began moving eastward out to sea and snows gradually began to end.

The "Blizzard of '88" presented some aspects that differed in many ways from the notion of a "classical" Northeast snowstorm. For example, many other cases of major Northeast snowstorms shown in chapters 3 and 4 (volume I) are preceded by an outbreak of very cold air across the eastern states, usually in association with an intense cell of high pressure over New York or New England. However, no such air mass was entrenched over the northeast United States prior to the blizzard. Rather, colder air moved into both regions at the same time as the cyclone was intensifying off the southern New England coast. Therefore, precipitation started as rain and later changed to snow over a large section of the Northeast. A north–south frontal zone to the north of the cyclone is also an unusual feature that undoubtedly played a major role in focusing the heaviest snowfall along and to the west of the frontal zone. This feature and its attendant temperature contrast are unique aspects of this case that are replicated by few other storms. Selected photographs taken during the storm are shown in Plates 9-1a–d.

c. 11–14 February 1899: "The great Arctic outbreak and Blizzard of 1899"

A 2-week period of exceptionally cold weather in what weather historian David Ludlum (1970) describes as "the greatest Arctic outbreak in history" culminated in an exceptional blizzard that affected the entire East Coast from Florida and the Gulf coast to Maine. The

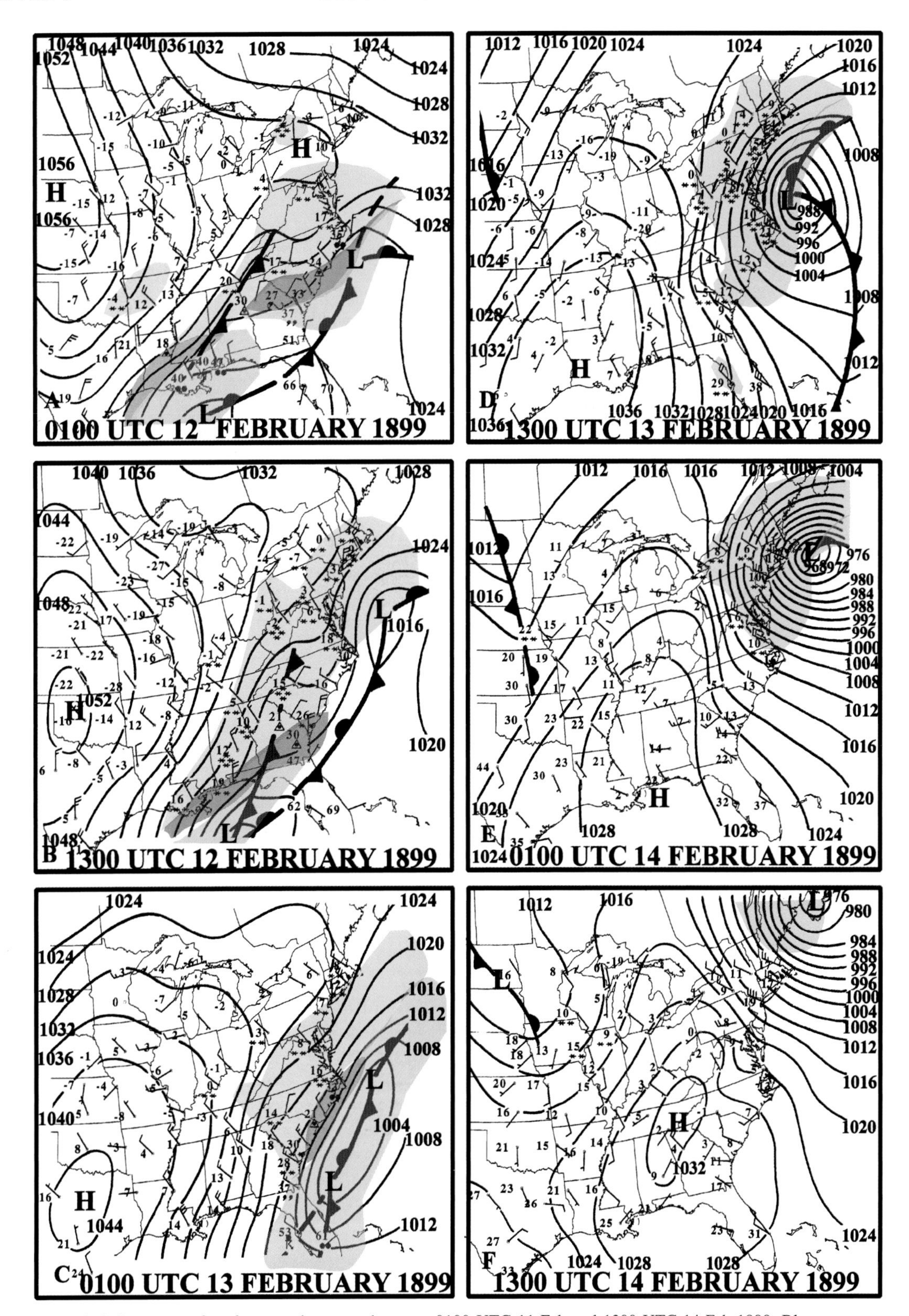

FIG. 9-6. Sequence of surface weather maps between 0100 UTC 11 Feb and 1300 UTC 14 Feb 1899. Blue, green, and violet shading represents snow, rain, and freezing rain/sleet. See Fig. 9-3 for details.

PLATE. 9-2a. The White House, Washington, DC, 15 Feb 1899 (from Ambrose 1993, p. 18).

cold-air outbreak caused temperatures to fall to 0°F (−18°C) along the beaches of the Gulf coast and ice flowed from the mouth of the Mississippi River into the Gulf of Mexico (Kocin et al. 1988). All-time record minimum temperatures were established in 12 states and the District of Columbia. A series of snowstorms accompanied the cold weather leaving a blanket of snow over much of the eastern United States that intensified the effects of the cold. As the final and most significant cold air mass spread over the central and eastern United States on 10–11 February 1899, a major storm developed along the leading edge of the cold pool and evolved into one of the most intense blizzards ever to affect the Gulf and East Coasts. Near-blizzard conditions occurred in such unlikely sites as New Orleans, Louisiana; Mobile, Alabama; and Pensacola, Florida. More than a foot (30 cm) of snow fell in a swath from the Carolinas to Maine (Fig. 9-4), accompanied by high winds and temperatures close to 0°F (−18°C) and snow extended throughout much of the Gulf coast and as far south as central Florida (Fig. 9-5). Illustrations of daily snowfall for the 5-day period surrounding the blizzard are provided in Brooks (1914). Much of the following discussion is derived from Kocin et al. (1988).

A sequence of surface weather maps documents the evolution of the storm from late on 11 February 1899 through 14 February 1899 (Fig. 9-6). At 0100 UTC 12 February (Fig. 9-6a), a massive surface anticyclone was located over much of the central United States with a 1058-hPa center over Wyoming. The cold front at the leading edge of this cold outbreak had passed off the Texas coast and was followed by northerly winds and subzero temperatures (<−18°C) as far south as north-central Texas and northwestern Arkansas. A weak cyclone and associated coastal front were located along the middle Atlantic coast with light snow and bitterly cold temperatures just above 0°F (−18°C) across Virginia and Maryland and mixed precipitation from the Carolinas to Georgia. A separate region of low pressure was located in the Gulf of Mexico along the leading

PLATE 9-2b. Music Academy, 9th and D St., Washington, DC, Feb 1899 (from Ambrose 1993, p. 15).

edge of the massive arctic outbreak with light rain, ice pellets, and snow from the Texas coast to northwestern Florida.

By the morning of 12 February, the coldest temperatures on record were being experienced across the central and southern plains states as the anticyclone continued to drift southward. The low pressure system over the Gulf of Mexico moved eastward overnight. Cold air poured into the Gulf coast; the cities of New Orleans, Mobile, and Pensacola experienced near-blizzard conditions with snow, strong northerly winds, and rapidly falling temperatures. Heavy snow extended northward from central Alabama and northern Georgia into the western Carolinas. A weak storm system that had developed the previous day along the Carolina coast had progressed northeastward to off the Maryland coast and was associated with a significant snowfall of 4–8 in. (10–20 cm) from Virginia to southern New England. The Gulf of Mexico low pressure system moved east-northeastward during the day and appeared to redevelop off the Southeast coast by evening. Snowfall expanded northeastward from eastern Georgia to southern New

PLATE 9-2c. Harlem, NY, 14 Feb 1899 (from Ambrose 1994, p. 19).

England with the heaviest snow falling from western South Carolina into North Carolina and southeastern Virginia. In the following 24 h, this storm would develop rapidly and blizzard conditions would be experienced along much of the eastern seaboard.

By Monday morning 13 February (Fig. 9-6c), the cyclone began to intensify explosively and was located east of the North Carolina coast with a central sea level pressure estimated at nearly 984 hPa. Snowfall rates increased significantly overnight from the Carolinas northward into New England as north to northeasterly winds increased along the coast. Near-zero visibilities in wind-blown heavy snow, gale force winds, and temperatures below 10°F (−12°C) occurred over a wide region from South Carolina to New England. By sunrise, more than 10 in. (25 cm) covered portions of South and North Carolina. Across Florida, rain changed to snow, leaving the heaviest accumulation on record [1.9 in. (5 cm)] at Jacksonville, the only measurable snow ever recorded at Tampa [0.1 in. (0.25 cm)], and snowflakes were observed as far south as Fort Myers. Prior to a rare snowfall in southern Florida during January 1977, this was the farthest south snow had ever been observed along the East Coast.

The cold air that had swept across the Southeast overnight left many all-time low temperature records across the region, including −9°F (−23°C) at Atlanta, Georgia, and −2°F (−19°F) at Tallahassee, Florida. During the day, the cyclone continued to deepen rapidly as it tracked northeastward and by evening (Fig. 9-6d) was located just southeast of Nantucket, Massachusetts, with a central pressure estimated at 966 hPa. Blizzard conditions raged along the middle Atlantic and New England states all day and 20 in. (50 cm) of new snow piled up in Washington and Baltimore, with more than 30 in. (75 cm) on the ground, the deepest snow cover on record (Plate 9-2a). As shown in Fig. 9-4, the heaviest snow fell from central North Carolina northeastward through Virginia, Maryland, and southern New Jersey, where 20-in. (50 cm) amounts were common, with Cape May,

Selected Snowfalls: 1900-1910

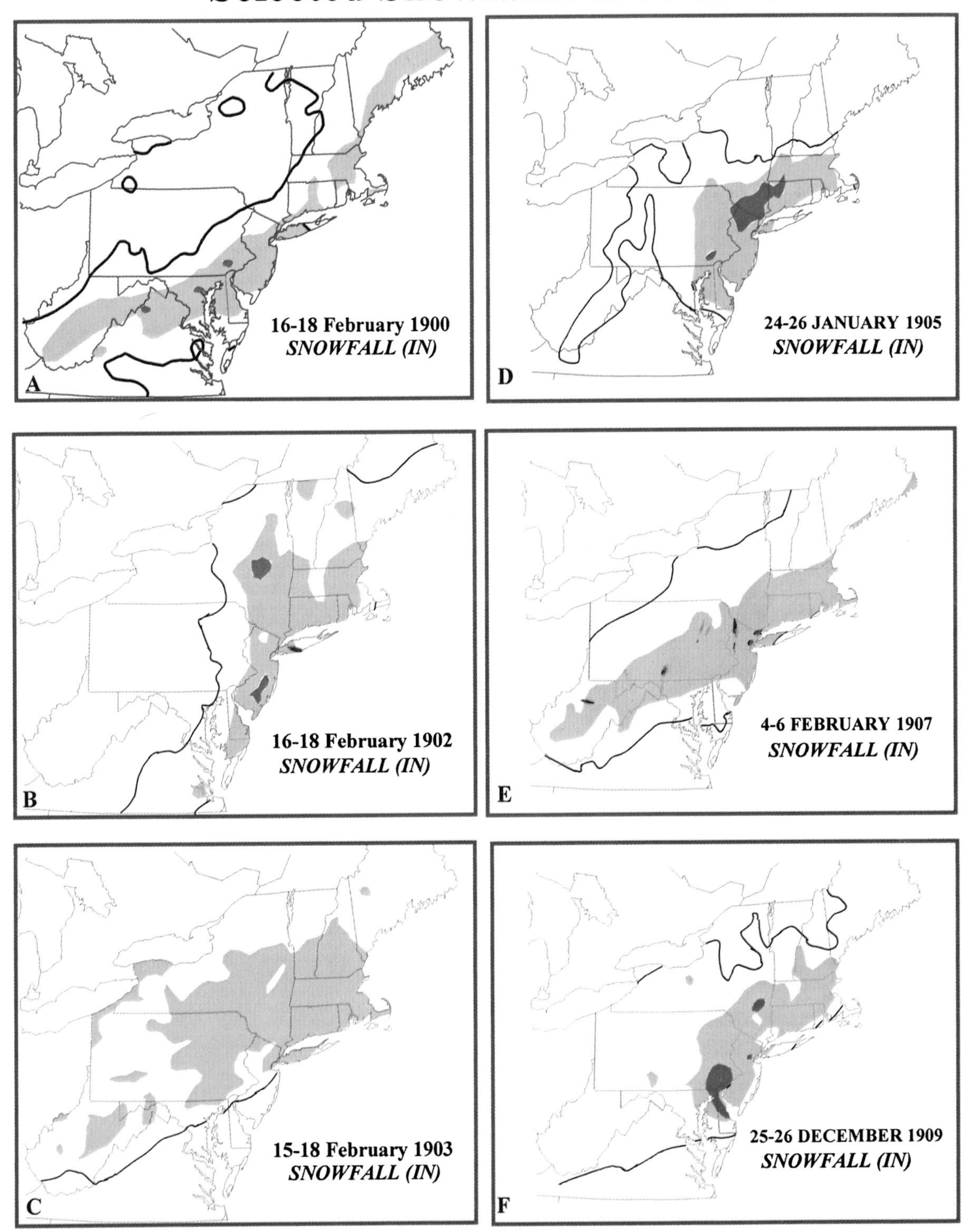

FIG. 9-7. Total snowfall for six cases, 1900–10: (a) 16–18 Feb 1900, (b) 16–18 Feb 1902, (c) 15–18 Feb 1903, (d) 24–26 Jan 1905, (e) 4–6 Feb 1907, and (f) 25–26 Dec 1909. Shading represents increments of 10 in. (25 cm) for amounts of 10 in. (25 cm) and greater. The solid line represents snowfall of 4 in. (10 cm) or greater.

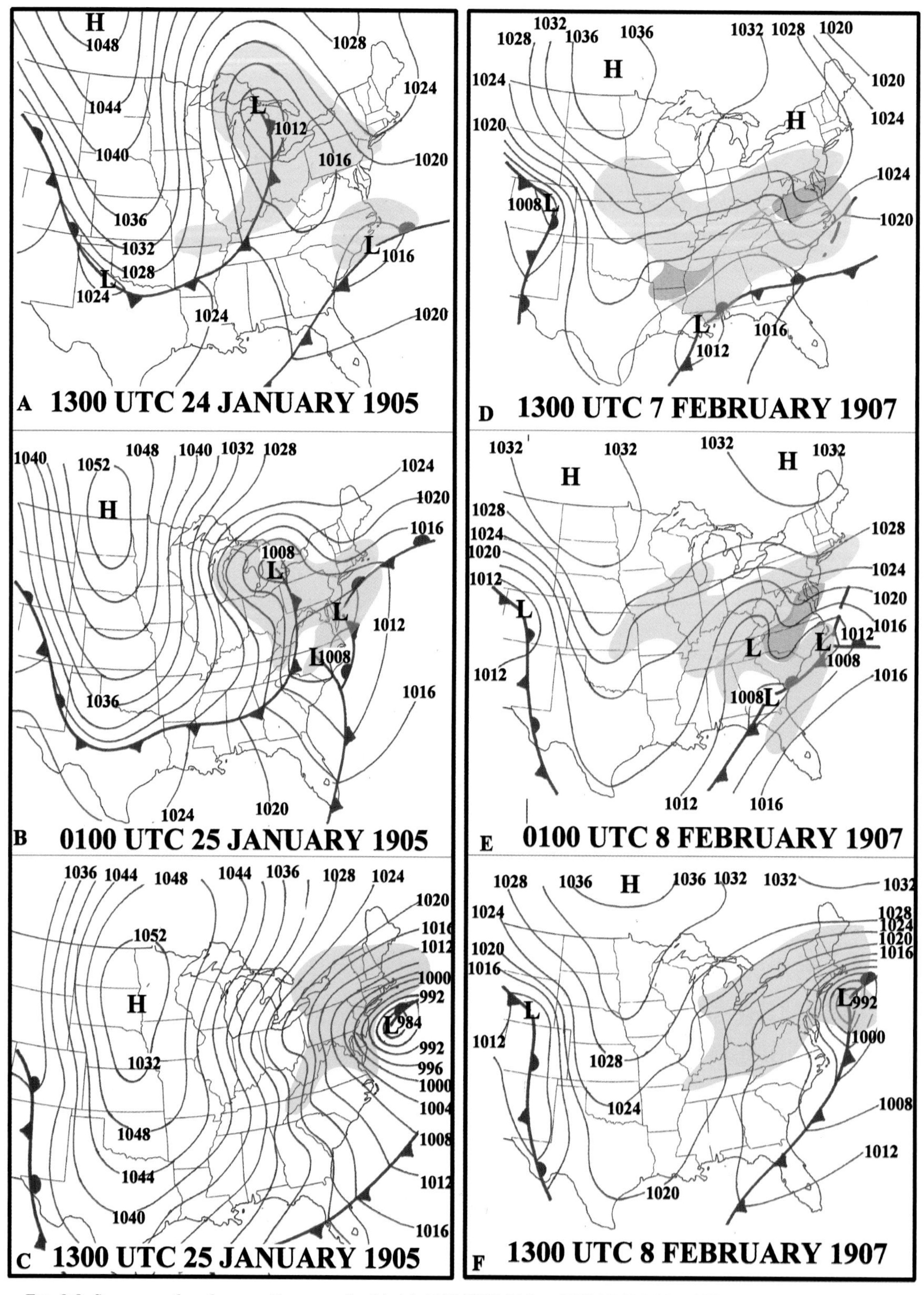

FIG. 9-8. Sequence of surface weather maps for (a)–(c) 1300 UTC 24 Jan–1300 UTC 25 Jan 1905 and (d)–(f) 1300 UTC 7 Feb–1300 UTC 8 Feb 1907. Blue, green, and violet shading represents snow, rain, and freezing rain/sleet.

PLATE 9-3. Philadelphia street scene, 26 Dec 1909 (Library Company of Philadelphia, provided by Jan Nese from Nese and Schwartz 2002, p. 65).

New Jersey, reporting the greatest amount with 34 in. (86 cm). Overnight, the storm center raced northeastward, and by morning the snow ended across all of New England except for Maine (Fig. 9-6f). Rapidly moderating conditions following the storm reduced Washington's record snow cover to a mere trace only a week later. This massive blizzard and cold wave signaled the winter's dramatic last gasp.

Weather forecasts made during the course of the storm were remarkably accurate according to a summary published in the February 1899 edition of the *Monthly Weather Review* (Garriott 1899). The Daily Weather Map, issued on the morning of 12 February, gave the following forecast: "Heavy snow is indicated this afternoon and to-night from northern Virginia northward through Eastern New York and southern New England, and the snow will continue during Monday in New England." According to the Times Union of Albany, New York, "It is seldom that the Weather Bureau fails in predicting a big storm, and it had been more successful this year." According to the *Boston Herald,* "The Weather Bureau is entitled to distinguished consideration for its services. . . It foretold the widespread disturbance with remarkable accuracy." Selected photographs taken during the storm are shown in Plates 9-2a–c.

2. Selected snowstorms: 1900–50

In this section, analyses and discussions of some of the most significant snowstorms of the first half of the 20th century are presented. Snowfall and surface analyses were prepared for more than 50 cases and several cases from each decade are discussed and described in varying detail throughout this section. These storms include the "Christmas Snowstorm" of 1909, an explosive blizzard in March 1914, Washington's "Knickerbocker" snowstorm of January 1922, New York City's "Big Snow" of December 1947, and concludes with the "Appalachian Storm" of November 1950.

The 1950 storm, one that produced very heavy snows in the Ohio Valley and Appalachians but did not produce significant snowfall in the Northeast urban corridor, is included here, not only because it also had a significant impact in the northeast United States, but also because rarely has any storm displayed an overall intensity and incomparable combination of severe weather elements as this storm.

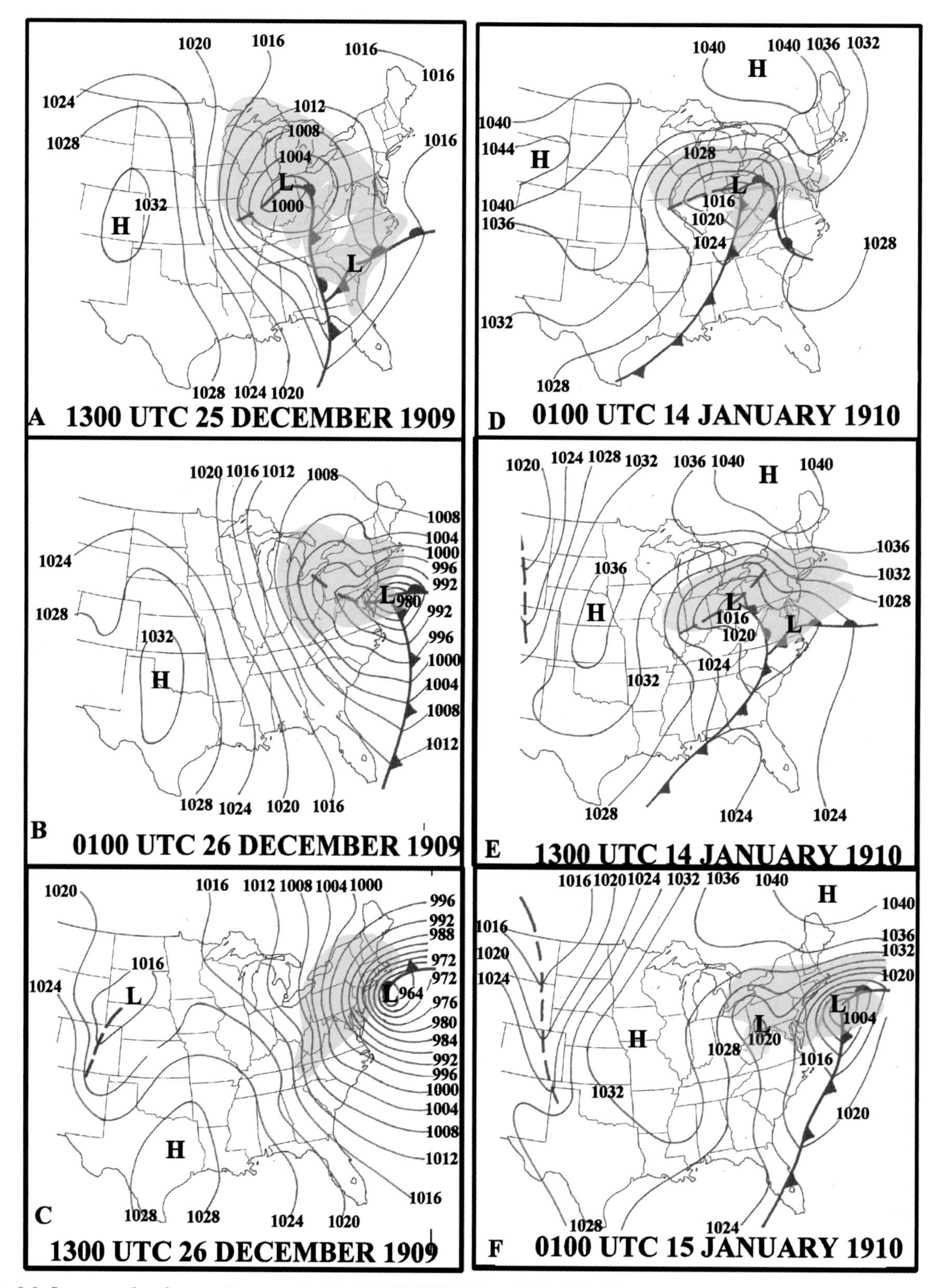

FIG. 9-9. Sequence of surface weather maps for (a)–(c) 1300 UTC 25 Dec–1300 UTC 26 December 1909 and (d)–(f) 0100 UTC 14 Jan–0100 UTC 15 Jan 1910. Blue, green, and violet shading represents snow, rain, and freezing rain/sleet.

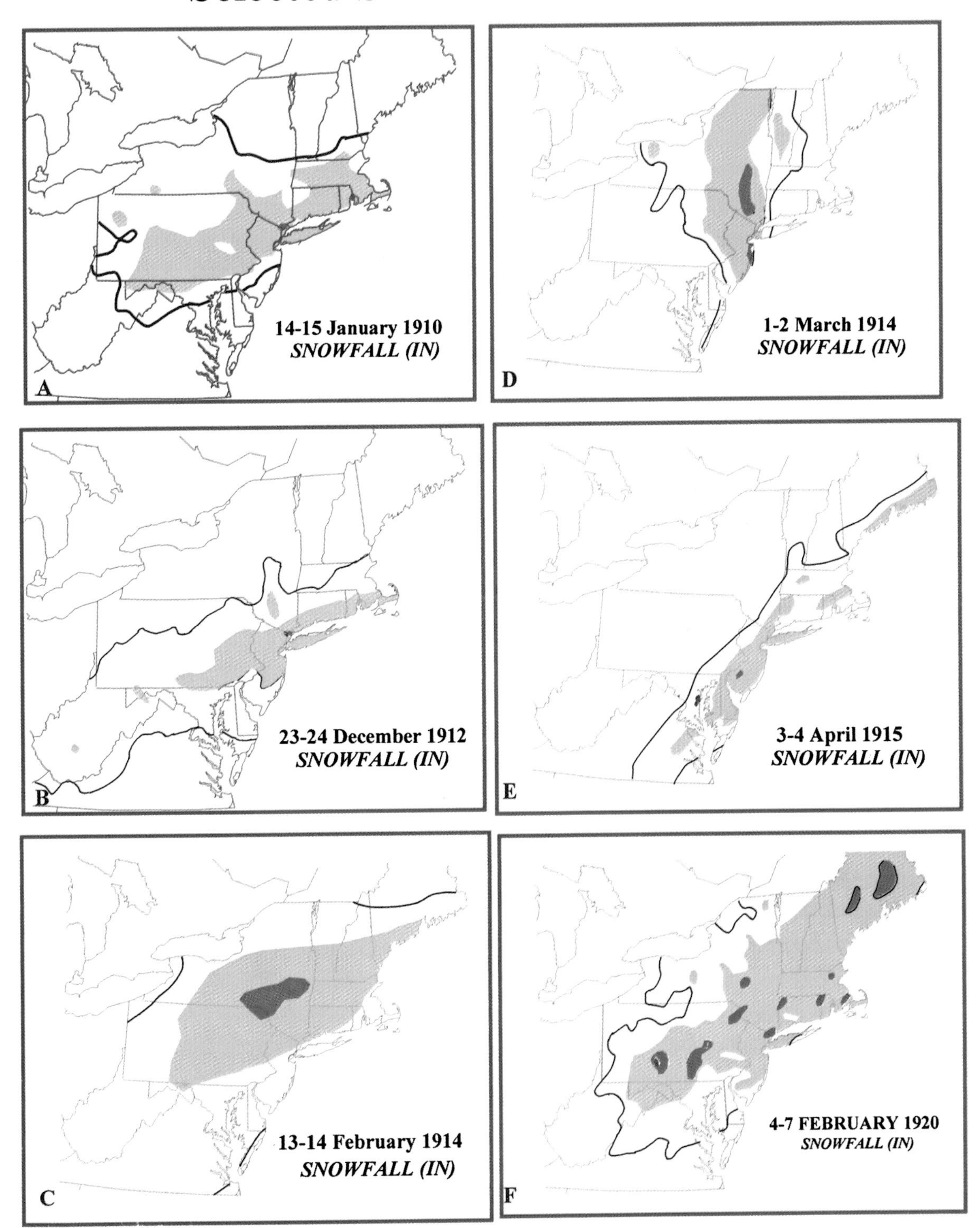

FIG. 9-10. Total snowfall for six cases, 1910–20: (a) 14–15 Jan 1910, (b) 23–24 Dec 1912, (c) 13–14 Feb 1914, (d) 1–2 Mar 1914, (e) 3–4 Apr 1915, and (f) 4–7 Feb 1920. Shading represents increments of 10 in. (25 cm) for amounts of 10 in. (25 cm) and greater. The solid line represents snowfall of 4 in. (10 cm) or greater.

PLATE 9-4. New York City, Broadway, 14 Jan 1910 (from Ambrose 1994, p. 27).

a. 1900–10

The first decade of the 20th century was marked by numerous significant snowstorms throughout the Northeast. Major snowstorms occurred in mid-February during three of the first four years of the 20th century (Figs. 9-7a–c), including a storm that occurred exactly 100 years to the day prior to the Presidents' Day snowstorm of 16–17 February 2003. Two major snowfalls within a week of each other occurred on 16–17 February 1902 and 21–23 February 1902. The first storm was a rapidly deepening cyclone that dropped up to 2 ft (60 cm) of snow in southern New Jersey and eastern New York. The second storm left a narrow band of heavy snow across northern Pennsylvania (where up to 33 in. of snow fell), southeastern New York, and extreme southern New England. A major snowstorm on 2–3 January 1904 left 10–15 in. (25–38 cm) of snow from New York City to Boston, including 23 in. (58 cm) in New London, Connecticut.

Two snowstorms in early January 1905 were followed by a blizzard on 24–26 January that left up to 30 in. (75 cm) of snow in southeastern New York (Fig. 9-7d). The surface analyses for this storm system (Figs. 9-8a–c) suggest that a merger (see volume I, chapter 4) of a fairly weak coastal cyclone with a separate weather system associated with a massive outbreak of cold air with a 1055-hPa anticyclone led to the explosive development of a storm system early on 25 January 1905. The storm left a large swath of snow greater than 10 in. (25 cm) from eastern Maryland northward into southern New England with a large area of +20 in. (+50 cm) snows across northern New Jersey, southeastern New York, and Connecticut.

The snowstorm of 4–6 February 1907 left heavy snow accumulations of 10 in. (25 cm) from Washington to Boston (Fig. 9-7e). This storm is a classic "nor'easter" or "Gulf of Mexico–Atlantic coastal development" type storm (see volume I, chapter 3) with low pressure developing in the Gulf of Mexico and moving northeastward along the Atlantic coast while a strong anticyclone dominated the northern half of the nation. This is an example of a storm system that produced a large area of heavy snow with a rapidly deepening surface low of the New England coast (Figs. 9-8d–f). The storm of 23–24 January 1908 left up to 22 in. (55 cm) of snow in southern New Jersey while the

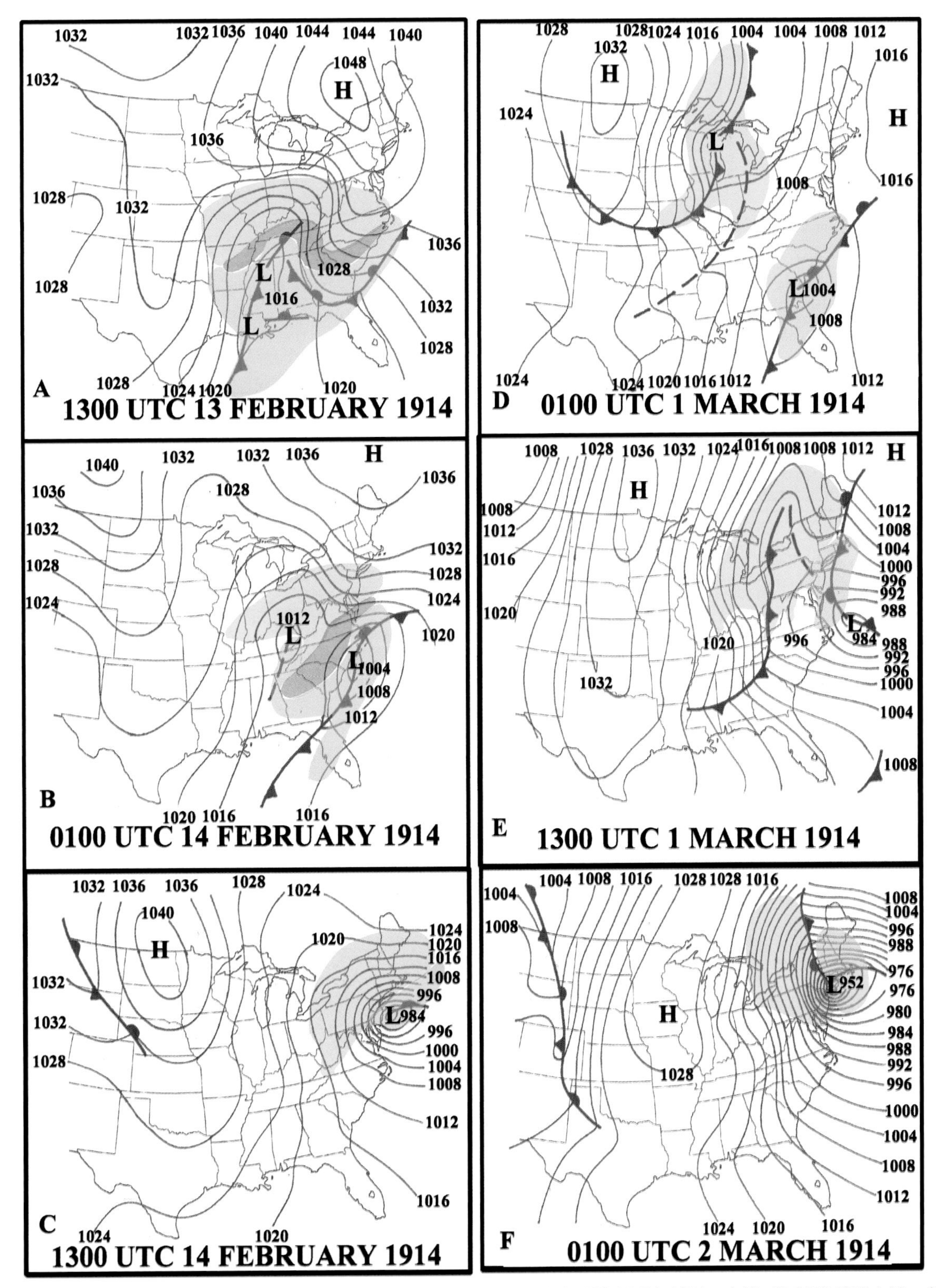

FIG. 9-11. Sequence of surface weather maps for (a)–(c) 1300 UTC 13 Feb–1300 UTC 14 Feb 1914 and (d)–(f) 0100 UTC 1 Mar–0100 2 Mar 1914. Blue, green, and violet shading represents snow, rain, and freezing rain/sleet.

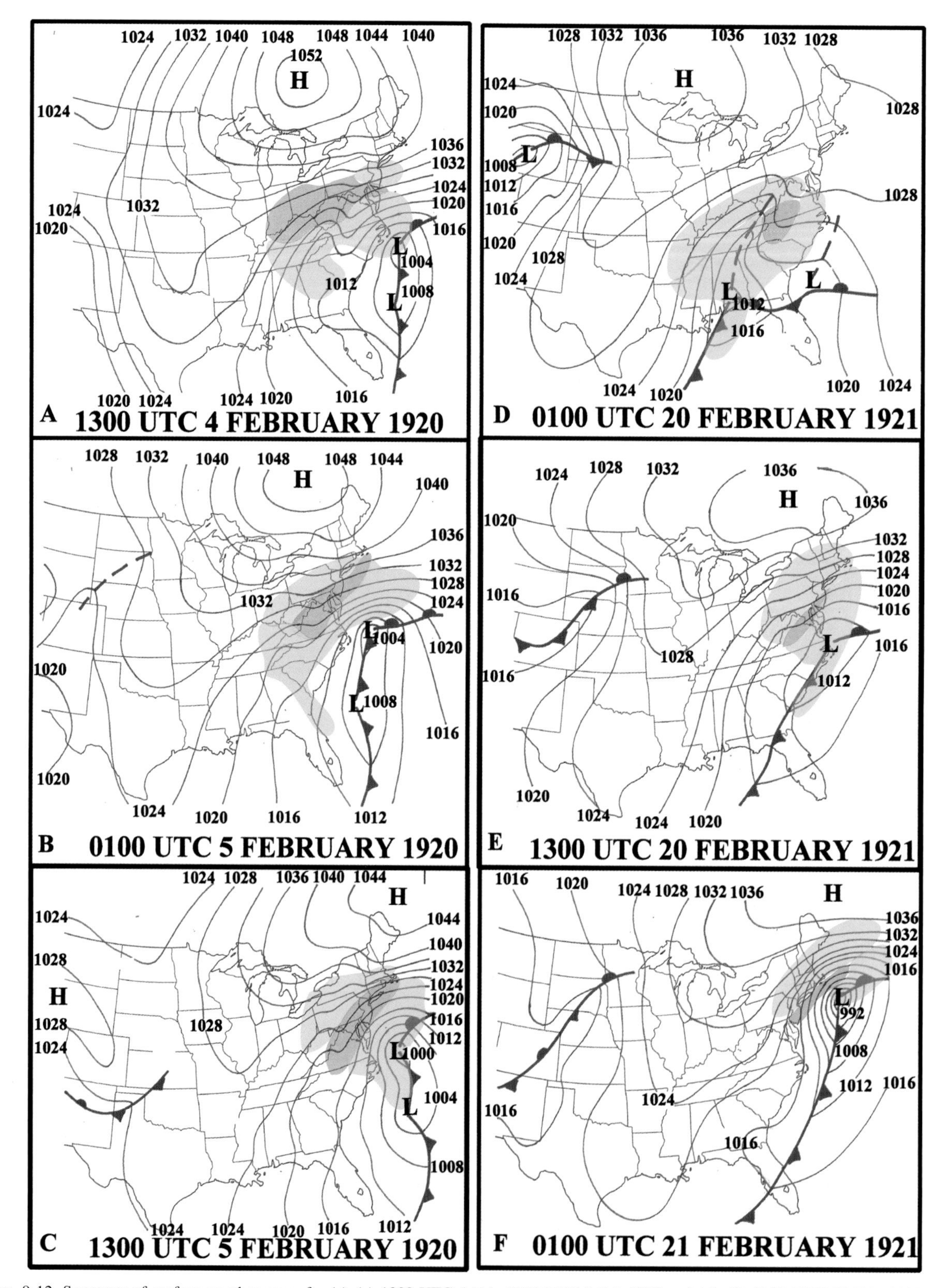

FIG. 9-12. Sequence of surface weather maps for (a)–(c) 1300 UTC 4 Feb–1300 UTC 5 Feb 1920 and (d)–(f) 0100 UTC 20 Feb–0100 UTC 21 Feb 1921. Blue, green, and violet shading represents snow, rain, and freezing rain/sleet.

PLATE 9-5. Seashore damage in Rockaway Beach, NY, 21 Feb 1920.

December 1908 storm buried Virginia with 17 in. (43 cm) at Richmond and nearly a foot (30 cm) of snow in Washington. President William Howard Taft was inaugurated as President on 3 March 1909 during a surprise snowstorm that left 10 in. (25 cm) in Washington and Baltimore, and up to 15 in. (38 cm) in surrounding areas.

The year 1909 ended up with a snowstorm on Christmas night and the following day that left more than 20 in. of snow (50 cm) in Philadelphia and other portions of eastern Pennsylvania and Delaware (Fig. 9-7f). This storm was described in the December 1909 *Monthly Weather Review* as one that "takes rank among the notably severe and destructive winter storms that have visited the North Atlantic States during the last half century. . . ." This storm produced the heaviest snows on record for much of Delaware and eastern Pennsylvania at the time, including 21 in. (53 cm) at Philadelphia (see Plate 9-3). The snow began in Philadelphia early on Christmas Day and continued all day with little wind and no drifting. By 2000 local time (Fig. 9-9b), only 5 in. (13 cm) of snow had fallen, but the storm suddenly gained in strength, and snowfall rates and winds increased. Overnight, the heavy wet snow was drifted badly by increasing north to northwesterly winds before ending by around 0800 LT on 26 December.

The storm was not preceded by the ominous signatures of other serious winter storms, including a large anticyclone poised over or north of the Northeast, or of a developing storm in the Gulf of Mexico. Rather, high pressure drifted eastward and weakened off the mid-Atlantic coast and a modest storm system moved northeastward from northern Missouri to Indiana with the hint of a secondary development over South Carolina (Fig. 9-9a). This secondary low pressure system then developed explosively (Figs. 9-9a–c), resulting in the lowest pressure ever observed at Cape May, New Jersey, (28.57 in.) to date. The *Monthly Weather Review* article

Selected Snowfalls: 1920-1930

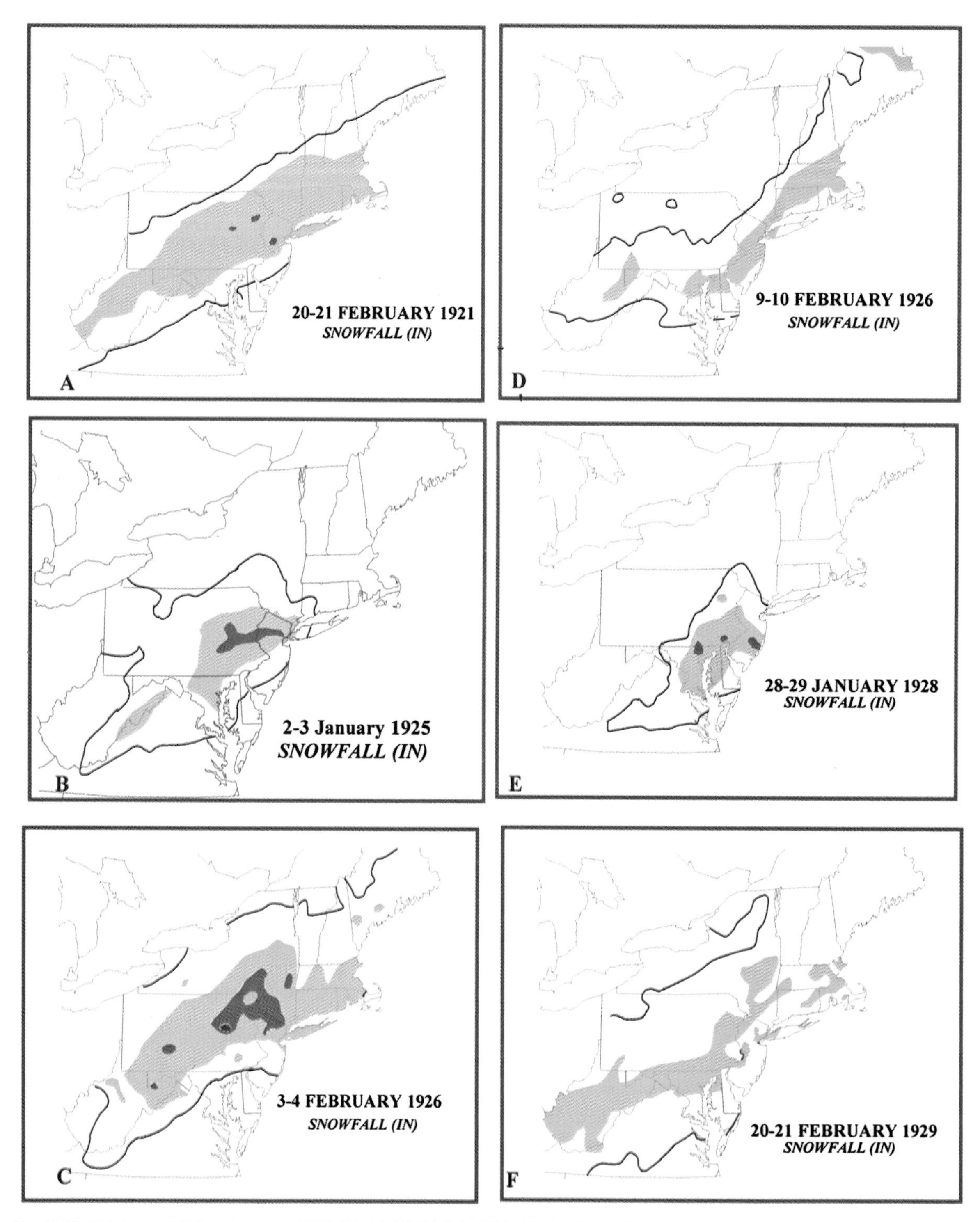

FIG. 9-13. Total snowfall for six cases: 1920–30: (a) 20–21 Feb 1921, (b) 2–3 Jan 1925, (c) 3–4 Feb 1926, (d) 9–10 Feb 1926, (e) 28–29 Jan 1928, and (f) 20–21 Feb 1929. Shading represents increments of 10 in. (25 cm) for amounts of 10 in. (25 cm) and greater. The solid line represents snowfall of 4 in. (10 cm) or greater.

PLATE 9-6a. The Knickerbocker Theater, Washington, DC, Jan 1922 (from Ambrose 1993, p. 24).

compares this storm to the 1888 blizzard and notes some similarities as well as the significant difference that this storm continued to move rapidly while the 1888 storm stalled off southern New England. It appears that this storm qualitatively resembles a recent storm on Christmas Day 2002, when a secondary low pressure system along the East Coast developed explosively without the presence of a strong surface anticyclone, in response to an amplifying upper-level jet–trough system. In the December 2002 case, the storm developed explosively just south of Long Island, while the heaviest snow fell in central New York. In this case, the low pressure system deepened farther to the south, east of the southern New Jersey, coastline before moving to the east, with heaviest snows falling to the west and northwest of the surface low, in Delaware and eastern Pennsylvania. While winds were not unusually strong with a storm of such low central pressure (possibly due to the lack of a significant surface high), the maximum winds along the New England coast coincided with the highest tides. In Boston, tides were said to be the highest in over 50 years, causing much coastal flooding. Winds gusted to 45 mi h^{-1} (20 m s^{-1}) in Boston, with 12 in. (30 cm) of snow, while New York City reported 10 in. of snow, accompanied by winds that gusted to 58 mi h^{-1} (26 m s^{-1}).

b. 1910–20

Less than 3 weeks after the 1909 Christmas snowstorm, another major snowfall on 14–15 January affected the mid-Atlantic states and southern New England with 10 in. (25 cm) of snow in Philadelphia and nearly 15 in. (37 cm) in New York City (Fig. 9-10a) (see Plate 9-4). The occurrence of a primary and secondary low pressure system is an example of "Atlantic coastal redevelopment" (see volume I, chapter 3), in which a primary surface low pressure system crosses the Ohio Valley and then a secondary low develops along the Virginia coastline, intensifies, and becomes the predominant cyclone (Figs. 9-9d–f). High pressure

PLATE 9-6b. Sidewalk in Washington, DC following the Jan 1922 Knickerbocker Snowstorm (Washington Division, D.C. Public Library; from Ambrose 1993, p. 29).

north of New York and New England provides the cold air for snowfall. At the end of 1910, another major snowfall in early December left around 10 in. (25 cm) of snow between Washington and Philadelphia. Two years later, in December 1912, another snowstorm just prior to Christmas left a band of 12–18 in. (30–45 cm) of snow across eastern Pennsylvania, central and northern New Jersey, New York City, and southern New England (Fig. 9-10b) as low pressure developed in the Gulf to Mexico and moved rapidly northeastward off the North Carolina coastline, another classic nor'easter.

During a 3-week period in February and March 1914, two severe blizzards enveloped portions of the Northeast (Figs. 9-10c and 9-10d). The first storm, on 13–14 February, followed a major arctic outbreak and produced snow, sleet, and high winds along the coast and 1–2 ft of snow (30–60 cm) across interior sections of the Northeast. A major component of this storm was a huge 1048-hPa anticyclone that moved across the northern United States and then drifted eastward across eastern Canada as low pressure deepened rapidly along the middle Atlantic coastline (Figs. 9-11a–c). The low pressure system developed near the Gulf coast, moved northeastward into the Tennessee Valley, and weakened as a strong secondary low pressure system developed along the Southeast coast and intensified rapidly as it moved near the Atlantic coast. The continued movement of the anticyclone to the east and the near-coastal track of the cyclone ensured a changeover from snow to sleet and freezing rain that reduced snowfall totals from Washington to Boston, placing this storm in a category termed interior snowstorm as described in chapter 5 of volume I.

The blizzard of 1–2 March 1914 was associated with one of the deepest extratropical cyclones to affect the United States during the 20th century. Sea level pressure fell to 28.38 in. (962 hPa) in New York City and 28.25 in. (957 hPa) at New Haven, Conneticut, the lowest pressures recorded at these locations. An examination of the data report from Bridgehampton, New York, indicates a pressure as low as 28.10 in. (952 hPa) on 1 March, one of the lowest pressures ever measured in the United States during an extratropical cyclone.

This storm has been described as the worst since the blizzard of 1888 in New York City as heavy rain changed to heavy, wind-driven snow. Fourteen inches (35 cm) of snow fell in New York as winds blew at gale to hurricane force. The area of heaviest snowfall was restricted to the west of the deepening surface low as it moved from the Virginia coast to a position near Long Island late on 1 March with heaviest snows falling across New Jersey and eastern New York with up to 30 in. reported in northeastern New Jersey (Fig. 9-10d).

Surface analyses (Figs. 9-11d–f) indicate an unusual development of the storm as a fairly innocuous low pressure system developing along the Southeast coast interacts with a major outbreak of arctic air and becomes a huge storm system, reminiscent of the kind of mergers (see volume I, chapter 4) noted with some other extreme storm developments, including the Blizzard of 1888 and the Ohio Valley blizzard of January 1978 (Salmon and Smith 1979; see volume I). The surface map at 2000 LT 1 March (Fig. 9-11f) represents the storm at its peak as the central pressure of the storm has plunged to 952 hPa, with New York City reporting blinding snow and winds gusting as high as 84 mi h^{-1} (38 m s^{-1}).

The spring snowstorm of 3 April 1915 left 19 in. (48 cm) of snow in Philadelphia, the city's fourth greatest snowfall on record, and up to 21 in. (53 cm) in southern New Jersey (Fig. 9-10e; see chapter 12). Two snowstorms in February 1916 and one during early March 1916 contributed to large seasonal snowfall totals during the winter of 1915/16 (see Fig. 2-3, volume 1). The snowfall of 2–3 February 1916 left up to a foot (30 cm) of snow from Hartford to Boston, while the storm of 11–13 February 1916 left 15 in. (38 cm) of snow in Boston. The winter of 1916/17 was dominated by two snowstorms: one on 15–16 December that left up to a foot of snow (30 cm) in New York City, and a second storm system on March 1917 that left up to 2 ft of snow (60 cm) on Long Island and eastern Massachusetts.

c. 1920–30

The decade of the 1920s began with some extremely harsh winter weather conditions. The Hartford February 1920 weather summary reported that the month of February brought some the harshest winter conditions ever experienced. A severe ice/sleet/snowstorm in late January was quickly followed by a more severe ice/sleet/snowstorm on 4–7 February that left deep accumulations of snow and slush over much of the Northeast. This storm left a widespread blanket of deep snows from northern Virginia to Maine (Fig. 9-10f) and the severity of the storm was compounded by its long duration, heavy amounts of sleet and freezing rain, and the resulting large water content of the snow/sleet/rain mass (Plate 9-5). The main features of this storm (Figs. 9-12a–c) were a very slow-moving low pressure system along the Atlantic coast interacting with a sprawling anticyclone (1054 hPa) over southern Canada (the synoptic setting is similar to the Knickerbocker snowstorm that had occurred two years later; described in the next few pages). High winds and coastal flooding compounded the severity of this storm (Plate 9-5).

A widespread snowstorm on 20–21 February 1921 produced more than a foot (30 cm) of snow from northern Virginia through much of southern New England (Fig. 9-13a). Surface analyses for this storm (Figs. 9-12d–f) show another classic nor'easter with a strong surface anticyclone drifting eastward along the U.S.–Canadian border while low pressure moves from the Gulf coast to the North Carolina coastline and then northeastward south of southern New England. Increasing northeasterly winds gusting to 57 mi h^{-1} (25 m s^{-1}) at New York City and 72 mi h^{-1} (32 m s^{-1}) at Nantucket caused extensive drifting of the snow, especially on the evening of 20 February, as the initially weak surface low deepened rapidly to the south of the strong Canadian anticyclone.

27–29 JANUARY 1922: "THE KNICKERBOCKER SNOWSTORM"

The heaviest snowstorm ever recorded in Washington, D.C., occurred exactly 150 years following the Washington–Jefferson snowstorm of 1772, a storm that produced similar amounts of snow across the middle Atlantic states. The 1922 "Knickerbocker" storm is known for its tragic consequences because it contributed to the greatest disaster in the city's history. The storm dropped 28 in. (71 cm) of snow, leading to the collapse of the roof of the Knickerbocker Theater during the evening of 28 January (Plate 9-6a), where hundreds of people were at the movies, one of the few activities left unaffected in a city that was immobilized by the record snowfall. The collapse of the theater roof led to the loss of approximately 100 lives.

Snow began at 1620 LT on 27 January and reached a depth of 9 in. (23 cm) by midnight. By 0800 LT on the morning of 28 January, the snow depth had reached 18 in. (46 cm), and most street and rail traffic came to a halt during the day. All government offices closed and both the Senate and the House adjourned by early afternoon when only a few members showed up. By 1400 LT, the snow depth reached 25 in. (63 cm), 26 in. (66 cm) by 1930 LT, and 28 in. (71 cm) by midnight. The roof of the theater collapsed at approximately 2100 LT. While the snow depth increased by only 3 in. after 1400 LT, the water equivalent of the snow totaled 0.69 in., indicating that the snow was settling while heavy snow continued to fall. Light snow continued from around 2100 until 0030 LT 29 January. Measurements in and around the city ranged from 24 in. (61 cm) to as much as 38 in. (97 cm), according to local reports, with a water equivalent of 3.02 in. (7.6 cm) at the Washington city office (Plates 9-6a,b).

The snowfall was produced by a slow-moving coastal storm that developed as a weak low pressure system off

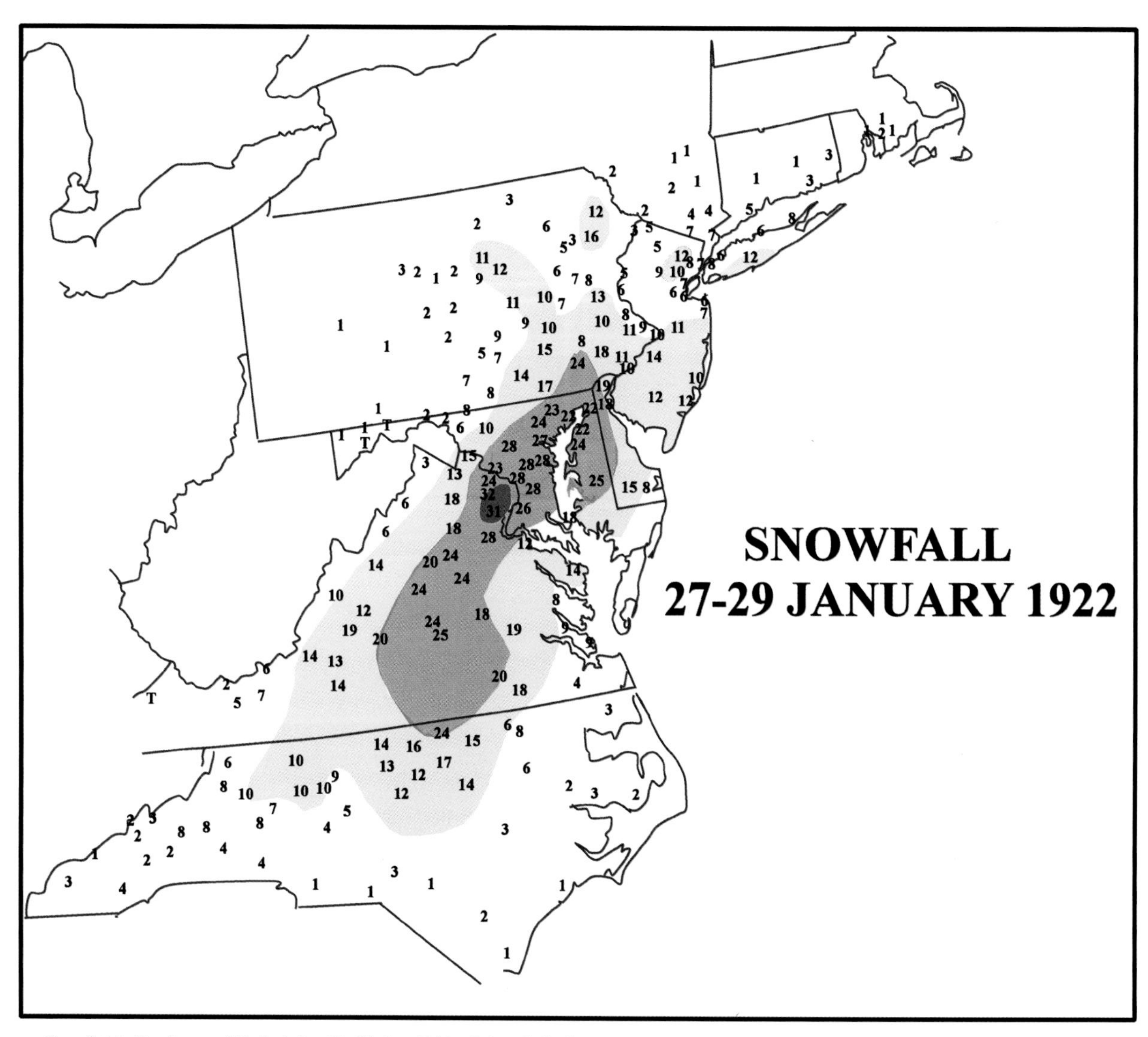

FIG. 9-14. Total snowfall (in.) for 27–29 Jan 1922. Colored shading represents increments of 10 in. (25 cm) for amounts of 10 in. (25 cm) and greater.

the Florida coastline (Day and Fergusson 1922; Mook 1956) and moved slowly northeastward along a coastal front before turning more to the east as it passed east of the Virginia–Maryland coastline. Heaviest snows developed in the south and middle Atlantic states, with the greatest accumulations of 50 cm or more primarily across Virginia and Maryland (Fig. 9-14). The following locations reported their greatest snowfalls of record at the time: Baltimore [26.5 in. (66 cm)], Lynchburg, Virginia [20.2 in. (51 cm)], and Richmond, Virginia [19.1 in. (48 cm)]. High winds were reported near the coast and produced serious drifting, particularly in Delaware and New Jersey, with sections of Norfolk submerged by coastal flooding, but winds were moderate in the Washington–Baltimore region, averaging 10–20 mi h^{-1} (5–10 m s^{-1}), limiting the amount of drifting in the metropolitan areas. More than a foot of snow (30 cm) fell over north-central North Carolina, much of central and eastern Virginia, Maryland (except the panhandle), Delaware, southeastern Pennsylvania, and central and southern New Jersey. Heaviest snows remained south of New York and New England.

The surface analyses in Fig. 9-15 show a fairly straightforward evolution of the storm. A large, nearly stationary surface high pressure system over southeastern Canada to the north of New York and New England provided the source of cold air for the snowfall event. This high pressure system expanded and moved slowly northwestward and appeared to be relatively stationary as the storm was developing off the southeast United States coast. A low pressure system developed just east of the Georgia coast on 26 January (Fig. 9-15a) and deepened slowly as it moved northeastward along a developing coastal front just east of the Southeast coast. By early Friday morning, 27 January (Fig. 9-15b), the storm center had moved to a position just east of Wil-

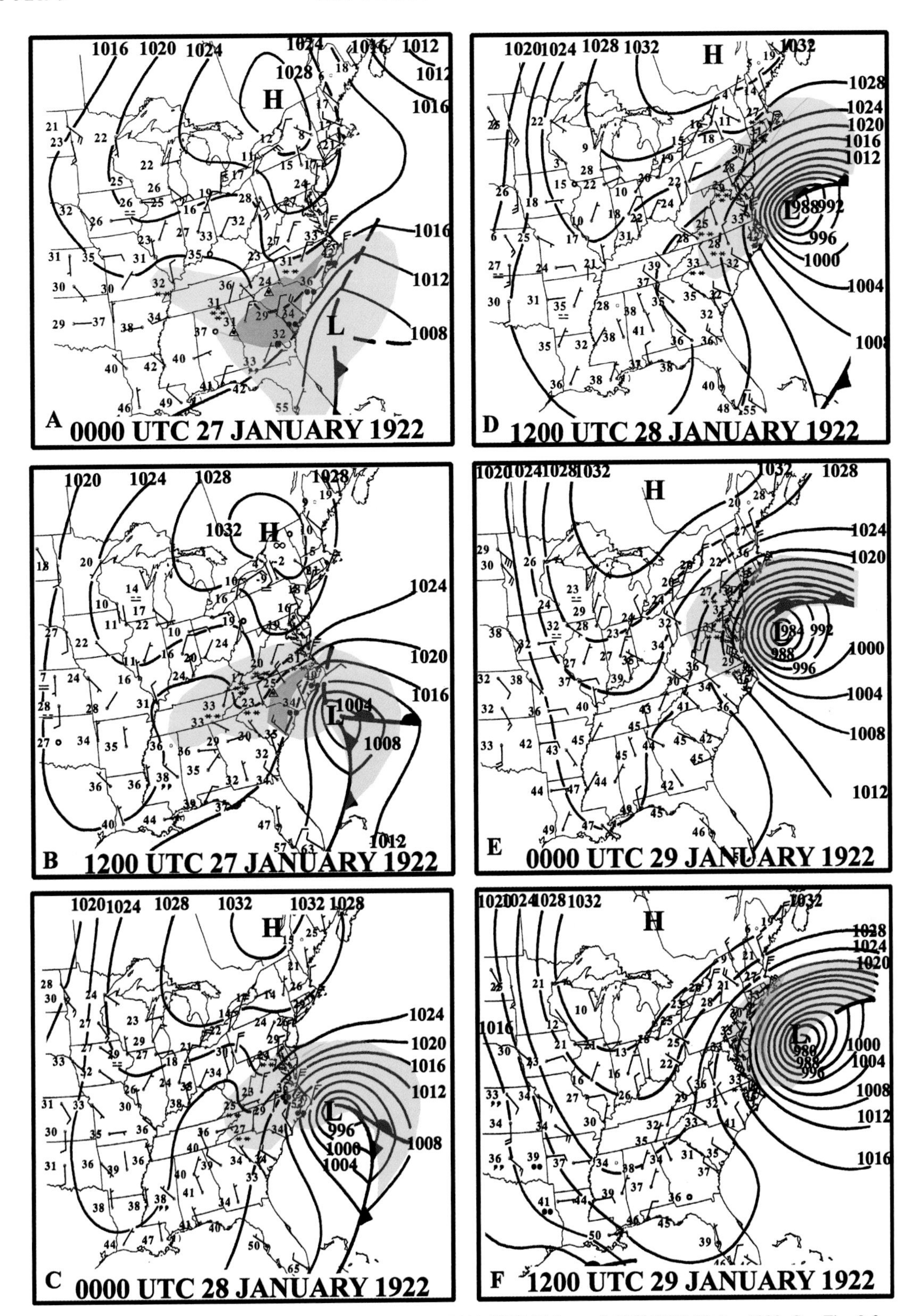

Fig. 9-15. Sequence of surface weather maps between 0000 UTC 27 Jan and 1200 UTC 29 Jan 1922. See Fig. 9-3 for details.

mington, North Carolina, and snow, some of it falling heavily, was occurring across eastern Tennessee, interior North Carolina, and southern Virginia.

The storm continued to deepen during the day as gale force winds began to pound the coasts of Virginia and Maryland, and the snow changed to rain across southeastern Virginia. Snow spread into the nation's capital by late afternoon, and by evening, the surface low deepened to 992 hPa just east of Cape Hatteras, North Carolina (Fig. 9-15e). During the next 24 h, the storm moved very slowly and began to turn more to the east. Therefore, heavy snow enveloped much of central and northern Virginia, Washington, Maryland, southeastern Pennsylvania, and southern New Jersey for the next 24 h, and its northward movement slowed appreciably (Figs. 9-15d,e). As the storm drifted more to the east than the north at 0000 and 1200 UTC 29 January, the snow spread northward into New York City and Long Island but barely affected extreme southern New England. Snows finally ended along the coast early on 29 January.

The forecast issued early on 27 January, about 8 h before the onset of snow in Washington, and as it appeared on the Daily Weather Map, stated that "The outlook is for snow to-night and Saturday in southern New England, southern New York, the Middle Atlantic States and West Virginia, and this afternoon and to-night in North Carolina, extreme northern South Carolina, and the extreme east portions of Tennessee and Kentucky. . . . On the middle Atlantic coast increasing northeast winds, probably becoming gales; snow to-night and Saturday." By early Saturday morning, with 18 in. (46 cm) on the ground in Washington and snow still falling, and with 4 in. (10 cm) at Philadelphia, by 10 a.m., the forecast read "The outlook is for snow to-night and probably Sunday in New England and southern and central New York, and this afternoon and to-night in the Middle Atlantic States and parts of North Carolina." According to the *Washington Post,* "The storm had been predicted by the weather bureau . . . weather bureau officials having predicted further snowfall today and possibly Sunday." Furthermore, according to reports in the *Washington Post* on Saturday morning, "When the weather forecasters left their work last night (Friday) they were satisfied that the storm was merely getting its stride; that the worst is yet to come. . . . Dr. Preston C. Day, who has been over at the weather bureau from the time they first began to keep records, said last night, unofficially, that the storm is in all probability the heaviest snowstorm seen in Washington since 1917. Going back even farther than that, Dr. Day compared the storm that began yesterday with the celebrated Blizzard of 1899."

Once forecasters saw how crippling the storm was in Washington, they forecast those conditions to translate northward toward New York and New England, but those conditions never materialized (the 28 January *Washington Star* reported that in New York, "The Storm didn't come up to expectations"). As the low pressure system passed east of the North Carolina coast, it turned more to the east, sparing the northeast United States north of New Jersey the bulk of the snow, yielding only 6 (15) to 12 in. (30 cm) in the New York City area and completely missing the Boston metropolitan area.

The winter of 1922/23 was characterized by many significant snowfalls and was a particularly snowy winter (see volume I, chapter 2, Fig. 2-3). New York City reported five separate snowfalls exceeding 6 in. (15 cm) during January and February while a powerful cyclone produced heavy snow across interior and northern portions of the Northeast in early March. A powerful spring storm system brought heavy snow and thunder to the Northeast on April Fools Day 1924 while January 1925 was greeted by a very heavy snowfall in eastern Pennsylvania and northern New Jersey on 2–3 January, leaving up to 27 in. (66 cm) of snow (Fig. 9-13b).

Two snowstorms within a week of each other in February 1926 (Figs. 9-13c,d) crippled many portions of the Northeast. The first snowstorm, on 3–4 February, left more than 20 in. (50 cm) of snow across a large portion of northeastern Pennsylvania, northwestern New Jersey, and southeastern New York while leaving a large area of a foot of snow or greater from northeastern West Virginia to southern New England. New York City and Boston were hard hit. Surface analyses (Figs. 9-16a–c) show a large, slow-moving low pressure system that moved northeastward along the Atlantic coast and deepened to around 970 hPa off southern New England early on 4 February with gale force winds gusting to 77 mi h^{-1} (34 m s^{-1}) at Nantucket. Less than a week later, a second significant snowfall produced its heaviest snowfall across the Northeast urban corridor from Washington (Plate 9-7) to Boston. Most of the snowfall amounts were 10–15 in. (25–38 cm) across a smaller area than the 3–4 February snowfall. Surface analyses (Figs. 9-16d–f) show that the snowfall was associated with a fairly disorganized storm system that suddenly congealed into a rapidly deepening cyclone late on the night of 9 February and the morning of 10 February. A primary surface low drifted eastward from the Midwest to near Pittsburgh, Pennsylvania, while a secondary low developed off the mid-Atlantic coast and deepened rapidly. At 0800 LT, 10 February, heavy snows were reported from Baltimore, Philadelphia, New York, and Boston as the surface low had intensified rapidly overnight, with an estimated pressure of 970 hPa south of New England. Heavy snows fell for a period of 12 h or less as the storm moved rapidly northeastward, while winds gusted to 66 mi h^{-1} (30 m s^{-1}) at Nantucket.

In early December 1926, a major snowfall left more than a foot of snow (30 cm) in Boston while another significant snowfall affected a large portion of the northern and interior portions of the Northeast on 19–20 February 1927. Another major snowfall blanketed the mid-Atlantic states on 28–29 January 1928 with up to 30 in. (75 cm) of snow reported in Maryland (see Fig.

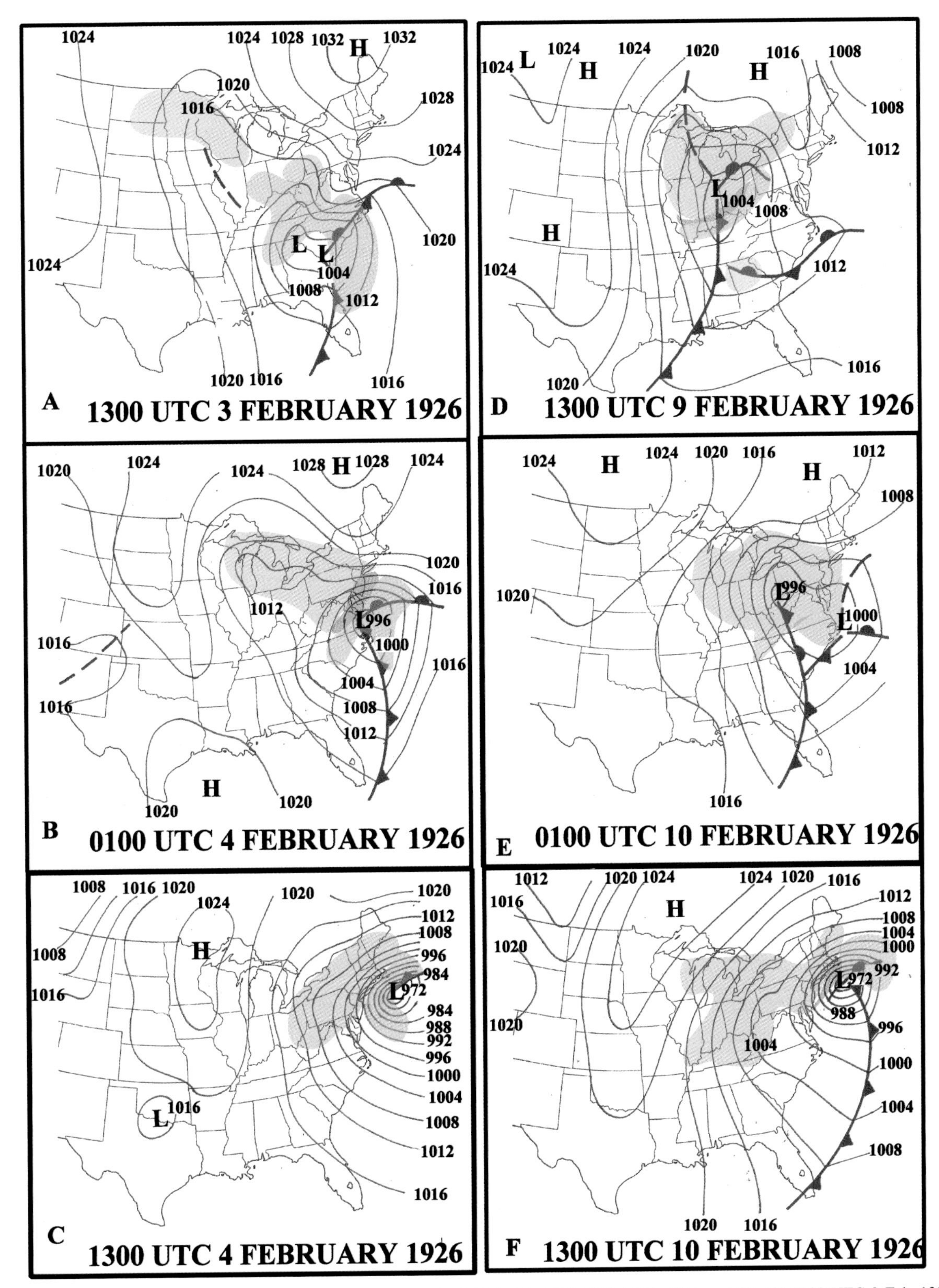

FIG. 9-16. Sequence of surface weather maps for (a)–(c) 1300 UTC 3 Feb–1300 UTC 4 Feb 1926 and (d)–(f) 1300 UTC 9 Feb–1300 UTC 10 Feb 1926. Blue, green, and violet shading represents snow, rain, and freezing rain/sleet.

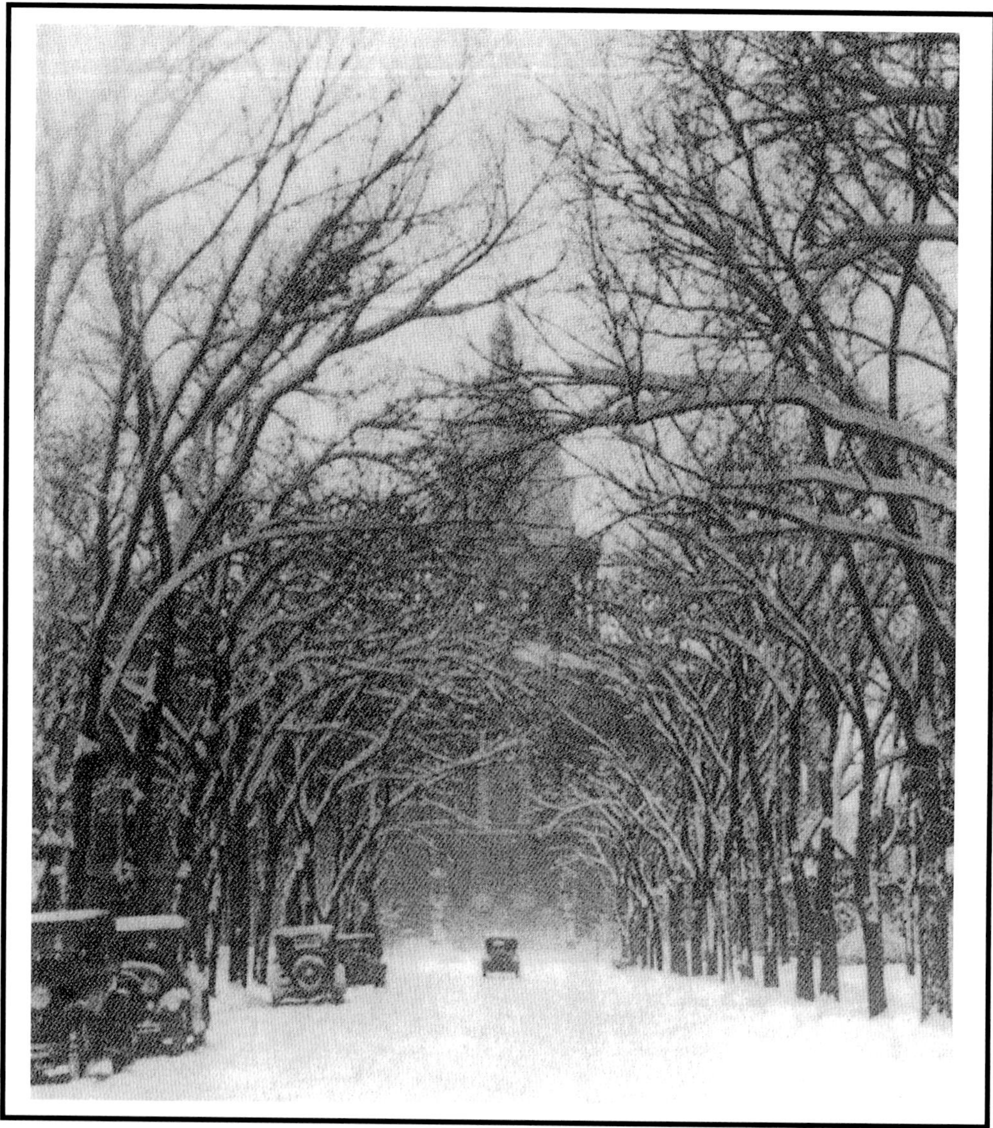

PLATE 9-7. Street scene in Washington, DC, 10 Feb 1926 (from Ambrose 1993, p. 88).

9-13e). This storm appeared to develop as an interaction or "merger" (see volume I, chapter 4) between an initially weak coastal low pressure system and a separate system frontal system/cold outbreak dropping southeastward across the Midwest. A final major snowstorm of the 1920s occurred on 20–21 February 1929 and left 8–12 in. (20–30 cm) along the urban corridor from Washington to Boston (Fig. 9-13f).

d. 1930–39

A notable snowstorm affected Washington, Maryland, and Virginia on 29–30 January 1930, leaving a foot (30 cm) of snow in Washington. This storm system was a rapidly moving cyclone that formed in the Gulf of Mexico and then moved east-northeastward off the North Carolina coast without curving northeastward, sparing much of the Northeast. While one of the deepest East Coast storms (Fig. 9-18a,b,c) of the 20th century produced more rain than snow in the Northeast urban corridor on 6–7 March 1932, the snowstorm of 16–17 December 1932 left a foot (30 cm) of snow in Washington and Baltimore The track of the surface low for this case is similar to that observed during the January 1930 storm but significant snow affected northern portions of the middle Atlantic states and southern New England. A fast-moving snowstorm on 10–11 February 1933 left 8–12 in. (20–30 cm) of snow from Baltimore to Boston.

Bitterly cold periods in December 1933 and February 1934, some of the coldest weather of the 20th century in the Northeast, were also accompanied by heavy snowfalls. A significant snowfall just after Christmas 1933 left heavy snow from New Jersey to New England with

Selected Snowfalls: 1930-1940

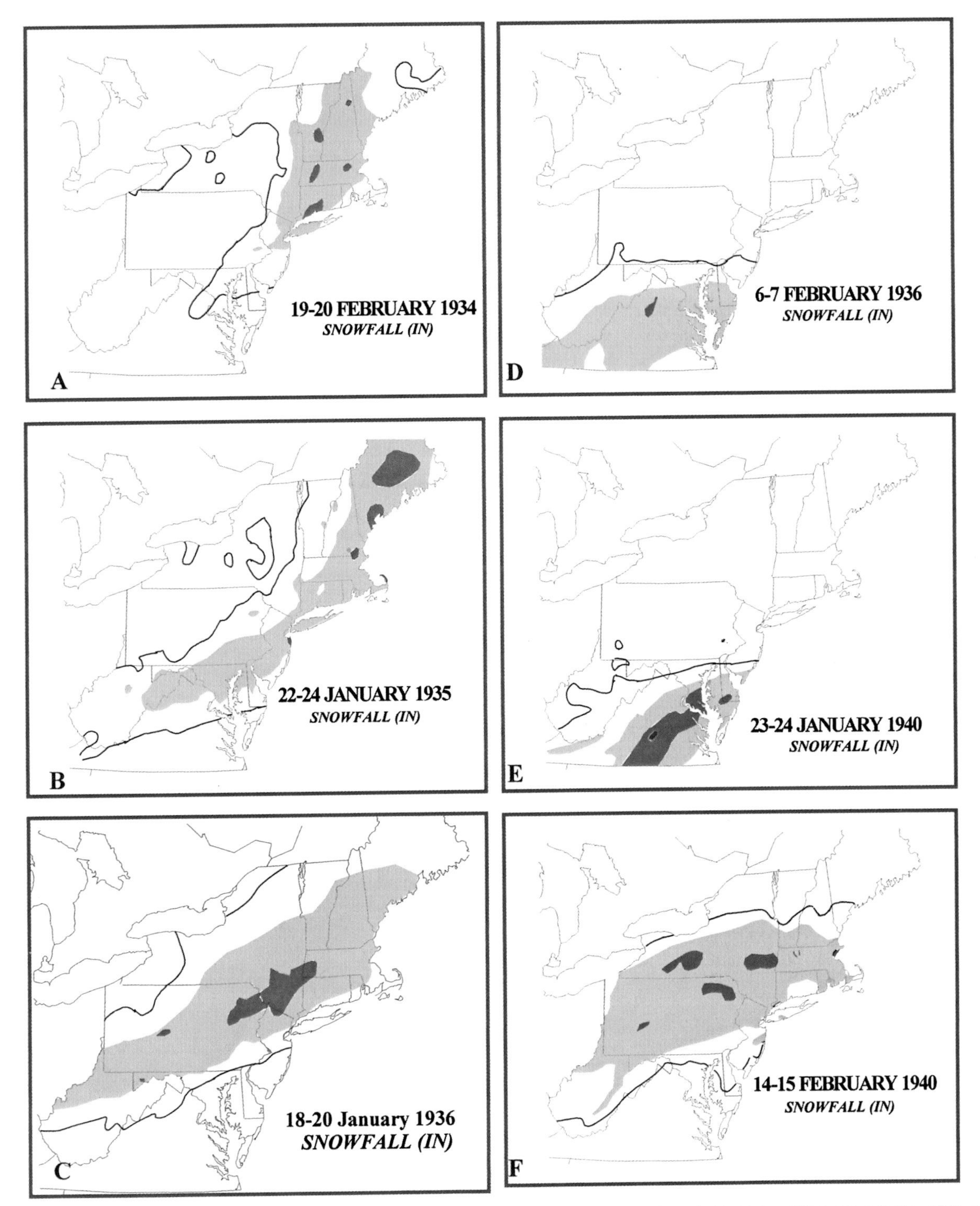

FIG. 9-17. Total snowfall for six cases, 1930–40: (a) 19–20 Feb 1934, (b) 22–24 Jan 1935, (c) 18–20 Jan 1936, (d) 6–7 Feb 1936, (e) 23–24 Jan 1940, and (f) 14–15 Feb 1940. Shading represents increments of 10 in. (25 cm) for amounts of 10 in. (25 cm) and greater. The solid line represents snowfall of 4 in. (10 cm) or greater.

13 in. (32 cm) in Boston and 10 in. (25 cm) in New York City, and was followed by temperatures that fell to −14°F (−26°C) in Boston. February 1934 was one of the coldest months ever recorded, which included the record-setting cold outbreak of 9 February, during which all-time low temperature records were shattered in Boston [−18°F (−28°C)], New York City [−15°F (−26°C)], and Philadelphia [−11°F (−24°C)]. During the same month, three major snowstorms also occurred. Snowstorms on 1–2 February and 26–27 February each produced large areas of deep snowfall [6–12 in. (15–30 cm) generally between Washington and Boston]. A separate storm on 19–20 February was a major blizzard (Fig. 9-17a).

The blizzard of 19–20 February 1934 stands out as the most severe of the three snowstorms during February 1934 and was one of the most severe blizzards of the first half of the 20th century. Snowfall was heaviest across Connecticut and Massachusetts where 15–25 in. (38–63 cm) of snow were common (Fig. 9-17a). The surface low developed within an inverted trough–frontal boundary along the East Coast and the surface low then developed explosively off the southern New England coast (Figs. 9-18d–f). High winds and falling temperatures added to the storm's impact as the storm deepened over New England. Newspaper accounts of the storm described its effects from New York City to Boston and especially across Connecticut as paralyzing. It was not the "classic" snowstorm in which the surface low develops to the south of the large high pressure system over New England. However, it has some of the signatures of those few storms that develop explosively and have perhaps some of the greatest impact—reminiscent of the development of the blizzards of March 1888 and 1914 (see Figs. 9-3 and 9-11).

Another widespread snowstorm battered the entire region from Washington to Boston on 22–24 January 1935, leaving 12–18 in. (30–45 cm) of snow in the cities and over 20 in. (50 cm) in Maine (Fig. 9-17b, Plate 9-8). This storm system was slow to develop as a complex series of low pressure systems consolidated along a frontal boundary across the southeastern United States on 21–22 January, with snow spreading from the Gulf coast to the middle Atlantic states. Snowfall rates intensified on 23 January as a low pressure consolidated off the Carolinas coastline and moved northeastward over the next 24 h (Figs. 9-19a–c).

The winter of 1935/36 was particularly severe, with several outbreaks of bitterly cold air and significant snowstorms, and was followed by catastrophic river flooding in March 1936 as heavy rain fell on melting snows. A vast snowstorm on 18–20 January 1936 produced heavy snow and sleet from New York City to Boston, leaving 8–12-in. (20–30 cm) accumulations as snow mixed with or changed to sleet, while 1–2 ft (30–60 cm) amounts were common across large portions of Pennsylvania, New York, and New England (Fig. 9-17c). In early February, a major snowstorm affected Virginia and Maryland, including 14 in. (35 cm) in Washington (Plate 9-9) and more than 10 in. of snow across much of Virginia (Fig. 9-17d). Surface analyses show that this storm developed over the Gulf of Mexico and followed a similar track to the January 1930 snowstorm, from the Gulf coast to North Carolina, and then east-northeastward over the Atlantic Ocean (Figs. 9-19d–f). This storm produced the heaviest snows in the middle Atlantic states while mostly avoiding New York and New England. The late 1930s did not see many significant snowstorms but an early season snow in November 1938 left a foot (30 cm) of snow in New York and New England around Thanksgiving.

e. 1940–49

The decade of the 1940s was dominated by many interesting and severe Northeast snowstorms, including New York City's greatest snowfall. January and February 1940 were bitterly cold with major snowstorms. One of the greatest Southeast snowstorms affected Georgia through Maryland on 23–24 January and was one of the heaviest snowstorms of record in Virginia, where up to 32 in. of snow (79 cm) was measured (Fig. 9-17e). This storm followed a track that extended from the Gulf of Mexico northeastward across northern Florida to a position off the South and North Carolina coast before drifting northeastward into the western Atlantic without affecting much of the northeast United States north of southern New Jersey (Figs. 9-20a–c).

The "Valentine's Day" snowstorm of February 1940 was an explosively developing storm that resulted in blizzard conditions from Pennsylvania, into New York, and to New England, and was one of the most memorable snowstorms on record in Boston because of the heavy snows and high winds. Snowfall from this storm was widespread (Fig. 9-17f), with more than 10 in. (25 cm) of snow covering much of Pennsylvania, the southern two-thirds of New York, and all of southern New England. The snowstorm was a rapidly developing cyclone that crossed the Ohio Valley and the Appalachians and then redeveloped explosively as it crossed the Virginia coastline early on 14 February 1940, reaching a central pressure of 965 hPa or lower (Figs. 9-20d–f).

Snowstorms in late February and early March 1941 left heavy accumulations in the middle Atlantic states and New England. One storm, on 28 February–1 March 1941, left up to 16 in. (40 cm) of snow at Atlantic City and paralyzed Philadelphia with up to a foot (30 cm) of snow (Fig. 9-21a). Another storm about a week later left the entire Washington to Boston corridor with more than 10 in. (25 cm) of snow (Fig. 9-21b), including 18 in. (45 cm) in New York City. Surface analyses of this storm show a slow-moving low pressure system over the southeast United States with a "center jump" (see volume I, chapter 3) as the surface low redeveloped

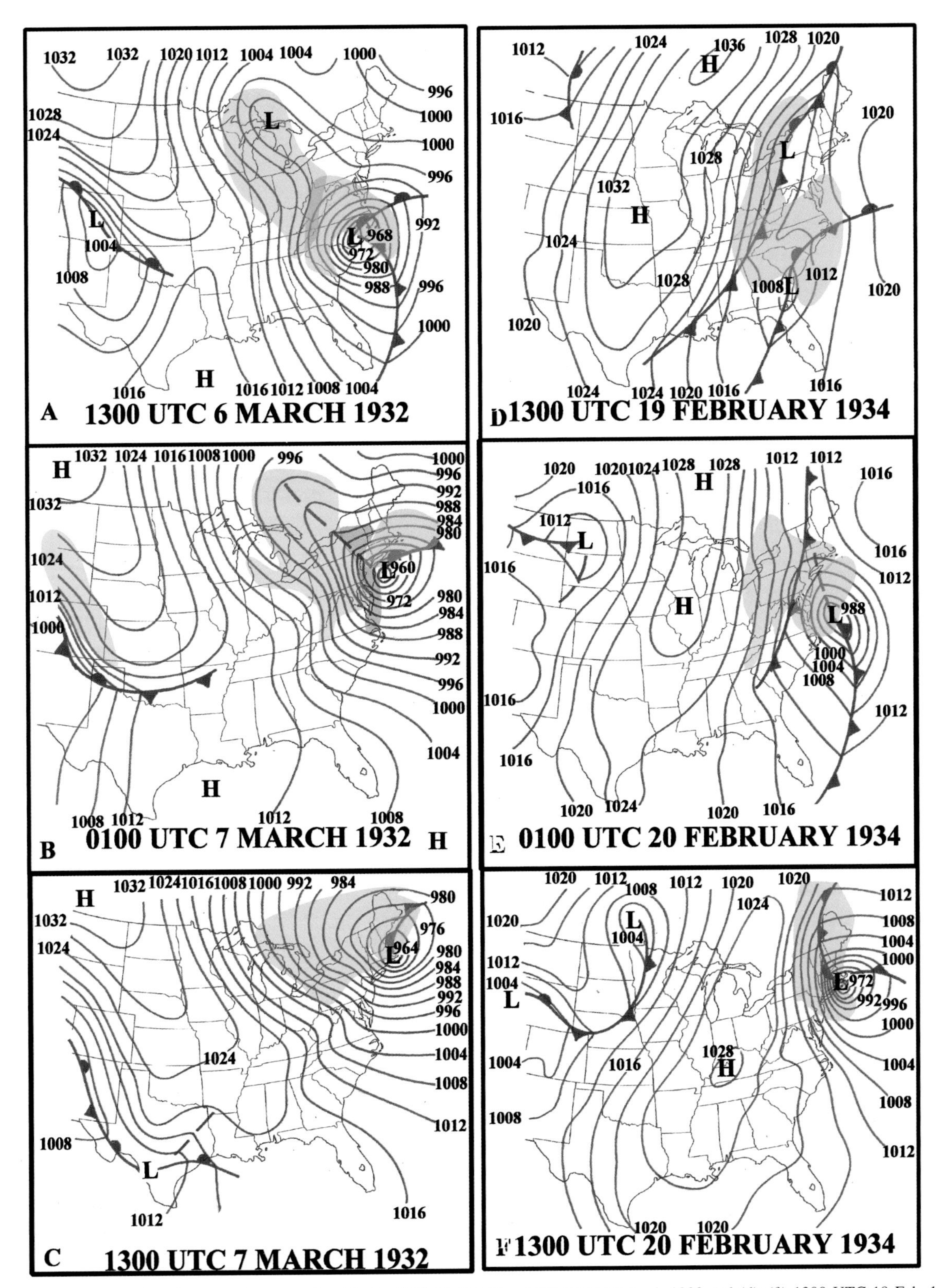

FIG. 9-18. Sequence of surface weather maps for (a)–(c) 1300 UTC 6 Mar–1300 UTC 7 March 1932 and (d)–(f) 1300 UTC 19 Feb–1300 UTC 20 Feb 1934. Blue, green, and violet shading represents snow, rain, and freezing rain/sleet.

PLATE 9-8. Clearing a sidewalk in New York City, 23 Jan 1935 (from Ambrose 1994, p. 44).

northeastward near the middle Atlantic coastline as very heavy snows moved northeastward (Figs. 9-22a–c).

One of the most anomalous snowstorms of the 20th century occurred on 28–29 March 1942 (see volume I, chapter 6), when a developing coastal low produced a surprise 11 in. (27 cm) of snow in Washington, 22 in. (54 cm) in Baltimore, and up to 3 ft (90 cm) of snow across central Pennsylvania. A series of snowstorms in late January 1943 left more than a foot (30 cm) of snow in southern New England. A snowstorm on 11–12 February 1944 left 8–12 in. (20–30 cm) of snow from New York City to Boston. The winter of 1944/45 was also quite severe with a major snowstorm on 14–16 January with heaviest accumulations of a foot or more (30 cm and greater) in eastern Pennsylvania, New York, and central and southern New England (Fig. 9-21c). Another storm on 8–9 February was severe in southern New England, leaving more than 14 in. (35 cm) of snow in Hartford; Providence, Rhode Island; and Boston. A third major snowstorm occurred in 1945 on 19–21 December with heaviest accumulations of 10–20 in. (25–50 cm) from northern New Jersey to eastern New England.

Two major snowstorms dominated 1947 with the first storm, on 21–22 February, overshadowed by New York City's greatest snowfall during December 1947. The February 1947 snowstorm was a widespread snowfall with 10–12 in. of snow (25–30 cm) or greater from Washington to Boston and affected a large portion of the Northeast (Fig. 9-21d). This storm can be considered a classic Northeast snowstorm as low pressure developed near the Gulf coast as cold, high pressure nosed southward across the northeast United States (Figs. 9-24d–f). The storm intensified as it moved northeastward along the East Coast, passing west of Cape Hatteras late on 20 February but remaining south of New England by the morning of 21 February (Fig. 9-22d) as the central pressure of the low fell near 970 hPa.

1) 26–27 DECEMBER 1947: NEW YORK CITY'S "BIG SNOW"

At 0320 LT 26 December 1947, snow began falling in New York City and by the time it ended nearly 24 h later, 26.4 in. (66 cm) had fallen at Central Park, 5 in. (13 cm) more than had fallen officially during the "Blizzard of '88." At 0700 LT, nearly 2 in. were measured (5 cm); by 1015 LT, 7 in. (18 cm) were measured; 1300 LT, 12 in. (25 cm); and by 1900 LT, 24 in.

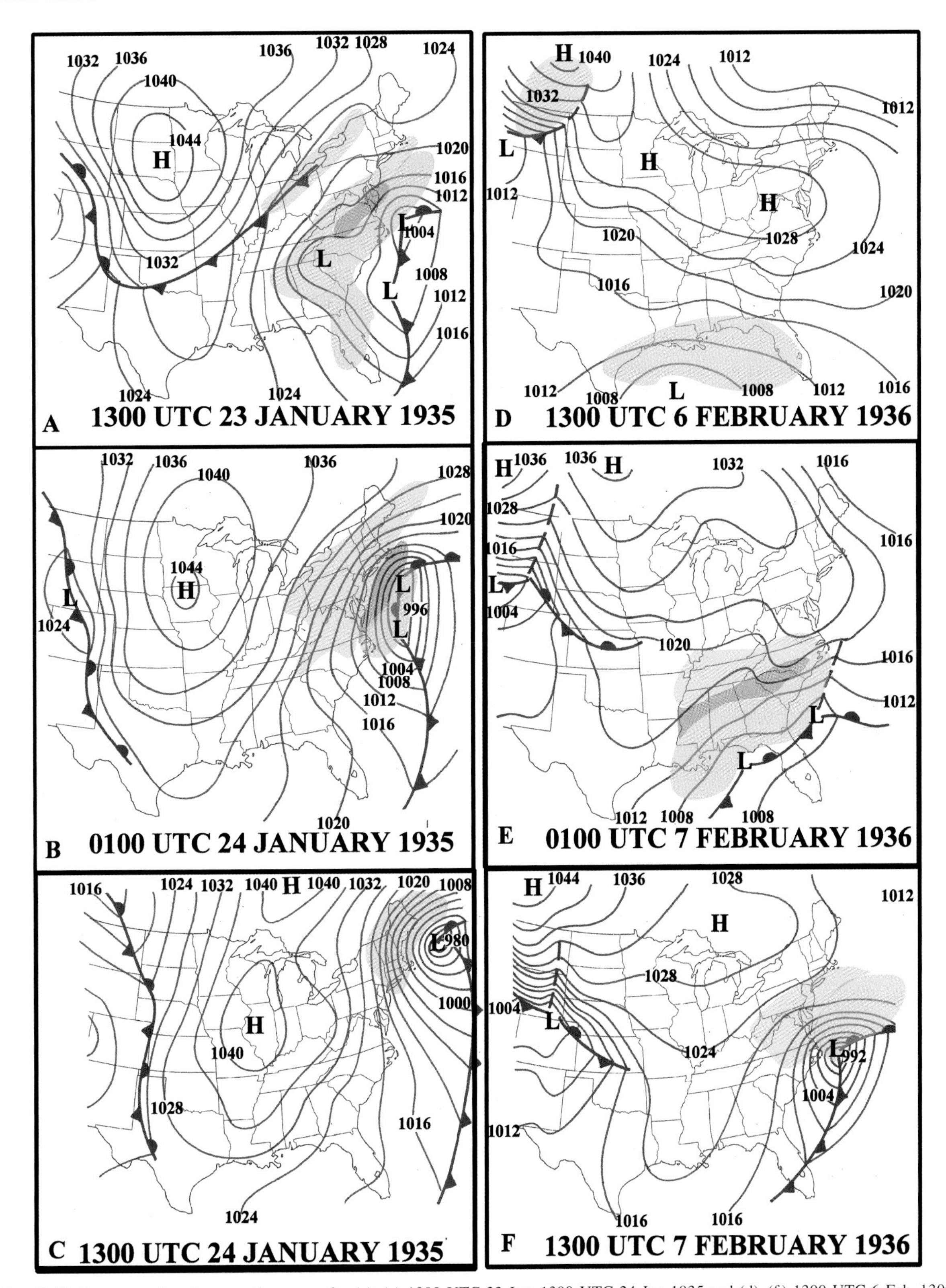

FIG. 9-19. Sequence of surface weather maps for (a)–(c) 1300 UTC 23 Jan–1300 UTC 24 Jan 1935 and (d)–(f) 1300 UTC 6 Feb–1300 UTC 7 Feb 1936. Blue, green, and violet shading represents snow, rain, and freezing rain/sleet.

PLATE 9-9. Streets in Washington, DC, 7 Feb 1936 (from Ambrose 1993, p. 41).

(50 cm). Eleven sites within New York City reported anywhere from 23 (58) to over 28 in. (71 cm), and several suburban locations in southeastern New York and northern New Jersey reported as much as 32 in. (80 cm; Fig. 9-23, Plate 9-10). Since many commuters were stranded, hotel accommodations were exhausted and many sought shelter in movie houses, armories, and other public places. Thousands of automobiles were stranded throughout the city and 24,000 extra workers were hired to shovel the snow. A 4-day ban on automobile traffic in the city was ordered by the mayor to clear the streets. The economic losses were estimated in the millions of dollars. By 4 January 1948, more than a week following the storm, all train, bus, subway, elevated, street car, and airplane service was restored, but only little more than half of the city's streets were cleared for traffic (U.S. Weather Bureau 1947).

The greatest snowstorm in New York City's history is another example of a weather forecast gone terribly wrong for what turned out to be a storm that rivaled

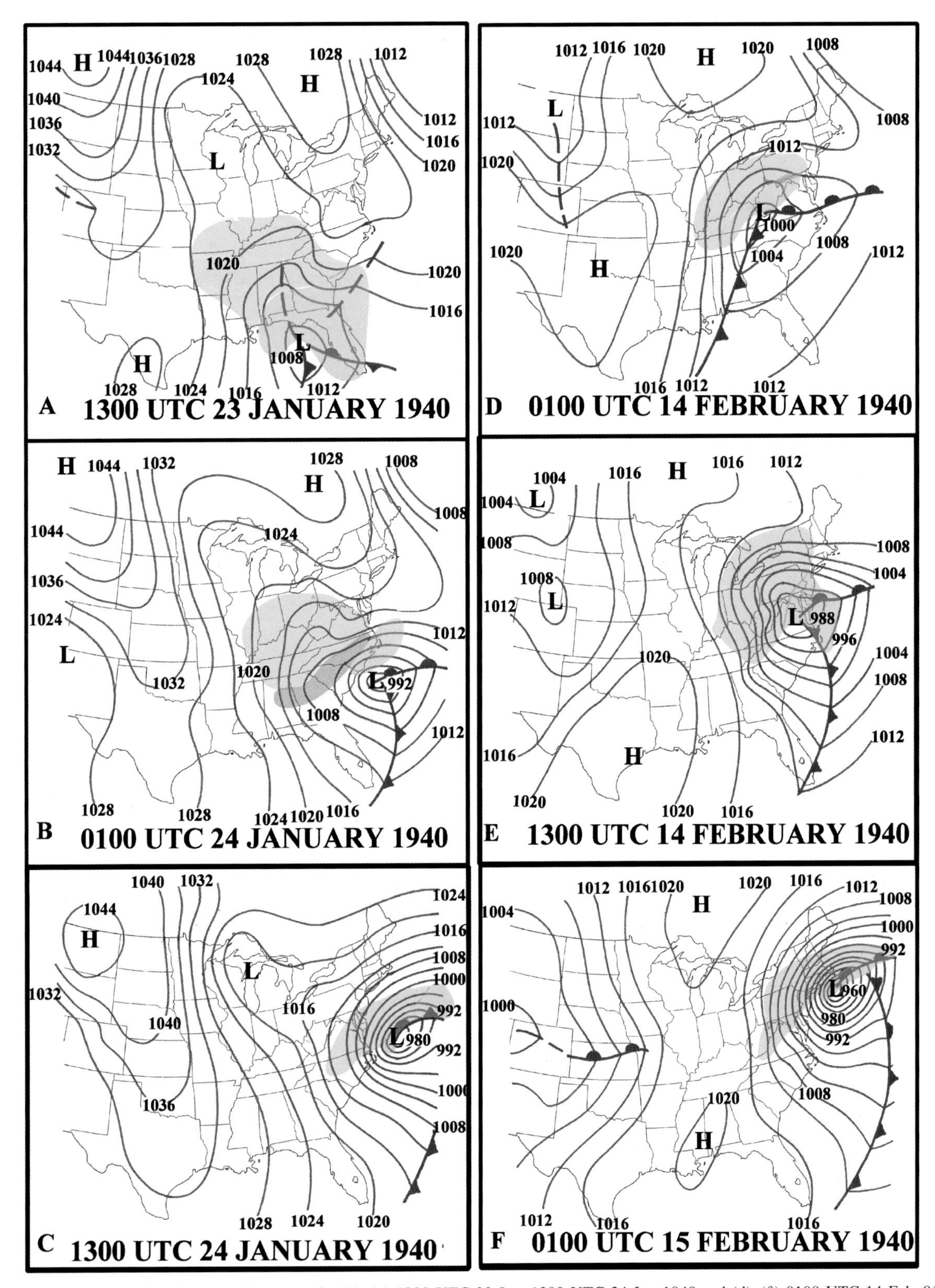

FIG. 9-20. Sequence of surface weather maps for (a)–(c) 1300 UTC 23 Jan–1300 UTC 24 Jan 1940 and (d)–(f) 0100 UTC 14 Feb–0100 UTC 15 Feb 1940. Blue, green, and violet shading represents snow, rain, and freezing rain/sleet.

Selected Snowfalls: 1940-1950

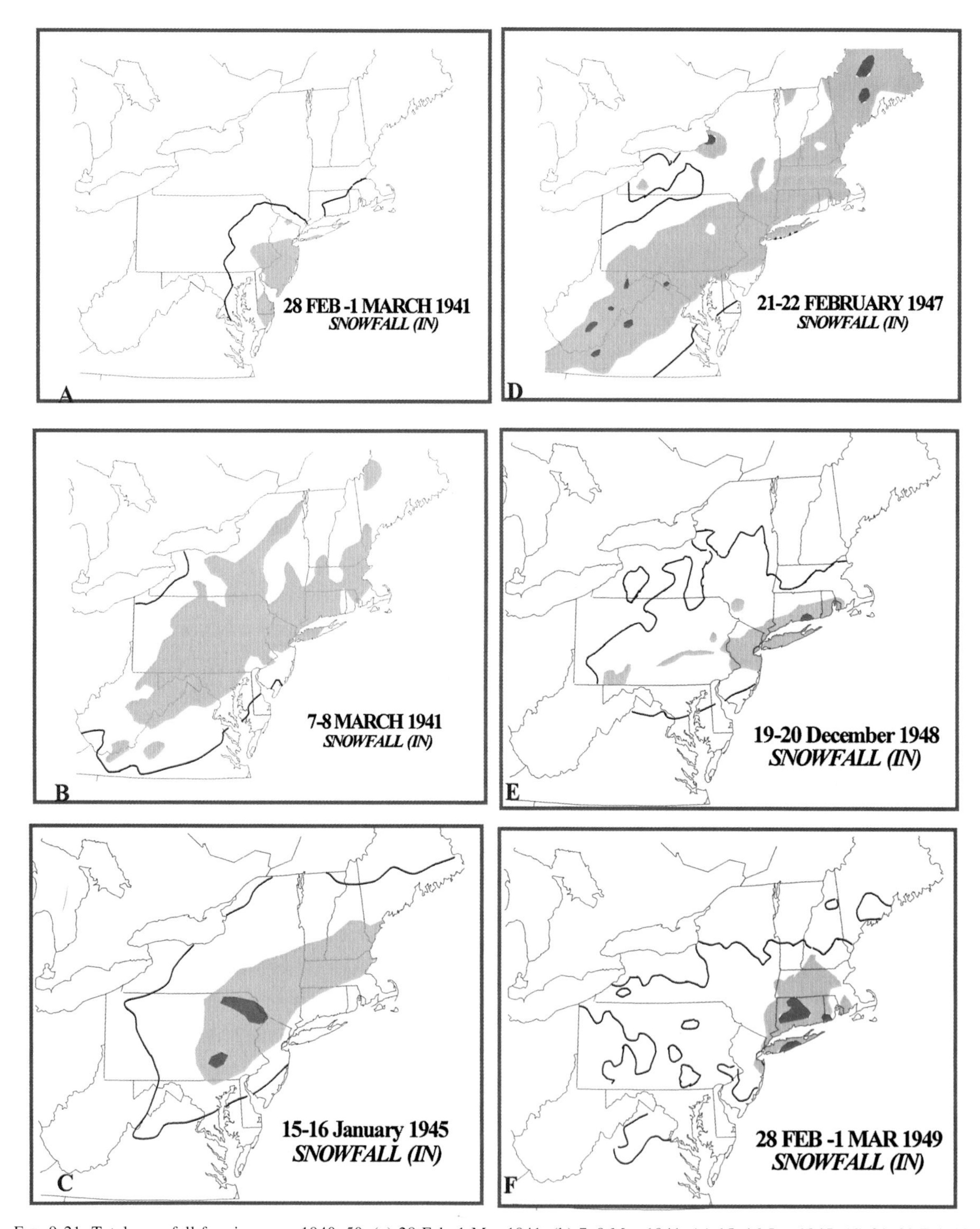

FIG. 9-21. Total snowfall for six cases, 1940–50: (a) 28 Feb–1 Mar 1941, (b) 7–8 Mar 1941, (c) 15–16 Jan 1945, (d) 21–22 Feb 1947, (e) 19–20 Dec 1948, and (f) 28 Feb–1 Mar 1949. Shading represents increments of 10 in. (25 cm) for amounts of 10 in. (25 cm) and greater. The solid line represents snowfall of 4 in. (10 cm) or greater.

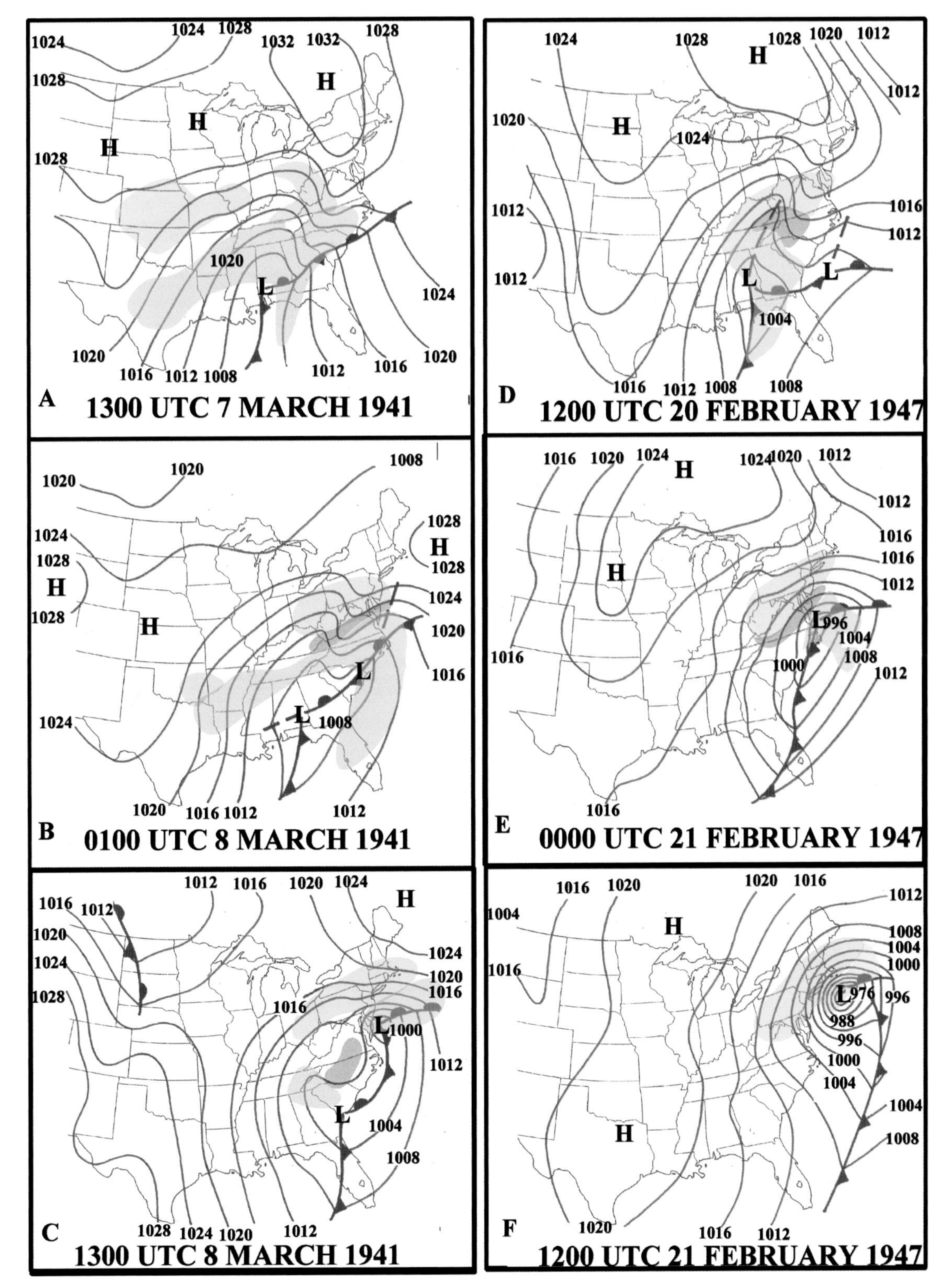

FIG. 9-22. Sequence of surface weather maps for (a)–(c) 1300 UTC 7 Mar–1300 UTC 8 Mar 1941 and (d)–(f) 1200 UTC 20 Feb–1200 UTC 21 Feb 1947. Blue, green, and violet shading represents snow, rain, and freezing rain/sleet.

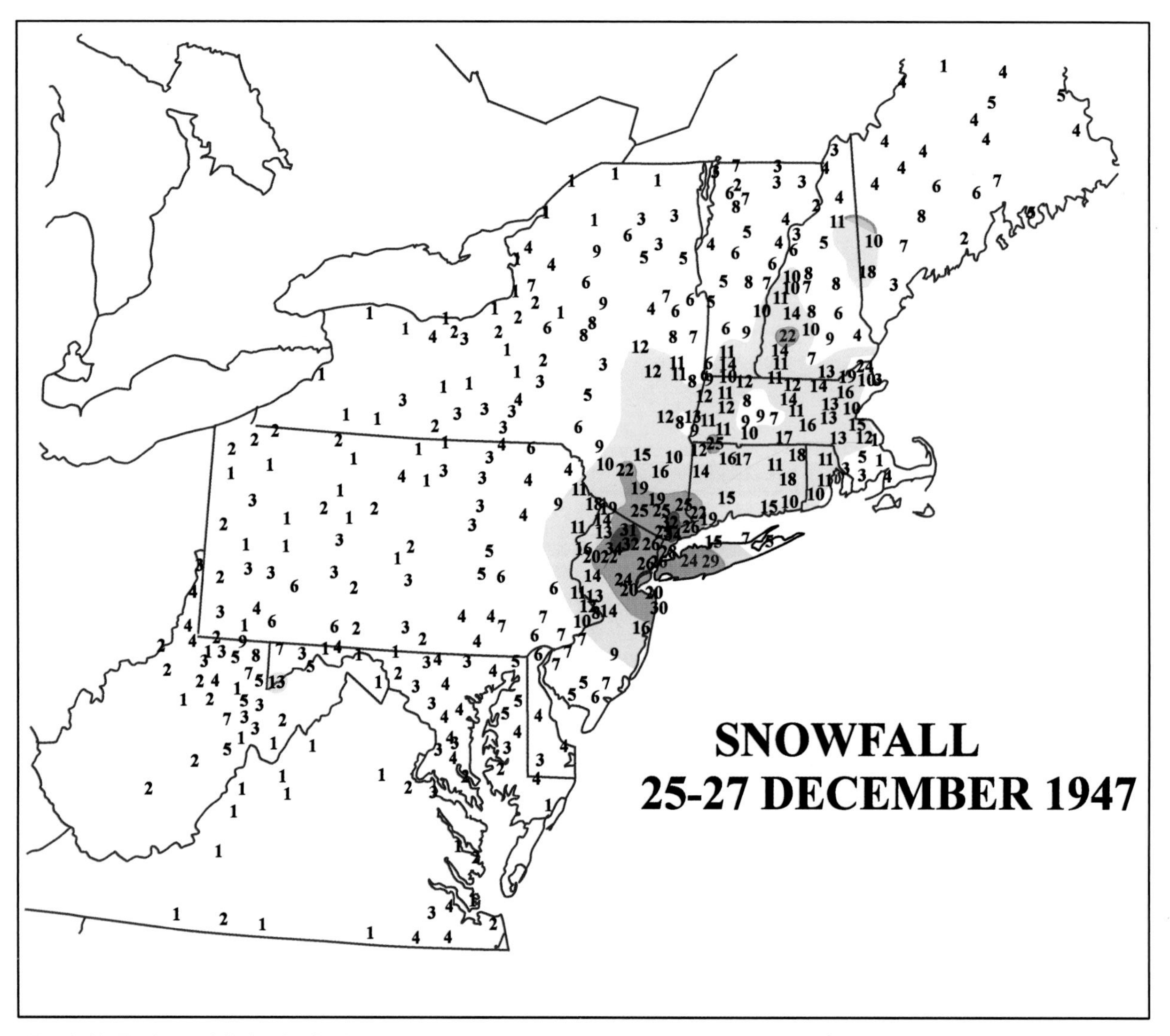

FIG. 9-23. Total snowfall (in.) for 25–27 Dec 1947. Colored shading represents increments of 10 in. (25 cm) for amounts of 10 in. (25 cm) and greater.

the impact of the Blizzard of 1888 in New York City and surrounding areas. The 0130 LT 26 December 1947 forecast for eastern and southeastern New York found on the early morning Daily Weather Map indicated that 2–4 in. (5–10 cm) of snow was likely as a small storm system moved northeastward along the Atlantic coast. A summary of special bulletins issued by the local Weather Bureau office in New York City illustrates the uncertainty that marred the forecast of this storm:

December 25

9:30 p.m. Tonight: cloudy with some snow possible toward morning. Friday—Cloudy with occasional snow ending during the afternoon followed by partial clearing.

9:50 p.m. Cloudy with snow towards morning ending during Friday afternoon, lowest temperature near 25. This is a borderline situation and the amount and time of precipitation depends on the movement of storm centered at 7:30 this evening off the Carolina coast. There is 1 chance in 4 that snow may be heavy and last most of the day, and 1 chance in 4 that no or little snow may fall.

December 26

4:30 a.m. Today. Snow ending during the afternoon. Tonight. Clearing.

10:30 a.m. About 7 in. of snow has fallen up to this time. Indications are for snow to continue to late afternoon or evening and accumulate 8 to 10 in.

Noon. Heavy snow will continue until 4:30 p.m. today ending completely at 7:30 p.m. Snow amount near 15 in. Strong northerly winds will cause drifting tonight.

PLATE 9-10. Bethpage, Long Island, New York. Louis Uccellini's brothers Charles and Walter and father Louis D. Uccellini after digging out from the 26–27 Dec 1947 storm. Note the windswept roof, an indication of the strong winds that accompanied this storm.

The area of snows in excess of 10 in. (25 cm) extended from extreme eastern Pennsylvania and central New Jersey northward into southeastern New York and southern New England (Fig. 9-23). Heaviest snows of 20 in. (50 cm) to greater than 30 in. (75 cm) were confined to northeastern New Jersey, extreme southeastern New York, the western half of Long Island, and southwestern Connecticut. Twelve-hourly surface analyses (Fig. 9-24) show that the storm developed along a frontal boundary near the Gulf and Southeast coasts with a weak inverted trough extending northward along the Appalachians and a developing coastal front off the Southeast coast. High pressure remained stationary over the Northeast early in the period and drifted northward as the coastal low moved toward the middle Atlantic coast. By midafternoon on Christmas Day (Fig. 9-24c), a 1004-hPa surface low is analyzed east of Charleston, South Carolina, with rain and snow across South and North Carolina. Low pressure was also moving southward toward the Great Lakes states with high pressure anchored over New England. In the next 12 h, the surface low moved northward to a position east of the Virginia coast by 0630 UTC 26 December (Fig. 9-24d) and deepened slowly as snow spread into northern Virginia, Maryland, southeastern Pennsylvania, and southern New Jersey. At this time, mostly light to moderate snows were falling in these locations, but snow would not begin in New York City for the next few hours.

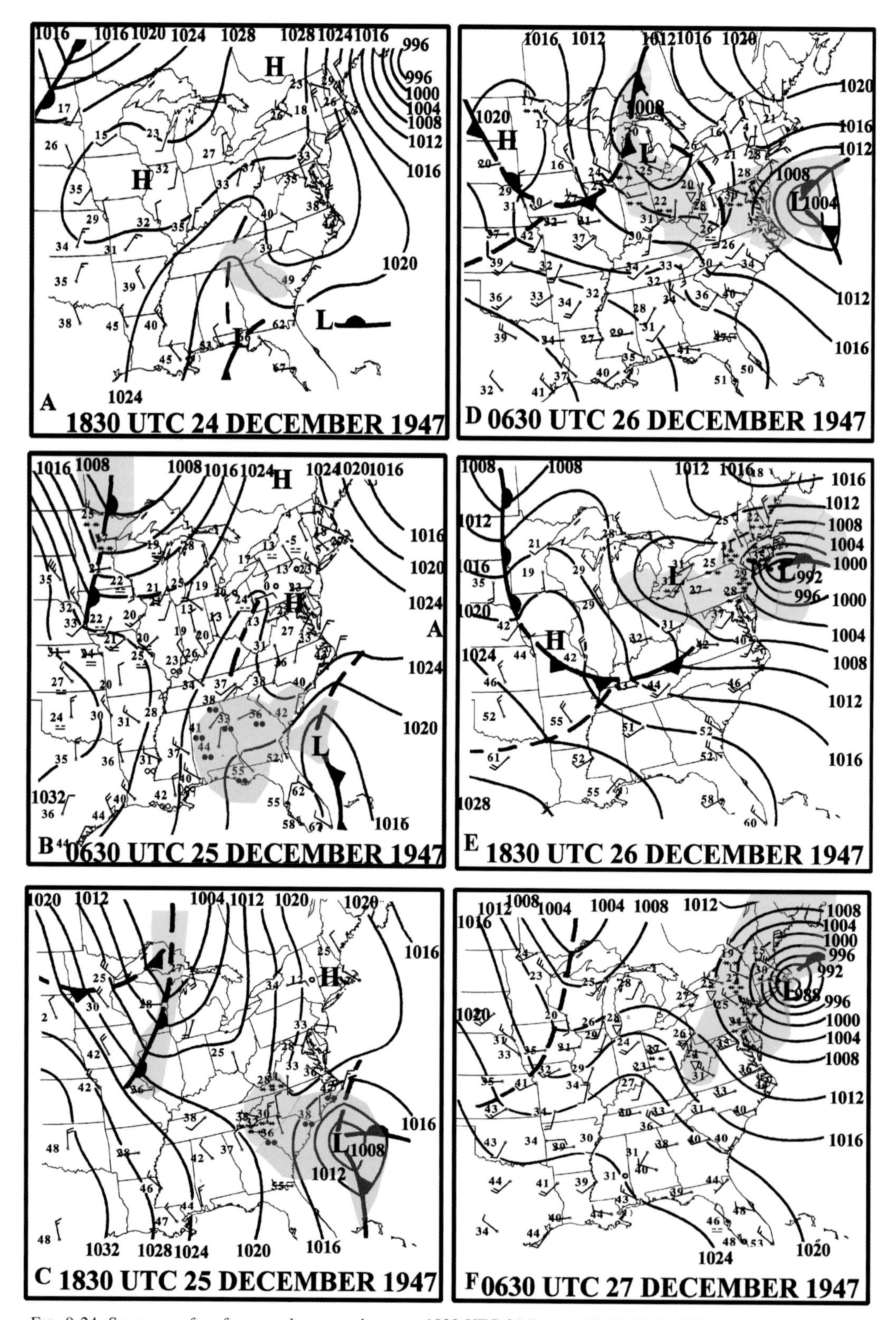

FIG. 9-24. Sequence of surface weather maps between 1830 UTC 24 Dec and 0630 UTC 27 Dec 1947. See Fig. 9-3 for details.

500 hPa HEIGHTS AND VORTICITY, 400 hPa WINDS

FIG. 9-25. Twelve-hourly analyses of 500-hPa geopotential height and upper-level wind fields for 0300 UTC 25 Dec–1500 UTC 27 Dec 1947. Analyses include locations of geopotential height maxima (H) or minima (L), contours of geopotential height (solid, at 60-m intervals, 522 = 5220 m), locations of 500-hPa absolute vorticity maxima (yellow/orange/brown areas beginning at 16×10^{-5} s^{-1}; intervals of 4×10^{-5} s^{-1}, and 400-hPa wind speeds exceeding 30 m s^{-1} (at 10 m s^{-1} intervals; alternate blue/white shading).

PLATE 9-11a. 42d St. between 7th and 8th Avenues, New York City (from Ambrose 1994, p. 49).

Within the next 12 h, the surface low deepened more rapidly and continued its northward course but began to slow as it neared southern New England as a 990-hPa low by 1830 UTC 26 December (Fig. 9-24e).

It was during this period of more rapid intensification that a mesoscale inverted pressure trough developed to the northwest of the surface cyclone center. This feature, shown in Fig. 6-10 (volume I, chapter 6), was first observed over New Jersey and then extended from central Long Island into southwestern Connecticut and southeastern New York. An analysis of the surface weather data indicates that the locally excessive snowfall amounts over northeastern New Jersey, southeastern New York, and southwestern Connecticut (Fig. 9-23) were associated with and located immediately to the west of an inverted sea level pressure trough extending north of the surface low pressure center. Pressure tendencies indicate that the trough developed across a relatively narrow area across western New England and separated northeasterly winds east of the trough axis from northwesterly winds in the region experiencing the greatest snowfall. This mesoscale feature appears to be a major factor in the excessive snowfall amounts and has been a recurrent feature for other heavy snowfalls that were not well predicted (see volume I, chapter 6). While the surface low continued to intensify and drift slowly eastward toward Nantucket Island (Fig. 9-24f), snowfall amounts across the rest of New England were clearly not as great as those across the New York City area (Fig. 9-23).

This case is the first major snowstorm examined in the monograph in which upper-air observations were available and points to their importance in forecasting the evolution of cyclones, especially for this case. This new dataset may have provided an indication that this storm had a "1 in 4 chance of developing into a major snowstorm," as forecasters surmised, and reveals that the cyclone appears to develop as the result of a merging or "phasing" of two upper-level short-wave troughs, one associated with the small coastal cyclone off the Southeast coast, and the other an amplifying or "digging" short-wave trough over the upper Midwest that propagated southeastward and interacted with the Southeast storm (see the 500-hPa analyses in Fig. 9-25). More details on trough mergers are discussed in chapter 4 of volume I. The interaction led to a rapidly developing cyclone that moved northward toward eastern Long Island and then drifted eastward to off the

PLATE 9-11b. 106th St., Central Park West, New York City, 27 Dec 1947 (from Ambrose 1994, p. 50).

southern New England coastline. Such interactions, involving an amplifying upstream short-wave trough and an incipient cyclone off the Southeast coast, appear to have produced some of the most exceptional, sudden, and dramatic cyclonic developments during the past century. Selected photographs taken during the storm are shown in Plates 9-11a–c.

New York City's greatest snowfall was followed less than a year later by another snowstorm that left up to 19.6 in. (50 cm) of snow on 19–20 December 1948, with up to 24 in. (60 cm) at New London, Connecticut (Fig. 9-21e). A complex storm system on 28 February–1 March 1949 left up to 10 in. of snow (25 cm) in New York City and Boston but up to 2 ft of snow (60 cm) across Long Island and Connecticut (Fig. 9-21f).

2) 25–27 NOVEMBER 1950: "THE APPALACHIAN STORM"

The "Appalachian Storm" of November 1950 is the greatest snowstorm on record for the western Appalachians and was accompanied by the coldest November

PLATE 9-11c. Park Avenue looking northeast along 79th St. (from Ambrose 1994, p. 53).

temperatures on record for portions of the South and Midwest (Smith 1950; Bristor 1951). Although this is not a traditional "Northeast snowstorm," since the heaviest snows fell far to the west of the Northeast coastline, it is included here because it represents perhaps the greatest combination of extreme atmospheric elements ever seen in the eastern United States. We feel that this storm is the benchmark against which all other major storms of the 20th century could be compared, as will shortly be described. The overused title "storm of the century" could be aptly used here. Rather, the storm *for* the century might be a more appropriate moniker.

As an interesting historical footnote, the Appalachian Storm served as the test case for the first numerical models using the newly engineered "store-foreward" computers (Phillips 1951; Charney and Phillips 1953; Phillips 2000). While the results of these first numerical experiments might appear crude compared to the sophisticated numerical weather prediction techniques in operational forecasting today, the success in simulating

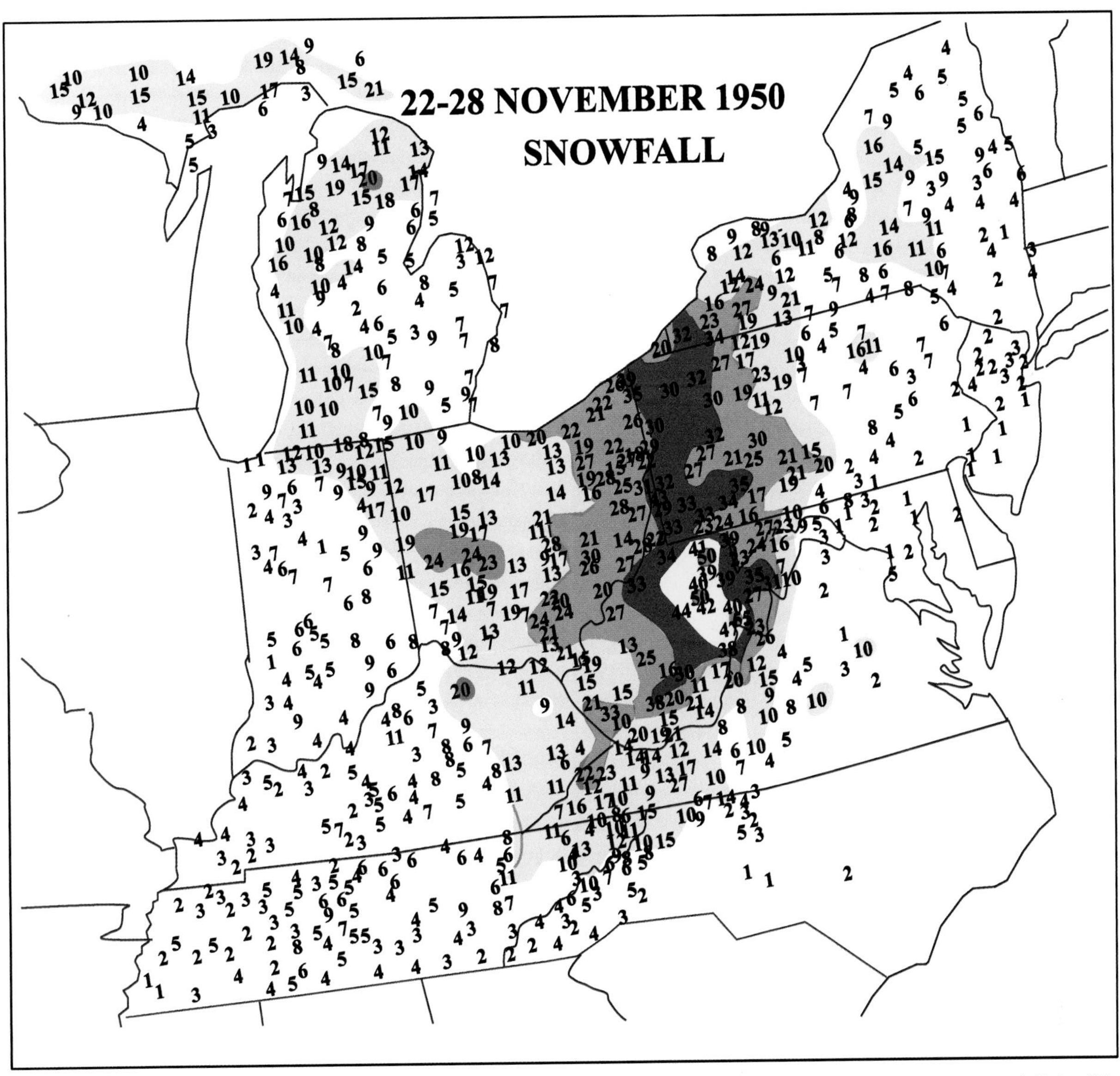

FIG. 9-26. Total snowfall (in.) for 22–28 Nov 1950. Colored shading represents increments of 10 in. (25 cm) for amounts of 10 in. (25 cm) and greater.

the general low-level and upper-level features of this massive storm provided the impetus for the concept of applying centralized model guidance as the basis for the forecast process that was successfully applied during the last half of the 20th century (volume I, chapter 8).

Snowfall totals of 20 in. (50 cm) or greater were common throughout a wide portion of West Virginia, western Pennsylvania, and eastern Ohio (Fig. 9-26) with 30 in. (75 cm) or greater in many of these locations. Snowfall totals included 29.5 in. (74 cm) at Elkins, West Virginia; 33.3 in. (85 cm) at Parkersburg, West Virginia; and 25–30 in. (62–75 cm) were common in the cities of Pittsburgh, Pennsylvania, and Cleveland and Youngstown, Ohio. The heaviest snow fell in West Virginia, where 40–50 in. (100–125 cm) or more were reported at many sites, including 62 in. (157 cm) at Coburn Creek, over a period of several days (Fig. 9-26). The heavy snow was accompanied by high winds and temperatures that fell to near 0°F (−18°C), producing life-threatening conditions for many days. In Dayton, Ohio, where "only" 11.2 in. (28 cm) of snow fell, bitter cold and 30 m s^{-1} winds created the worst blizzard on record.

Some of the coldest temperatures ever experienced during November were common in the Midwest and South with 0°F (−18°C) temperatures reaching central Tennessee and northern Georgia, including −26°F (−32°C) at Mount Mitchell, North Carolina; −23°F (−31°C) at Pellston, Michigan; −1°F (−19°C) at Nashville, Tennessee; and 3°F (−16°C) at Atlanta, Georgia, to name just a few.

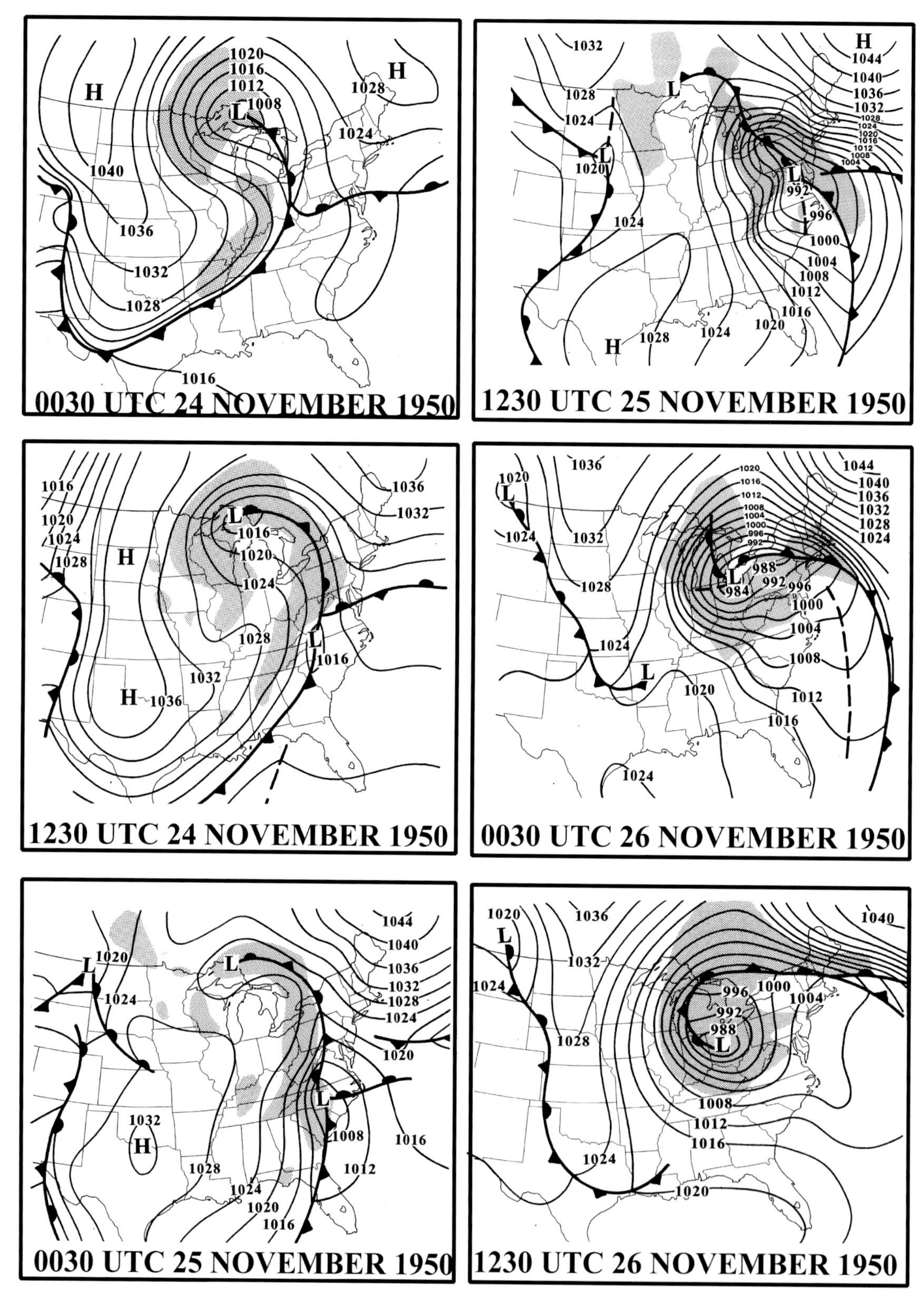

FIG. 9-27. Sequence of surface weather maps between 0030 UTC 24 Nov and 1230 UTC 26 Nov 1950, including fronts and isobars, and shading represents current precipitation.

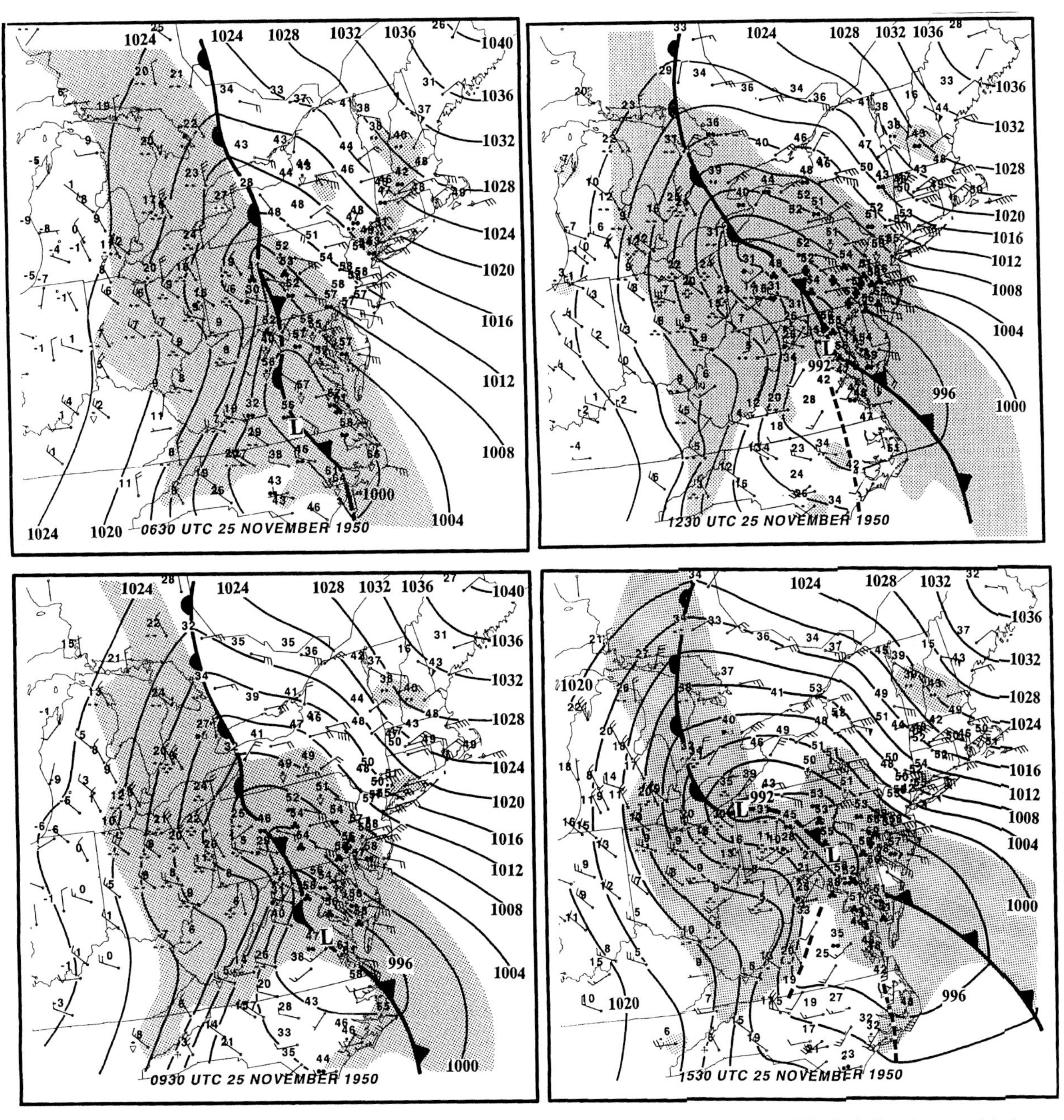

FIG. 9-28a. Detailed 3-hourly surface analyses between 0630 UTC 25 Nov and 0330 UTC 26 Nov 1950, including fronts and isobars (hPa at 4-hPa intervals, station symbols, temperature °F, winds, current weather). Shading represents current precipitation.

Easterly gales with hurricane force gusts prevailed across much of the northeastern United States and coincided with high tide, producing destructive coastal flooding. Wind gusts in excess of 100 mi h^{-1} (50 m s^{-1}) were recorded in Newark, New Jersey (108 mi h^{-1}); Hartford, Conneticut; and Concord, New Hampshire, and wind speeds of hurricane force were experienced not only along the coast, but over much of the interior of the Northeast as well. The high wind speeds were generated by a strengthening pressure gradient associated with a nearly stationary 1045–1050-hPa anticyclone north of New England and by the intense cyclogenesis moving north and northwestward over the middle Atlantic states toward the Ohio Valley.

The sequence of 12-hourly surface charts (Fig. 9-27) shows a strong cold front crossing the Ohio Valley, lower Tennessee Valley, and Texas at 0030 UTC 24 November), extending southwestward from a low pressure system over the Great Lakes. In the following 12 h (Fig. 9-27b), the low over the Great Lakes remained stationary and weakened while a new 1015-hPa low center developed along the cold front in western North Carolina with snow falling in the cold air over the Appalachians, west of the front. By 0030 UTC 25 Novem-

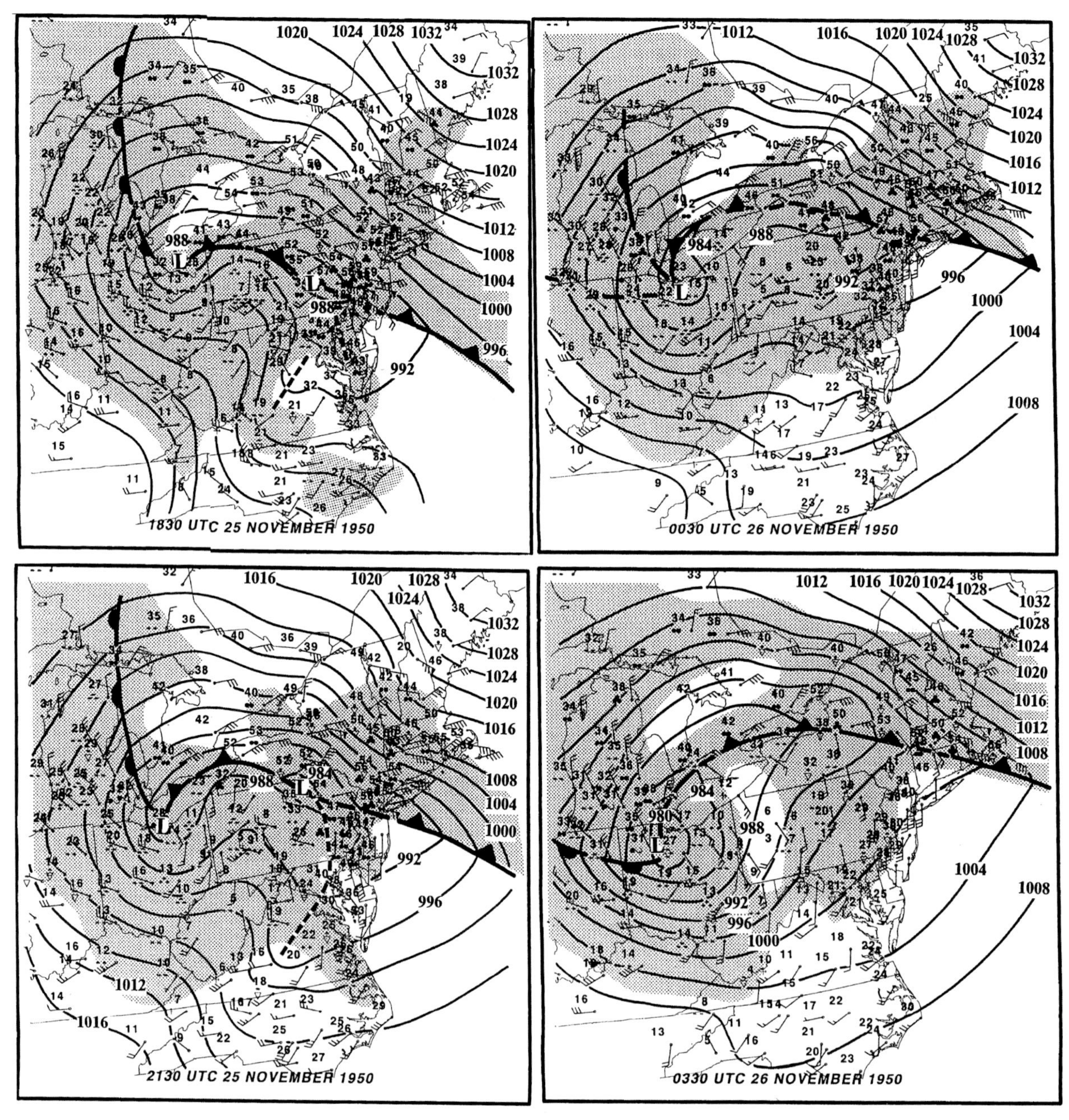

FIG. 9-28b. Detailed 3-hourly surface analyses between 0630 UTC 25 Nov and 0330 UTC 26 Nov 1950.

ber (Fig. 9-27c), the low pressure system over the western Carolinas had briefly dropped southward and then began to move northward as a 1004-hPa low along the nearly stationary cold frontal boundary with snows increasing in intensity along the Appalachians. Note that at the same time this storm was beginning to intensify, high pressure north of Maine also strengthened from 1030 hPa to nearly 1050 hPa in only 24 h. Over the next 24 h (Figs. 9-27d,e), the storm increased rapidly in intensity and moved, or redeveloped, to the north, then to the northwest, and then to the west, reaching a position over northern Ohio by 0030 UTC 26 November 1950 with a central pressure near 980 hPa. During this period of intensification, the heaviest snows fell across the northern Appalachians of West Virginia, western Pennsylvania, and southwestern New York, and also spread westward across Ohio, Michigan, and Indiana.

As the storm spiraled into Ohio, various frontal boundaries surrounding the cyclone created exceptionally complex weather patterns that can be best described by examining detailed surface analyses of the storm, shown at 6-hourly intervals in Figs. 9-28a and 9-28b. A few of the most spectacular meteorological anomalies that best typify this sequence of charts include the oc-

500 hPa HEIGHTS AND VORTICITY, 400 hPa WINDS

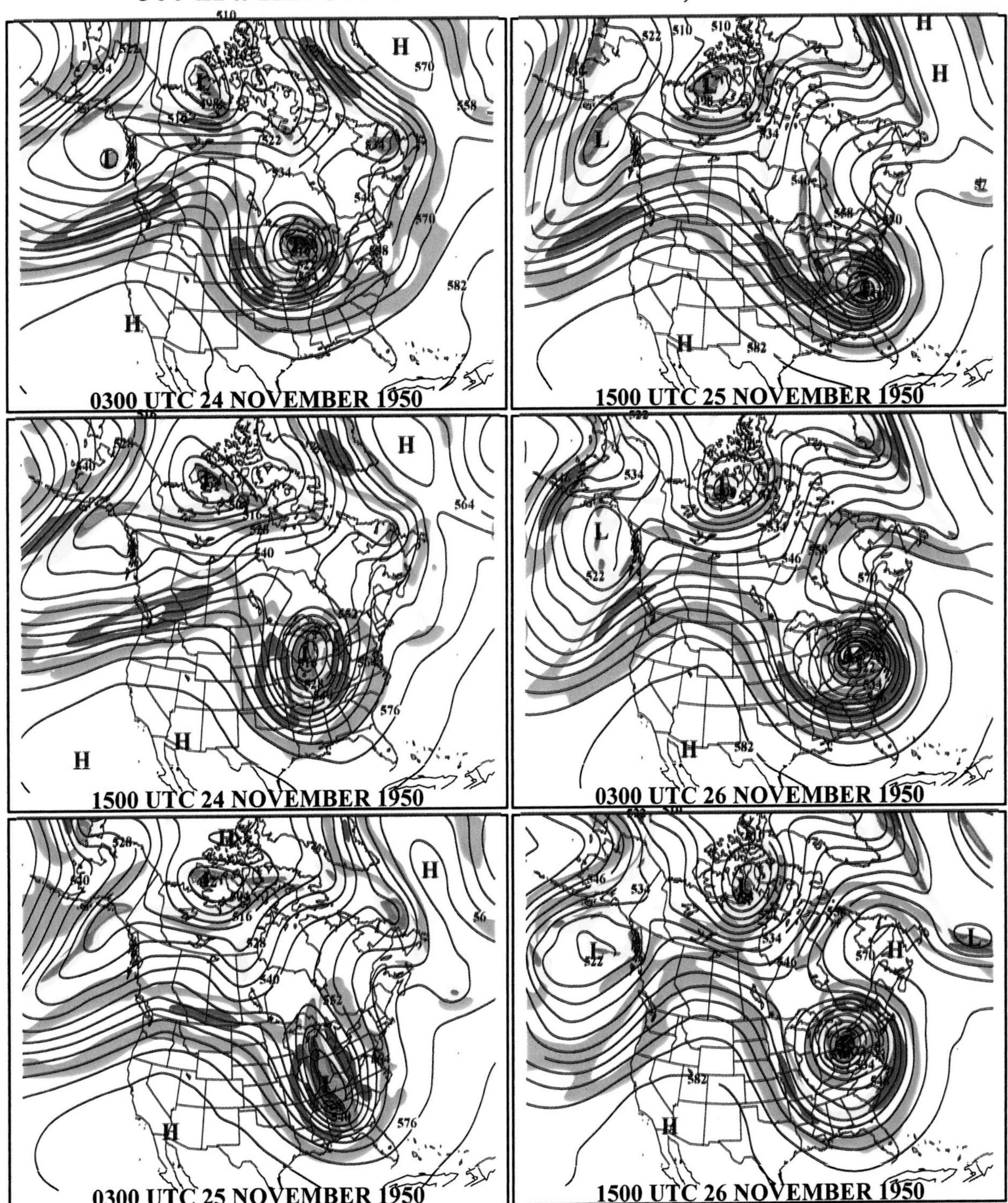

FIG. 9-29. Sequence of 500-hPa charts between 0300 UTC 24 Nov and 1500 UTC 26 Nov 1950. See Fig. 9-25 for details.

currence of blizzard conditions at Pittsburgh, with heavy snow and rapidly falling temperatures [from 21°F (−6°C) at 1230 UTC 25 November to 9°F (−13°C) by 1830 UTC] as winds shifted from the northwest to a *southerly* direction. This occurred as the storm was redeveloping to the north and west of Pittsburgh and the coldest air moved into western Pennsylvania from the south and west. Meanwhile, north of the cyclone, heavy

PLATE 9-12a. Webster Ave., Pittsburgh, PA, 25 Nov 1950 (from Bristor 1951, p. 10).

rains, high winds, and much warmer temperatures contrasted with the arctic conditions farther south. At 1830 UTC 25 November, the time that Pittsburgh was experiencing heavy snow and 9°F (−6°C) temperatures, Buffalo, New York, was a balmy 54°F (12°C) with occasional heavy rain and hurricane force wind gusts, which were common at this time across much of New York, New Jersey, and southern New England. In Detroit, Michigan, heavy snow changed to rain after 1830 UTC as a warm front north of the developing storm moved westward, raising temperatures from the low teens (°F; −11°C) at 1230 UTC to the mid- and upper 30°s F (3°–5°C) by 0030 UTC 26 November as the boundary crossed the city late on November 25. The passage of the warm front was followed only a few hours later by the passage of the cold front from the east as the storm center spun into Ohio (Fig. 9-28b). Temperatures across Detroit, northern Ohio, northwestern Pennsylvania, and western New York dropped rapidly as winds shifted into an east-to-southeasterly direction behind the front. Along the western fringes of the storm, Indianapolis, Indiana, started the morning of 25 November with clear skies and −1°F (−19°C) temperatures. As the day progressed, skies grew cloudy from the east as the storm, initially too far east to have any impact, moved toward Ohio. By nightfall, the blizzard would reach into central Indiana, where even Indianapolis reeled from 7 in. (18 cm) of snow and high winds.

The 500-hPa charts (Fig. 9-29) show that this amazing storm was accompanied by a dramatic amplification of an upper-level trough over the east-central United States on 24 November and evolved into a huge upper-level vortex. The amplified trough developed a northwest to southeast "negative tilt," which contributed to the northwestward track of the surface low from the Appalachians to the Great Lakes. Selected photographs taken during the storm are shown in Plates 9-12a–c.

f. Summary of snowstorms: 1900–50

Many memorable snowstorms occurred during the first half of the 20th century. While most of these storms can only be defined by their surface weather characteristics prior to the advent of operational upper-level observations during the late 1940s, all storms exhibited many of the features shown by 30 major snowstorms during the second half of the 20th century, as discussed in chapters 3 and 4 of volume I.

PLATE 9-12b. A parking lot, LaGuardia Airport, New York City, 25 Nov 1950 (from Bristor 1951).

Classic nor'easters, with surface low pressure systems developing over the Gulf of Mexico and following a path along the southeast U.S. coast to near Cape Hatteras, North Carolina, were observed with many heavy snowfalls. In some instances, low pressure veered eastward, sparing New England, while Virginia, Maryland, Delaware, and Washington bore the brunt of the heaviest snows. Examples include storms in January 1930, December 1932, February 1936, and January 1940. In other instances, low pressure centers reached Cape Hatteras and then continued northeastward off the southern New England coastline, producing heaviest snows from Pennsylvania and New York northeastward into New England, as in February 1907, February 1921, and February 1947. The location of surface anticyclones, some of which revealed high central pressures, provided the cold air for precipitation to reach the ground as snow. Examples of slow-moving storm systems following a track from the Gulf of Mexico northeastward along the East Coast and interacting with strong anticyclones include the destructive storms of February 1920 and the Knickerbocker storm of January 1922.

Other snowstorms exhibited a primary low pressure system moving toward the eastern United States followed by the development of a secondary low pressure system, as discussed in chapter 3 of volume I. Significant examples include December 1909, January 1910, February 1914, and 9–10 February 1926. In many cases, the presence of a significant anticyclone poised north of New York and New England appears to play a major role in providing cold air for the snowfall and helping produce the baroclinic zones in which heavy snow developed. A few exceptions are noted, especially December 1909, in which heavy snow occurred without the presence of a strong anticyclone.

Some of the most spectacular cyclogenesis occurred in conjunction with initially weak cyclones that interacted with significant outbreaks of cold air, possible surface reflections of trough "mergers" (volume I, chapter 4). Some of the most memorable snowstorms fall into this category, including the March blizzard of 1888, the snowstorm of 24–25 January 1905, a sudden snowstorm in March 1909 (not shown) that affected the inauguration of President William Howard Taft, the blizzards of March 1914 and February 1934, and the New York City's heaviest snowfall in December 1947.

The review of the first 50 years of the 20th century is consistent with chapter 2 of volume I, which provides

PLATE 9-12c. Digging out in Pittsburgh, PA, 27 Nov 1950.

a climatological review of the second half of the century, in that the first 50 years were marked by at least 30 storms similar to those described in chapters 3 and 4 (volume I), illustrating that these storms were relatively rare events, but possibly more frequent than during the second half of the century, when seasonal snowfalls appeared to diminish (see volume I, chapter 2). Episodes of major snowstorms are noted throughout the first decade of the century, the mid- to late 1910s, the early and middle 1920s (especially February 1926), the middle 1930s, and several winters during the 1940s, separated by seasons of relatively low activity (such as the early and late 1910s, the early and late 1930s, and parts of the early 1940s), pointing toward the episodic characteristic of these snowstorms. Out of the 30 storms singled out for the first 50 years of the century, as discussed in section 2 of this chapter, more than half (16) occurred during February, reconfirming a belief (see volume I, chapter 2) that February is the month of the "big snows" in the Northeast urban corridor, at least during much of the 20th century.

The advent of upper-level data during the late 1940s provides evidence that the linkage of surface cyclones and cold anticyclones to characteristic upper-level signatures of upper trough–ridge patterns and their associated upper-level jet streaks (see volume I, chapter 4) applies to three cases in the late 1940s (not shown), as they did to many cases during the second half of the 20th century. These surface and upper-level signatures are shown in detail for 30 cases between 1950 and 2000 from chapters 3 and 4 of volume I, as well as the recent Presidents' Day Storm of February 2003, providing the basis for the following chapter.

Chapter 10

THIRTY-TWO SELECTED SNOWSTORMS: 1950–2003

This chapter systematically illustrates the organization and evolution of the 30 snowstorms documented in chapters 3 and 4 of volume I and the recent February 2003 and December 2003 snowstorms, focusing on the period prior to and during the development of heavy snowfall along the northeastern coast of the United States. Each case study includes a total snowfall chart and 12-hourly sequences of surface weather maps, 850-hPa charts, 500-hPa geopotential height analyses with 400-hPa wind analyses superimposed, and 250-hPa geopotential height and wind analyses. Infrared satellite image sequences are also displayed for all cases beginning with 19–20 January 1978.

Each case begins with a brief summary of the storm's major effects on the Northeast urban corridor, as well as a brief summary of the weather regime in which the storm developed, including the phases of El Niño–Southern Oscillation (ENSO) and the North Atlantic Oscillation (NAO). The snowfall charts that follow (Fig. 10.*n*-1, where $n = 1 \ldots, 32$) were constructed from daily snowfall measurements provided in the National Climatic Data Center publication *Climatological Data*. Some urban measurements are listed in Tables 10.*n*-1, including the storm's classification under the Northeast Snowfall Impact Scale [NESIS; see volume I, chapter 8.4; also see Kocin and Uccellini (2004)]. Accumulations in excess of 10, 20, and 30 in. (25, 50, and 75 cm) are contoured and shaded and amounts for selected locations are noted in the text.

The surface charts depict fronts, high and low pressure centers, isobars, and precipitation (snow, mixed precipitation, and rain; Figs. 10.*n*-2). The sequence of maps is ordered so that the top-right panel captures the surface low pressure center along the East Coast, when many of the storms were deepening rapidly. A discussion of the surface charts for each case focuses on the coastal frontogenesis, the intensification, track, and duration of the cyclone, the anticyclone, and cold-air damming east of the Appalachians.

The evolutions of the lower and upper troposphere are described with analyses provided through a reanalysis procedure described earlier, using the T254, 64-level version of the National Centers for Environmental Prediction (NCEP) Global Forecast System (GFS). The 850-hPa charts display geopotential height contours, isotherms, and analyses of wind speed and direction (Figs. 10.*n*-3). The 850-hPa low track and intensity, the evolution of the lower-tropospheric baroclinic zone, and the development of low-level wind maxima, or jet streaks, are highlighted in the accompanying discussions.

The upper troposphere is depicted by a map sequence that displays 500-hPa geopotential height contours, isotachs of 400-hPa wind speeds, and the locations of cyclonic absolute vorticity maxima at 500 hPa (Figs. 10.*n*-4). The wind analyses at 400 hPa were selected to illustrate the evolution of polar jet streaks that influence the development of the storms, recognizing the possibility that maximum wind speeds may occur between standard pressure levels. The 250-hPa analyses of the geopotential heights and isotachs are presented to represent a more complete picture of the polar jet streaks and the evolution of subtropical jet streaks prior to and during the evolution of the storm (Figs. 10.*n*-5).

These fields are examined to address how upper-level troughs and jet streaks evolve in space and time and how these features influence the surface cyclogenesis and associated precipitation patterns. A general overview of the large-scale circulation patterns across the United States, Canada, and surrounding oceans prior to cyclogenesis and a discussion of ENSO and the NAO are presented to identify any recognizable patterns that precede or accompany the major snowstorms.

Various aspects of the troughs and their flanking ridges, most notably the cyclonic vorticity and vorticity advection, amplitude, wavelength, trough axis orientation, and the development of a closed circulation at 500 hPa, are highlighted in relation to the surface cyclogenesis and the development of precipitation. Changes in the strength and location of upper-level wind maxima are also discussed. The shortening wavelength of the trough–ridge system and the increasing magnitudes of jet streaks, which have been discussed by Palmén and Newton (1969) and noted by Mullen (1983) and Uccellini et al. (1984) as being associated with nonlinear processes influential in the development of cyclones, are also reviewed. The associations between diffluent flow, negatively tilted troughs, and rapidly deepening cyclones are covered, since these factors point to the importance of along-stream variations in the upper-level height (mass) and wind fields in the development of the storm systems. The discussions of upper-level features also include mention of the trough and confluence zones over southeastern Canada and the northeastern United States, and their associated jet streaks.

Twelve-hourly Geostationary Operational Environmental Satellite (GOES-East) infrared imagery, corresponding closely to the times of the surface and upper-air analyses, are presented for each case since 1978 to depict the evolution of cloud systems during cyclogenesis (Figs. 10.*n*-6).

Coauthor Louis Uccellini near his home in Bethpage, Long Island, New York, during the 18–20 Mar 1956 storm.

1. 18–20 March 1956

a. General remarks

- The first snowstorm of the 30-case sample did not affect a very large area, but the heaviest snow occurred across the densely populated sections of eastern Pennsylvania, New Jersey, southeastern New York, and southern New England. Local newspaper accounts indicate that this storm was poorly predicted, with local forecasts only calling for flurries the day the snow began in New York City. As a result, many travelers became stranded on the roads. An early account of the storm can be found in Mook and Norquest (1956). This system followed a more widespread snowstorm on 15–16 March that produced similar snow amounts in the interior of New York and New England (see chapter 11.1), ending a period of relatively snow-free winters across the Northeast urban corridor since 1950.
- Several regions had snow accumulations exceeding 10 in. (25 cm): southeastern Pennsylvania, New Jersey (except the extreme south), southeastern New York, Connecticut, Rhode Island, Massachusetts (except northwest), and southeastern New Hampshire.
- Several regions also had snow accumulations exceeding 20 in. (50 cm): scattered areas of northern New Jersey, southeastern New York, Connecticut, and Massachusetts.

TABLE 10.1-1. Snowfall amounts for urban and selected locations for 18–20 Mar 1956. Also included is the Northeast Snowfall Impact scale (NESIS; see volume I, chapter 8.4).

NESIS = 2.23 (category 1)	
Urban center snowfall amounts	
Washington, D.C.–National Airport	1.7 in. (4 cm)
Baltimore, MD	5.5 in. (14 cm)
Philadelphia, PA	8.7 in. (22 cm)
New York, NY–Central Park	13.5 in. (34 cm)
Boston, MA	13.3 in. (34 cm)
Other selected snowfall amounts	
Babylon, NY	25.6 in. (65 cm)
Newark, NJ	18.2 in. (46 cm)
Providence, RI	14.7 in. (37 cm)
Hartford, CT	14.0 in. (36 cm)
Trenton, NJ	12.2 in. (31 cm)

b. Surface analyses

- Cold air for the snowfall was provided by a weak, but intensifying, anticyclone well north of New York, which developed behind the major storm of 15–16 March. Sea level pressures at the center of this anticyclone rose from 1021 hPa at 0300 UTC 18 March to 1033 hPa by 1500 UTC 19 March.
- Cold-air damming was observed at 1530 UTC 18 March, as evidenced by an inverted sea level pressure ridge from southeastern New York to northern Virginia. The damming occurred at the same time that heavier snowfall was beginning to develop over the middle Atlantic states. Warm frontogenesis is analyzed along the mid-Atlantic coast.
- The primary low pressure center tracked southeastward from the northern plains states to the Ohio Valley on 16–17 March in the wake of the intense storm that had advanced northeastward along the East Coast on 15–16 March (Fig. 11.1-5; see chapter 11.1). The cyclone was weak, as indicated by the small sea level pressure gradients in the vicinity of the low center, and was associated with only scattered light precipitation.
- A secondary cyclone formed over Virginia during the early afternoon of 18 March and deepened slowly. The secondary development was not especially pronounced, even though the primary cyclone dissipated over West Virginia in only about 6 h. Thundershowers formed across Virginia and North Carolina, however, and steady snowfall developed from Maryland to New York City, as the cyclone crossed the Virginia coast and moved over the Atlantic Ocean.
- Heavy snowfall commenced late on 18 March and continued into 19 March from Pennsylvania and New Jersey into southern New England, as the storm drifted slowly northeastward and gradually deepened to 1000 hPa. The central sea level pressure dropped slowly, only 8 hPa over the 30-h period ending at 1500 UTC 19 March.
- Northeasterly winds increased on 19 March as the sea level pressure gradient north of the cyclone center increased.
- Snowfall gradually ended across southern New England late on 19 March as the cyclone continued drifting slowly east-northeastward at an average speed of 10 m s^{-1}.

c. 850-hPa analyses

- Northwesterly flow dominated the northeastern United States prior to this storm, with the 0°C isotherm located over the southeastern United States on 17 March.
- The 850-hPa low center, like its sea level counterpart, did not intensify rapidly. The only deepening occurred east of the New Jersey coast between 0300 and 1500 UTC 19 March, when the geopotential height minimum fell 60 m. This 12-h period of intensification occurred following the commencement of sea level deepening.
- The region of maximum temperature gradient, including the 0°C isotherm, was initially displaced several hundred kilometers from the 850-hPa low center on 17–18 March. The 0°C isotherm became nearly collocated with the cyclone center by 0300 UTC 19 March as the sea level cyclone deepened east of Maryland.
- A pattern of enhanced 850-hPa temperature advec-

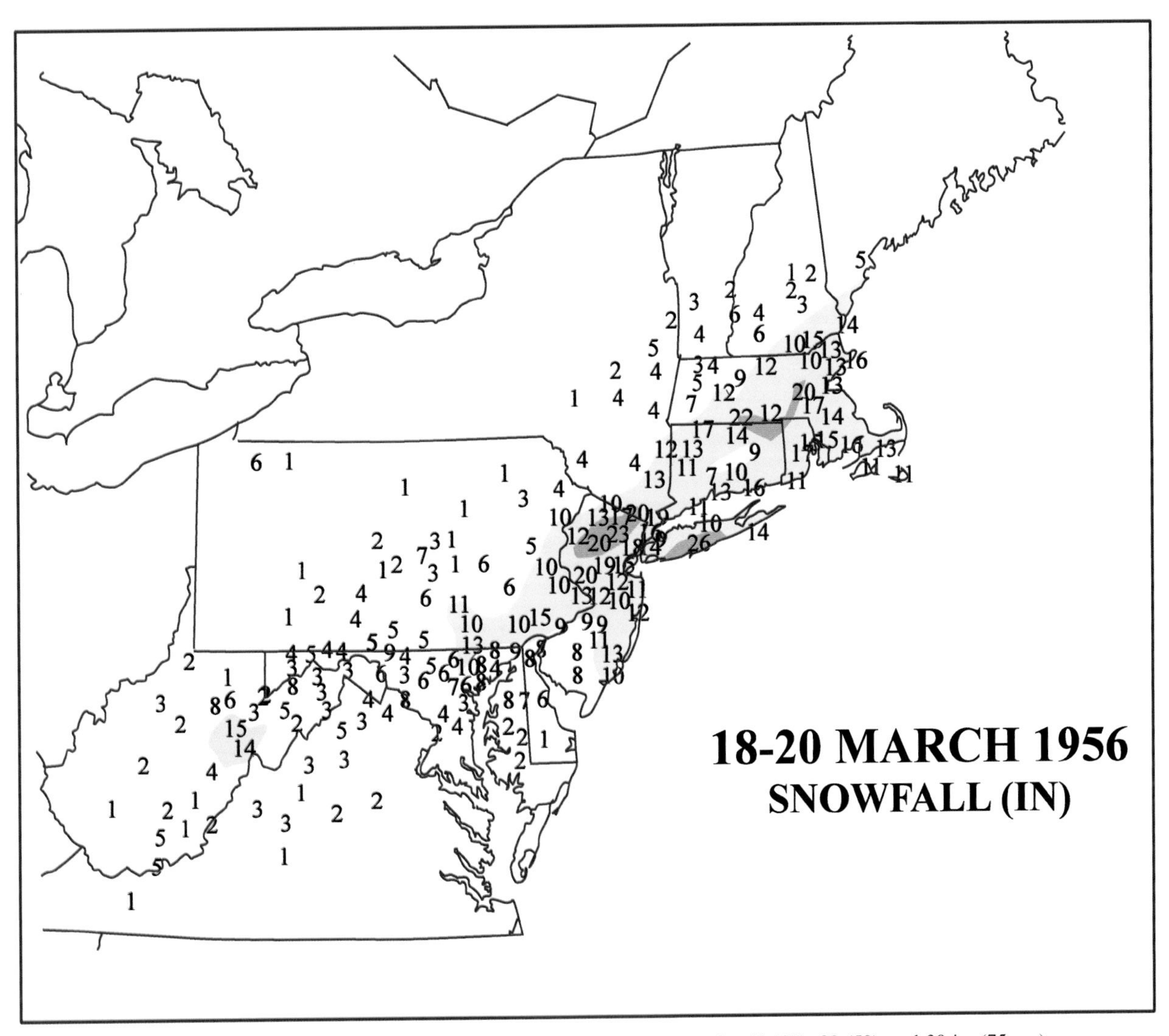

FIG. 10.1-1. Snowfall (in.) for 18–20 Mar 1956. Shading contours are for 10 (25), 20 (50), and 30 in. (75 cm).

tions was established by 0300 UTC 18 March, as the sea level cyclone propagated through the Ohio Valley. At this time, strong cold-air advection can be inferred over the eastern plains states and the middle Mississippi Valley. By 1500 UTC 18 March, vigorous cold-air advections west of the 850-hPa low center and warm-air advections east of the low were quite evident. An S-shaped isotherm pattern resulted near the East Coast late on 18 March and early on 19 March.

- Southerly 850-hPa winds over North Carolina and Virginia increased to 15–20 m s^{-1} at 1500 UTC 18 March, as snowfall intensified across the middle Atlantic states and convection developed in Virginia and North Carolina. By 0300 and 1500 UTC 19 March, a 20–25 m s^{-1} low-level jet (LLJ) developed north of the 850-hPa low center. This easterly LLJ augmented the moisture transport toward the region of heavy snowfall in New Jersey, New York City, and points north.
- As the 850-hPa winds increased along the East Coast between 0300 and 1500 UTC 19 March, there is evidence of cross-contour flow, indicating an important ageostrophic component as the LLJ developed to the north and east of the 850-hPa circulation.

d. 500-hPa geopotential height and 400-hPa wind analyses

- The precyclogenetic environment over the United States and Canada before 19 March was dominated by a region of low geopotential heights over Quebec and eastern Canada, and a ridge over the extreme western United States and western Canada. A pronounced ribbon of large geopotential height gradients extended southeastward from western Canada through the Plains states to the southeastern United States.

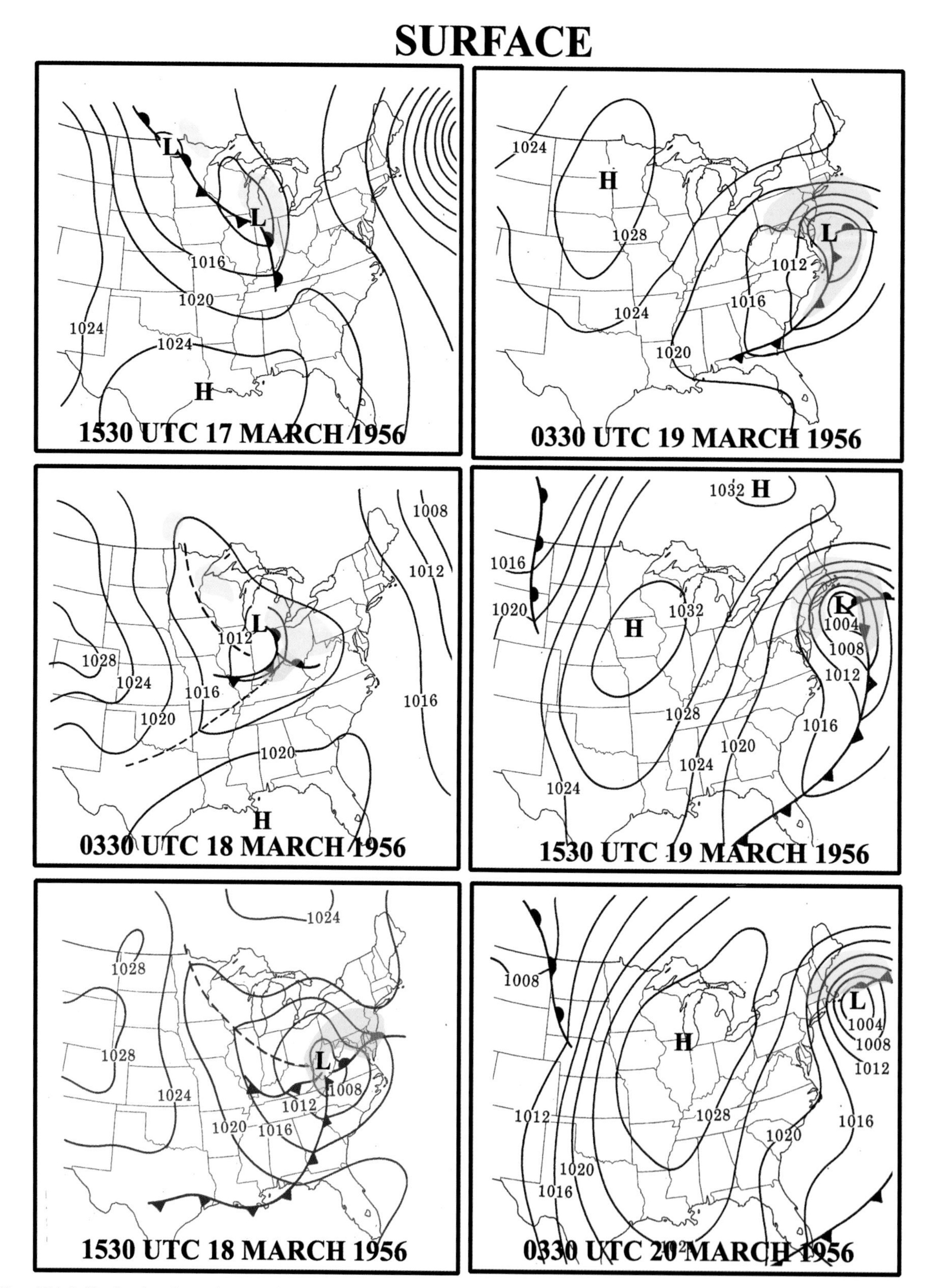

FIG. 10.1-2. Twelve-hourly surface weather analyses for 1530 UTC 17 Mar–0330 UTC 20 Mar 1956. Maps include surface high and low pressure centers and fronts. Shading indicates blue, snow; violet, mixed precipitation; and green, rain. Solid lines are isobars (4-hPa intervals), and dashed lines represent axes of surface troughs not considered to be fronts.

850 hPa HEIGHTS, WINDS, TEMPERATURE

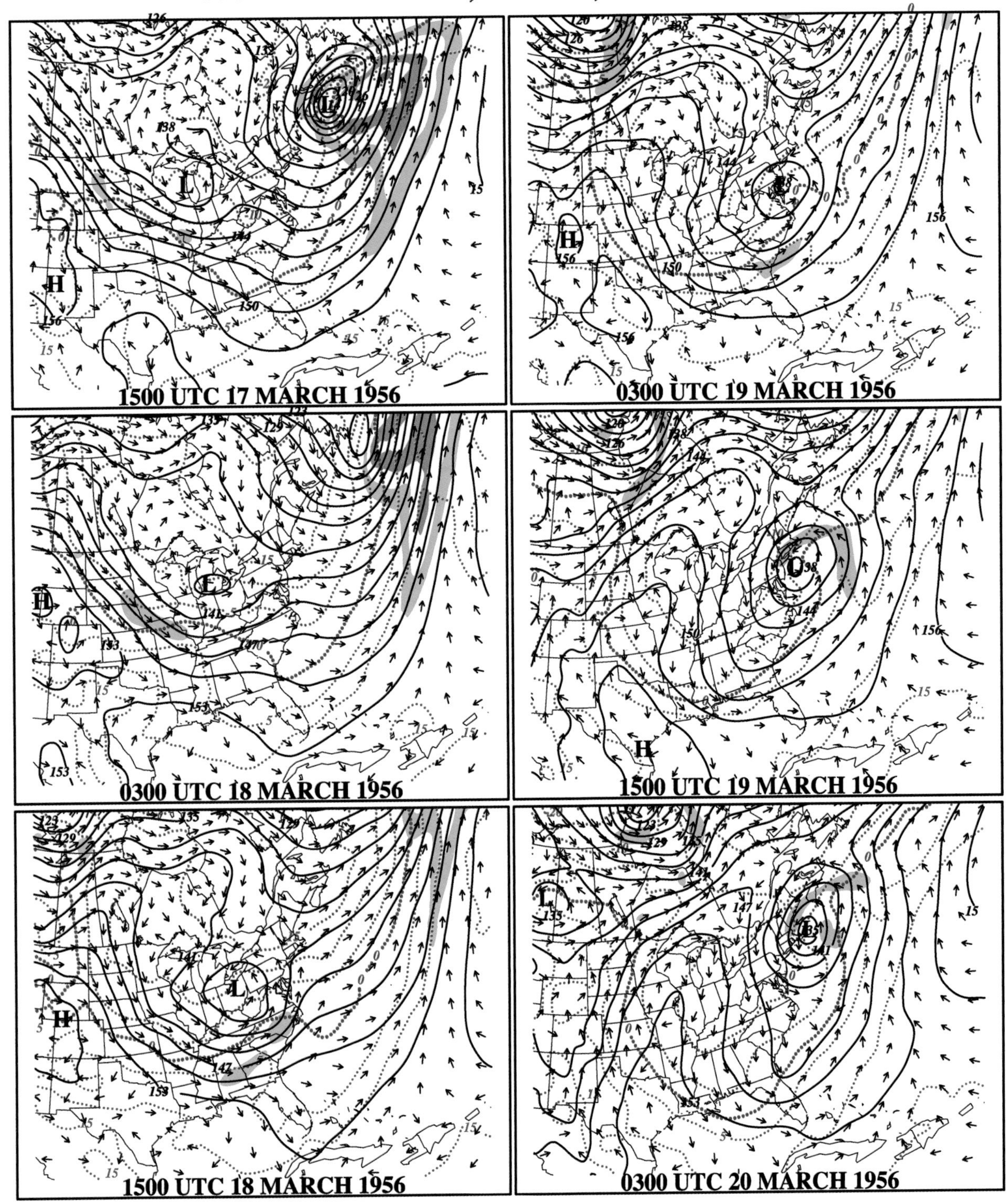

FIG. 10.1-3. Twelve-hourly 850-hPa analyses for 1500 UTC 17 Mar–0300 UTC 20 Mar 1956. Analyses include contours of geopotential height (solid, at 30-m intervals; 156 = 1560 m), isotherms (dotted, °C, in 5°C intervals; blue, 0° and less; red, 5° and greater), and intervals of wind speed greater than 20 m s^{-1} (at 5 m s^{-1} intervals; alternating blue/white shading; red shading, 40 m s^{-1}).

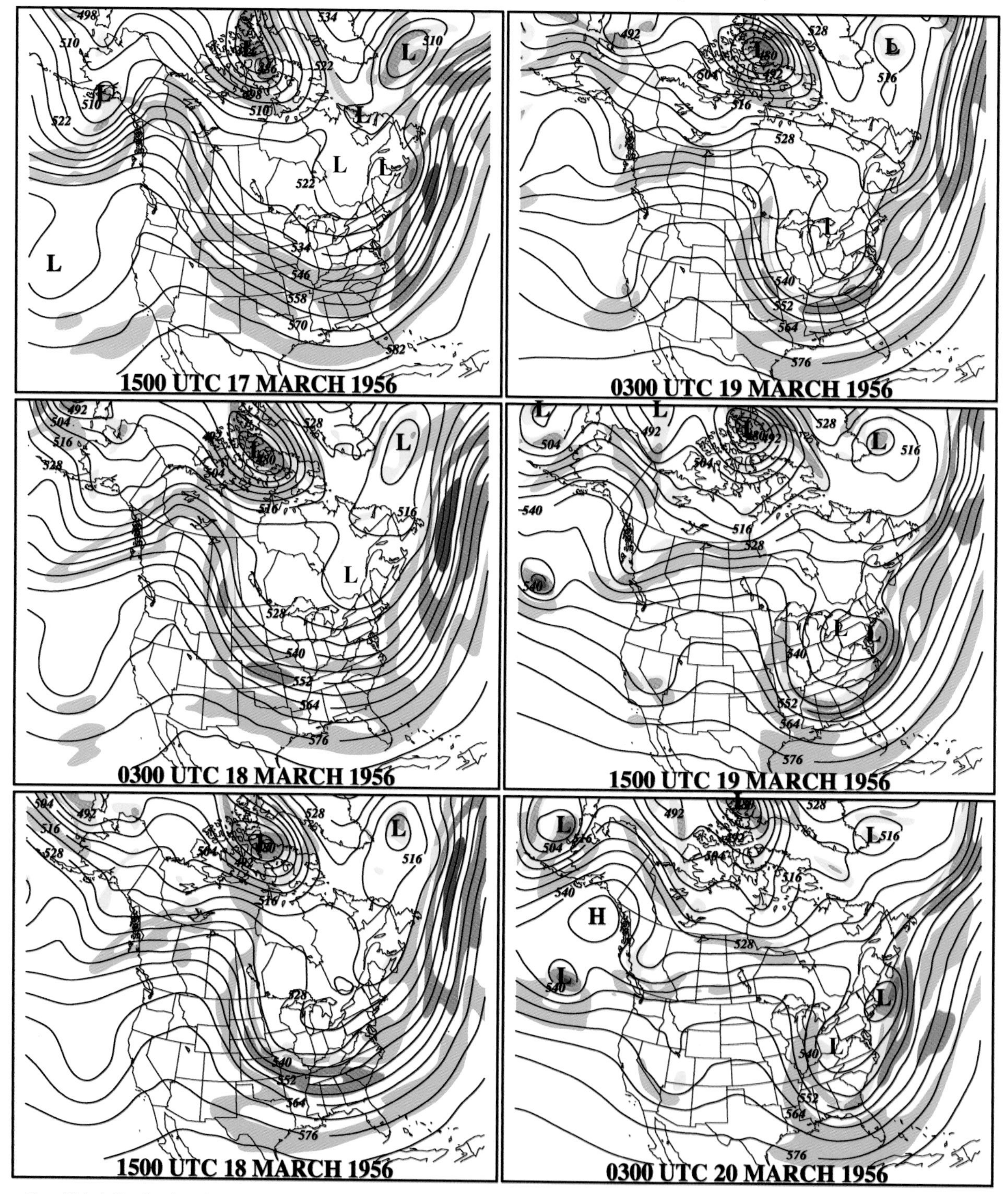

FIG. 10.1-4. Twelve-hourly analyses of 500-hPa geopotential height and upper-level wind fields for 1500 UTC 17 Mar–0300 UTC 20 Mar 1956. Analyses include locations of geopotential height maxima (H) or minima (L), contours of geopotential height (solid, at 60-m intervals; 522 = 5220 m), locations of 500-hPa absolute vorticity maxima (yellow/orange/brown areas beginning at $16 \times 10^{-5}\ s^{-1}$; intervals of $4 \times 10^{-5}\ s^{-1}$), and 400-hPa wind speeds exceeding 30 m s^{-1} (at 10 m s^{-1} intervals; alternate blue/white shading).

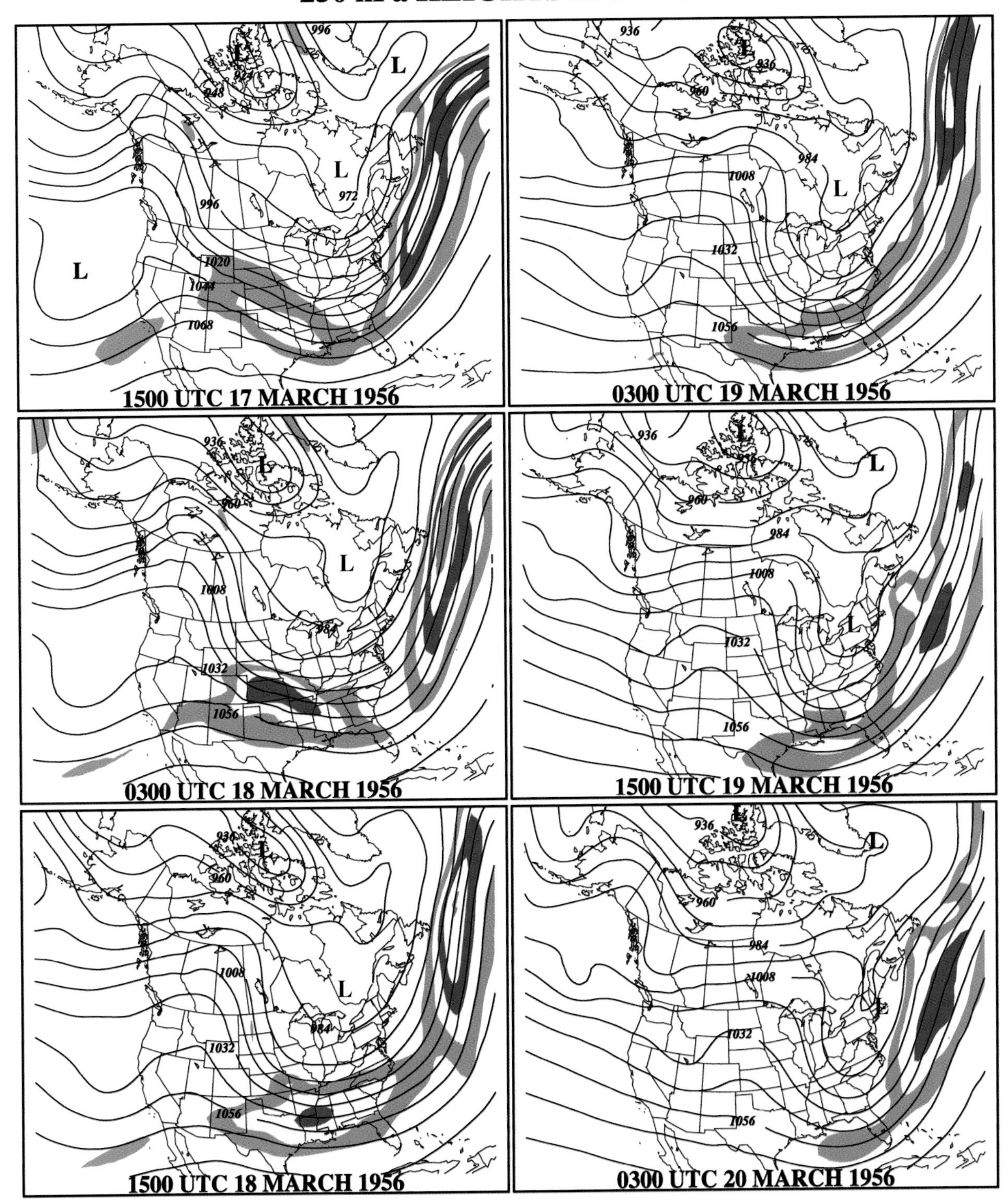

FIG. 10.1-5. Twelve-hourly analyses of 250-hPa geopotential height for 1500 UTC 17 Mar–0300 UTC 20 Mar 1956 (solid at 120-m intervals; 1032 = 10 320 m), with wind speeds exceeding 50 m s^{-1} (at 10 m s^{-1} intervals; alternate blue/white shading).

- A confluent pattern can be discerned in the geopotential height field from the central Great Lakes and Midwest, eastward across West Virginia to New York, especially at 1500 UTC 18 March. By 0300 UTC 19 March, this region had shifted to eastern New England eastward over the Atlantic. Rising sea level pressures in association with an anticyclone over eastern Canada were located beneath this confluence zone.
- The 500-hPa trough associated with the sea level cyclone initially appeared as a slight perturbation in the northwesterly flow just north of North Dakota at 1500 UTC 17 March. The trough propagated southeastward and then eastward in the following 24–36 h, reaching the Appalachian Mountains as the sea level cyclone neared the East Coast at 0300 UTC 19 March. During this period, the axis of the trough rotated from a northeast–southwest tilt to a north–south orientation.
- A region of cyclonic absolute vorticity at 500 hPa was associated with the southeastward-propagating trough. The vorticity maximum tracked from the Ohio Valley to the middle Atlantic coast between 1500 UTC 18 March and 1500 UTC 19 March. The developing surface low and outbreak of heavy snow were located downwind of the propagating maximum in a region of cyclonic vorticity advection.
- By 1500 UTC 19 March, a center of low 500-hPa geopotential heights (denoted by an L) developed over New Jersey, as the trough appeared to rotate northeastward about a preexisting 500-hPa low center near Lake Erie. In the following 12 h, the new low center deepened 60 m and evolved into a separate vortex off the New England coast. The development of this vortex coincided with a period in which the sea level cyclone did not appear to deepen, although it had deepened earlier.
- The 500-hPa trough exhibited a large increase in amplitude, especially at 1500 UTC 18 March and 0300 UTC 19 March. The amplification occurred as heavy snow developed across the middle Atlantic states and southern New England, and as the sea level cyclone deepened, although at only a modest rate.
- The increase of amplitude was especially apparent because of the increase in 500-hPa heights downwind of the trough axis, forming an amplifying ridge. The formation of the ridge at 1500 UTC 18 March and 0300 UTC 19 March occurred in conjunction with the development of strong warm-air advection at 850 hPa.
- By 0300 UTC 19 March, a wind maximum at 400 hPa extended off the coast of the northeastern United States, within confluent flow south of the trough over eastern Canada. The sea level anticyclone, low-level northwesterly flow, and cold-air advection were located beneath the entrance region of this jet streak, especially at 1500 UTC 18 March.
- Although gaps in wind coverage were common during the 1950s, two distinctly separate jet systems were analyzed from the central to the southern United States throughout the course of this event.
- At 400 hPa, these jets appear to merge into one jet streak with a well-defined exit region off the mid-Atlantic coast by 0300 and 1500 UTC 19 March. The surface low, 850-hPa LLJ, and heavy snowfall all appear to develop within the exit region during this period.

e. 250-hPa geopotential height and wind analyses

- A subtropical jet (STJ) was apparent at 250 hPa across the southern United States. Wind speeds within this jet increased between 1500 UTC 17 March and 0300 UTC 18 March, with maximum velocities of 70 m s^{-1} at 0300 and 1500 UTC 18 March, prior to cyclogenesis along the East Coast.
- In this case, the STJ appears to remain south and east as the polar jet analyzed at 400 hPa appears to pass beneath it and propagate northeastward with the developing cyclone.

Fans leave the racetrack at Bowie, Maryland, 15 Feb 1958 (photo courtesy of Kevin Ambrose, *Blizzards and Snowstorms of Washington, D.C.*, Historical Enterprises, 1993).

2. 14–17 February 1958

a. General remarks

- The "Blizzard of '58" was a very widespread storm that produced snow accumulations in excess of 10 in. (25 cm) from Alabama to Maine. The storm occurred during one of the stormiest and snowiest winters of the late 20th century. Intense cold and high winds persisted after the snow ended, prolonging the severe effects of the storm.
- The following regions had snow accumulations exceeding 10 in. (25 cm): central Virginia, central Maryland, northern Delaware, eastern Pennsylvania, New Jersey, New York (except southwest and northwest), Connecticut, Rhode Island, Massachusetts, Vermont, New Hampshire, and sections of western Maine, West Virginia, western Virginia, and Maryland.
- The following regions had snow accumulations exceeding 20 in. (50 cm): eastern Pennsylvania, western and eastern New York, southern Vermont, eastern Massachusetts, and scattered areas of New Hampshire, Connecticut, New Jersey, and Maryland.

TABLE 10.2-1. Snowfall amounts for urban and selected locations for 14–17 Feb 1958; see Table 10.1-1 for details.

NESIS = 5.98 (category 3)	
Urban center snowfall amounts	
Washington, D.C.–National Airport	14.4 in. (37 cm)
Baltimore, MD	15.5 in. (39 cm)
Philadelphia, PA	13.0 in. (33 cm)
New York, NY–La Guardia Airport	10.1 in. (26 cm)
Boston, MA	19.4 in. (49 cm)
Other selected snowfall amounts	
Callicoon, NY	38.7 in. (98 cm)
Rochester, NY	30.6 in. (78 cm)
Syracuse, NY	26.5 in. (67 cm)
Albany, NY	18.3 in. (46 cm)
New Haven, CT	17.2 in. (44 cm)
Allentown, PA	15.8 in. (40 cm)
Worcester, MA	13.6 in. (35 cm)
Newark, NJ	13.3 in. (34 cm)
Trenton, NJ	13.0 in. (33 cm)

b. Surface analyses

- As one anticyclone moved off the East Coast on 14 February, cold air for the snowfall was reinforced by another large anticyclone (1035 hPa) moving southward from the Canadian plains toward the northern plains states. A lobe of high pressure extended eastward from this anticyclone across Ontario and into southern Quebec.
- Cold-air damming was observed over the southeastern United States on 15 February, as an inverted ridge of sea level pressure developed from North Carolina to Georgia. Coastal frontogenesis occurred immediately off the Southeast coast at the same time and coincided with the development of moderate to heavy precipitation across Georgia and the Carolinas.
- The sea level cyclone followed a path from the western Gulf of Mexico to Georgia, then northeastward up the Atlantic coast. The cyclone center appeared to jump or redevelop northeastward when it reached South Carolina at 1800 UTC 15 February. A new center formed over eastern North Carolina and reached southeastern Virginia by 0000 UTC 16 February. The average propagation rate of the storm was 17 m s^{-1}, but during this 6-h period, the rate increased to nearly 25 m s^{-1}. The forward rate of motion slowed to about 10 m s^{-1} as the cyclone neared southern New England on 16 February.
- A weak, inverted sea level pressure trough extending northward from the cyclone center during 14–15 February was accompanied by light snowfall. This feature reflected an upper-level trough and vorticity maximum that was separate from the upper-level system associated with the cyclone. The inverted sea level trough lost its identity late on 15 February as it became incorporated into the expanding circulation about the deepening cyclone.
- Sea level deepening occurred nearly continuously from 14 to 16 February as the cyclone moved from the western Gulf of Mexico to the New England coast. During this period, the central pressure fell from 1004 to 970 hPa. Deepening rates exceeding -3 hPa (3 h)$^{-1}$ prevailed between 1200 UTC 15 February and 1200 UTC 16 February. Heavy precipitation was observed along the East Coast during this period of rapid intensification.
- Concurrent with the rapid deepening of the sea level cyclone, strong northeasterly winds developed north of the storm center along the middle Atlantic and southern New England coasts. As the cyclone moved off the New England coast on 16–17 February, bitterly cold air followed in its wake. Temperatures fell from $-10°$ to $-20°$C across the northeastern United States on 17–18 February.

c. 850-hPa analyses

- As the cyclone was developing over Texas and Louisiana on 14 February, cold air dominated the lower troposphere over the central and eastern United States, with the 0°C isotherm suppressed over the southern United States. At the same time, strong northwesterly flow behind an intense cyclone east of Maine was associated with 850-hPa temperatures between $-10°$ and $-20°$C in the northeastern states. The northwesterly flow was located beneath a confluent upper-level geopotential height pattern over the northeastern United States.
- The 850-hPa low center formed over Louisiana by 0000 UTC 15 February and deepened during the next 48 h. Maximum deepening occurred as the center moved from Alabama to southeastern Virginia [-90 m (12 h)$^{-1}$] and then to near Long Island [-150 m (12

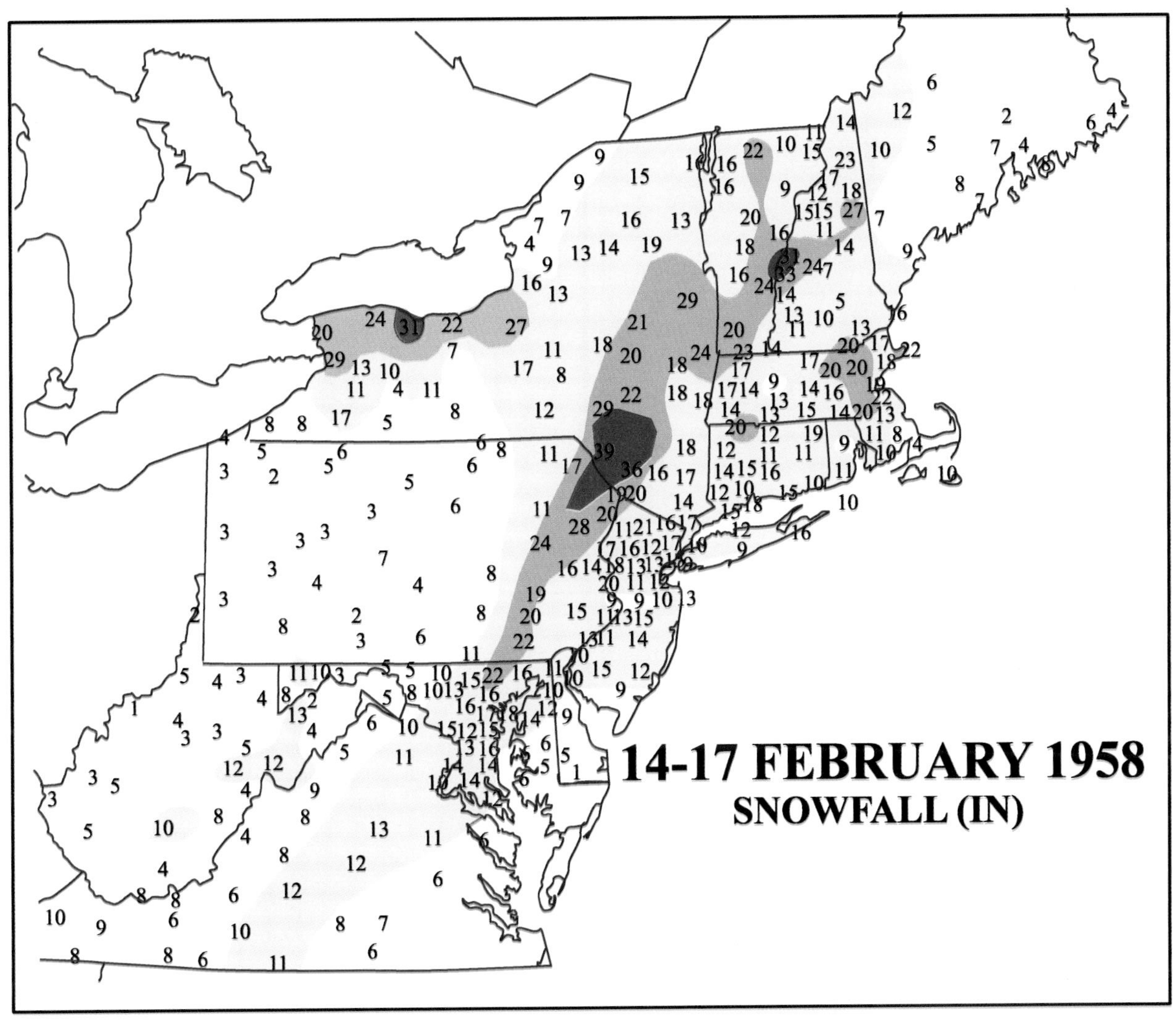

FIG. 10.2-1. Snowfall (in.) for 14–17 Feb 1958. See Fig. 10.1-1 for details.

h)$^{-1}$] between 1200 UTC 15 February and 1200 UTC 16 February. This period coincided with the rapid intensification of the sea level cyclone. The 850-hPa center then deepened only 30 m as it moved to near Boston, Massachusetts, by 0000 UTC 17 February, an indication of the occlusion of the cyclone.

- The 850-hPa low evolved along a band of isotherms over the southern United States, where the shape of the isotherm pattern took on the characteristic "S" shape by 0000 UTC 16 February as the temperature gradient increased between 14 and 15 February. The development of the low center at 0000 UTC 15 February was accompanied by a strong surge of cold-air advection west of the low over Texas. In the next 12 h, this cold-air surge was followed by the development of a pronounced southerly LLJ over the southeastern United States. The development of this jet coincided with an outbreak of moderate to heavy precipitation over the southeastern states and the formation of the coastal front.
- As the LLJ developed in the Southeast and then farther northward along the East Coast between 0000 UTC 15 February and 1200 UTC 16 February, distinct cross-contour flow is noted, indicating a significant ageostrophic component to the wind field as the winds increased along the axis of the LLJ.
- The enhanced cold-air advection west of the low center and warm-air advection east of the low center created an S-shaped isotherm pattern over the eastern United States on 15 February.
- The 0°C isotherm was nearly collocated with the 850-hPa low center until the cyclone reached New England at 0000 UTC 17 February, at which time the center was located in colder air (another indication of occlusion). The 0°C isotherm appeared to represent the rain–snow demarcation rather well for this case, since

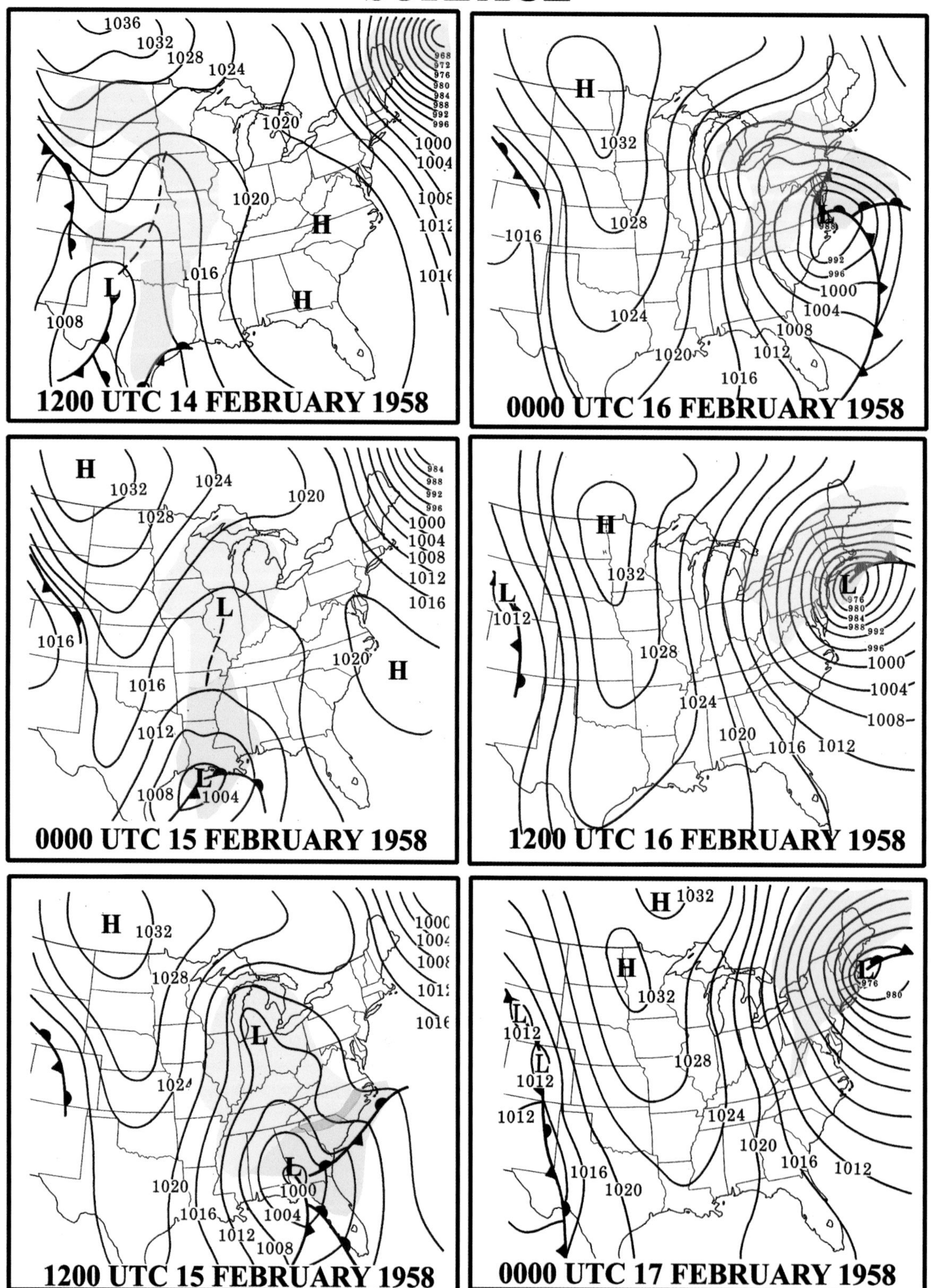

FIG. 10.2-2. Twelve-hourly surface weather analyses for 1200 UTC 14 Feb–0000 UTC 17 Feb 1958. See Fig. 10.1-2 for details.

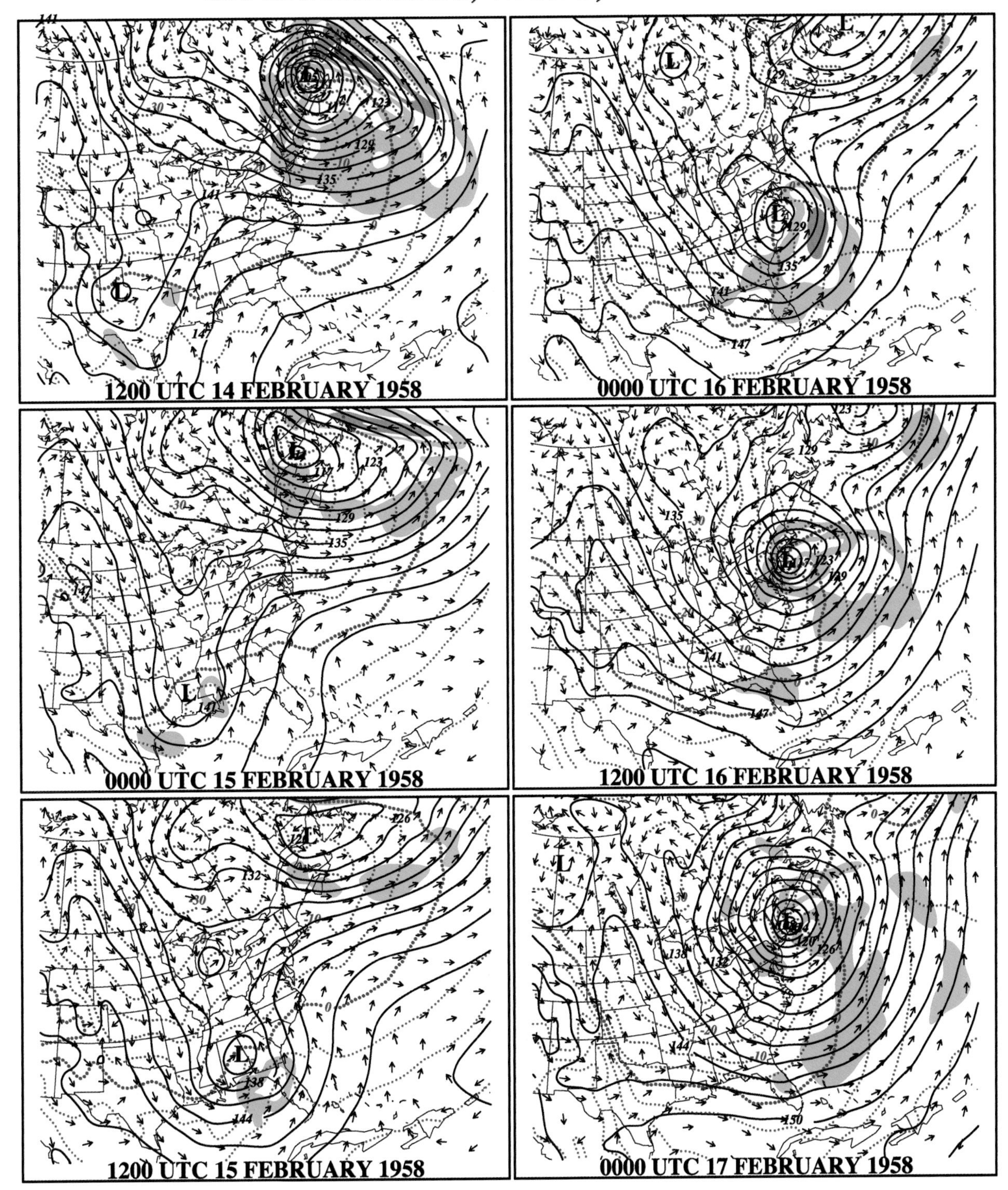

FIG. 10.2-3. Twelve-hourly 850-hPa analyses for 1200 UTC 14 Feb–0000 UTC 17 Feb 1958. See Fig. 10.1-3 for details.

500 hPa HEIGHTS AND VORTICITY, 400 hPa WINDS

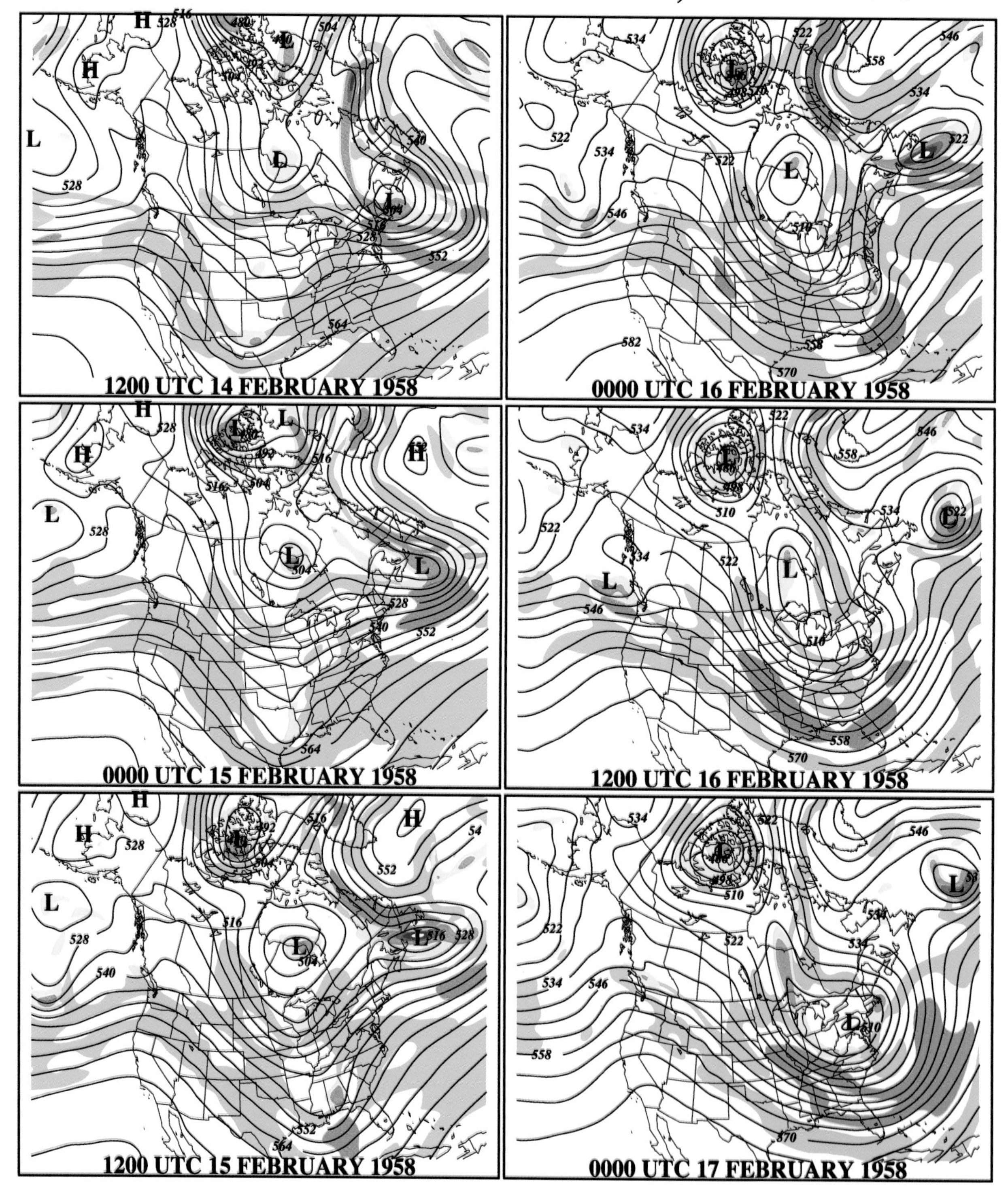

FIG. 10.2-4. Twelve-hourly analyses of 500-hPa geopotential heights and upper-level wind fields for 1200 UTC 14 Feb–0000 UTC 17 Feb 1958. See Fig. 10.1-4 for details.

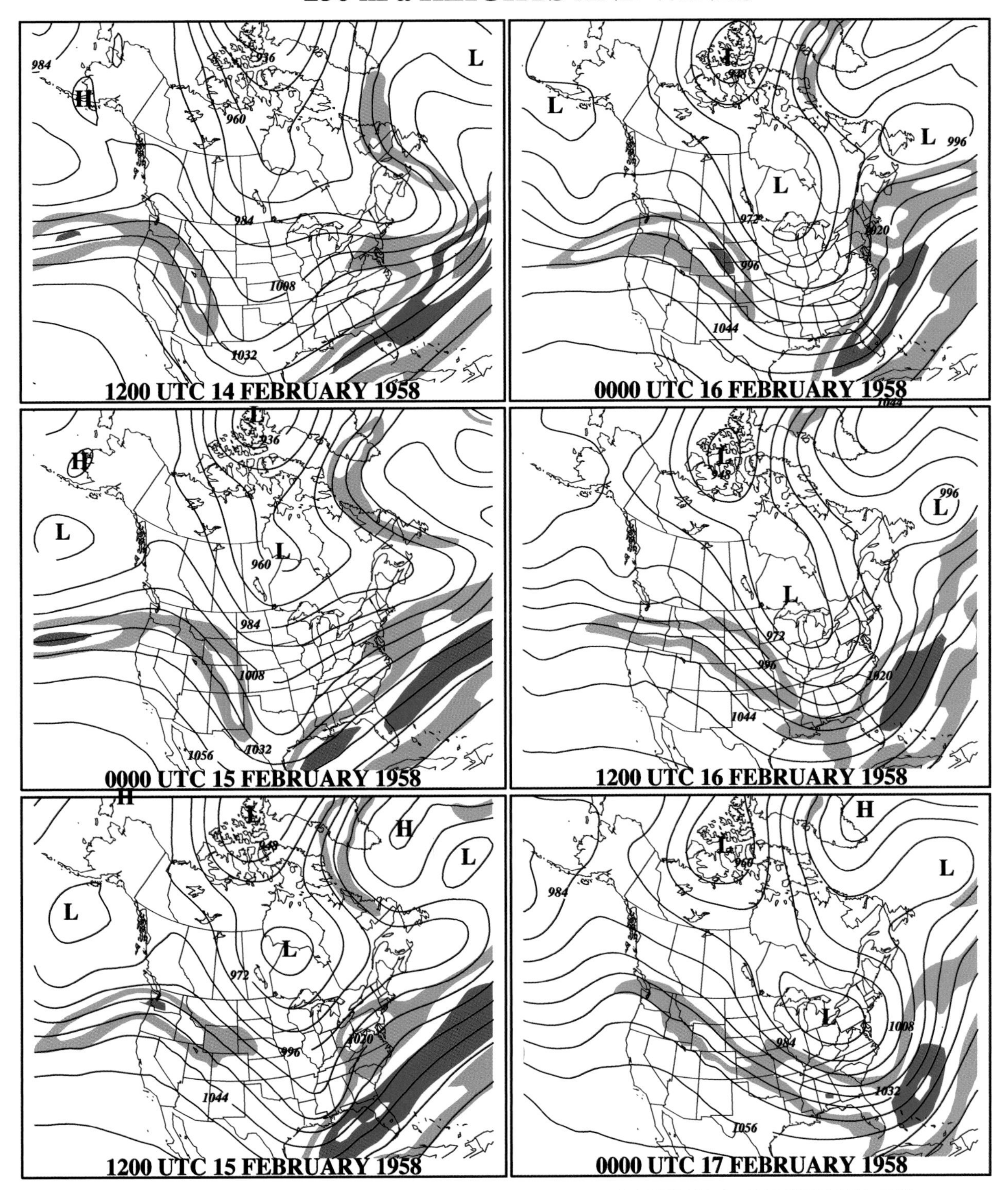

FIG. 10.2-5. Twelve-hourly analyses of 250-hPa geopotential height for 1200 UTC 14 Feb–0000 UTC 17 Feb 1958. See Fig. 10.1-5 for details.

it progressed no farther northward than the Delaware–Maryland coast (cf. with snowfall chart).

- By 1200 UTC 16 February, a complex easterly wind maximum nearing 30 m s^{-1} was established north of the 850-hPa low center as it deepened rapidly and was directed toward the regions experiencing heavy snowfall.

d. 500-hPa geopotential height and 400-hPa wind analyses

- This blizzard and the following March snowstorm occurred during a winter marked by a strong El Niño signal (Fig. 2-19, volume I) and negative NAO. A lengthy period of negative NAO began during the middle of January 1958 and persisted until mid-April without break. The storm developed in a period when the daily NAO values were becoming more negative and then grew less negative once the storm occurred. Possible evidence for the negative NAO may be observed in the 500-hPa charts by the emerging cutoff upper ridge developing near southern Greenland from 14 through 17 February.
- The precyclogenetic period, prior to 15 February, featured an upper-level ridge across the western United States and Canada, an intense cyclonic circulation over extreme southeastern Canada, a separate cyclonic vortex near Hudson Bay, and a closed anticyclonic circulation south of Greenland. The greatest geopotential height gradients were located over the United States, where several jet streams were observed. A trough was propagating from the southwestern United States into the south-central states.
- The surface high pressure ridge and cold temperatures in the northeastern United States were located beneath a distinct region of confluent geopotential heights over the northeastern United States upwind of a deep 500-hPa trough over southeastern Canada on 14–15 February.
- The intensifying sea level cyclone was associated with an upper-level trough that tracked eastward across the southern United States and then rotated northeastward along the east coast on 16 February.
- A separate upper-level trough associated with the sea level inverted trough over the Ohio Valley rotated about the Hudson Bay vortex between 1200 UTC 14 February and 1200 UTC 15 February. It then appeared to merge or phase with the trough moving northeastward along the East Coast at 0000 UTC 16 February.
- The deepening cyclone and heavy precipitation were associated with cyclonic vorticity advection downwind of a vorticity maximum crossing the southern and eastern United States on 14–16 February. The sea level inverted trough over the Ohio Valley on 14–15 February was associated with a separate vorticity maximum that probably merged with the other vorticity field by 0000 UTC 16 February.
- As the trough associated with the cyclone moved northeastward along the Atlantic seaboard and merged with the other trough swinging around the vortex near Hudson Bay, geopotential heights fell rapidly along the East Coast between 1200 UTC 15 February and 0000 UTC 17 February. During this period, the sea level cyclone deepened rapidly and a major shift in the upper-level circulation pattern occurred, as the Hudson Bay vortex became reestablished over the northeastern United States.
- The sea level cyclone underwent its greatest intensification after 1200 UTC 15 February within the diffluent region downwind of the upper-level trough as the trough axis neared the East Coast and changed its orientation from north–south to northwest–southeast (a negative tilt), and as the 500-hPa low center either propagated, or more likely redeveloped over the northeastern United States.
- Increases in the amplitude of the trough and the downstream ridge during cyclogenesis were small relative to the other cases. The half-wavelength between the trough and downstream ridge decreased substantially, especially between 1200 UTC 15 February and 1200 UTC 16 February.
- Serious gaps in wind observation coverage plagued this case, especially over the western United States, but the event featured a variety of jet systems that were captured by the reanalysis and influenced the development of the cyclone.
- A distinct entrance region into a strong polar jet was in evidence at 1200 UTC 14 February and 0000 UTC 15 February across the northeastern United States to the south of an intense vortex at 500 hPa over extreme southeastern Canada. The cold-air outbreak across the eastern United States occurred within the confluent entrance region of this jet.
- A separate jet (which was difficult to analyze because of missing wind reports) was analyzed nonetheless west of and near the upper trough propagating through the south-central states on 14–15 February. Rapid cyclogenesis appeared to commence within the exit region of this jet as the jet reached the base of the deepening upper-level trough along the middle Atlantic coast between 1200 UTC 15 February and 0000 UTC 16 February.

e. 250-hPa geopotential height and wind analyses

- A subtropical jet with velocities exceeding 70–80 m s^{-1} was observed over the southeastern United States through 1200 UTC 15 February, when its axis was located south of the developing surface low.
- By 1200 UTC 15 February, a separate 60 m s^{-1} jet had developed from the eastern Ohio Valley to the Atlantic coast at the crest of an amplifying upper-level

ridge. This jet formed immediately prior to rapid cyclogenesis along the East Coast as the precipitation area spread northeastward with the developing coast low.
- The amplification of the separate jet over the expanding region of precipitation and amplifying ridge appears to strengthen (or reinforce) the entrance region of this jet system over the middle Atlantic states during the beginning phases of the snowstorm.

A U.S. Air Force helicopter assists some of the nearly 800 stranded travelers at the Morgantown exit of the Pennsylvania Turnpike, Mar 1958 (photo courtesy of Temple University Libraries, Urban Archives, Philadelphia).

3. 18–21 March 1958

a. General remarks

- This late winter snowstorm was one of the more unusual cases of the sample, for which elevation played a very significant role in the snowfall distribution. Although snow amounts at the official reporting sites in the five largest cities did not reach 12 in. (30 cm), some of the Washington, Baltimore, Philadelphia, and New York City suburbs received from 14 (35) to more than 20 in. (50 cm). While many local forecasters from Washington to Boston assumed that precipitation would fall primarily as rain, heavy, wet snow with temperatures near or slightly above freezing resulted in tremendous destruction of power lines and trees. In parts of eastern Pennsylvania and northern Maryland, this was the greatest snowfall on record. Further descriptions of the storm can be found in Sanderson and Mason (1958).
- The following regions reported snow accumulations exceeding 10 in. (25 cm): northern Virginia, central Maryland, northern Delaware, eastern Pennsylvania, New Jersey (except near the coast), southeastern New York, western Connecticut, and scattered areas of eastern West Virginia, western Pennsylvania, Rhode Island, east-central Massachusetts, New Hampshire, and Maine.
- The following regions reported snow accumulations exceeding 20 in. (50 cm): sections of north-central Maryland, eastern Pennsylvania, western New Jersey, and scattered areas of northern Virginia, northern Delaware, southeastern New York, Massachusetts, and New Hampshire.

TABLE 10.3-1. Snowfall amounts for urban and selected locations for 18–21 Mar 1958, see Table 10.1-1 for details.

NESIS = 3.92 (category 2)	
Urban center snowfall amounts	
Washington, D.C.–National Airport	4.8 in. (12 cm)
Baltimore, MD	8.4 in. (21 cm)
Philadelphia, PA	11.4 in. (29 cm)
New York, NY–Central Park	11.7 in. (30 cm)
Boston, MA	6.7 in. (17 cm)
Other selected snowfall amounts	
Morgantown, PA	50.0 in. (127 cm)
Mount Airy, MD	33.0 in. (84 cm)
Allentown, PA	20.3 in. (51 cm)
Wilmington, DE	19.0 in. (48 cm)
Worcester, MA	18.8 in. (48 cm)
Trenton, NJ	17.8 in. (45 cm)
Newark, NJ	14.8 in. (38 cm)

b. Surface analyses

- Cold air (or, more appropriately, cool air) for this event was provided by a large anticyclone centered over extreme northern Canada, with a weak ridge of high pressure extending south-southeastward into the northeastern United States. On 19 March, late winter sunshine helped to boost surface temperatures in this air mass to as high as 10°C in Maine. Where clouds and precipitation dominated, however, temperatures generally ranged between 0° and 2°C.
- Weak cold-air damming was observed, especially at 0000 UTC 19 March, when an inverted high pressure ridge extended from New Jersey to South Carolina. At the same time, coastal frontogenesis, which was more pronounced in the wind field than in the temperature field, was occurring off the Carolina coast.
- The most striking characteristic of the sea level cyclone was its slow movement, which averaged only 6 m s^{-1} during the 60-h period shown in the analyses. The cyclone headed northeastward along the Southeast coast at a rate of approximately 10 m s^{-1} on 19 March, but then slowed to 5 m s^{-1} or less on 20–21 March off the New Jersey and southern New England coasts. The low pressure center then drifted ashore over Rhode Island on 21 March. The slow movement of the cyclone resulted in a 3-day period of steady precipitation, with melted totals greater than 10 cm over portions of Maryland and Pennsylvania.
- No redevelopment occurred in this case, at least during the period studied.
- The cyclone was initially rather weak, as indicated by the small pressure gradients surrounding the low center off the Southeast coast on 19 March. It deepened continuously through 0000 UTC 21 March, becoming a rather intense storm on 20 March off the middle Atlantic coast. The greatest deepening occurred between 0000 UTC 20 March and 0000 UTC 21 March, when the central pressure fell from 997 to 976 hPa.
- The heaviest precipitation occurred from late on 19 March through early on 21 March, as the cyclone underwent its greatest intensification. Strong northeasterly winds developed from the middle Atlantic coast to the southern New England coast during the same period.

c. 850-hPa analyses

- A weak 850-hPa anticyclone was located over New England just prior to the storm. Northwesterly flow had prevailed over the northeastern United States on 17–18 March, but was confined to southeastern Canada by 19 March.
- While the 850-hPa low center was located on the Alabama–Georgia border at 0000 UTC 19 March, the remnants of a weakening short-wave trough nearing the East Coast brought a separate region of relatively low geopotential heights, large temperature gradients, and 10 m s^{-1} southeasterly winds to the Carolina coast. These features remained near the Carolina coastline at 1200 UTC 19 March and were related to

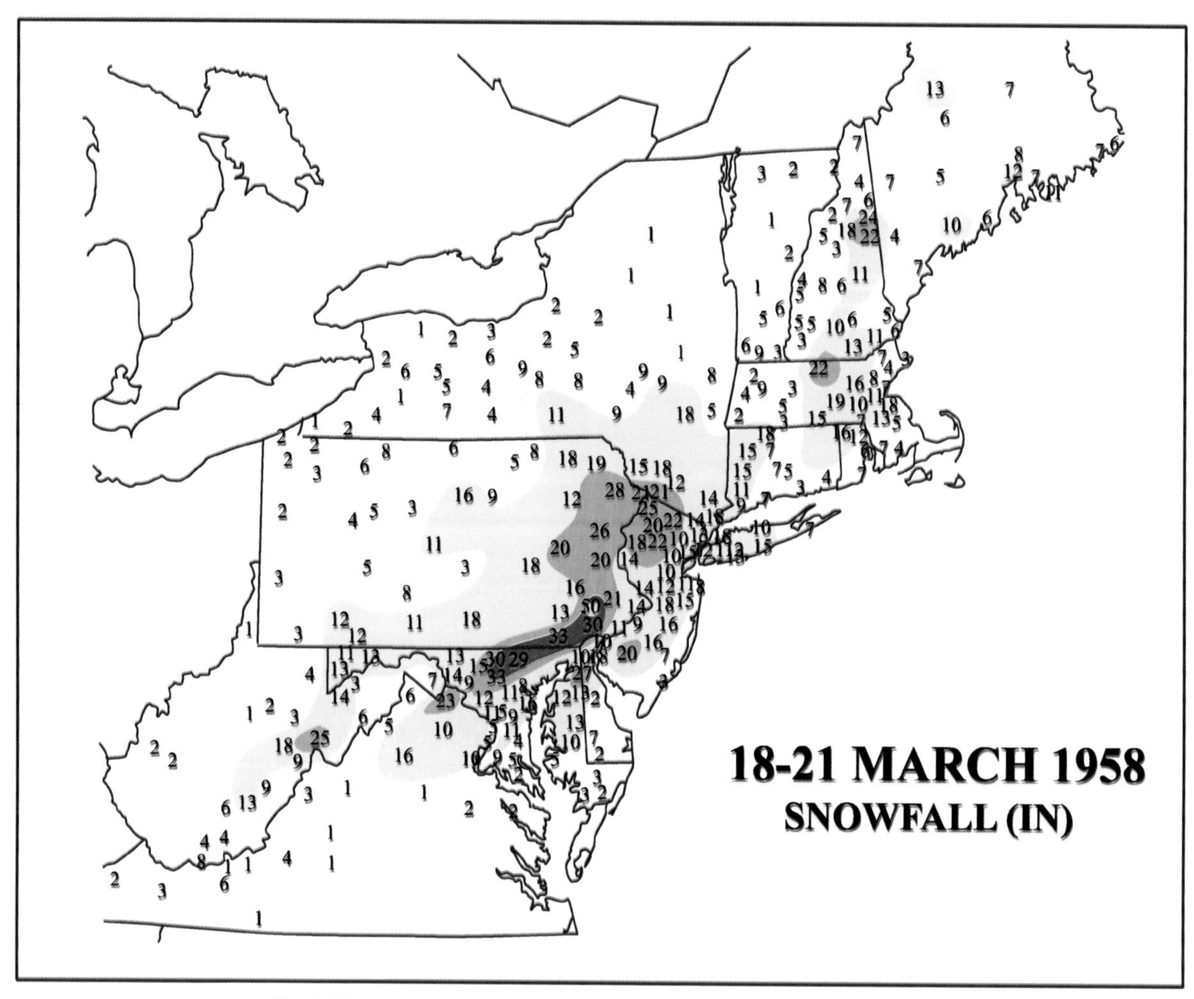

FIG. 10.3-1. Snowfall (in.) for 18–21 Mar 1958. See Fig. 10.1-1 for details.

the light precipitation occurring along the East Coast during this period.

- By 0000 UTC 20 March, a well-defined and strengthening 850-hPa center was in evidence over northeastern North Carolina as the sea level cyclone was beginning to intensify offshore. This center deepened rapidly during the following 24 h as it moved to near Long Island.
- The 850-hPa low developed along a preexisting band of isotherms that extended from the Gulf of Mexico to the North Carolina and Virginia coasts on 19 March. The low center was located slightly to the warm side of the 0°C isotherm through 1200 UTC 20 March, then was located to the cold side while the cyclone was occluding on 21 March. The 0°C isotherm roughly coincided with the demarcation between heavy and light snowfall amounts.
- Strong easterly winds developed north of the low center prior to the period of rapid intensification, with wind speeds greater than 25 m s^{-1} extending toward the middle Atlantic coast by 0000 UTC 20 March with some hint of cross-contour flow at this time.
- By 1200 UTC 20 March and 0000 UTC 21 March, a distinct (low-level jet) LLJ is observed north of the 850-hPa low center with 35 m s^{-1} speeds over New York City at 1200 UTC 20 March and 0000 UTC 21 March. The development of heavy precipitation at these times suggests that the easterly LLJ increased the moisture transport into the region of heavy precipitation.

d. 500-hPa geopotential height and 400-hPa wind analyses

- This storm developed during the same El Niño winter and during the same continuous 4-month period of negative North Atlantic Oscillation (NAO) values as did the February 1958 storm. An examination of daily values of the NAO also shows that the storm occurred during a week-and-a-half period in which the negative values were becoming even more negative. The presence of an upper ridge over Greenland at 500 hPa during the development of the storm is consistent with a negative NAO and the ridge appears to merge with

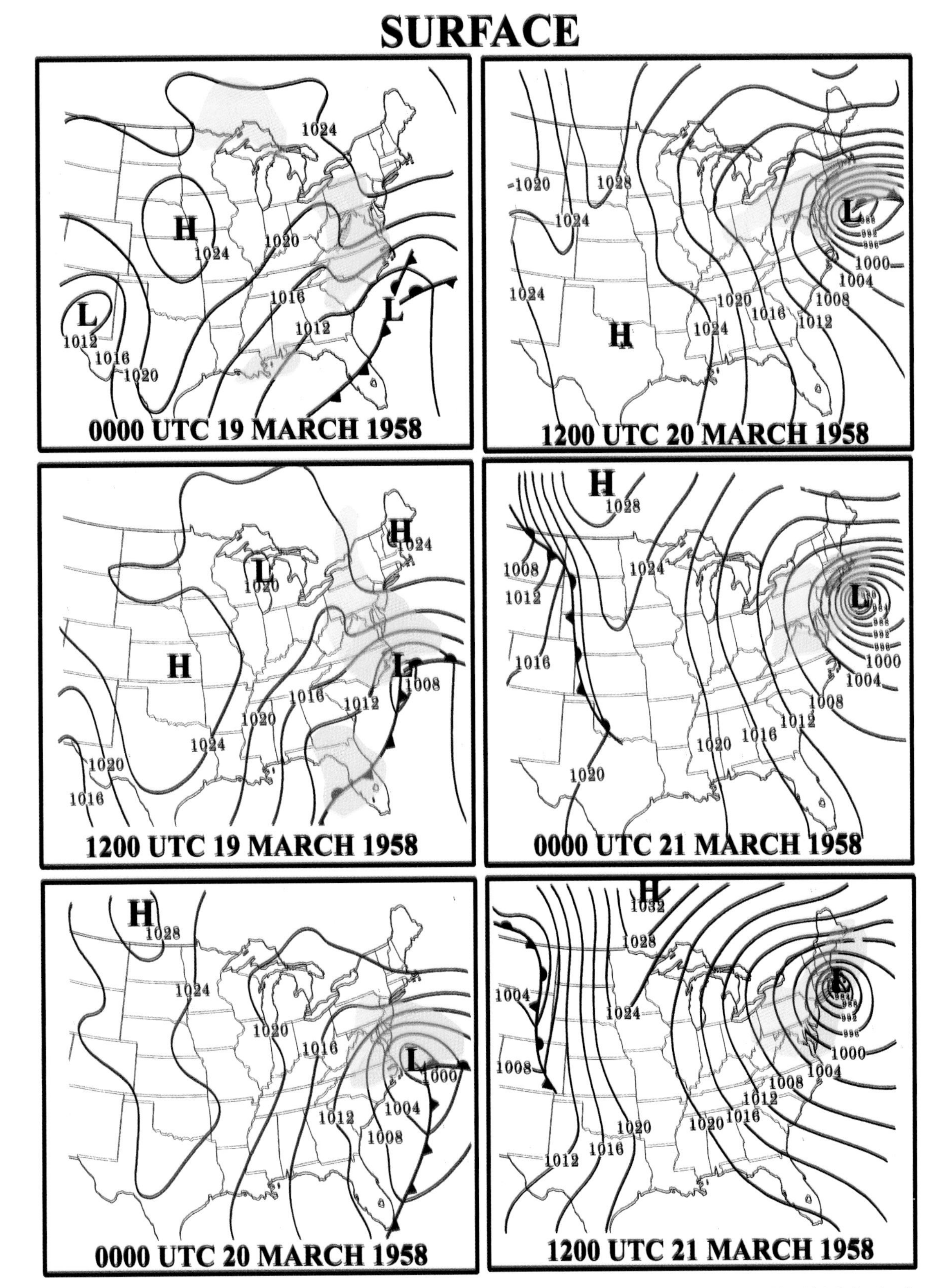

FIG. 10.3-2. Twelve-hourly surface weather analyses for 0000 UTC 19 Mar–1200 UTC 21 Mar 1958. See Fig. 10.1-2 for details.

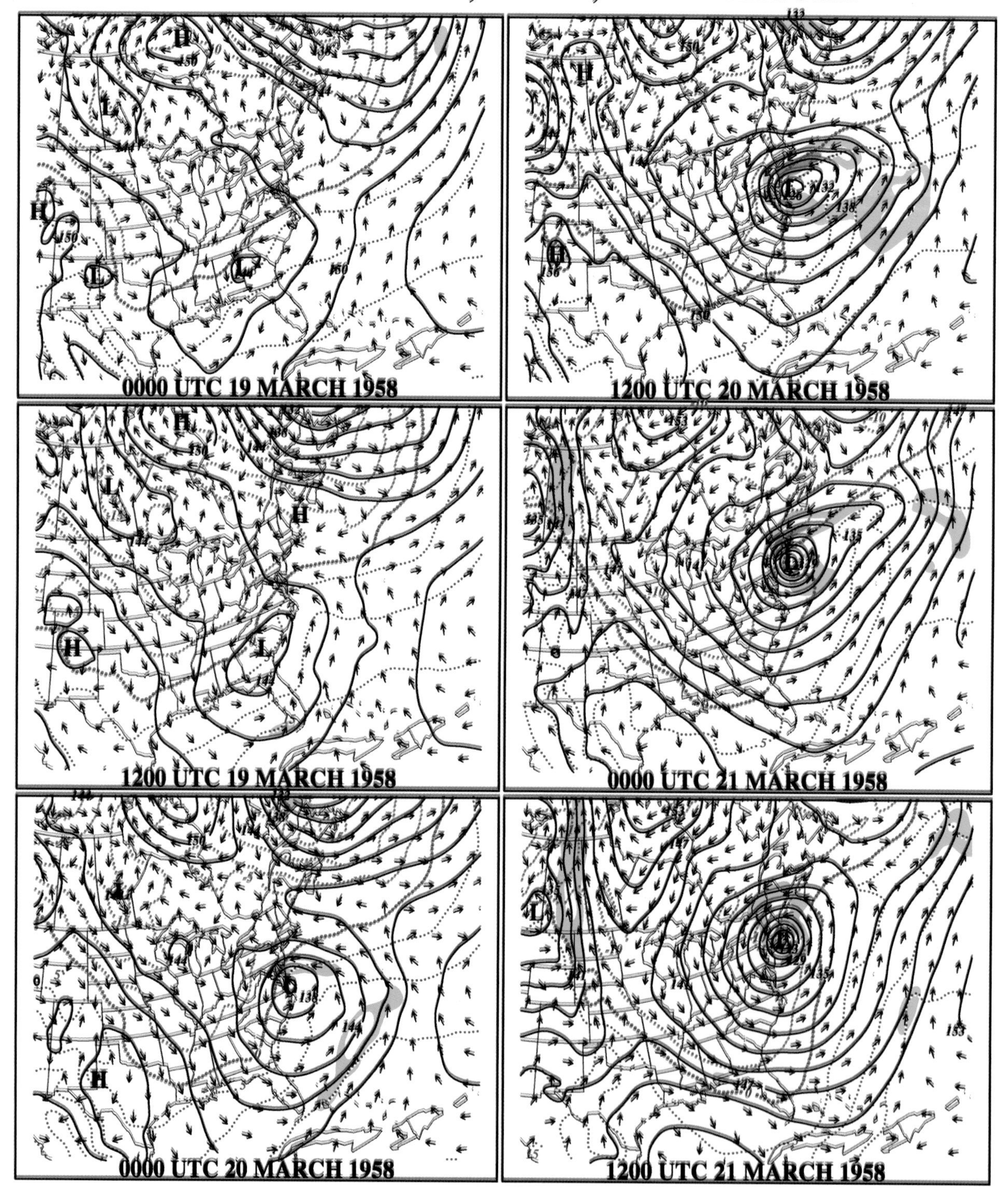

FIG. 10.3-3. Twelve-hourly 850-hPa analyses for 0000 UTC 19 Mar–1200 UTC 21 Mar 1958. See Fig. 10.1-3 for details.

FIG. 10.3-4. Twelve-hourly analyses of 500-hPa geopotential heights and upper-level wind fields for 0000 UTC 19 Mar–1200 UTC 21 Mar 1958. See Fig. 10.1-4 for details.

FIG. 10.3-5. Twelve-hourly analyses of 250-hPa geopotential height for 0000 UTC 19 Mar–1200 UTC 21 Mar 1958. See Fig. 10.1-5 for details.

an amplifying ridge near Hudson Bay on 19 March, with the main axis of the ridge appearing to move toward the west as the southern trough undercuts the area of high geopotential heights.

- The precyclogenetic environment was characterized by a trough that bisected the central and eastern United States, with a pronounced ridge over Hudson Bay and a deep cyclonic vortex east of Labrador. The ridge and vortex over eastern Canada moved little during the 60-h study period, indicating that the flow regime was generally blocked, consistent with the slow movement of this storm.
- The surface anticyclone over northern Canada and its associated high pressure ridge extending southeastward into New England on 19–20 March were associated with the stationary 500-hPa anticyclone near Hudson Bay and the 500-hPa cyclonic vortex over extreme eastern Canada.
- A 500-hPa trough of small amplitude and short wavelength weakened as it reached the East Coast. By 0000 UTC 19 March, this feature was nearly unidentifiable in the height field, but was still associated with a weak sea level cyclone off South Carolina and light precipitation along the East Coast.
- A broad trough with a weakly defined axis was located over the central and eastern half of the United States at 0000 UTC 19 March. During the following 48 h, this trough amplified slowly and evolved into a large closed cyclonic circulation over the northeastern United States. The vortex first developed over the upper Midwest at 1200 UTC 19 March, then redeveloped near the East Coast by 0000 UTC 21 March as the sea level cyclone deepened rapidly. The vortex intensified considerably by 1200 UTC 21 March, as it became nearly vertical with the surface system.
- As the broad trough amplified on 19–20 March, several poorly defined, shorter-wavelength trough features rotated about the 500-hPa low center located over the upper Midwest. While these features were difficult to distinguish in the geopotential height field, they were revealed by the height tendencies (not shown) and were associated with distinct cyclonic vorticity maxima. Two of these short-wave features propagated toward the East Coast during 1200 UTC 19 and 20 March before merging near Long Island at 0000 UTC 21 March. The rapid development of the sea level cyclone near the East Coast on 20 March occurred downwind of these propagating features, in a region of cyclonic vorticity advection.
- Rapid sea level deepening commenced by 0000 UTC 20 March as a region of upper-level diffluence, which had developed downwind of the trough axis, crossed the coastline.
- The half-wavelength between the trough axis and the downstream ridge decreased rapidly during the period of most rapid cyclogenesis, especially between 1200 UTC 19 March and 1200 UTC 20 March.
- The upper-level wind field in this reanalysis remained poorly defined through 0000 UTC 20 March, owing perhaps to a lack of data near the core of the jets. Nevertheless, a distinct polar jet is evident near the base of the trough by 0000 UTC 20 March, which then slowly propagates northeast along the East Coast by 0000 UTC 21 March. The rapid surface cyclogenesis occurs within the exit region of this jet.
- A separate jet streak is evident in the confluent zone just off the middle Atlantic coast at 1200 UTC 19 March and shifts northward with the confluent zone during the remainder of the event, merging with a separate confluent zone associated with a cutoff low in eastern Canada. The entrance region of this jet remains collocated with the expanding precipitation shield between 1200 UTC 19 March and 0000 UTC 20 March.

e. 250-hPa geopotential height and wind analyses

- The evolution of the 250-hPa height field illustrates a gradual reinforcement of a blocked pattern over North America. This occurred as the trough over the central United States slowly deepened between 0000 UTC 19 March and 1200 UTC 21 March as a ridge remained over southern Canada and the downstream trough slowly moved eastward and deepened during the same period. This evolution is consistent with the slow movement of the surface cyclone along the East Coast on 20–21 March.
- A subtropical jet, whose axis crossed Florida, appeared to either weaken or move off the coast when maximum wind speeds dropped from 70 to 50 m s^{-1} between 0000 UTC 19 March and 0000 UTC 20 March. Meanwhile, a polar jet streak across the middle Atlantic states contained maximum wind speeds of 50 m s^{-1} near 300 hPa.
- A polar jet streak extending along the middle Atlantic states slowly intensified even up to the 250-hPa level on 20–21 March, with maximum wind speeds exceeding 50 m s^{-1} coincident with the period of maximum surface cyclogenesis.

The Supreme Court, Washington D.C., on 4 Mar 1960 (photo courtesy of Kevin Ambrose, *Blizzards and Snowstorms of Washington, D.C.*, Historical Enterprises, 1993).

Coauthor Louis Uccellini near his home in Bethpage, Long Island, New York, after the Mar 1960 storm.

4. 2–5 March 1960

a. General remarks

- This widespread late winter storm was especially fierce in eastern New England. Severe blizzard conditions occurred in eastern Massachusetts, as snow accumulations exceeded 20–30 in. (50–75 cm) and near-hurricane force winds battered the coast.
- The following regions reported snow accumulations exceeding 10 in. (25 cm): eastern West Virginia, northern and western Virginia, much of Pennsylvania, parts of Maryland and Delaware, southeastern and western New York, Connecticut, Rhode Island, Massachusetts, southern Vermont, New Hampshire, and Maine.
- The following regions reported snow accumulations exceeding 20 in. (50 cm): eastern Massachusetts, Rhode Island, and scattered areas of northern New Jersey, southeastern New York, Connecticut, and New Hampshire.

TABLE 10.4-1. Snowfall amounts for urban and selected locations for 2–5 Mar 1960; see Table 10.1-1 for details.

NESIS = 7.63 (category 4)	
Urban center snowfall amounts	
Washington, D.C.–National Airport	7.9 in. (20 cm)
Baltimore, MD	10.4 in. (26 cm)
Philadelphia, PA	8.4 in. (21 cm)
New York, NY–La Guardia Airport	15.5 in. (39 cm)
Boston, MA	19.8 in. (50 cm)
Other selected snowfall amounts	
Nantucket, MA	31.3 in. (79 cm)
Milton, MA–Blue Hill Observatory	30.3 in. (77 cm)
Worcester, MA	22.1 in. (56 cm)
Scranton, PA	18.0 in. (46 cm)
Providence, RI	17.7 in. (45 cm)
Roanoke, VA	17.4 in. (44 cm)
Allentown, PA	14.4 in. (37 cm)
Hartford, CT	13.0 in. (33 cm)
Pittsburgh, PA	12.8 in. (33 cm)

b. Surface analyses

- A large, cold anticyclone preceded the storm. The axis of highest sea level pressure extended from the Great Lakes states to the middle and south Atlantic coasts on 2–3 March. A pronounced inverted high pressure ridge along the south Atlantic coast indicated cold-air damming, especially on 2 March. Coastal frontogenesis occurred off the Southeast coast on 2 March, prior to the development of a secondary sea level cyclone.
- The primary low pressure center moved from the Gulf coast to the Ohio Valley on 2–3 March. Moderate to heavy precipitation fell across the Tennessee and Ohio Valleys in association with this system. The low deepened 8 hPa in the 12 h ending at 0000 UTC 3 March, prior to the development of the secondary low, then deepened slowly and erratically before finally dissipating during 3 March. The primary low was evident for 15 h after the onset of secondary cyclogenesis.
- The secondary low pressure center formed near the South Carolina coast between 0000 and 0600 UTC 3 March. The development occurred along a pronounced coastal front off the Southeast coast, as cold air wedged as far south as northern Florida.
- The secondary low center deepened 45 hPa in the 30-h period from 0600 UTC 3 March to 1200 UTC 4 March, with the central pressure falling to 960 hPa off New England. Heavy snowfall and high winds were widespread across the middle Atlantic states and southern New England during this phase of rapid development.
- The movement of the storm center from the Carolinas to off the New Jersey coast on 3 March was quite rapid, averaging greater than 20 m s^{-1}. It subsequently slowed to less than 10 m s^{-1} off the New England coast on 4 March, as the lowest pressure was reached. The deceleration of the cyclone near New England was attended by strong northeasterly flow, prolonging the snowfall and enhancing accumulations across southeastern New England.

c. 850-hPa analyses

- Northwesterly flow maintained 850-hPa temperatures of less than −10°C north of Virginia through 1200 UTC 2 March.
- The 850-hPa low center propagated east-northeastward from the southern plains states to the East Coast on 2–4 March. Its deepening rate increased from −60 m (12 h)$^{-1}$ in the 12-h period ending at 0000 UTC 3 March to −90 m (12 h)$^{-1}$ by both 1200 UTC 3 March and 0000 UTC 4 March, and to −150 m (12 h)$^{-1}$ by 1200 UTC 4 March. The increase in the 850-hPa deepening rate coincided with the rapid intensification of the secondary surface low after 0000 UTC 3 March.
- The elongation of the 850-hPa low at 1200 UTC 3 March depicts two centers, with the southeast extension over Virginia reflecting the rapid development of the secondary surface low.
- A concentrated isotherm ribbon moved from the southern United States on 2 March northward to the middle Atlantic states by 3 March. The thermal gradient intensified by 1200 UTC 3 March, as the secondary surface low developed. The 0°C isotherm was generally collocated with the 850-hPa low center at 0000 and 1200 UTC 3 March.
- Strong cold and warm-air advection accompanied the 850-hPa low center at 0000 UTC 3 March, producing an S-shaped isotherm pattern over the eastern United States. The development of this pattern coincided with the onset of rapid deepening of both the 850-hPa and

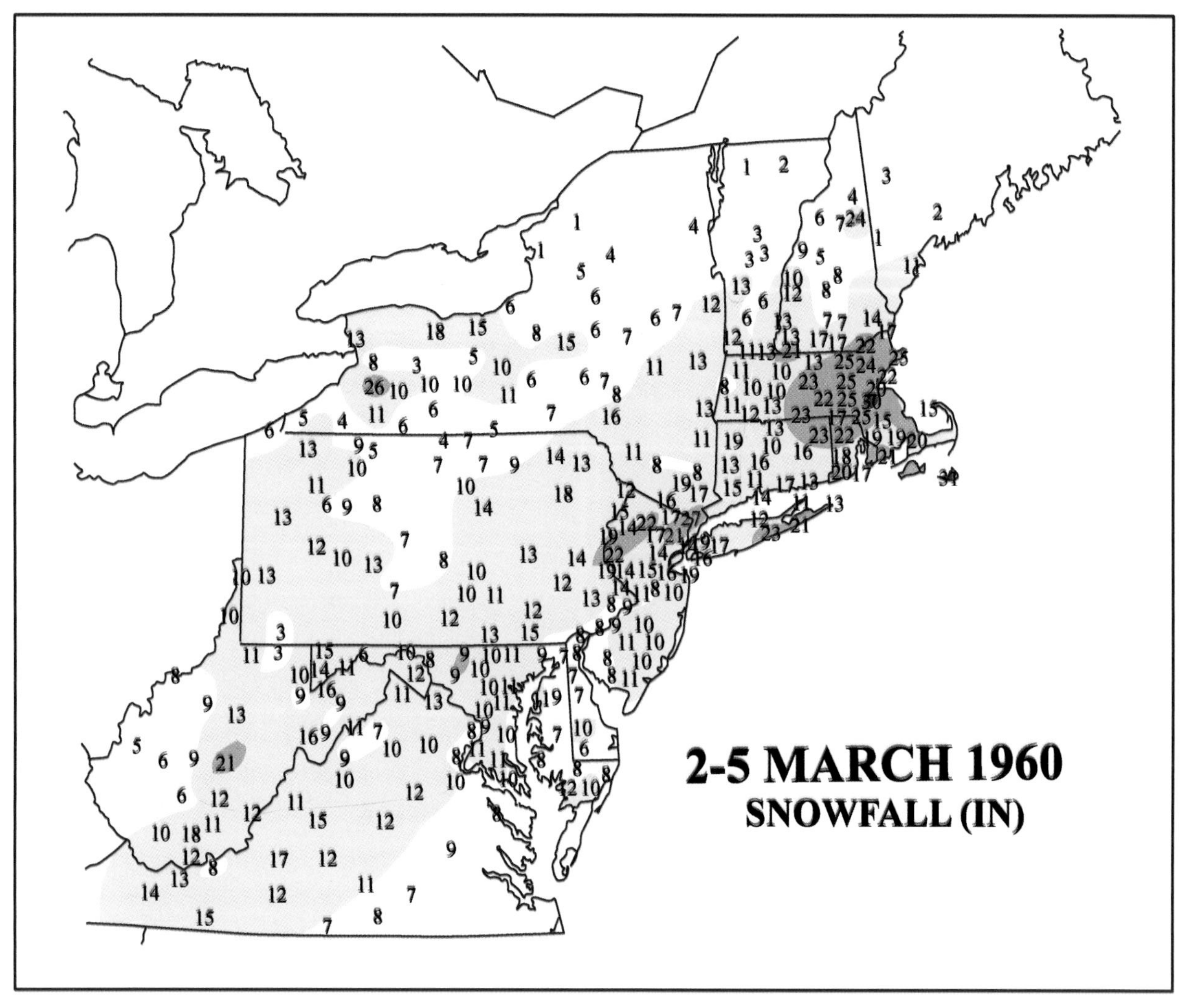

FIG. 10.4-1. Snowfall (in.) for 2–5 Mar 1960. See Fig. 10.1-1 for details.

surface lows. The 850-hPa low was located just west of the apex of the thermal ridge.

- By 0000 UTC 3 March, a southerly LLJ was observed extending from the Louisiana Gulf coast toward Alabama, Georgia, and South Carolina, with wind speeds exceeding 30 m s^{-1}, prior to rapid secondary development. The jet was associated with strong warm-air advection across the southeastern United States with moderate to heavy precipitation extending from Tennessee to southwestern Virginia, immediately downwind of this jet (see corresponding surface charts).
- Wind speeds in the core of the LLJ continued to increase to a magnitude greater than 30 m s^{-1} off the North Carolina coast by 1200 UTC 3 March with a noticeable cross-contour ageostrophic component coincident with the period of rapid secondary cyclogenesis off the East Coast.
- Heavy snow developed along the middle Atlantic coast by 1200 UTC 3 March as the secondary low formed, local geopotential height falls intensified, the temperature gradient increased, and a low-level east-to-southeasterly jet formed north of the 850-hPa low center.
- To the north of the 850-hPa low, an easterly LLJ with speeds exceeding 25 m s^{-1} enhanced moisture transports from New Jersey to southern New England by 0000 UTC 4 March. Winds within the now northeasterly LLJ increased to greater than 35 m s^{-1} across eastern New England by 1200 UTC 4 March, coinciding with heavy snowfall.
- The enhanced temperature gradient combined with increasing low-level winds to intensify the warm-air advection and moisture transport into the developing precipitation area covering the middle Atlantic states.

d. 500-hPa geopotential height and 400-hPa wind analyses

- This major snowstorm developed during a very stormy and snowy March for the middle Atlantic states

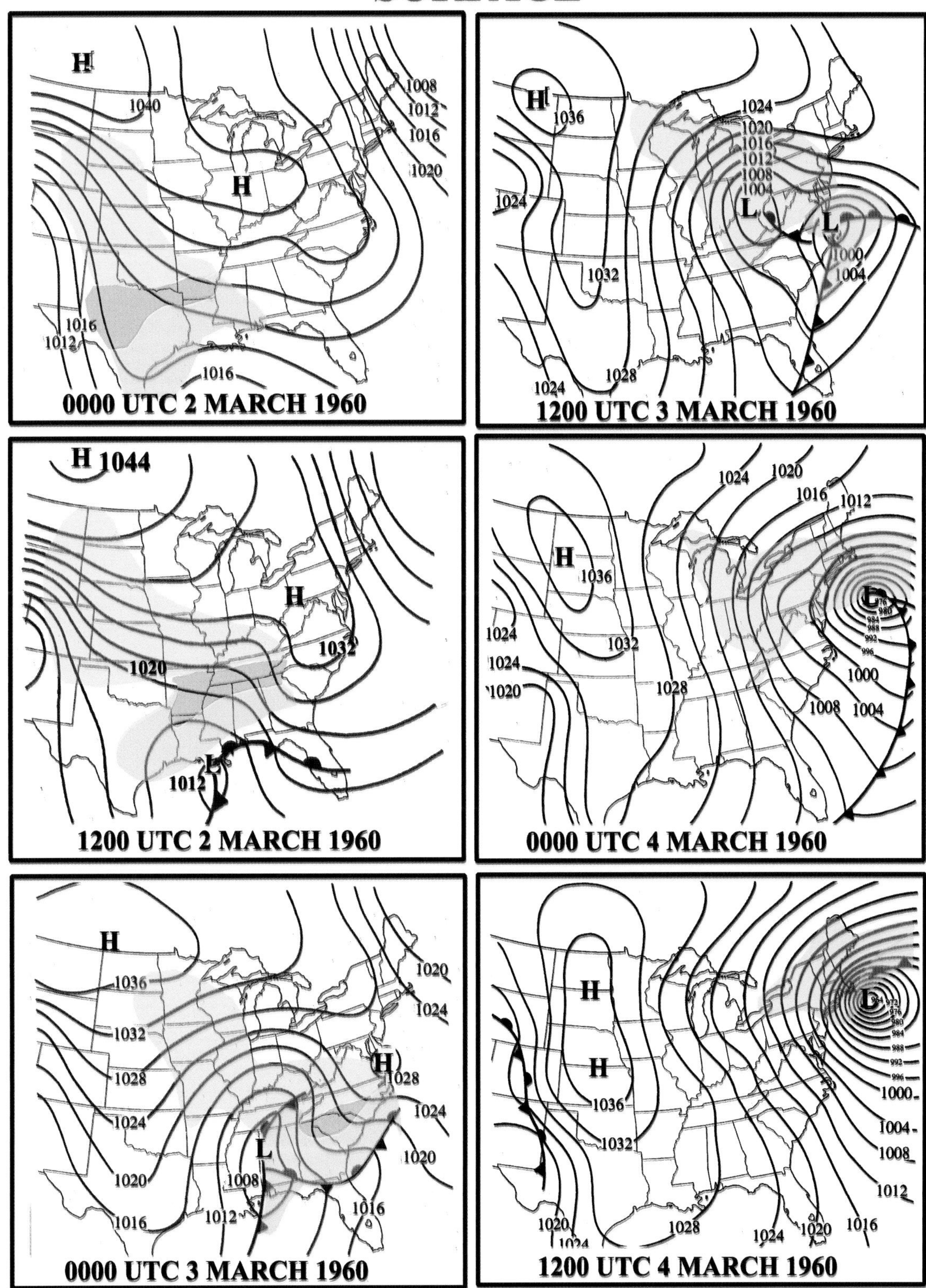

FIG. 10.4-2. Twelve-hourly surface weather analyses for 0000 UTC 2 Mar–1200 UTC 4 Mar 1960. See Fig. 10.1-2 for details.

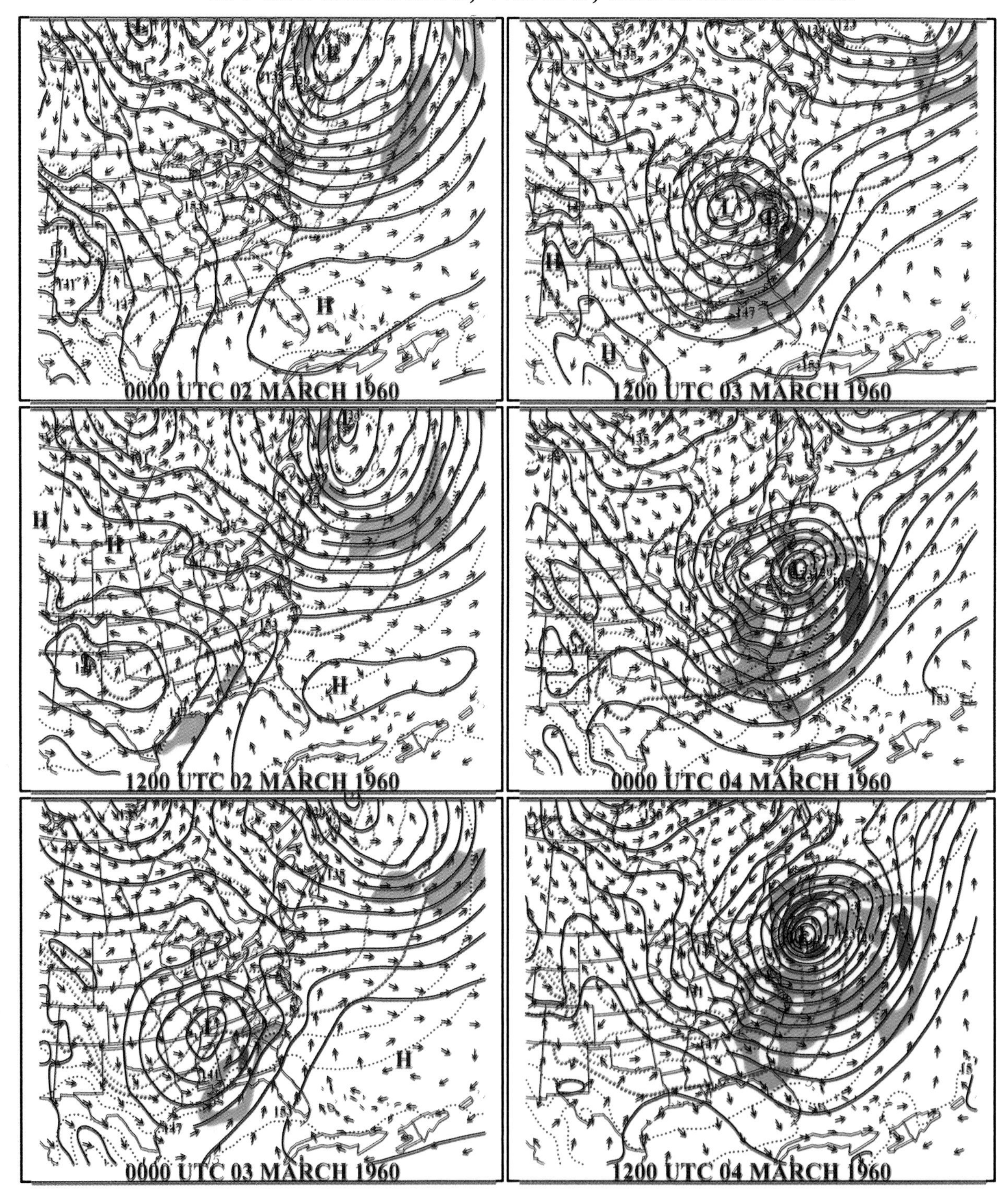

FIG. 10.4-3. Twelve-hourly 850-hPa analyses for 0000 UTC 2 Mar–1200 UTC 4 Mar 1960. See Fig. 10.1-3 for details.

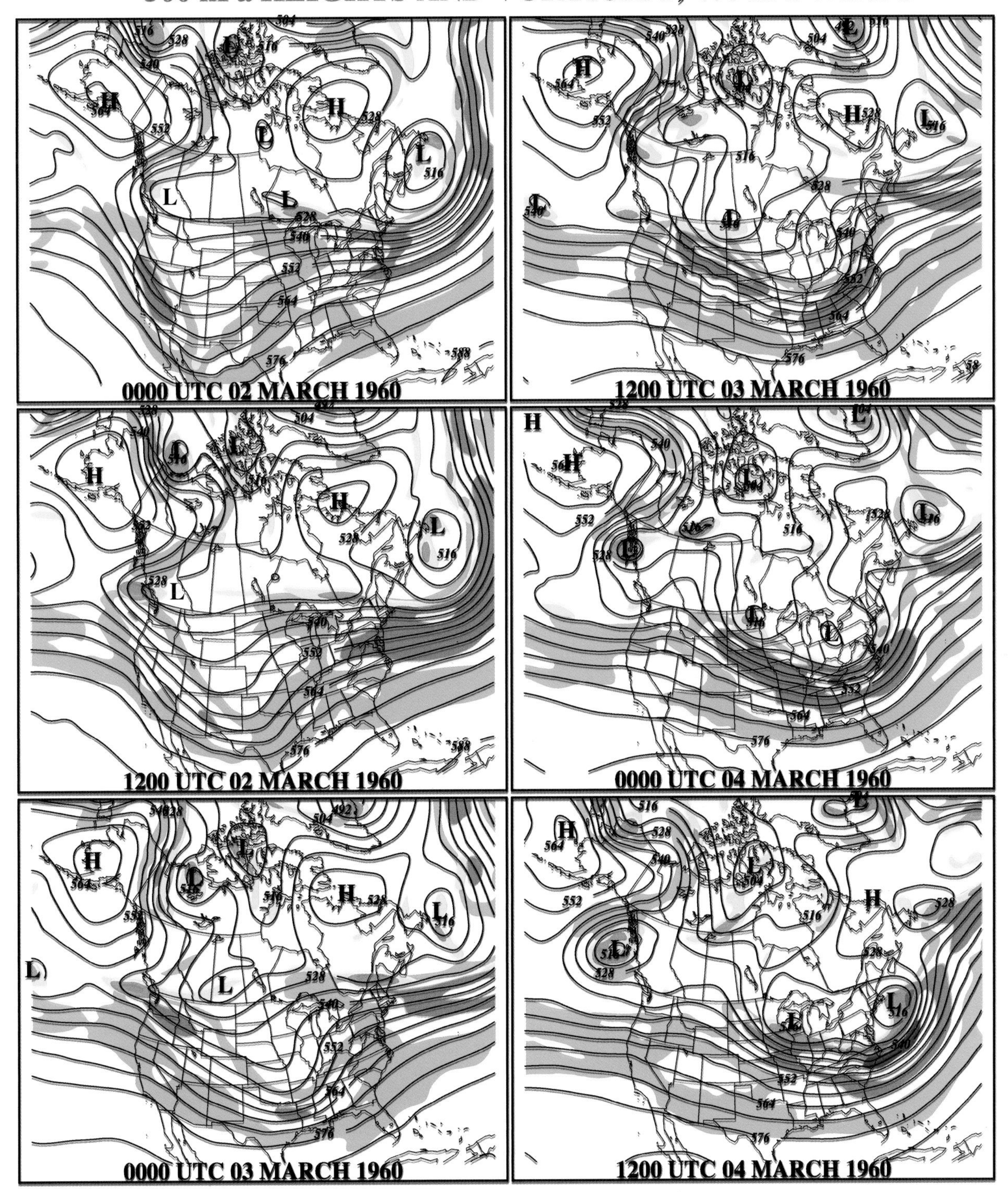

FIG. 10.4-4. Twelve-hourly analyses of 500-hPa geopotential heights and upper-level wind fields for 0000 UTC 2 Mar–1200 UTC 4 Mar 1960. See Fig. 10.1-4 for details.

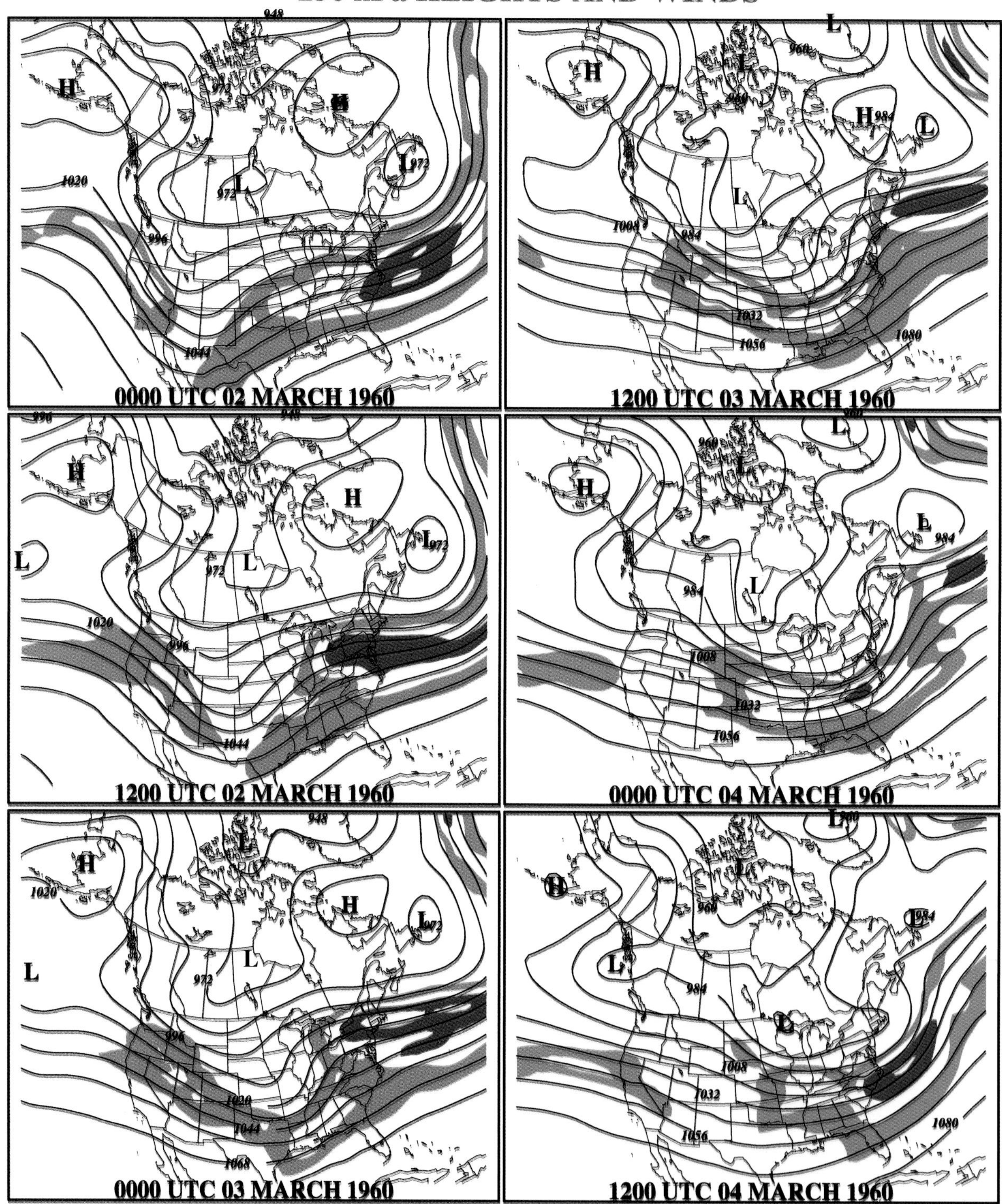

FIG. 10.4-5. Twelve-hourly analyses of 250-hPa geopotential height for 0000 UTC 2 Mar–1200 UTC 4 Mar 1960. See Fig. 10.1-5 for details.

and New England. However, it appears to occur during a winter in which ENSO (El Niño and La Niña) had little influence (Fig. 2–19, Volume I). However, the storm did occur during a winter in which a nearly persistent negative NAO was in place from the first week of January through the beginning of April. Immediately prior to the storm's development, daily values of the NAO were clearly negative, but trending toward less negative. As the storm was developing on 2–5 March, daily values decreased. The 500-hPa signal for the presence of a negative NAO may be the cutoff ridge observed over extreme northern Quebec and southern Greenland and the blocking pattern evident across all of eastern Canada.

- Strong geopotential height gradients and westerly flow dominated the precyclogenetic period across the United States prior to 2 March. Weak geopotential height gradients and large, nearly stationary vortices were observed over Canada, including a closed anticyclonic circulation between Quebec and Greenland, and cyclonic circulations over central and southeastern Canada.
- A 500-hPa trough exiting southeastern Canada on 2 March produced a significant confluent flow regime across the northeastern United States. This pattern was associated with strong northwesterly flow and cold-air advection at low levels.
- The 500-hPa trough associated with the sea level cyclone extended from the United States–Canada border into Mexico at 0000 UTC 2 March, then propagated to the east coast by the end of 3 March. A closed 500-hPa center formed during rapid cyclogenesis late on 3 March. It then deepened rapidly on 4 March, as the surface low moved slowly off the New England coast.
- The trough axis became oriented from northwest to southeast (a negative tilt) during 2 March, as the primary cyclone deepened. The geopotential height gradient at the base of the trough increased as the secondary low pressure center developed and intensified along the East Coast on 3–4 March. This secondary cyclogenesis occurred beneath a pronounced diffluence region downwind of the 500-hPa trough.
- The cyclone and heavy snowfall developed downwind of a propagating and intensifying cyclonic vorticity maximum, in the region of strong vorticity advection.
- Although the surface cyclonic development was explosive in this case, changes in the amplitude of the 500-hPa trough and downstream ridge were relatively small. There was some increase in amplitude of the trough and downstream ridge over the northern United States during secondary cyclogenesis on 3 March. Little increase in the amplitude of the trough and upstream ridge was observed prior to cyclogenesis.
- Although changes in amplitude were small, decreases in the half-wavelength between the trough and downstream ridge were substantial. They appeared to occur by 1200 UTC 2 March, during primary cyclogenesis, and by 0000 UTC 4 March, during secondary cyclogenesis, an apparent consequence of the "self-development" processes described earlier in volume I.
- This case is marked by a distinct dual-jet-streak pattern, with one jet extending from the Great Lakes eastward off the Atlantic coast between 2 and 3 March in the confluent height field noted earlier and the other more southerly jet streak associated with the distinct trough that contributed to the East Coast cyclogenesis after 0000 UTC 3 March.
- Wind speeds exceeding 60 m s^{-1} were analyzed within the confluent region upstream of the trough axis over southeastern Canada through 0000 UTC 3 March. The confluent entrance region of this polar jet was located above the cold high pressure ridge that became entrenched over the northern United States by 0000 UTC 3 March. The anticyclonic side of the entrance region of this jet also coincided with the expanding region of precipitation along the East Coast.
- Wind speeds associated with the second noticeable jet streak near the base of the trough increased to over 70 m s^{-1} by 0000 UTC 4 March, as secondary cyclogenesis commenced near the Carolina coast and then moved northeastward. The secondary cyclogenesis occurred along the East Coast in the diffluent exit region of this intensifying upper-level jet streak located over the southeastern United States on 3 March, as the trough also maintained a slightly negative tilt. The ageostrophic LLJ at 850 hPa (located earlier) intensifying along the Carolina coast at 1200 UTC 3 March also occurs within the diffluent exit region of this jet streak system.

e. 250-hPa geopotential height and wind analyses

- The 250-hPa level is marked by a distinct westerly flow regime across the entire country with a trough propagating from the southwestern United States at 0000 UTC 2 March with little amplification.
- A subtropical jet (STJ) is apparent at 0000 and 1200 UTC 2 March extending from Texas to the East Coast. This STJ appears to remain on the east side of the trough axis during the event.
- A separate wind maximum developed over the Ohio Valley (within the downstream ridge) by 1200 UTC 2 March and amplified to greater than 80 m s^{-1} by 0000 UTC 3 March. This wind maximum develops as the confluent zone is enhanced in the entrance region of the 400-hPa polar jet and as precipitation spread northward into the Ohio Valley with an inverted surface trough in Tennessee on 2 March. The expanding vertical extent of the polar jet increased the along-stream wind variation in the entrance region of this jet and is consistent with an enhanced pattern of ascent associated with the heavy snowfall that developed north of the inverted trough.

Park Avenue and 37th Street, New York City, 12 Dec 1960 (photo reprinted courtesy of the *New York Times*).

5. 10–13 December 1960

a. General remarks

- This early season storm was the first of three big snowstorms during the 1960/61 winter season. It was accompanied by wind gusts of up to 42 m s^{-1} at Block Island, Rhode Island, and 38 m s^{-1} at Nantucket, Massachusetts. Temperatures falling below −7°C and high winds created blizzard conditions across the region from of the middle Atlantic to New England.
- The following regions reported snow accumulations exceeding 10 in. (25 cm): West Virginia panhandle, extreme northern Virginia, Maryland (except for the lower eastern shore), parts of Delaware, southern and eastern Pennsylvania, New Jersey, southeastern New York, Connecticut, Rhode Island, Massachusetts (except the extreme west), extreme southeastern Vermont, southern New Hampshire, and coastal Maine.
- The following regions reported snow accumulations exceeding 20 in. (50 cm): scattered locations in northern New Jersey and eastern Massachusetts.

b. Surface analyses

- An Arctic front passed through the northeastern states on 10–11 December and was followed by much colder weather as rising sea level pressures extended eastward into New England from an anticyclone north of Minnesota (1035 hPa). Surface temperatures remained in the high 10s and low 20s (°F; −8° to −5°C) in New Jersey, New York, and New England on 11 December.
- As the surface high pressure system extended eastward into New England, the inverted sea level pressure ridge along the East Coast on 11 December indicates that cold-air damming was occurring immediately prior to the formation of a coastal front and secondary cyclogenesis late on 11 December.
- The primary low developed in the western Gulf of Mexico by 1200 UTC 10 December. It propagated northward to Oklahoma, then eastward across Kentucky to West Virginia by 0000 UTC 12 December. The low center deepened slowly from 1012 to 1002 hPa during this period. This system was associated with light to moderate precipitation amounts, with snow accumulations of generally less than 12 cm across the plains states, Midwest, and Ohio Valley states. The primary low was observed for only 6 h after the onset of secondary cyclogenesis at 1800 UTC 11 December.
- The secondary low formed over South Carolina along a rapidly evolving coastal front. The low moved off the Atlantic coast after passing near Cape Hatteras, North Carolina, at 0000 UTC 12 December, advancing northeastward at approximately 16 m s^{-1}. The central sea level pressure dropped 27 hPa during this one 12-h period and reached a minimum of 966 hPa by 0000 UTC 13 December. Heavy snow fell from Virginia to New England as the secondary cyclone propagated from eastern North Carolina to a position south of Nantucket, Massachusetts, between 0000 and 1200 UTC 12 December.
- Sea level pressure gradients north of the low center intensified rapidly as the cyclone deepened on 12 December. Strong north-northeasterly winds were reported along the coastline, contributing to the blizzard conditions observed with this storm.
- Extreme cold followed the storm, with surface temperatures across the eastern United States remaining 10°C below seasonal levels for the following few days.

TABLE 10.5-1. Snowfall amounts for urban and selected locations for 10–13 Dec 1960; see Table 10.1-1 for details.

NESIS = 4.47 (category 3)	
Urban center snowfall amounts	
Washington, D.C.–National Airport	8.5 in. (22 cm)
Baltimore, MD	14.1 in. (36 cm)
Philadelphia, PA	14.6 in. (37 cm)
New York, NY–The Battery	17.0 in. (44 cm)
Boston, MA	13.0 in. (33 cm)
Other selected snowfall amounts	
Newark, NJ	20.4 in. (52 cm)
Trenton, NJ	16.6 in. (42 cm)
Portland, ME	14.9 in. (38 cm)
Hartford, CT	13.4 in. (34 cm)
Providence, RI	11.2 in. (28 cm)

c. 850-hPa analyses

- Strong lower-tropospheric cold advection was observed across the northeastern United States through 1200 UTC 11 December in association with rising sea level pressures beneath a region of pronounced upper-level confluence.
- The 850-hPa low center intensified and its circulation expanded on 10 December, as the primary surface low deepened over Oklahoma. The 850-hPa low strengthened slowly during 11 December as it tracked from Oklahoma to West Virginia. It then either redeveloped along the coast and deepened explosively by 1200 UTC 12 December, which is coincident with the explosive development phase of secondary sea level cyclogenesis.
- Prior to 0000 UTC 12 December, the largest low-level temperature gradients were concentrated in the cold air over the northeastern United States, while the thermal gradient across the southeastern United States remained fairly weak. The temperature gradient south of the 0°C isotherm increased during 11 December as warm advection, and a southerly 20–25 m s^{-1} low-

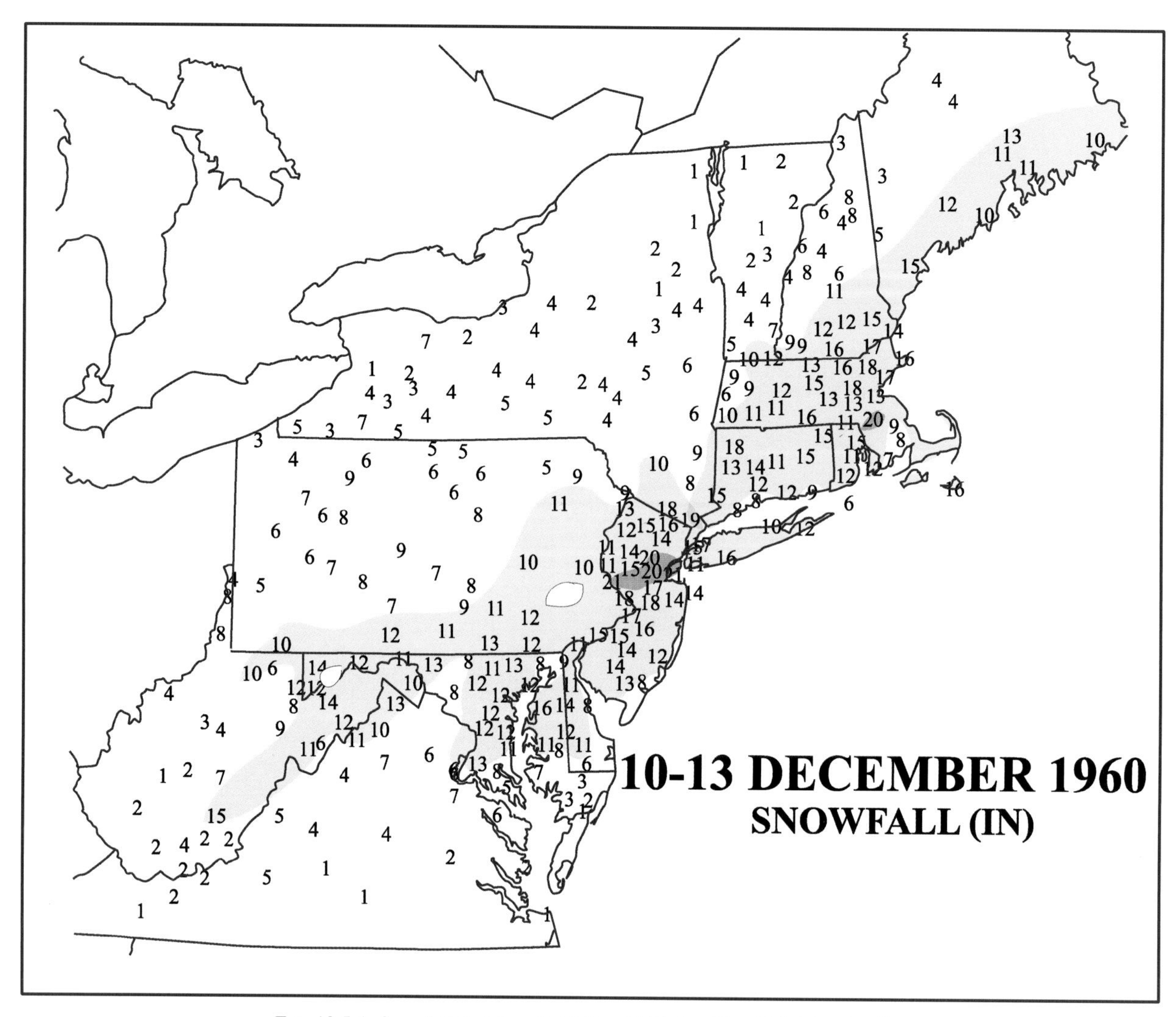

FIG. 10.5-1. Snowfall (in.) for 10–13 Dec 1960. See Fig. 10.1-1 for details.

level jet (LLJ), formed over the southeastern United States. The 0°C isotherm remained nearly stationary across West Virginia and Virginia prior to and during secondary cyclogenesis over North Carolina, despite significant warm-air advection between 1200 UTC 11 December and 0000 UTC 12 December.

- Very cold air plunged southward from central Canada as two upper-level troughs merged during 12 December.
- An S-shaped isotherm pattern became established along the East Coast by 0000 UTC 12 December as secondary cyclogenesis commenced. The S-shaped signature developed as strong warming occurred along the southeast coast, while temperatures remained relatively constant in the northeastern United States.
- Strong south-to-southwesterly low-level winds developed across the southeastern United States beneath the diffluent exit region of the upper-level jet system crossing the Gulf states on 11 December. The high wind speeds enhanced the moisture transport into the precipitation area, which expanded over the eastern United States by 0000 UTC 12 December.
- Strong southeasterly winds were observed at 0000 UTC 12 December in advance of the 850-hPa low center marking a distinct LLJ just east of the developing 850-hPa circulation, with wind speeds exceeding 40 m s^{-1} by 1200 UTC 12 December.
- Noticeable cross-contour flow at the 850-hPa level marked the rapid increase in wind speeds in the core of the LLJ at 0000 and 1200 UTC 12 December. The cross-contour flow coincided with the rapid surface cyclogenesis and heavy snowfall expanding northeastward along the East Coast between 0000 and 1200 UTC 12 December.
- Large easterly wind components were found north of the 850-hPa center from 0000 UTC 11 December through 1200 UTC 12 December, and were associated with moderate to heavy precipitation at each map time.

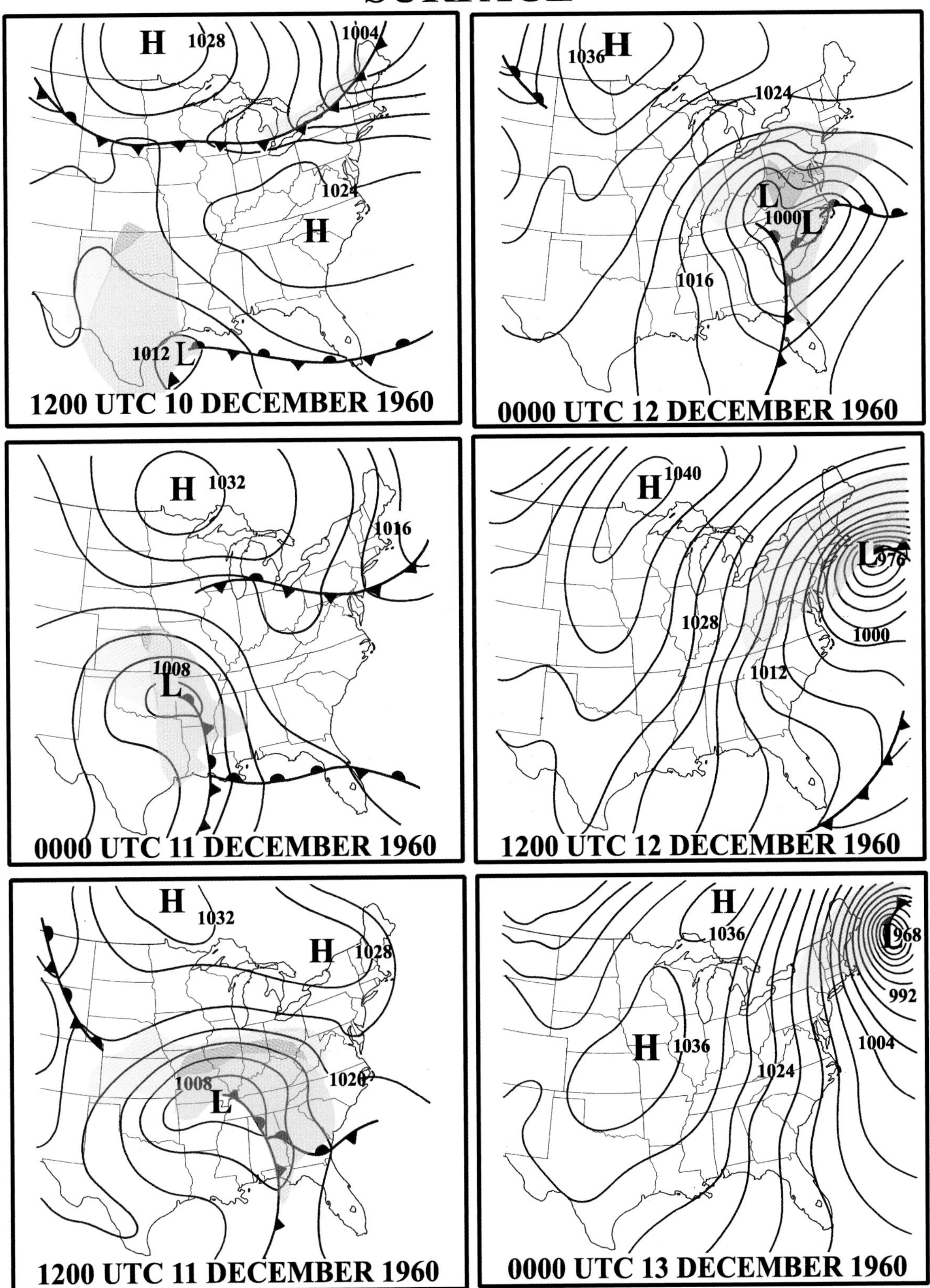

FIG. 10.5-2. Twelve-hourly surface weather analyses for 1200 UTC 10 Dec–0000 UTC 13 Dec 1960. See Fig. 10.1-2 for details.

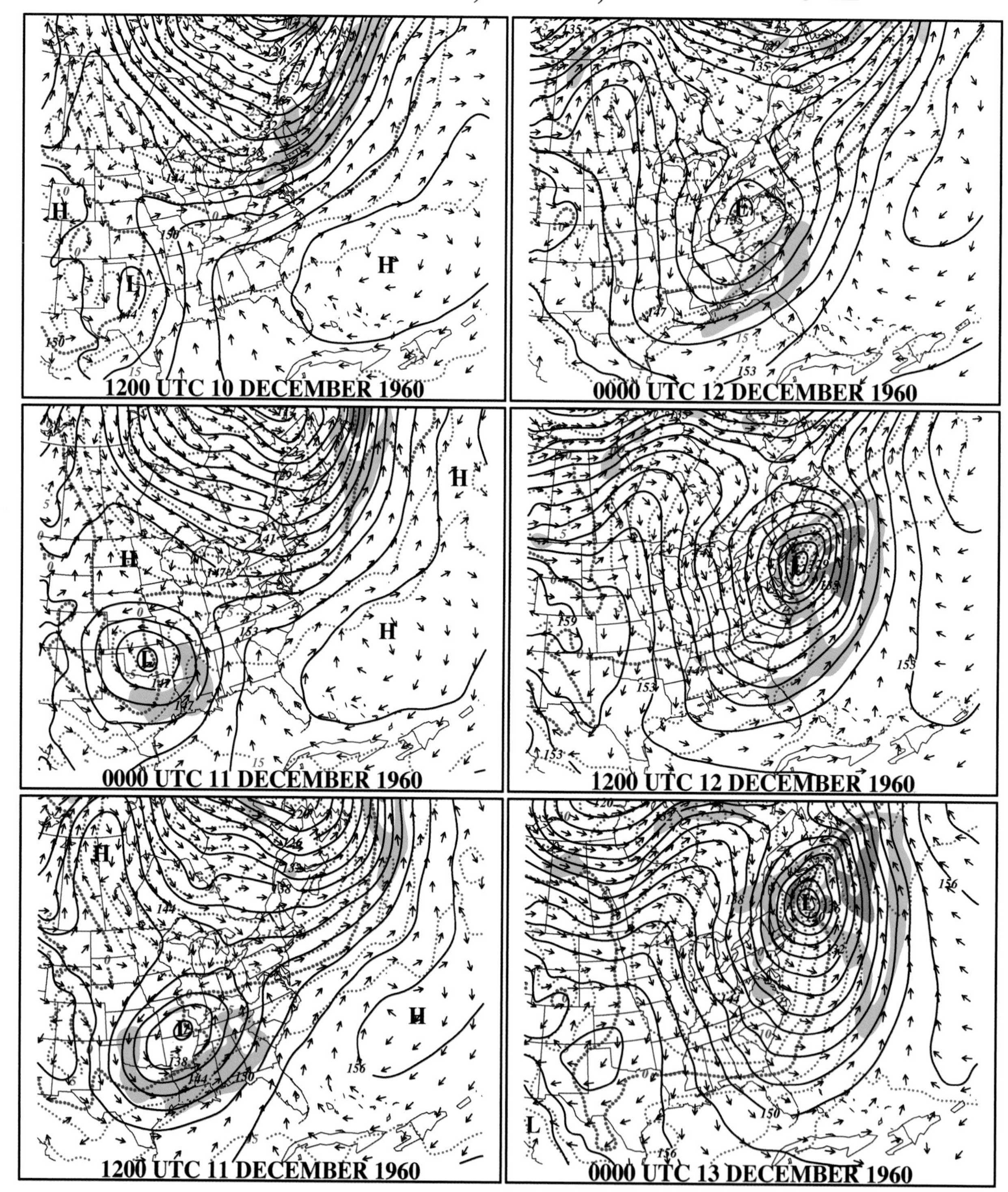

FIG. 10.5-3. Twelve-hourly 850-hPa analyses for 1200 UTC 10 Dec–0000 UTC 13 Dec 1960. See Fig. 10.1-3 for details.

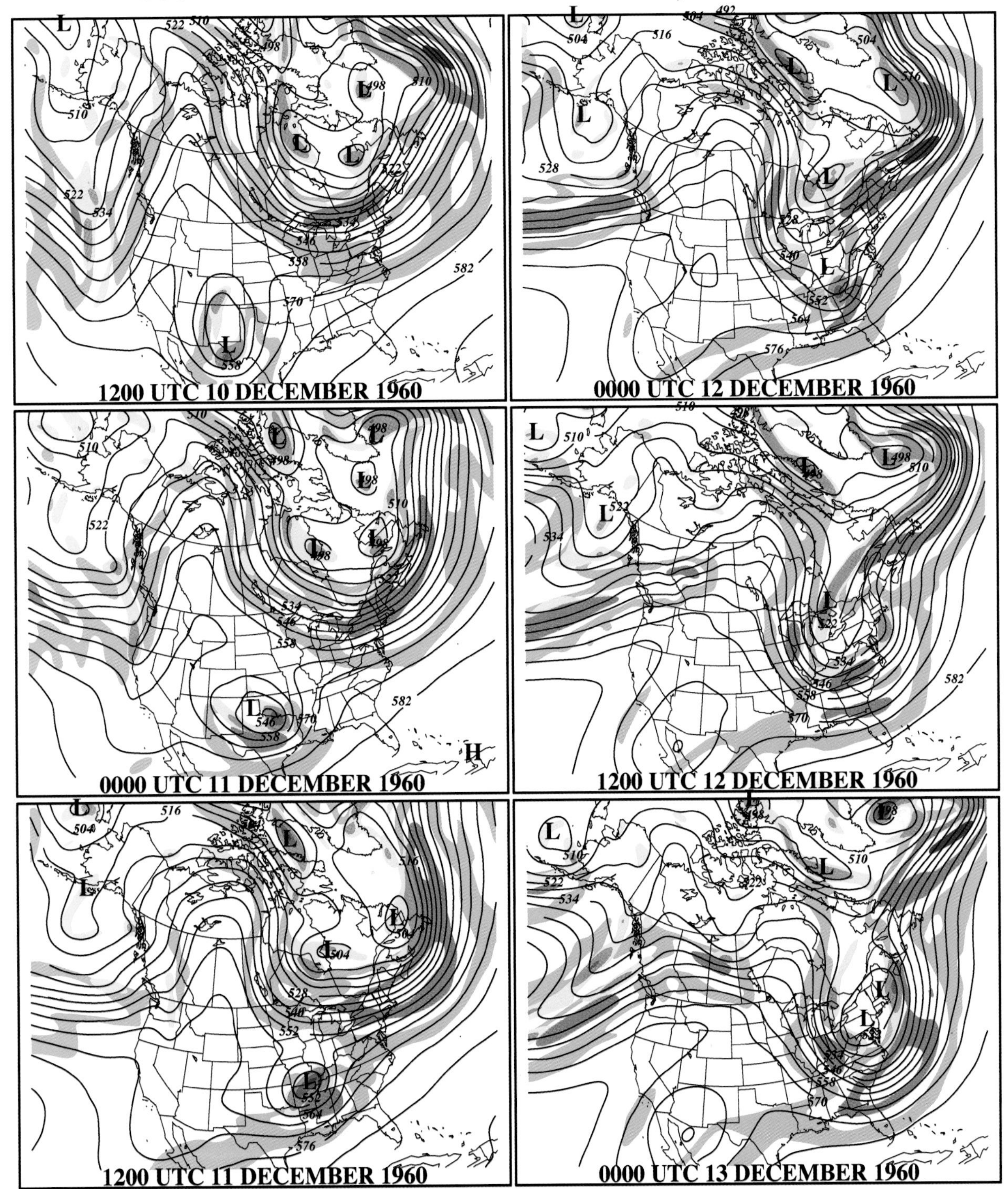

FIG. 10.5-4. Twelve-hourly analyses of 500-hPa geopotential heights and upper-level wind fields for 1200 UTC 10 Dec–0000 UTC 13 Dec 1960. See Fig. 10.1-4 for details.

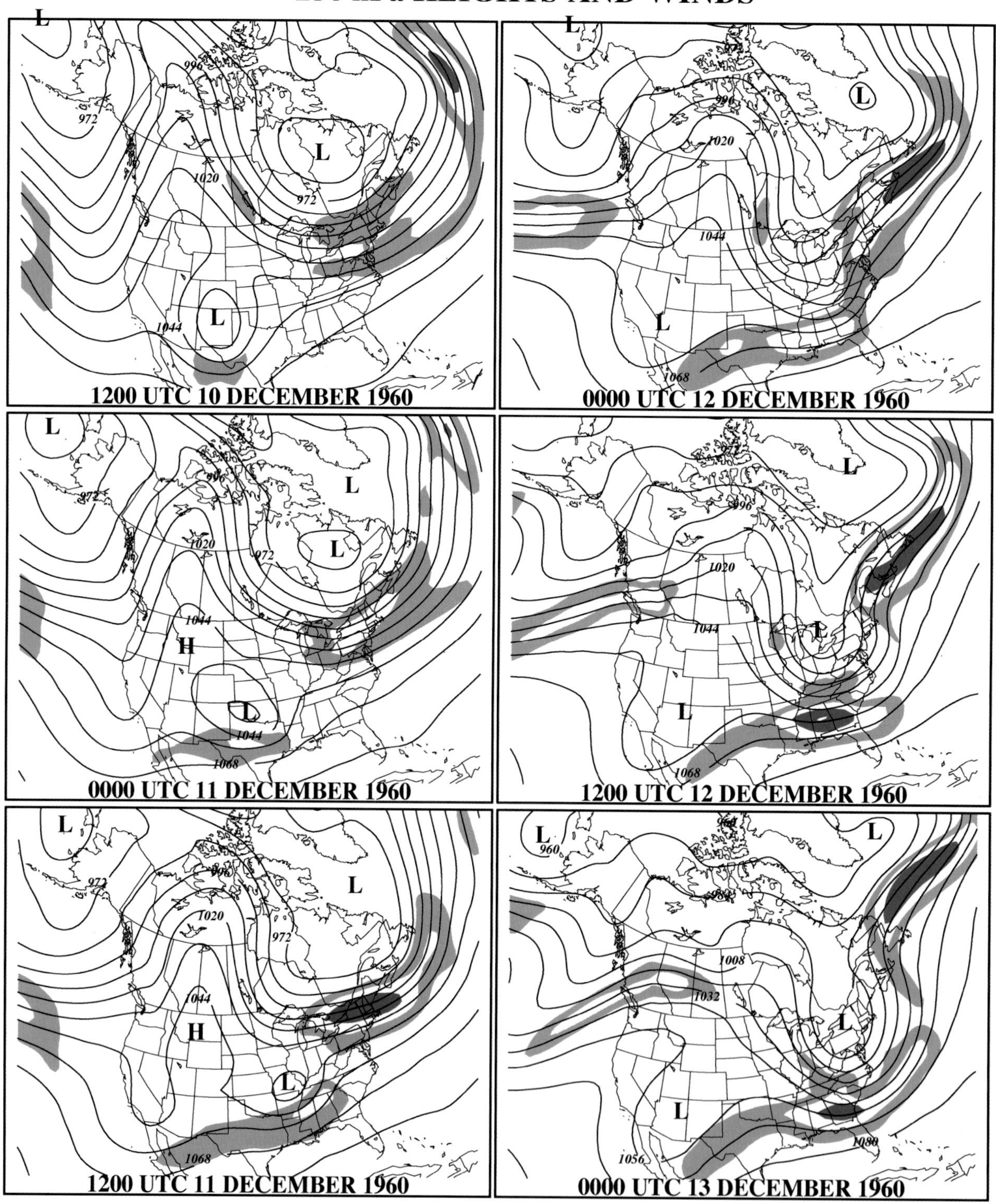

FIG. 10.5-5. Twelve-hourly analyses of 250-hPa geopotential height for 1200 UTC 10 Dec–0000 UTC 13 Dec 1960. See Fig. 10.1-5 for details.

d. 500-hPa geopotential height and 400-hPa wind analyses

- This early season snowstorm developed at the end of a very weak La Niña that lasted no longer than four months. It also developed during a winter in which the North Atlantic Oscillation (NAO) also does not exhibit a strong signal, except during December, when it is clearly positive. This storm developed during one such positive NAO period that began on 1 December and then became slightly negative for about a week following the storm. The 500-hPa analyses show that the Northeast storm developed during a period in which a deep upper low was found over southern Greenland, which is not a characteristic of a negative NAO.
- An intense vortex was evident over eastern Canada prior to and during the early stages of East Coast cyclogenesis. Cyclonic flow extended from central Canada across the northeastern United States on 10 and 11 December. The strong cyclonic flow combined with a ridge near the southeast coast to produce pronounced confluence over the eastern United States between 1200 UTC 10 December and 1200 UTC 11 December. Very cold air and high pressure at the surface extended across the northeastern United States beneath this confluence zone.
- A high-amplitude ridge was located over the western United States and Canada, upstream of the closed 500-hPa trough in the southwestern United States that was associated with the primary surface low pressure center.
- The evolution of the sea level cyclone appears to have been linked to the propagation of the southwestern trough as it moved to the east on 12 December. As the cyclone neared the East Coast, however, it also interacted with a separate trough that propagated southeastward from Canada. This trough propagated southeastward from extreme western Canada at 1200 UTC 10 December to Minnesota and Wisconsin at 0000 UTC 12 December.
- The southwestern 500-hPa trough was characterized by closed contours through 1200 UTC 11 December, but "opened up" as it neared the East Coast, became negatively tilted by 1200 UTC 12 December, and appeared to merge with another trough propagating southward from central Canada. This apparent interaction occurred as the secondary cyclone deepened rapidly along the East Coast.
- These troughs were marked by distinct cyclonic vorticity maxima. Secondary cyclogenesis and the development of heavy snowfall along the East Coast were heavily influenced by the cyclonic vorticity advections associated with these several features, but the vorticity maximum associated with the trough that originated over the southwestern United States appears to have exerted the most important direct influence.
- The merged troughs formed a deep closed-contour low center over the northeastern United States by 0000 UTC 13 December.
- Amplitude changes were difficult to assess due to the transformations from vortex to open trough to vortex, but the Canadian trough that propagated southward and the downstream ridge amplified sharply after 0000 UTC 12 December, as rapid cyclogenesis continued off the Atlantic coast.
- The half-wavelength between the southwestern 500-hPa closed low and the downstream ridge decreased prior to primary cyclogenesis between 1200 UTC 10 December and 0000 UTC 11 December. It also appeared to decrease during the rapid secondary cyclogenesis on 12 December.
- By 1200 UTC 11 December, two distinct and laterally coupled jet streaks are clearly evident: one in the confluent zone over New England and the other associated with the southwestern trough moving toward the East Coast.
- Intense polar jet streaks located off the northeast coast increased to greater than 70 m s^{-1} by 0000 UTC 12 December in association with highly confluent flow above the cold sea level high pressure ridge that extended across this region. The entrance region of the polar jet coincided with the expanding precipitation area as the secondary cyclone formed and began to deepen rapidly along the East Coast.
- The jet streak south of the initially closed trough over the southwestern United States was marked by increasing wind speeds, with a maximum wind in this jet increasing to greater than 50 m s^{-1} at the 400-hPa level by 1200 UTC 11 December. The increases in wind speed and areal coverage occurred as the primary low and heavy precipitation developed in the southern United States.
- By 0000 UTC 12 December, the coupled jet streak pattern is clearly observed along the East Coast, with the rapidly developing surface cyclone and ageostrophic LLJ found beneath the exit region of the southern jet and the expanding area of heavy precipitation found in the right entrance region of the northern jet streak. This laterally coupled jet streak pattern persisted through 0000 UTC 13 December as the storm moved northeastward off the Maine Coast.

e. 250-hPa geopotential height and wind analyses

- The 250-hPa height analyses reflect the broad trough over eastern Canada and the slow-moving trough over the southwestern United States on 1200 UTC 10 December. These two troughs evolved and merged together by 0000 UTC 12 December as the cyclone rapidly developed off the East Coast between 0000 and 1200 UTC 12 December, in a manner similarly described for the 500-hPa level.
- The dual-jet- streak pattern is clearly evident from

0000 UTC 11 December to 1200 UTC 12 December. The northern jet streak represents the upper-level extension of the polar jet in the confluent region over New England and increased to greater than 70 m s^{-1}. The southern jet had subtropical characteristics, remaining located in the southeast United States, and increased to nearly 80 m s^{-1} by 1200 UTC 12 December over Alabama as the storm moved northeastward.

- The rapid surface cyclogenesis and heavy snow occurred in the left exit region of the southern jet streak and right entrance region of the northern jet steak, most noticeably for a 24-h period between 1200 UTC 11 December and 1200 UTC 12 December.

President-elect and Mrs. John F. Kennedy on the way to the Inaugural Gala, 19 Jan 1961 (photo courtesy of the John F. Kennedy Library and Museum, and AP/Wide World Photos).

6. 19–20 January 1961

a. General remarks

- The "Kennedy Inaugural Snowstorm" occurred on the eve of John F. Kennedy's presidential inauguration in Washington, D.C., and was the second of three major East Coast winter storms during the 1960/61 season. The area of heaviest snowfall was similar in location to that of the December 1960 storm. Blizzard or near-blizzard conditions developed across the northeastern United States as the cyclone deepened rapidly offshore.
- The following regions had snow accumulations exceeding 10 in. (25 cm): northern Maryland and Delaware, parts of West Virginia and Virginia, southeastern Pennsylvania, New Jersey, southeastern New York, Connecticut, Rhode Island, Massachusetts (except for the northwest corner), and southern sections of New Hampshire and Maine.
- The following regions had snow accumulations exceeding 20 in. (50 cm): scattered parts of eastern Pennsylvania, northern New Jersey, southeastern New York, northwestern Connecticut, northeastern Massachusetts, and southern New Hampshire.

TABLE 10.6-1. Snowfall amounts for urban and selected locations for 19–20 Jan 1961; see Table 10.1-1 for details.

NESIS = 3.47 (category 2)	
Urban center snowfall amounts	
Washington, D.C.–National Airport	7.7 in. (20 cm)
Baltimore, MD	8.4 in. (21 cm)
Philadelphia, PA	13.2 in. (34 cm)
New York, NY–Central Park	9.9 in. (25 cm)
Boston, MA	12.3 in. (31 cm)
Other selected snowfall amounts	
Harrisburg, PA	18.7 in. (47 cm)
Worcester, MA	18.7 in. (47 cm)
Nantucket, MA	16.0 in. (41 cm)
Hartford, CT	14.2 in. (36 cm)
Newark, NJ	13.7 in. (35 cm)

b. Surface analyses

- A front ushered in cold air for the heavy snow event across New England and the middle Atlantic states on 18 January. This cold air was associated with high pressure north of the Great Lakes.
- There was no evidence of cold-air damming from New England into the middle Atlantic states immediately prior to cyclogenesis on 19 January. No inverted sea level pressure ridge was observed, and surface temperatures did not show the characteristic wedge of cold air along the eastern slopes of the Appalachians.
- The surface low appeared to follow a relatively straight path, even across the Appalachian range, as it moved toward the Atlantic coast on 19 January. Although no separate secondary cyclone was observed, the surface low did have a tendency to "jump" or reform slightly farther to the east as it crossed the Appalachian Mountains.
- The cyclone propagated rapidly, covering approximately 700–800 km every 12 h. The forward movement of the storm may have slowed slightly near the New England coast.
- There was no evidence of coastal frontogenesis, which could be related, in part, to both the lack of cold-air damming and the speed with which this system developed and moved across the East Coast.
- The cyclone deepened rapidly from 1200 UTC 19 January through 0000 UTC 21 January. During this time, the central pressure fell 43 hPa over one 24-h period. The lowest sea level pressure was analyzed at 964 hPa east of New England late on 20 January. The intensification late on 19 January and early on 20 January was accompanied by a large increase in the pressure gradients surrounding the cyclone center and an increase in wind speeds along the coast.
- The rapid intensification of the surface low center coincided with an expansion of the precipitation shield on 19 January. Virtually no precipitation was occurring at 0000 UTC 19 January, with light snow breaking out across the Ohio Valley by 1200 UTC 19 January. Precipitation rates increased dramatically over the following 12–24 h from the middle Atlantic states northeastward into New England.
- Cold air was reinforced following the storm as a cold anticyclone plunged south-southeastward into the plains states late on 20 January, a pattern that was repeated numerous times in the following 2 weeks.

c. 850-hPa analyses

- Northwesterly flow and cold advection were established over the northeastern United States between 1200 UTC 18 January and 1200 UTC 19 January, beneath a region of upper-level confluence. The 0°C isotherm remained virtually stationary near the Virginia–North Carolina border through 19 January.
- The 850-hPa low commenced deepening between 0000 and 1200 UTC 19 January [-90 m (12 h)$^{-1}$] over the Ohio Valley. It intensified most rapidly between 0000 and 1200 UTC 20 January [-180 m (12 h)$^{-1}$], at the same time that the surface low deepened by 24 hPa along the East Coast.
- The initial development of the 850-hPa low over the central plains states on 18 January was characterized by strong cold-air advection west of the center. A northerly 25 m s^{-1} low-level jet (LLJ) was located beneath the entrance region of an upper-level jet across the southern plains states (see upper-level wind analyses) by 1200 UTC 19 January. East of the 850-hPa low, only weak southwesterly flow and slight

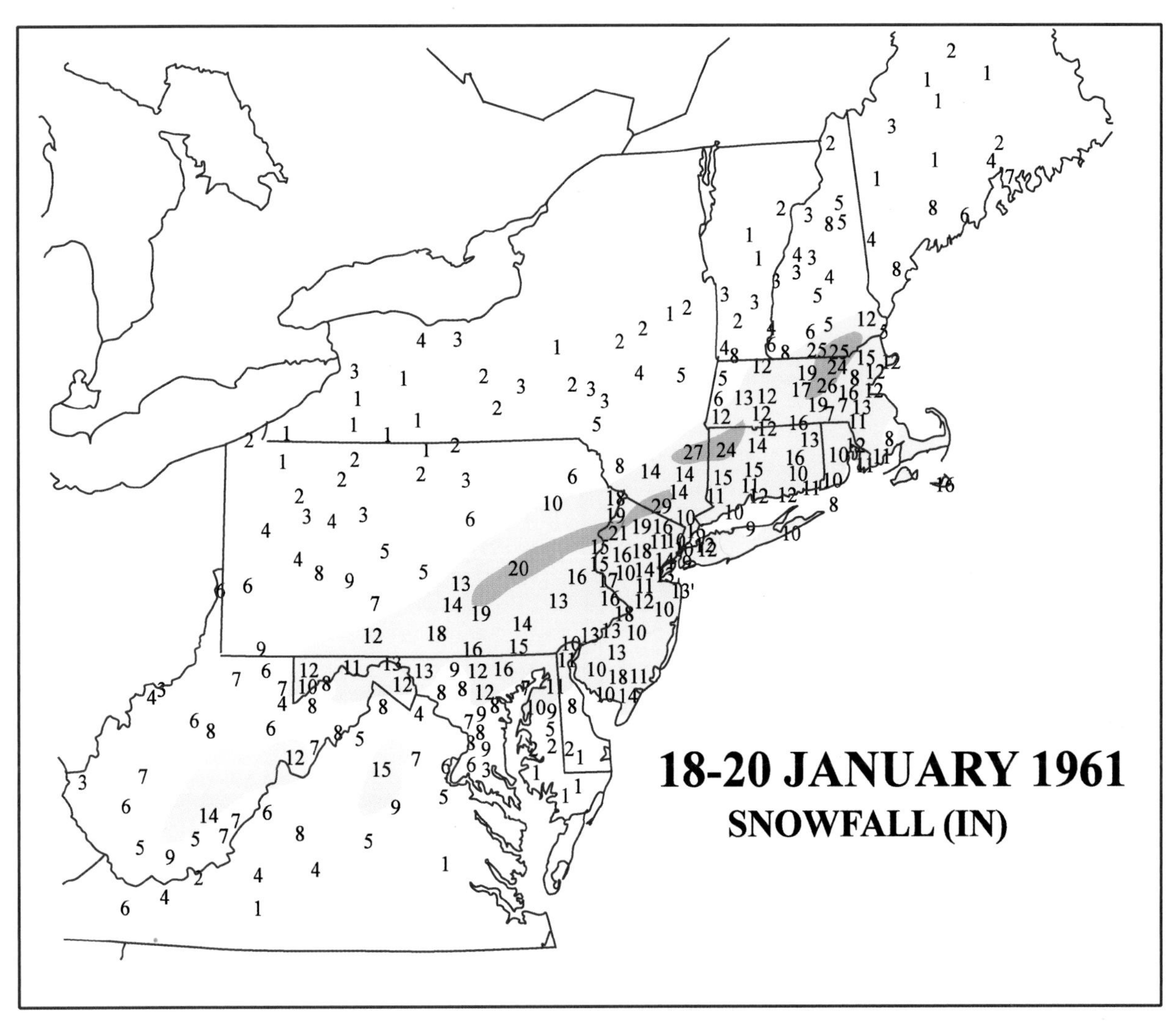

FIG. 10.6-1. Snowfall (in.) for 18–20 Jan 1961. See Fig. 10.1-1 for details.

warm-air advection were evident prior to 1200 UTC 19 January, when little precipitation was reported.

- Increasing warm-air advection east of the deepening 850-hPa low at 1200 UTC 19 January and 0000 UTC 20 January was associated with the widespread outbreak of precipitation. The strong cold-air advection behind the low combined with the developing warm-air advection ahead of it to produce an S-shaped isotherm pattern along the East Coast by 0000 UTC 20 January. This coincided with the start of the cyclone's explosive deepening stage. Temperature gradients were concentrated around the 850-hPa low at this time, with the 0°C isotherm nearly collocated with the low center.
- An easterly to southeasterly wind component developed to the north and northeast of the 850-hPa center after 1200 UTC 19 January as the 850-hPa and surface lows began to deepen. By 1200 UTC 20 January, wind speeds had increased by more than 15 m s^{-1} north of the low center to nearly 30 m s^{-1}, directed toward the expanding area of precipitation, probably contributing to the intensification of snowfall rates during this phase of the storm's development.
- A southerly LLJ with maximum winds exceeding 30 m s^{-1} developed along the East Coast by 0000 UTC 20 January. The wind speeds east and north of the intensifying low-level circulation increased dramatically thereafter, with 40 m s^{-1} wind speeds analyzed northeast of the storm center by 0000 UTC 21 January.

d. 500-hPa geopotential height and 400-hPa wind analyses

- January 1961 is characterized by a neutral El Niño–Southern Oscillation (ENSO) pattern, indicating that neither El Niño or La Niña had any effect during this period. Like the preceding storm the prior month, the NAO is characterized by daily values fluctuating slightly above and below zero, also indicating a weak

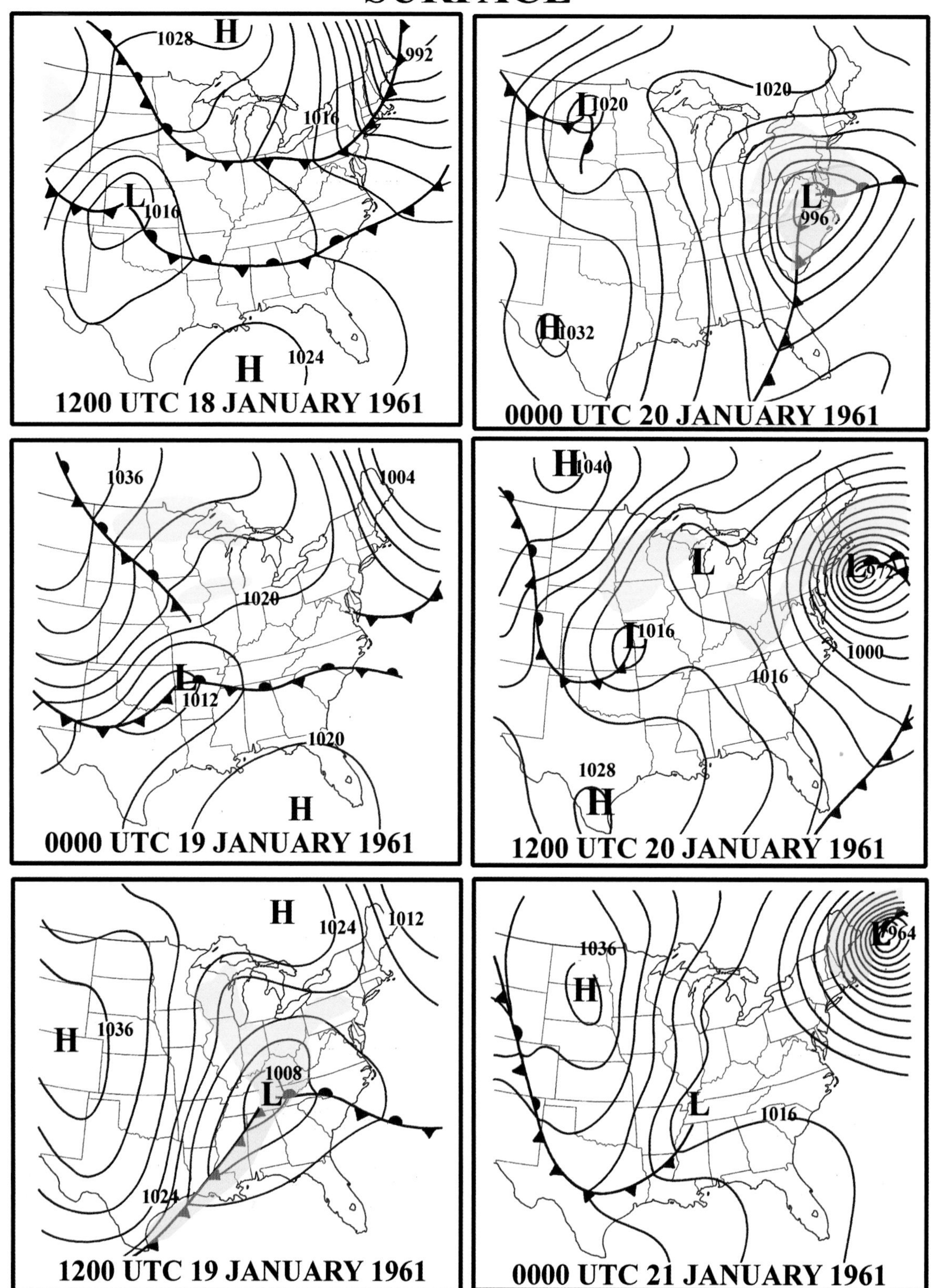

FIG. 10.6-2. Twelve-hourly surface weather analyses for 1200 UTC 18 Jan–0000 UTC 21 Jan 1961. See Fig. 10.1-2 for details.

850 hPa HEIGHTS, WINDS, TEMPERATURE

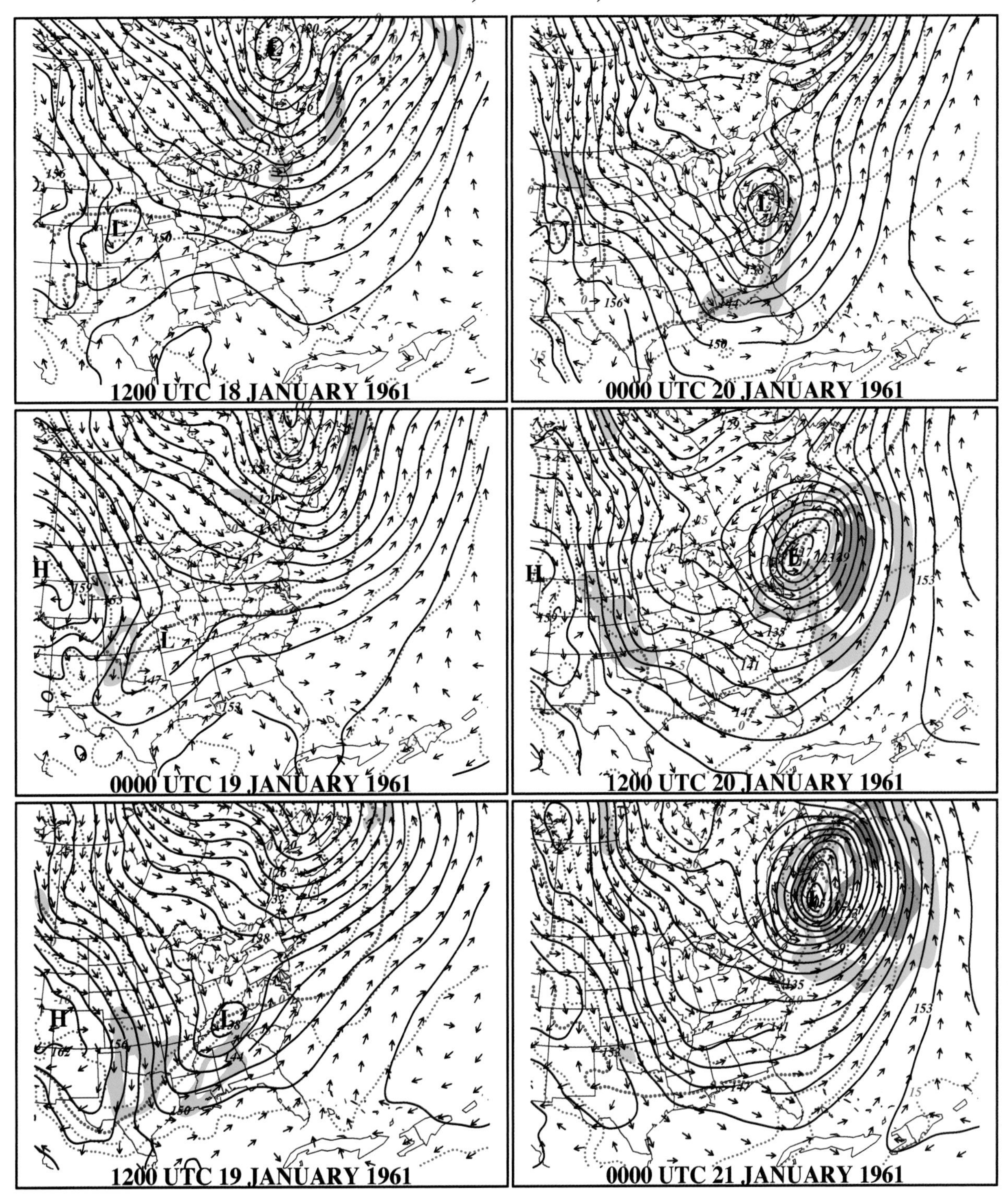

FIG. 10.6-3. Twelve-hourly 850-hPa analyses for 1200 UTC 18 Jan–0000 UTC 21 Jan 1961. See Fig. 10.1-3 for details.

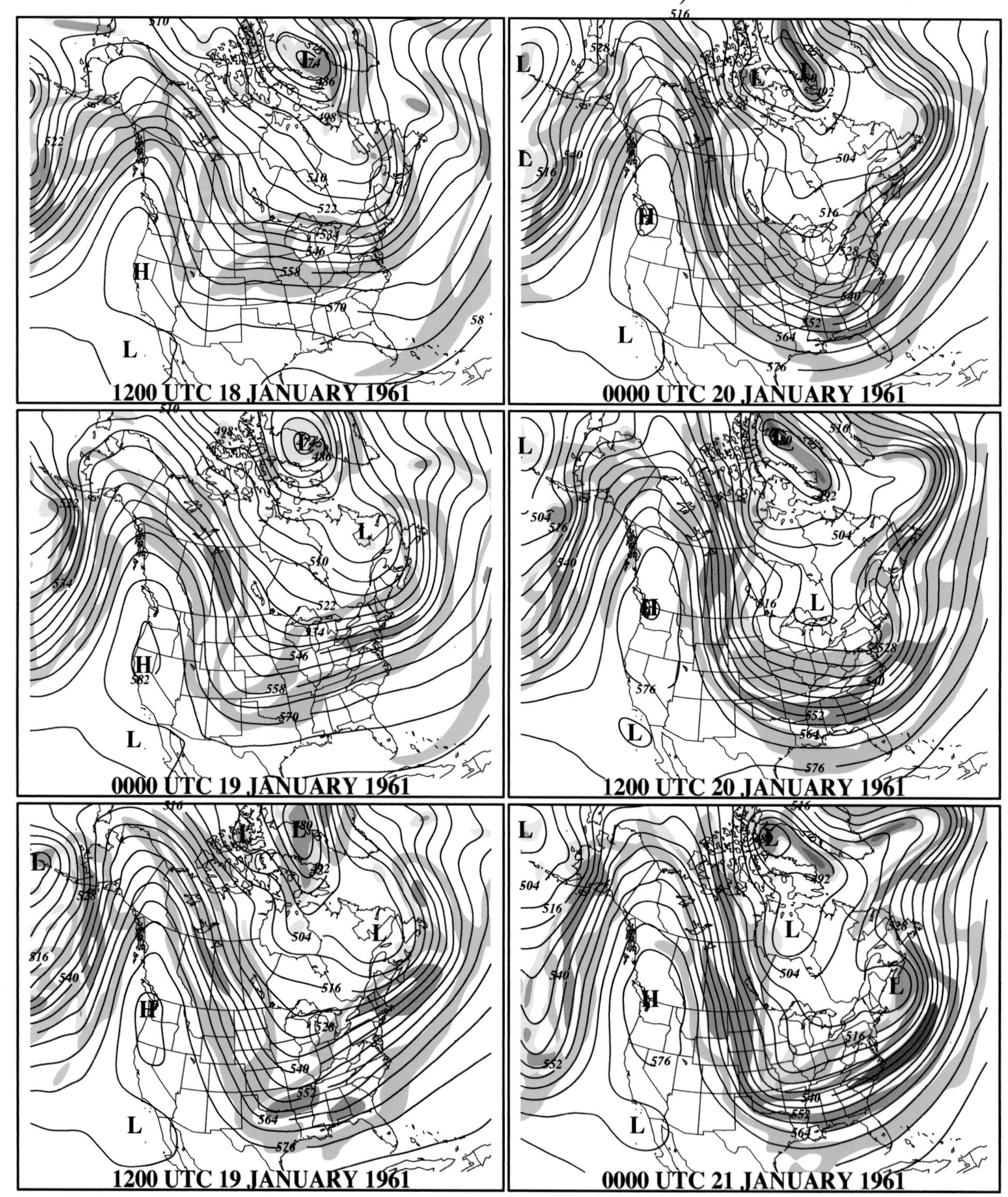

FIG. 10.6-4. Twelve-hourly analyses of 500-hPa geopotential heights and upper-level wind fields for 1200 UTC 18 Jan–0000 UTC 21 Jan 1961. See Fig. 10.1-4 for details.

250 hPa HEIGHTS AND WINDS

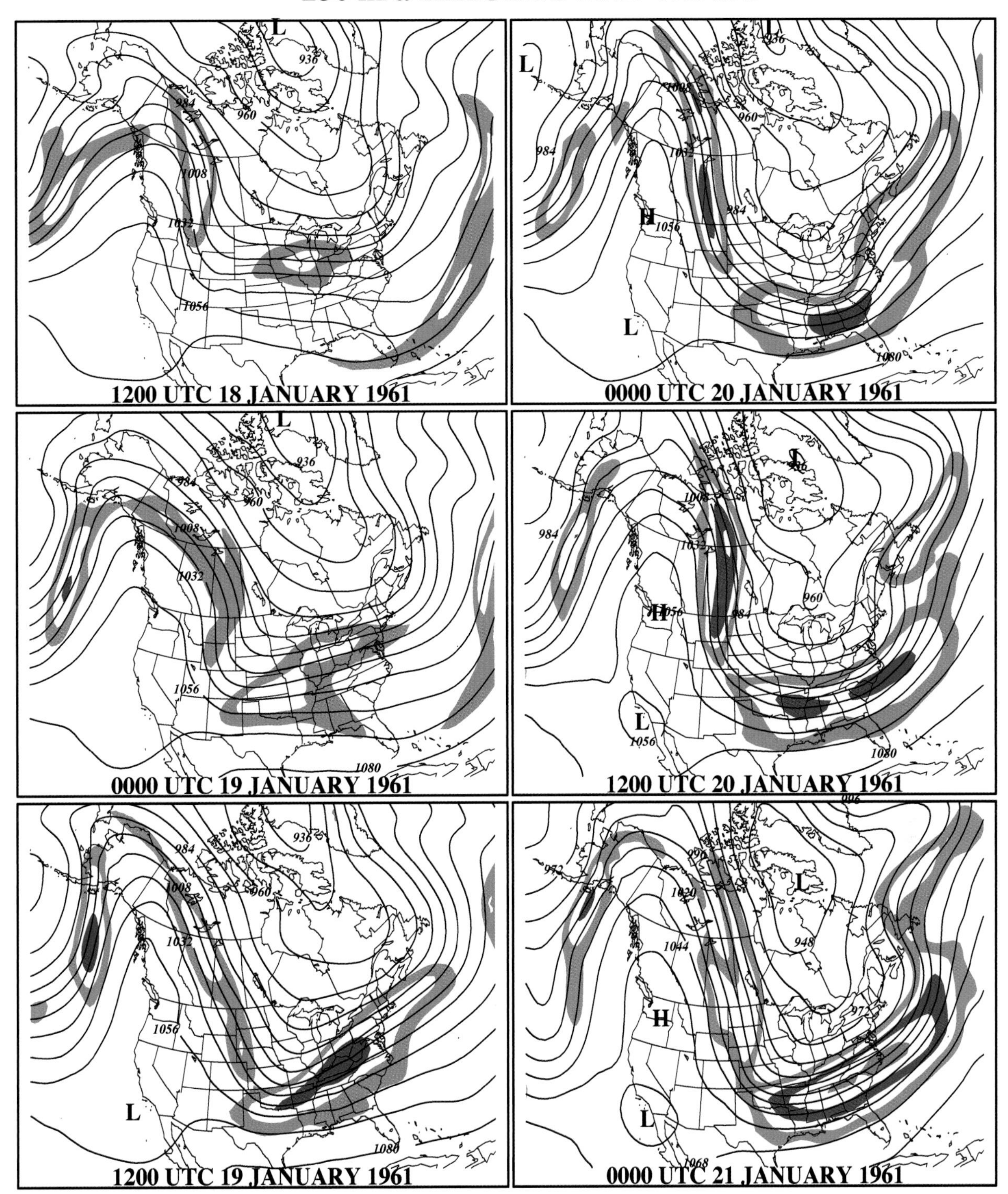

FIG. 10.6-5. Twelve-hourly analyses of 250-hPa geopotential height for 1200 UTC 18 Jan–0000 UTC 21 Jan 1961. See Fig. 10.1-5 for details.

NAO signal as well. However, daily values of NAO are slightly negative during the evolution of this storm and remain so for a few days following the storm.

- The precyclogenetic period featured a high-amplitude ridge anchored over the west coast of North America. The trough that was associated with the cyclone was located downstream of this ridge, embedded in a belt of enhanced geopotential height gradients extending from central Canada across the northern United States north of a ridge over the Gulf of Mexico.
- A slow-moving trough in eastern Canada was associated with a confluent geopotential height pattern across the Great Lakes and New England during 19 January. This upper-level confluent flow pattern was located above the surface anticyclone north of the Great Lakes that was supplying cold air to the northeastern United States.
- The 500-hPa trough associated with the cyclone developed over the central United States and propagated eastward on 19 and 20 January and evolved into a negatively tilted trough and deep closed center by 0000 UTC 21 January, as the surface low intensified rapidly off the New England coast. The rapid development of the cyclone and precipitation shield occurred to the east of an amplifying cyclonic vorticity center, in the region of strong vorticity advection. This feature propagated from the Rocky Mountain states to Oklahoma, then up the Tennessee Valley, and finally off the middle Atlantic coast on 18–20 January. The geopotential height gradient at the base of the trough increased at 0000 UTC 20 January, concurrent with the onset of sea level development.
- As this system approached the East Coast, diffluence increased and the trough axis assumed a northwest–southeast orientation (a negative tilt) by 1200 UTC 20 January. The sea level cyclone was deepening rapidly off southeastern New England at this time.
- The amplitude of the trough over the central United States and the ridge upstream along the West Coast increased dramatically in the 24-h period ending at 1200 UTC 19 January, as the surface low moved from Kansas to Tennessee. The amplitude increased slowly thereafter. The increase of amplitude was associated with the development of an upper-level anticyclone along the West Coast.
- As the surface low began to deepen on its trek from Tennessee to the Atlantic Ocean between 1200 UTC 19 January and 1200 UTC 20 January, the 500-hPa geopotential heights east of the cyclone rose significantly. The rising heights produced a well-defined ridge extending into New England by 1200 UTC 20 January. The increase in the amplitude of the trough and this downstream ridge occurred during the period when the surface low intensified rapidly.
- The half-wavelength between the trough and its downstream ridge over the eastern United States decreased after 0000 UTC 19 January, especially during the period of rapid cyclogenesis on 20 January.
- A distinct dual-jet pattern developed between 0000 UTC 19 January and 0000 UTC 20 January. The development of the dual pattern coincided with the rapid surface cyclogenesis and heavy snow occurring in the left-exit region of the southern jet and the right-entrance region of the northern jet, especially between 0000 and 1200 UTC 20 January.
- Prior to cyclogenesis, confluent flow and a polar jet streak extended from the Great Lakes across New Jersey and offshore between 1200 UTC 18 January and 1200 UTC 19 January, with the wind maximum increasing to greater than 60 m s^{-1} just east of New England by 1200 UTC 19 January. A cold surface high pressure ridge built from the Great Lakes region into New England during this period.
- The southern jet streak became more evident near the base of the deepening trough between 0000 and 1200 UTC 19 January, and extended eastward and increased in magnitude to greater than 60 m s^{-1} over the southeast United States by 0000 UTC 20 January.
- The 850-hPa LLJ, a rapidly expanding area of heavy snowfall, and rapid surface cyclogenesis all occurred within the well-defined diffluent exit region of this jet streak and the right-entrance region of the northern jet streak with a lateral coupling focused along the middle Atlantic coastline by 0000 UTC 20 January.
- Wind speeds at 400 hPa increased to greater than 60–70 m s^{-1} along the broad southern end of the large-scale trough that existed over the eastern two-thirds of the country by 0000 UTC 21 January.

e. 250-hPa geopotential height and wind analyses

- The broad West Coast ridge and developing trough in the central United States, with confluent flow over the northeast, are clearly evident during the precyclogenetic period between 1200 UTC 18 January and 1200 UTC 19 January.
- The dual-jet-streak pattern slowly evolved at this level by 0000 UTC 20 January, with the most notable feature being the enhancement of the southern jet, with subtropical characteristics between 0000 UTC 20 January and 0000 UTC 21 January. As the storm system moved northeastward and the large-scale trough amplified, covering the eastern two-thirds of the United States, maximum winds within the subtropical jet increased to greater than 90 m s^{-1}.

Snowbound cars on the Belt Parkway in Brooklyn, 5 Feb 1961 (photo courtesy of Kevin Ambrose, *Great Blizzards of New York City*, Historical Enterprises, 1994).

7. 2–5 February 1961

a. General remarks

- The third major snowstorm of the 1960/61 winter season occurred at the end of a prolonged cold spell across the northeastern United States. It produced near-record snow cover in the major metropolitan areas since snow fell on unmelted accumulations from the previous storms. This storm also produced paralyzing gale to hurricane force winds on the coast, with wind gusts of 43 m s^{-1} at Blue Hill Observatory in Milton, Massachusetts, and 41 m s^{-1} at New York City (La Guardia Airport). Temperatures rose to near freezing during the storm, producing heavy, wet, wind-driven snow accumulations along the coast that quickly hardened and froze as temperatures fell back into the 20s (°F) during the last segment of the storm.
- The following regions reported snow accumulations exceeding 10 in. (25 cm): panhandle of West Virginia, parts of northern Virginia and northern Maryland, Pennsylvania, central and southern New York, central and northern New Jersey, Connecticut, Rhode Island, Massachusetts, southern Vermont, and New Hampshire.
- The following regions reported snow accumulations exceeding 20 in. (50 cm): northern Pennsylvania, central New York, extreme southeastern New York, Long Island, northern New Jersey, and scattered portions of New England.

TABLE 10.7-1. Snowfall amounts for urban and selected locations for 2–5 Feb 1961; see Table 10.1-1 for details.

NESIS = 6.24 (category 4)	
Urban center snowfall amounts	
Washington, D.C.–National Airport	8.3 in. (21 cm)
Baltimore, MD	10.7 in. (27 cm)
Philadelphia, PA	10.3 in. (26 cm)
New York, NY–Kennedy Airport	24.0 in. (61 cm)
New York, NY–La Guardia Airport	19.0 in. (48 cm)
Boston, MA	14.4 in. (37 cm)
Other selected snowfall amounts	
Cortland, NY	40.0 in. (102 cm)
Newark, NJ	22.6 in. (57 cm)
Worcester, MA	18.8 in. (46 cm)
Providence, RI	18.3 in. (45 cm)
Allentown, PA	17.3 in. (44 cm)
Nantucket, MA	14.4 in. (37 cm)
New Haven, CT	14.0 in. (36 cm)

b. Surface analyses

- A large, very cold anticyclone (1048 hPa) north of the Great Lakes preceded the storm, with record low temperatures observed in the northeastern United States on 2 February. The anticyclone remained north of New York and New England prior to and during the snowstorm, serving as a continuous source of cold air for the snowfall in the Northeast.
- Cold-air damming occurred on 2–3 February, with a distinct inverted ridge of high pressure extending from New York to northern Florida at 0000 and 1200 UTC 3 February.
- A coastal front developed off the southeast United States on 3 February, immediately prior to the development of the secondary cyclone.
- The primary low pressure center intensified only slowly as it propagated through the Tennessee Valley on 2 February, then filled as it reached the Ohio Valley on 3 February. The central pressure of this system never fell below 1010 hPa. The primary low remained a separate entity for 9 h following the onset of secondary cyclogenesis along the East Coast on 3 February.
- The secondary low formed east of South Carolina shortly after 1200 UTC 3 February and deepened rapidly off the middle Atlantic coast on 4 February. Intensification occurred primarily between 1200 UTC 3 February and 1200 UTC 4 February. The central sea level pressure fell from 1008 hPa near Norfolk, Virginia, at 0000 UTC 4 February to 992 hPa off the New Jersey coast 12 h later. The secondary center moved northeastward parallel to the coast through 1200 UTC 4 February, then turned to the east during the next 12-h period.
- Large sea level pressure gradients formed to the north of the rapidly deepening secondary low on 4 February as heavy snow and strong northeasterly winds developed from northern Virginia to southern New England.
- The primary and secondary lows were both relatively slow movers, averaging only 12–13 m s^{-1}.

c. 850-hPa analyses

- An 850-hPa cutoff anticyclone located north of the Great Lakes was responsible for northwesterly flow and cold advection over the Northeast and middle Atlantic states on 2 February. The cold-air advection was located beneath a large region of upper-level confluence.
- The 850-hPa low center located over the central United States on 2 February deepened slowly until 0000 UTC 3 February. This system propagated eastward and weakened slightly by 0000 UTC 4 February, then deepened explosively in the 12 h ending at 1200 UTC 4 February [-240 m (12 h)$^{-1}$] as the surface cyclone was intensifying rapidly off the middle Atlantic coast. Dual 850-hPa low centers at 0000 UTC 4 February (one over the Ohio Valley and one along the Carolina coast) reflected the coastal redevelopment of the surface low center. Local height falls amplified markedly by 1200 UTC 4 February concurrent with the rapid deepening of the 850-hPa low center along the coast.
- The 850-hPa low evolved along a low-level baroclinic

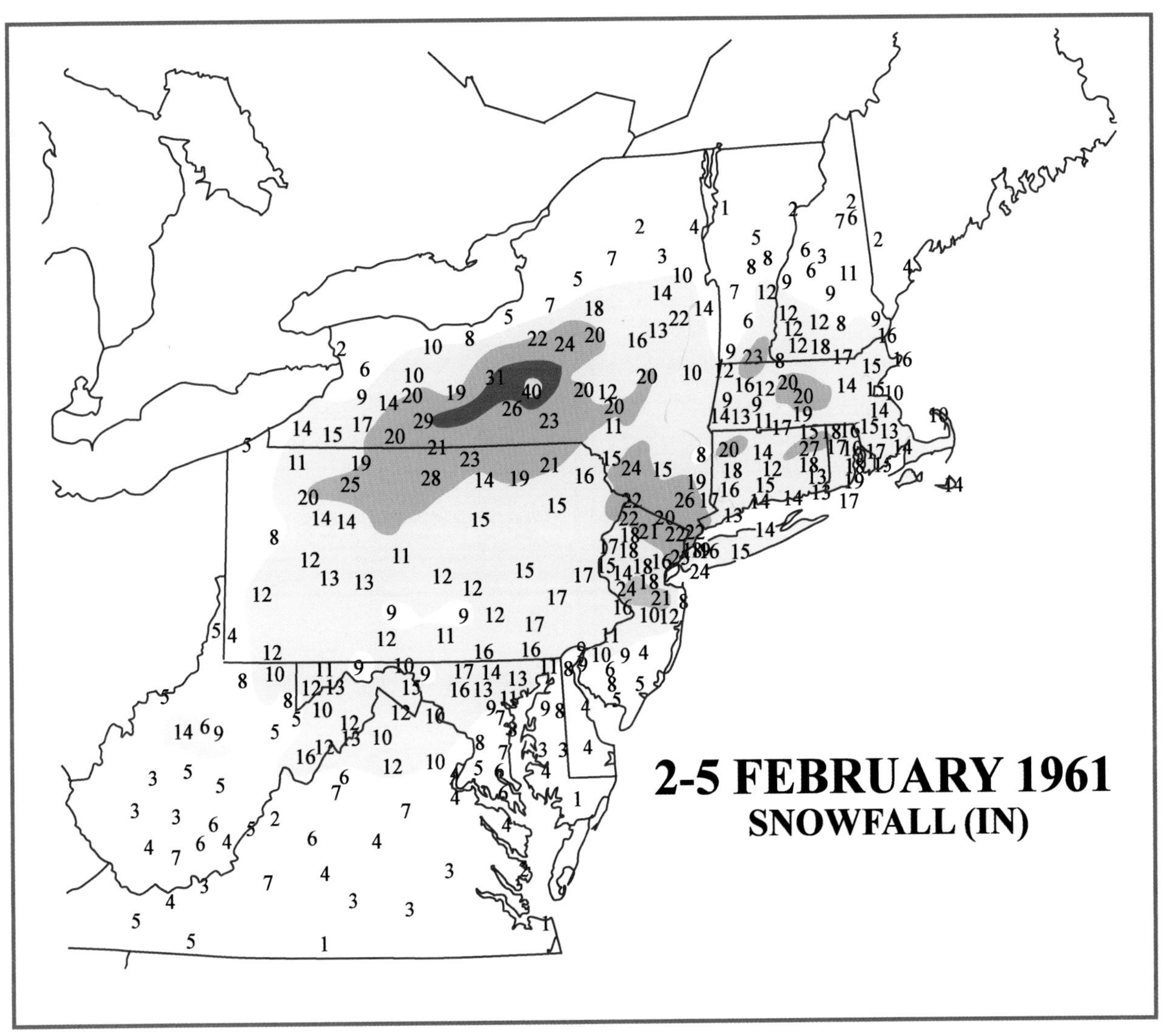

FIG. 10.7-1. Snowfall (in.) for 2–5 Feb 1961. See Fig. 10.1-1 for details.

zone that extended from the central plains states to the middle Atlantic coast on 2–3 February. At 1200 UTC 2 February, the largest temperature gradients were oriented from west to east on the cold side of the 0°C isotherm. A pattern of cold advection west of the low and warm advection east of the low became quite evident by 0000 UTC 3 February. This established the S-shaped isotherm pattern along the East Coast during 3 February, as the secondary cyclone was forming. The strong warm-air advection was close to the low center at 0000 UTC 3 February, but shifted to the coast during 3 February as coastal frontogenesis and cyclogenesis occurred along the Southeast coast.

- A southeasterly low-level jet (LLJ) with wind speeds exceeding 25 m s^{-1} formed over the middle Atlantic states by 0000 UTC 4 February, in conjunction with secondary cyclogenesis, as the local height falls shifted from the Midwest to the East Coast. This jet appeared to enhance the moisture transport into the developing area of heavy precipitation along the East Coast.
- Significant cross-contour flow fed into the southeasterly LLJ along and just east of the Carolina coast by 0000 UTC 4 February as rapid cyclogenesis and heavy snowfall commenced immediately along the coast.
- Strong winds greater than 30 m s^{-1} circled the rapidly expanding circulation on 4 February, with the easterly LLJ located just north of the deepening 850-hPa center directed toward the region of heaviest snowfall at 1200 UTC 4 February.

d. 500-hPa geopotential height and 400-hPa wind analyses

- As with the previous two storms, the El Niño–Southern Oscillation (ENSO) is near neutral. During January through mid-February, the North Atlantic Os-

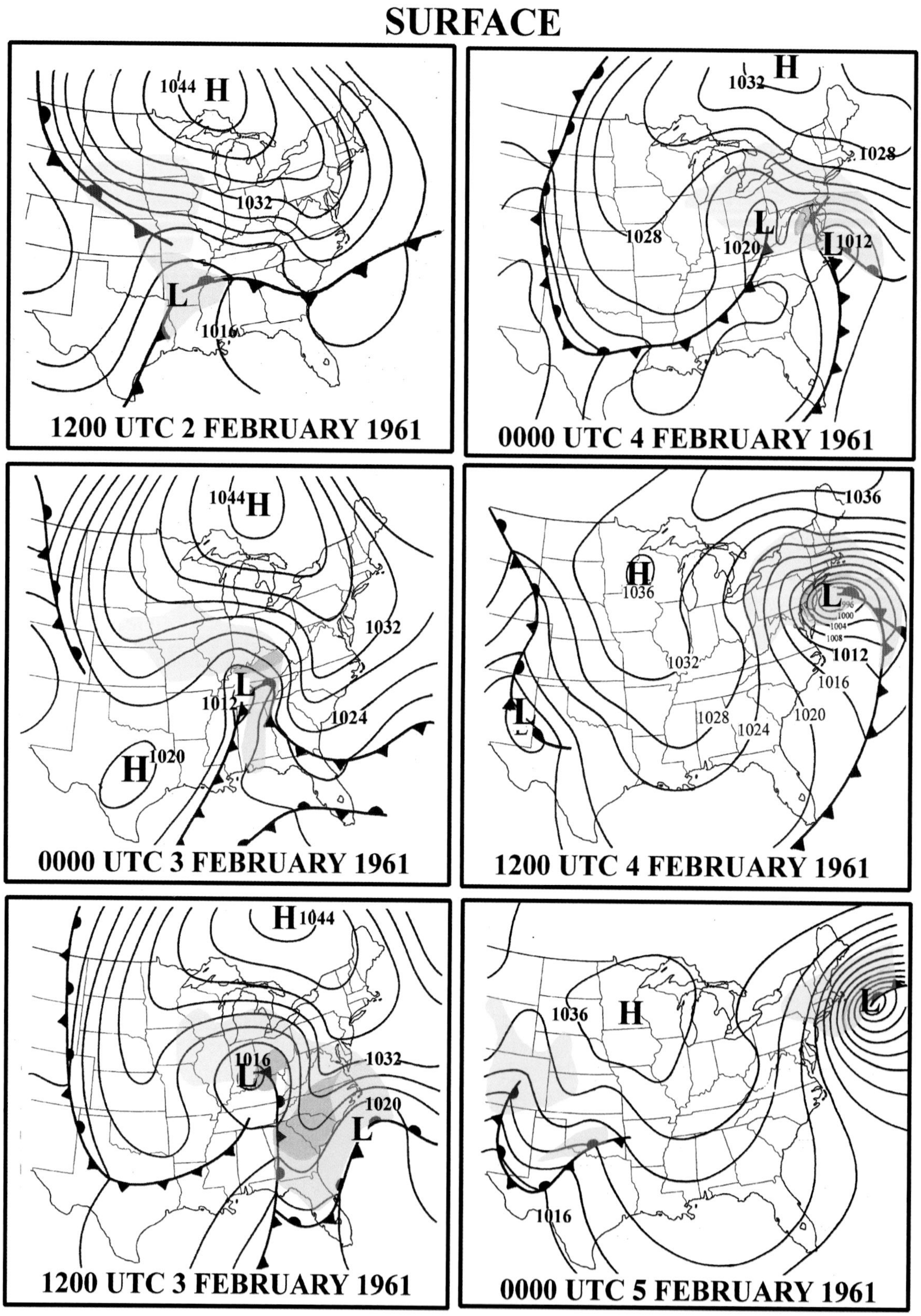

SURFACE

1200 UTC 2 FEBRUARY 1961

0000 UTC 4 FEBRUARY 1961

0000 UTC 3 FEBRUARY 1961

1200 UTC 4 FEBRUARY 1961

1200 UTC 3 FEBRUARY 1961

0000 UTC 5 FEBRUARY 1961

FIG. 10.7-2. Twelve-hourly surface weather analyses for 1200 UTC 2 Feb–0000 UTC 5 Feb 1961. See Fig. 10.1-2 for details.

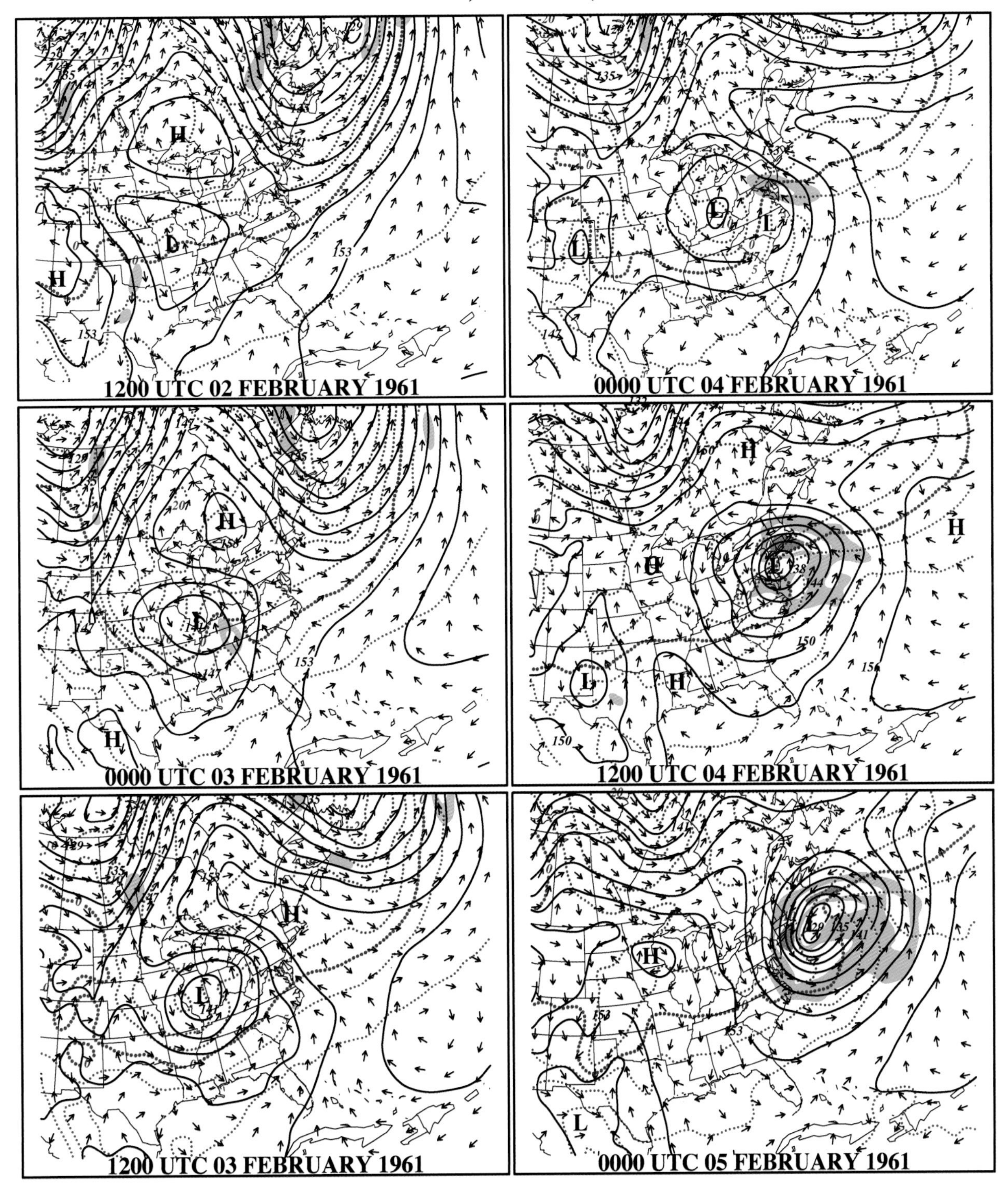

FIG. 10.7-3. Twelve-hourly 850-hPa analyses for 1200 UTC 2 Feb–0000 UTC 5 Feb 1961. See Fig. 10.1-3 for details.

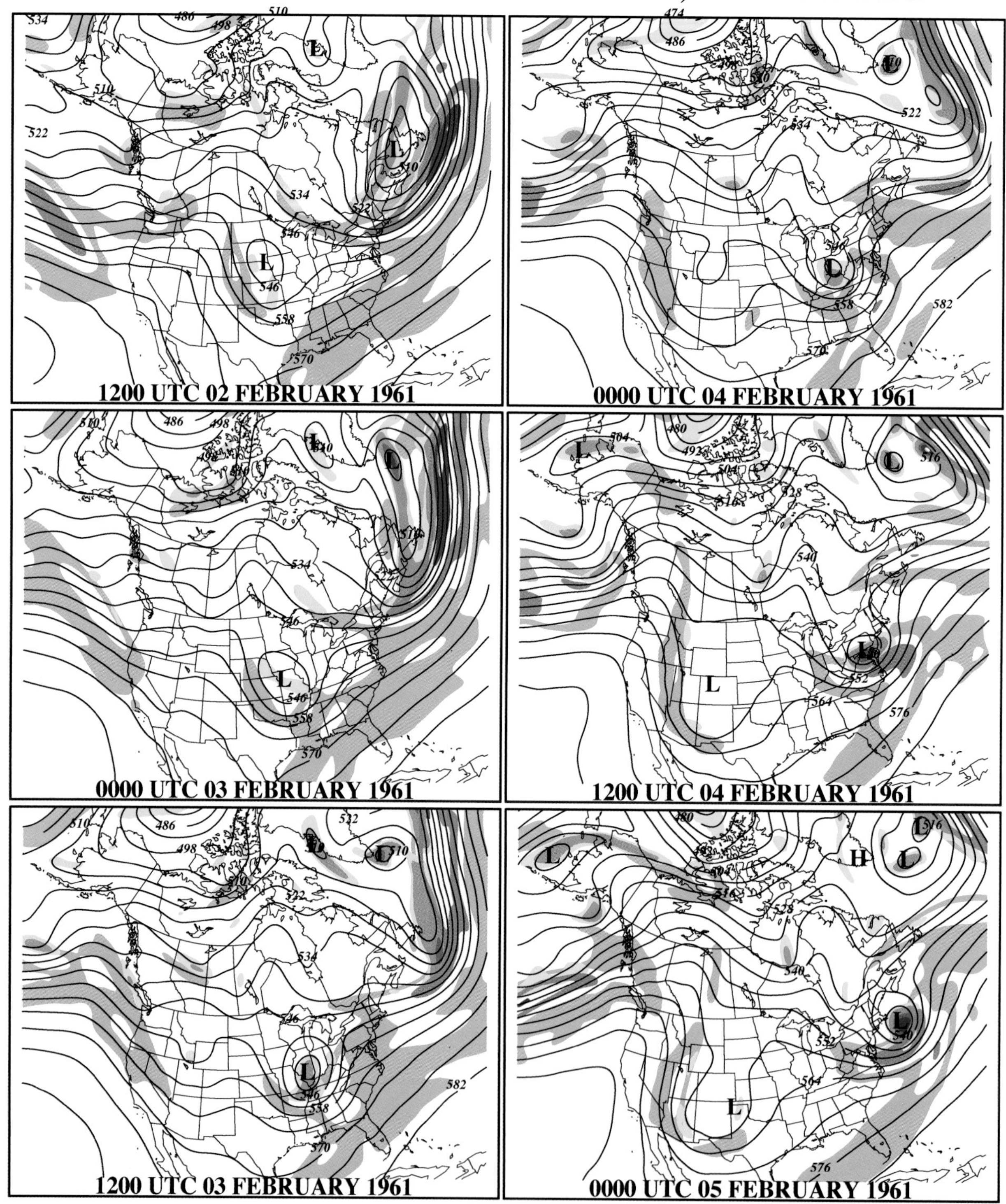

FIG. 10.7-4. Twelve-hourly analyses of 500-hPa geopotential heights and upper-level wind fields for 1200 UTC 2 Feb–0000 UTC 5 Feb 1961. See Fig. 10.1-4 for details.

250 hPa HEIGHTS AND WINDS

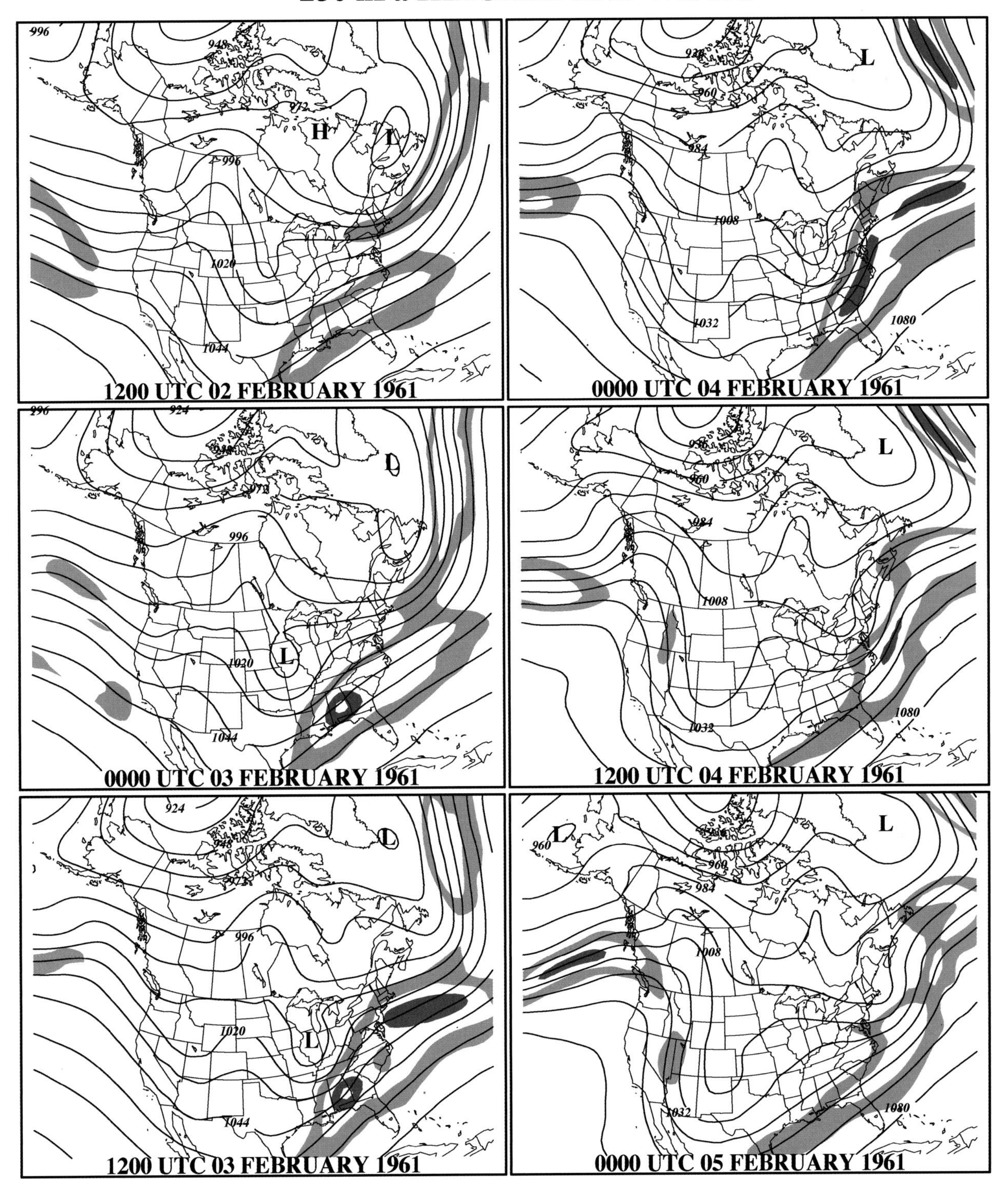

FIG. 10.7-5. Twelve-hourly analyses of 250-hPa geopotential height for 1200 UTC 2 Feb–0000 UTC 5 Feb 1961. See Fig. 10.1-5 for details.

cillation (NAO) fluctuated very slightly above and below zero, with this storm occurring during a week in which the NAO is very slightly negative. Analyses of 500-hPa data hint at a weak upper ridge over southern Greenland, but as with the two other snowstorms, there appears to be little relationship between the storm and the influences of ENSO or the NAO.

- The precyclogenetic environment over the northeastern United States was dominated by a large cyclonic circulation that drifted eastward across southeastern Canada. This vortex was associated with a region of pronounced confluence over southeastern Canada, the Great Lakes region, and the northeastern United States, directly above the cold surface high that extended southeastward from the Great Lakes region during this period.
- The sea level cyclone was associated with a closed 500-hPa low that drifted slowly eastward from the central plains states to the East Coast from 2 to 4 February. This circulation was collocated with a well-defined cyclonic vorticity maximum and was associated with strong vorticity advections downwind of the trough axis.
- The closed cyclonic circulation was evident over the plains states prior to any major development at the surface. It deepened by 120 m between 0000 UTC 4 February and 0000 UTC 5 February, as the secondary cyclone developed rapidly off the East Coast. Geopotential height gradients increased at the base of the trough during this period.
- Intense sea level development occurred between 0000 and 1200 UTC 4 February, as diffluence downwind of the trough axis crossed the coastline and the trough axis acquired a northwest–southeast orientation (a negative tilt), while the downstream ridge over New England became better defined.
- This storm was associated with an upper-level trough and downstream ridge of relatively small amplitude and short half-wavelength, as compared with other cases. A small increase in amplitude between the trough and downstream ridge occurred as the sea level cyclone deepened after 0000 UTC 4 February. The half-wavelength between the trough and downstream ridge decreased between 1200 UTC 3 February and 1200 UTC 4 February, as the trough approached the East Coast and the surface cyclone underwent explosive development.
- This case was marked by a dual-jet pattern with a northern jet extending from the Great Lakes region to the North Atlantic Ocean during the precyclogenetic period and a southern jet associated with the trough that moved slowly from west to east and was located along the East Coast by 4 February.
- The entrance region of a 60–70 m s^{-1} polar jet within the confluent region over the northeastern United States coincided with the cold surface ridge that extended from New England to the middle Atlantic states on 2 and 3 February.
- The more southern jet with maximum wind speeds approaching 50 m s^{-1} propagated toward the base of the trough by 0000 UTC 3 February as the primary low developed just east of the trough axis and precipitation expanded into the Ohio Valley. At 1200 UTC 3 February, the axis of maximum wind speeds shifted east of the base of the trough, concurrent with an increase in the geopotential height gradients there. This jet appears to represent a merger of a polar and subtropical jet system, which is more evident at the 250-hPa level.
- The rapid development of the secondary cyclone along the coast, the distinct LLJ, and the expanding area of heavy snowfall all occurred in the diffluent exit region of the jet streaks located near the base of the trough after 0000 UTC 4 February.

e. 250-hPa geopotential height and wind analyses

- This case is marked by a general west-to-east flow pattern with the major trough over southeastern Canada and the developing trough over the central United States clearly evident in the precyclogenetic period before 0000 UTC 3 February.
- A strong subtropical jet exceeding 80 m s^{-1} extended from Texas to the Southeast coast during the entire storm period, although maximum winds within the jet diminished after 1200 UTC 4 February as the system moved off the coast.
- The northern jet system extending from the New Jersey coast eastward also amplified on 3 February, likely in response to the expanding area of heavy precipitation as the storm system began to develop just off the East Coast.

A student walks home from school, 13 Jan 1964 (courtesy of *Washington Weather: The Weather Sourcebook for the D.C. Area*, Historical Enterprises, 2002).

8. 11–14 January 1964

a. General remarks

- This system was a large, slow-moving storm that produced severe winter weather through much of the central and eastern United States. Blizzard conditions prevailed throughout the middle Atlantic states and southern New England, as temperatures fell below −7°C and wind speeds increased to greater than gale force. The heaviest snows fell across Pennsylvania, where amounts of 16–24 in. (40–60 cm) were common.
- The following regions reported snow accumulations exceeding 10 in. (25 cm): central and northern West Virginia, northern Maryland, the northern tip of Virginia, Pennsylvania (except the extreme northwest and southeast), New Jersey, the southeastern half of New York, portions of Connecticut and Massachusetts, Rhode Island, southern Vermont, southern New Hampshire, and the southern tip of Maine.
- The following regions reported snow accumulations exceeding 20 in. (50 cm): portions of Pennsylvania and south-central New York.

TABLE 10.8-1. Snowfall amounts for urban and selected locations for 11–14 Jan 1964.; see Table 10.1-1 for details.

NESIS = 5.74 (category 3)	
Urban center snowfall amounts	
Washington, D.C.–Dulles Airport	10.2 in. (26 cm)
Washington, D.C.–National Airport	8.5 in. (22 cm)
Baltimore, MD	9.9 in. (25 cm)
Philadelphia, PA	7.2 in. (18 cm)
New York, NY–Central Park	12.5 in. (32 cm)
Boston, MA	9.2 in. (23 cm)
Other selected snowfall amounts	
Williamsport, PA	24.1 in. (61 cm)
Scranton, PA	21.1 in. (54 cm)
Nantucket, MA	19.2 in. (49 cm)
Harrisburg, PA	18.1 in. (46 cm)
Pittsburgh, PA	15.6 in. (40 cm)
Hempstead, NY	14.7 in. (37 cm)

b. Surface analyses

- A large, intense anticyclone was located near the Minnesota–Canada border at 0000 UTC 12 January with a central pressure of 1046 hPa. Ridge axes extended south from the northern plains to Texas and southeast across the Great Lakes, then southward along the Atlantic coast. Cold temperatures at or below 20°F (−6°C) were established across the northeastern states on 11–12 January as sea level pressures rose over the middle Atlantic states and New England. This high pressure ridge remained north of New York during the storm period, serving as a continuous source of cold air along the coast.
- Cold-air damming was in evidence along much of the East Coast at 1200 UTC 12 January and 0000 UTC 13 January. The characteristic inverted ridge of high pressure extended from the middle Atlantic states into the southeastern states.
- A slowly intensifying primary low pressure center tracked from Kansas across Tennessee to Kentucky before weakening over western Virginia by 1200 UTC 13 January, 24 h after the onset of secondary cyclogenesis along the East Coast. The primary low was accompanied by significant snowfall from Missouri and Iowa up the Ohio Valley into Pennsylvania, with accumulations exceeding 10 in. (25 cm).
- A secondary cyclone developed as a series of low pressure centers over the southeastern United States consolidated into one cyclone center off the North Carolina coast late on 12 January. The secondary cyclogenesis commenced before 1200 UTC 12 January during a period that featured coastal frontogenesis, cold-air damming, and the development of moderate to heavy precipitation west of the coastal front.
- The secondary cyclone had an erratic history of intensification. The low deepened rapidly between 1200 UTC 12 January and 0000 UTC 13 January, did not deepen at all early on 13 January, then deepened rapidly again later on 13 January off the southern New England coast; coincident with this behavior was the intermittent nature of the snowfall in New York City and Long Island, with the heaviest snowfall not developing until after 1800 UTC 13 January, well after it was forecast.
- Both the primary and secondary low pressure centers moved at a relatively slow rate, averaging 9–12 m s^{-1}. The slow movement and erratic intensification created many problems for forecasters concerned with the onset time and duration of the snowfall throughout the Northeast.

c. 850-hPa analyses

- Northwesterly flow and cold-air advection were located beneath a region of upper-level confluence over the northeastern United States between 1200 UTC 11 January and 1200 UTC 12 January. The cold northwesterly flow across New England persisted until 1200 UTC 13 January.
- A well-defined 850-hPa low center drifted eastward from Kansas to West Virginia and deepened very slowly from 11 through 13 January. A secondary 850-hPa low began to form near the South Carolina–Georgia border at 1200 UTC 12 January. This center deepened at a more rapid pace than did the primary low, as it propagated northeastward along the East Coast. Between 1200 UTC 13 January and 0000 UTC 14 January, the two 850-hPa low centers consolidated into one strong center near the coast, just south of New York City.
- The 0°C isotherm was displaced to the south and east

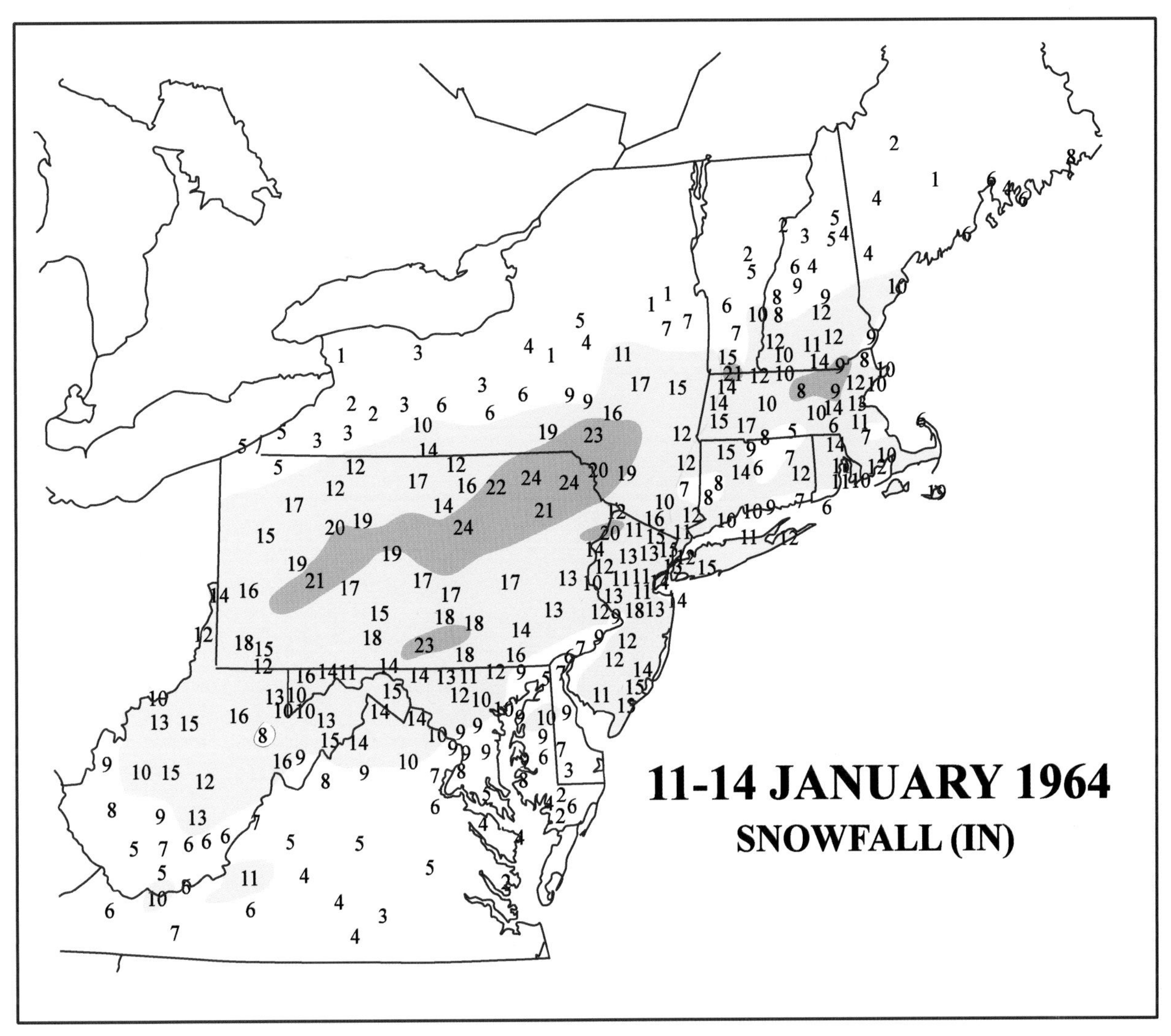

FIG. 10.8-1. Snowfall (in.) for 11–14 Jan 1964. See Fig. 10.1-1 for details.

of the initial 850-hPa low on 11–12 January and was later located to the north of the secondary 850-hPa center at 0000 UTC 13 January. The east-to-west orientation of the 850-hPa isotherms at 1200 UTC 11 January evolved into an S-shaped pattern by 0000 UTC 12 January as cold air was advected southward behind the 850-hPa low and warm air was drawn northward ahead of it.

- The temperature gradients increased along the middle Atlantic coast during 12 January with the onset of secondary cyclogenesis, coastal frontogenesis, and the formation of the low-level jet (LLJ).
- The development of a secondary 850-hPa low center during 12 January was associated with the formation of a 20–25 m s^{-1} south-to-southeasterly LLJ along the Florida to South Carolina coast. These developments coincided with the secondary sea level cyclogenesis and coastal frontogenesis off the Southeast coast, and the outbreak of moderate to heavy precipitation across Georgia and South Carolina. Nevertheless, the evolution of the LLJ along the East Coast after 1200 UTC 12 January was not straightforward, with no strong jet analyzed (as in other cases) to the north of the circulation centers.
- An easterly LLJ was established to the north of both the primary and secondary 850-hPa low centers on 12 and 13 January, with wind speeds exceeding 20 m s^{-1} in the Ohio Valley at 0000 UTC 13 January and barely exceeding 20 m s^{-1} from the New Jersey coast to Pennsylvania.
- The development of a coastal height-fall center and LLJ over the 12-h period ending at 1200 UTC 12 January with a noticeable sharp turn in the wind field to a southeasterly direction is similar to that which occurred in the February 1979 "Presidents' Day Storm" [Bosart (1981); Uccellini et al. (1984, 1987); see chapter 11.18]. The LLJ and height-fall center formed at a considerable distance from the primary

FIG. 10.8-2. Twelve-hourly surface weather analyses for 1200 UTC 11 Jan–0000 UTC 14 Jan 1964. See Fig. 10.1-2 for details.

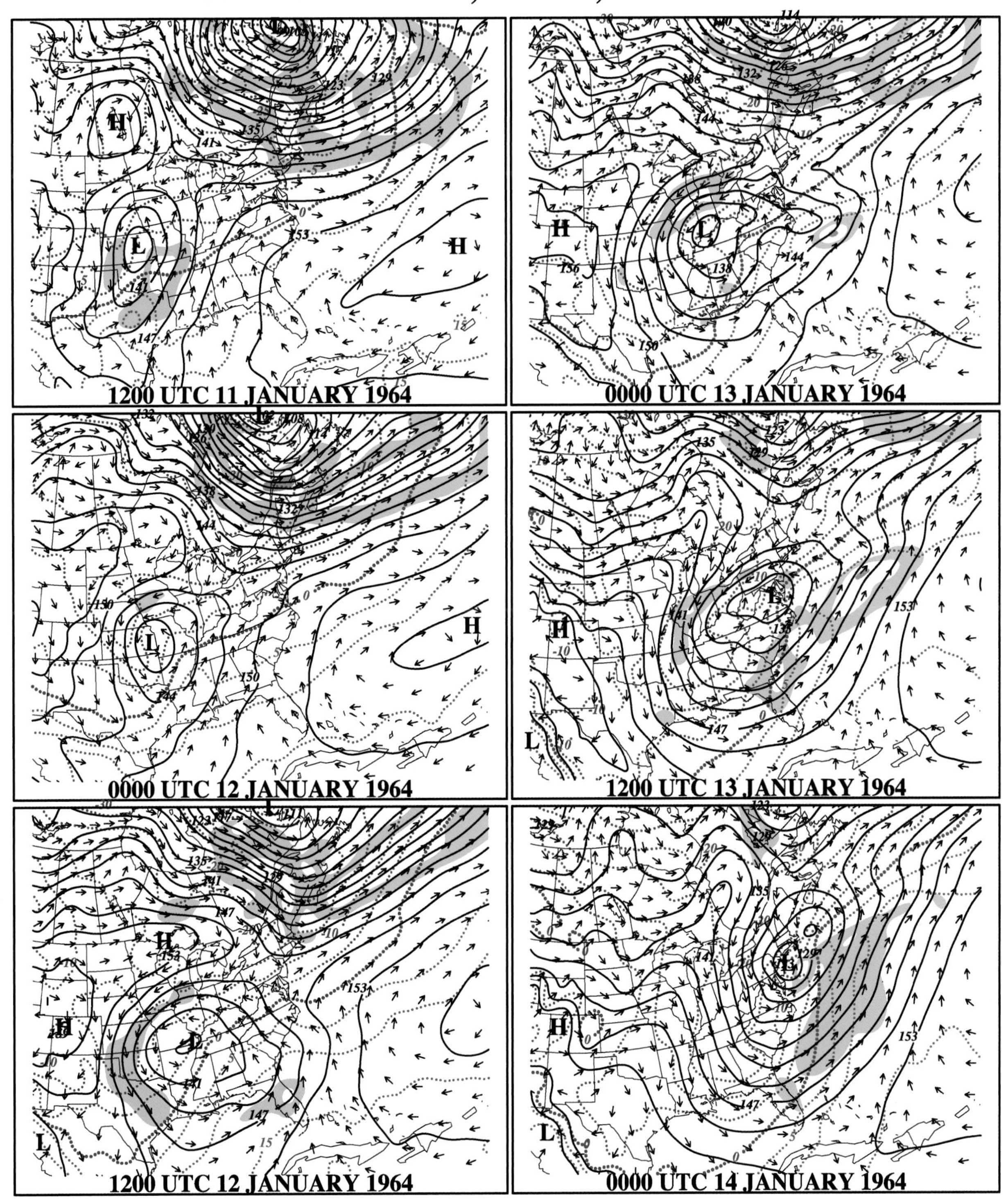

FIG. 10.8-3. Twelve-hourly 850-hPa analyses for 1200 UTC 11 Jan–0000 UTC 14 Jan 1964. See Fig. 10.1-3 for details.

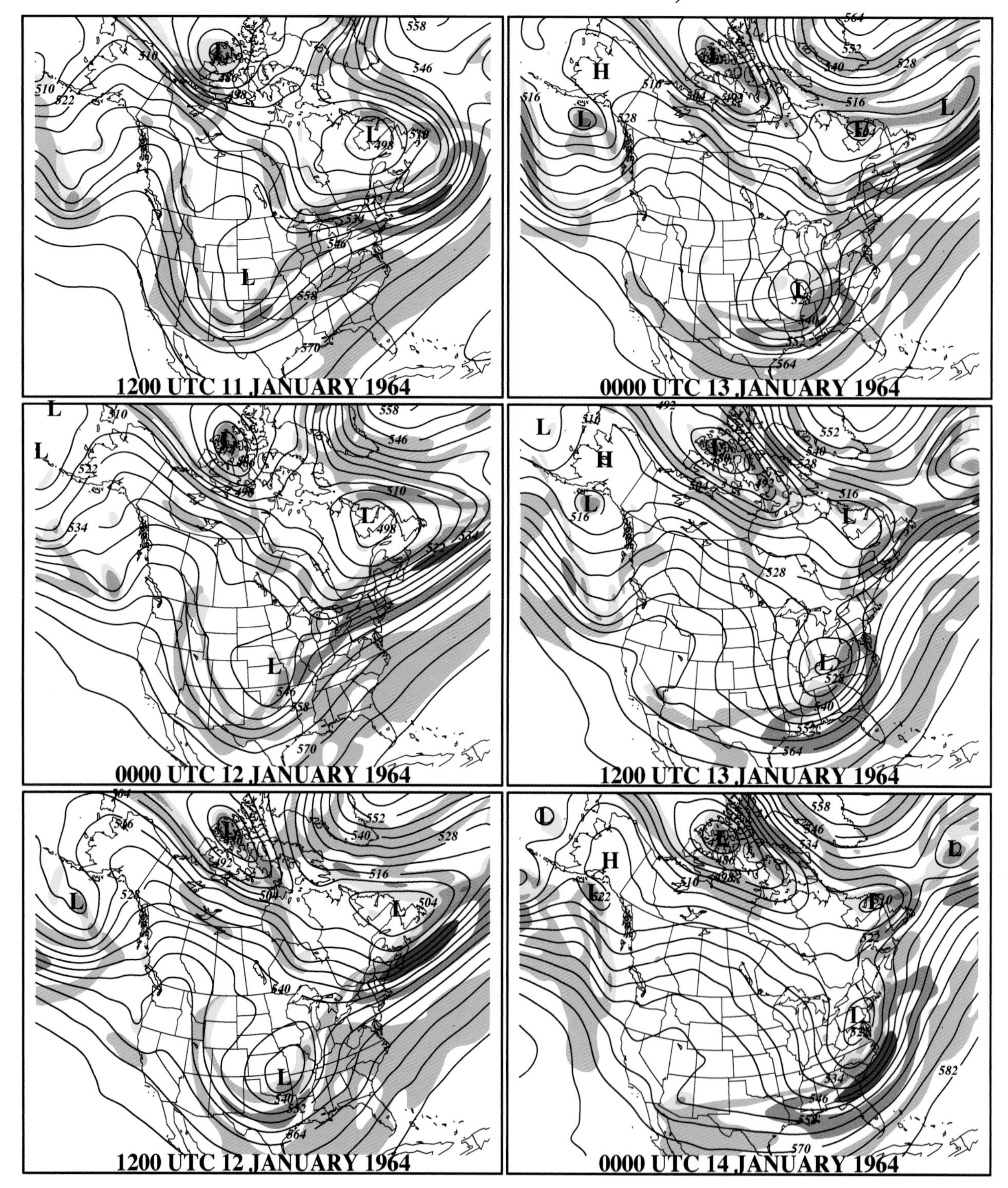

FIG. 10.8-4. Twelve-hourly analyses of 500-hPa geopotential heights and upper-level wind fields for 1200 UTC 11 Jan–0000 UTC 14 Jan 1964. See Fig. 10.1-4 for details.

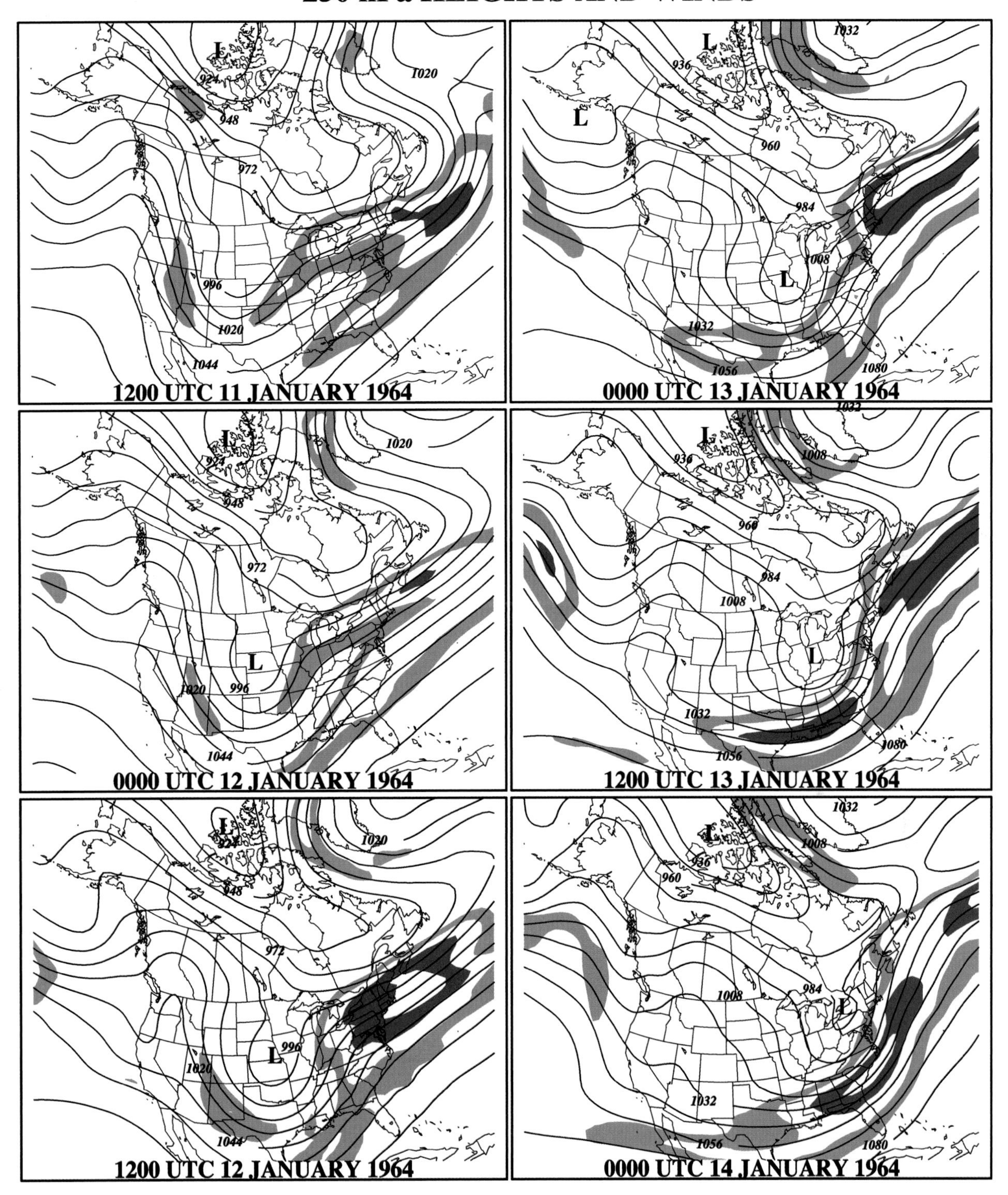

FIG. 10.8-5. Twelve-hourly analyses of 250-hPa geopotential height for 1200 UTC 11 Jan–0000 UTC 14 Jan 1964. See Fig. 10.1-5 for details.

850-hPa low located over the central United States at 1200 UTC 12 January. These features developed at the 850-hPa level while a subtropical jet strengthened at 200 hPa, cold-air damming occurred east of the Appalachian Mountains, and coastal frontogenesis and heavy precipitation were developing near the Southeast coast. The LLJ increased the thermal and moisture advections across the middle Atlantic states and acted to enhance the moisture transport toward the expanding precipitation area observed in the Southeast on 1200 UTC 12 January.

d. 500-hPa geopotential height and 400-hPa wind analyses

- This snowstorm developed during a winter that was similar to the winter of 1957/58. It was a winter during which a strong El Niño was present (although it would quickly evolve into a La Niña pattern by March) and was also a winter dominated by negative values of the North Atlantic Oscillation (NAO). The snowstorm developed in a period during which daily values of the NAO were negative and dropping sharply. At 500 hPa, evidence of the negative NAO may be found in the westward expansion of a large upper ridge to a position just east of Greenland over the Atlantic Ocean.
- An intense anticyclonic circulation over Greenland and a strong cyclonic vortex over eastern Canada established a region of confluent geopotential heights across the northeastern United States from 1200 UTC 11 January through 1200 UTC 13 January. This confluent flow regime was located above the rising sea level pressures that accompanied a surge of cold air from the Great Lakes to the middle Atlantic coast on 12 January.
- A separate trough was located in the central United States on 11 and 12 January, downstream from a high-amplitude ridge over the western United States. This trough evolved into a rather intense 500-hPa circulation by 0000 UTC 13 January as the primary surface cyclone moved into the Ohio Valley and the primary surface cyclone moved into the Ohio Valley, with the vortex sustaining its intensity during the remainder of the storm. The large-amplitude trough–vortex system drifted across the central United States as the primary surface low developed in Kansas and moved slowly to the Ohio Valley on 11 and 12 January. The circulation center deepened from 5300 m at 1200 UTC 12 January to 5120 m by 0000 UTC 14 January, as the secondary surface low formed along the East Coast.
- An intense area of cyclonic vorticity was collocated with the vortex in association with cyclonic wind shear at the base of the trough. The sea level cyclone developed off the East Coast in a region of cyclonic vorticity advection into a diffluent region east of the vortex, especially on 13 January.
- The amplitude of the trough and its flanking ridges did not appear to change significantly immediately prior to or during secondary cyclogenesis. The amplitude of the trough and downstream ridge increased only slightly throughout the study period.
- The half-wavelength between the trough and downstream ridge decreased only slightly, primarily between 1200 UTC 11 January and 0000 UTC 13 January, when secondary surface cyclogenesis was under way.
- A dual-jet pattern slowly evolved for this case but is clearly established by 0000 UTC 13 January with a northern jet in the confluent zone over the Great Lakes to the Atlantic Ocean during the precyclogenetic period and a southern jet associated with this distinct trough moving toward the coast.
- Missing wind reports over the northeastern United States at 1200 UTC 11 January and 0000 UTC 12 January were indicative of the strong winds associated with a highly confluent polar jet. During this period, high sea level pressure and cold air began to surge into the northeastern United States beneath the confluent entrance region of this jet..
- A polar jet with maximum wind speeds of 50–60 m s^{-1} propagated into the base of the trough over the east-central United States by 0000 UTC 13 January, as the development of the secondary cyclone was initiated near the East Coast. Analyzed winds in the base of the trough at 0000 and 1200 UTC 13 January increased to near 70 m s^{-1} while secondary cyclogenesis was in progress just along the East Coast.
- The secondary low developed in the diffluent exit region of this southern jet system extending to just east of the trough axis and in the right-entrance region of the confluent jet over New England at 0000 and 1200 UTC 13 January.

e. 250-hPa geopotential height and wind analyses

- The broad trough over southeastern Canada and the developing trough in the central United States (downwind of the West Coast ridge) are clearly evident.
- Wind speeds at 200 and 250 hPa increased significantly over the middle Atlantic states and New England in the 12-h period ending at 1200 UTC 12 January, strengthening to greater than 80 m s^{-1}. This feature developed and amplified at the crest of the downstream ridge as a coastal front, secondary cyclone, and heavy precipitation were developing over the southeastern United States. The amplification of this upper-level and the low-level jet along the Southeast coast (see the 850-hPa maps), in conjunction with the development of heavy precipitation in Georgia and South Carolina at 1200 UTC 12 January, is similar to the sequence observed in the February 1979 Presidents' Day Storm [Uccellini et al. (1984, 1987); see

chapter 11.18]. This sequence is also evident in other cases throughout the sample.

- An intensifying subtropical jet, with maximum winds exceeding 70 m s^{-1} by 1200 UTC 13 January and 80 m s^{-1} by 0000 UTC 14 January, was located over the southeastern United States. The amplification occurred as the surface low intensified and propagated northeastward along the East Coast on 13 January.

Hamilton Street in Hyattsville, Maryland, during a blizzard, 30 Jan 1966 (photo courtesy of Kevin Ambrose, *Blizzards and Snowstorms of Washington, D.C.*, Historical Enterprises, 1993).

9. 29–31 January 1966

a. General remarks

- This storm is referred to as the Blizzard of '66. The storm was the third and most severe in a series of three snowstorms that occurred over a 10-day period along the middle Atlantic coast. Heavy snow fell from North Carolina northward to New York and combined with temperatures below −10°C and wind gusts in excess of 25 m s^{-1} to create widespread blizzard conditions.
- The axis of heaviest snowfall extended from southwest to northeast across Virginia, but shifted to a more north–south orientation north of Virginia. Maximum snowfall was displaced west of the immediate coast from New Jersey to New England, sparing the major metropolitan areas from Philadelphia across much of southern New England the storm's worst effects.
- Cyclone snowfall combined with "lake effect" snow over parts of New York state to produce record accumulations of 60–100 in. (150–250 cm) immediately south and east of Lake Ontario.
- The following regions had snow accumulations exceeding 10 in. (25 cm): parts of North Carolina, central Virginia, parts of West Virginia, Maryland (except the extreme west), Delaware, northwestern New Jersey, central and eastern Pennsylvania, New York (except the northeast, southeast, and parts of the southwest), and scattered locations throughout New England.
- The following region had snow accumulations exceeding 20 in. (50 cm): central New York.

b. Surface analyses

- The cyclone developed along the advancing edge of a record setting cold air mass associated with a 1055-hPa anticyclone over northern Canada.
- Cold-air damming was observed early on 29 January in conjunction with an inverted high pressure ridge near the Southeast coast. Concurrently, a coastal front developed just offshore.
- The surface low pressure center initially moved along the Gulf coast at a rate of about 15 m s^{-1} before turning northeastward along the Southeast coast on 29 January. The low deepened slowly as it propagated northeastward along the coastal front through the afternoon of 29 January. The center tended to "jump" or redevelop northeastward along the coastal front near the Carolina coast late on 29 January. It advanced at 25 m s^{-1} as it crossed eastern North Carolina by 0000 UTC 30 January.
- Once the cyclone center passed eastern North Carolina, it took a more northerly track late on 29 January, deepening explosively from 0000 UTC 30 January until 1800 UTC 30 January as the central pressure fell from 997 to 970 hPa.
- Very heavy snow fell from Virginia to New York during this stage of rapid development. The subsequent track of the storm took it inland across extreme eastern New Jersey and New York City, then up the Hudson Valley of New York. To the west of this path, temperatures never rose much above −10°C, and north-northwest surface winds strengthened, creating blizzard conditions from central Virginia to central New York.
- A wedge of drier and warmer air raced up the coast near and to the east of the low center. Temperatures approached 5°C in New York City and southern New England, cutting snow accumulations in these regions. Snowfall ceased or changed to rain for a 6–9-h period as the low passed to the west before changing back to snow with falling temperatures and southerly winds just before 1800 UTC 30 January.
- Intense sea level pressure gradients formed, especially to the west and north of the low center, prolonging high winds even after the low pressure system had passed.

TABLE 10.9-1. Snowfall amounts for urban and selected locations for 29–31 Jan 1966; see Table 10.1-1 for details.

NESIS = 5.05 (category 3)	
Urban center snowfall amounts	
Washington, D.C.–National Airport	13.8 in. (35 cm)
Baltimore, MD	12.1 in. (31 cm)
Philadelphia, PA	8.3 in. (21 cm)
New York, NY–Central Park	6.8 in. (17 cm)
Boston, MA	6.3 in. (16 cm)
Other selected snowfall amounts*	
Syracuse, NY	39.0 in. (99 cm)
Rochester, NY	26.7 in. (68 cm)
Binghamton, NY	20.1 in. (51 cm)
Roanoke, VA	12.3 in. (31 cm)
Harrisburg, PA	12.2 in. (31 cm)

* Snowfall amounts for the New York cities are accumulations through 31 Jan.

c. 850-hPa analyses

- A strong northwesterly flow of very cold air was maintained across much of the north-central and northeastern United States through 1200 UTC 29 January, suppressing the 0°C isotherm southward to the Gulf states. Most of the northern states were covered by 850-hPa temperatures below −20°C.
- The 850-hPa low intensified only slightly through 0000 UTC 29 January, then deepened at a rate of −60 m (12 h)$^{-1}$ over the 24-h period ending at 0000 UTC 30 January, as the surface low reached North Carolina. During the following 12-h period, the surface low intensified explosively and the 850-hPa center also deepened rapidly [−150 m (12 h)$^{-1}$]. A slower rate of intensification [−60 m (12 h)$^{-1}$] was then observed in the 12-h period ending at 0000 UTC 31 January, with a major vortex covering the northeastern quadrant of the United States by this time.

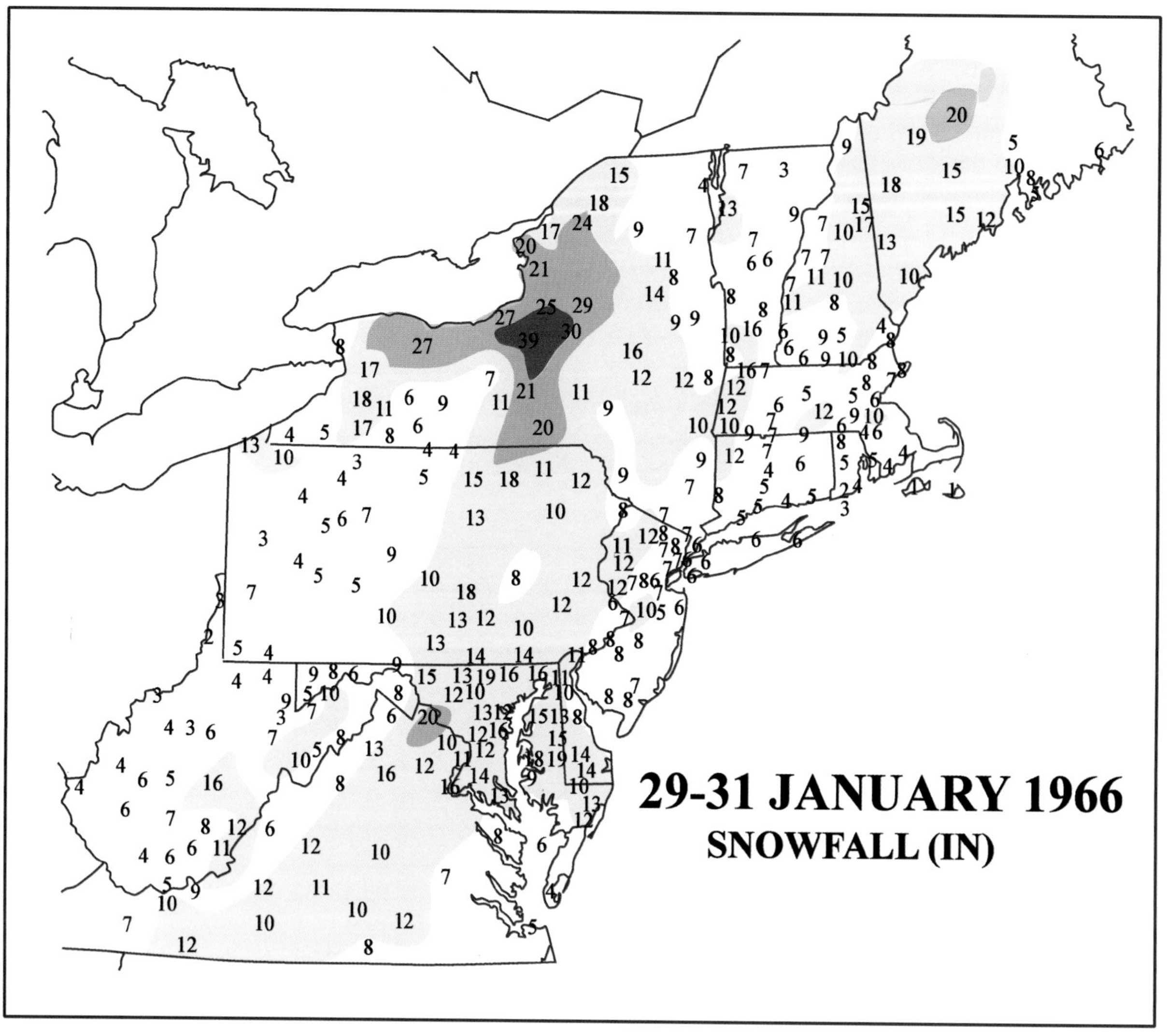

FIG. 10.9-1. Snowfall (in.) for 29–31 Jan 1966. See Fig. 10.1-1 for details.

- The 850-hPa low developed along the southern edge of the intense baroclinic zone that extended from the southern plains to the middle Atlantic coast. The pronounced temperature gradient was maintained over the middle Atlantic states within a confluent upper-tropospheric geopotential height pattern on 28 and 29 January. The east–west orientation of the thermal field along the East Coast on 28–29 January changed to north–south on 30 January. This change in orientation occurred during the rapid development phase of the storm and was marked by a 24°C temperature difference between Washington, D.C. (−24°C), and New York City (0°C) at 1200 UTC 30 January. The 0°C isotherm was located near the 850-hPa low center, except during the occluding stage of the cyclone late on 30 January, when the center was found in the colder air over New England.
- An S-shaped isotherm pattern did not become clearly evident until 1200 UTC 30 January, during the period of explosive cyclogenesis. This pattern emerged as strong warm-air advection developed along the middle Atlantic coast by 0000 UTC 30 January, and across the northeastern states at 1200 UTC 3 January, with cold-air advection quite evident to the west and the south of the developing vortex.
- On 28–29 January, during the early stages of cyclogenesis along the Gulf coast, a southerly 20 m s^{-1} low-level jet (LLJ) was directed toward areas of heavy precipitation in advance of the 850-hPa low. A very strong southeasterly jet developed east of the 850-hPa low after 0000 UTC 30 January, in association with the rapid 850-hPa deepening. The area of maximum winds slowly moved around the developing vortex with wind speeds of 30–40 m s^{-1} along the southern New England coast by 1200 UTC 30 January. This strong low-level flow was directed toward the region of heavy snowfall, indicating that the vertical motions and moisture transport associated with the LLJ were

SURFACE

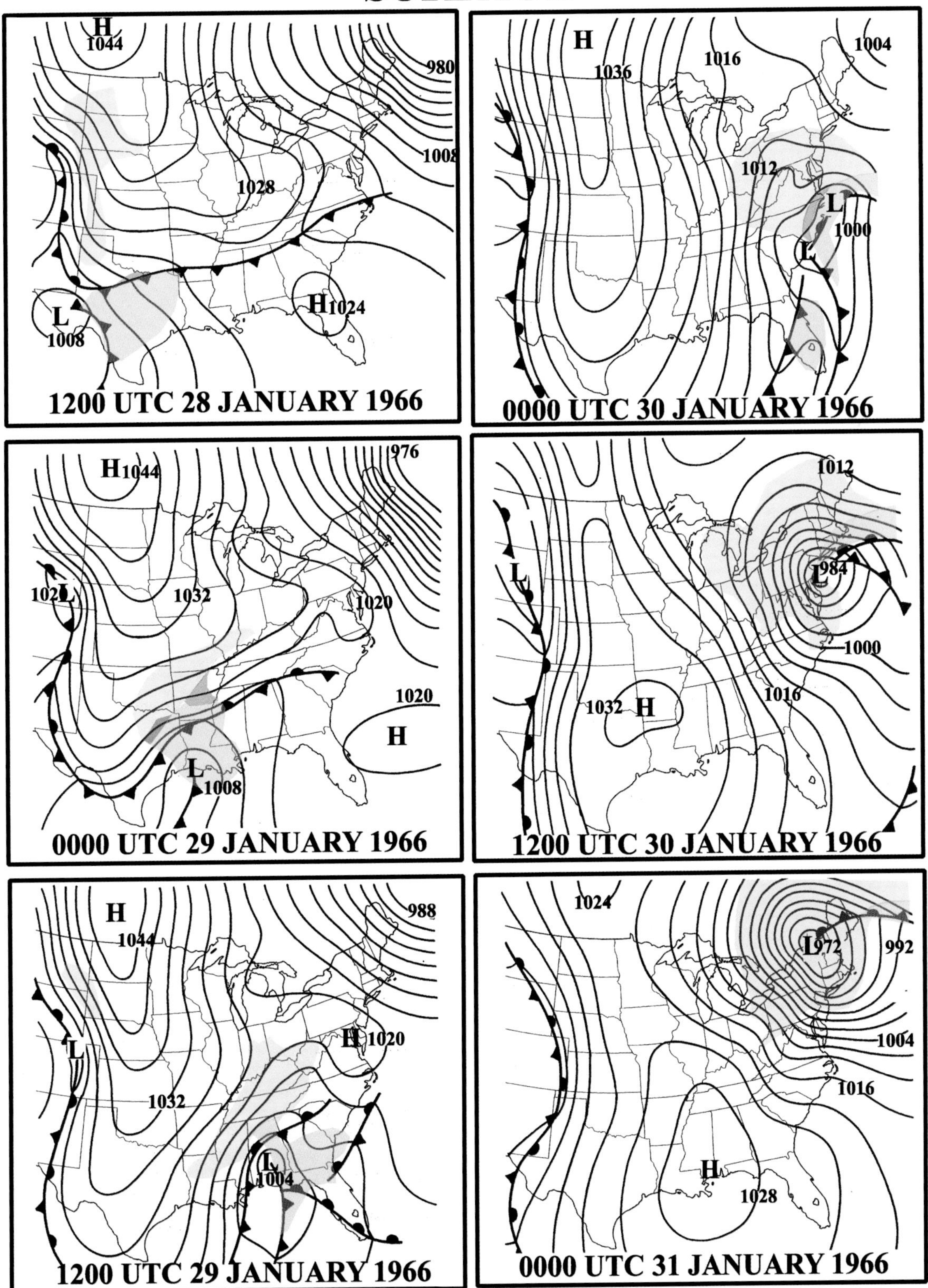

FIG. 10.9-2. Twelve-hourly surface weather analyses for 1200 UTC 28 Jan–0000 UTC 31 Jan 1966. See Fig. 10.1-2 for details.

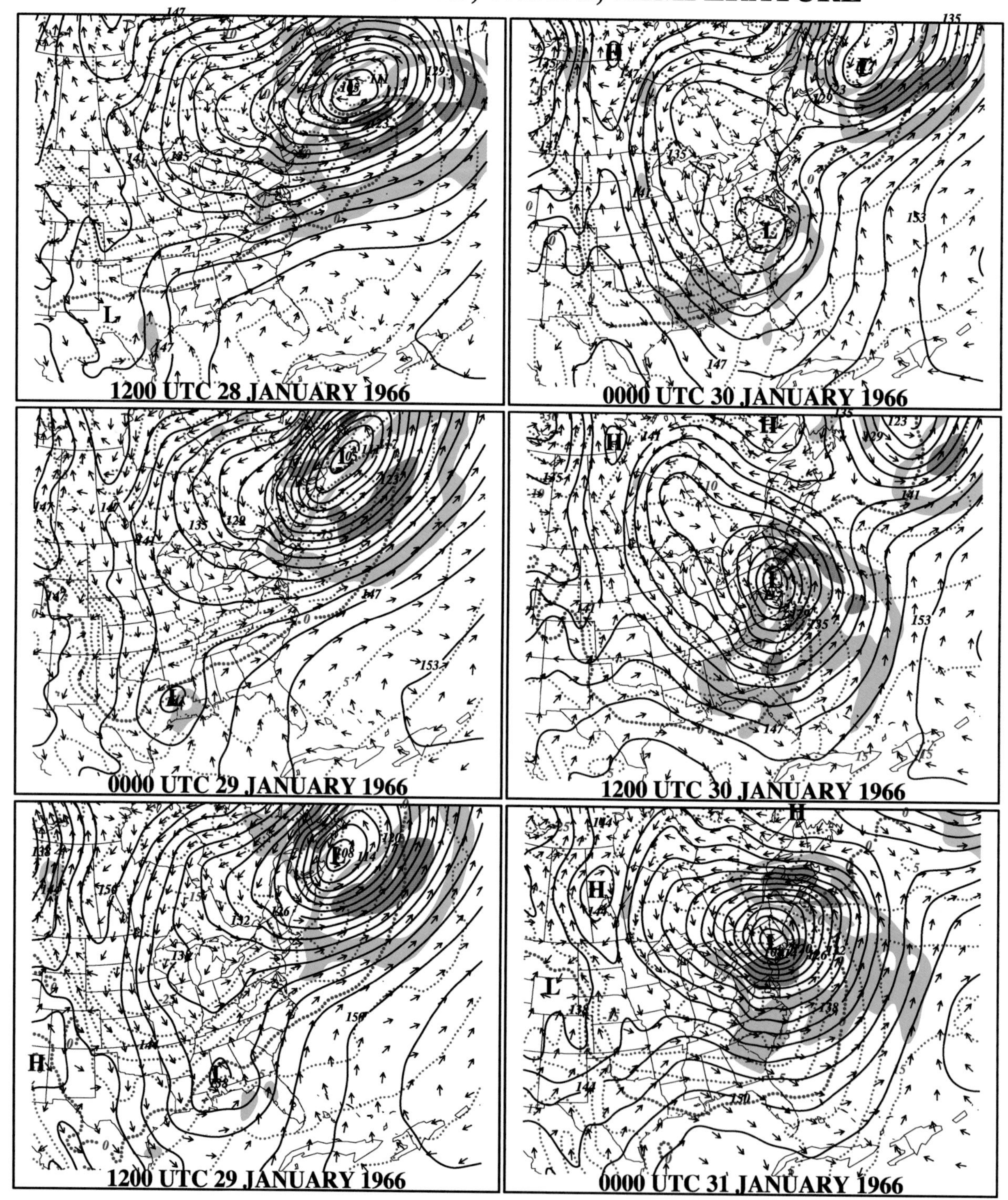

FIG. 10.9-3. Twelve-hourly 850-hPa analyses for 1200 UTC 28 Jan–0000 UTC 31 Jan 1966. See Fig. 10.1-3 for details.

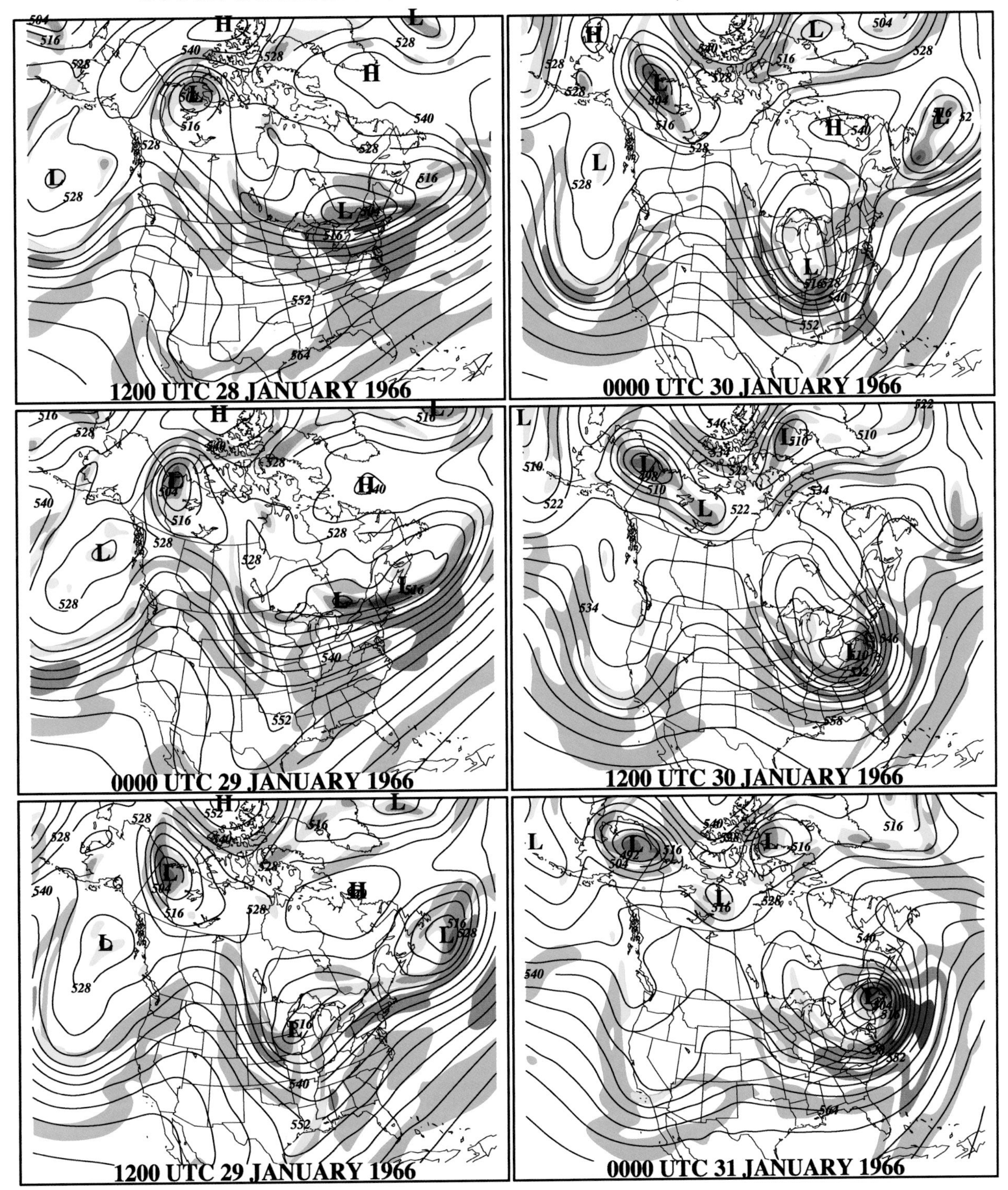

FIG. 10.9-4. Twelve-hourly analyses of 500-hPa geopotential heights and upper-level wind fields for 1200 UTC 28 Jan–0000 UTC 31 Jan 1966. See Fig. 10.1-4 for details.

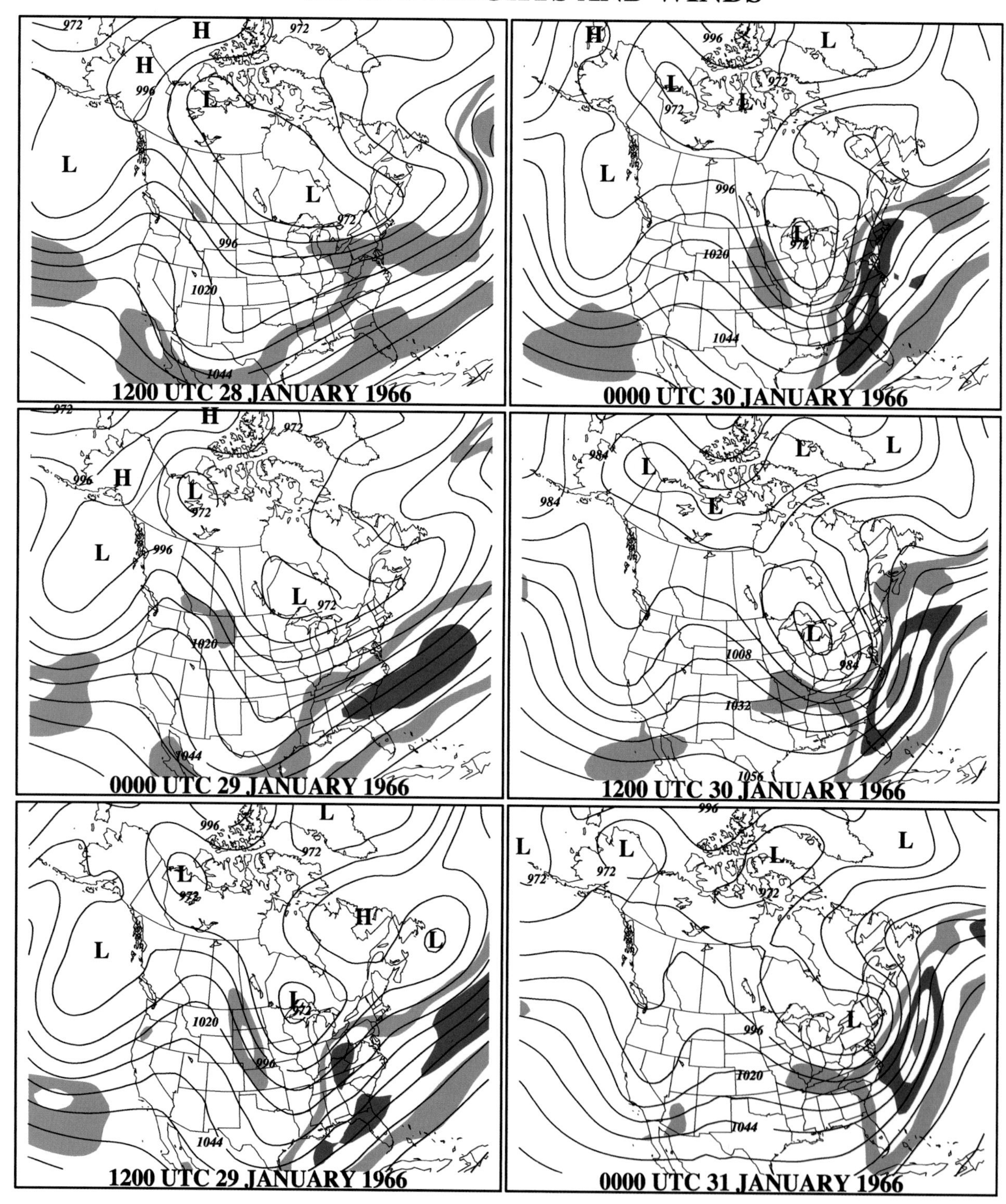

FIG. 10.9-5. Twelve-hourly analyses of 250-hPa geopotential height for 1200 UTC 28 Jan–0000 UTC 31 Jan 1966. See Fig. 10.1-5 for details.

important factors in the precipitation field expanding northward into western New York on 30 January.

d. 500-hPa geopotential height and 400-hPa wind analyses

- Prior to rapid cyclogenesis on 30 January, an amplifying ridge moved inland from the west coast as a trough propagated out of the southwestern United States. An anticyclonic circulation between Quebec and Greenland moved slowly westward, with an elongated cyclonic circulation extending from the Great Lakes eastward beyond Nova Scotia.
- This case was characterized by a major reorientation of the upper-level features. The propagating trough from the southwestern United States merged with a separate trough that propagated and amplified downwind of the ridge axis over the northwestern United States. This separate trough was marked by a distinct upper-level jet streak and an amplifying cyclonic vorticity maximum that tracked southeastward from Saskatchewan to Iowa between 1200 UTC 28 January and 1200 UTC 29 January.
- The merger of these features was associated with the splitting of the elongated closed low near the Canadian border into two separate vortices by 1200 UTC 29 January. One center moved eastward off the Newfoundland coast on 29 January. The other center, located near Lake Superior at 1200 UTC 29 January, rotated from an east–west to a north–south orientation by 0000 UTC 30 January, with sharply rising heights analyzed over the East Coast by 0000 UTC 30 January.
- The surface low deepened explosively on 30 January as the vorticity maximum that had dropped southeastward from the northern plains on 28–29 January neared the East Coast. The merger of the two troughs and the increasing cyclonic wind shear yielded one vorticity maximum near the North Carolina–Virginia coast by 1200 UTC 30 January.
- The rotation of the Great Lakes trough and its subsequent merger with the trough from the southwestern United States occurred as the amplitude of the combined trough system and upstream ridge increased substantially between 1200 UTC 28 January and 0000 UTC 30 January. The amplitude of the trough and the downstream ridge over the northeastern United States increased after 1200 UTC 29 January, as the surface low moved to the southeast coast, then intensified rapidly.
- The maximum intensification of the surface low occurred as the diffluent region downwind of the trough became better defined at 0000 UTC 30 January and the trough axis became oriented from northwest to southeast (a negative tilt) by 1200 UTC 30 January, with the surface low approaching New York City. This region also contained the strongest vorticity advection downwind of the diffluent trough axis.
- The vortex deepened significantly between 1200 UTC 30 January and 0000 UTC 31 January. The half-wavelength of the trough and downstream ridge axes decreased significantly as the surface low developed along the East Coast and heavy precipitation expanded across the Northeast.
- This case, like many others, was marked by a dual-jet pattern with a northern jet extending from the northeast United States eastward across the Atlantic Ocean during the precyclogenetic period (prior to 0000 UTC 30 January) and a complex southern jet system associated with the deepening trough that continued to amplify throughout the cyclogenetic period between 0000 UTC 30 January and 0000 UTC 31 January.
- Unlike other cases, these jets did not appear to become laterally coupled as the northern jet moved far offshore on 30 January while the southern jet moved northeastward along the East Coast by 0000 UTC 31 January.
- A 60 m s^{-1} polar jet streak was located over the northeastern United States on 28 January, immediately south of the elongated 500-hPa vortex. The extension of cold air across the eastern states prior to cyclogenesis occurred in the confluent entrance region of this jet.
- An intensifying polar jet streak with maximum wind speeds increasing from 40 to 60 m s^{-1} propagated southeastward from southwestern Canada to the plains states between 1200 UTC 28 January and 1200 UTC 29 January as the West Coast ridge amplified and the Great Lakes vortex began to rotate into a north–south orientation
- The strongest upper-level wind speeds developed and were consistently located downstream of the major trough axis as the entire wind field increased at the base of the trough as it amplified by 1200 UTC 30 January. The rapid cyclogenesis occurred in the diffluent exit region of the complex jet system that became better defined along the Southeast coast by 1200 UTC 30 January, and then propagated northeastward along the Atlantic coast by 0000 UTC 31 January as wind speeds increased to greater than 80 m s^{-1} at 400 hPa.
- By the time the intense jet system amplified over the East Coast, the upper-level ridge rapidly amplified as far north as southeastern Canada, forcing warmer air through much of New England east of the storm center as the entrance region of the northern jet moved far to the east of the southeastern Canadian coast.

e. 250-hPa geopotential and wind analyses

- The geopotential height field clearly depicts the major trough over southeastern Canada and confluent flow over the Great Lakes region that marked the cold, precyclogenetic environment in the north-central and northeastern United States. The height field also clearly shows the amplifying trough into the central and eastern United States by 30 January.

- An interesting wind feature was the development of a 70 m s^{-1} jet streak at 250 hPa across the middle Atlantic and New England states by 1200 UTC 29 January (wind speeds exceeded 80 m s^{-1} at 300 hPa). This jet developed near the crest of the downstream ridge as the southern and northern troughs merged in the central United States prior to rapid cyclogenesis on the East Coast and before heavy snowfall enveloped the middle Atlantic region by 0000 UTC 30 January.
- The increasing wind speeds to greater than 70 m s^{-1} over the Ohio Valley and 80 m s^{-1} over the mid-Atlantic coast and Southeast at 0000 UTC 30 January occurred as the heavy precipitation spread into these regions within the amplifying ridge downstream of the developing storm system.
- Maximum wind speeds associated with the subtropical jet extending across the southeastern United States at 250 hPa increased from 70 to greater than 90 m s^{-1} between 1200 UTC 28 January and 1200 UTC 30 January, during cyclogenesis. A coastal front developed along the Southeast coast and precipitation expanded across the Gulf coast region concurrent with the early period of increasing wind speeds.
- The strong winds along and over the western Atlantic at 400 and 250 hPa at 0000 UTC 31 January indicates a likely merger of the diffluent polar jet and the overlying subtropical jet during this period as the intense trough and very cold underlying air moved up along the East Coast, another indication of just how intense the dynamical pattern was for this storm.

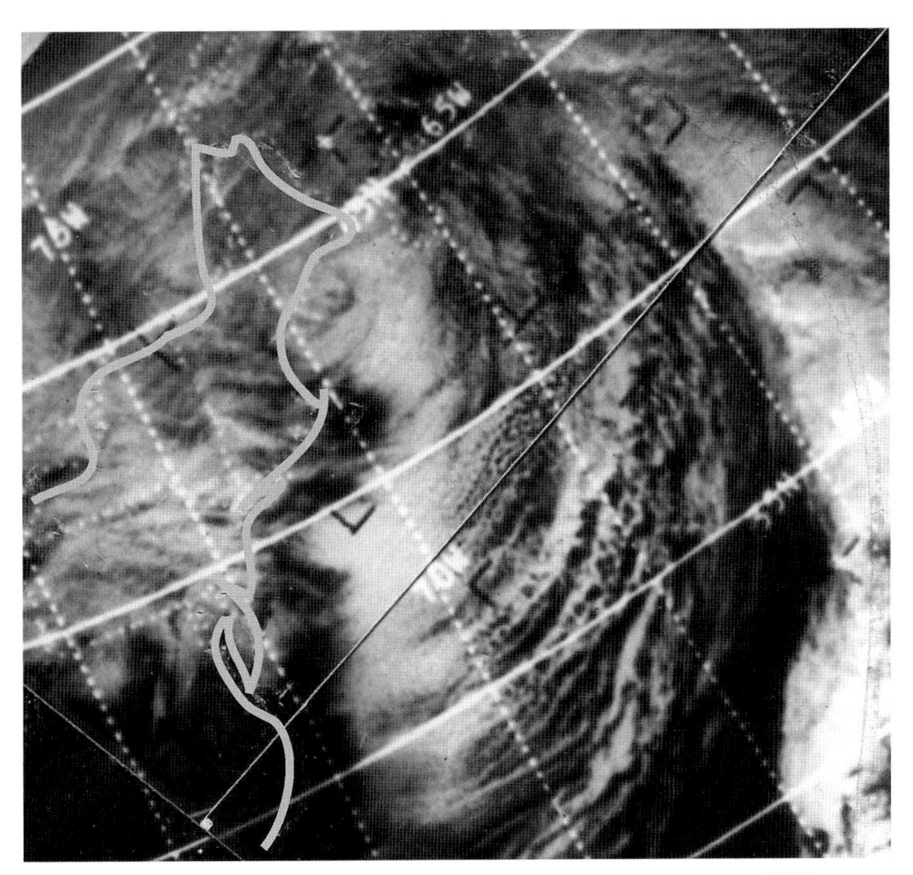

Essa-3 visible satellite photo on 25 Dec 1966 (image courtesy of NOAA/NESDIS).

10. 23–25 December 1966

a. General remarks

- This case is notable as a Christmas Eve snowstorm that deposited heavy snow over a wide area extending from the southern plains states to New England. An intriguing aspect of this storm was the numerous reports of thunderstorms with heavy snow from the middle Atlantic states to New England. Snowfall along the coast was reduced by a changeover to sleet during the afternoon of 24 December.
- The following regions reported snow accumulations exceeding 10 in. (25 cm): portions of central and northern Virginia, parts of Maryland and West Virginia, northern Delaware, eastern Pennsylvania, western New Jersey, eastern New York, and western New England.
- The following region reported snow accumulations exceeding 20 in. (50 cm): parts of eastern New York.

b. Surface analyses

- A strong high pressure cell located over the plains states was associated with a large mass of cold air covering the central and eastern United States on 22 and 23 December.
- Cold-air damming, as indicated by a weak inverted sea level pressure ridge, was observed over the middle Atlantic states between 0000 and 1200 UTC 24 December.
- Coastal frontogenesis was not observed along the Southeast coast, given the existence of a preexisting frontal zone located along the Carolina coast.
- The surface low developed along the edge of the cold-air outbreak and moved eastward to the South Carolina coast by the morning of 24 December. The center deepened slowly during this period, while advancing at a pace of 15 m s^{-1}. The cyclone then moved northeastward just off the East Coast and deepened rapidly during the ensuing 24-h period. The central pressure decreased 22 hPa in 24 h, reaching 978 hPa by the time the low was located along the Maine coast at 1200 UTC 25 December.
- The surface low exhibited a "center jump" along the preexisting frontal boundary near the Carolina coast around 1200 UTC 24 December, as the storm began to accelerate northeastward along the East Coast.

c. 850-hPa analyses

- A major vortex over southeastern Canada provided west-to-northwesterly flow over the north-central and northeastern United States, sustaining cold air in the lower troposphere into 24 December.
- The 0°C isotherm, oriented on an east–west axis, extended from a low over the Texas–Oklahoma border to southern Virginia at 1200 UTC 24 December.
- The 850-hPa low center deepened only slightly as it moved across the southern United States on 23 December, but intensified more rapidly after 0000 UTC 24 December as it neared the East Coast. The 12-hourly central height changes were -60 m (12 h)$^{-1}$ at 1200 UTC 24 December, -90 m (12 h)$^{-1}$ at 0000 UTC 25 December, and -120 m (12 h)$^{-1}$ at 1200 UTC 25 December.
- Strong cold-air advection west of the low center preceded the development of warm-air advection east of the center. The cold-air advection was observed by 0000 UTC 23 December, as strong winds developed nearly perpendicular to the band of isotherms that stretched across Texas. A strong northerly jet to the rear of the 850-hPa low dominated early cyclogenesis across the southern United States on 23 December. Otherwise, weak flow was observed near the low center prior to 24 December. Significant warm-air advection was not seen until 0000 UTC 24 December as the 850-hPa low propagated slowly eastward and southerly winds increased along the Southeast coast.
- The 850-hPa low propagated along the narrow and intense east–west temperature gradient. The isotherms evolved into an S-shaped configuration between 0000 and 1200 UTC 24 December as the warm-air advection pattern strengthened east of the low. The 0°C isotherm ran through the 850-hPa low center until the cyclone occluded. By 1200 UTC 25 December, the low center was located in colder air over New England.
- Wind speeds increased near and to the southeast of the low center by 1200 UTC 24 December as it neared the East Coast and began to intensify. A low-level jet (LLJ) developed to the south and east of the low as the surface cyclone was beginning a 24-h period of rapid deepening. Easterly flow strengthened north of the low by 0000 UTC 25 December, reaching 20–25 m s^{-1}, during rapid 850-hPa intensification, and was directed toward the area of heavy snowfall developing in Pennsylvania and New York.

TABLE 10.10-1. Snowfall amounts for urban and selected locations for 23–25 Dec 1966; see Table 10.1-1 for details.

NESIS = 3.79 (category 2)	
Urban center snowfall amounts	
Washington, D.C.–Dulles Airport	9.2 in. (23 cm)
Baltimore, MD	8.5 in. (22 cm)
Philadelphia, PA	12.7 in. (32 cm)
New York, NY–Central Park	7.1 in. (18 cm)
Boston, MA	5.7 in. (14 cm)
Other selected snowfall amounts	
Albany, NY	18.7 in. (47 cm)
Burlington, VT	14.9 in. (38 cm)
Allentown, PA	13.3 in. (34 cm)
Wilmington, DE	12.5 in. (32 cm)
Trenton, NJ	11.7 in. (30 cm)

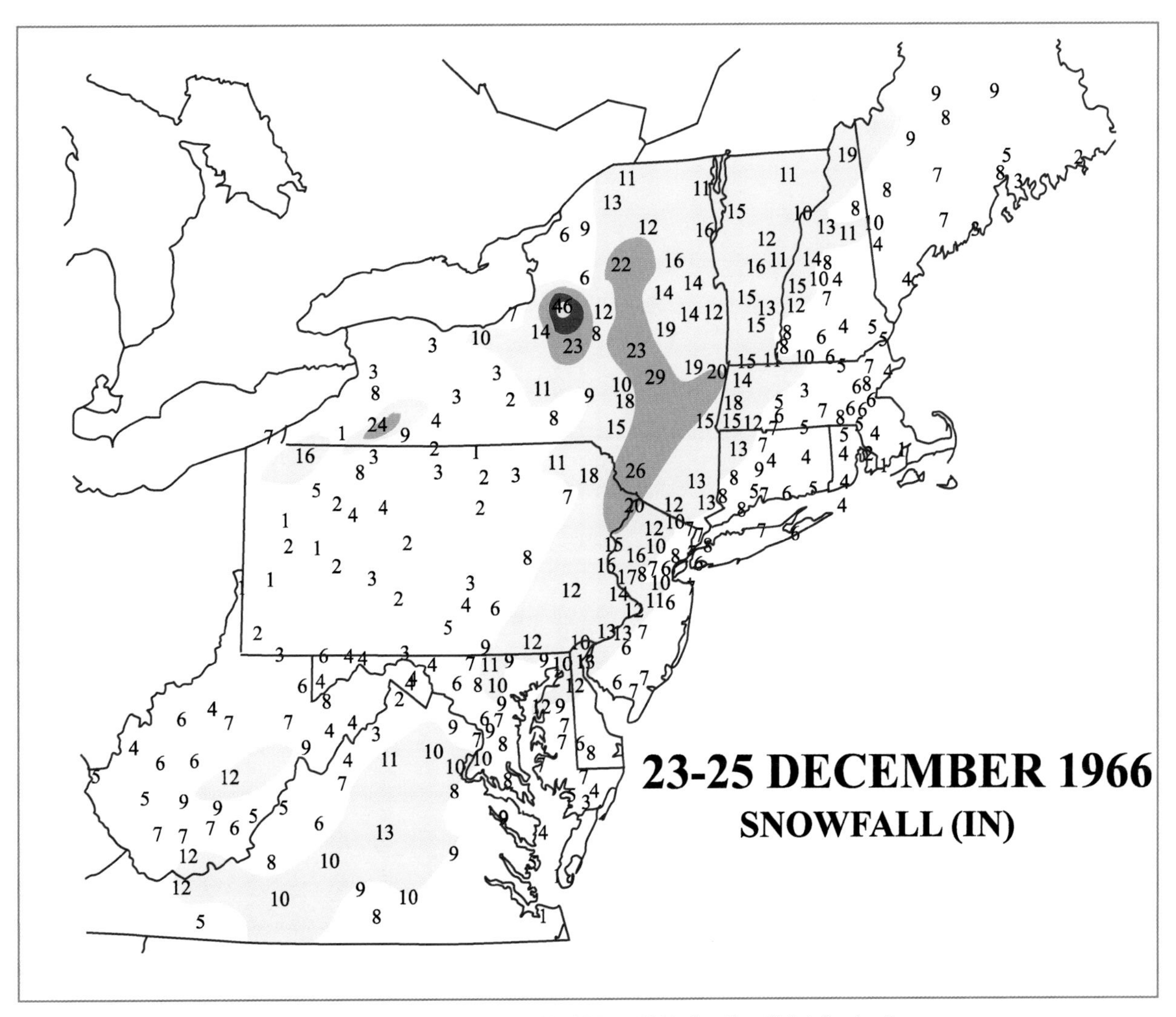

FIG. 10.10-1. Snowfall (in.) for 23–25 Dec 1966. See Fig. 10.1-1 for details.

d. 500-hPa geopotential height and 400-hPa wind analyses

- This storm was one of several snowstorms during the snowy winter of 1966/67 in the Northeast (Fig. 2–4, volume I). This storm occurred during a period when the effects of the El Niño–Southern Oscillation (ENSO) were relatively small, with a weak La Niña, and during a month when the North Atlantic Oscillation (NAO) was mainly positive. The storm did occur during a brief period when the NAO became decidedly negative for about a week. The 500-hPa analysis shows a trough over southern Greenland, which is not consistent with a negative NAO; a blocking pattern over eastern Canada, with a cutoff low near the U.S. border; and a developing cutoff ridge over northern Quebec, which is consistent with the negative NAO observed for this particular period.
- The precyclogenetic environment featured a ridge located just inland from the West Coast and a trough extending from Nebraska to New Mexico early on 23 December. A closed cyclonic circulation was located just north of Lake Huron with a trough axis extending eastward toward New England. An amplifying ridge over northeastern Canada was later associated with the development of a closed anticyclone that drifted slowly westward across northern Quebec.
- The juxtaposition of the Great Lakes vortex and the trough in the plains states produced confluence across the northern and eastern United States. The confluent region is observed above the surface high pressure ridge building eastward into the middle Atlantic states from the anticyclone center located over the plains states on 23 December.
- The development of the surface low occurred in conjunction with the eastward propagation of the trough initially over the central plains states at 0000 UTC 23 December as it interacted with the cyclonic vortex over the Great Lakes by 1200 UTC 24 December.
- A cyclonic vorticity maximum propagated northeast-

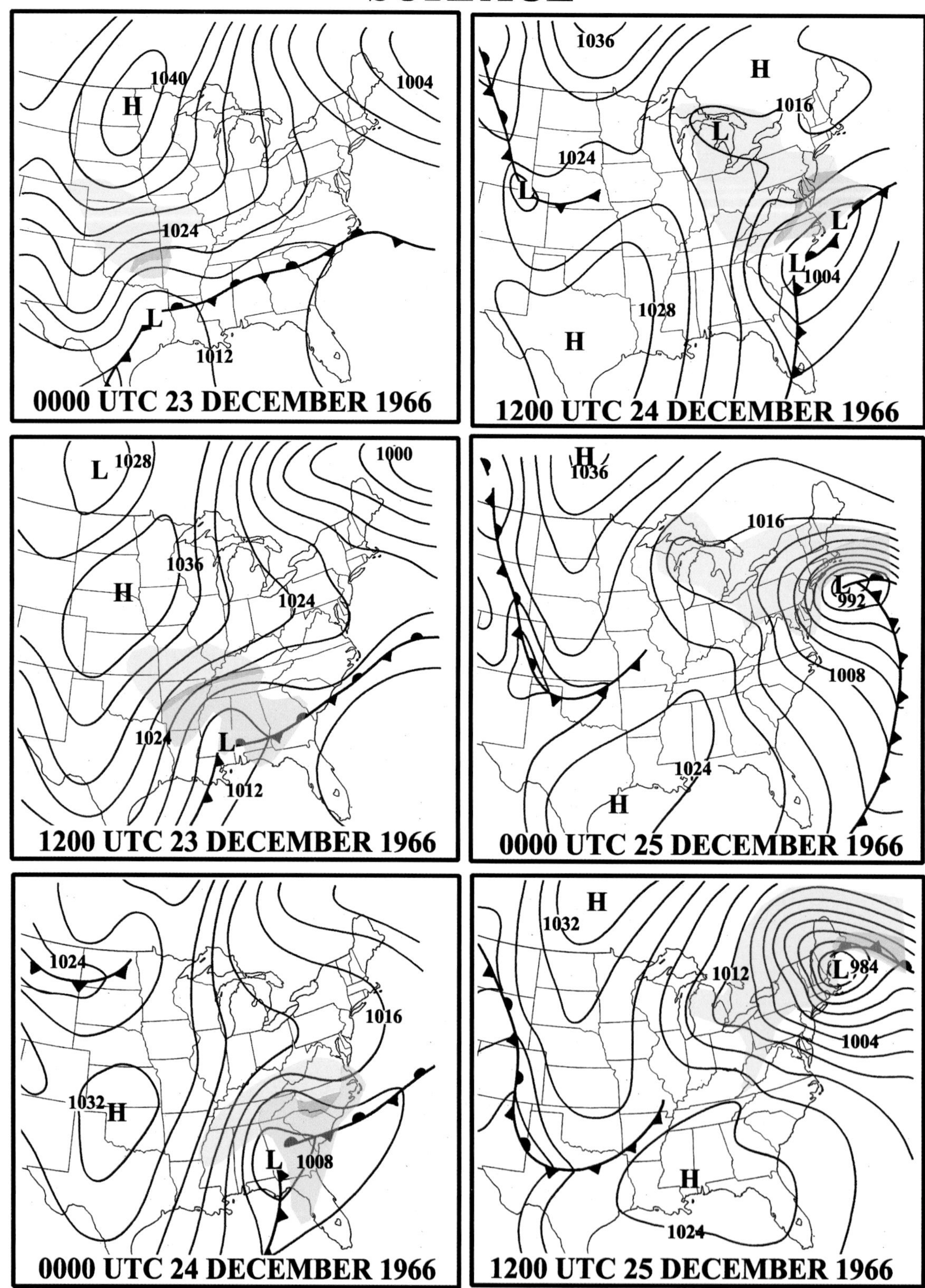

FIG. 10.10-2. Twelve-hourly surface weather analyses for 0000 UTC 23 Dec–1200 UTC 25 Dec 1966. See Fig. 10.1-2 for details.

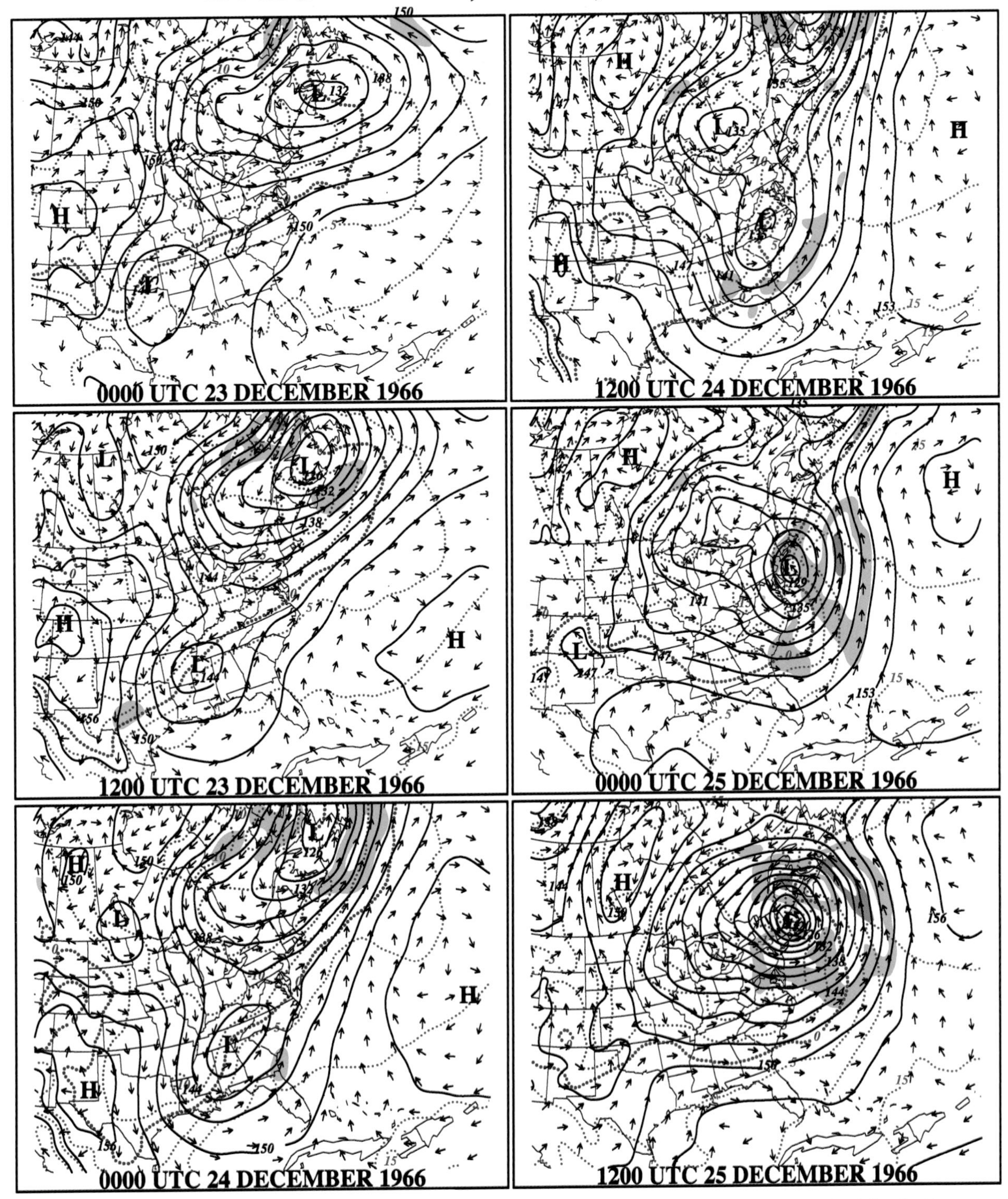

FIG. 10.10-3. Twelve-hourly 850-hPa analyses for 0000 UTC 23 Dec–1200 UTC 25 Dec 1966. See Fig. 10.1-3 for details.

500 hPa HEIGHTS AND VORTICITY, 400 hPa WINDS

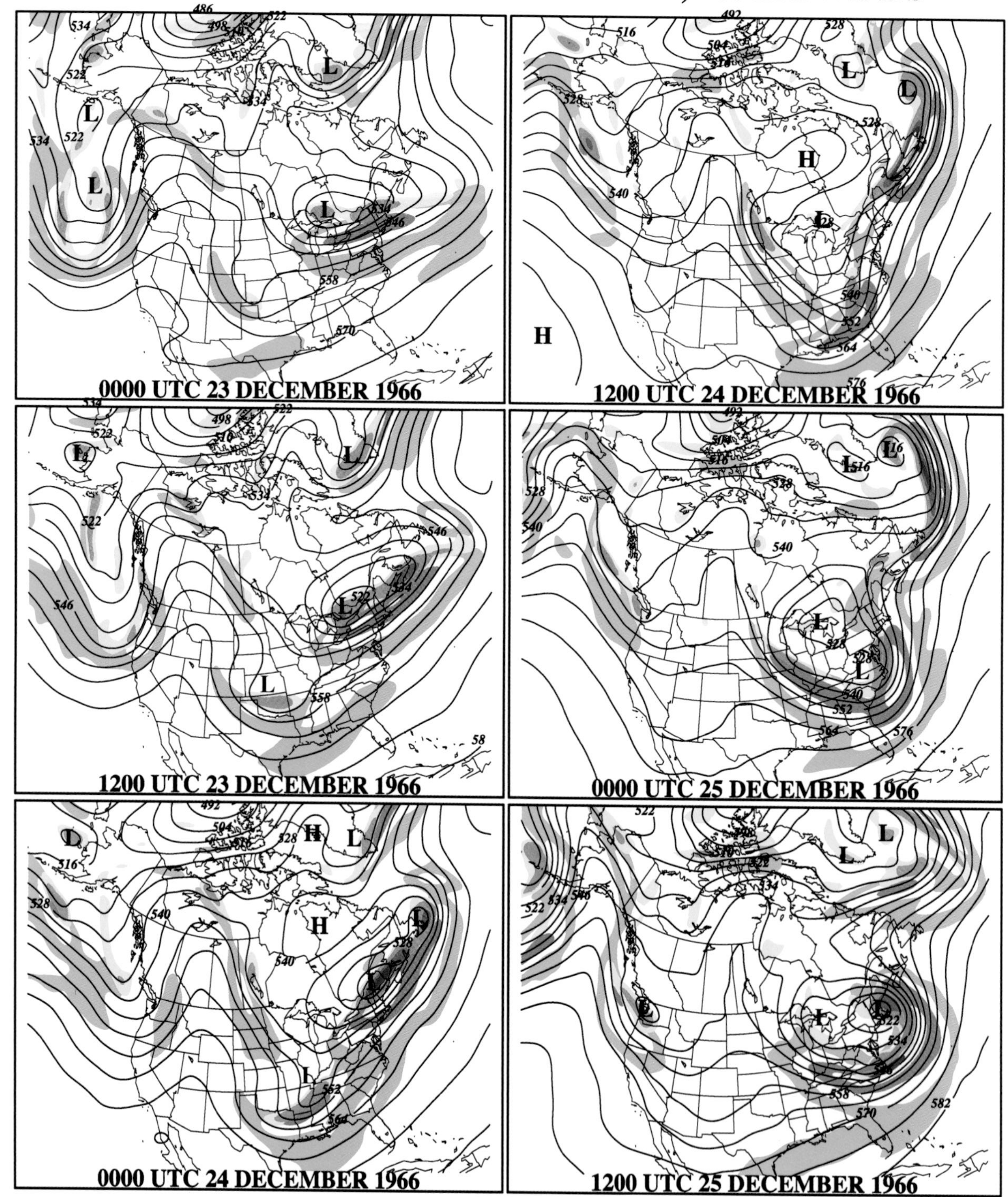

FIG. 10.10-4. Twelve-hourly analyses of 500-hPa geopotential heights and upper-level wind fields for 0000 UTC 23 Dec–1200 UTC 25 Dec 1966. See Fig. 10.1-4 for details.

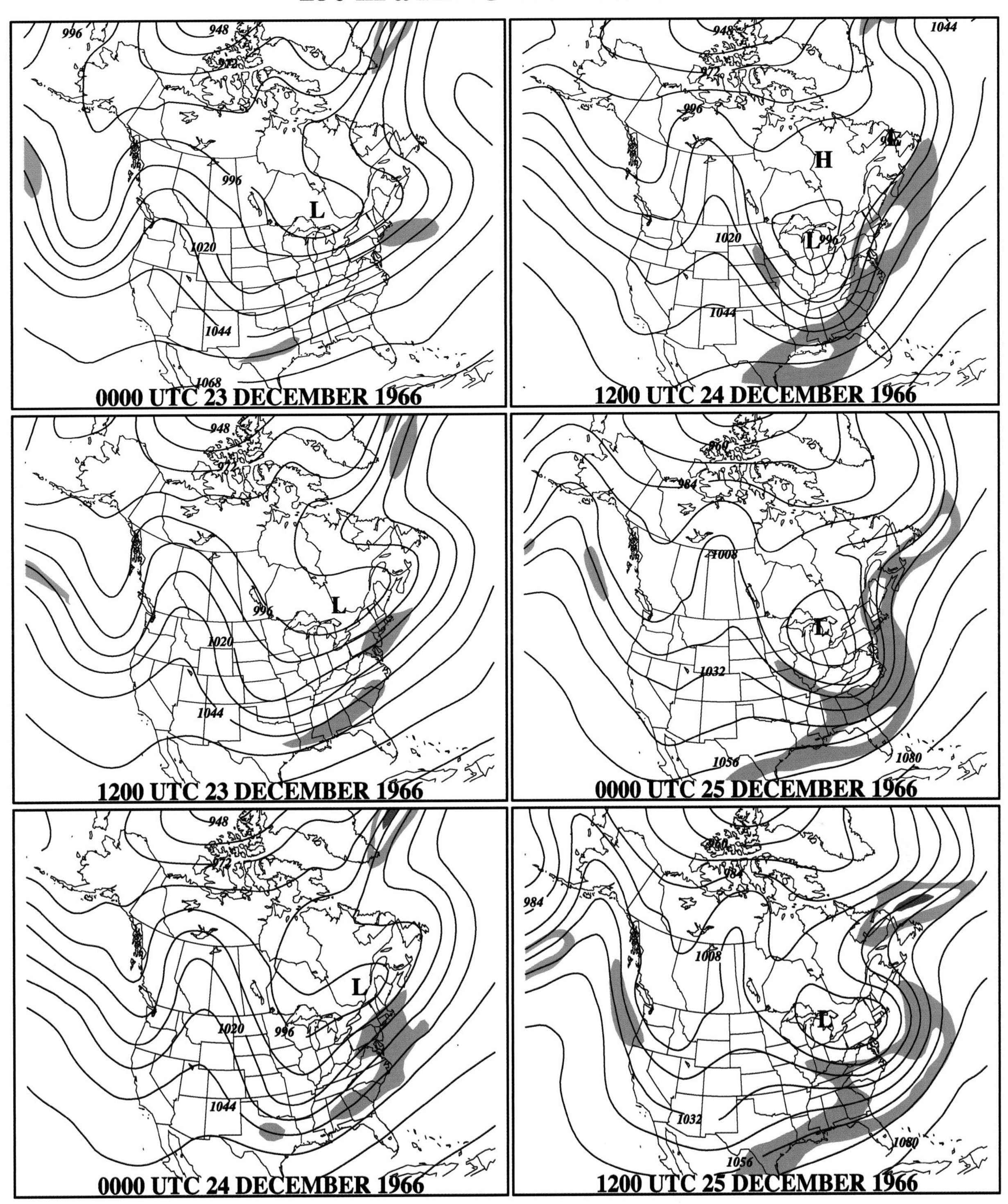

FIG. 10.10-5. Twelve-hourly analyses of 250-hPa geopotential height for 0000 UTC 23 Dec–1200 UTC 25 Dec 1966. See Fig. 10.1-5 for details.

ward along the East Coast as the cyclone deepened on 24–25 December.

- The vortex over the Great Lakes on 23 December split into two separate trough systems early on 24 December. One trough moved eastward from Newfoundland and the other became established over the central Great Lakes, similar to the January 1966 case. As this occurred, the trough over the plains states merged with the trough system forming over Michigan, becoming negatively tilted by 0000 UTC 25 December.
- As the 500-hPa low center remained over Michigan, a new center developed over Maryland at 0000 UTC 25 December, coincident with the rapid deepening of the surface low off the East Coast. This new center then pivoted counterclockwise around the Michigan low and was located over New England at 1200 UTC 25 December, deepening rapidly by -120 m $(12\ h)^{-1}$ while the surface low continued to intensify off the New England coast.
- The sea level cyclone intensified beneath a region of increasing diffluence east of the trough axis, which attained a northwest–southeast orientation (a negative tilt) by 0000 UTC 25 December.
- The amplitude of the upstream ridge over the western United States and the trough over the central plains increased sharply by 0000 UTC 24 December. Meanwhile, the trough drifted eastward to the lower Mississippi Valley, prior to the intensification of the surface low along the East Coast. The amplitude of the trough and downstream ridge increased markedly over the 12-h period ending at 0000 UTC 24 December, prior to East Coast cyclogenesis, and increased moderately thereafter as the cyclone deepened.
- The half-wavelength between the trough and downstream ridge decreased slightly between 0000 UTC 24 December and 0000 UTC 25 December during surface cyclogenesis and as the precipitation shield spread northeastward along the East Coast.
- This case was marked by the distinct dual-jet pattern, with a northern jet extending from the Great Lakes to New England and the Atlantic coast during the precyclogenetic period and a separate, distinct jet streak associated with the southern trough that influenced the rapid cyclogenesis on 24 December.
- A 60 m s^{-1} polar jet streak was observed over the northeastern United States on 23 December within the confluent region south of the vortex over the Great Lakes. The confluent entrance region of this jet was associated with rising sea level pressures and colder temperatures extending into the middle Atlantic states.
- The southern jet amplified near and east of the trough in the central United States by 0000 UTC 24 December, with wind speeds at the 400-hPa level increasing to greater than 50 m s^{-1}.
- As this jet–trough system approached the East Coast, the winds continued to increase to greater than 60 m s^{-1} by 0000 UTC 25 December. The heavy snowfall, LLJ, and rapid cyclogenesis along the East Coast all developed within the diffluent exit region of this jet streak and the right-entrance region of the northern jet on 24 December.

e. 250-hPa geopotential height and wind analyses

- The geopotential height fields in the 250-hPa analyses clearly depict the separate trough systems over southern Canada and the central United States slowly merging by 0000 UTC 24 December with the western portion of the Canadian trough merging with the central United States trough, and amplifying by 1200 UTC 24 December.
- The analyses illustrate the dual-jet pattern, noted earlier at 400 hPa, through 1200 UTC 24 December and 0000 UTC 25 December. The wind speeds increased in the northern jet from Virginia to southern New England as the precipitation shield expanded northeastward and the surface low began to deepen rapidly along the coast.
- A subtropical jet at 250 hPa, located downwind of the trough over the central United States, had maximum winds of 50 m s^{-1}, which then increased to greater than 60 m s^{-1} by 1200 UTC 24 December These speeds were less than those observed in many other cases.

Times Square, New York City, 7 Feb 1967 (photo reprinted courtesy of the *New York Times*).

11. 5–7 February 1967

a. General remarks

- Snowfall was produced from two separate low pressure systems. The first cyclone produced a narrow band of snow across the northern middle Atlantic states and southern New England, with accumulations of generally less than 4 in. (10 cm). It also ushered very cold air into New England and the middle Atlantic states. The second storm produced heavy snowfall rates, but for a relatively short duration. As this storm moved rapidly northeastward along the East Coast, blizzard conditions developed across the middle Atlantic states and southern New England.
- The following regions reported snow accumulations exceeding 10 in. (25 cm): northern Virginia, northeastern West Virginia, central and northern Maryland, northern Delaware, southeastern Pennsylvania, parts of western Pennsylvania, most of New Jersey, extreme southeastern New York, Connecticut, Rhode Island, central and eastern Massachusetts (except Cape Cod), southeastern New Hampshire, and coastal Maine.
- The following region reported snow accumulations exceeding 20 in. (50 cm): scattered locations in northern New Jersey and southeastern New York.

TABLE 10.11-1. Snowfall amounts for urban and selected locations for 5–7 Feb 1967; see Table 10.1-1 for details.

NESIS = 3.82 (category 3)	
Urban center snowfall amounts	
Washington, D.C.–Dulles Airport	11.8 in. (30 cm)
Baltimore, MD	10.6 in. (27 cm)
Philadelphia, PA	9.9 in. (25 cm)
New York, NY–Central Park	15.2 in. (39 cm)
Boston, MA	9.5 in. (24 cm)
Other selected snowfall amounts	
Newark, NJ	16.5 in. (42 cm)
Bridgeport, CT	14.9 in. (38 cm)
Worcester, MA	14.4 in. (37 cm)
Trenton, NJ	13.8 in. (35 cm)
Allentown, PA	13.0 in. (33 cm)

b. Surface analyses

- Two cyclones were responsible for the heavy snowfall. The first surface low moved from southern Iowa across northern Kentucky to northern Virginia on 5–6 February. It propagated along a cold front that moved slowly southward into the middle Atlantic states by 6 February. The surface low was weak and filled as it neared the East Coast, but still produced 2–4 in. (5–10 cm) accumulations from Pennsylvania into New York City and Long Island.
- Following the first cyclone, an anticyclone (1040 hPa) moved across southern Ontario into Quebec, accompanied by bitterly cold temperatures (less than −20°C). A weak inverted sea level pressure ridge west of the middle Atlantic coast at 0000 and 1200 UTC 7 February is suggestive of cold-air damming.
- Coastal frontogenesis was not observed in this case.
- The second low, which produced the heaviest snow, formed along the same cold front as it reached the Gulf coast on 6 February. This cyclone moved very rapidly, propagating from northwestern Florida at 0000 UTC 7 February to a position off the Virginia coast 12 h later. The low center appeared to "jump" from central South Carolina to the Virginia coast between 0900 and 1200 UTC 7 February. It then raced northeastward to a position off the coast of Maine by 0000 UTC 8 February, covering more than 1000 km in 12 h.
- An area of precipitation expanded quickly to the east and north as the surface low moved rapidly up the Atlantic coast on 6–7 February. The heaviest precipitation was initially concentrated just north of the cyclone center, but was later observed approximately 200 km northwest of the low center. Heavy snow generally fell for only 12 h or less due to the storm's rapid movement. Temperatures ranging between −7° and −15°C combined with gale force winds to create blizzard conditions across the middle Atlantic and southern New England states on 7 February.
- The cyclone deepened slowly as it tracked across the southeastern United States between 0000 and 0600 UTC 7 February. Deepening rates increased to nearly -1 hPa h^{-1} after 1200 UTC 7 February, as the storm center moved over the Atlantic Ocean. Although the deepening rates exhibited by this system were not as spectacular as those in many other cases, the storm's rapid northeastward progression produced local pressure falls across southeastern Canada that were as large as those observed in slower-moving cyclones that had intensified more vigorously.

c. 850-hPa analyses

- Separate 850-hPa low centers accompanied the two surface low pressure systems in this case. The first 850-hPa low weakened as it moved up the Ohio Valley and into Pennsylvania by 1200 UTC 6 February. Its subsequent passage off the East Coast was followed by the southward displacement of the 850-hPa baroclinic zone to the middle Atlantic region. The passage of this frontal zone to the south established cold air across the northeastern United States for the second system.
- The second and more significant 850-hPa low center developed in Texas and Louisiana at 1200 UTC 6 February along the southern edge of a narrow, well-defined baroclinic zone that extended from Texas across the lower Ohio Valley to the middle Atlantic states. A 20–25 m s^{-1} north-to-northeasterly jet accompanied the development of the 850-hPa low over Texas on 6 February. Strong cold-air advection occurred with this jet west of the newly formed center.

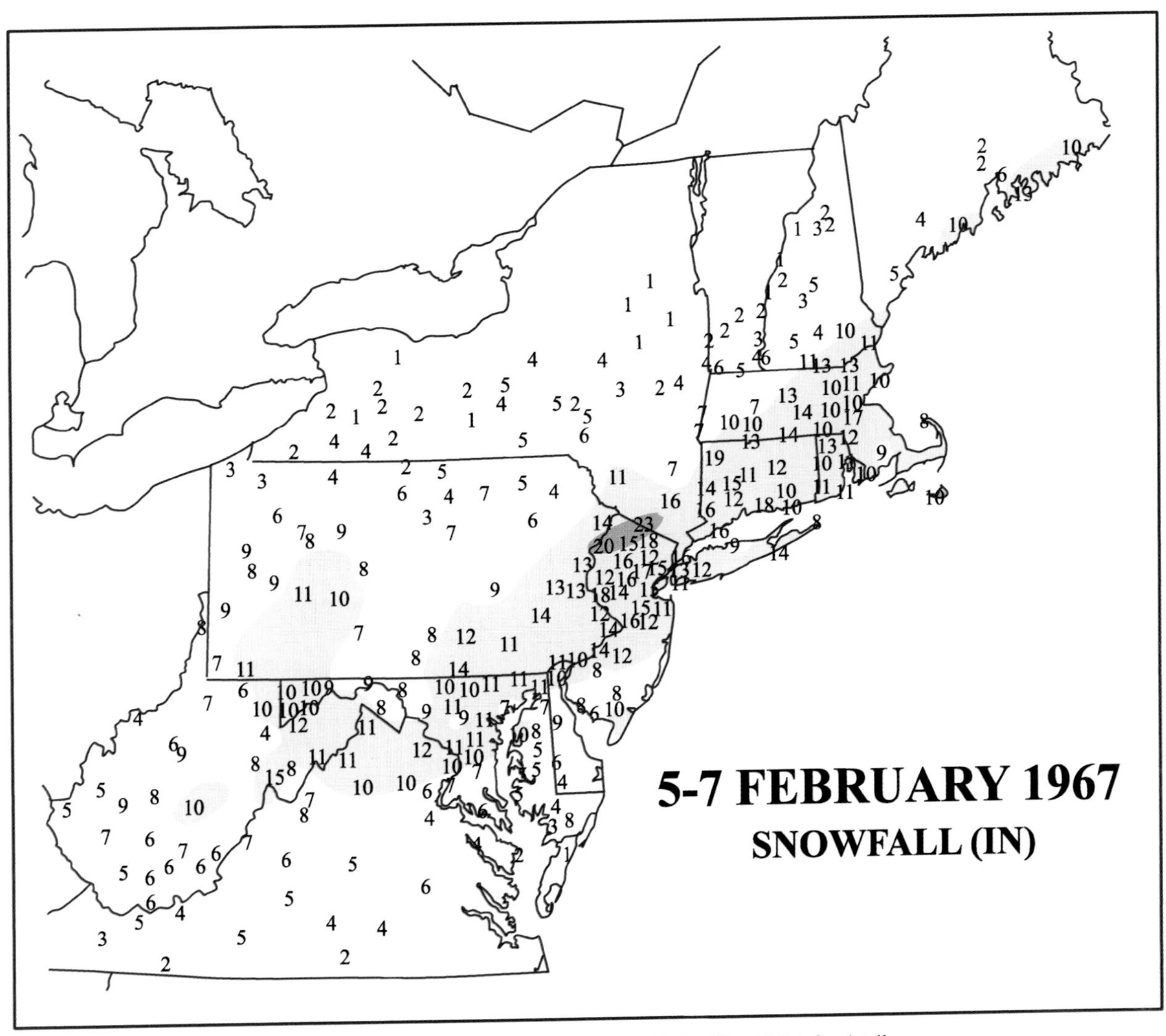

FIG. 10.11-1. Snowfall (in.) for 5–7 Feb 1967. See Fig. 10.1-1 for details.

- The low deepened slowly as it propagated northeastward to the middle Atlantic coast by 1200 UTC 7 February. It intensified more rapidly once it was well off the coast later on 7 February.
- The low center tracked along the narrow, intense band of isotherms that had been established prior to cyclogenesis. Only a hint of the S-shaped isotherm pattern appeared along the East Coast at 1200 UTC 7 February, a signature that was much better defined in many of the other cases.
- A 25 m s^{-1} southwesterly low-level jet (LLJ) formed in southern Georgia and northern Florida by 0000 UTC 7 February. By 1200 UTC 7 February, a greater than 30 m s^{-1} southerly jet (LLJ) is analyzed over eastern North Carolina, as wind speeds generally increased with the deepening 850-hPa low along the Atlantic coast. The strengthening circulation enhanced the warm-air advection and moisture transport over the middle Atlantic states while moderate to heavy snowfall was developing in the region.
- The LLJ continued to amplify to near 50 m s^{-1} on 8 February as the surface and 850-hPa low intensified and raced to the northeast. The strongest winds remained over the Atlantic Ocean east and south of the circulation center for this case.

d. 500-hPa geopotential height and 400-hPa wind analyses

- This storm also occurred during the snowy winter of 1966/67, which was marked by a weak La Niña. Although much of January 1967 was dominated by a negative North Atlantic Oscillation (NAO), the index reversed through much of February and March 1967, when the NAO was positive. Daily values of the NAO

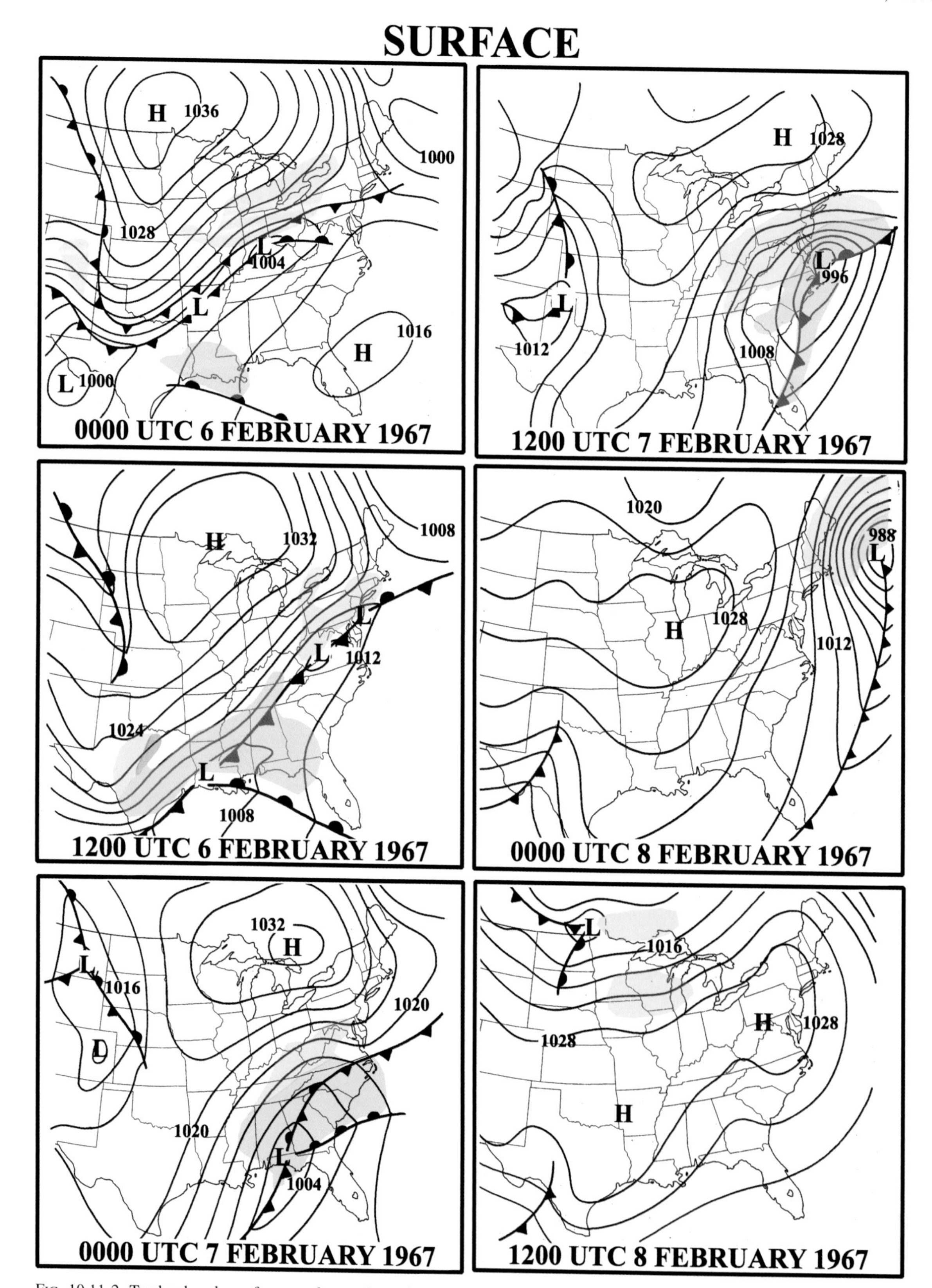

FIG. 10.11-2. Twelve-hourly surface weather analyses for 0000 UTC 6 Feb–1200 UTC 8 Feb 1967. See Fig. 10.1-1 for details.

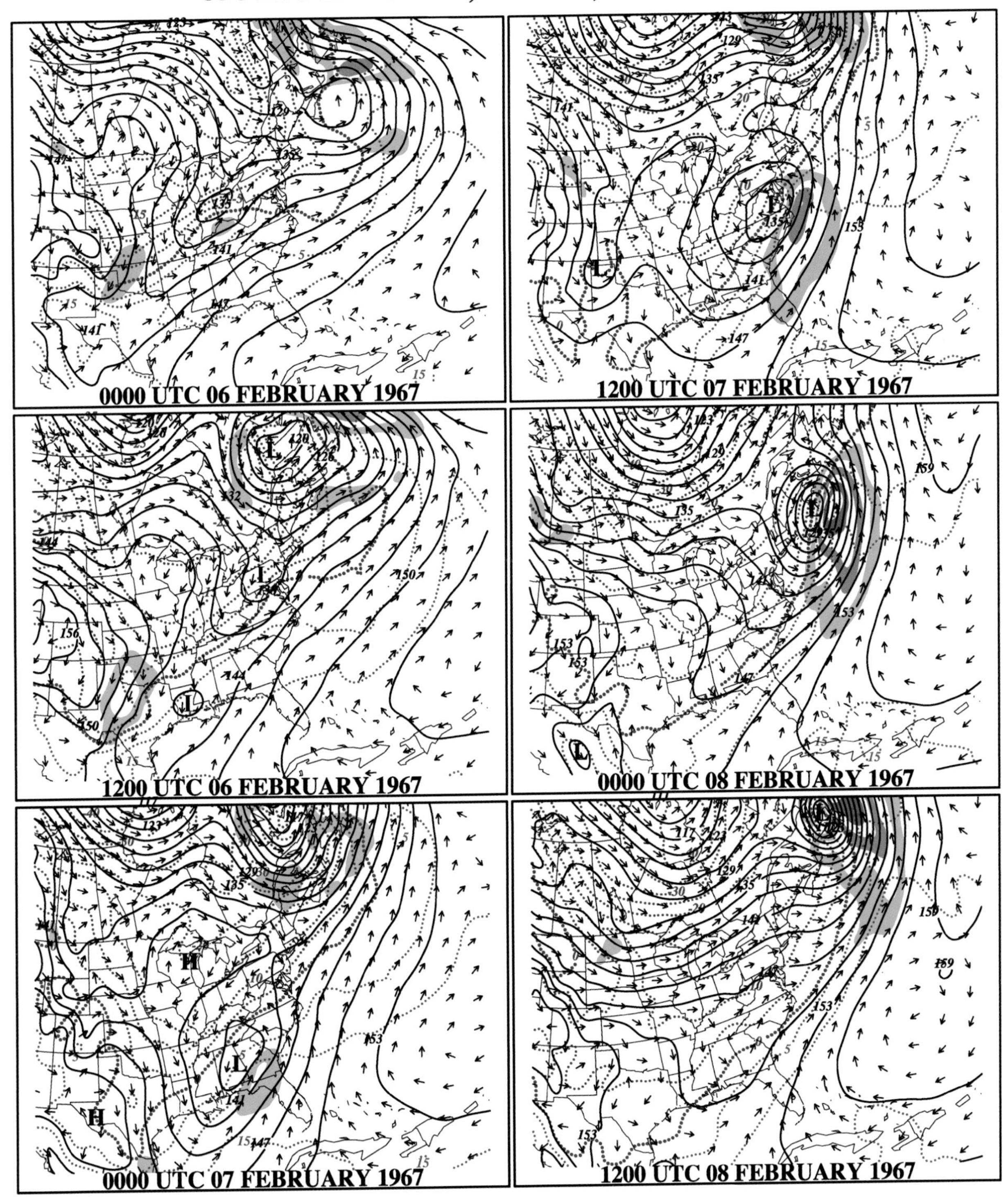

FIG. 10.11-3. Twelve-hourly 850-hPa analyses for 0000 UTC 6 Feb–1200 UTC 8 Feb 1967. See Fig. 10.1-3 for details.

500 hPa HEIGHTS AND VORTICITY, 400 hPa WINDS

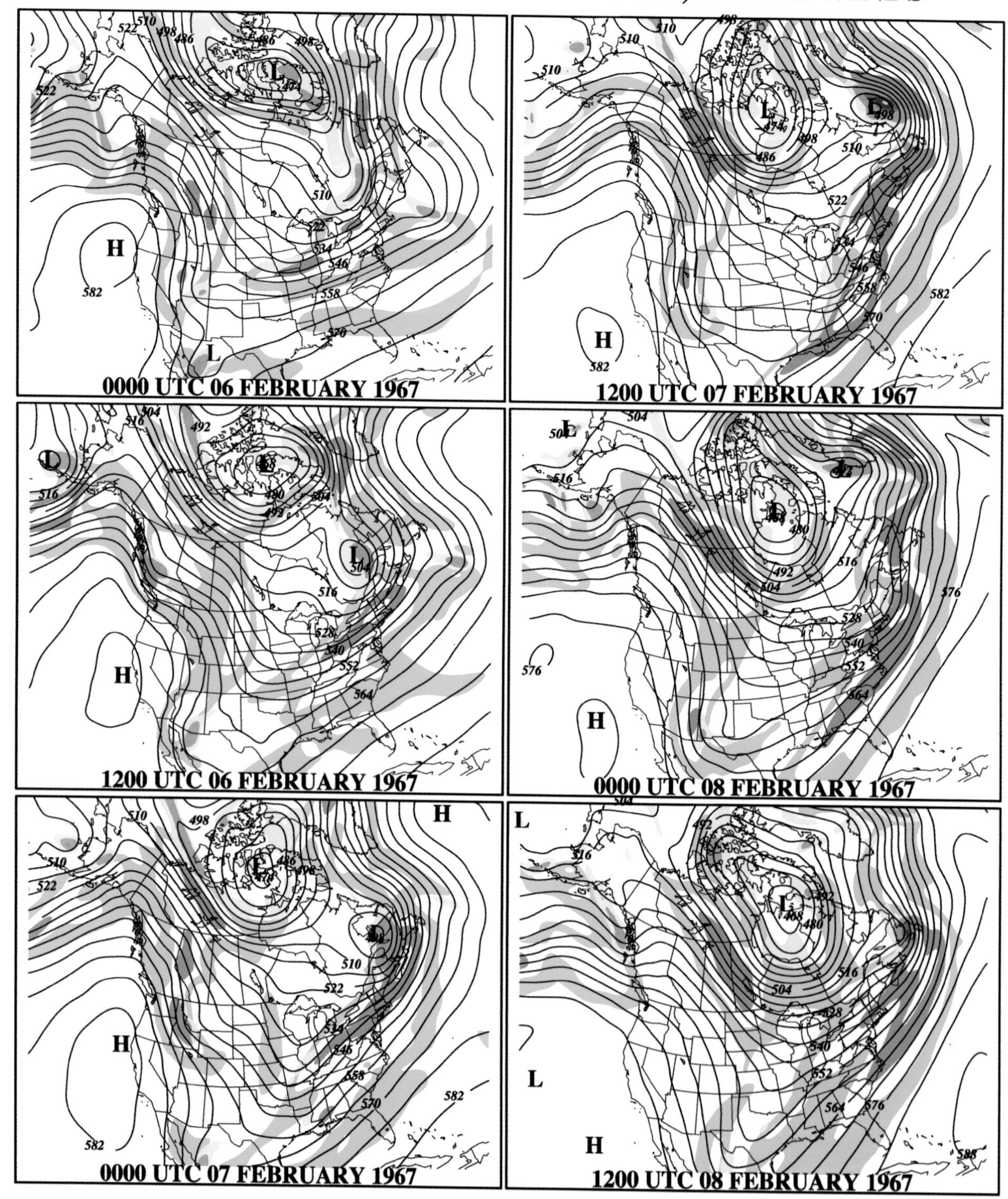

FIG. 10.11-4. Twelve-hourly analyses of 500-hPa geopotential heights and upper-level wind fields for 0000 UTC 6 Feb–1200 UTC 8 Feb 1967. See Fig. 10.1-4 for details.

250 hPa HEIGHTS AND WINDS

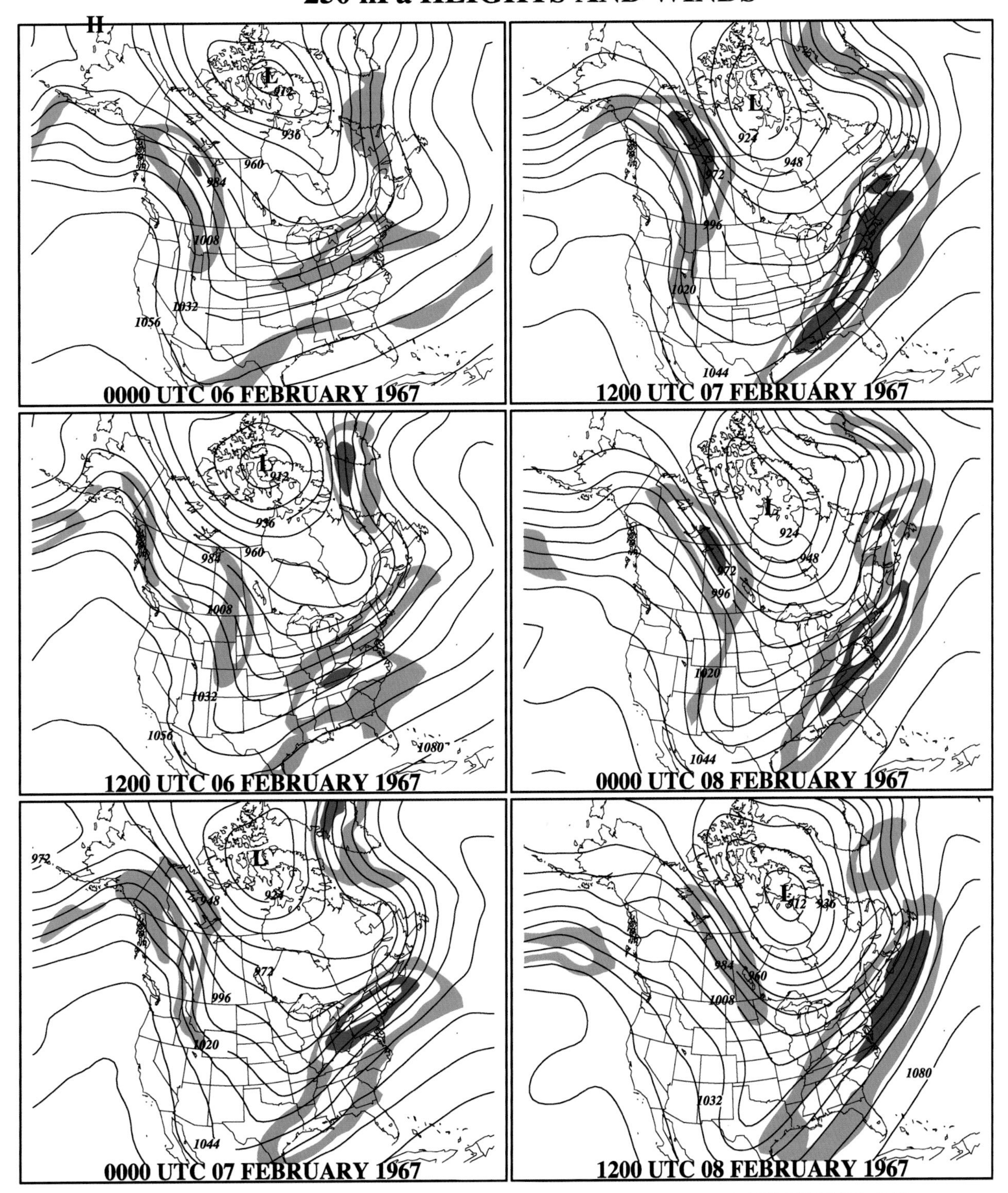

FIG. 10.11-5. Twelve-hourly analyses of 250-hPa geopotential height for 0000 UTC 6 Feb–1200 UTC 8 Feb 1967. See Fig. 10.1-5 for details.

show that the storm developed during the very beginning of the 2-month period in which the NAO tended positive, but actually occurred during a 2-day period in which the NAO remained slightly negative.

- A deep trough propagating slowly across eastern Canada on 6 February produced highly confluent flow over southeastern Canada and the northeastern United States. An anticyclone with very cold surface temperatures moved into New England and toward the middle Atlantic states on 6–7 February beneath the confluent upper-level flow.
- The first weakening cyclone that propagated to the south of New York City at 1200 UTC 6 February was associated with a weak trough over the Ohio Valley. Although this trough was not easy to identify in the geopotential height field within the strongly confluent flow over the Ohio Valley and New England, there was an area of cyclonic vorticity advection over the Ohio Valley at 0000 UTC 6 February and over the middle Atlantic states at 1200 UTC 6 February.
- A trough over northern Mexico and another trough propagating southeastward downwind of a high-amplitude ridge along the Pacific coast set the stage for the second, more intense cyclone on 6–7 February. As the trough became better defined over the central United States at 0000 UTC 7 February, a separate vorticity maximum became apparent over Louisiana. The trough advanced rapidly to southeastern Canada by 1200 UTC 8 February, while the sea level cyclone intensified along the East Coast.
- No closed cyclonic circulation developed at 500 hPa and no distinct minimum in the geopotential height field was observed until 1200 UTC 8 February over extreme eastern Canada. Unlike most of the other cases, this system did not appear to have a marked diffluent region downstream of the trough axis until it reached eastern Canada, even though the trough developed a slight northwest–southeast (negative) tilt after 1200 UTC 7 February. The absence of this feature may be related to the rapid acceleration of the entire storm system after 0000 UTC 7 February. The geopotential height gradients increased at the base of the trough after 0000 UTC 7 February, during the period of rapid cyclogenesis.
- Large increases in amplitude were associated with geopotential height rises near the East Coast downstream of the trough at 0000 and 1200 UTC 7 February, as the cyclone was developing along the Southeast and middle Atlantic coasts. There was little change in the half-wavelength between the trough and downstream ridge until the cyclone was east of Newfoundland on 8 February.
- There is evidence of the dual-jet pattern that marks many of these cases with a northern jet streak extending into New England from the Ohio Valley by 1200 UTC 6 February and 0000 UTC 7 February, and with a developing southern jet near 70 m s^{-1} approaching the East Coast by 1200 UTC 7 February.
- A 60 m s^{-1} polar jet streak was analyzed within the confluent region across the Ohio Valley and New England from 1200 UTC 5 February through 1200 UTC 6 February. This jet accompanied the first, weak surface low and was associated with the outbreak of cold air to the rear of this cyclone. Furthermore, the first band of snow appears to develop in the right entrance region of the jet streak on 6 February.
- During 7 February, the surface low was located in the exit region of the 60 m s^{-1} jet to the south of the trough and the entrance region of the jet streak over New England and southeastern Canada, a pattern that is similar to that found in other cases. Nevertheless, the exit region of the southern jet is not marked by diffluence and significant along-stream wind variations as is the case in many of the other cases in the sample.

e. 250-hPa geopotential height and wind analyses

- This case is marked by a broad trough that nearly extends across much of Canada and the United States, reflecting the very cold air that accompanied this storm, which penetrated far into the south-central and middle Atlantic states.
- A short-wave trough is analyzed, extending to southeastern Canada by 0000 UTC 7 February, as is a weakly defined trough moving rapidly from the south-central United States into New England by 0000 UTC 8 February.
- Large increases in the upper-level wind to greater than 80 m s^{-1} are observed in the confluent region over Pennsylvania and New York by 0000 UTC 7 February, as the precipitation shield expanded rapidly to the northeast on 7 February and cyclogenesis commenced in the Southeast.
- Wind speeds within the subtropical jet over the southeastern United States increased by greater than 20 m s^{-1} during the 24-h period ending at 1200 UTC 6 February, as precipitation developed across the Gulf coast. Wind speeds in the subtropical jet continued to increase to greater than 90 m s^{-1} along the East Coast by 0000 UTC 8 February, as the cyclone moved rapidly northeastward into Canada and continued to intensify.

A pedestrian fights the wind in Manhattan during the Lindsay Storm, Feb 1969 (source is NOAA Historical Photograph Collection, photo courtesy of Donald Sutherland's digital photography library, www.geocities.com/donsutherland1/snow1.html).

12. 8–10 February 1969

a. General remarks

- This cyclone is known as the "Lindsay Storm" in New York City since Mayor John Lindsay ran into political misfortune after sections of the city remained unplowed for a week following the snowfall. The storm was poorly predicted in New York City as forecasters first thought the precipitation would fall primarily as rain. Even when it appeared that snow would predominate on 9 February, predicted snowfall amounts persistently lagged the actual totals, as snow continued to fall long past the time it was forecast to cease. The rapid development and deceleration of the storm brought paralyzing snow and increasing winds from northern New Jersey through most of New England.
- The following regions reported snow accumulations exceeding 10 in. (25 cm): northern New Jersey, southeastern New York (except eastern Long Island), most of Connecticut, central and northern Rhode Island, Massachusetts (except the southeast), Vermont (except the northwest), New Hampshire, and southern and central Maine.
- The following regions reported snow accumulations exceeding 20 in. (50 cm): parts of the New York City and Boston metropolitan areas, western Connecticut, western and eastern Massachusetts, southern Vermont, northern Rhode Island, eastern New Hampshire, and southern Maine.

TABLE 10.12-1. Snowfall amounts for urban and selected locations for 8–10 Feb 1969; see Table 10.1-1 for details.

NESIS = 3.34 (category 2)	
Urban center snowfall amounts	
Washington, D.C.–Dulles Airport	5.0 in. (13 cm)
Baltimore, MD	3.0 in. (8 cm)
Philadelphia, PA	2.9 in. (7 cm)
New York, NY–John F. Kennedy Airport	20.2 in. (51 cm)
Boston, MA	11.1 in. (28 cm)
Other selected snowfall amounts	
Bedford, MA	25.0 in. (64 cm)
Portland, ME	21.5 in. (55 cm)
Bridgeport, CT	17.7 in. (45 cm)
Hartford, CT	15.8 in. (40 cm)
Worcester, MA	15.6 in. (40 cm)

b. Surface analyses

- Bitterly cold air neither preceded nor followed the passage of this storm. This case was unusual in that no large, cold surface anticyclone was present over Ontario, Quebec, or northern New England immediately prior to East Coast cyclogenesis. Instead, a rather meager 1022-hPa anticyclone drifted off New England by 0000 UTC 9 February, resulting in southeasterly winds along the coast late on 8 February and temperatures slightly above freezing. The winds backed to a northeasterly direction along the Atlantic coast as the secondary low developed rapidly and moved northeastward off the Virginia coast on 9 February.
- Very weak cold-air damming was observed along the middle Atlantic coast on 8 February as the moderately cold high pressure system crossed New England.
- The primary surface low propagated east-northeastward from Oklahoma to Kentucky in the 24-h period ending at 0000 UTC 9 February, covering approximately 1000 km. Little change in the central sea level pressure occurred during this period, but moderate to heavy rain was observed north of the low from Missouri to Ohio on 8 February. This low weakened rapidly as a secondary center developed along the East Coast between 0000 and 1200 UTC 9 February. The primary cyclone disappeared within 9 h of the commencement of secondary cyclogenesis.
- The secondary low pressure system formed over Georgia after 1800 UTC 8 February as heavy rains developed in the Carolinas. Mixed snow and rain spread across the middle Atlantic states between 0000 and 0600 UTC 9 February, with wet snow predominating in northern Virginia, central Maryland, eastern Pennsylvania, and New Jersey. Generally light to moderate precipitation amounts were observed in these locations. Heavy snow developed from New Jersey northward by 1200 UTC 9 February.
- The secondary low developed inland over Georgia along the warm front of the primary low pressure system. Although there was some tendency for this front to extend itself to the Southeast coast where small land–sea air temperature differences were observed, it cannot strictly be called a coastal front.
- The low pressure center moved northeastward from the North Carolina coast to near Long Island by 0000 UTC 10 February, then eastward off Cape Cod by 1200 UTC 10 February. The low deepened 32 hPa over an 18-h period, reaching 970 hPa by 0000 UTC 10 February just east of Long Island.
- The forward motion of the secondary low slowed dramatically on 9 February, with the center advancing only 200 km in 12 h. The storm's slow movement prolonged the snowfall from New York City across much of New England.

c. 850-hPa analyses

- Very weak cold-air advection was observed on 8 February across the northeastern United States prior to East Coast cyclogenesis. Lower-tropospheric temperatures ranged between 0° and −10°C from Virginia to Maine.
- The 850-hPa low supporting the development of the primary low was located at the edge of an intense temperature gradient over the southern plains states

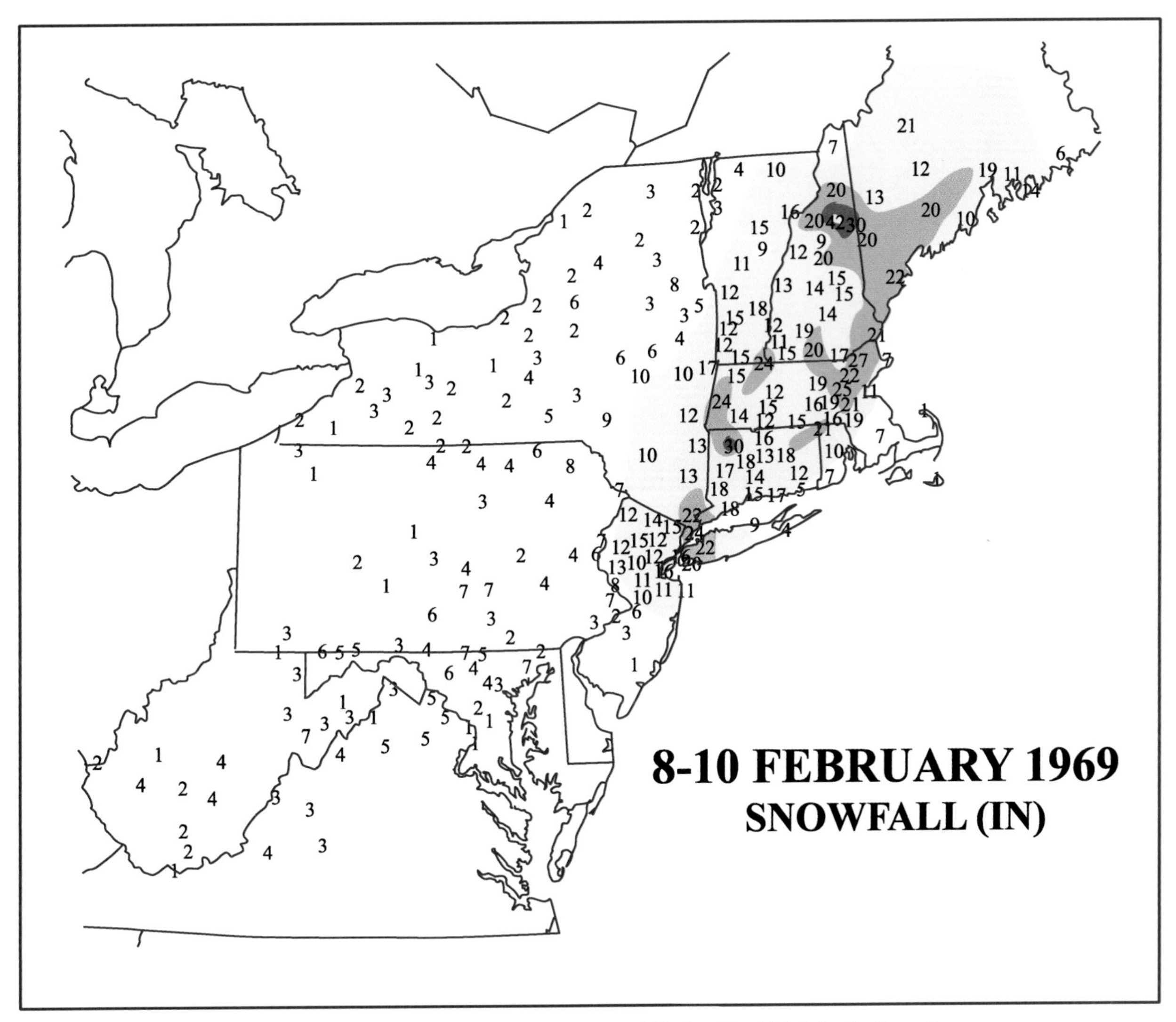

FIG. 10.12-1. Snowfall (in.) for 9–10 Feb 1969. See Fig. 10.1-1 for details.

at 0000 UTC 8 February, which weakened as the low drifted eastward toward the Ohio Valley in the following 24 h. A tongue of warm air was located northeast and east of the 850-hPa center on 8 February, yielding rain in the Ohio Valley.

- A separate 850-hPa low began deepening along the East Coast after 0000 UTC 9 February as the primary surface low was crossing the Ohio Valley. The 850-hPa deepening rate reached a maximum of −150 m (per 12 h) between 1200 UTC 9 February and 0000 UTC 10 February along the coast, concurrent with a 20-hPa intensification of the secondary surface low off the middle Atlantic coast.
- An elongated region of height falls from Ohio to South Carolina at 0000 UTC 9 February suggested secondary 850-hPa low development over the southeastern United States. Secondary development toward North Carolina was also supported by the large temperature gradient between Georgia and Virginia, well south and east of the main 850-hPa low center.
- A pattern of strong cold-air advection behind the 850-hPa low emerged over Texas and Oklahoma by 1200 UTC 8 February, 12 h prior to the onset of secondary cyclogenesis along the East Coast. Southwesterly winds increased over Louisiana, Mississippi, and Alabama, initiating the stronger warm-air advection that later attended the early stages of secondary cyclogenesis along the Atlantic coast at 0000 UTC 9 February. The warm advection was greatest in the region with enhanced geopotential height falls and where precipitation was breaking out across the Carolinas.
- The isotherms began to exhibit an S shape by 1200 UTC 9 February, at the start of rapid cyclogenesis off the Virginia coast. The 0°C isotherm was consistently located near the 850-hPa low center.
- A strong low-level jet (LLJ) with winds exceeding 25

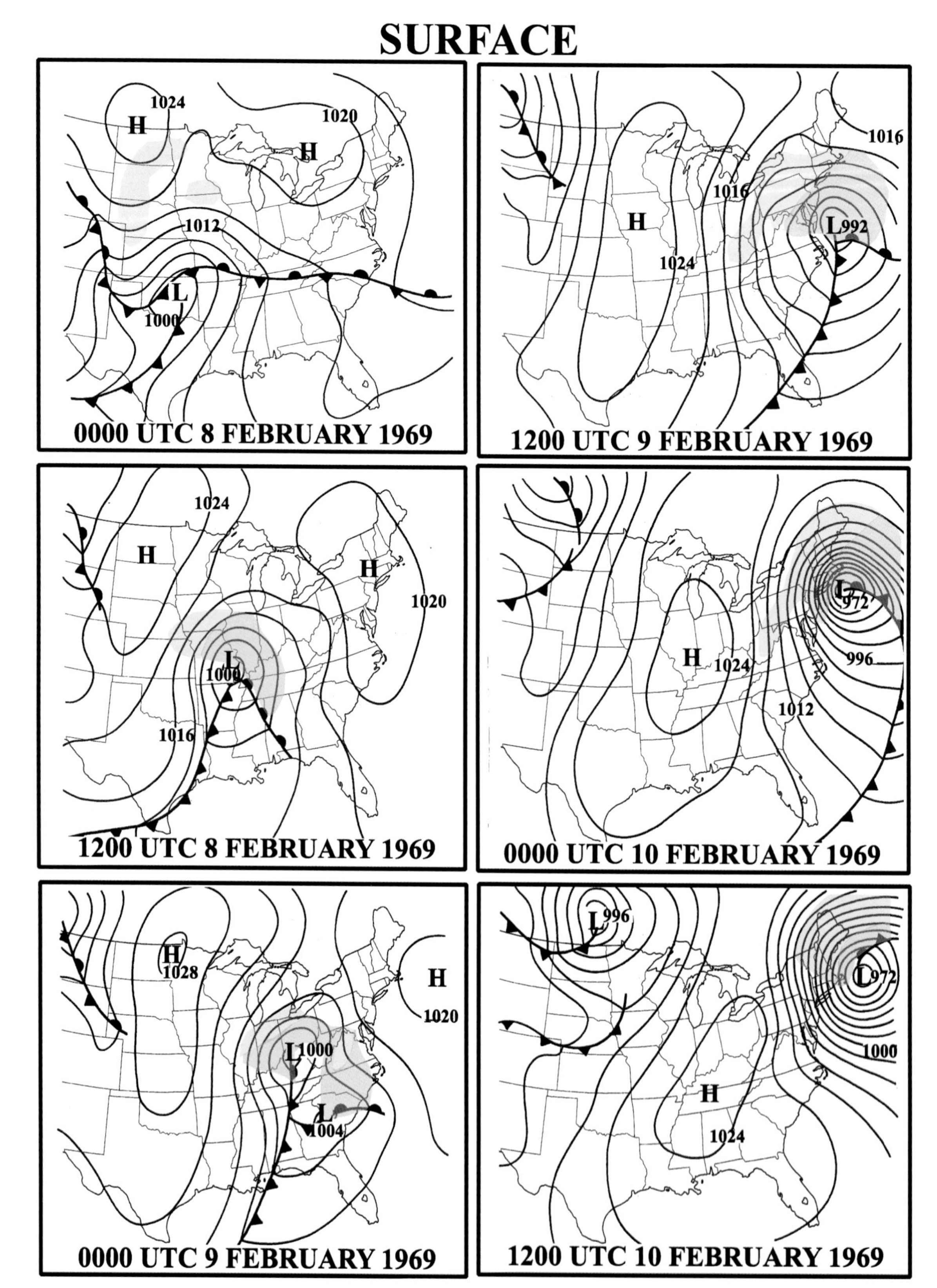

FIG. 10.12-2. Twelve-hourly surface weather analyses for 0000 UTC 8 Feb–1200 UTC 10 Feb 1969. See Fig. 10.1-2 for details.

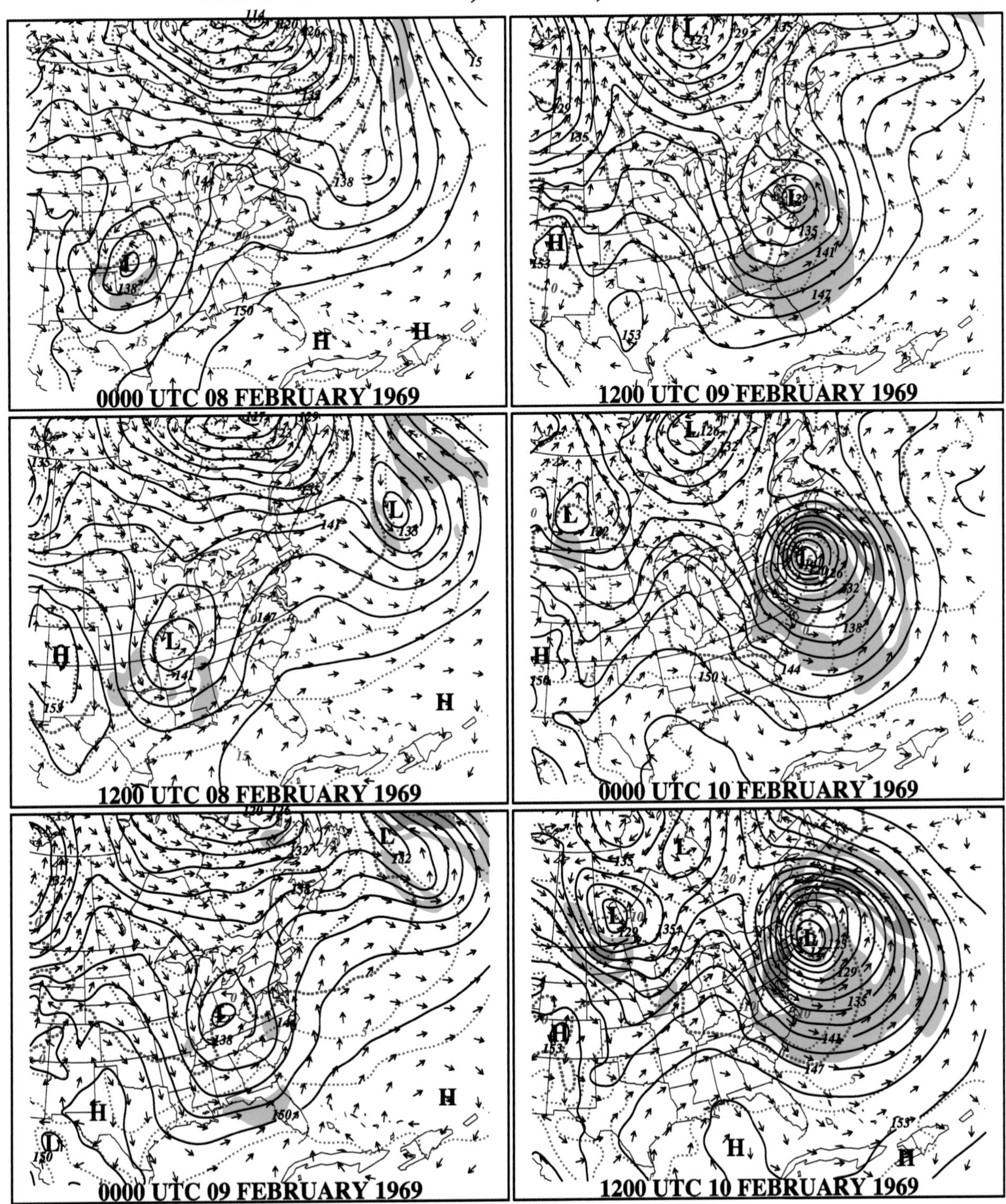

FIG. 10.12-3. Twelve-hourly 850-hPa analyses for 0000 UTC 8 Feb–1200 UTC 10 Feb 1969. See Fig. 10.1-3 for details.

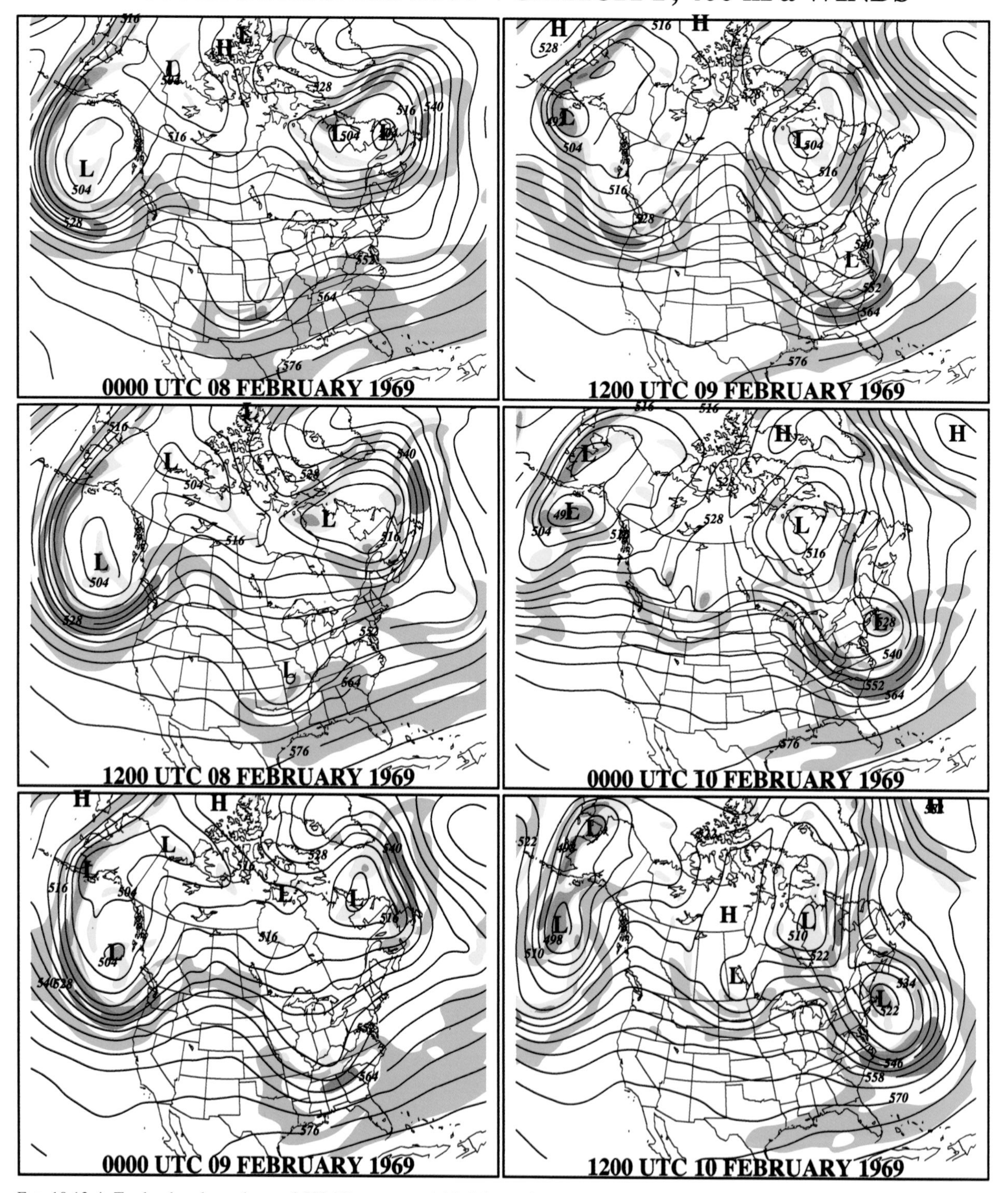

FIG. 10.12-4. Twelve-hourly analyses of 500-hPa geopotential heights and upper-level wind fields for 0000 UTC 8 Feb–1200 UTC 10 Feb 1969. See Fig. 10.1-4 for details.

FIG. 10.12-5. Twelve-hourly analyses of 250-hPa geopotential height for 0000 UTC 8 Feb 1969 through 1200 UTC 10 Feb 1969. See Fig. 10.1-5 for details.

$m\ s^{-1}$ developed south and east of the deepening 850-hPa low by 1200 UTC 9 February. There is some indication of cross-contour flow as the LLJ developed rapidly and then appeared to wrap itself around the intensifying vortex by 0000 UTC 10 February with maximum winds exceeding 35 $m\ s^{-1}$.

- Easterly winds in the vicinity of the 850-hPa low strengthened markedly during the 12-h period ending at 0000 UTC 10 February, when rapid surface deepening was in progress and heavy precipitation was spreading across New York and New England, just downwind of this wind maximum. Wind speeds increased to greater than 40 $m\ s^{-1}$ in the easterly LLJ by 1200 UTC 10 February as the storm reached its maximum intensity off the New England coast.

d. 500-hPa geopotential height and 400-hPa wind analyses

- The winter of 1968/69 combined both a significant El Niño and negative North Atlantic Oscillation (NAO), similar to that which occurred in the winters of 1957/58, 1963/64, and 1965/66. The magnitude of this El Niño (Fig. 2–19, volume I) was smaller than the strong El Niños of 1957/58, 1972/73, 1982/83, 1986/87, 1991/92, and 1997/98. Daily values of the NAO entered a negative phase in mid-December 1968 and remained consistently negative throughout the rest of the winter before going into a positive period in the last days of March. This storm occurred as negative values of the NAO were decreasing daily before reaching a nadir around 15–18 February. The 500-hPa analyses show an upper ridge over south-central Greenland, but the center of the upper ridge appears to be much farther southeast, over the central Atlantic.
- Prior to the major storm along the East Coast, a large cyclonic vortex drifted northward over eastern Canada on 8 February. This trough seems to stall and extend southward into the Great Lakes by 1200 UTC 9 February. Weak confluence was briefly observed over the northeastern United States at 1200 UTC 8 February but eroded over the next 24 h as a weak surface high drifted eastward into New England on 8 February. The absence of confluence over the northeastern United States represents a marked deviation from the precyclogenetic conditions of the other major snowstorms.
- The trough associated with the developing storm was located in the south-central United States by 0000 UTC 8 February, immediately downstream of an amplifying ridge in the northwestern United States and southwestern Canada. The trough propagated eastward from the central plains as the primary cyclone moved into the Ohio Valley on 8 February, reaching the middle Atlantic coast as the secondary cyclone developed along the East Coast on 9 February.
- A distinct vorticity maximum was evident at the base of the trough, where cyclonic curvature and shear were maximized. The primary and secondary cyclogenesis both occurred downstream of this vorticity maximum beneath the region of large vorticity advections.
- Between 0000 UTC 8 February and 0000 UTC 9 February, there was little change in the configuration of the geopotential height field that defined the trough as it propagated toward the East Coast. However, between 1200 UTC 9 February and 1200 UTC 10 February, geopotential height values within the trough dropped significantly, with a closed 500-hPa low developing by 0000 UTC 10 February. Furthermore, this trough seemed to develop a negative tilt and merge with the Canadian trough extending southward into the Great Lakes by 1200 UTC 9 February. These changes occurred during the period of rapid intensification of the surface cyclone along the East Coast.
- Diffluence was observed downwind of the trough at all times. The rapid development of the secondary cyclone between 0000 and 1200 UTC 9 February coincided with the passage of the diffluent height contours across the coastline.
- The amplitude of the trough over the northeastern United States and the downstream ridge increased, especially between 0000 UTC 9 February and 0000 UTC 10 February, as the ridge surged northward into eastern Canada during rapid cyclogenesis. The half-wavelength decreased between the trough and the downstream ridge off the East Coast, especially by 1200 UTC 9 February as precipitation spread across the Northeast and the cyclone began to deepen rapidly.
- Unlike other cases, the confluent entrance region of a northern jet remained well to the north, with the jet axis bisecting southeastern Canada. This lack of a confluent jet is consistent with the lack of colder air preceding this storm.
- A separate jet upstream of the trough over the southern plains appeared to propagate with the trough on 8 February, then merged with a jet system extending from the Great Lakes by 1200 UTC 9 February. Wind speeds at 400 hPa increased to greater than 50 $m\ s^{-1}$ near and upstream of the trough axis between 0000 UTC 8 February and 0000 UTC 9 February, prior to the East Coast surface development.
- The secondary surface low and its associated precipitation shield and the rapid accelerations associated with the LLJ all developed in the diffluent exit region of this jet system as it amplified at the base of the trough while propagating toward the East Coast between 0000 and 1200 UTC 9 February. Wind speeds continued to increase to greater than 60 $m\ s^{-1}$ at the base of the trough at 1200 UTC 9 February.

e. 250-hPa geopotential height and wind analyses

- The geopotential height fields depict a general west to east flow with uniform gradients in the precyclogenetic environment. The trough in the middle United States is the dominant feature through 0000 UTC 9 February.

- The merging of this trough with the Canadian trough extending south to the Great Lakes region is clearly evident by 1200 UTC 9 February.
- A subtropical jet was observed at 250 hPa downstream of the trough, which was approaching the East Coast between 0000 UTC 8 February and 0000 UTC 9 February. The subtropical jet was similar to those observed in other cases, and amplified to 80 m s^{-1} as it propagated across the southern United States at 1200 UTC 9 February.
- While there is some evidence of a separate jet system in the Ohio Valley on 8 February as the primary low moved slowly eastward, this feature is relatively weak compared to other cases and reflects the lack of colder air observed prior to this storm.

Snowdrifts cover fences and cars in West Peabody, Massachusetts (*Boston Globe* photo from Weatherwise Magazine, 1969, p. 235).

13. 22–28 February 1969

a. General remarks

- This was an unusual storm due to its slow movement, long duration, moderate intensity, erratic intensification, lack of large thermal contrast at the surface, and chaotic upper-level geopotential height patterns. The storm produced excessive amounts of snow across New England with accumulations of greater than 30 in. (75 cm) across large sections of eastern Massachusetts, New Hampshire, and Maine, with 40 in.+ (100 cm) snows observed in a large area from New Hampshire to Maine.
- The following regions reported snow accumulations exceeding 10 in. (25 cm): central and northern Connecticut and Rhode Island, central and eastern Massachusetts (except Cape Cod and the Islands), eastern Vermont. New Hampshire, and most of Maine.
- The following regions reported snow accumulations exceeding 20 in. (50 cm): eastern Massachusetts, northeastern Vermont, New Hampshire, and most of Maine.

TABLE 10.13-1. Snowfall amounts for urban and selected locations for 22–28 Feb 1969; see Table 10.1-1 for details.

NESIS = 4.01 (category 3)	
Urban center snowfall amounts	
Philadelphia, PA	1.9 in. (5 cm)
New York, NY–La Guardia Airport	1.7 in. (4 cm)
Boston, MA	26.3 in. (67 cm)
Other selected snowfall amounts	
Mount Washington, NH	97.8 in. (248 cm)
Pinkham Notch, NH	77.3 in. (196 cm)
Long Falls Dam, ME	56.0 in. (142 cm)
Old Town, ME	43.6 in. (111 cm)
Rockport, MA	39.0 in. (99 cm)
Portland, MA	26.9 in. (68 cm)

b. Surface analyses

- A large anticyclone was anchored over Hudson Bay prior to and during cyclogenesis (1040 hPa). The air immediately along the East Coast during the storm period was maritime in origin, however, with temperatures generally between −5° and 5°C.
- A ridge of high pressure extending down the East Coast prior to cyclogenesis reflected the cold-air damming on 22 and 23 February.
- The development of a primary surface low was ill defined. An inverted trough initially in the Ohio Valley slowly evolved into a weak low pressure center, which meandered northeastward over the eastern Great Lakes on 23–24 February. This system exhibited no frontal structure and was attended by only light precipitation.
- A second inverted trough and coastal front developed along the Southeast coast by 1200 UTC 23 February. A weak cyclone then formed along the coastal front. The surface low deepened slowly [-4 hPa (12 h)$^{-1}$] as it propagated from near the Virginia capes to a position east of New Jersey between 0000 and 1200 UTC 24 February. The low then intensified more rapidly in the following 12-h period off the southern New England coast as its central pressure fell 10 hPa to 994 hPa by 0000 UTC 25 February. Heavy snowfall developed over southeastern New England during this period. Little deepening occurred thereafter. The surface low drifted very slowly after 1200 UTC 24 February, moving only 250 km in 24 h.
- The cyclone meandered near the southeastern New England coast from 25 to 28 February, prolonging the heavy snowfall in eastern New England.
- Sea level pressure gradients to the north and northeast of the surface low were moderately intense, with winds approaching gale force only along the New England coast.

c. 850-hPa analyses

- The 850-hPa charts provide evidence for blocked flow with a trough over the Great Lakes and over the Atlantic Ocean separated by a strong ridge along the coast, capped off by a significant anticyclone over south-central Canada on 23–24 February.
- Cold-air advection associated with northwesterly flow was not observed at 850 hPa over the northeastern United States prior to this storm.
- Two separate 850-hPa low centers were observed in this case. The first and weaker low moved northward from Kentucky to Michigan on 23 February. This system was associated with the inverted trough, surface low, and light precipitation over the Ohio Valley. The low was located in a weak and poorly defined 850-hPa temperature gradient. It propagated far to the north and west of the 0°C isotherm and did not deepen appreciably.
- A second 850-hPa low formed by 1200 UTC 23 February along a band of somewhat enhanced temperature gradients near the Carolina coast. This center was associated with the sea level cyclone that moved northeastward along the Atlantic seaboard. The 850-hPa low deepened slowly at first, as it drifted up the East Coast. More vigorous deepening ensued after 1200 UTC 24 February coinciding with the more rapid development of the surface low.
- The initial development of the 850-hPa lows in the Great Lakes region and then along the Carolina coast was not marked by the strong cold-air advection west of the center that characterized many of the other cases. A weak S-shaped isotherm pattern developed along the East Coast by 1200 UTC 24 February as warm-

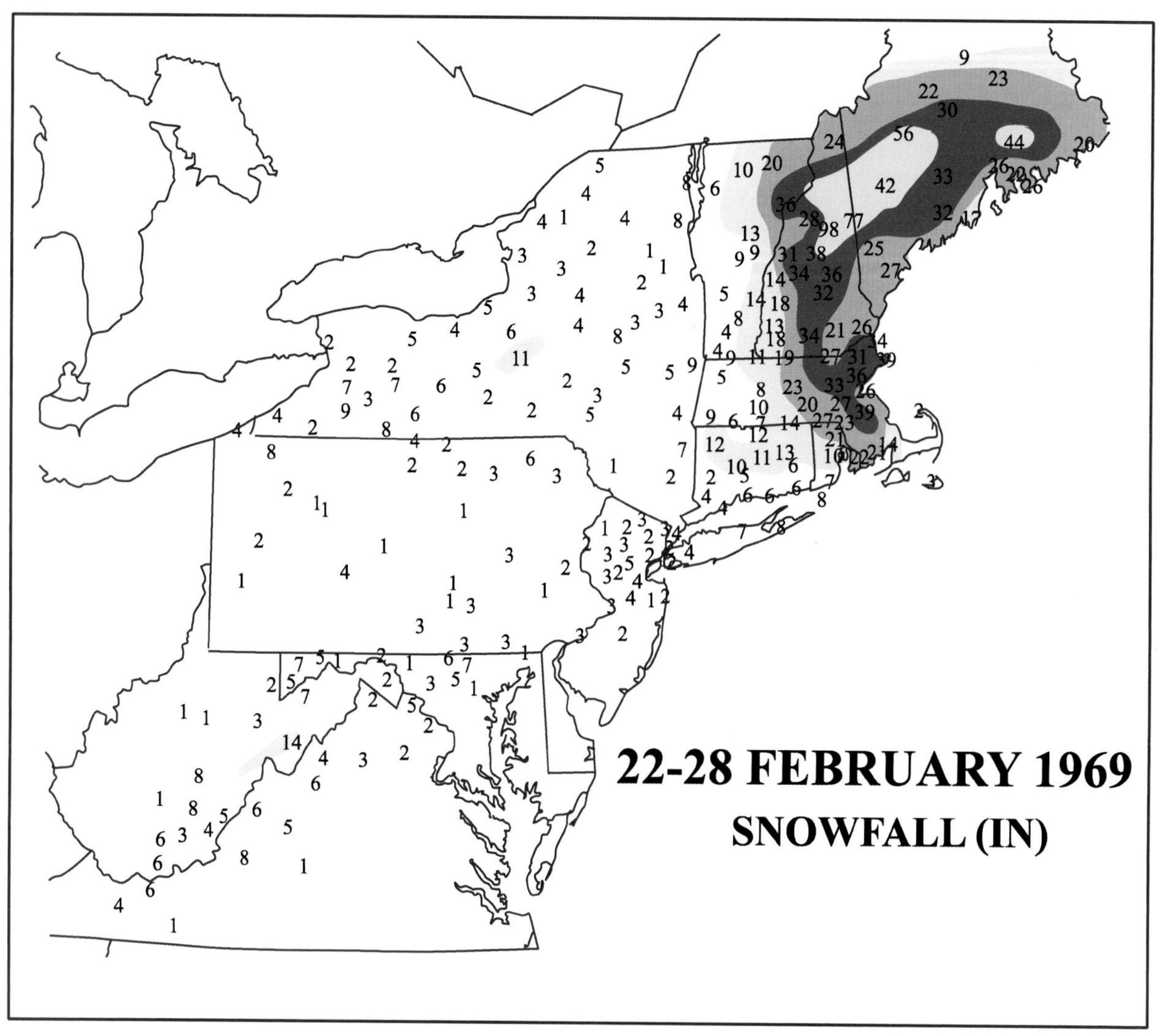

FIG. 10.13-1. Snowfall (in.) for 22–28 Feb 1969. See Fig. 10.1-1 for details.

air advection increased along the Northeast coast. The low initially formed on the warm side of the 0°C isotherm, but was located closer to the 0°C line during cyclogenesis near New England late on 24 February and on 25 February.

- The cyclone evolved in the most diffuse 850-hPa temperature field of the survey. However, a very narrow, yet significant, 850-hPa temperature gradient formed over New England by 1200 UTC 24 February as the cyclone was beginning to intensify and heavy snow was falling in southeastern New England.
- Southeasterly to easterly winds and their associated moisture transports strengthened to the north of the coastal low center on 24 and 25 February as the system intensified and heavy snow fell across eastern New England. The wind speeds increased to 30 m s^{-1} in the low-level jet (LLJ) by 1200 UTC 26 February, directed toward the heaviest snows in Maine and New Hampshire. Heaviest snows fell within an area in which an easterly LLJ remained directed overhead for at least 36 h.

d. 500-hPa geopotential height and 400-hPa wind analyses

- As with the previous storm during the same month, this storm occurred during a winter dominated by El Niño conditions and a negative North Atlantic Oscillation (NAO). Daily values of the NAO reached a nadir on 15–18 February, then rose before the storm developed. Negative NAO values then leveled off during the evolution of the storm. Despite the occurrence of a negative NAO, 500-hPa analyses are not suggestive of a strong cutoff upper ridge near Greenland, characteristic of other cases that occurred during a negative NAO.
- The precyclogenetic geopotential height field at 500

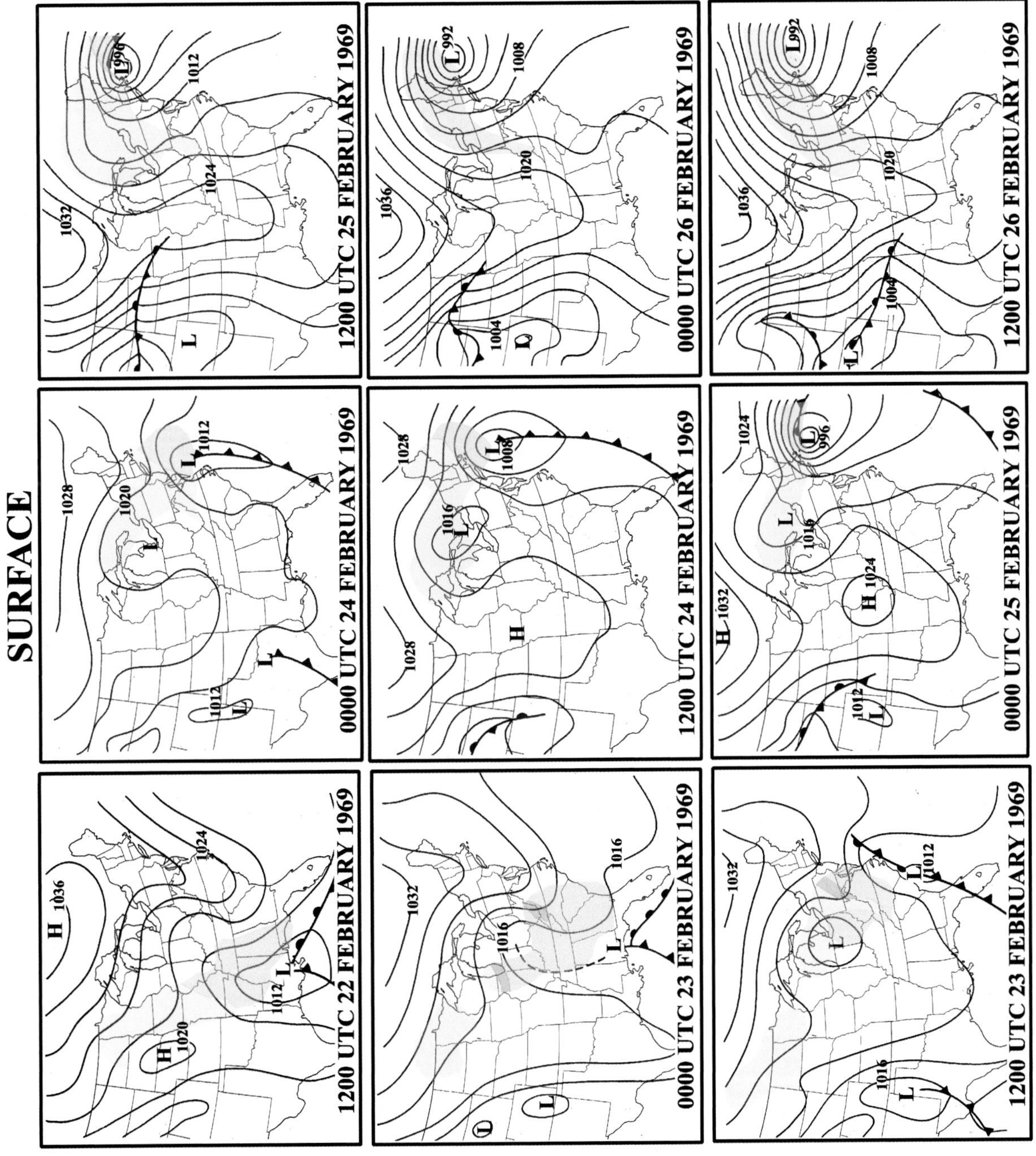

FIG. 10.13-2. Twelve-hourly surface weather analyses for 1200 UTC 22 Feb–1200 UTC 26 Feb 1969. See Fig. 10.1-2 for details.

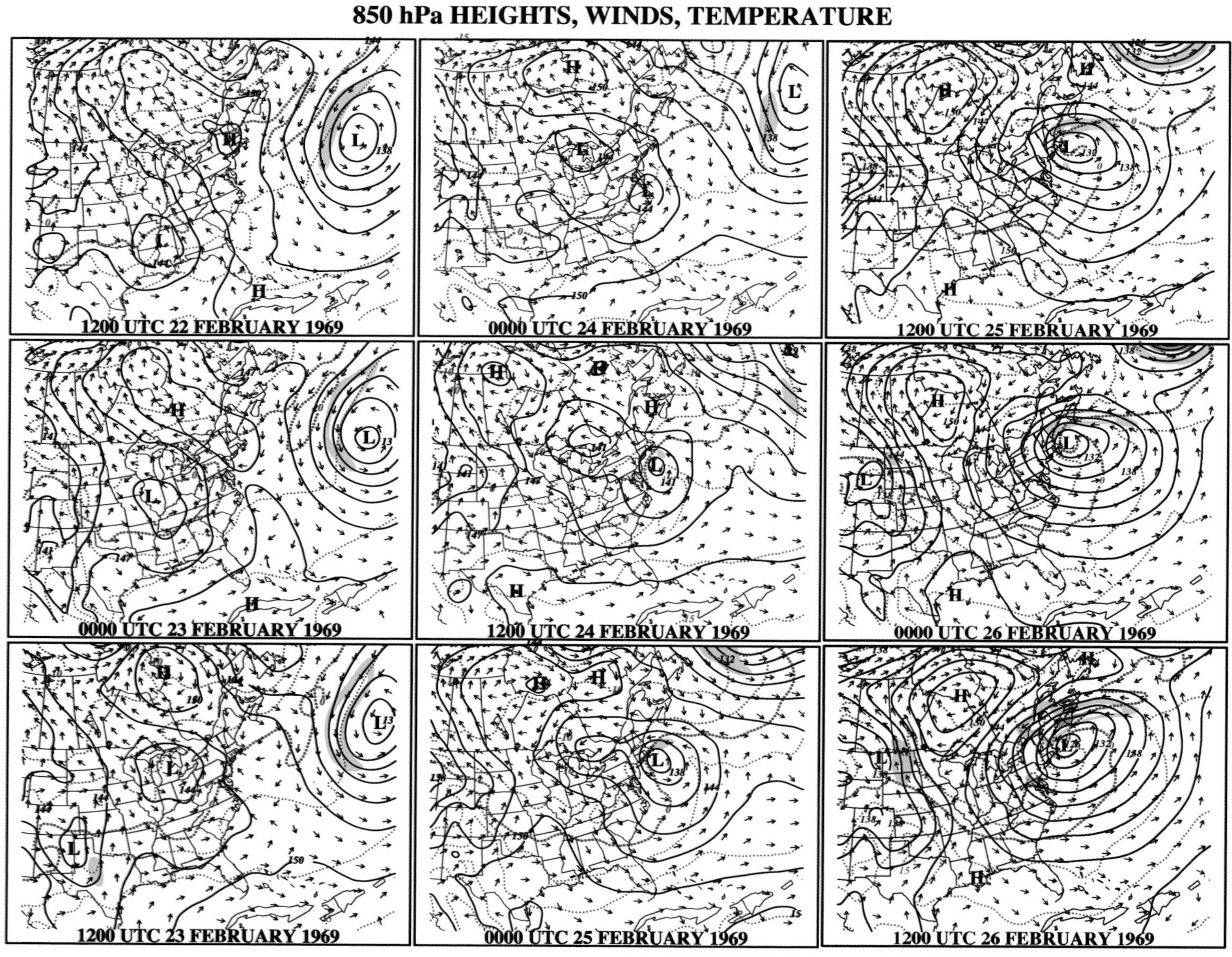

FIG. 10.13-3. Twelve-hourly 850-hPa analyses for 1200 UTC 22 Feb–1200 UTC 26 Feb 1969. See Fig. 10.1-3 for details.

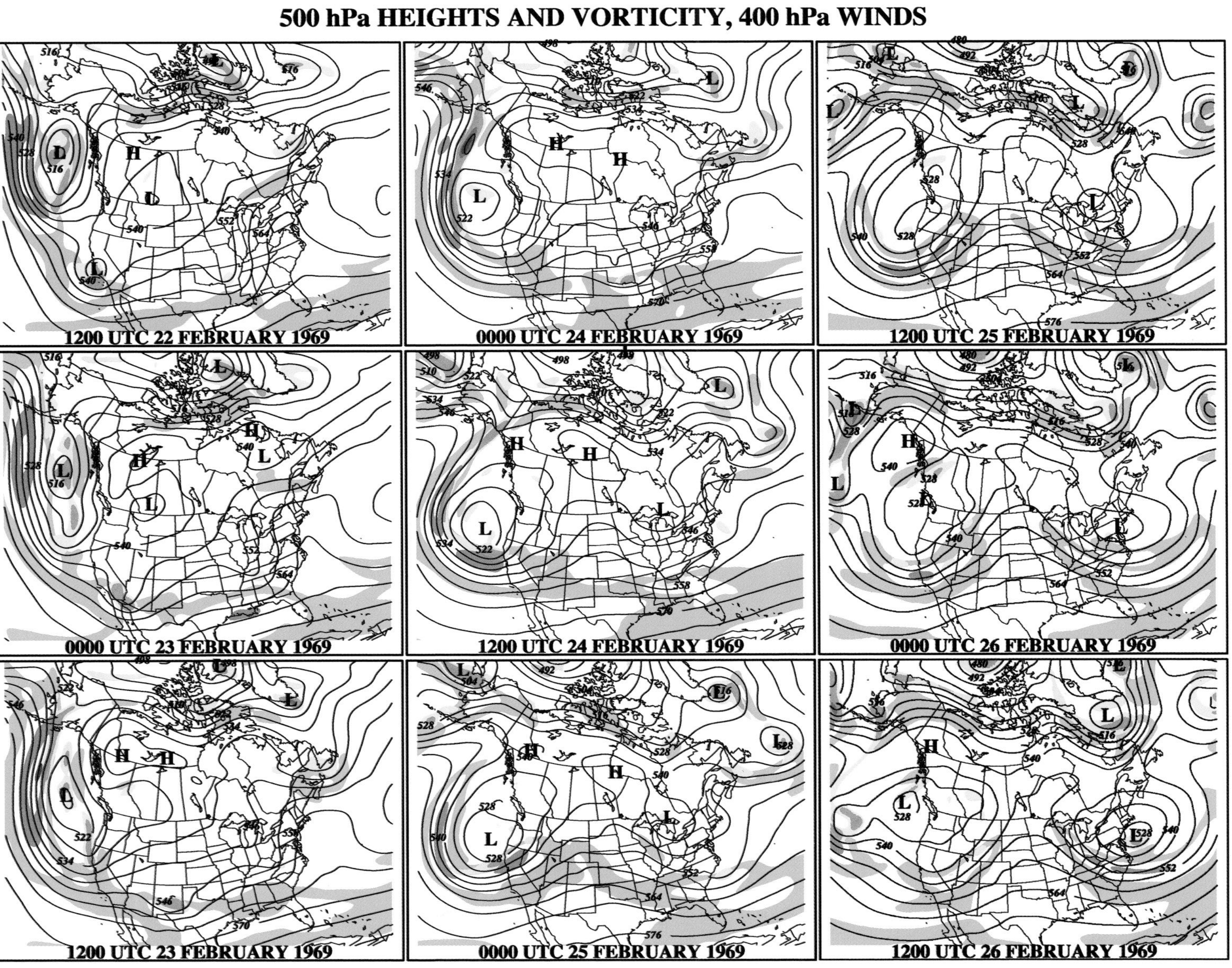

FIG. 10.13-4. Twelve-hourly analyses of 500-hPa geopotential heights and upper-level wind fields for 1200 UTC 22 Feb–1200 UTC 26 Feb 1969. See Fig. 10.1-4 for details.

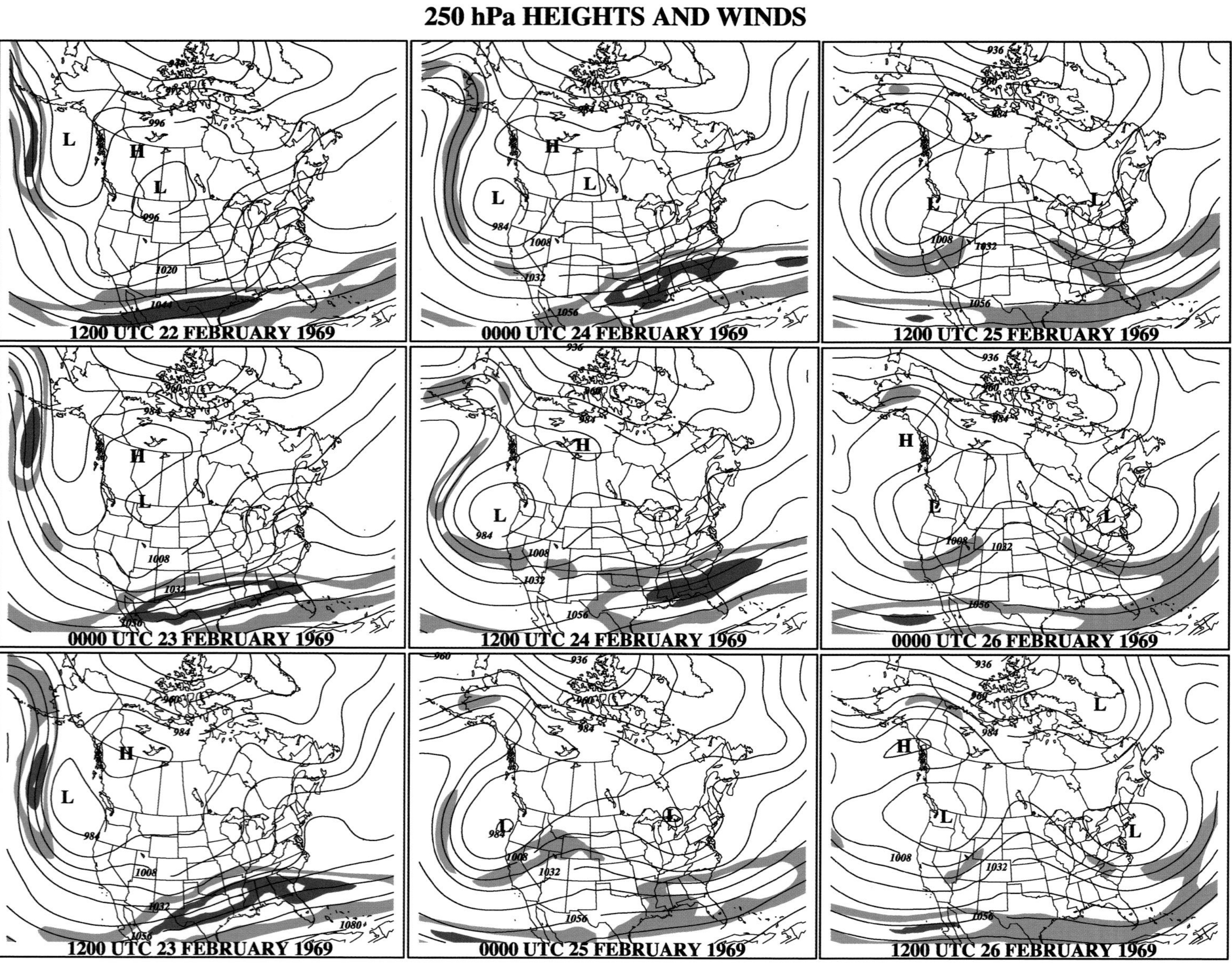

FIG. 10.13-5. Twelve-hourly analyses of 250-hPa geopotential height for 1200 UTC 22 Feb–1200 UTC 26 Feb 1969. See Fig. 10.1-5 for details.

hPa did not exhibit the coherent trough–ridge patterns found in the other cases. An unusually large number of trough and ridge features can be identified across North America on 22–23 February, suggestive of a blocking pattern that could account for the slow movement of the storm.

- Troughs over the south-central United States and southwestern United States at 1200 UTC 22 February later influenced the growth of the East Coast storm. The 500-hPa chart also showed a trough off the West Coast, closed cyclonic circulations north of Montana and near Bermuda, and a ridge along the East Coast. A trough in southeastern Canada produced confluent flow, which was associated with the large surface anticyclone located over eastern Canada on 23 February.
- A trough in the Ohio Valley was associated with the inverted surface trough, a weak cyclone, and light precipitation across the eastern United States on 23 February. A second trough in the southwestern United States propagated to the East Coast by 24 February. It appeared to merge with the Ohio Valley trough by 0000 UTC 25 February, and may have influenced the more rapid development of the storm near New England on 24–25 February.
- Two separate cyclonic vorticity maxima were observed in connection with the two troughs over the central and eastern United States between 1200 UTC 23 February and 1200 UTC 24 February. The cyclonic vorticity maximum initially associated with the weak Ohio Valley trough first tracked eastward, then turned southeastward toward the middle Atlantic coast by 26 February. The second vorticity maximum was associated with the separate trough–jet system that crossed the southern United States between 22 and 25 February, located off the North Carolina coast by 0000 UTC 25 February.
- After 1200 UTC 24 February, the chaotic regime of small-amplitude, high-wavenumber features evolved into a simpler, more organized lower-wavenumber pattern with troughs and ridges of increasing amplitude that take on blocking characteristics. A major trough remained off the West Coast with a ridge over central North America and a deep, closed cyclonic vortex along the Atlantic seaboard, reflecting the development of the surface cyclone near the East Coast, with a downstream ridge extending into southeastern Canada and a cutoff low farther downstream over the North Atlantic Ocean.
- A pattern of weak diffluence was evident as early as 1200 UTC 23 February. A more pronounced pattern of diffluence was observed by 0000 and 1200 UTC 25 February, following the only 12-h period in which the surface low intensified rapidly.
- Over the 36-h period between 0000 UTC 23 February and 1200 UTC 24 February, during which the surface low deepened and moved toward New England, the amplitude between the troughs along the coast and the downstream ridge appeared to decrease, a deviation from the other cases examined. The half-wavelength, however, also decreased during this time, a factor that is consistent with the other cases.
- The passage of a trough across the southeastern United States on 24–25 February was followed by a general increase in amplitude as a longer-wavelength trough–ridge pattern and a probable blocking situation became established over the United States. The emergence of the low-wavenumber regime was associated with the slow movement of the storm off the coast for the following 3–4 days.
- This case is marked by the absence of a confluent jet streak in the Great Lakes to New England area during the precyclogenetic period. The southern jet streak is a dominant factor throughout the entire period.
- A jet marked by wind speeds of less than 50 m s^{-1} at 300 hPa was located in the confluent region south of a small cutoff low over eastern Canada on 23 February. This jet probably influenced the large surface anticyclone over eastern Canada, although this pattern was relatively weak and poorly defined as compared to the other cases.
- Although the 500-hPa geopotential height fields were chaotic, clearly defined upper-level jet streaks extended from northern Mexico to the Southeast coast from 23 to 25 February. These features were observed over the region of largest height gradients along the periphery of the chaotic trough pattern. Wind speeds generally exceeded 40 m s^{-1} at 400 hPa for the entire observing period. The cyclonic wind shear associated with each jet streak embedded within this flow regime contributed to the separate vorticity maxima observed over the southern United States prior to cyclogenesis. The evolution of the various jet streaks and troughs into one coherent, negatively tilted system by 1200 UTC 25 February is analyzed for this case, to the east of the coastline, and well south of the developing low.

e. 250-hPa geopotential height and wind analyses

- The upper troposphere was marked by a general west to east flow regime, with the short-wave troughs noted at 500 hPa still evident at this level. The height gradients are maximized along a line from near the Gulf coast and the Atlantic coast off South Carolina and Georgia.
- A strong subtropical jet (STJ) is located within this region. The STJ amplified to speeds greater than 80 m s^{-1} on 23–24 February and slowly decreased in magnitude thereafter.
- This is no evidence of any northern jet at this level prior to, during, or after the surface .cyclogenesisi along the East Coast, an attribute that is quite different from many other major East Coast snowstorms.

East End Ave, New York City, Dec 1969 (photo courtesy of Kevin Ambrose, *Great Blizzards of New York City*, Historical Enterprises, 1994).

14. 25–28 December 1969

a. General remarks

- This storm was a near miss (see chapter 5, volume I) for the large cities of the northeastern United States as heavy snow turned to rain (and back to snow in many areas). The heaviest snow fell immediately north and west of the coastal plain. This system is one of the heaviest snowstorms on record for eastern and northern New York. Accumulations of greater than 20 in. (50 cm) covered a wide area of central and eastern New York into northwestern New England.
- The following regions reported snow accumulations exceeding 10 in. (25 cm): western Virginia, western Maryland, portions of West Virginia, central and eastern Pennsylvania (except the extreme southeast), central and eastern New York (except the coast), Connecticut (except the coast), central and western Massachusetts, Vermont, New Hampshire, and western Maine.
- The following regions reported snow accumulations exceeding 20 in. (50 cm): northeastern Pennsylvania, much of eastern New York, and Vermont.

TABLE 10.14-1. Snowfall amounts for urban and selected locations for 25–28 Dec 1969; see Table 10.1-1 for details.

NESIS = 5.19 (category 3)	
Urban center snowfall amounts	
Washington, D.C.–Dulles Airport	12.1 in. (31 cm)
Baltimore, MD	6.1 in. (15 cm)
Philadelphia, PA	5.2 in. (13 cm)
New York, NY–La Guardia Airport	7.4 in. (19 cm)
Boston, MA	4.2 in (11 cm)
Other selected snowfall amounts	
Burlington, VT	29.8 in. (76 cm)
Albany, NY	26.4 in. (67 cm)
Binghamton, NY	21.9 in. (56 cm)
Williamsport, PA	17.2 in. (44 cm)
Roanoke, VA	16.4 in. (42 cm)
Hartford, CT	15.2 in. (39 cm)
Harrisburg, PA	12.9 in. (33 cm)

b. Surface analyses

- An anticyclone (1033 hPa) accompanied by cold surface temperatures was centered north of New England prior to 26 December. The high center drifted slowly northeastward as cyclogenesis commenced along the coast.
- Cold-air damming and coastal frontogenesis were prominent along the East Coast on 25 December and along the New England coast on 26 December.
- The surface low developed over Texas, well to the south of an occluding cyclone near the Minnesota–Canada border on 24–25 December. The low crossed the Gulf coast states, then moved northeastward along the coastal front near the East Coast on 25–26 December. Precipitation spread rapidly northeastward with heavy snow falling along the Appalachian Mountains by late on 25 December. The storm track's close proximity to the coastline combined with the location of the surface anticyclone over extreme eastern Canada acted to draw warmer air inland, resulting in snow changing to rain over coastal areas. Central New England experienced a severe ice storm.
- The surface low moved rapidly northeastward along the East Coast between 0000 and 1200 UTC 26 December, then slowed considerably as it reached the New Jersey shore. It then moved very slowly along the New England coast for the following 24–36 h, advancing at only 5–10 m s^{-1}.
- Rapid intensification occurred during two periods. Between 0000 and 1200 UTC 26 December, the storm's central pressure fell 12 hPa in 12 h as it moved from Georgia to New Jersey. During this period, heavy precipitation spread quickly across the middle Atlantic states into New England. The following 12 h were marked by little intensification as the surface low moved very slowly from New Jersey to just east of Long Island. The period between 0000 and 1200 UTC 27 December brought the second surge of rapid deepening, as the central pressure fell another 12 hPa to 976 hPa, as the cyclone drifted toward the Massachusetts coast. Snowfall rates increased to the west of the storm center during this second period of intensification, producing significant snow across Long Island, a rare occurrence even in this type of storm. The cyclone continued to drift slowly east-northeastward in the following 24 h.
- The slow movement of the storm center after 1200 UTC 26 December contributed to the very heavy snowfall in northern New York and Vermont. Considerable warming also occurred over northern and eastern New England as persistent onshore winds raised temperatures well above freezing. The warm air from the Atlantic and a wedge of cold air remaining over western New England created a strong frontal boundary across central New England on 27 December, and was an important factor in creating the conditions for the ice storm in central New England.

c. 850-hPa analyses

- Only weak northwesterly flow and cold-air advection were observed across the northeastern United States on 24 December, as the colder air was already established over the northeast prior to the snowstorm. The 850-hPa temperatures ranged from 0°C over North Carolina to −15°C across northern New York and New England.
- An 850-hPa low center was located near the U.S.–Canadian border at 0000 UTC 25 December in association with the occluding surface low over the

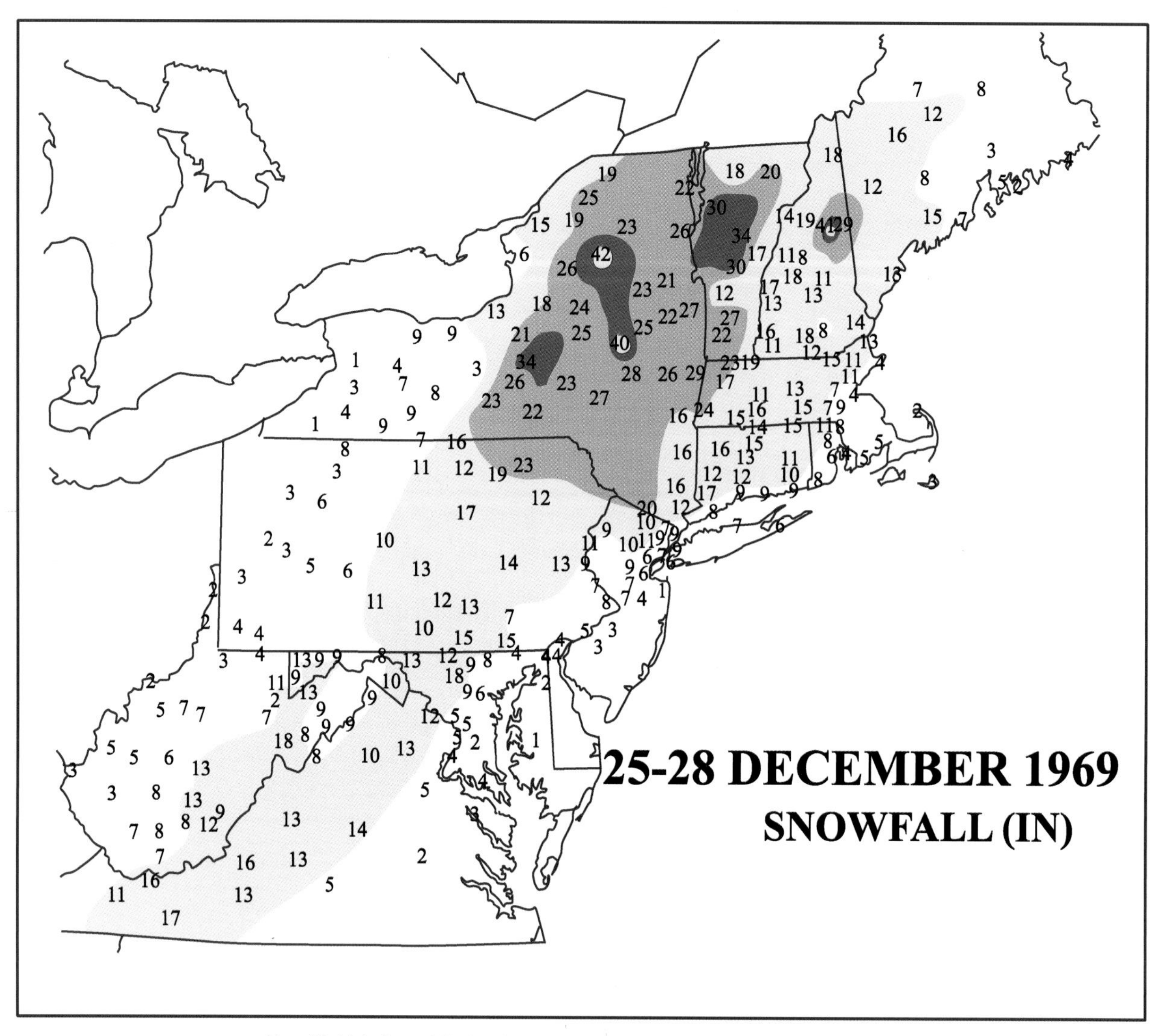

FIG. 10.14-1. Snowfall (in.) for 25–28 Dec 1969. See Fig. 10.1-1 for details.

north-central United States. The cyclonic circulation around this center produced warm-air advection across the Great Lakes region through 0000 UTC 26 December.

- The 850-hPa low center that was associated with the major storm drifted eastward from the south-central to the southeastern United States on 25 December. The low center deepened very slowly as the surface low crossed the Gulf coast. It then experienced two 12-h periods of rapid deepening, concurrent with an intensification period of the surface low. The first period occurred between 0000 and 1200 UTC 26 December (-90 m per 12 h), with the second period occurring between 0000 and 1200 UTC 27 December (-90 m per 12 h). During the 12-h interval separating these periods, the 850-hPa low center deepened at a rate of only -30 m $(12\text{ h})^{-1}$.
- The 850-hPa low developed along an east–west thermal ribbon extending across the southern United States. Cold-air advection behind the low increased across Texas by 1200 UTC 25 December while strong warm-air advection east of the low coincided with the rapid outbreak of moderate to heavy precipitation. These amplifying advections resulted in an S-shaped isotherm pattern along the coast by 1200 UTC 26 December. This pattern gradually became oriented north to south from the Carolinas to Maine on 27 December, as the low moved over Massachusetts and warmer air moved over Long Island and coastal New England, changing the snow to rain in that region.
- A southerly low-level jet (LLJ) exceeding 25 m s^{-1} developed over Mississippi and Alabama by 1200 UTC 25 December and shifted to the Southeast coast by 0000 UTC 26 December, immediately prior to the first period of rapid deepening along the Atlantic coast. Coastal frontogenesis occurred and precipitation broke out across the southeastern states during this period. The jet formed as the rapidly advancing

SURFACE

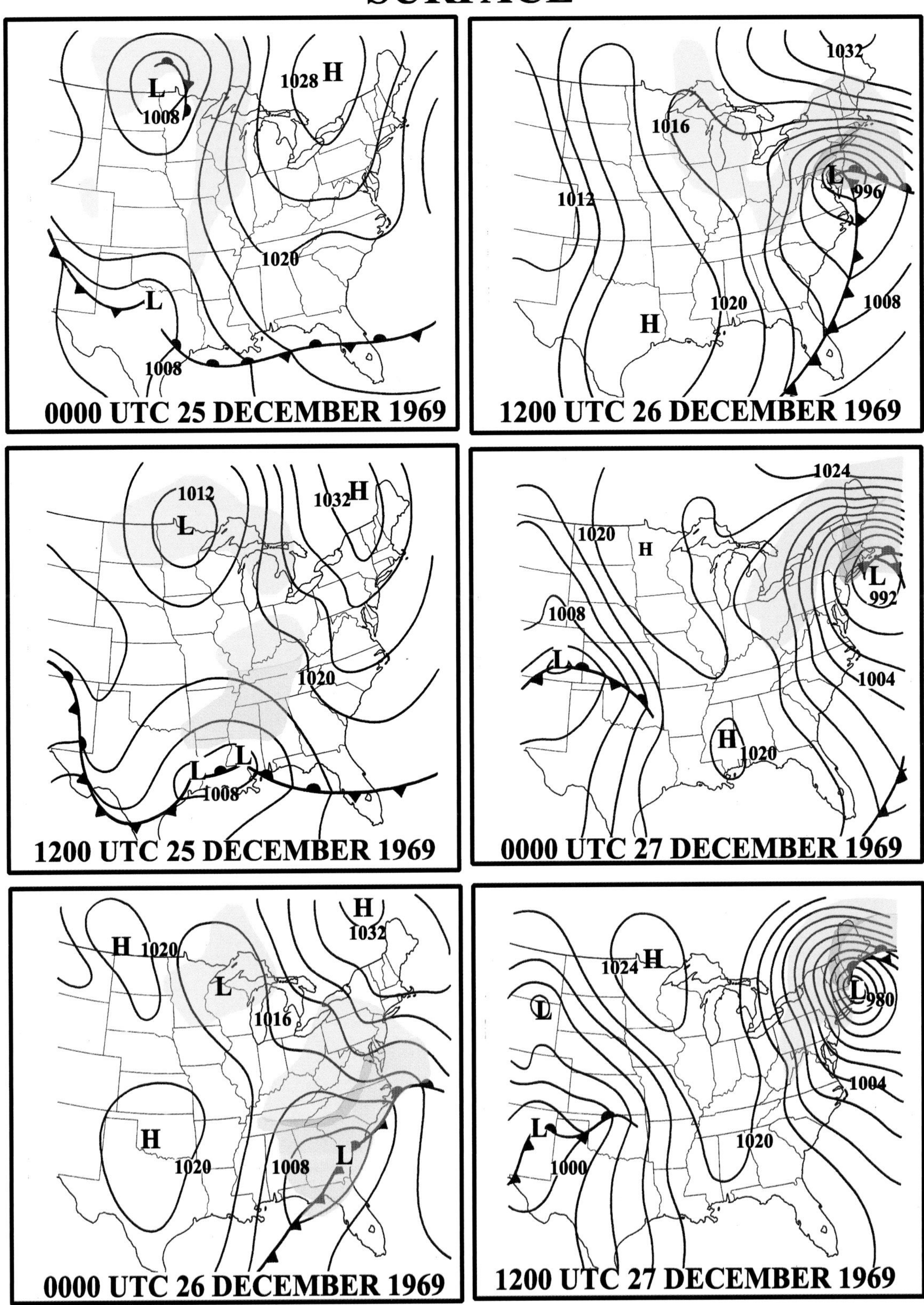

FIG. 10.14-2. Twelve-hourly surface weather analyses for 0000 UTC 25 Dec–1200 UTC 27 Dec 1969. See Fig. 10.1-2 for details.

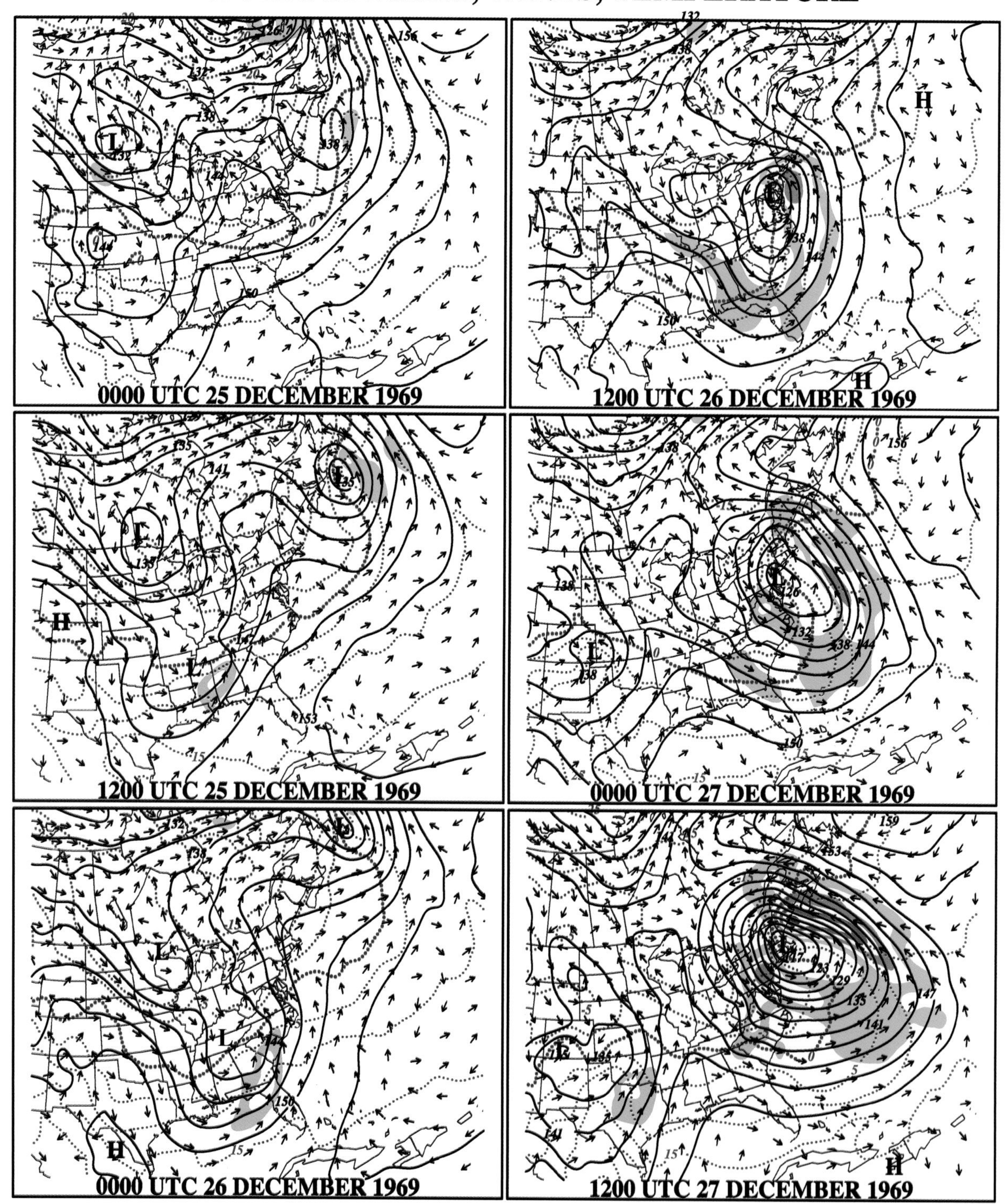

FIG. 10.14-3. Twelve-hourly 850-hPa analyses for 0000 UTC 25 Dec–1200 UTC 27 Dec 1969. See Fig. 10.1-3 for details.

500 hPa HEIGHTS AND VORTICITY, 400 hPa WINDS

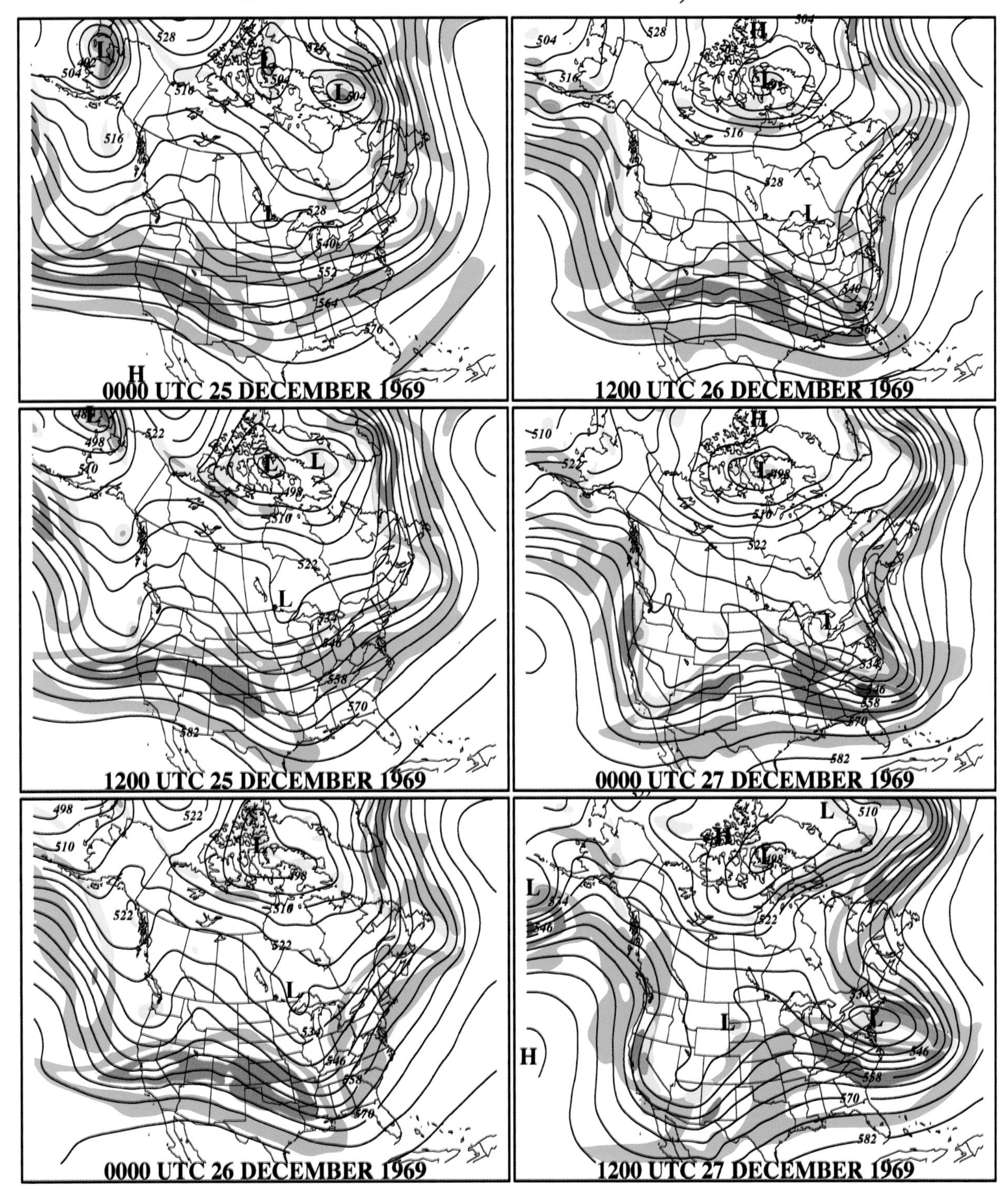

FIG. 10.14-4. Twelve-hourly analyses of 500-hPa geopotential heights and upper-level wind fields for 0000 UTC 25 Dec–1200 UTC 27 Dec 1969. See Fig. 10.1-4 for details.

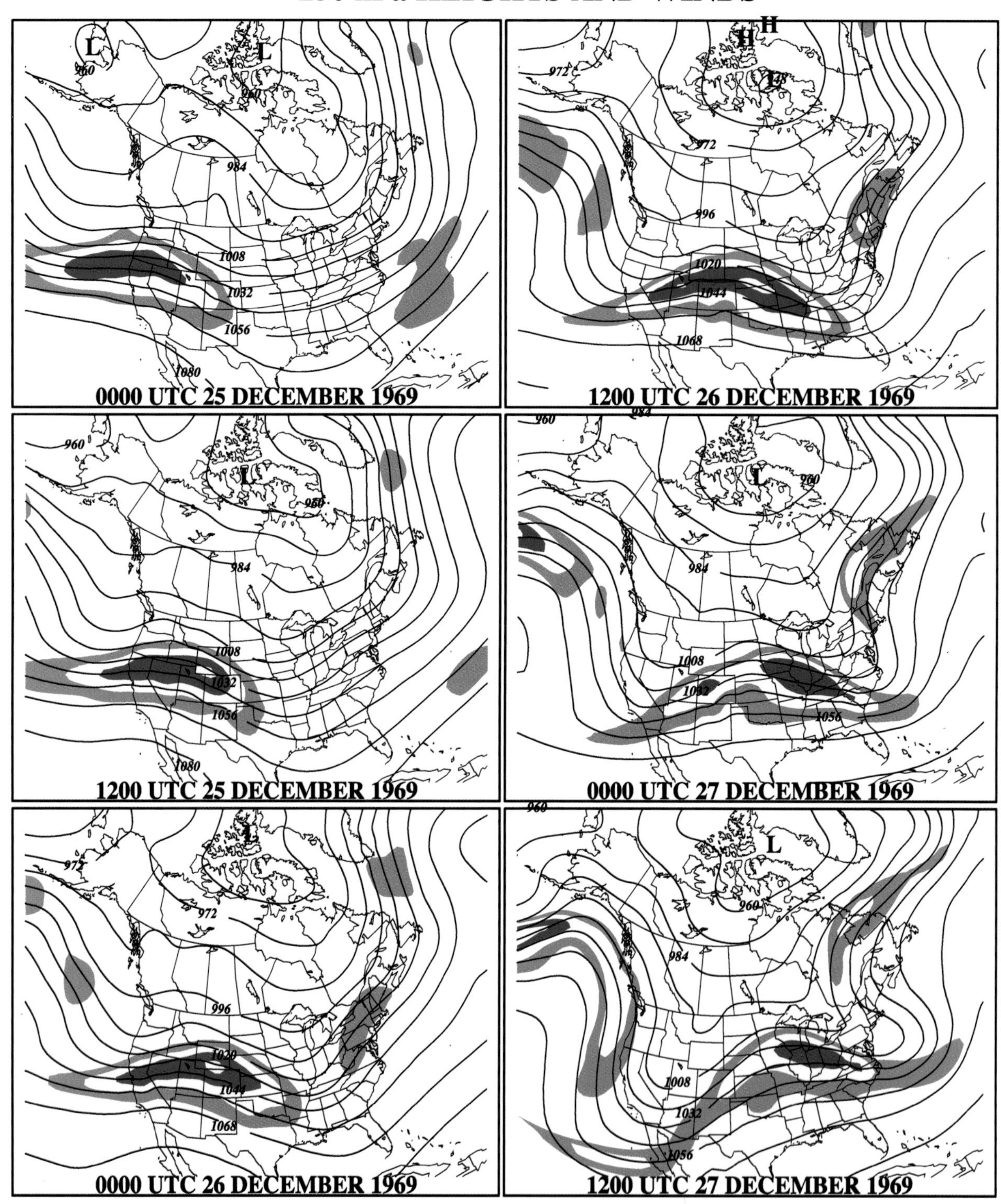

FIG. 10.14-5. Twelve-hourly analyses of 250-hPa geopotential height for 0000 UTC 25 Dec–1200 UTC 27 Dec 1969. See Fig. 10.1-5 for details.

and deepening 850-hPa low encountered a slower-moving ridge to its east. As a result, the geopotential height gradients increased along the Southeast coast.

- The LLJ continued to intensify to greater than 30 m s^{-1} and extended on a north–south axis toward Long Island and southern New England as the 850-hPa low intensified over the middle Atlantic coast at 1200 UTC 26 December. The enhanced warm-air advection associated with this jet system quickly changed the snow to rain and freezing rain east of the storm center.
- The LLJ began to wrap around the north of the low by 1200 UTC 26 December, as the cyclone was undergoing its first period of rapid deepening. The southerly and easterly low-level jets enhanced the moisture transports into regions of heavy precipitation at this time. By 1200 UTC 27 December, the LLJ took on a southeasterly orientation and increased to over 55 m s^{-1} (the highest analyzed for all cases) off the New England coast as the entire system underwent the second period of rapid deepening.

d. 500-hPa geopotential height and 400-hPa analyses

- The winter of 1969/70 was a continuation of the long-period El Niño that was continuing from the previous winter. This storm occurred during a period in which daily values of the North Atlantic Oscillation (NAO) fluctuated between slightly negative and slightly positive. The 3-day period in which this storm evolved was marked by an NAO that tended to be negative. Immediately following the occurrence of the storm, the NAO went into a pronounced negative phase for much of January 1970, a very cold month for the Northeast.
- The period prior to 26 December was marked by nearly zonal flow across much of the southern United States with no high-amplitude features present anywhere over North America on 24–25 December. This pattern is consistent with the generally colder air that was already in place across the north-central and northeastern United States by 25 December.
- A trough over eastern Canada did not contain the geopotential height minimum, closed cyclonic circulation, or intense vorticity gradients that marked many of the other cases. The flow over New England and the Great Lakes was not characterized by the degree of confluence seen in the precyclogenetic periods of many other cases. A weakly defined pattern of confluent geopotential heights was found across southern Ontario and Quebec at 0000 and 1200 UTC 25 December, above the surface anticyclone. A better-defined pattern of confluence appeared across extreme eastern Canada at 1200 UTC 26 December and 0000 UTC 27 December as the surface high drifted toward Newfoundland.
- The cyclone's development was linked to the amplification of a trough that propagated eastward from the central United States after 0000 UTC 25 December to the southeast coast by 1200 UTC 26 December. The amplifying trough seemed to pivot about an ill-defined cyclonic vortex over the Great Lakes region on 26 December. This feature was associated with the occluded, dying surface low pressure system over the northern United States.
- The cyclonic vorticity maximum and vorticity advection patterns became much better defined by 1200 UTC 26 December as the trough and jet streak system amplified over the south-central and southeast United States.
- As the 500-hPa trough deepened over the southeastern United States by 1200 UTC 26 December, it became oriented from northwest to southeast (a negative tilt), with a diffluent pattern appearing downwind of the trough axis. These changes in trough structure occurred as the surface low moved rapidly toward New England and went through its first period of significant deepening. By 1200 UTC 27 December, the 500-hPa trough deepened into a closed circulation with easterly flow extending up the 400-hPa level over New England.
- Diffluence was observed over New England between the trough and the downstream ridge. This coincided with the second period of rapid deepening, as the surface low moved very slowly along the New England coast.
- The cyclogenetic period from 26 to 27 December was characterized by a general increase of amplitude of the trough–ridge pattern across the United States. A large increase in amplitude occurred between the trough over the eastern United States and the downstream ridge over New England and southeastern Canada on 26 December.
- The half-wavelength of the trough and downstream ridge shortened during cyclogenesis, especially between 0000 UTC 26 December and 0000 UTC 27 December.
- This case was marked by a dual-jet-streak pattern, like many of the previous cases. However, by 0000 UTC 26 December, the northern jet became oriented on a north–south axis with no confluent entrance region over the Northeast to maintain the cold air along the coast.
- The more southern jet streak propagated eastward with the deepening trough, was located over the south-central United States, and had wind speeds greater than 60 m s^{-1} by 1200 UTC 26 December.
- The initial development of the surface low and the 850-hPa LLJ appeared to occur beneath the exit region of the jet system diving southeastward toward the Gulf coast, and distinct entrance region of the north–south-oriented jet extending from Georgia to New England at 0000 UTC 26 December. However, the surface low

and LLJ quickly moved northeastward along the coast the following 12 h and were far removed from the exit region of the southern jet by 1200 UTC 26 December, indicating that lateral coupling was not a factor in this case.

- The changeover to rain and freezing rain along the Northeast coast on 26 December is consistent with several "near miss" cases (volume I, chapter 5) in which the 500-hPa trough undergoes amplification and attains a negative tilt far to the south and west of the East Coast (Mississippi and Alabama by 0000 UTC 26 December). This pattern, combined with lack of a confluent entrance region of a northern jet over New England, allowed warmer air to flood across the Northeast urban corridor and southern New England. Snow changed to rain by 1200 UTC 26 December as far north as New York City, and changed to rain throughout New England after 1200 UTC 26 December.

e. 250-hPa geopotential height and wind analyses

- The geopotential height analysis clearly reflects the general westerly flow into the central United States during the precyclogenetic period. The deepening trough over the south-central and southeast United States attained a negative tilt even up to this level by 0000 UTC 26 December.
- The most distinctive feature is the strong upper-level jet that propagates into the south-central United States upwind of the deepening trough prior to the onset of cyclogenesis. The separate, southwest-to-northeast-oriented jet amplified downwind of the trough axis as the cyclone developed and moved northeastward along the East Coast by 1200 UTC 26 December.
- The northern jet has the characteristics of an "outflow" jet emanating from the region of rapid cyclogenesis along the coast and very heavy snowfall occurring over the more inland areas of the Northeast.

A woman shovels snow in Fair Lawn, New Jersey, 18 Feb 1972 (photo by Edward Hausner).

15. 18–20 February 1972

a. General remarks

- This storm was one of the few to pose a threat of heavy snow in the Northeast urban corridor during the early and middle 1970s and another near miss for the major cities as the heaviest snow fell immediately to their west and north. Strong easterly winds and rough seas caused significant damage along the middle Atlantic and New England coasts.
- The following regions reported snow accumulations exceeding 10 in. (25 cm): eastern West Virginia, northern Virginia, central and northern Maryland, Pennsylvania (except the west and southeast), northwestern New Jersey, New York (except the extreme west and coastal regions), central and northern Connecticut, Massachusetts (except the southeast), Maine, and parts of New Hampshire and Vermont.
- The following regions reported snow accumulations exceeding 20 in. (50 cm): north-central Pennsylvania, central New York, and parts of West Virginia.

TABLE 10.15-1. Snowfall amounts for urban and selected locations for 18–20 Feb 1972; see Table 10.1-1 for details.

NESIS = 4.19 (category 3)	
Urban center snowfall amounts	
Washington, D.C.–Dulles Airport	10.0 in. (25 cm)
Baltimore, MD	3.2 in. (8 cm)
Philadelphia, PA	3.7 in. (9 cm)
New York, NY–La Guardia Airport	6.3 in. (16 cm)
Boston, MA	6.3 in. (16 cm)
Other selected snowfall amounts	
Binghamton, NY	24.4 in. (62 cm)
Williamsport, PA	22.8 in. (58 cm)
Syracuse, NY	20.0 in. (51 cm)
Portland, ME	15.4 in. (39 cm)
Worcester, MA	15.2 in. (39 cm)
Harrisburg, PA	13.0 in. (33 cm)
Albany, NY	12.1 in. (31 cm)

b. Surface analyses

- An anticyclone (1035 hPa) was located to the northeast of New England on 18 February, prior to East Coast cyclogenesis. Its eastward location, in concert with the surface low's track close to the shoreline, produced more of an east than northeasterly flow along the coast. The resultant warming ensured that a significant portion of the precipitation would fall as rain within the Northeast urban corridor.
- Cold-air damming and coastal frontogenesis were observed along the Southeast coast on 18 February, prior to the coastal cyclogenesis.
- The primary low was a slow-moving, intense storm over the upper Midwest. It filled 10 hPa in 24 h between 0000 UTC 18 February and 0000 UTC 19 February as it moved over the Great Lakes. Despite the obvious dissipation of the primary low, it remained evident for 27 h after the onset of secondary cyclogenesis.
- A secondary low developed over southern Georgia after 1200 UTC 18 February, 1600 km south of the primary center. This separation represents the largest distance between primary and secondary low centers in the sample. The secondary low deepened rapidly after 0300 UTC 19 February, with the central sea level pressure falling 25 hPa in 12 h, reaching 975 hPa over New Jersey by 1500 UTC 19 February.
- The secondary low propagated northeastward at an average rate of 15 m s^{-1} along the Southeast coast. The forward motion slowed along the middle Atlantic and New England coast on 19 February. By 1800 UTC 19 February, the cyclone developed dual centers as it moved slowly and erratically off the New Jersey and New England coasts.
- An intense pressure gradient developed to the northeast of the rapidly developing storm on 19 February and was associated with very strong winds gusting to 30–35 m s^{-1}, producing a strong easterly fetch of air toward the middle Atlantic and southern New England shorelines. Convergent winds associated with an inverted trough extending northward from the secondary low may have been instrumental in the development of heavy snowfall over central Pennsylvania, western Maryland, and northern Virginia on 19 February.

c. 850-hPa analyses

- As in the previous case, the 24-h period prior to East Coast cyclogenesis did not feature the northwesterly flow and cold-air advection over the Great Lakes and New England that characterized most of the other cases. A small cyclonic system moving eastward over the western Atlantic on 17 February contributed to a pattern of weak cold-air advection in the middle Atlantic states. The 850-hPa 0°C isotherm was displaced to the North Carolina–South Carolina border by 0000 UTC 18 February.
- An intense 850-hPa low center drifted slowly eastward over the upper Great Lakes region through 0000 UTC 19 February, yielding a distinct regime of southerly winds and warm-air advection from the middle Atlantic states to south-central Canada. This system was located well to the north of the 0°C isotherm. It weakened rapidly in the 24-h period after 1200 UTC 18 February as the largest temperature gradient progressed southward away from the low center.
- The circular low center over the upper Great Lakes became elongated on a north–south axis during 18 February as cold-air advection west of the trough drove the 0°C isotherm southward to the Gulf coast. A new center developed over the Carolinas by 0000 UTC 19 February, near the 0°C isotherm. The Great Lakes low center disappeared by 1200 UTC 19 Feb-

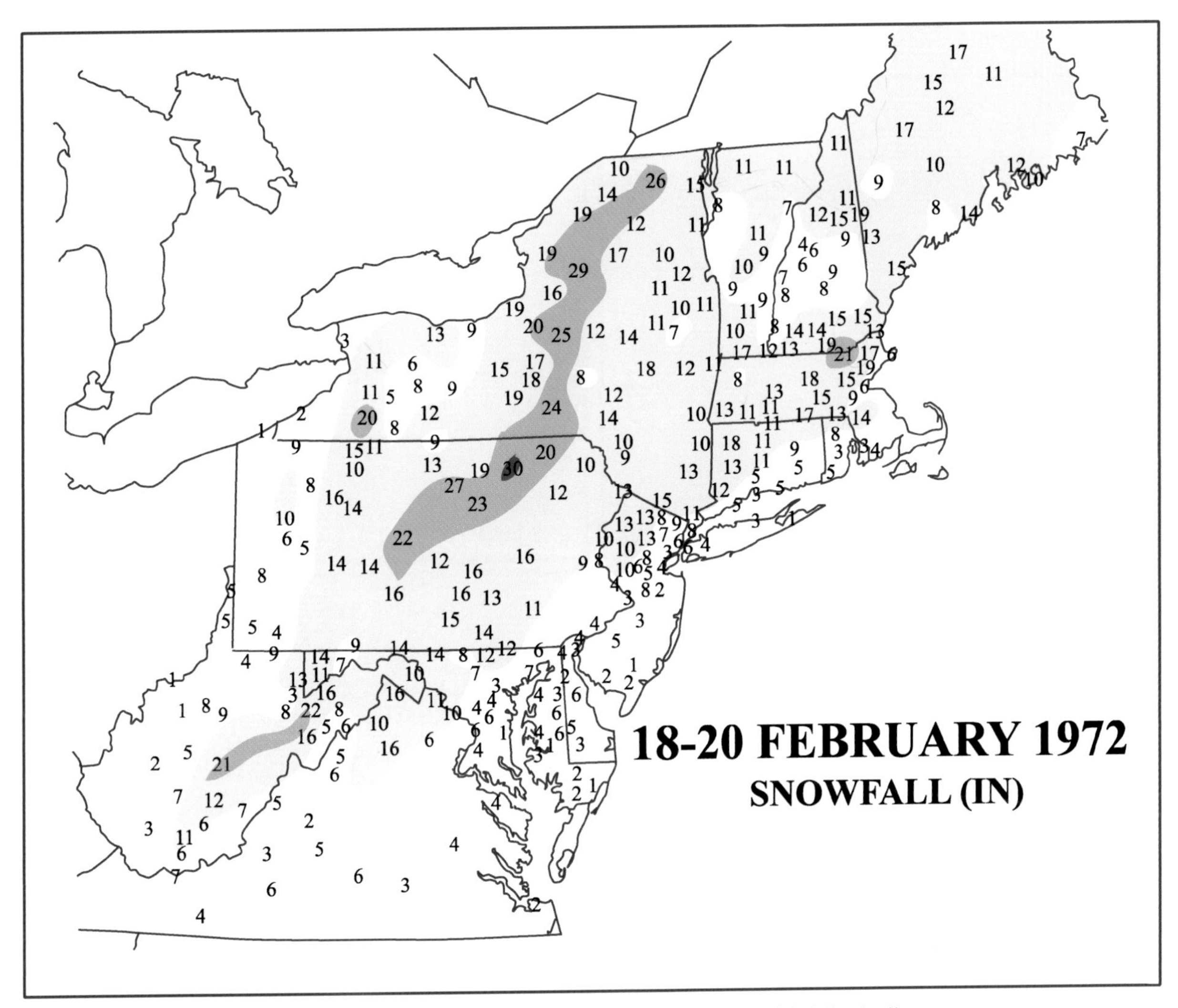

FIG. 10.15-1. Snowfall (in.) for 18–20 Feb 1972. See Fig. 10.1-1 for details.

ruary as the secondary 850-hPa circulation intensified rapidly along the East Coast.

- A northerly jet with wind speeds exceeding 25 m s^{-1} was established west of the low centers prior to secondary cyclogenesis, with cold-air advection occurring throughout the south-central and southeastern United States.
- The secondary 850-hPa low deepened 150 m in the 12-h period ending at 1200 UTC 19 February and 90 m in the following 12-h period.
- The secondary development began concurrent with an increase in warm-air advection along the East Coast by 0000 UTC 19 February. The 850-hPa low developed near the inflection point of a developing S-shaped isotherm pattern. The temperature gradient intensified along the East Coast by 1200 UTC 19 February during a period when the surface and 850-hPa lows deepened rapidly.
- The 850-hPa circulation pattern that developed by 1200 UTC 19 February is marked by a general increase of winds with a distinct and intense southeasterly-to-easterly jet forming to the north of the rapidly deepening low between 0000 and 1200 UTC 19 February. The jet attained wind speeds greater than 40 m s^{-1} by 0000 UTC 20 February and was directed toward and coincided with the outbreak of heavy snow and rain in the middle Atlantic states and New England. This suggests that the increasing moisture inflow associated with the intensifying low-level circulation was an important factor in the expanding area of heavy precipitation on 19 February.

d. 500-hPa geopotential height and 400-hPa wind analyses

- This storm developed during a period when La Niña conditions were transitioning into one of the strongest

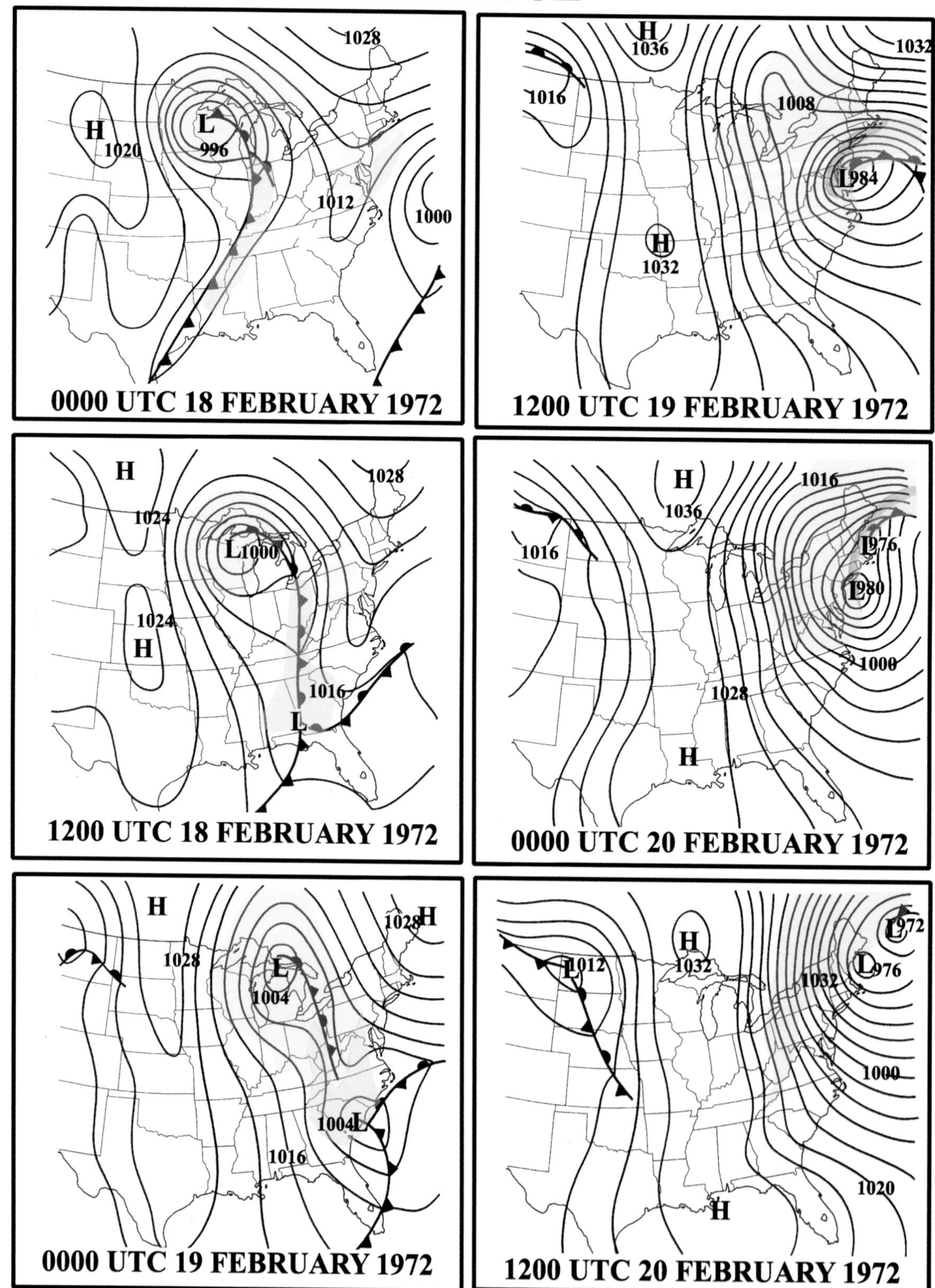

FIG. 10.15-2. Twelve-hourly surface weather analyses for 0000 UTC 18 Feb–1200 UTC 20 Feb 1972. See Fig. 10.1-2 for details.

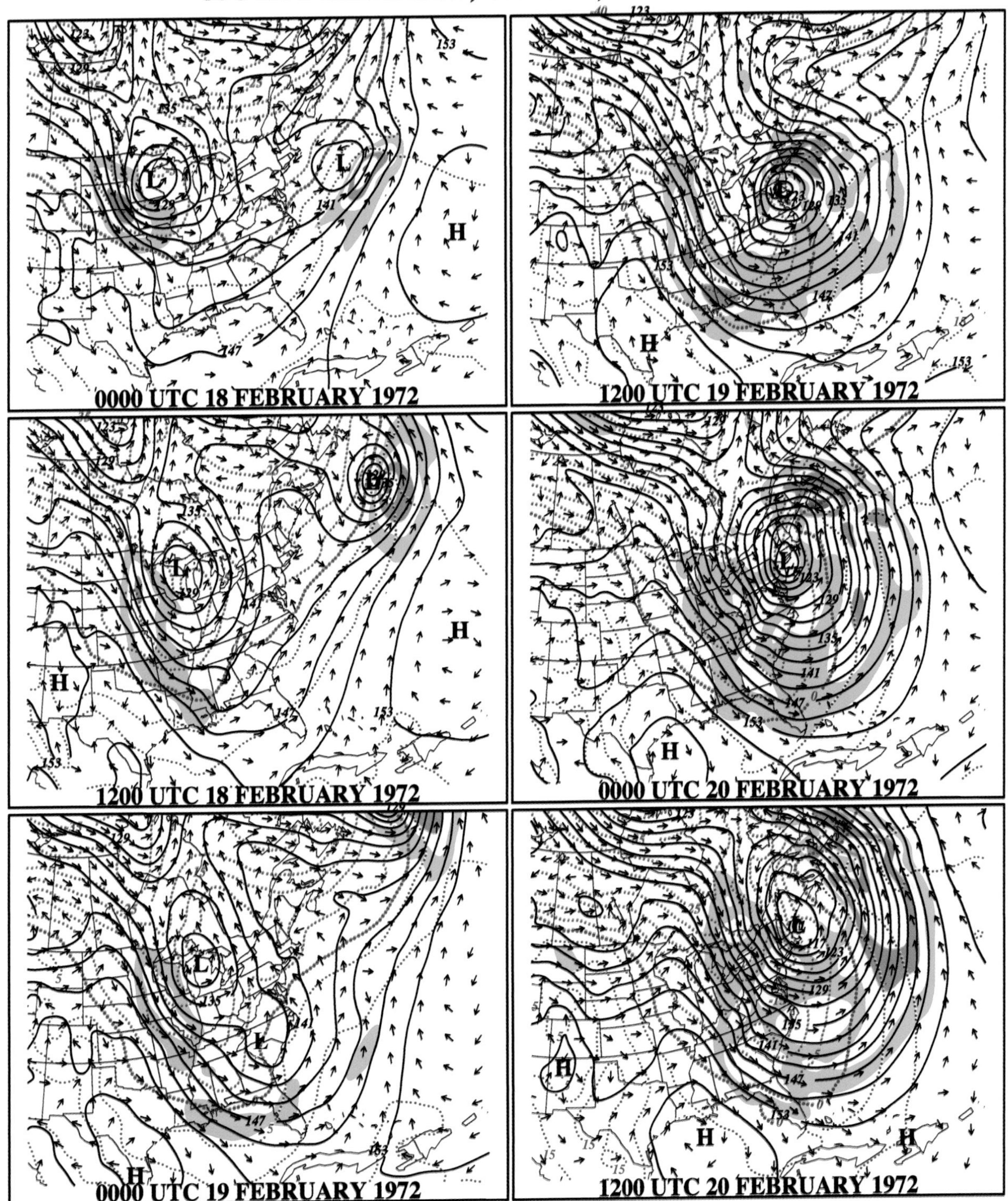

FIG. 10.15-3. Twelve-hourly 850-hPa analyses for 0000 UTC 18 Feb–1200 UTC 20 Feb 1972. See Fig. 10.1-3 for details.

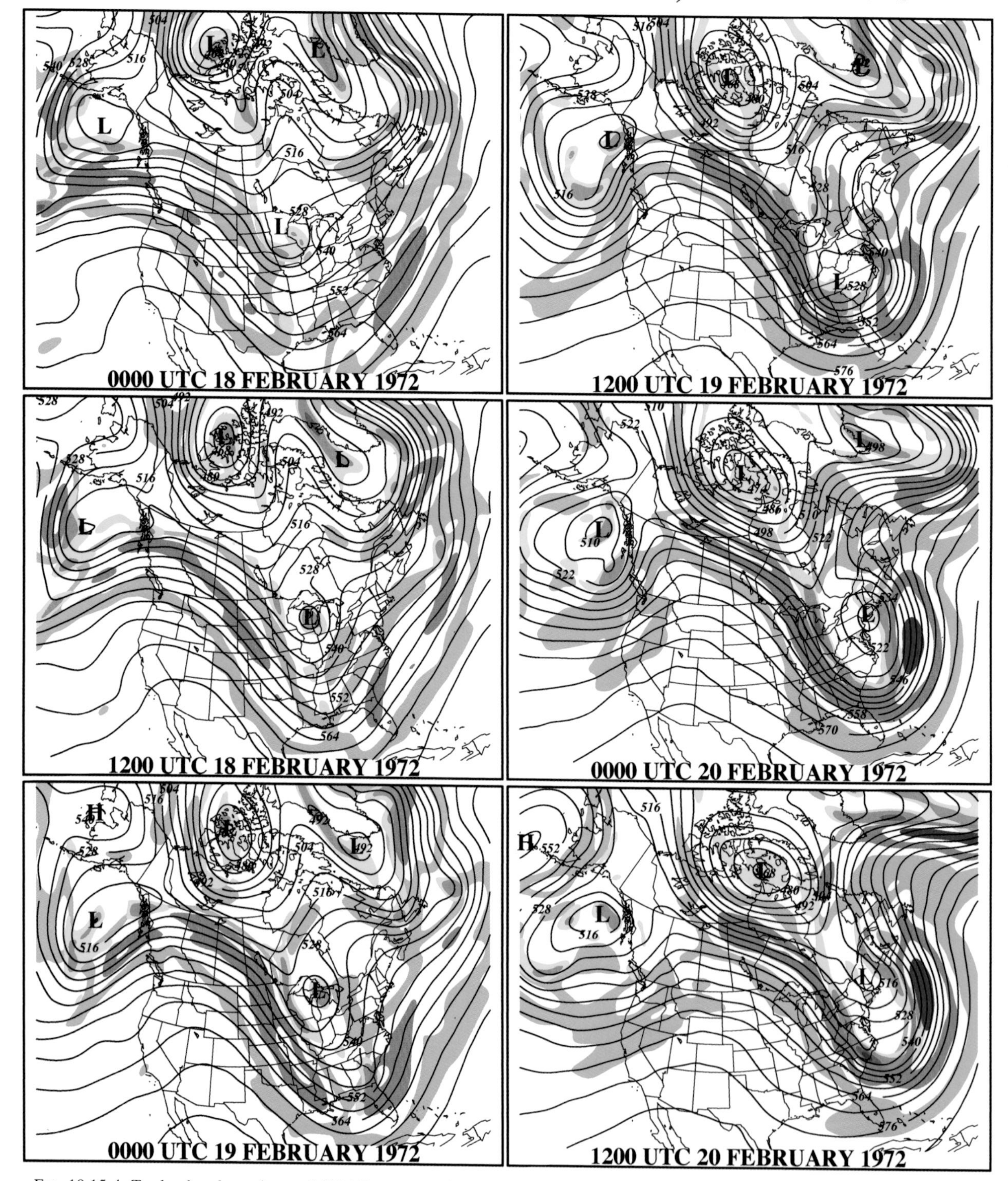

FIG. 10.15-4. Twelve-hourly analyses of 500-hPa geopotential heights and upper-level wind fields for 0000 UTC 18 Feb–1200 UTC 20 Feb 1972. See Fig. 10.1-4 for details.

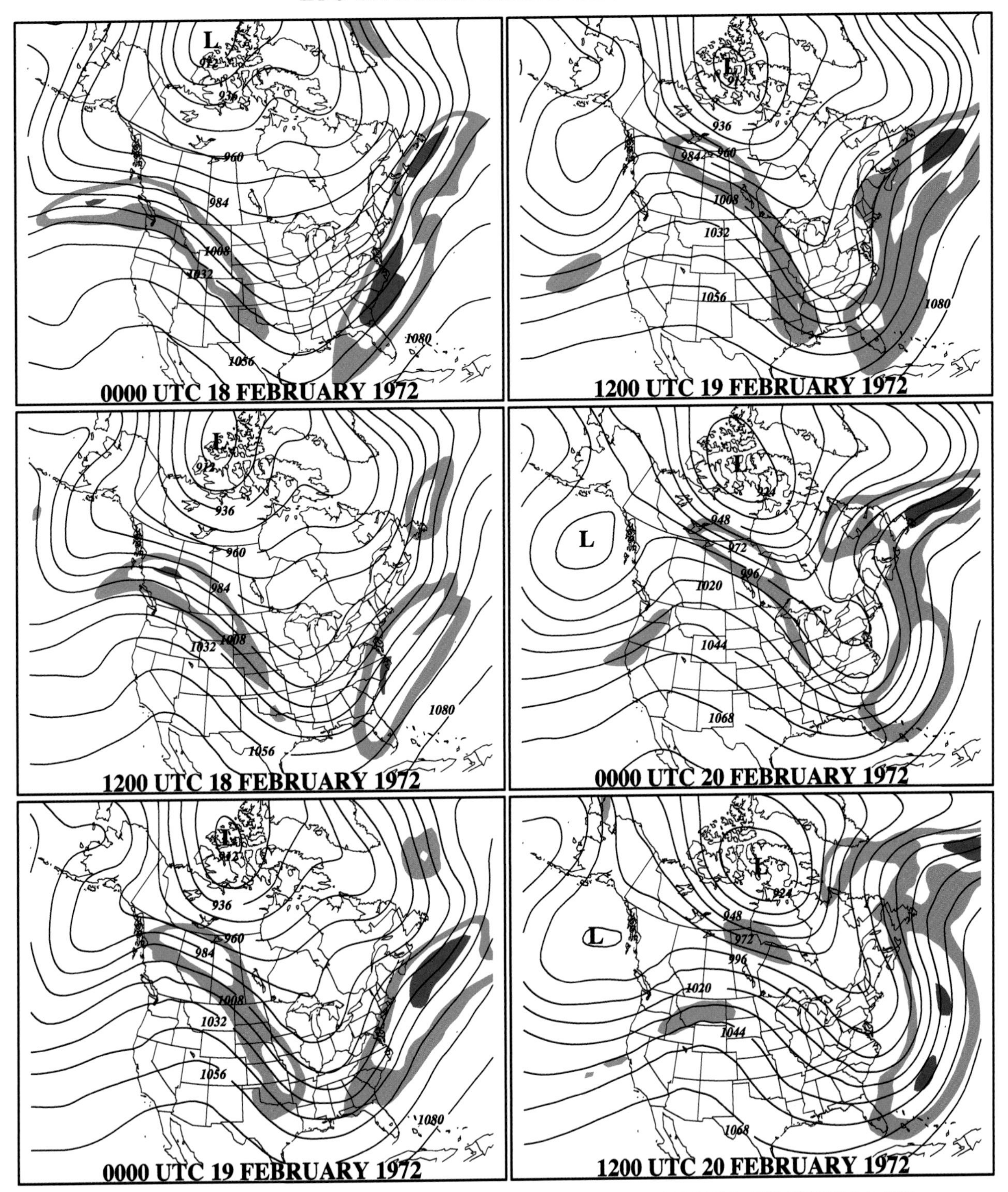

FIG. 10.15-5. Twelve-hourly analyses of 250-hPa geopotential height for 0000 UTC 18 Feb–1200 UTC 20 Feb 1972. See Fig. 10.1-5 for details.

El Niños observed in the last half of the 20th century (Fig. 2-19, volume I). The winter was characterized by a slightly positive phase of the North Atlantic Oscillation (NAO), following four consecutive winters in which a negative NAO was dominant. Daily values of the NAO during this storm period were decidedly positive.

- A trough over eastern Canada was associated with confluence across the Maritime Provinces on 18–19 February. This confluence region was associated with the surface anticyclone over eastern Canada that funneled relatively cold air down the East Coast. The Canadian trough and its associated confluence zone continued moving eastward over the Atlantic Ocean during the following 2 days.
- By 0000 UTC 19 February, the confluent area was located far to the north and east of New England. The withdrawal of the confluence zone from the northeastern United States was consistent with the retreat of cold air along the coast and the changeover from snow to rain that occurred in many places on 19 February.
- A slowly amplifying ridge was established over the western United States on 18 February. The trough that was later linked to the major East Coast storm amplified over the central United States downstream of the ridge on 18 and early 19 February, prior to cyclogenesis. This trough continued to amplify through 1200 UTC 19 February as it propagated toward the East Coast.
- A 500-hPa geopotential height minimum over the western Great Lakes on 18 February was nearly collocated with the weakening primary surface low. This feature disappeared by 1200 UTC 19 February as a new cyclonic vortex deepened rapidly over the southeastern United States in association with the secondary cyclogenesis along the middle Atlantic coast.
- The cyclone over the Great Lakes was associated with a cyclonic vorticity maximum that weakened by 19 February. The developing cyclone along the East Coast was associated with a separate increasing and intense vorticity maximum located over Georgia on 19 February that later swept northeastward along the Southeast and middle Atlantic coasts on 19 and 20 February. This vorticity maximum intensified as the curvature increased near the base of the amplifying trough and the cyclonic shear associated with the jet strengthened, especially between 1200 UTC 18 February and 1200 UTC 19 February.
- The rapid development of the secondary cyclone occurred beneath an increasingly diffluent geopotential height pattern downwind of the trough between 0000 and 1200 UTC 19 February, as the trough axis became oriented from northwest to southeast (a negative tilt). Geopotential height gradients at the base of the trough increased at 0000 UTC 19 February, immediately prior to the rapid secondary cyclogenesis.
- The amplitude of the trough and downstream ridge increased between 0000 UTC 18 February and 1200 UTC 19 February. There was also a shortening of the half-wavelength between the trough and downstream ridge during the secondary development, especially between 0000 UTC 19 February and 0000 UTC 20 February.
- As in the December 1969 case, this storm does not display an interaction of a northern and a southern jet system that would maintain colder air in the Northeast while a storm system developed and propagated northeastward along the East Coast.
- A 50–60 m s^{-1} jet streak is observed in the confluent region over southeastern Canada on 18 February, but it remained too far north and east to have much influence. The lack of cold-air advection at 850 hPa over the Northeast during the precyclogenetic period is consistent with this observation.
- The surface low that developed explosively near the East Coast on 19 February and evolved into two centers off the Maine coast by 0000 UTC 30 February appears to be located in the left-exit region of the north–south-oriented jet streak east of the trough axis.
- The strong south–north-oriented jet east of the trough axis that became negatively tilted over the Southeast by 0000 UTC 19 February is consistent with the changeover from snow to rain along the coast, as with other "near miss" cases described in chapter 11.

e. 250-hPa geopotential height and wind analyses

- The height analyses at this level illustrate the general westerly flow in the initial precyclogenetic period that rapidly evolved into the deep, negatively tilted trough by 19 February.
- While there is a hint of a dual-jet pattern east of the trough axis at 1200 UTC 19 February and 0000 UTC 20 February, the lateral coupling is not entirely obvious in this analysis, probably due to the lack of observations off the coast for this case.

Cars covered in snow in New York City, Jan 1978 (source is *The Weather Book*, Little, Brown and Co., courtesy of Donald Sutherland's digital photography library, www.geocities.com/donsutherland1/snow1.html).

16. 19–21 January 1978

a. General remarks

- For many of the urban centers that span the northeastern United States, this was the most debilitating snowstorm since 1969. This storm was the last in a series of three storms (one on 13–14 January, the next on 17–18 January; see chapters 11.1 and 11.3) during a week that produced a variety of winter weather conditions across the Northeast. Along the coast, snowfall was underforecast since a predicted changeover from snow to rain either did not occur or took place after significant accumulations had already occurred. Boston, Massachusetts, set its 24-h snowfall record with this storm, only to have it broken 2 weeks later on 6–7 February 1978 (next section). The storm was accompanied by wind gusts exceeding 20 m s^{-1} from New Jersey to the New England coast.
- The following regions reported snow accumulations exceeding 10 in. (25 cm): sections of West Virginia, western and northern Virginia and Maryland, much of Pennsylvania, central and northern New Jersey, and much of New York and New England.
- The following regions reported snow accumulations exceeding 20 in. (50 cm): portions of West Virginia, western Maryland, central and northeastern New York, eastern Massachusetts, and northern Rhode Island.

TABLE 10.16-1. Snowfall amounts for urban and selected locations for 19–21 Jan 1978; see Table 10.1-1 for details.

NESIS = 5.90 (category 3)	
Urban center snowfall amounts	
Washington, D.C.–Dulles Airport	7.5 in. (19 cm)
Baltimore, MD	5.6 in. (14 cm)
Philadelphia, PA	13.2 in. (34 cm)
New York, NY–John F. Kennedy Airport	14.2 in. (36 cm)
Boston, MA	21.5 in. (54 cm)
Other selected snowfall amounts	
Syracuse, NY	18.9 in. (48 cm)
Charleston, WV	18.4 in. (47 cm)
Newark, NJ	17.8 in. (45 cm)
Bridgeport, CT	16.7 in. (42 cm)
Hartford, CT	15.5 in. (39 cm)
Pittsburgh, PA	14.0 in. (36 cm)

b. Surface analyses

- At the beginning of the analysis, a surface low off the Maine coast at 0000 UTC 19 January is associated with the second in a series of three major snowstorms. This storm was responsible for heavy snow accumulations across inland sections of the northeastern United States on 17–18 January (see chapter 11.1). Rising sea level pressures and colder air followed the passage of this storm as a large, cold anticyclone (1046 hPa) over the northern plains states on 19 January built eastward toward northern New England.
- On 19 January, cold-air damming and an inverted sea level pressure ridge developed east of the Appalachian Mountains. At the same time, an inverted trough and coastal front developed near the Carolina coast.
- The cyclone formed over the Gulf of Mexico on 18 January and moved northeastward across the eastern Gulf of Mexico and along the East Coast on 19–20 January. The low propagated at about 15–20 m s^{-1} and tended to redevelop or "jump" northeastward along the coastal front near the Carolina coast between 0000 and 1200 UTC 20 January.
- This storm exhibited rather erratic intensification. After forming in the Gulf of Mexico, the central pressure of the cyclone vacillated as it moved northeastward. It deepened rapidly for only a brief period off the New Jersey coast near 1800 UTC 20 January, at which time the center reached its lowest pressure of 995 hPa.
- Although the minimum sea level pressure was not particularly low relative to the other cases, the large areal extent of the cyclone's circulation and its interaction with the Canadian anticyclone to the north produced a widespread snowstorm accompanied by gale force winds along the Atlantic coast. Winds increased along the middle Atlantic coast on 20 January as the distance between the anticyclone and the advancing surface low decreased.
- The heaviest precipitation fell hundreds of kilometers in advance of the surface low center, especially from 0000 through 1200 UTC 20 January, as the inverted trough and coastal front became established along the Carolina coast.

c. 850-hPa analyses

- West-to-northwesterly flow to the rear of the cyclone moving northeastward past the Canadian Maritime Provinces at 0000 UTC 19 January maintained low-level cold-air advection across the northeastern United States through 0000 UTC 20 January. The cold air moved into New England and the middle Atlantic states as an 850-hPa low developed along the Gulf coast.
- This 850-hPa low developed in a broad southwest–northeast baroclinic zone extending across the southeastern United States early on 19 January. The cyclonic circulation moved northeastward along the isotherm ribbon and expanded in areal extent during 19 January, covering much of the eastern United States by 20 January.
- The 850-hPa low deepened only slowly as it propagated from the Texas coast to southern New England. The greatest intensification (−60 m per 12 h) occurred between 1200 UTC 19 January and 0000 UTC 20 January, although the surface low deepened only slightly during the same period. The small 850-hPa

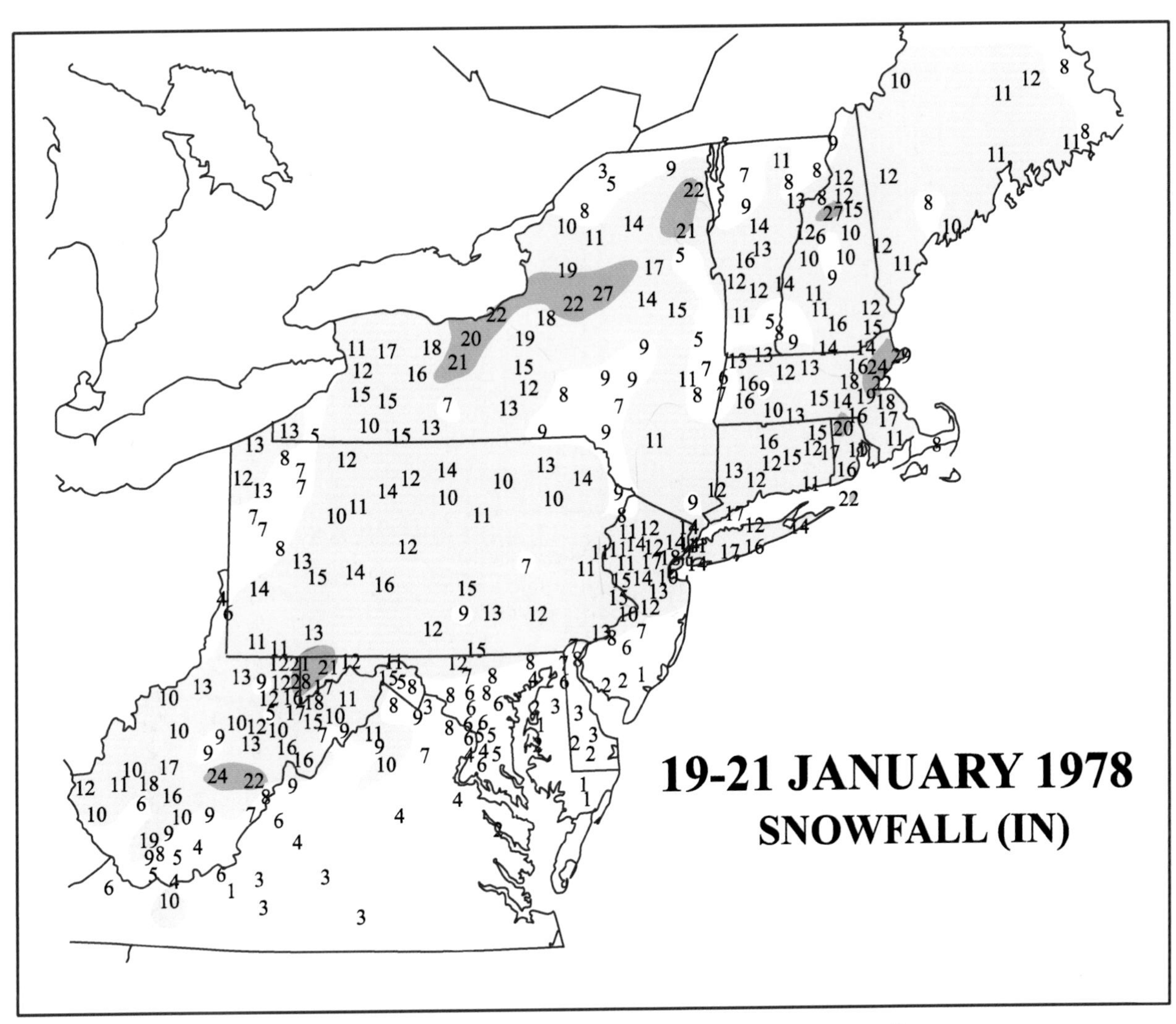

FIG. 10.16-1. Snowfall (in.) for 19–21 Jan 1978. See Fig. 10.1-1 for details.

deepening rates are consistent with those of the surface low for this system.

- As the cyclone propagated northeastward, the 850-hPa temperature gradient increased in advance of the low, especially on 20 January, as heavy snow spread across the middle Atlantic states and southern New England. The S-shaped isotherm pattern observed in many other storms was not as pronounced in this case since the cold-air advection to the rear of the cyclone remained weak, especially after 0000 UTC 20 January. The low developed on the warm side of the 0°C isotherm on 18 January, but tracked closer to the 0°C isotherm by the time it reached Virginia on 20 January.
- A southerly low-level jet (LLJ) developed in the northern Gulf of Mexico and across Florida by 1200 UTC 19 January, immediately upwind of an outbreak of moderate to heavy precipitation over the southeastern United States. The LLJ continued to intensify east of the slow-moving 850-hPa low on 20 January and attained speeds greater than 35 m s^{-1} by 1200 UTC 20 January.
- Easterly winds strengthened to the north of the low center as it intensified slightly and moved from Louisiana to Virginia between 1200 UTC 19 January and 1200 UTC 20 January. The southerly LLJ at 1200 UTC 20 January and the easterly LLJ at 0000 UTC 21 January were both directed toward the region of heaviest precipitation.

d. 500-hPa geopotential height and 400-hPa wind analyses

- January 1978 was a very stormy month, and occurred during an extended El Niño that lasted from mid-1976 through February 1978. However, while the North Atlantic Oscillation (NAO) had some significant negative periods during the winter, it was clearly positive

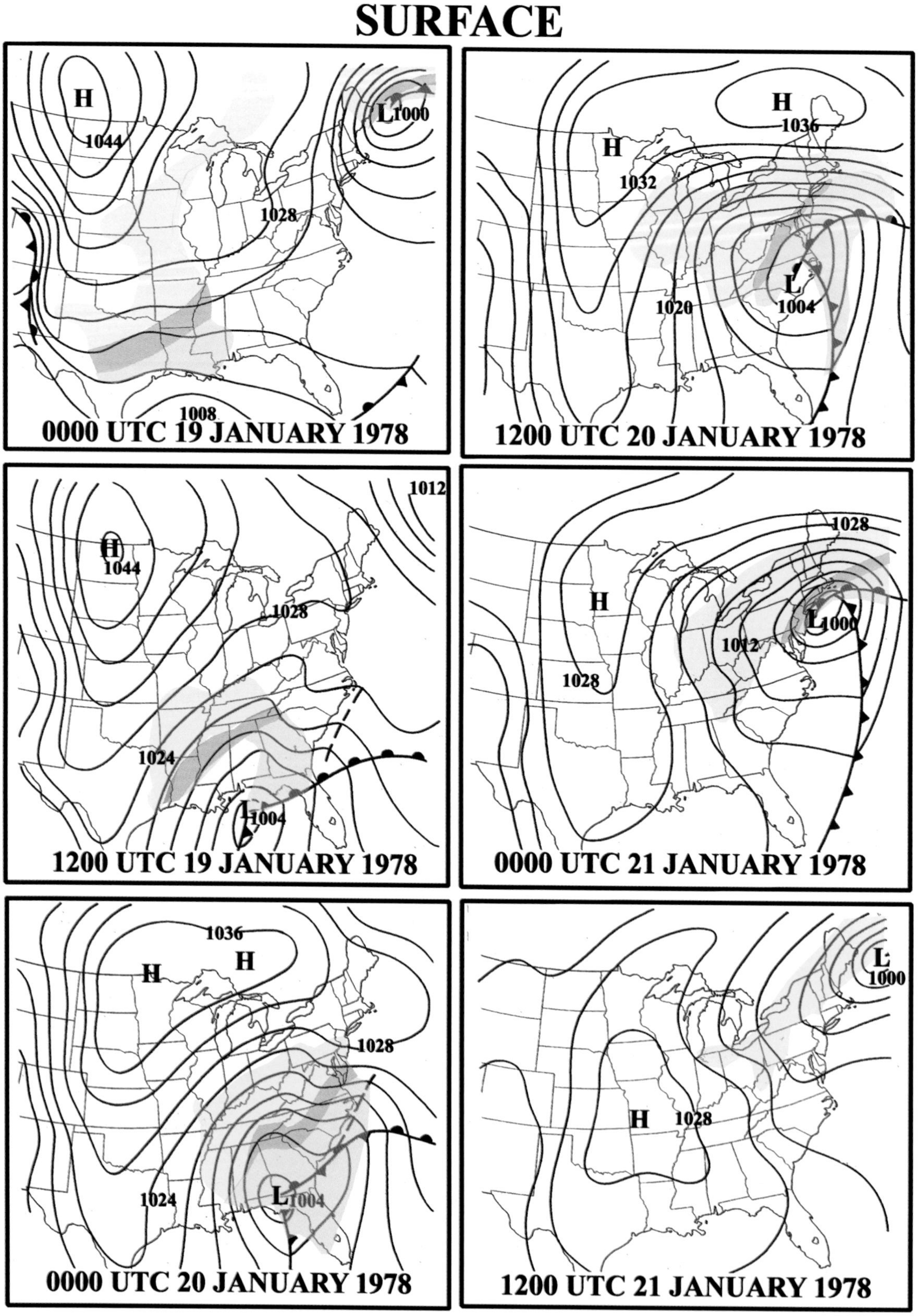

SURFACE

0000 UTC 19 JANUARY 1978

1200 UTC 20 JANUARY 1978

1200 UTC 19 JANUARY 1978

0000 UTC 21 JANUARY 1978

0000 UTC 20 JANUARY 1978

1200 UTC 21 JANUARY 1978

FIG. 10.16-2. Twelve-hourly surface weather analyses for 0000 UTC 19 Jan–1200 UTC 21 Jan 1978. See Fig. 10.1-2 for details.

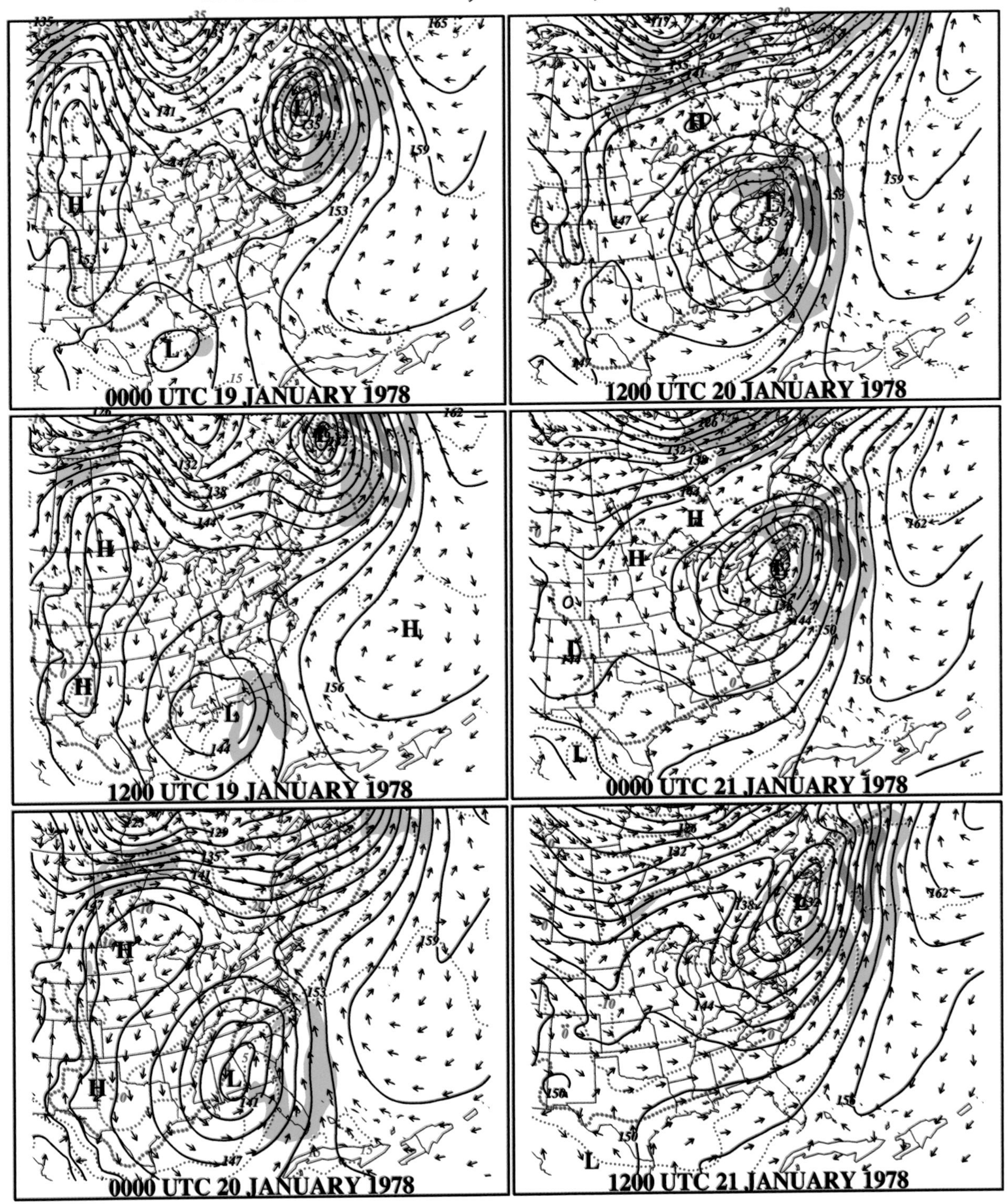

FIG. 10.16-3. Twelve-hourly 850-hPa analyses for 0000 UTC 19 Jan–1200 UTC 21 Jan 1978. See Fig. 10.1-3 for details.

500 hPa HEIGHTS AND VORTICITY, 400 hPa WINDS

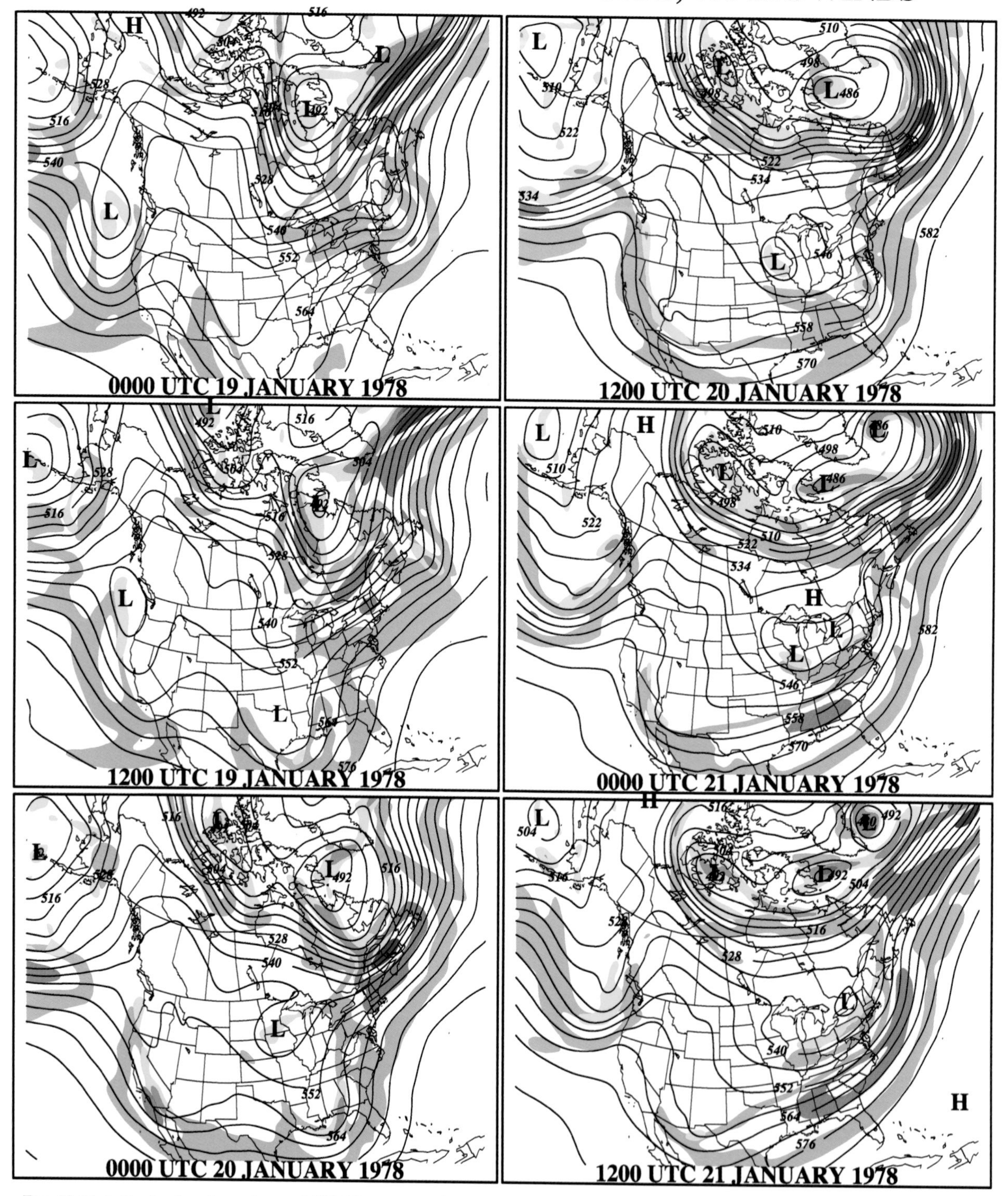

FIG. 10.16-4. Twelve-hourly analyses of 500-hPa geopotential heights and upper-level wind fields for 0000 UTC 19 Jan–1200 UTC 21 Jan 1978. See Fig. 10.1-4 for details.

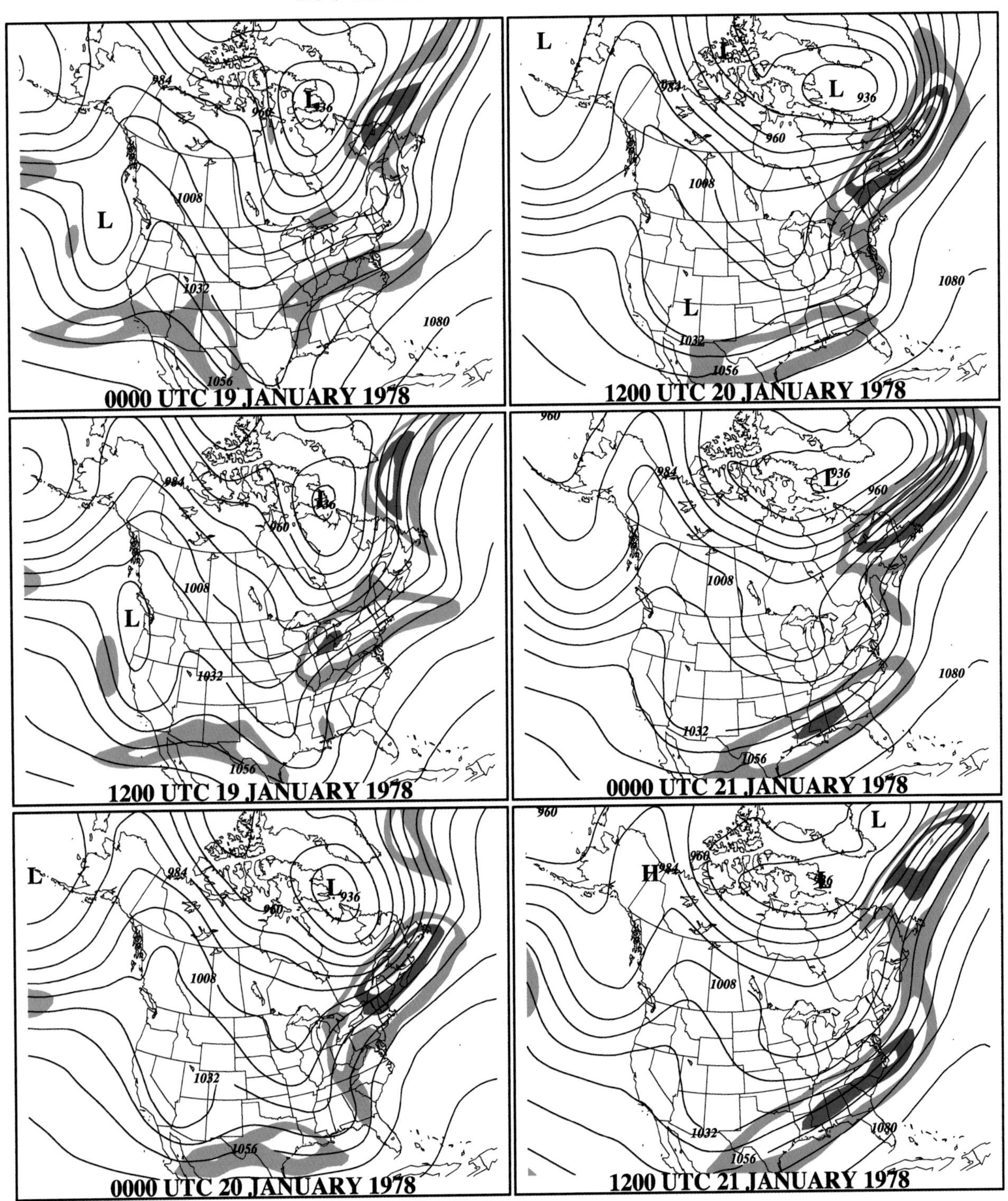

FIG. 10.16-5. Twelve-hourly analyses of 250-hPa geopotential height for 0000 UTC 19 Jan–1200 UTC 21 Jan 1978. See Fig. 10.1-5 for details.

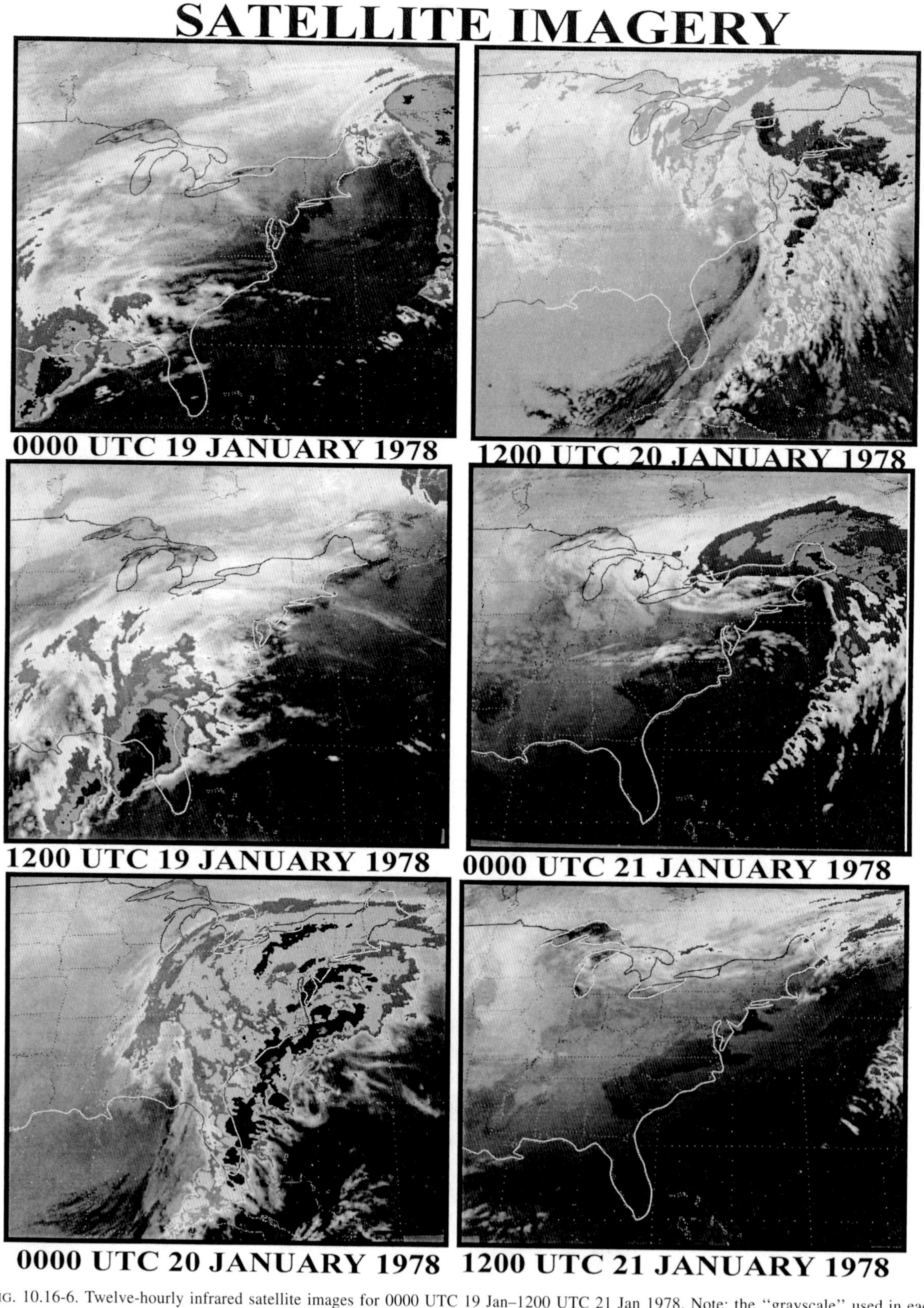

SATELLITE IMAGERY

0000 UTC 19 JANUARY 1978 **1200 UTC 20 JANUARY 1978**

1200 UTC 19 JANUARY 1978 **0000 UTC 21 JANUARY 1978**

0000 UTC 20 JANUARY 1978 **1200 UTC 21 JANUARY 1978**

FIG. 10.16-6. Twelve-hourly infrared satellite images for 0000 UTC 19 Jan–1200 UTC 21 Jan 1978. Note: the "grayscale" used in each image provides a means of using black, white, and shades of gray to distinguish intervals of cloud-top temperature. The grayscale at 0000 and 1200 UTC 20 Jan differs from the scale at the other times.

in the daily values as this storm and several others were developing in January 1978. Oddly enough, there is some evidence of an upper ridge near central and northern Greenland during the days in which the storm developed, which would suggest a negative NAO.

- As the surface cyclone was forming in the Gulf of Mexico on 18 January, two separate troughs encircled an upper-level vortex over northern Quebec. One trough was associated with the cyclone that had brought heavy snowfall to sections of the northeastern United States on 17–18 January and was located over Newfoundland by 1200 UTC 19 January. This trough was followed by a significant increase in geopotential heights over the western Atlantic on 19 January. The second trough extended from the vortex over northern Quebec to Nebraska at 0000 UTC 19 January and later split into two separate trough features. The southern part developed into a cyclonic circulation over the midwestern states while the northern portion pulled away and propagated across eastern Canada on 19 January.
- The combination of rising geopotential heights off the East Coast and the passage of a trough across eastern Canada resulted in a strongly confluent height field near the U.S.–Canada border between 0000 and 1200 UTC 20 January. This pattern was observed as the surface anticyclone built eastward across southern Quebec and cold-air damming developed east of the Appalachian Mountains.
- The trough that was associated with the Gulf cyclone propagated across the southern United States on 18 and 19 January, then moved northeastward along the east coast on 20 January. This trough rotated about the trough developing over the Midwest. Both troughs were associated with well-defined cyclonic vorticity maxima that maintained their individual identities through 1200 UTC 20 January.
- A closed 500-hPa circulation did not develop in association with the cyclone tracking northeastward to New England. The amplitude of the trough and downstream ridge along the Atlantic coast decreased slightly between 1200 UTC 19 January to 0000 UTC 21 January. This is only one of a few cases in which the amplitude decreased.
- The trough axis became negatively tilted (oriented from northwest to southeast) over the northern Gulf of Mexico by 0000 UTC 20 January. The half-wavelength between the trough and downstream ridge decreased significantly by 1200 UTC 20 January, but the flow appeared to become only slightly diffluent east of the trough axis between 1200 UTC 20 January and 1200 UTC 21 January. During this period, the surface low deepened by only 4 hPa, although the heaviest snow was falling across the northeastern United States. The lack of a strongly diffluent height pattern may explain the small surface deepening rate observed in this storm. Additionally, the tendency for the trough to attain a negative tilt across the southern United States in this case is consistent with the changeover from snow to mixed precipitation that kept most snowfall amounts less than 10 in. (25 cm) across Virginia and Maryland.
- While the jet pattern is not as distinct in this case as in others, a dual-jet pattern does emerge during East Coast cyclogenesis (0000 UTC 20 January) with a northern jet over extreme southeastern Canada and a southern jet over the southeast United States.
- The northern jet streak amplified from 50 to 70 m s^{-1} in the ridge over New England by 0000 UTC 20 January within a region of strong confluence. This jet streak continued to amplify over southeastern Canada, attaining maximum winds greater than 80 m s^{-1} by 1200 UTC 20 January. The cold surface anticyclone ridged southeastward into New England and toward the middle Atlantic states beneath the confluent entrance region of this jet.
- The precipitation area to the north and northeast of the storm center also developed and expanded in the entrance region of this amplifying jet as it crossed New England between 1200 UTC 19 January and 1200 UTC 20 January.
- The surface low and LLJ developed in the exit region of the southern jet streak propagating around the base of the trough in the southeastern United States on 20 January with one well-defined jet streak with maximum winds exceeding 40 m s^{-1} analyzed off the Southeast coast by 1200 UTC 20 January. The jet pattern at 1200 UTC 20 January is similar to many of the other cases examined here, but the magnitude of the northern jet is clearly dominating that of the southern jet.

e. 250-geopotential height and wind analyses

- The combination of the intense vortex over northeastern Canada and the confluent flow pattern over New York to New England is evident during the precyclogenetic period, as is the separate deepening trough over the southeast United States by 0000 UTC 20 January.
- The dual-jet pattern is more apparent at this level than at 400 hPa. The northern jet amplified to greater than 100 m s^{-1} over southeastern Canada by 1200 UTC 20 January with a well-defined entrance region over the Ohio Valley at 0000 UTC 20 January and at 1200 UTC 20 January. The region of precipitation raced northward in the entrance region of the amplifying jet during this period.
- The southern jet propagating toward the base of the amplifying trough and the exit region along the East Coast are both well defined at 1200 UTC 20 January and 0000 UTC 21 January.
- The surface low, southerly LLJ, and area of heavy precipitation all appeared to become focused in the

region between the right-entrance region of the northern jet streak and the left-exit region of the southern jet, especially at 1200 UTC 20 January.

f. Infrared satellite imagery sequence

- The initial cyclone development in the Gulf of Mexico was characterized by a cold cloud-top core, as indicated by the dark enhancement south of Louisiana at 0000 UTC 19 January. This area, which may have contained intense convective elements, moved over Florida by 1200 UTC 19 January.
- The rapid northeastward expansion of the high cloud shield on 19 and 20 January extended over 1000 km north of the surface low pressure center. This expansion occurred in the entrance region of the amplifying jet streak over the northeastern United States, especially by 0000 UTC 20 January 1978. The sharp northern edge of the high cloud mass at this time corresponds with the axis of the 100 m s^{-1} upper-level jet analyzed at 250 hPa just north of New England.
- A clear area marked by a region free of upper-level clouds developed along the Gulf coast at 0000 UTC 20 January and moved into North Carolina by 1200 UTC 20 January. The clear wedge corresponds to the axis of the jet located downwind of the upper-level trough (see 400- and 250-hPa winds). The surface low was located very close to the eastern edge of the dry slot or along the western edge of the cloud area.
- The upper cloud mass developed a distinct comma shape by 0000 UTC 21 January as the northern part of the cloud area pivoted northward while the comma tail swept to the east.
- The heaviest precipitation was located within the anticyclonically curved cloud region marking the warm conveyor belt (see volume I, chapter 4.4) through 1200 UTC 20 January, but then shifted toward the southern and western edges of the comma head, or cold conveyor belt (see chapter 4.4), as the low moved toward New England. It is difficult to distinguish the separate airstreams from the infrared images.

Hempstead, Long Island, New York, 6 Feb 1978 (photo reprinted courtesy of *Newsday*).

17. 5–7 February 1978

a. General remarks

- Hurricane force winds and record-breaking snowfall made this storm one of the more intense to occur during the 20th century across parts of the northeastern United States.
- Many meteorologists and weather enthusiasts in the Northeast, especially New England, have expressed their feelings that this storm was a "defining moment" in their careers and weather interests.
- A small area of 50 or more in. (125 cm) of snowfall was reported in northern Rhode Island (not depicted in Fig. 11.17–1). The cyclone was forecast remarkably well several days in advance by operational numerical forecast models (Brown and Olson 1978). Despite these accurate predictions, many people were stranded on the roads in the New York City area, because the onset of heavy snow occurred slightly later than predicted during the Monday morning rush hour, resulting in thousands of people trapped in their cars commuting to and from work from New York City to Boston. People were generally skeptical of the warnings issued by operational weather forecasters following a series of inaccurate forecasts of winter weather during the preceding month.
- The most severely affected regions were Long Island, Connecticut, Rhode Island, and Massachusetts, where businesses and schools were shut down for a week or more. This storm is a remarkable example of sudden and rapid cyclonic development associated with a major trough amplification at 500 hPa.
- The following regions reported snow accumulations exceeding 10 in. (25 cm): eastern Maryland; Delaware; eastern Pennsylvania; New Jersey; southeastern, northeastern, and portions of western New York; Connecticut; Rhode Island; Massachusetts; central and southern Vermont; New Hampshire; and Maine.
- The following regions reported snow accumulations exceeding 20 in. (50 cm): sections of northeastern Pennsylvania, northern New Jersey, western and southeastern New York, Connecticut, Rhode Island, Massachusetts, southern Vermont, and parts of New Hampshire and Maine.

TABLE 10.17-1. Snowfall amounts for urban and selected locations for 5–7 Feb 1978; see Table 10.1-1 for details.

NESIS = 6.25 (category 4)	
Urban center snowfall amounts	
Washington, D.C.–National Airport	2.2 in. (6 cm)
Baltimore, MD	9.1 in. (23 cm)
Philadelphia, PA	14.1 in. (36 cm)
New York, NY–Central Park	17.7 in. (45 cm)
Boston, MA	27.1 in. (69 cm)
Other selected snowfall amounts	
Woonsocket, RI	38.0 in. (97 cm)
Rockport, MA	32.5 in. (83 cm)
Providence, RI	28.6 in. (73 cm)
Rochester, NY	25.8 in. (66 cm)
Riverhead, NY	25.0 in. (64 cm)
Worcester, MA	20.2 in. (51 cm)
Hartford, CT	16.9 in. (43 cm)
Trenton, NJ	16.1 in. (41 cm)
Wilmington, DE	14.5 in. (37 cm)

b. Surface analyses

- The cyclone developed in a regime of unusually high sea level pressure and very cold temperatures. A 1055-hPa anticyclone drifted across central Canada prior to and during the storm, with two high pressure ridges dominating the precyclogenetic period on 5–6 February. One ridge axis extended southward through the center of the country and the other extended from New England to the southeastern United States.
- Weak cold air damming was indicated by the inverted high pressure ridge along the East Coast on 5 February. Despite the wedge of very cold air along the Atlantic seaboard, no coastal frontogenesis was observed with this case.
- A weak 1025-hPa low crossing the Great Lakes states on 5 February produced only light snows and provided few clues that major cyclogenesis would occur the following day. This low dissipated early on 6 February following the development of a secondary cyclone east of North Carolina.
- The secondary low developed in a data-void region, but appeared to deepen at rates exceeding −3 hPa per 3 h between 0600 UTC 6 February and 0000 UTC 7 February as it moved northward toward Long Island. Since the cyclone developed in an environment dominated by very cold high pressure, it deepened to "only" 984 hPa at its peak intensity. Nevertheless the pressure gradient to the north and west of the low center was among the largest in the 50-yr sample, producing gale to hurricane force northeasterly winds that blew the heavy snow, reducing visibilities to near zero over a large area from New England to New Jersey and eastern Pennsylvania.
- The surface low propagated at approximately 10–12 m s^{-1} as it tracked northward from a position well off the North Carolina coast to just south of Long Island on 6 February. The cyclone then may have performed a small counterclockwise loop before it drifted very slowly eastward just south of New England on 7 February. This slow movement prolonged the heavy snowfall from Long Island to eastern New England.
- The heaviest snowfall occurred to the north and northeast of the surface low, as is often the case. Heavy snowfall, however, was also reported to the west and southwest of the center over Maryland and Delaware late on 6 February as the precipitation rotated southwestward around the intensifying storm.

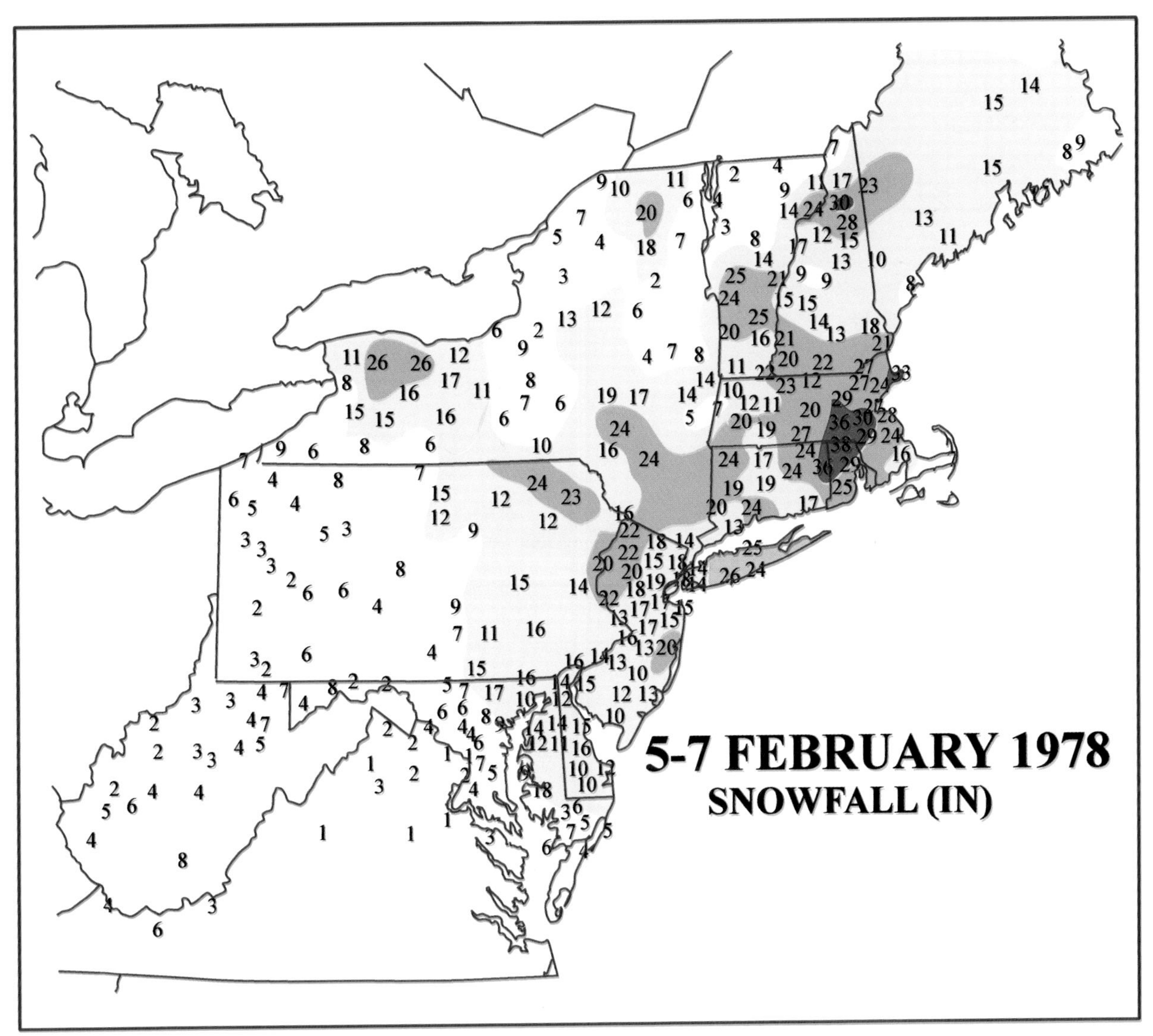

FIG. 10.17-1. Snowfall (in.) for 5–7 Feb 1978. See Fig. 10.1-1 for details.

c. 850-hPa analyses

- Northwesterly flow and weak cold-air advection preceded the storm as cold air had already become established over the north-central and northeast United States by 0000 UTC 5 February. The 850-hPa temperatures generally ranged between −5° and −15°C over the northeastern United States.
- A weak 850-hPa low dropped southeastward across the Great Lakes on 5 February and deepened very gradually. A new 850-hPa low center formed just off the Virginia coast by 1200 UTC 6 February. The redevelopment of the 850-hPa low was first evident at 0000 UTC 6 February when the largest height falls were located in the Carolinas, well south of the primary center, which was positioned near Buffalo, New York. Meanwhile, east-to-southeasterly flow is evident just off the Carolina coast. The secondary low deepened rapidly (−150 m per 12 h) as it moved northward close to the New Jersey coast by 0000 UTC 7 February, then drifted to the east without any further deepening.
- A very pronounced pattern of cold-air advection was evident to the west of the 850-hPa low center prior to and during the development of the secondary low. The −20°C isotherm was driven from Manitoba to North Carolina on 5–7 February. A pattern of warm-air advection did not develop until 0000 UTC 6 February, over eastern North Carolina and Virginia and then northeastward toward Long Island, as the secondary cyclone was beginning to form over the Atlantic Ocean. An S-shaped isotherm pattern evolved along the East Coast by 1200 UTC 6 February, reflecting the strong cold- and warm-air advections surrounding the 850-hPa low during secondary cyclogenesis.

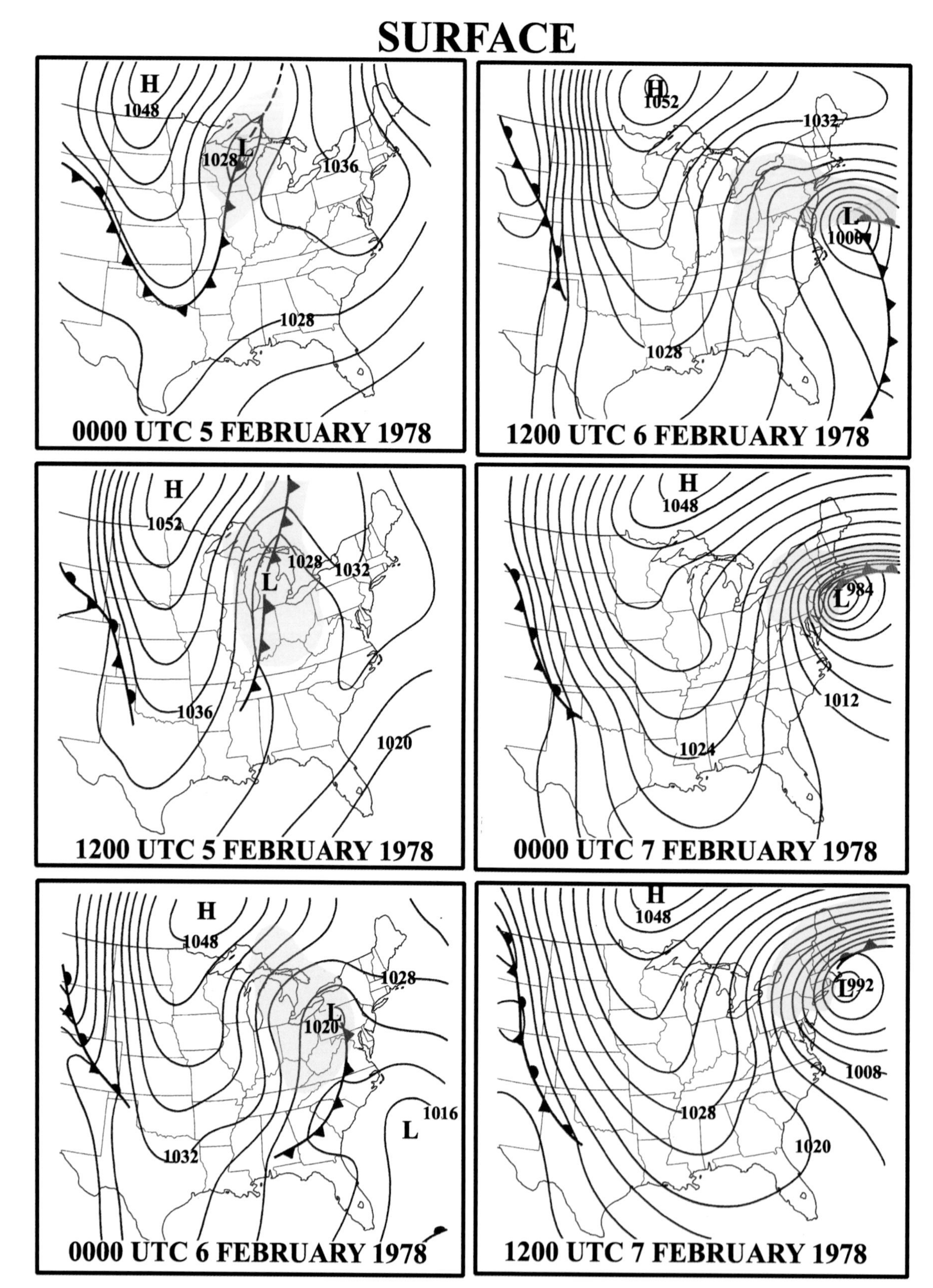

FIG. 10.17-2. Twelve-hourly surface weather analyses for 0000 UTC 5 Feb–1200 UTC 7 Feb 1978. See Fig. 10.1-2 for details.

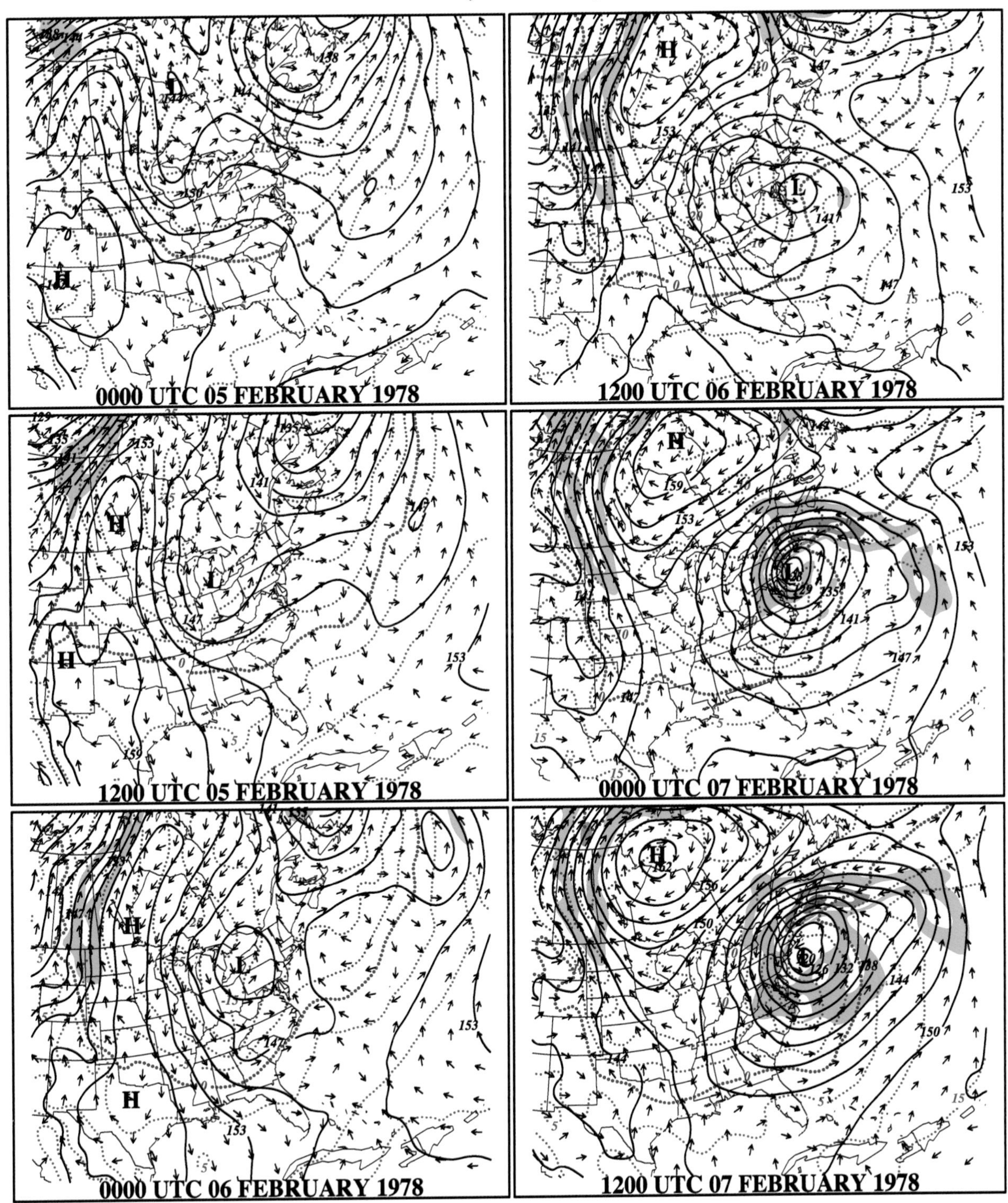

FIG. 10.17-3. Twelve-hourly 850-hPa analyses for 0000 UTC 5 Feb–1200 UTC 7 Feb 1978. See Fig. 10.1-3 for details.

500 hPa HEIGHTS AND VORTICITY, 400 hPa WINDS

FIG. 10.17-4. Twelve-hourly analyses of 500-hPa geopotential heights and upper-level wind fields for 0000 UTC 5 Feb–1200 UTC 7 Feb 1978. See Fig. 10.1-4 for details.

250 hPa HEIGHTS AND WINDS

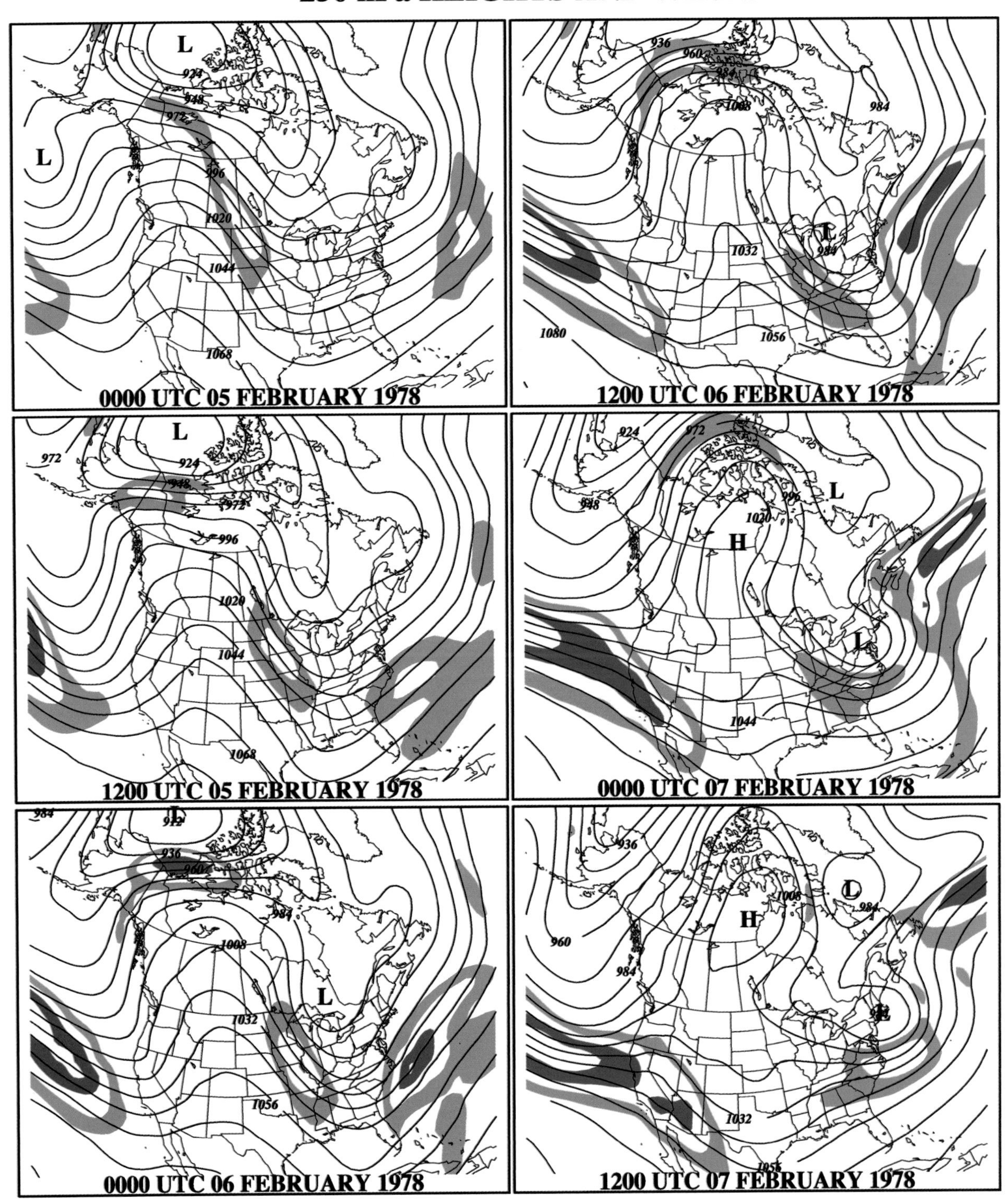

FIG. 10.17-5. Twelve-hourly analyses of 250-hPa geopotential height for 0000 UTC 5 Feb–1200 UTC 7 Feb 1978. See Fig. 10.1-5 for details.

(a) **SATELLITE IMAGERY**

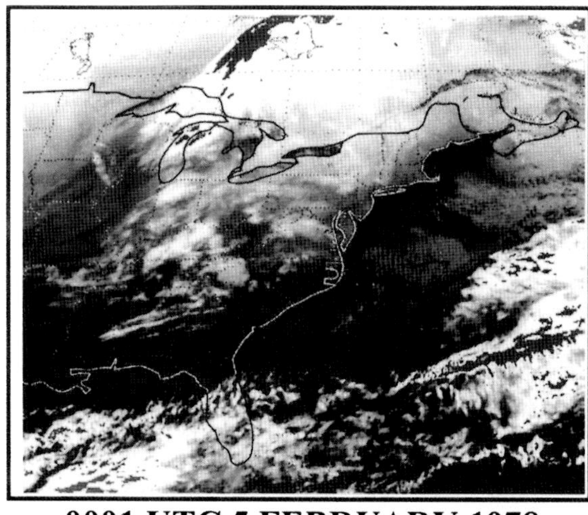

0001 UTC 5 FEBRUARY 1978

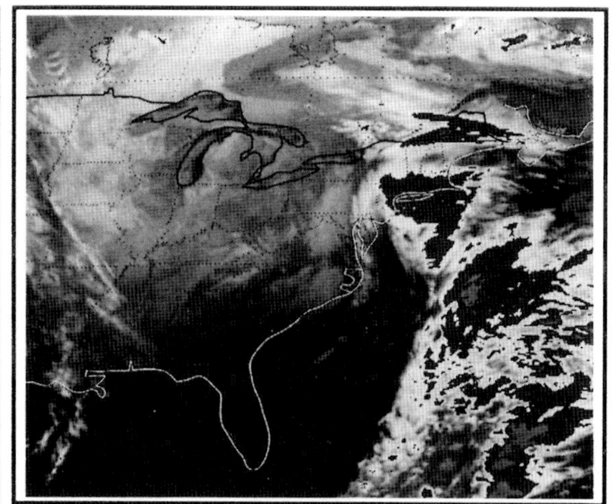

1200 UTC 6 FEBRUARY 1978

1200 UTC 5 FEBRUARY 1978

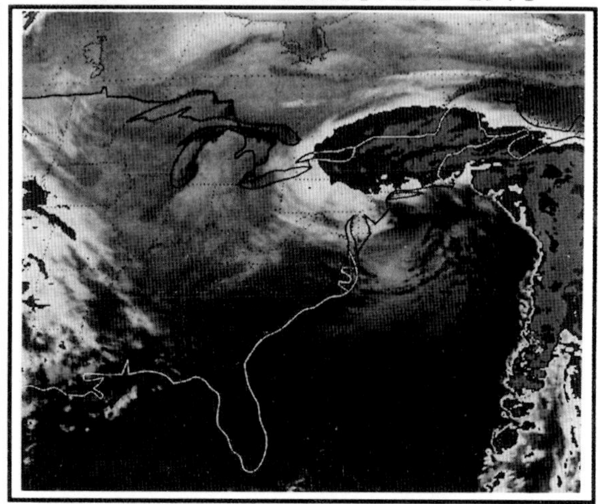

0001 UTC 7 FEBRUARY 1978

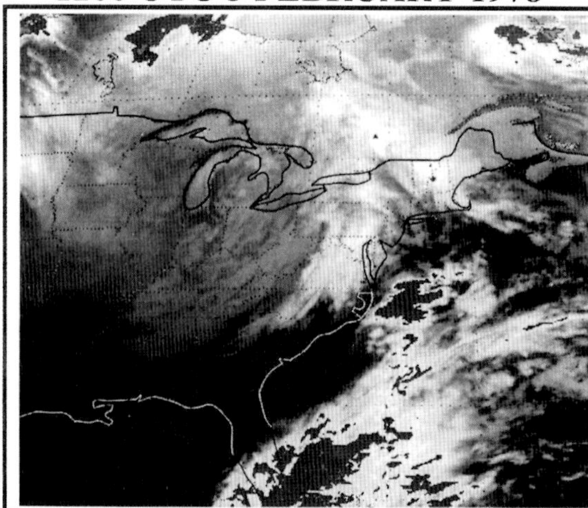

0001 UTC 6 FEBRUARY 1978

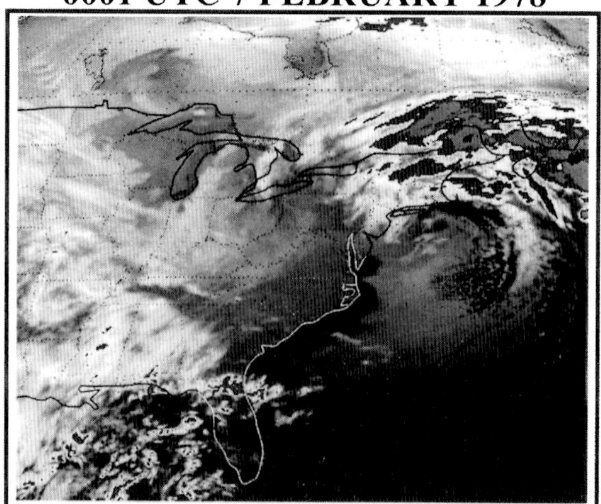

1200 UTC 7 FEBRUARY 1978

FIG. 10.17-6a. Twelve-hourly infrared satellite images for 0000 UTC 5 Feb–1230 UTC 7 Feb 1978. See Fig. 10.16-6 for details.

(b)

FIG. 10.17-6b. Visible satellite image for 1430 UTC 7 Feb 1978 showing eyelike cloud-free region.

- The analyses do not capture a strong low-level jet off the East Coast (in a data-void region) at 0000 and 1200 UTC 6 February. However, a strong easterly low-level jet (LLJ) with velocities exceeding 40 m s^{-1} developed to the north of the cyclone center in conjunction with the rapid deepening between 1200 UTC 6 February and 0000 UTC 7 February. The development of this jet coincided with the rapid northward and westward expansion of snowfall over the middle Atlantic and New England states. The moisture transports and ascending motions associated with the LLJ were apparently important elements in the development of heavy snowfall over the northeastern United States.

d. 500-hPa geopotential height and 400-hPa wind analyses

- As with January 1978, modest El Niño conditions existed through February as well. Unlike January 1978, the North Atlantic Oscillation (NAO) underwent a noticeable change with significant negative NAO values computed throughout the month. This storm occurred as daily values of the NAO were becoming increasingly negative. A weak but stationary upper ridge in the 500-hPa analyses near Greenland is consistent with the negative phase of the NAO.

- The precyclogenetic environment at 500 hPa included a relatively stationary ridge over Greenland, a closed cyclonic circulation over southeastern Canada, a trough across south-central Canada and the Great Lakes region, and a ridge over western North America on 5 February. The vortex over southeastern Canada was associated with a confluent geopotential height pattern over New England that was located above the 850-hPa cold-air advection area and sea level high pressure ridge.
- The East Coast cyclone developed while the flow regime across North America underwent a remarkable transformation. The 500-hPa pattern was characterized by nearly zonal flow prior to 5 February, by strongly meridional flow on 5–6 February, and by large, symmetrical vortices on 7 February. The upper-level ridge over the West amplified on 5–6 February and evolved into a closed anticyclonic circulation in central Canada by 7 February. Concurrently, the trough in southcentral Canada amplified and dropped southeastward, evolving into a large cyclonic circulation over the eastern United States as the surface low intensified off the middle Atlantic coast. This is an excellent example of a deepening or "digging" trough system (see volume I, chapter 4.1a).
- The growing cyclonic circulation was accompanied by an amplifying cyclonic vorticity maximum that propagated southeastward from the Great Lakes region on 5 February to the Virginia coast on 6 February. This vorticity maximum appeared to be associated with the cyclonic curvature and increasing cyclonic wind shears associated with a jet streak propagating southeastward over the central United States.
- A diffluent geopotential height pattern developed downstream of the trough between 1200 UTC 5 February and 1200 UTC 6 February. The trough axis rotated dramatically between 0000 UTC 6 February and 0000 UTC 7 February, the period during which the rapid sea level development took place.
- Very pronounced amplitude changes also occurred during the cyclogenetic period. The amplitude of the trough and upstream ridge increased greatly from 0000 UTC 5 February through 0000 UTC 7 February. The latitudinal separation of the 5520-m contour at the ridge and trough axes grew from 14° to greater than 30° latitude (1500–3000 km). The amplitude of the trough and downstream ridge increased most rapidly between 1200 UTC 5 February and 0000 UTC 7 February, during the period of rapid cyclogenesis.
- The half-wavelength between the trough and the upstream ridge increased prior to East Coast cyclogenesis on 6 February. The half-wavelength between the trough and the downstream ridge decreased during 6–7 February while cyclogenesis was under way off the coast.
- The intense cyclonic and anticyclonic upper-level vortices established a blocking pattern over the eastern half of North America. The large cyclonic vortex drifted from the New England coast into eastern Canada, where it meandered for several weeks following the storm. This resulted in persistent cold and dry weather over the northeastern United States for the remainder of February 1978.
- The analysis for this case is not marked by a well-defined dual-jet pattern at 400 hPa. Although there is a well-defined confluent region over northern New England and southeast Canada on 5 February, the entrance region of a northern jet does not become apparent until 0000 UTC 7 February,

e. 250-hPa geopotential height and wind analyses

- The most dominant jet streak for this case is a polar jet streak with winds of 50 m s^{-1} positioned upwind of the trough axis in the central United States prior to cyclogenesis. The strongest winds remained just west of the diffluent portion of the trough as the system moved toward the East Coast on 5 and 6 February. The cyclone development commenced beneath the diffluence exit region of this jet streak downwind of the trough axis, where cyclonic vorticity advection and upper-level divergence were maximized.
- The entire polar jet system appears to wrap around the middle-tropospheric vortex by 1200 UTC 7 February, illustrating again the intense nature of this storm.
- The amplifying ridge in the western United States and deepening trough in the central, then eastern, United States is clearly evident in the upper troposphere. The diffluent trough becomes negatively tilted to an almost west-to-east axis by 1200 UTC 7 February.
- The jet pattern is not straightforward in these analyses. The jet streak off the East Coast on 5 February represents the lower portion of a subtropical jet that is propagating east-northeastward during the precyclogenetic period.
- The wind maximum that is evident east of the trough axis at 1200 UTC 6 February amplifies to greater than 80 m s^{-1} at 0000 UTC 7 February during intense surface cyclogenesis and concurrent development of heavy precipitation.
- The polar jet diving southeastward into the central and southeast United States is evident throughout the precyclogenetic period, with the dual-jet pattern becoming apparent by 1200 UTC 6 February. At this time, cyclogenesis has commenced east of the North Carolina coastline within the left-exit region of the southern jet and the general entrance region of the northern jet.

f. Infrared satellite imagery sequence

- A diffuse cloud area moved from the Great Lakes toward the eastern United States within the diffluent exit region of the trough–polar jet system prior to 1200 UTC 5 February.
- The 0001 UTC 6 February satellite image depicts the first stage of major cyclogenesis. The cloud mass over the eastern United States was more distinct, with a region of cold cloud tops forming off the North Carolina coast. This developing cloud mass was located east of the upper-level trough axis and was associated with the initial cyclogenesis. An expanding region of cold cloud tops off Florida appeared to be associated with the 70–80 m s^{-1} subtropical jet analyzed at 250 hPa over the Atlantic Ocean.
- By 1200 UTC 6 February, the cloud feature off Florida had moved northeastward over the Atlantic Ocean. Its northern edge displayed the distinctly anticyclonic curvature characteristic of the warm conveyor belt described in chapter 4d of volume I. The cloud mass originally located off North Carolina expanded northward and then westward over the northeastern states by 0000 UTC 7 February, coinciding with the rapid development of the 40 m s^{-1} easterly LLJ and as the 500-hPa vortex developed along the Virginia coast. This cloud feature resembles the cold conveyor belt described in chapter 4d, and became the comma head within a cloud mass that developed the comma signature by 0001 UTC 6 February. The heaviest snow fell over the middle Atlantic states and southern New England on 6 and 7 February, along the southern edge of the coldest tops in the comma head.
- As the cyclone occluded off the New England coast, the cloud mass spiraled around the low center by 1200 UTC 7 February. The coldest cloud tops now had become disorganized and were located well to the north of the low center.
- A pronounced eyelike cloud-free region is observed in the visible image presented here (Fig. 10.17-6b) at 1430 UTC on the morning of 7 February, nearly collocated with the positions of the surface low and the 500-hPa vortex.

Wisconsin Avenue, Georgetown, Washington D.C., 19 Feb 1979 (copyright Fred Maroon).

18. 18–20 February 1979

a. General remarks

- This heavy snowstorm hit the middle Atlantic states with record snow amounts on the Presidents' Day holiday, shutting down the cities of Washington, Baltimore, Philadelphia, and New York with more than a foot (30 cm) to 20 in. (50 cm) of snow. Snowfall rates approached 12 cm h^{-1} on the morning of 19 February (Foster and Leffler 1979). The cyclone has been the subject of numerous studies (Bosart 1981; Bosart and Lin 1984; Uccellini et al. 1984, 1985, 1987; Atlas 1987) since the rapidly developing storm and heavy snowfall rates observed on 19 February were poorly forecast by operational numerical prediction models. The snowstorm occurred at the end of an unusually cold period that culminated in a massive cold-air outbreak on 17–18 February and the development of the "Presidents' Day Storm" (this storm is now termed the "Presidents' Day Storm I") since a *second* Presidents' Day snowstorm occurred in February 2003 (see chapter 10.31).
- The following regions reported snow accumulations exceeding 10 in. (25 cm): portions of West Virginia and Virginia (especially northern Virginia), Maryland, Delaware, southern Pennsylvania, most of New Jersey, and New York City.
- The following regions reported snow accumulations exceeding 20 in. (50 cm): parts of Maryland, Delaware, and New Jersey.

TABLE 10.18-1. Snowfall amounts for urban and selected locations for 18–20 Feb 1979; see Table 10.1-1 for details.

NESIS = 4.42 (category 3)	
Urban center snowfall amounts	
Washington, D.C.–National Airport	18.7 in. (47 cm)
Baltimore, MD	20.0 in. (51 cm)
Philadelphia, PA	14.3 in. (36 cm)
New York, NY–Central Park	12.7 in. (32 cm)
Other selected snowfall amounts	
Dover, DE	25.0 in. (64 cm)
Atlantic City, NJ	17.1 in. (43 cm)
Newark, NJ	16.6 in. (42 cm)
Wilmington, DE	16.5 in. (42 cm)
Harrisburg, PA	14.2 in. (36 cm)
Richmond, VA	10.9 in. (28 cm)

b. Surface analyses

- The cyclone occurred at the end of a prolonged 2–3-week period of bitterly cold air over the eastern half of the country, and was preceded by a massive anticyclone (1050 hPa) that established many record low temperatures across the eastern third of the United States. Following the President's Day Storm, temperatures moderated quickly and much of the snow melted within a week.
- Cold-air damming occurred to the lee of the Appalachian Mountains on 18 February, as indicated by the pronounced inverted sea level pressure ridge over the eastern United States at 1200 UTC 18 February and 0000 UTC 19 February. When snow began falling across the middle Atlantic states on 18 February, surface temperatures generally ranged between $-15°$ and $-10°$C.
- Two inverted sea level pressure troughs formed prior to the development of the cyclone. One trough formed from the Gulf coast to the Ohio Valley late on 17 February. The other appeared off the Southeast coast early on 18 February, in conjunction with the development of a pronounced coastal front. Heavy snow, freezing rain, and ice pellets broke out across the southeastern United States to the west of the developing coastal front–inverted trough.
- A primary low pressure center formed over Kentucky but had no associated frontal features. This initial low deepened to only 1024 hPa over the Ohio Valley and was associated with light to moderate snows just south of the Great Lakes on 18 February. The low dissipated as rapid secondary cyclogenesis commenced along the East Coast early on 19 February.
- The secondary cyclone formed along the coastal front off the Georgia–Florida coast by 1800 UTC 18 February and propagated northeastward to a position near the Maryland coast by 1200 UTC 19 February. A period of rapid deepening commenced around 0600 UTC 19 February. Heavy snow developed across the middle Atlantic states, with snowfall rates of 5–7.5 cm h^{-1} common from Washington to New York City. The low pressure center made an abrupt turn to the east after 1200 UTC 19 February, sparing New England significant accumulations. The cyclone propagated at an average rate of 15 m s^{-1}.
- The storm developed in a regime of very high sea level pressure and deepened to only 995 hPa [based on National Meteorological Center (NMC, now known as the National Centers for Environmental Prediction, NCEP) analyses; Bosart (1981) established a central pressure of 990 hPa by 1800 UTC 19 February] east of the Maryland coast. The heaviest snows fell across the middle Atlantic states on 19 February, when the deepening rates approached a maximum of -2 to -3 hPa h^{-1}.

c. 850-hPa analyses

- An 850-hPa ridge that was associated with the large surface anticyclone dominated the eastern United States prior to east coast cyclogenesis. Cold-air advection and northwesterly flow east of the ridge axis were located beneath a region of confluent geopotential heights at 500 hPa. The 850-hPa 0°C isotherm was

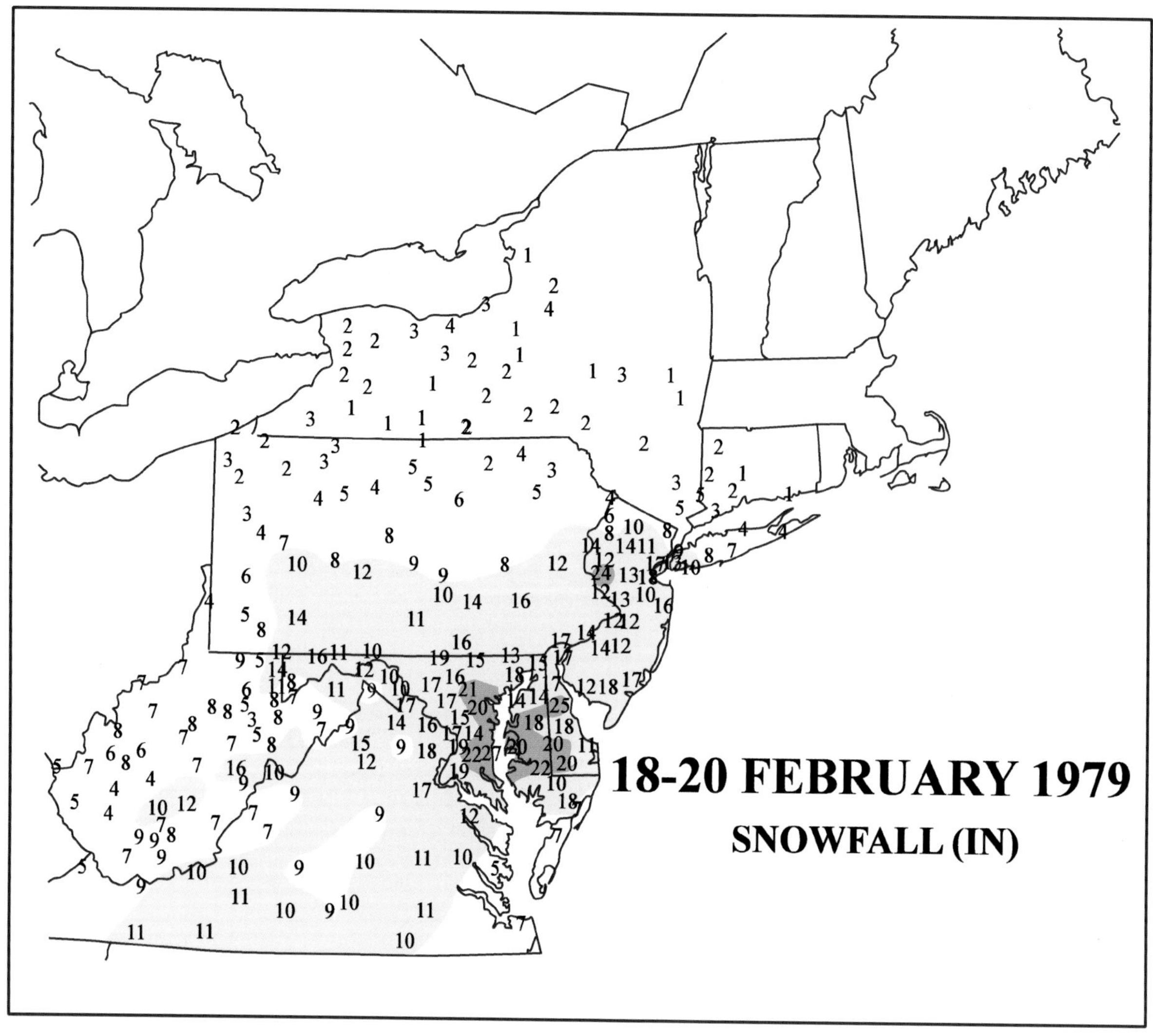

FIG. 10.18-1. Snowfall (in.) for 18–20 Feb 1979. See Fig. 10.1-1 for details.

forced as far south as South Carolina, with temperatures below −20°C common across the northeastern United States on 17 February.

- An 850-hPa low center formed over the Midwest at 0000 UTC 19 February, well north of the 0°C isotherm and the largest temperature gradient. A separate region of geopotential height falls developed across the southeastern United States at 1200 UTC 18 February and 0000 UTC 19 February. This height-fall center was located near the 0°C isotherm in a region where heavy precipitation and the coastal inverted trough were developing. The 850-hPa low appeared to redevelop over southeastern Virginia by 1200 UTC 19 February, probably in response to an amplification of the geopotential height-fall center as it moved northeastward with the deepening surface low. It may reflect a merger of the Ohio Valley low center with the amplifying coastal height-fall center.
- Detailed analyses by Bosart (1981) and Uccellini et al. (1984) show a southeasterly, ageostrophic low-level jet (LLJ) with winds of up to 35 m s^{-1} formed over Georgia and South Carolina by 1200 UTC 18 February. The LLJ appeared at the same time that the coastal front and heavy precipitation developed across the southeastern United States.
- More detailed case studies linked the development of the LLJ to a combination of interrelated factors. These included a geopotential height-fall center and related isallobaric wind, an indirect transverse circulation across the southeastern United States related to a subtropical jet stream and related vertical displacement of air parcels in a region of large 850-hPa temperature gradients, sensible and latent heating, and cold-air damming near the surface (see Uccellini et al. 1984, 1987). The LLJ in this case was shown to be an important factor in enhancing the mass divergence in the lower troposphere, focusing the sea level pressure falls along the Carolina coast that marked the initial de-

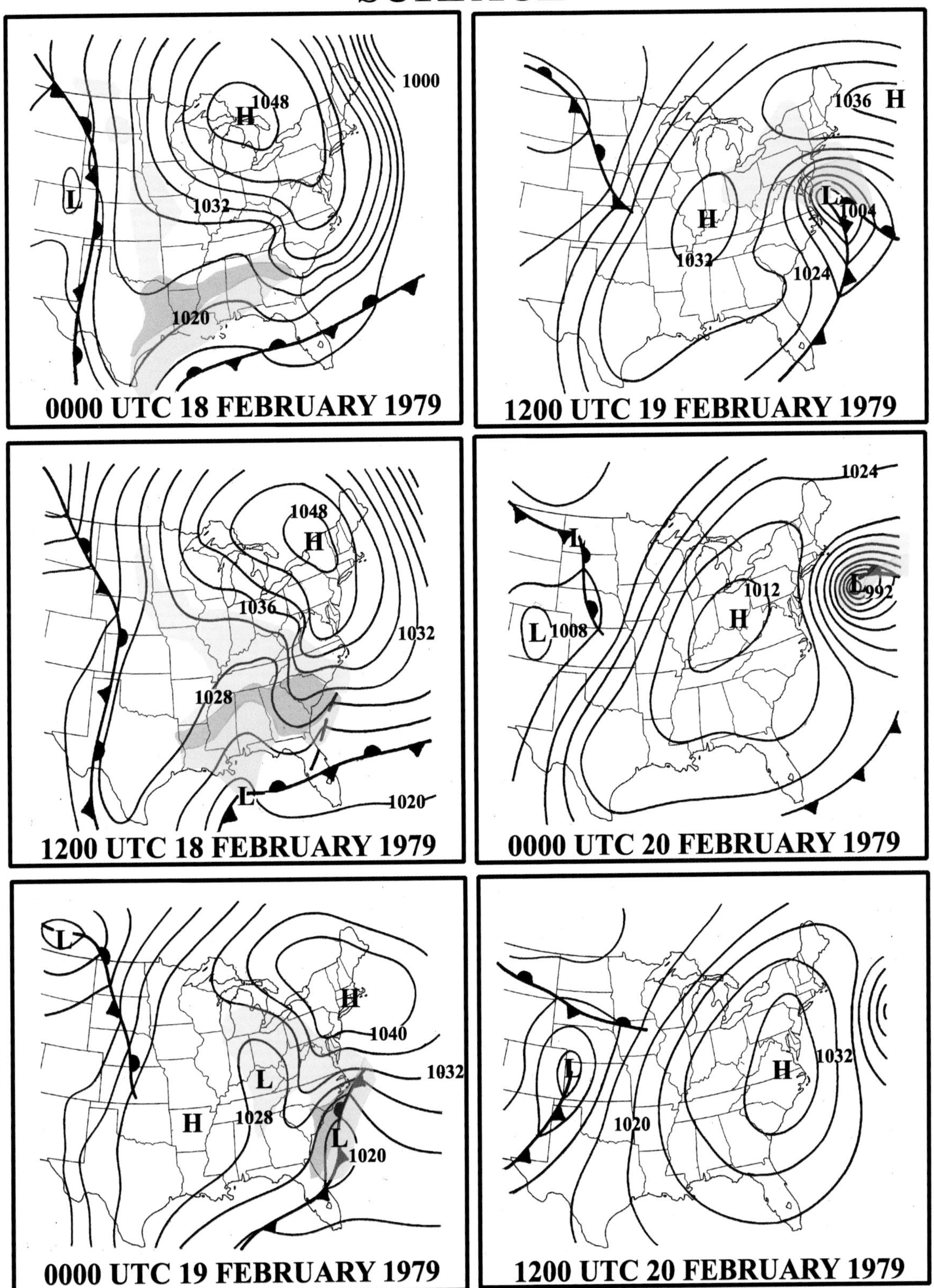

FIG. 10.18-2. Twelve-hourly surface weather analyses for 0000 UTC 18 Feb–1200 UTC 20 Feb 1979. See Fig. 10.1-2 for details.

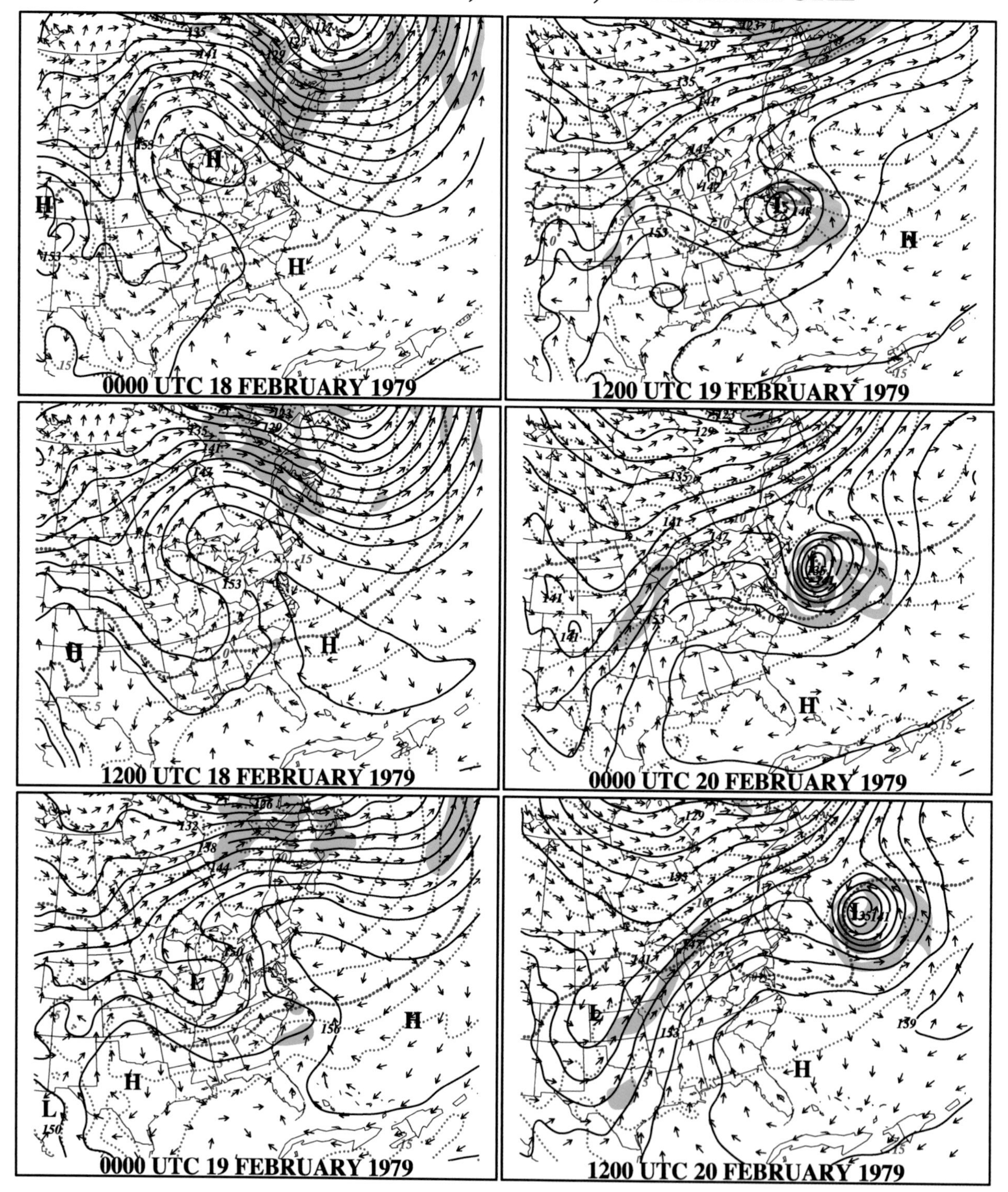

FIG. 10.18-3. Twelve-hourly 850-hPa analyses for 0000 UTC 18 Feb–1200 UTC 20 Feb 1979. See Fig. 10.1-3 for details.

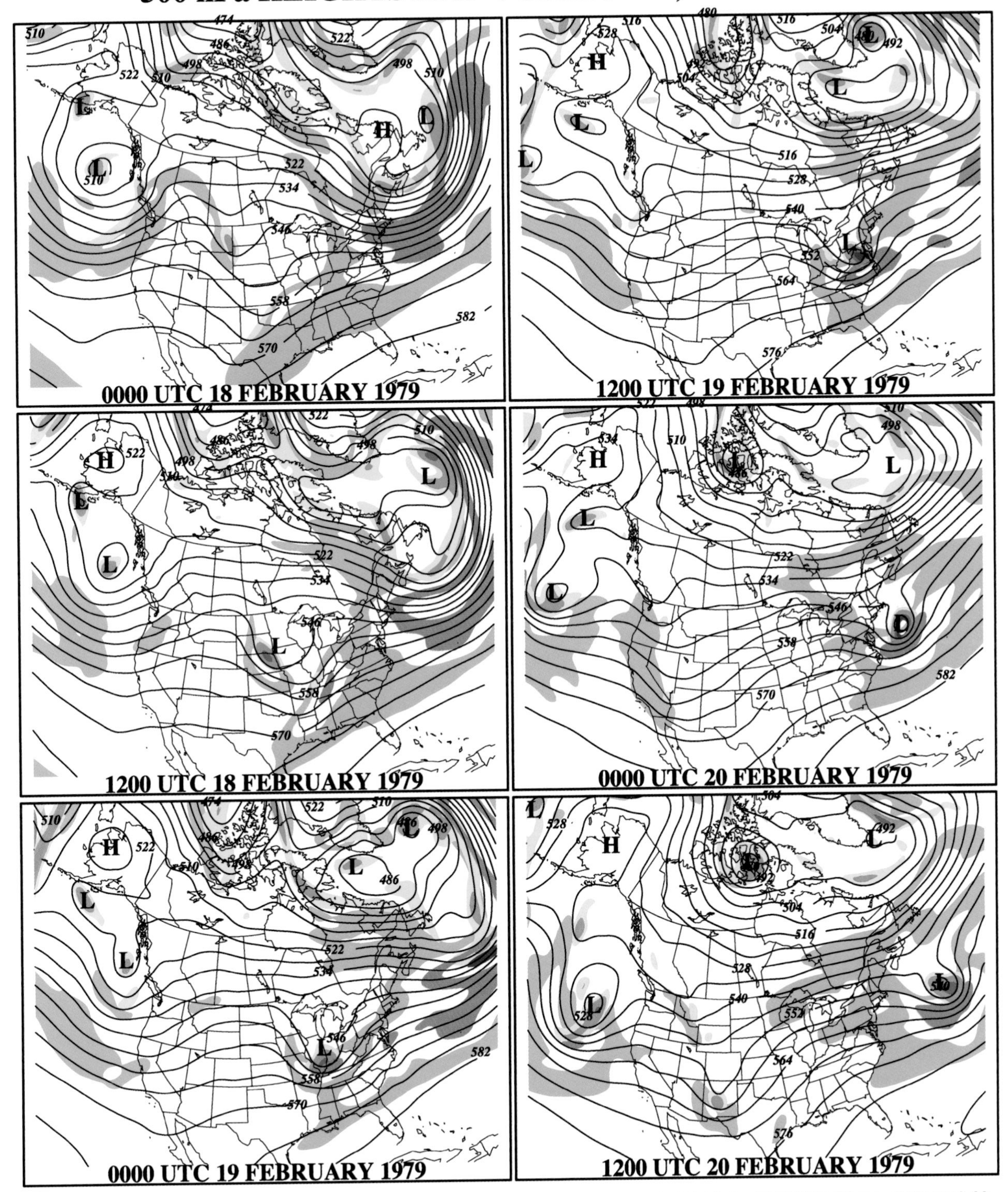

FIG. 10.18-4. Twelve-hourly analyses of 500-hPa geopotential heights and upper-level wind fields for 0000 UTC 18 Feb–1200 UTC 20 Feb 1979. See Fig. 10.1-4 for details.

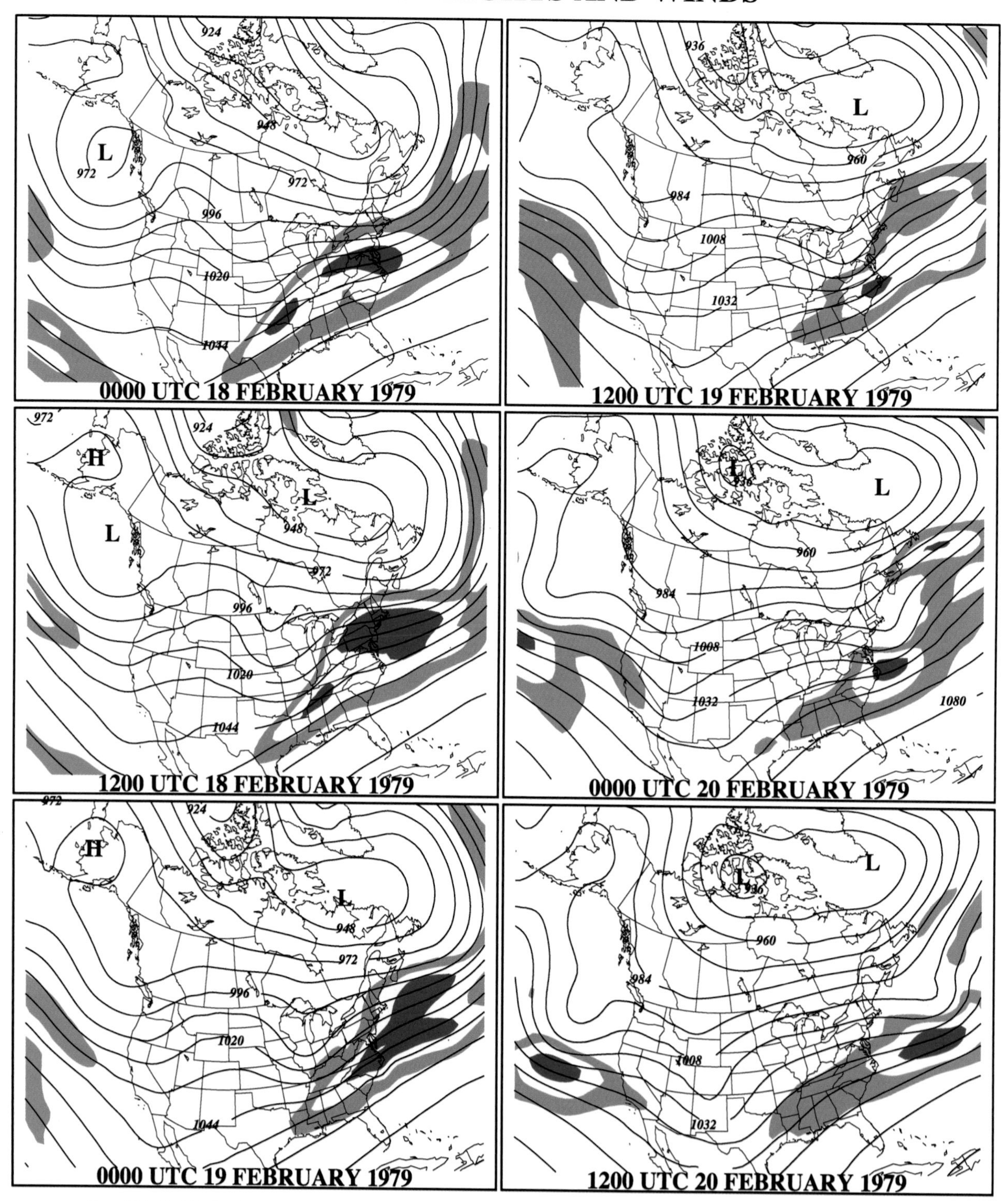

FIG. 10.18-5. Twelve-hourly analyses of 250-hPa geopotential height for 0000 UTC 18 Feb–1200 UTC 20 Feb 1979. See Fig. 10.1-5 for details.

SATELLITE IMAGERY

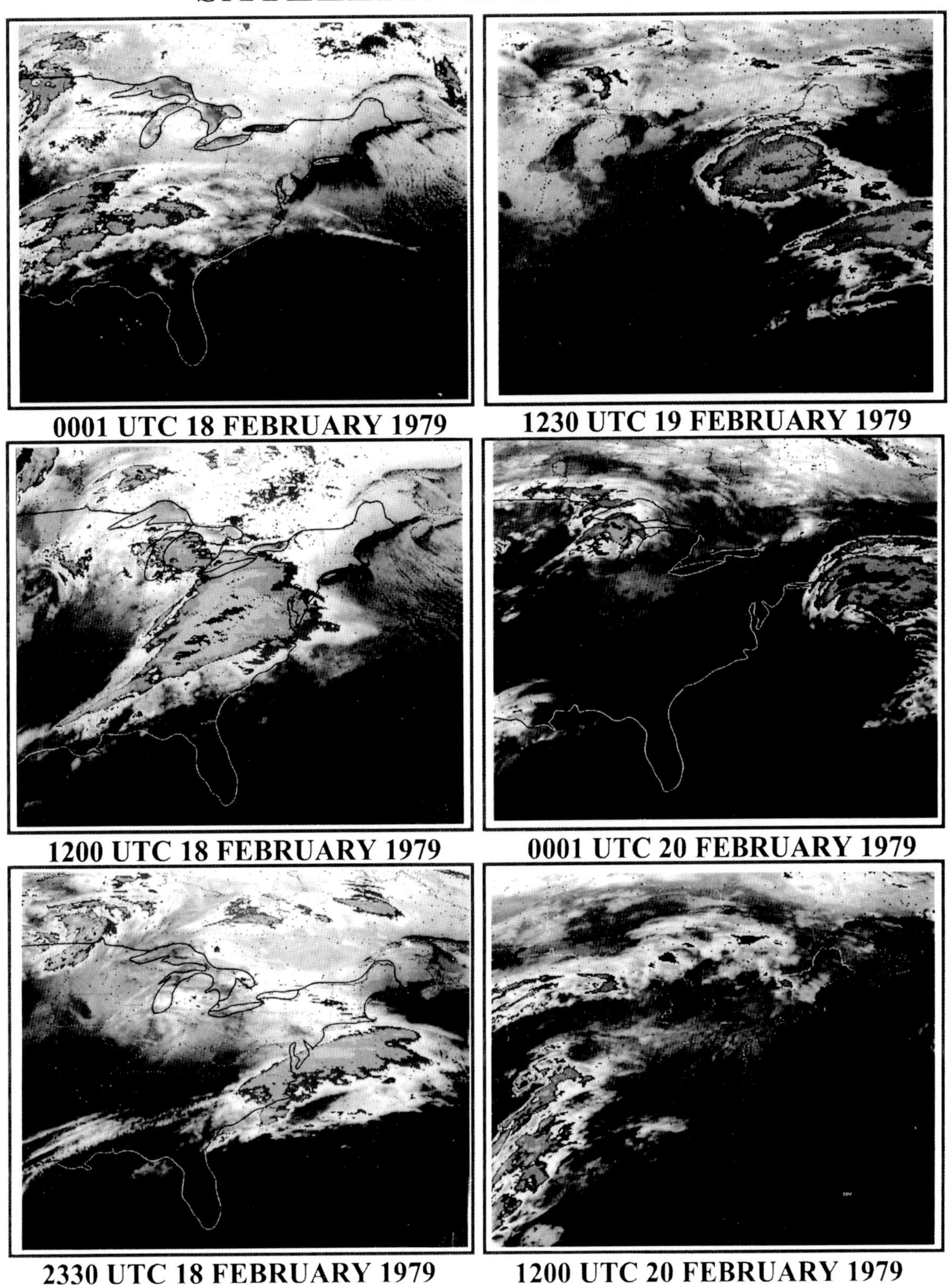

FIG. 10.18-6. Twelve-hourly infrared satellite images for 0001 UTC 18 Feb–1200 UTC 20 Feb 1979. See Fig. 10.16-6 for details.

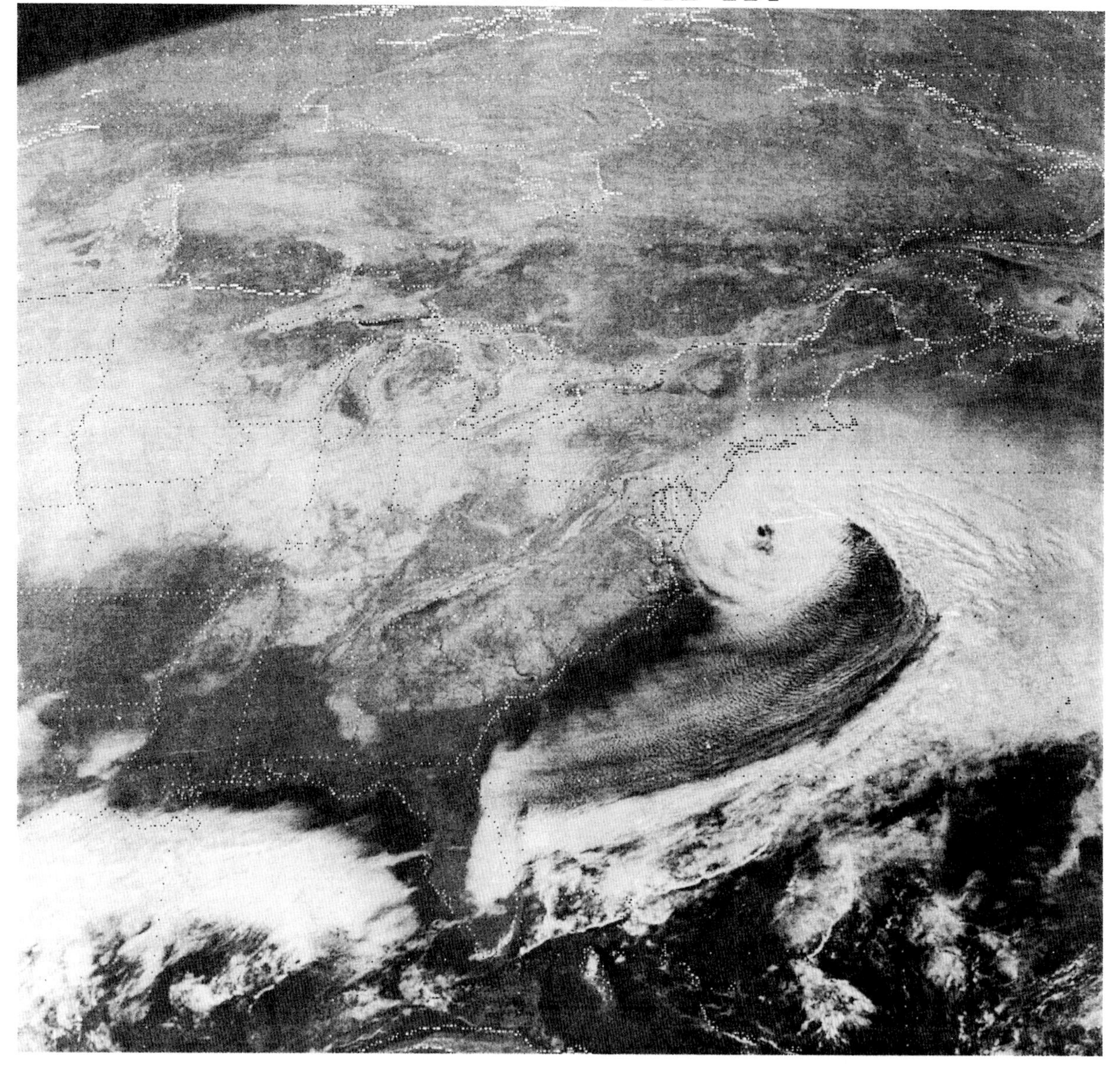

FIG. 10.18-6b. Visible Satellite image for 1830 UTC 19 Feb 1979 showing eyelike cloud-free region.

velopment of a coastal inverted trough by 0000 UTC 19 February.

- The LLJ shifted farther north along the East Coast and wrapped around the developing vortex off the Maryland–Virginia coast during the 24 h ending at 1200 UTC 19 February. The easterly LLJ was a significant factor in doubling the moisture transports into the region of heavy snowfall on 18 and 19 February (Uccellini et al. 1984). Strong warm-air advection was associated with the LLJ between 1200 UTC 18 February and 1200 UTC 19 February, helping to create an S-shaped isotherm pattern along the East Coast, within which the surface cyclone was deepening rapidly.

d. 500-hPa geopotential height and 400-hPa wind analyses

- Weak El Niño conditions began during late 1978 and continued through much of 1979, including February. A distinct period of negative NAO daily values began in early January 1979 and remained negative until they reversed sign on 19 February, the date of the Presidents' Day Storm, and went into a distinct positive phase. Much of February 1979 was exceptionally cold in the northeast United States, consistent with the negative phase, but quickly moderated once this storm passed. Therefore, this storm developed during a transitional period in which the negative phase of the NAO was weakening rapidly. An upper ridge east of central and northern Greenland, possibly consistent with a negative NAO, is seen in the 500-hPa analyses. The ridge appears to weaken by 1200 UTC 20 February.

- A distinct pattern of confluent geopotential heights across the northeastern United States on 17 and 18 February was associated with the movement of a very cold anticyclone from the Great Lakes region to New England. The circulation across the remainder of the United States and southern Canada was dominated by a nearly zonal flow. A low-amplitude trough developed in and then crossed the central United States from west to east on 18 February prior to east coast cyclogenesis.
- Separate troughs, identified by distinct cyclonic vorticity maxima north of Montana and over New Mexico at 1200 UTC 17 February (not shown), appeared to merge into one trough over the Mississippi Valley by 1200 UTC 18 February. Uccellini et al. (1984, 1985) showed that a tropopause fold associated with this trough extruded stratospheric air with high potential vorticity down to the 700-hPa level by 0000 UTC 19 February. The trough and its associated vorticity maximum propagated eastward, crossing the East Coast as the surface low deepened rapidly after 1200 UTC 19 February. No closed cyclonic circulation was observed at 500 hPa up until this time.
- Diffluence became increasingly evident downwind of the trough axis after 0000 UTC 18 February, coinciding with the formation of the inverted troughs and low pressure centers over the Ohio Valley and Atlantic coast. The orientation of the upper-level trough axis rotated slightly from north–south–southeast (a negative tilt) by 1200 UTC 19 February, with a pronounced pattern of diffluent geopotential heights along the East Coast as the secondary cyclone was deepening rapidly off Maryland.
- The amplitude of the trough and downstream ridge increased slightly between 1200 UTC 18 February and 0000 UTC 19 February, but actually appeared to decrease slightly in the 12-h period ending at 1200 UTC 19 February, while the secondary cyclone was intensifying rapidly off the middle Atlantic coast. Between 0000 UTC 18 February and 1200 UTC 19 February, the distance between the trough axis and the downstream ridge decreased.
- The precipitation area expanded on 18 and 19 February as the half-wavelength between trough and downstream ridge decreased and rapid cyclogenesis commenced off the East Coast.
- Details on the evolution of the upper-level features can be found in Uccellini et al. (1984, 1985). See Bosart (1981) and Bosart and Lin (1984) for surface and quasigeostrophic analyses.
- As described in Uccellini et al. (1984, 1987) and Whitaker et al. (1988), this case is influenced by a complex interaction of a northern polar jet, a southern subtropical jet, and a separate polar jet system associated with the deepening trough–tropopause fold in the central United States on 19 February.
- A polar jet attained speeds of up to 70 m s^{-1} at 400 hPa near the New England coast by 1200 UTC 18 February. This jet was embedded within the confluent region to the rear of the upper-level trough moving away from eastern Canada and was located above the large, cold surface anticyclone positioned over the northeastern United States.
- A separate 40–50 m s^{-1} polar jet streak at 400 hPa propagated from the northern plains states to the East Coast between 0000 UTC 18 February and 1200 UTC 19 February. A tropopause fold was diagnosed along the axis of this jet (Uccellini et al. 1985). The rapid development of precipitation across the middle Atlantic states, the continued enhancement of the LLJ off the East Coast, and the explosive surface cyclogenesis occurred within the diffluent exit region of this jet after 0000 UTC 19 February, as the polar jet propagated toward the East Coast.
- See Uccellini et al. (1984, 1985, 1987) and Whitaker et al. (1988) for a more detailed description of the jet streaks that influenced the development of the Presidents' Day Storm.

e. 250-hPa geopotential height and wind analyses

- The geopotential height analyses depict the evolution of the trough over southeastern Canada, the confluent flow over the Northeast, and the separate trough over the central United States that later yielded the explosive East Coast cyclogenesis. The transformation of the bitterly cold weather to a warmer regime by 1200 UTC 20 February is apparent as the upper-level ridge over Greenland is replaced by a trough and the westerly flow across the United States has a uniform height gradient depicting the relaxation of the highly baroclinic regime that helped produce conditions favorable for the Presidents' Day Storm.
- Winds within the subtropical jet streak at 250 hPa strengthened near the crest of the ridge approaching the East Coast as heavy snow developed over the southeastern United States early on 18 February, prior to east coast cyclogenesis. Wind speeds increased from 60 m s^{-1} on 17 February to near 80 m s^{-1} by 1200 UTC 18 February as the flow developed a noticeable cross-contour component near the ridge crest. Increasing upper-level divergence associated with this subtropical jet contributed to the initial outbreak of precipitation in the southeastern United States (Uccellini et al. 1984) and provided a basis for a transverse ageostrophic cicculation pattern that was coupled to the development of the LLJ along the coast by 1200 UTC 19 February.

f. Infrared satellite imagery sequence

- The cyclone and heavy snowfall were associated with two separate cloud systems.

- One cold-top cloud mass streamed eastward along the axis of the subtropical jet between 0000 and 2330 UTC 18 February. The heaviest precipitation occurred along the southern edge of this cloud system, with heavy snow and freezing precipitation falling across Kentucky, Tennessee, Georgia, and the Carolinas on 18 February. These clouds moved off the East Coast and diminished in areal coverage as cyclogenesis commenced along the Carolina coast late on 18 February.
- A small comma-shaped cloud area located over Nebraska and Iowa at 1200 UTC 18 February traversed the eastern half of the country in the ensuing 24-h period. This cloud mass was located in the exit region of the polar jet during this entire period and expanded rapidly in the 6-h period between 0900 and 1500 UTC 19 February (see Uccellini et al. 1985). Much colder cloud-top temperatures blossomed from Virginia to Massachusetts at 1200 UTC 19 February as the surface cyclone deepened rapidly off the Virginia–Maryland coast, with very heavy snows in the middle Atlantic states. The coldest cloud-top temperatures were found to the north of the surface low. The heaviest snows were located from the coldest cloud-top area near New York City southward to the edge of the cold tops over northern Virginia at 1200 UTC 19 February. The rapid development of this feature resembles a cold conveyor belt (Carlson 1980) and represents the emerging head of a comma-shaped cloud mass, which became quite distinct over the Atlantic Ocean by 0001 UTC 20 February.
- A cloud-free "eye" developed over the Atlantic by 1800 UTC 19 February (see visible satellite image in Fig. 10.18-6b) after the storm went through a period of rapid sea level intensification.

New York City, 6 Apr 1982 (photo reprinted courtesy of the *New York Times*).

19. 5–7 April 1982

a. General remarks

- This unusual late season storm produced near-blizzard conditions over much of New York, and New England, including New York City and Boston. The rapidly intensifying cyclone spread heavy snow amounts across the Midwest and the Ohio Valley before reaching the northern middle Atlantic states and New England. The snow and cold temperatures postponed opening day of the Major League Baseball season in many cities. Thunderstorms with frequent lightning were reported in New York City during the heaviest snowfall. The storm was followed by one of the coldest air masses on record for April. The temperature at Boston, Massachusetts, remained near −10°C during the afternoon of 7 April. Operational numerical simulations were successful in forecasting this storm. [See Kaplan et al. (1982) for a mesoscale simulation of the secondary sea level development associated with this case.]
- The following regions reported snow accumulations exceeding 10 in. (25 cm): portions of northern and eastern Pennsylvania, northern New Jersey, southern New York, central and northern Connecticut and Rhode Island, Massachusetts (except the extreme southeast), southern Vermont and New Hampshire, and Maine.
- The following regions reported snow accumulations exceeding 20 in. (50 cm): scattered portions of eastern New York, southern Vermont, northeastern Massachusetts, and southeastern New Hampshire.

b. Surface analyses

- Relatively cold air moved into the northeastern United States behind an intense cyclone that propagated across eastern Canada on 4 April. Unseasonably cold temperatures were associated with an anticyclone (1033 hPa) located over central Canada at 1200 UTC 5 April. A sea level high pressure ridge extending southeastward from the anticyclone shifted from the Ohio Valley to the northeastern United States on 5 April, providing cold air for the snowstorm that developed the next day. Following the snowstorm, the anticyclone moved southward into the Midwest. Its effects combined with the snow cover to produce record low temperatures from the plains states to the East Coast on 6–8 April.
- Only weak cold-air damming was observed along the East Coast prior to and during cyclogenesis on 5–6 April.
- Coastal frontogenesis was evident late on 5 April and early on 6 April along the Carolina coast, but was not characterized by the large temperature contrasts seen in many of the other cases.

TABLE 10.19-1. Snowfall amounts for urban and selected locations for 5–7 Apr 1982; see Table 10.1-1 for details.

NESIS = 3.75 (category 2)	
Urban center snowfall amounts	
Philadelphia, PA	3.5 in. (9 cm)
New York, NY–Central Park	9.6 in. (24 cm)
Boston, MA	13.3 in. (34 cm)
Other selected snowfall amounts	
Grafton, NY	22.8 in. (58 cm)
Albany, NY	17.7 in. (45 cm)
Portland, ME	15.9 in. (40 cm)
Worcester, MA	15.0 in. (38 cm)
Hartford, CT	14.1 in. (36 cm)
Concord, NH	13.9 in. (35 cm)
Newark, NJ	12.8 in. (33 cm)
Allentown, PA	11.4 in. (29 cm)

- A primary low pressure system moved rapidly from the southern plains through the Ohio Valley to Pennsylvania between 1200 UTC 5 April and 1200 UTC 6 April. The center deepened 10 hPa in the 12-h period ending at 0900 UTC 6 April. The primary low produced moderate to heavy snows and strong winds from Chicago to Detroit and Cleveland as it moved through the Ohio Valley on 5–6 April. Although this case exhibited the most intense primary surface low pressure system of any in the sample, the primary low was very quickly absorbed into the circulation of the secondary coastal storm on 6 April.
- A weak low pressure center was observed on the coastal front near the Georgia coast by 0000 UTC 6 April. Explosive cyclogenesis commenced farther to the north, however, along the Virginia coast, between 0000 and 1200 UTC 6 April. The storm deepened at a rate of −1 to −3 hPa h^{-1}, with the central sea level pressure falling from 994 hPa at 0900 UTC 6 April to 968 hPa by 0000 UTC 7 April. The low center tracked east-northeastward from the Virginia–Maryland coast, passing south of Long Island and New England, then turned northeastward toward Nova Scotia on 7 April. The forward motion of the storm maintained a rate of about 15 m s^{-1} throughout this period.
- Over the northeastern United States, most of the snow fell between 1200 UTC 6 April and 1200 UTC 7 April. In many locations, snowfall rates of 1–2 in. (2.5–5 cm) per hour occurred in conjunction with north-northeasterly surface winds increasing to 15–20 m s^{-1} with higher gusts.

c. 850-hPa analyses

- Prior to East Coast cyclogenesis, low-level northwesterly flow behind the intense cyclone in eastern Canada generated strong cold-air advection across the northeastern United States on 4–5 April. The northwesterly flow and cold-air advection were located beneath a region of upper-level confluence. The strong

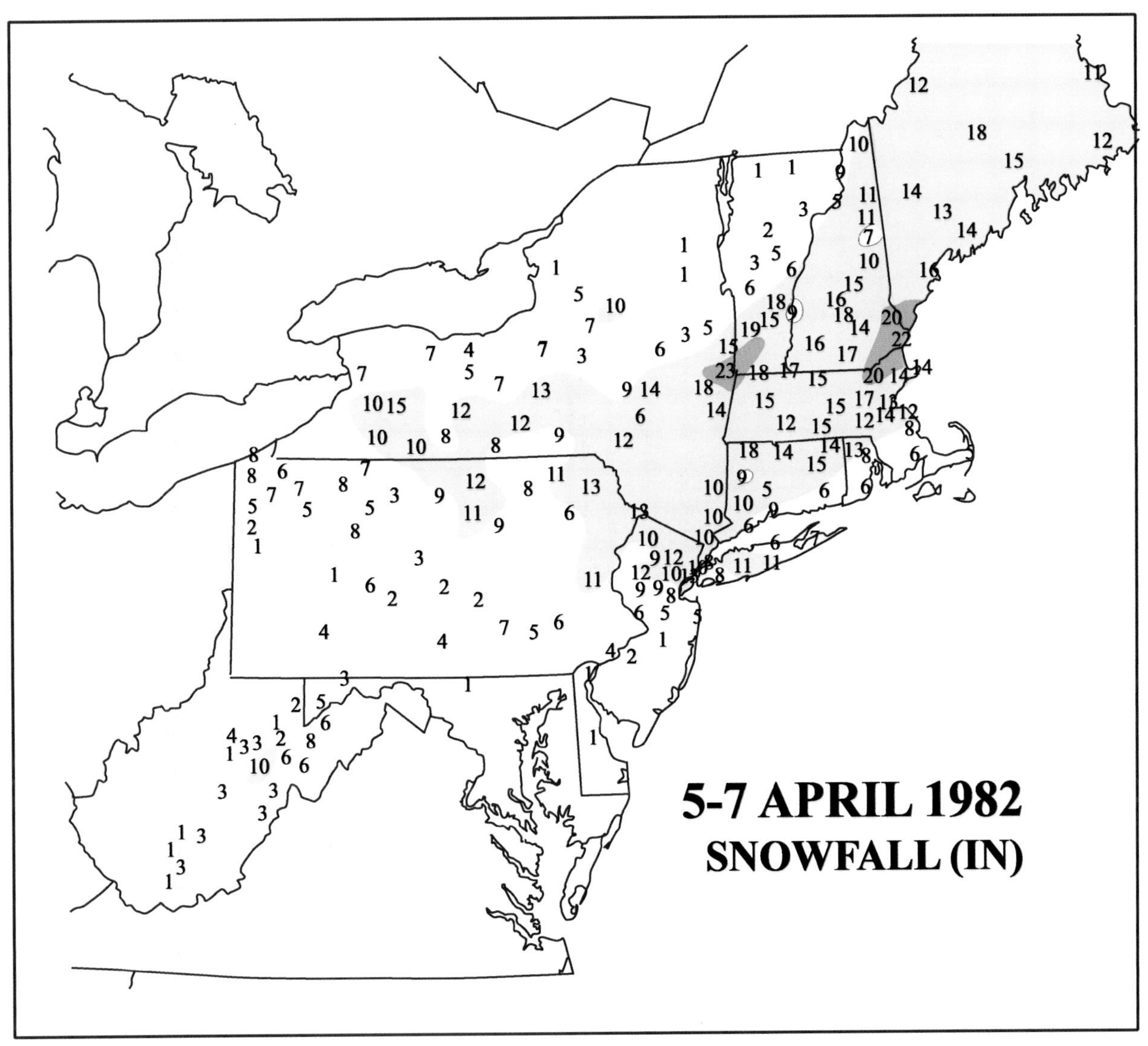

FIG. 10.19-1. Snowfall (in.) for 5–7 Apr 1982. See Fig. 10.1-1 for details.

cold-air advection helped to create a narrow, intense temperature gradient over the middle Atlantic states on 5 April.

- The 850-hPa low intensified at an accelerating rate as it moved from the central plains to near southern New England between 1200 UTC 5 April and 0000 UTC 7 April. It deepened at rates of −60 m per 12 h as it crossed the upper Mississippi Valley by 0000 UTC 6 April, −90 m per 12 h as it approached the East Coast by 1200 UTC 6 April, and −150 m per 12 h as it moved off the coast by 0000 UTC 7 April.
- A secondary 850-hPa geopotential height-fall center appeared over the southeastern United States at 0000 UTC 6 April, possibly reflecting the secondary surface low development. This feature was difficult to follow at 1200 UTC 6 April since the entire 850-hPa circulation had intensified over the eastern United States.
- This case was marked by very strong temperature gradients during cyclogenesis. Pronounced cold-air advection occurred behind the low, especially at 1200 UTC 5 April and 0000 UTC 6 April, with strong warm-air advection ahead of it, especially at 1200 UTC 6 April and 0000 UTC 7 April. The advection patterns intensified as the 850-hPa low moved toward the East Coast on 6 April, producing an S-shaped isotherm pattern within which the surface low rapidly intensified.
- The 0°C isotherm was generally located near the 850-hPa low center except by 1200 UTC 7 April, when the low was centered in the colder air as it occluded.
- A low-level jet (LLJ) appears to form off the East Coast by 1200 UTC 6 April with wind speeds exceeding 25 m s^{-1}. By 0000 UTC 7 April, this wind maximum appears to be rotating around the rapidly developing vortex and increasing to greater than 35 m s^{-1} by 1200 UTC 7 April north of the storm center.

SURFACE

0000 UTC 5 APRIL 1982

1200 UTC 6 APRIL 1982

1200 UTC 5 APRIL 1982

0000 UTC 7 APRIL 1982

0000 UTC 6 APRIL 1982

1200 UTC 7 APRIL 1982

FIG. 10.19-2. Twelve-hourly surface weather analyses for 0000 UTC 5 Apr–1200 UTC 7 Apr 1982. See Fig. 10.1-2 for details.

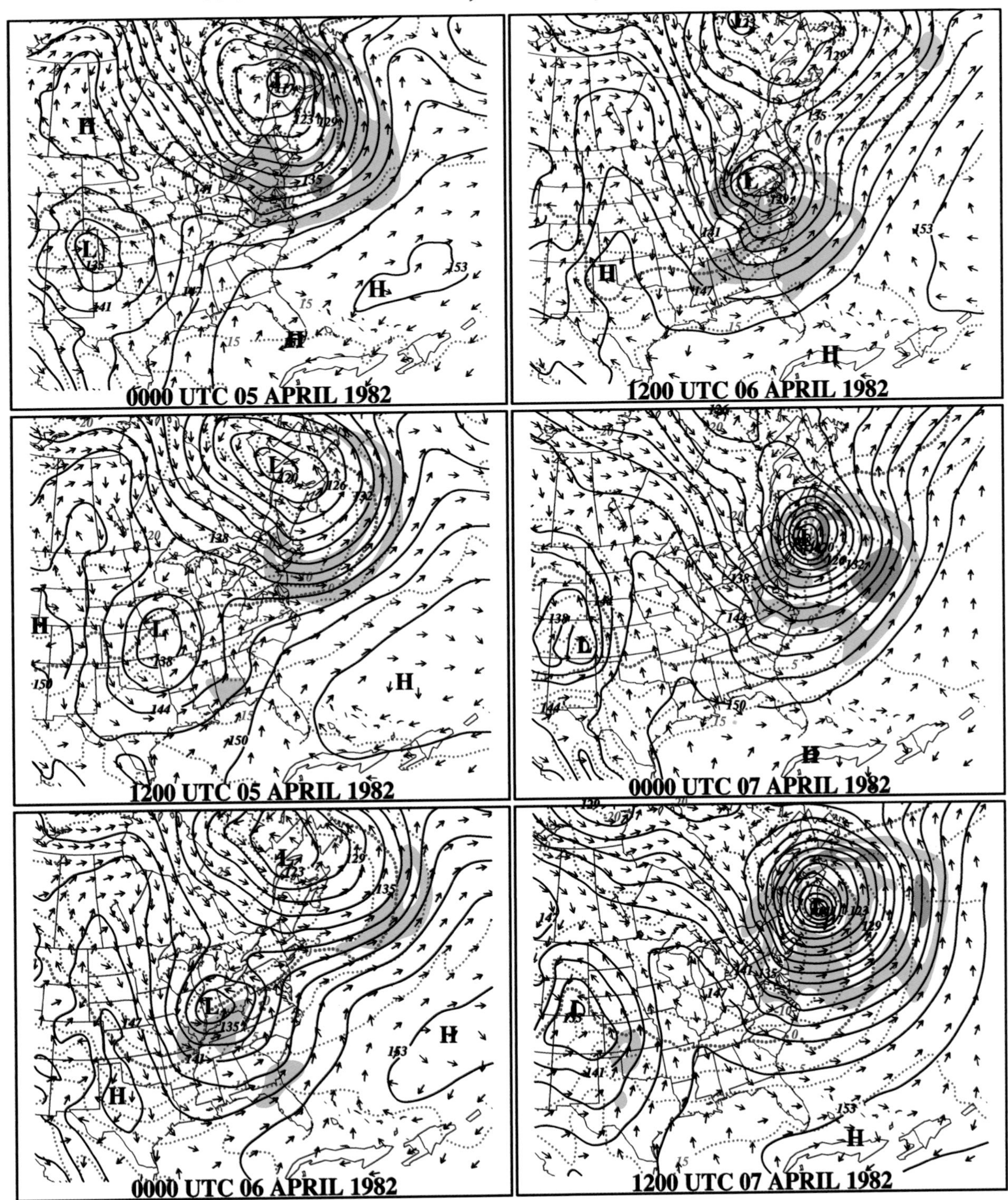

FIG. 10.19-3. Twelve-hourly 850-hPa analyses for 0000 UTC 5 Apr–1200 UTC 7 Apr 1982. See Fig. 10.1-3 for details.

500 hPa HEIGHTS AND VORTICITY, 400 hPa WINDS

FIG. 10.19-4. Twelve-hourly analyses of 500-hPa geopotential heights and upper-level wind fields for 0000 UTC 5 Apr–1200 UTC 7 Apr 1982. See Fig. 10.1-4 for details.

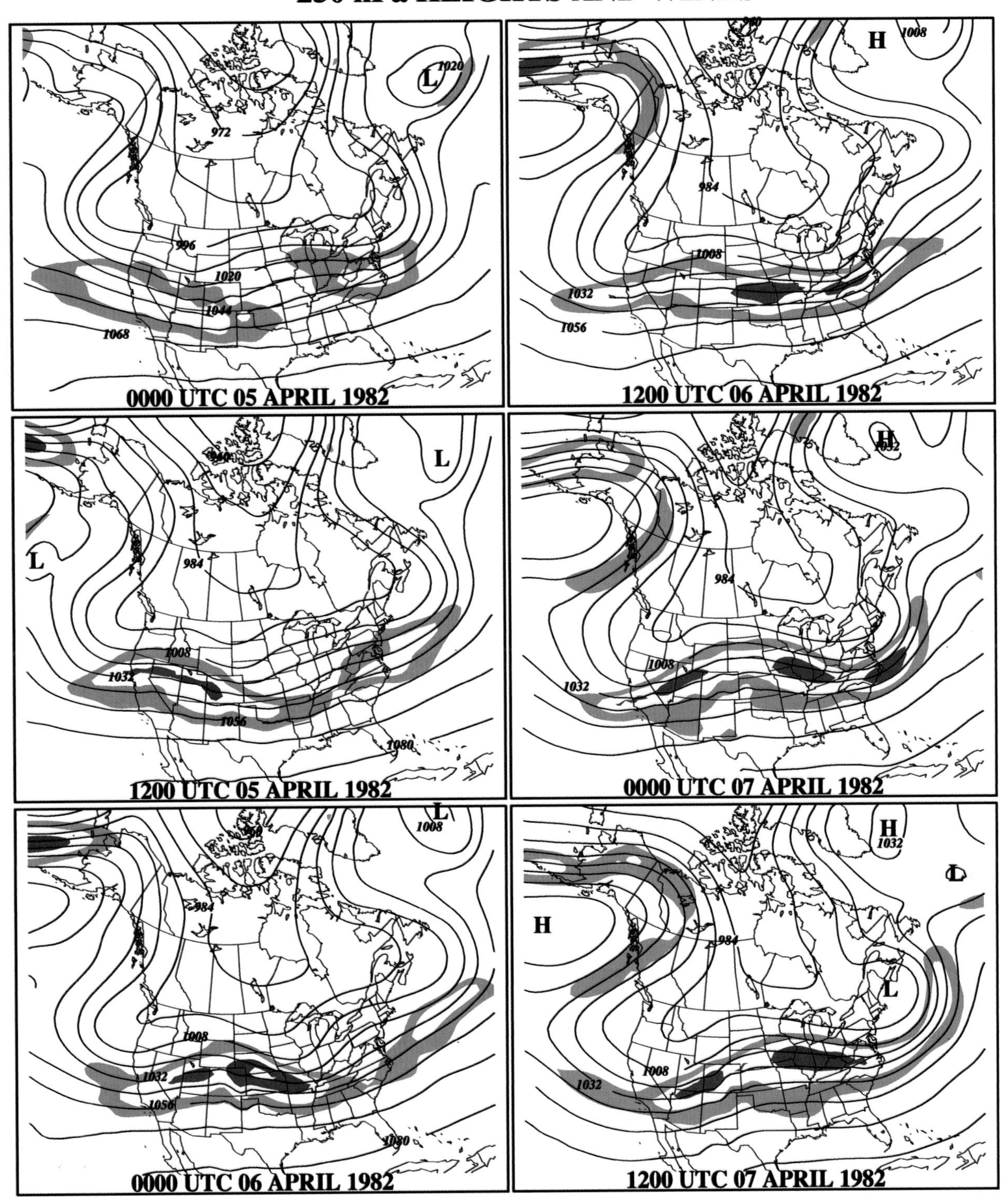

FIG. 10.19-5. Twelve-hourly analyses of 250-hPa geopotential height for 0000 UTC 5 Apr–1200 UTC 7 Apr 1982. See Fig. 10.1-5 for details.

SATELLITE IMAGERY

0000 UTC 5 APRIL 1982

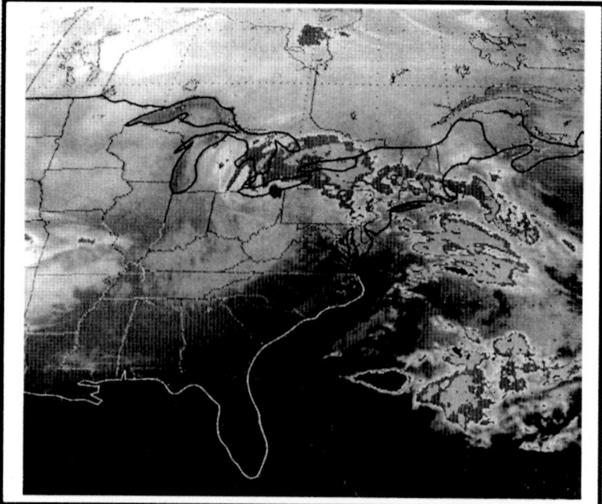

1200 UTC 6 APRIL 1982

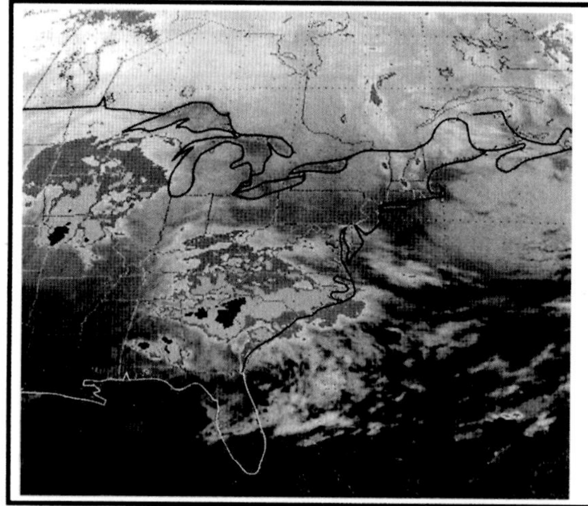

1200 UTC 5 APRIL 1982

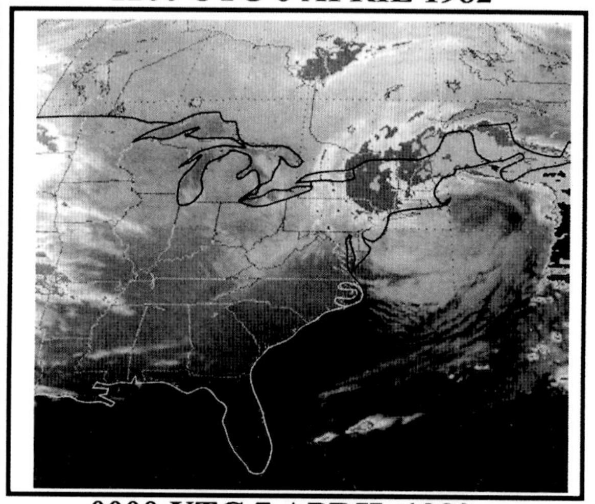

0000 UTC 7 APRIL 1982

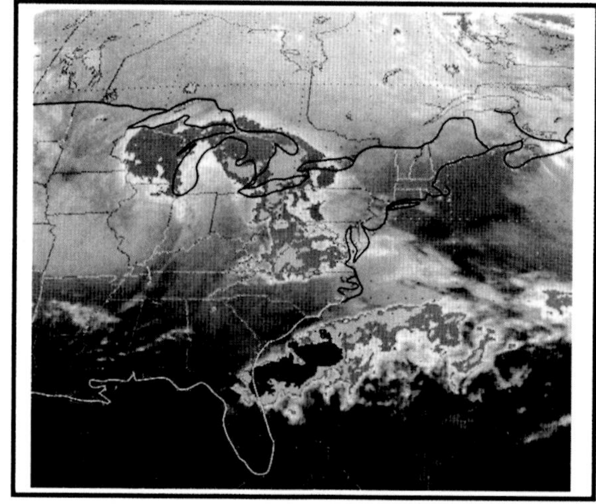

0030 UTC 6 APRIL 1982

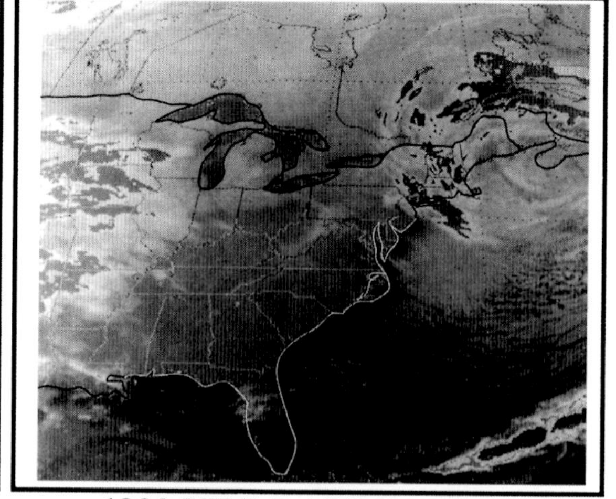

1200 UTC 7 APRIL 1982

FIG. 10.19-6. Twelve-hourly infrared satellite images for 0000 UTC 5 Apr–1200 UTC 7 Apr 1982. See Fig. 10.16-6 for more details.

- The 850-hPa wind speeds also increased dramatically south and west of the low between 0000 and 1200 UTC 6 April, as the system continued to amplify and the vortex strengthened off the New England coast.

d. 500-hPa geopotential height and 400-hPa wind analyses

- El Niño conditions existed in April 1982 and eventually went on to become one of the strongest El Niños of the late 20th century through the 1982–83 period. In early April, the North Atlantic Oscillation (NAO) underwent a transition from positive to negative. Daily NAO values were trending downward during the 3-day period in which this storm developed and intensified. A cutoff upper ridge developed near southern Greenland late on 6 April, a feature that is consistent with the negative phase of the NAO.
- Prior to 6 April, the precyclogenetic environment at 500 hPa was characterized by a developing anticyclonic circulation near Greenland and a high-amplitude trough over southeastern Canada. This configuration produced a confluent geopotential height pattern across the Great Lakes and northeastern United States. The surface anticyclone and cold air became entrenched over the Great Lakes region on 5–6 April beneath the confluent region.
- The trough associated with the cyclone propagated eastward with the strong westerly flow across the United States from 4 to 7 April. The trough system was small in amplitude as it crossed the western third of the country on 4–5 April, but amplified and formed a deepening cyclonic vortex at 500 hPa off the East Coast by 0000 UTC 7 April.
- The amplifying trough was associated with an intensifying cyclonic vorticity maximum as it crossed the United States. Increasing vorticity advection accompanied the rapid development of the surface low and the outbreak of heavy snowfall on 6 April.
- Geopotential height gradients increased at the base of the trough between 0000 and 1200 UTC 6 April as both the primary and secondary surface lows intensified. At the same time, diffluence became more evident downwind of the trough, as the trough axis became oriented from northwest to southeast (a negative tilt) by 1200 UTC 6 April. The surface low deepened rapidly and heavy precipitation developed as the diffluence pattern downwind of the negatively tilted 500-hPa trough became better defined and moved across the coast on 6 April.
- The amplitude between the trough and downstream ridge increased between 1200 UTC 5 April and 0000 UTC 7 April, although only a small increase in amplitude occurred during rapid cyclogenesis on 6 April. The half-wavelength between the trough and downstream ridge decreased throughout the 36-h period ending at 0000 UTC 7 April.
- This case is marked by a distinct dual-jet pattern with a northern jet system extending across the northeast United States on 4–5 April and a southern jet moving with the short-wave trough becoming well defined at 400 hPa by 0000 UTC 6 April.
- A polar jet stretched across the northeastern United States on 5–6 April, within the confluent region upwind of the trough system over southeastern Canada. This confluent jet pattern was located above the cold surface anticyclone over the Great Lakes region.
- A separate polar jet with maximum wind speeds exceeding 50 m s^{-1} at 400 hPa extended from the west coast into the base of the upper-level trough as the trough amplified over the central United States on 5 April. The increasing cyclonic vorticity maximum was located in the cyclonic exit region of the jet, at the base of the 500-hPa trough by 0000 UTC 6 April.
- Missing wind reports over the Carolinas at 1200 UTC 6 April indicate that wind speeds increased near and upwind of the base of the trough immediately before and during the rapid development of the secondary cyclone on 6 April. The analyses show wind speeds in the jet had increased to greater than 70 m s^{-1} at the 400-hPa level by 1200 UTC 6 April.
- The storm system developed rapidly within the diffluent exit region of the polar jet streak–trough system as it approached the East Coast between 0600 and 1200 UTC 6 April (as modeled by Kaplan et al. 1982) and the right-entrance region of the more northern polar jet extends east from the New England coast.

e. 250-hPa geopotential height and wind analyses

- The height analyses clearly show the general west-to-east flow concentrated over the central United States in the precyclogenetic period, with the amplification of the trough over the central United States by 6 April.
- The most intense winds at this level tend to remain south of the developing surface low. The 80 m s^{-1} jet maximum to the west of the amplifying trough axis at 0000 UTC 6 April represents an area where the polar jet is evident at 400 hPa and is an indication of the subtropical jet that is maximized near 200 hPa. These jet systems remain interconnected as they approach the East Coast with a well-defined exit region at 1200 UTC 6 April, the time of rapid cyclogenesis.

f. Infrared satellite imagery sequence

- A distinct cloud mass with cold cloud tops moved through the southeastern United States between 0000 UTC 5 April and 0000 UTC 6 April. These upper-level clouds were positioned well in advance of the amplifying upper-level trough crossing the United States, and may have been related to the wind maximum evident at 250 hPa passing off the Southeast coast during this period. The cold cloud region was associated with an area of light to moderate rainfall across the southeastern United States. By 0000 UTC 6 April, it was located near the weak surface low along the Southeast coast.

- A separate cloud mass extending from Missouri to Minnesota at 1200 UTC 5 April was associated with the upper-level trough and was located north of the surface low over Oklahoma. This cloud mass remained in the diffluent exit region of the polar jet as it propagated eastward and increased in magnitude at 400 hPa between 1200 UTC 5 April and 1200 UTC 6 April.
- The comma shape of the cloud mass was evident as early as 1200 UTC 5 April, in association with the development of the primary cyclone. There was no marked redevelopment of the comma head as the secondary cyclone deepened after 1200 UTC 6 April. The cloud mass consolidated into a more distinct comma-shaped cloud feature by 0000 UTC 7 April, however, as the surface low intensified rapidly off the East Coast. The heaviest snows fell along the southern edge of this cloud feature.
- The comma cloud evolved into the spiraling cloud mass, which characterized the occlusion stage of the cyclone by 1200 UTC 7 April. An eye was briefly visible late on 6 April as the cyclone was deepening rapidly south of New England.

Snarled traffic on the Northern State Parkway, 12 Feb 1983 (photo by Jack Balletti, courtesy of Kevin Ambrose, *Great Blizzards of New York City*, Historical Enterprises, 1994).

20. 10–12 February 1983

a. General remarks

- This snowstorm was one of many cyclones to affect the eastern United States during a winter that was unusually warm and stormy, but it was one of the few storms accompanied by temperatures cold enough for snowfall in the Northeast urban corridor. The heaviest snows from this storm were oriented near a line through Washington, Baltimore, Philadelphia, New York, and Boston. The 24-h snowfalls at Philadelphia, Harrisburg, and Allentown, Pennsylvania, and Hartford, Connecticut, were the greatest on record. For many other cities, this was one of the heaviest snowstorms on record. Accumulations reached 30 in. (75 cm) in parts of northern Virginia, western Maryland, and the panhandle of West Virginia. Given the widespread severe effects of this storm in the heavily populated Northeast urban corridor, Sanders and Bosart (1985a,b) named this the Megalopolitan Snowstorm. Thunderstorms were observed at many locations from Washington to New York City during the afternoon and evening of 11 February. Winds were not as crippling as in some other storms, but wind speeds were still high enough to create blizzard or near-blizzard conditions from eastern Pennsylvania and northern Delaware to Massachusetts.
- The following regions reported snow accumulations exceeding 10 in. (25 cm): West Virginia (except the extreme southwest and northwest), Virginia (except the extreme southwest and the Tidewater area), Maryland (except the extreme southeast), Delaware, the southeastern half of Pennsylvania, New Jersey, southeastern New York, Connecticut, Rhode Island, Massachusetts, and extreme southern New Hampshire.
- The following regions reported snow accumulations exceeding 20 in. (50 cm): northern Virginia, northeastern West Virginia, central and northern Maryland, southeastern Pennsylvania, central and northern New Jersey, southeastern New York, central Connecticut, and parts of Massachusetts.

TABLE 10.20-1. Snowfall amounts for urban and selected locations for 10–12 Feb 1983; see Table 10.1-1 for details.

NESIS = 6.28 (category 4)	
Urban center snowfall amounts	
Washington, D.C.–Dulles Airport	22.8 in. (58 cm)
Baltimore, MD	22.8 in. (58 cm)
Philadelphia, PA	21.3 in. (54 cm)
New York, NY–La Guardia Airport	22.0 in. (56 cm)
Boston, MA	13.5 in. (34 cm)
Other selected snowfall amounts	
Woodstock, VA	32.0 in. (81 cm)
Allentown, PA	25.2 in. (64 cm)
Harrisburg, PA	25.0 in. (64 cm)
Hartford, CT	21.0 in. (53 cm)
Roanoke, VA	18.6 in. (47 cm)
Richmond, VA	17.7 in. (45 cm)

b. Surface analyses

- A 1035–1040-hPa anticyclone drifted very slowly from eastern Ontario into southwestern Quebec while the cyclone was propagating across the southeastern United States on 10–11 February. This anticyclone brought one of the few colder than normal air masses to affect the United States during the entire winter.
- The distinct high pressure ridge from New York to North Carolina indicated cold-air damming on 10–11 February. Coastal frontogenesis was observed along the coast of the Carolinas prior to 1200 UTC 10 February.
- The surface low developed along the Gulf coast on 9 February and moved eastward to southern Georgia by 0000 UTC 11 February. It then propagated northeastward along the coastal front near the Southeast coast during 11 February, before turning more to the east-northeast after passing the Virginia coast at 0000 UTC 12 February. The low appeared to jump or redevelop from southern Georgia to near the South Carolina coast early on 11 February. Heavy snow developed in the mountains of North Carolina and in southern and western Virginia after 0000 UTC 11 February and spread through the urban centers of the Northeast coastal region during 11 February. The forward speed of the cyclone varied between 10 and 20 m s^{-1}.
- The storm deepened slowly and erratically as it propagated from Georgia to near southeastern Virginia on 10–11 February with the central sea level pressure remaining above 995 hPa. The cyclone deepened rapidly only when it had moved east of New England after 1200 UTC 12 February.
- Pressures never fell below 1012 hPa in the major cities of the Northeast, with oscillating pressure tendencies observed from Washington to Boston, Massachusetts, during the course of the storm. Evidence presented by Bosart and Sanders (1986) indicates that the erratic pressure tendencies and an outbreak of thunderstorms from Washington to New York City were associated with a gravity wave. Snowfall rates of 2–5 in. (5–12 cm) per hour were reported during the convective outbreak.
- The distance between the low and high pressure centers along the Atlantic seaboard decreased as the storm moved up the coast on 11 February. This increased the sea level pressure gradient north of the low center and strengthened the easterly flow from the Atlantic toward the region of maximum snowfall, and helped create blizzard or near-blizzard conditions along the Northeast coast from Delaware northward.

c. 850-hPa analyses

- Strong northwesterly flow and cold-air advection at 850 hPa covered the northeastern United States

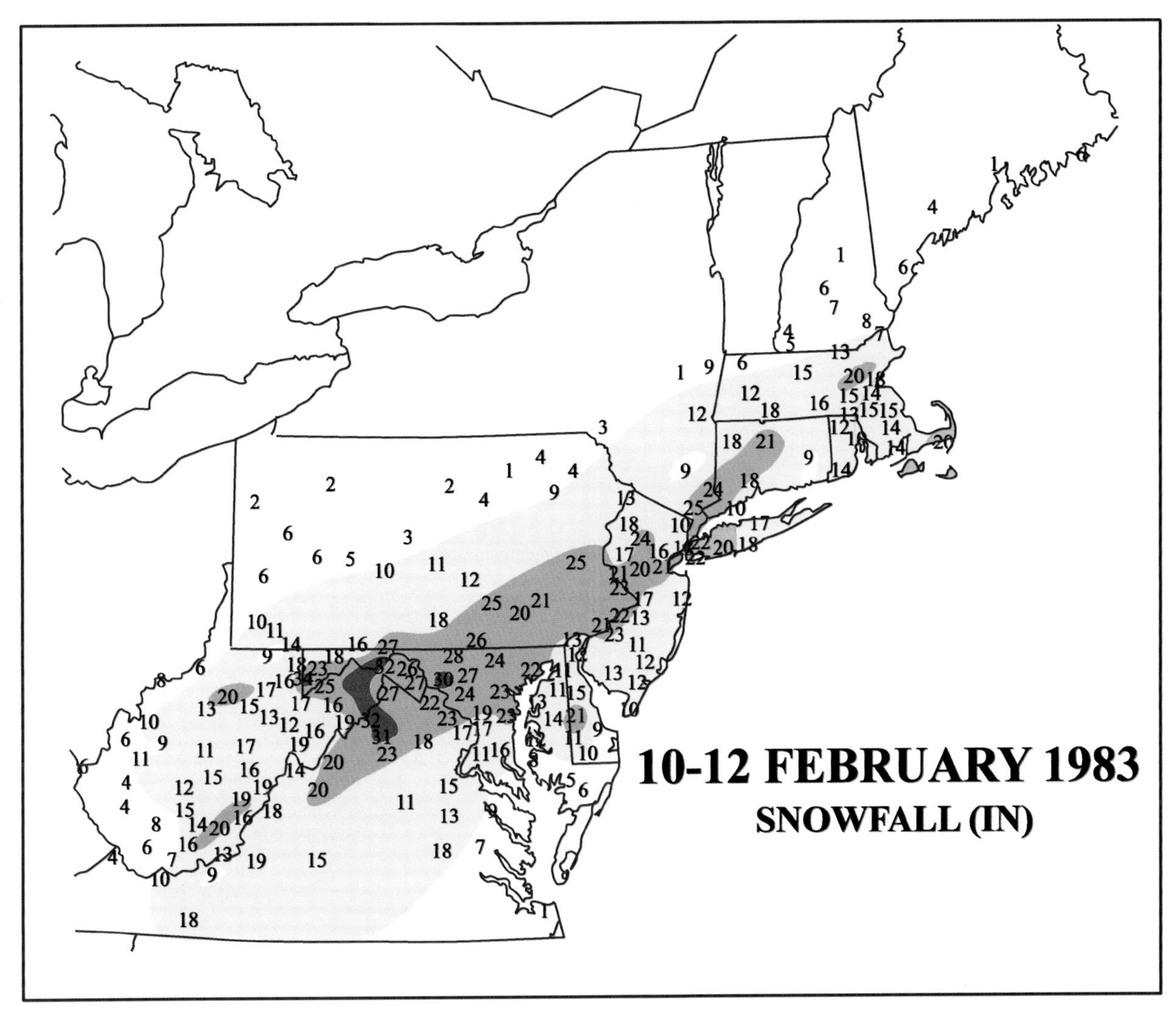

FIG. 10.20-1. Snowfall (in.) for 10–12 Feb 1983. See Fig. 10.1-1 for details.

through 10 February, beneath a region of confluent upper-level geopotential heights. A large 850-hPa temperature gradient was established along the East Coast, with temperatures ranging from 0°C over central North Carolina to −20°C over Maine.

- By 0000 UTC 11 February, cold air remained over New England while south-to-southeasterly winds and warm-air advection increased near the Southeast and middle Atlantic coasts, where 850-hPa heights were falling significantly. As a result, temperature gradients along the Southeast and middle Atlantic coasts increased sharply at 0000 UTC 11 February and 0000 UTC 12 February. At these times, moderate to heavy precipitation developed to the west of the coastal front, over the regions experiencing the largest warm-air advection at 850 hPa.
- The increasing temperature gradients were associated with the development of an S-shaped isotherm pattern along the East Coast between 1200 UTC 11 February and 0000 UTC 12 February. The 850-hPa low center was located on the warm side of the 0°C isotherm through 0000 UTC 11 February, but became collocated with the 0°C isotherm as it approached the Atlantic coast at 1200 UTC 11 February.
- The 850-hPa low alternately deepened and filled slightly between 0000 UTC 10 February and 1200 UTC 11 February. It then deepened slowly after 1200 UTC 11 February as it moved off the East Coast.
- Although this storm produced some of the deepest snow amounts of any of the cases examined, the low-level jets (LLJs) associated with it were only of moderate intensity. Easterly-to-southeastery winds to the north of the 850-hPa low center at 1200 UTC 11 February were analyzed at 25 m s^{-1}, although several key reports were missing. The increasing southeasterly to easterly winds along and just east of the coast acted to enhance the moisture transport into the region of heavy snowfall on 11–12 February. This effect was aided by the slow movement of the 850-hPa low, averaging less than 15 m s^{-1}, which maintained a strong

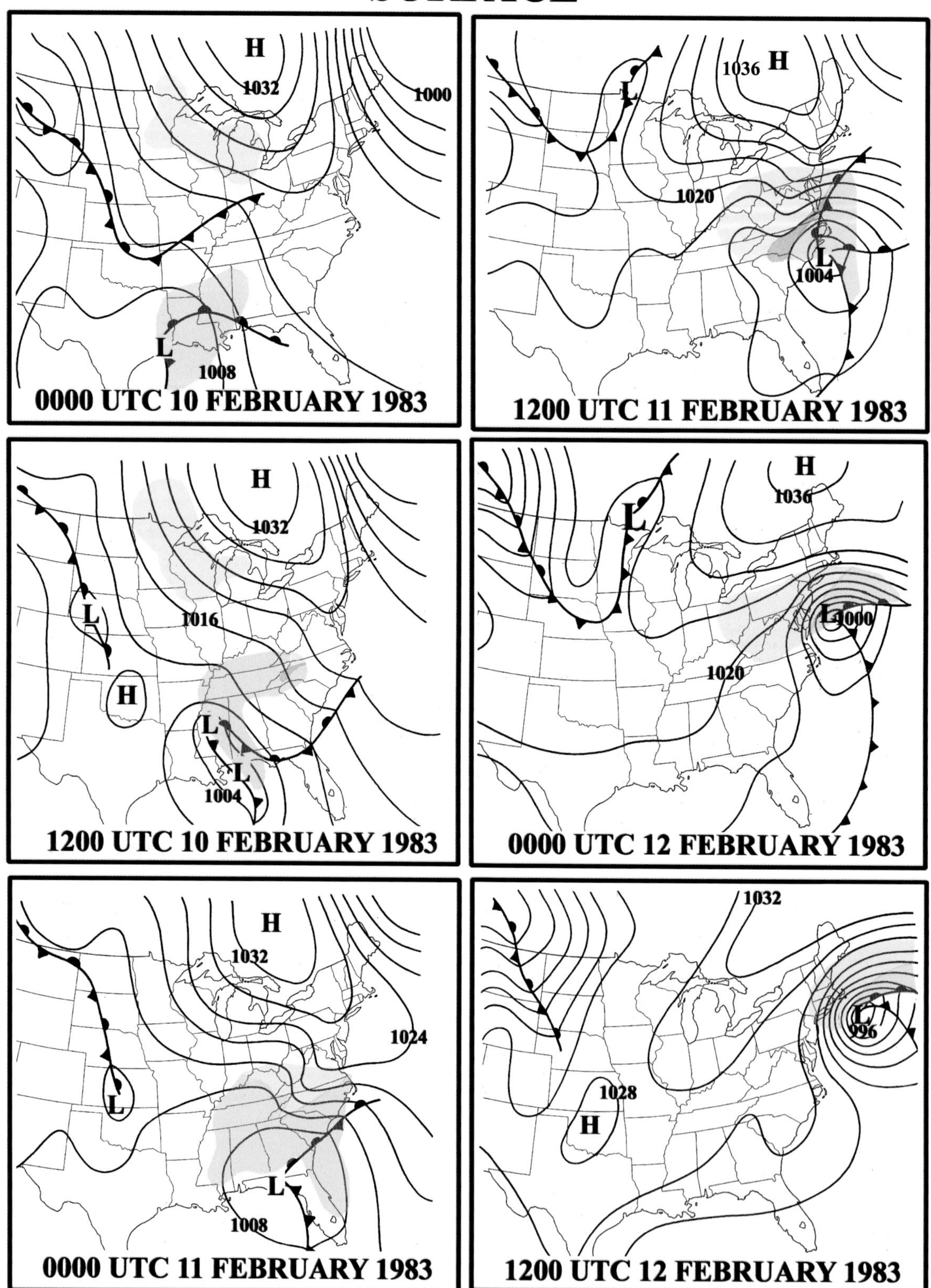

FIG. 10.20-2. Twelve-hourly surface weather analyses for 0000 UTC 10 Feb–1200 UTC 12 Feb 1983. See Fig. 10.1-2 for details.

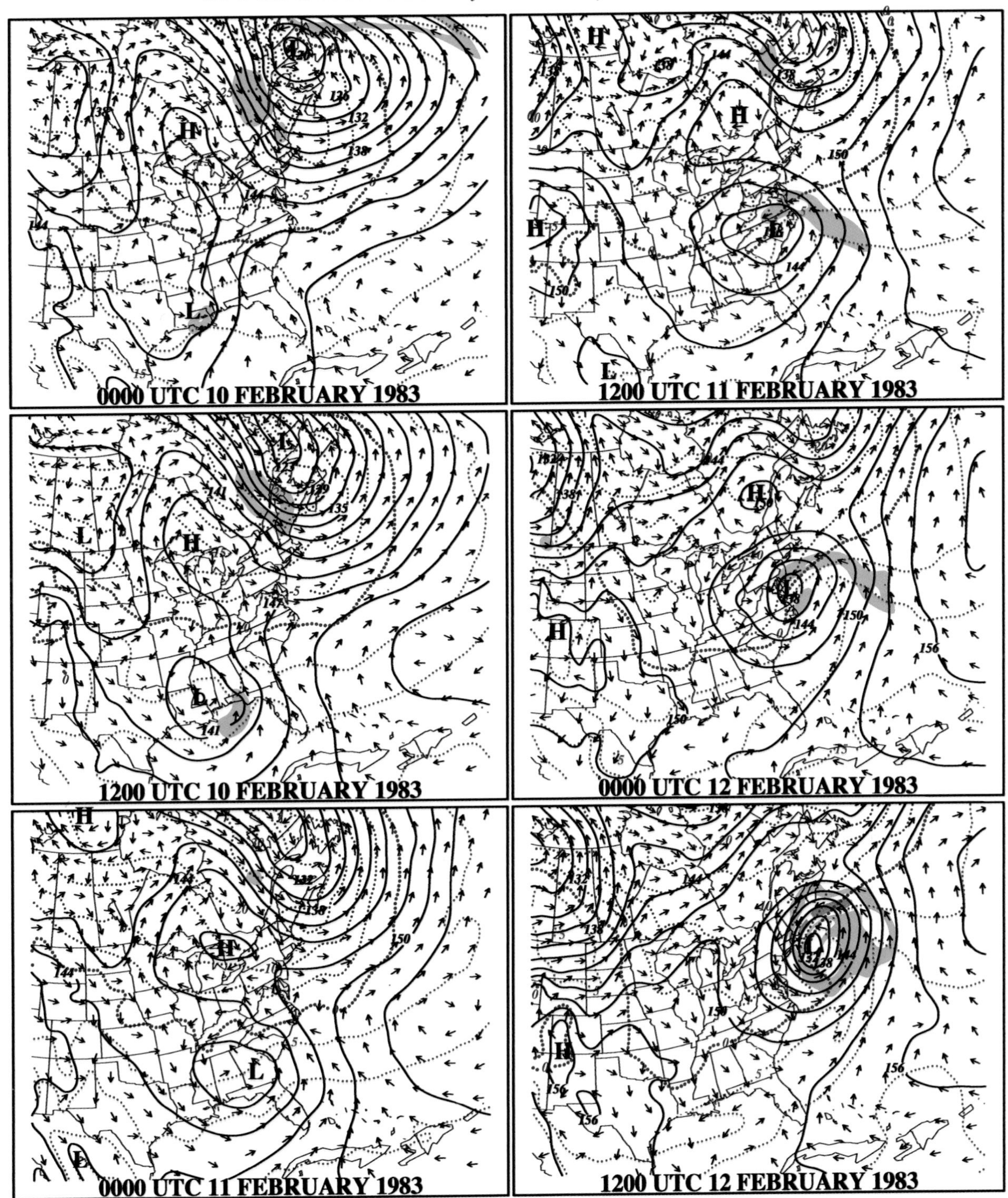

FIG. 10.20-3. Twelve-hourly 850-hPa analyses for 0000 UTC 10 Feb–1200 UTC 12 Feb 1983. See Fig. 10.1-3 for details.

500 hPa HEIGHTS AND VORTICITY, 400 hPa WINDS

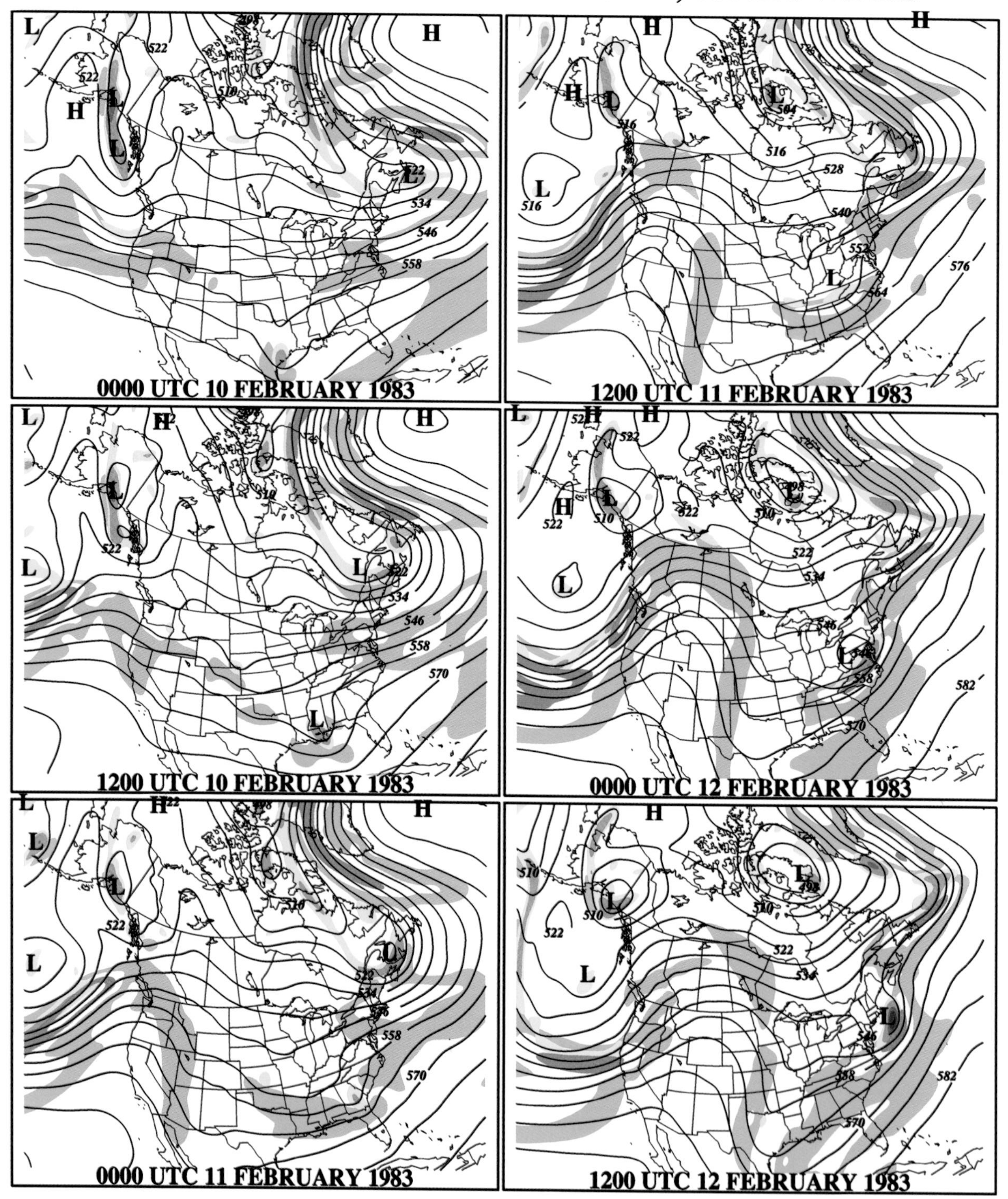

FIG. 10.20-4. Twelve-hourly analyses of 500-hPa geopotential heights and upper-level wind fields for 0000 UTC 10 Feb–1200 UTC 12 Feb 1983. See Fig. 10.1-4 for details.

250 hPa HEIGHTS AND WINDS

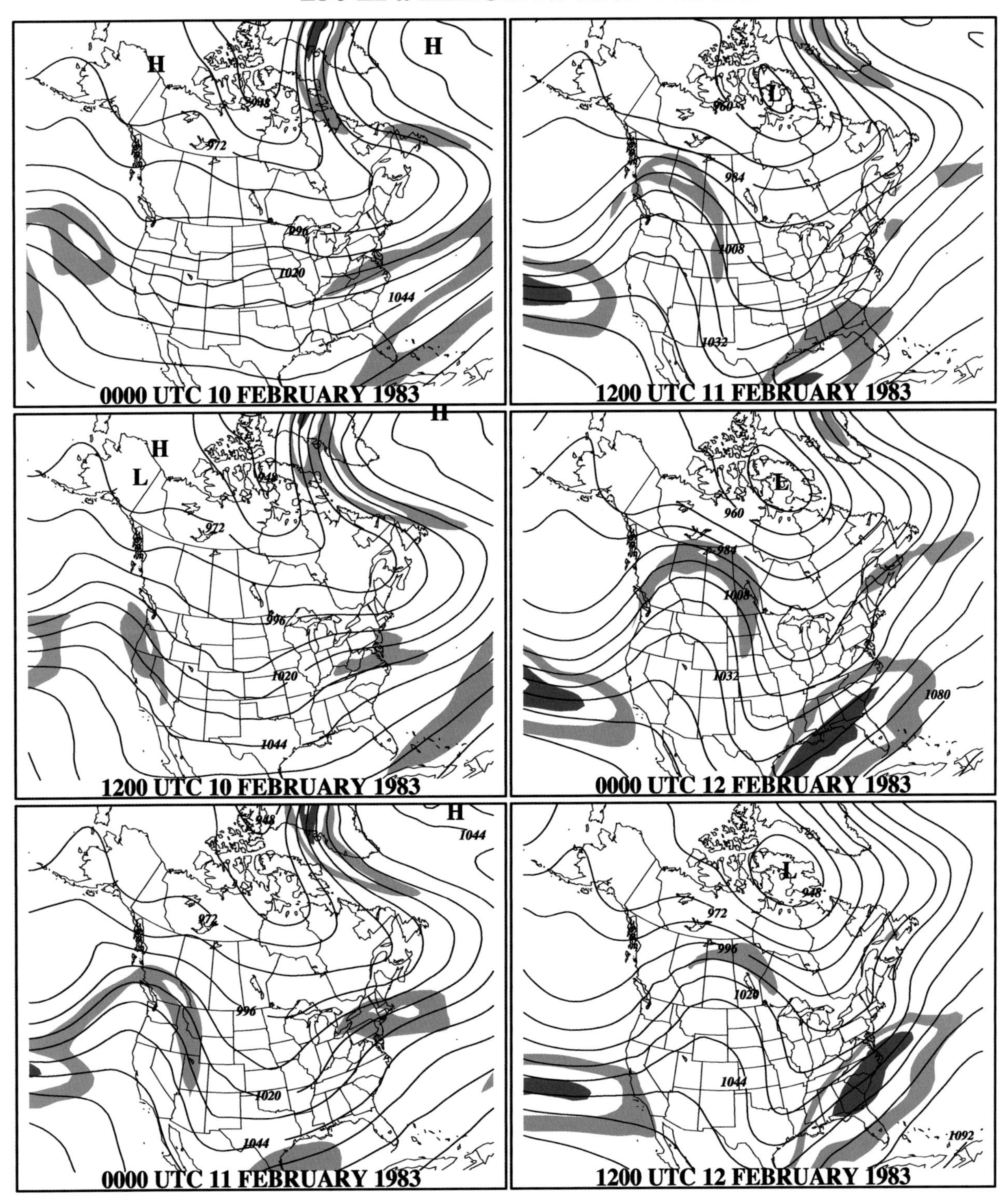

FIG. 10.20-5. Twelve-hourly analyses of 250-hPa geopotential height for 0000 UTC 10 Feb–1200 UTC 12 Feb 1983. See Fig. 10.1-5 for details.

SATELLITE IMAGERY

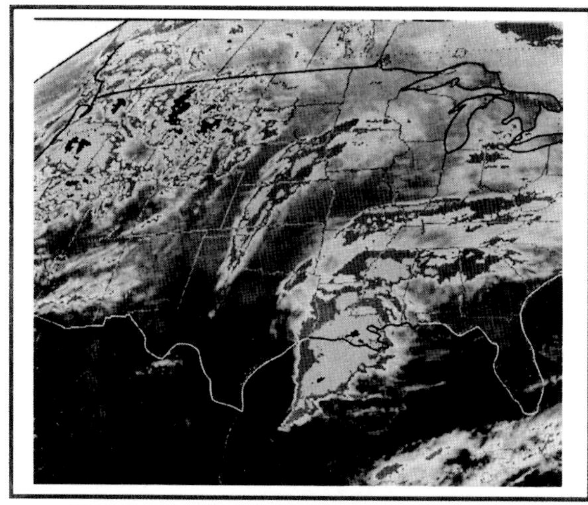

0000 UTC 10 FEBRUARY 1983

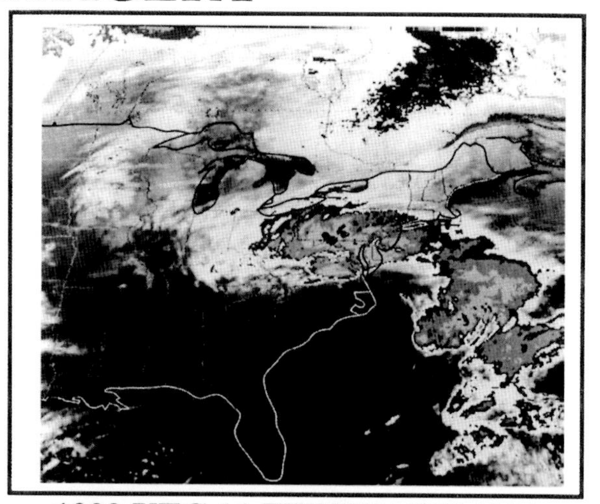

1300 UTC 11 FEBRUARY 1983

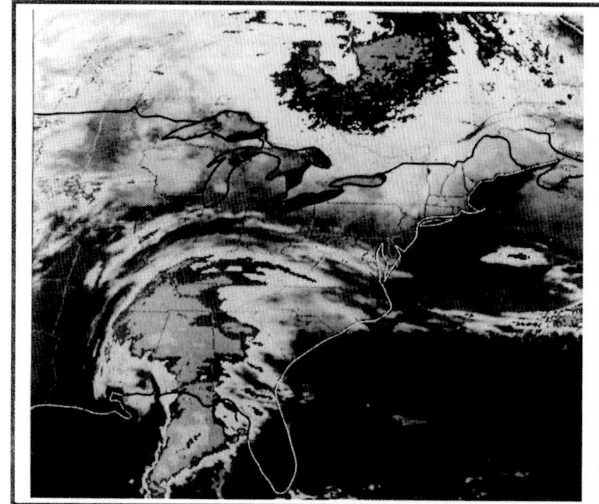

1200 UTC 10 FEBRUARY 1983

2300 UTC 11 FEBRUARY 1983

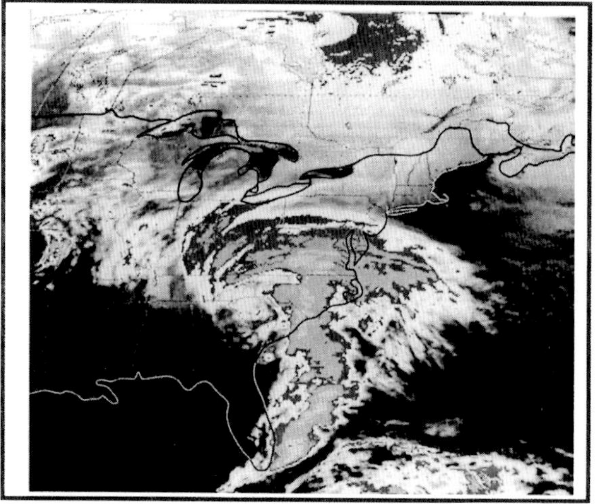

2330 UTC 10 FEBRUARY 1983

1230 UTC 12 FEBRUARY 1983

FIG. 10.20-6. Twelve-hourly infrared satellite images for 0000 UTC 10 Feb–1230 UTC 12 Feb 1983. See Fig. 10.16-6 for more details.

moisture flow into the middle Atlantic region for a protracted period.

- By 1200 UTC 12 February, strong winds circled the developing 850-hPa vortex, with easterly winds on the north side of the vortex exceeding 30 m s^{-1}, directed at the area of heaviest snowfall over eastern Massachusetts.

d. 500-hPa geopotential height and 400-hPa wind analyses

- This storm developed during one of the strongest El Niño's of the 20th century. It also occurred during one of the few periods in the months following October 1982 in which the North Atlantic Oscillation (NAO) was negative. A distinct negative NAO in the daily values began around 3 February and continued through much of February, reaching its lowest value near 10 February, then rising sharply through 12 February (but still remaining negative). The negative NAO was characterized by a large cutoff upper ridge at 500 hPa over the North Atlantic southeast of Greenland that moved eastward past Iceland by 12 February.
- Prior to cyclogenesis along the East Coast, an intense, nearly stationary 500-hPa anticyclone east of Greenland and a slow-moving trough across southeastern Canada dominated the circulation pattern. The trough in Canada combined with the general westerly flow across the northern United States to produce a pronounced confluent geopotential height pattern over the Great Lakes and northeastern United States on 10–11 February. The surface anticyclone and low-level cold-air advection were located beneath this confluence zone.
- A basically zonal flow pattern with several embedded weak troughs extended across the United States on 10 February. A ridge developed over the western United States on 11 February with two separate well-defined troughs developing east of the Rocky Mountains. One trough was associated with the East Coast storm on 11–12 February while the other produced a new low pressure center in the Gulf of Mexico on 12 February.
- The trough that accompanied the East Coast storm was associated with the merger of two distinct cyclonic vorticity maxima: one propagating across the Gulf states and the other moving east-southeastward across the central plains on 10 February. Between 0000 UTC 10 February and 0000 UTC 11 February, a pronounced pattern of diffluent geopotential heights developed immediately downwind of the trough axis over the southeastern United States. The surface low and heavy precipitation developed within this diffluence region in conjunction with cyclonic vorticity advection along the East Coast.
- The amplitude of the trough and downstream ridge increased only slightly through 1200 UTC 11 February. A closed cyclonic circulation appeared briefly over the middle Atlantic states at 0000 UTC 12 February.
- The half-wavelength between the trough and the downstream ridge decreased, especially between 1200 UTC 11 February and 0000 UTC 12 February, when heavy snow was falling across the middle Atlantic states and southern New England.
- More detailed diagnostic analyses of this case examining the influence of gravity wave phenomena, the possible roles of frontogenetical forcing and symmetric instability, and a detailed Doppler radar analysis can be found in Emanuel (1985), Bosart and Sanders (1986), and Sanders and Bosart (1985a,b). A description of the upper-level jet streaks and their associated vertical transverse circulations can be found in Uccellini and Kocin (1987) and is highlighted for this case in chapter 4 in volume I.
- The case is marked by two distinct jet systems: a northern jet located within the confluent zone preceding cyclogenesis and a southern jet associated with the trough moving toward the East Coast on 11 February.
- A 50 m s^{-1} polar jet located within the confluent flow over the northeastern United States coincided with the southeastward extension of the surface anticyclone over New England on 10–11 February. Uccellini and Kocin (1987) show that the direct transverse circulation in the confluent entrance region of this jet contributed to the low-level cold-air advection pattern in the northeastern United States that maintained the cold air that kept the precipitation as all snow in the Northeast urban corridor.
- Unlike many of the other cases, the jets propagating into the western portion of the trough that was associated with the East Coast storm were relatively weak (close to 50 m s^{-1}), especially before the merger of the two short-wave features at 0000 UTC 11 February.
- As shown by Uccellini and Kocin (1987), an indirect circulation in the exit region of the polar jet streak over the southeast United States at 1200 UTC 11 February accentuated the low-level warm-air advection and moisture transport toward the region of heavy snow over the middle Atlantic states. The developing cyclone and LLJ off the East Coast were both located within this diffluent exit region of this jet. The rising branches of the transverse circulations of both jet systems over New England combined to produce a sloped pattern of ascent collocated with the region of heavy snowfall that appeared to extend up toward the right-entrance region of the northern jet.
- By 1200 UTC 11 February and through 1200 UTC 12 February, the right-entrance region of the southern jet became laterally coupled.
- A jet streak with pronounced cross-contour flow propagated from the ridge crest in southwestern Canada toward the base of the second major trough over the southern United States by 0000 UTC 12 February. These features were associated with the development

of a new cyclone over the Gulf of Mexico late on 12 February.

e. 250-hPa geopotential height and wind analyses

- The height analyses clearly illustrate the pronounced trough extending into southeast Canada and general west-to-east flow that followed the precyclogenetic period for this case. The developing trough in the Southeast with increasing diffluence downwind of the trough axis on 10–11 February and enhanced confluence over the Northeast that mark other cases are evident by 0000 UTC 11 February.
- While there was an increase in the wind speeds over the Northeast at 0000 UTC 11 February as the precipitation shield expanded northward and the confluent flow increased, this change was rather modest as compared to other cases.
- The most significant jet feature at this level was the subtropical jet (STJ) that emerged from the Mexico–Texas border on 11 February and translated across the northern Gulf of Mexico to the Southeast coast by 1200 UTC 12 February. By 1200 UTC 11 February, the diffluent exit region of this STJ extended above the exit region of the polar jet propagating eastward, then northeastward along the East Coast during the same period.

f. Infrared satellite imagery sequence

- The shape of the cloud mass associated with the storm changed character during the life cycle of the cyclone. At 0000 UTC 10 February, the cloud mass was marked by a distinct western edge and a small, organized center over Louisiana, similar to Carlson's (1980) schematic for the initial stages of cyclone development. A distinct comma-shaped pattern emerged at 1200 and 2330 UTC 10 February as the surface low moved slowly eastward along the Gulf coast. The northern edge of the cloud shield was generally aligned with the axis of the confluent polar jet at 400 and 250 hPa over the northeast United States.
- At 2330 UTC 10 February, a comma-shaped cloud over the eastern United States was located downwind of the 500-hPa cyclonic vorticity maximum over Georgia. The larger cloud mass extended westward to a small comma-shaped cloud pattern over the Kansas–Missouri border, which was associated with a separate vorticity maximum located there.
- As the two cyclonic vorticity maxima merged into one over Tennessee by 1200 UTC 11 February, the cloud mass changed character again, stretching into a narrower east–west band of cold cloud tops in the 1300 UTC 11 February image. This band moved slowly northward into the northeastern United States by 2300 UTC 11 February, then drifted eastward off the New England coast by 1230 UTC 12 February, as heavy snow was ending across southeastern New England. The heaviest snow occurred along the southern edge of the high cloud tops associated with this band and was collocated with the axis of the 850-hPa LLJ from 11 to 12 February.
- The very cold cloud tops over New Jersey, southeastern New York, and New England at 2300 UTC 11 February reflect the embedded convection that was associated with the gravity wave and produced very heavy snowfall.
- The position of the surface cold front and low center corresponded well with the back edge of the enhanced cloud mass at 0000 and 1200 UTC 10 February. At later times, the surface low center was displaced south and west of the region of high cloud tops, as a wedge of dry air moved over the storm center from the southwest.

Jogger on Delacourt Circle, Central Park, Jan 1987 (copyright Alan Schein, courtesy of Kevin Ambrose, *Great Blizzards of New York City*, Historical Enterprises, 1994).

21. 21–23 January 1987

a. General remarks

- The winter of 1986/87 will be remembered for its many heavy snowstorms over the middle Atlantic states and southern New England in January and February.
- This storm yielded heavy snowfall from Alabama and Georgia to New York and portions of New England. New York City, Philadelphia, Baltimore, and Washington experienced their largest snowfall in nearly four years. Snow accumulations were underpredicted by many forecasters who anticipated a changeover from snow to rain. In New York City, the changeover did occur, but not until 8 to 12 in. (20–30 cm) of snow had already fallen. Some lightning and thunder were reported from Virginia to New York on 22 January.
- The following regions reported snow accumulations exceeding 10 in. (25 cm): Virginia (except the southeast), the southeastern half of West Virginia, Maryland (except the eastern shore), northern Delaware, central and eastern Pennsylvania, western and northern New Jersey, central and eastern New York, northwestern Connecticut, western Massachusetts, much of New Hampshire and Vermont, and scattered area of Maine.
- The following regions reported snow accumulations exceeding 20 in. (50 cm): scattered sections of Virginia, Pennsylvania, and New York.

TABLE 10.21-1. Snowfall amounts for urban and selected locations for 21–23 Jan 1987; see Table 10.1-1 for details.

NESIS = 4.93 (category 3)	
Urban center snowfall amounts	
Washington, D.C.–Dulles Airport	11.1 in. (28 cm)
Baltimore, MD	12.3 in. (31 cm)
Philadelphia, PA	8.8 in. (22 cm)
New York, NY–La Guardia Airport	11.3 in. (29 cm)
Boston, MA	5.3 in. (11 cm)
Other selected snowfall amounts	
Albany, NY	16.6 in. (42 cm)
Williamsport, PA	15.8 in. (40 cm)
Patuxent River, MD	13.0 in. (33 cm)
Lynchburg, VA	12.3 in. (31 cm)
Wilmington, DE	12.1 in. (31 cm)
Harrisburg, PA	11.4 in. (29 cm)
Allentown, PA	11.1 in. (28 cm)
Newark, NJ	11.0 in. (28 cm)
Hartford, CT	10.2 in. (26 cm)

b. Surface analyses

- The cold air for this snowstorm was associated with a relatively weak anticyclone (1022 hPa) that propagated from the southern plains northeastward across the middle Atlantic states and New England to eastern Canada. This differs markedly from the large, intense anticyclones (>1035 hPa) moving across southern Canada that characterized most of the other cases (see volume I, chapter 3.3).
- An inverted sea level high pressure ridge extending from New Jersey to North Carolina indicated cold-air damming at 1200 UTC 22 January. The appearance of this high pressure ridge coincided with the development of a pronounced coastal front parallel to the southeastern United States coast, along which the surface cyclone advanced.
- As the cyclone developed in the western Gulf of Mexico, it moved very slowly, averaging only 10 m s^{-1} between 1200 UTC 21 January and 0000 UTC 22 January. The low center then accelerated northeastward as it neared northern Florida, averaging 18–20 m s^{-1} in the 12-h period ending at 1200 UTC 22 January. As the cyclone reached northern Florida, a new center formed to its northeast along the coastal front off Georgia and South Carolina. Moderate to heavy snow spread from Georgia to Virginia and Maryland as the new center developed. The initial low was absorbed into the new circulation as the second low center began deepening rapidly.
- The low center moved northeastward along the coastal front through eastern North Carolina on 22 January, reaching a position just off the New Jersey coast by 0000 UTC 23 January. The center advanced at a rate of 18–20 m s^{-1} over the 12-h period ending at 0000 UTC 23 January. The track of the low remained close to the coastline, causing a changeover from snow to rain along the immediate coast of the middle Atlantic states. The cyclone then crossed central Long Island and central and eastern New England, reaching southern Maine by 1200 UTC 23 January. The forward speed of the storm slowed as it crossed New England, averaging only 12 m s^{-1}. The onshore path of the center brought a changeover from snow to rain from Delaware to New York City across eastern New England. Accumulations of greater than 10 cm generally preceded the changeover, except in extreme southeastern Massachusetts, where little snow fell.
- A 300-km-wide swath of 10-in. (25 cm) snowfall was oriented nearly parallel to the path of the surface low. The easternmost edge of the 10-in. (25 cm) accumulations was generally located between 30 and 160 km west of the low track.
- At first, the cyclone deepened slowly as it crossed the Gulf of Mexico, with the central pressure falling from 1007 hPa at 1200 UTC 21 January to 1003 hPa by 0000 UTC 22 January, and to 996 hPa by 1200 UTC 22 January. As the low redeveloped and accelerated along the coastal front and moved to a position east of New Jersey, it underwent rapid deepening, with the central pressure falling 20 hPa in only 12 h, reaching 975 hPa by 0000 UTC 23 January. The storm continued to intensify as it crossed Long Island and New England, attaining a minimum pressure of 966 hPa over Maine. The rapid development phase was ac-

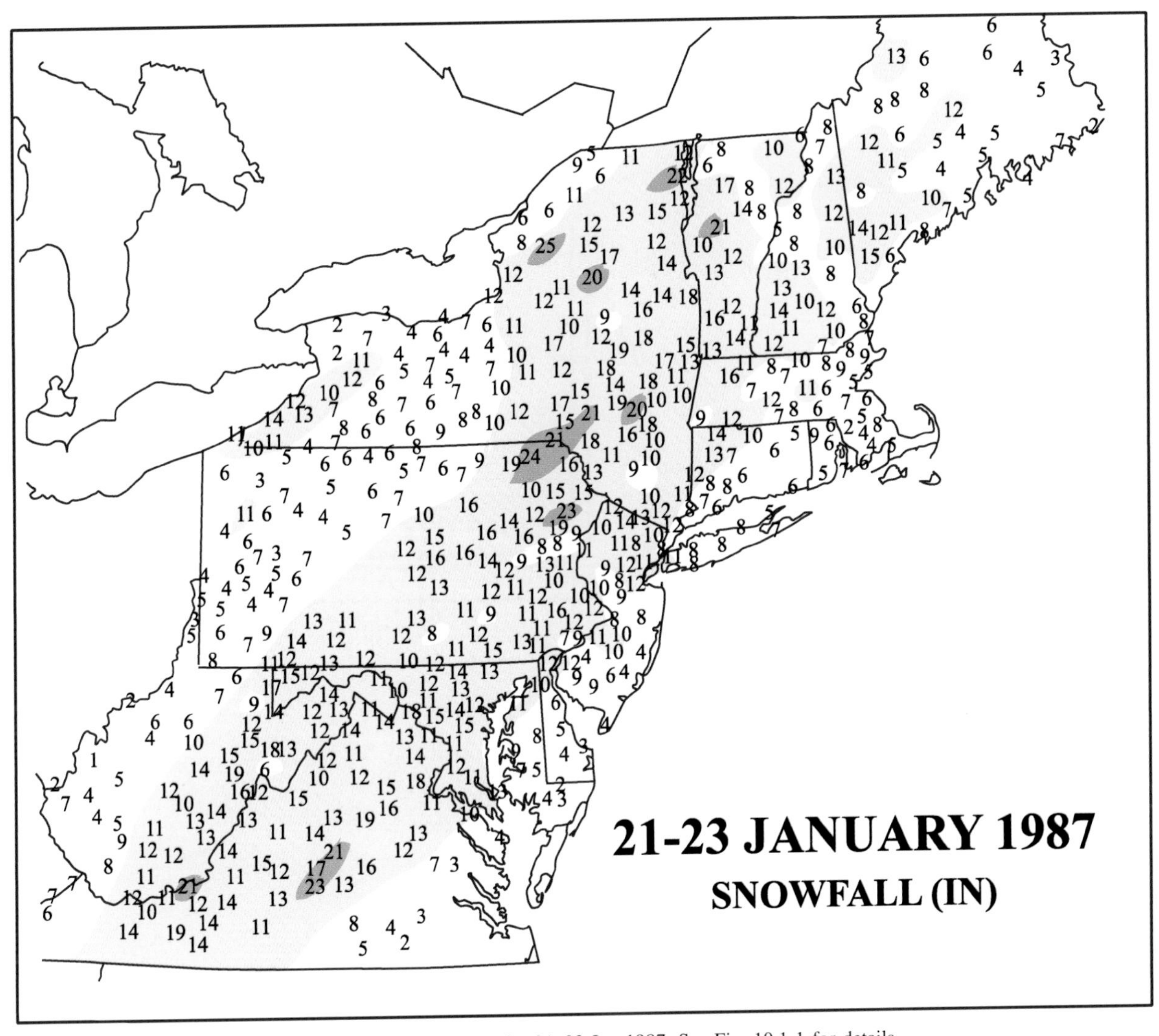

FIG. 10.21-1. Snowfall (in.) for 21–23 Jan 1987. See Fig. 10.1-1 for details.

companied by snowfall rates of 1–3 in. (2.5–7.5 cm) per hour from North Carolina to New England.

- A dissipating cold front ushered in the coldest air of the season in the wake of the storm, resulting in persistent winter weather conditions across the middle Atlantic states and New England, and setting up the middle Atlantic states for another major winter storm 2 days later.

c. 850-hPa analyses

- A wedge of cold 850-hPa temperatures (−10° to −15°C) advanced from the Great Lakes into the middle Atlantic states and New England on 21 January behind a weak upper-level trough crossing the northeastern United States and eastern Canada. This pool of cold air was located beneath a region of confluence upwind of the 500-hPa trough through 0000 UTC 22 January. Unlike many of the other cases described in this chapter, however, the cold-air advection was associated with moderate westerly flow at 850 hPa, not with strong northwesterly flow or with an intense surface anticyclone.
- The 0°C isotherm extended from the Texas coast across North Carolina on 21 January, prior to the development of the 850-hPa low in the southeastern United States. As the cyclone moved northeastward, the 0°C isotherm remained nearly stationary through 1200 UTC 22 January. It then buckled into the familiar S-shaped pattern by 0000 UTC 23 January, as warm-air advection increased in advance of the 850-hPa low center and cold-air advection increased to its rear. The increasing warm-air advection from 1200 UTC 22 January through 0000 UTC 23 January was accompanied by heavy snowfall rates from the Carolinas to New England.
- The 850-hPa low center deepened at an accelerating

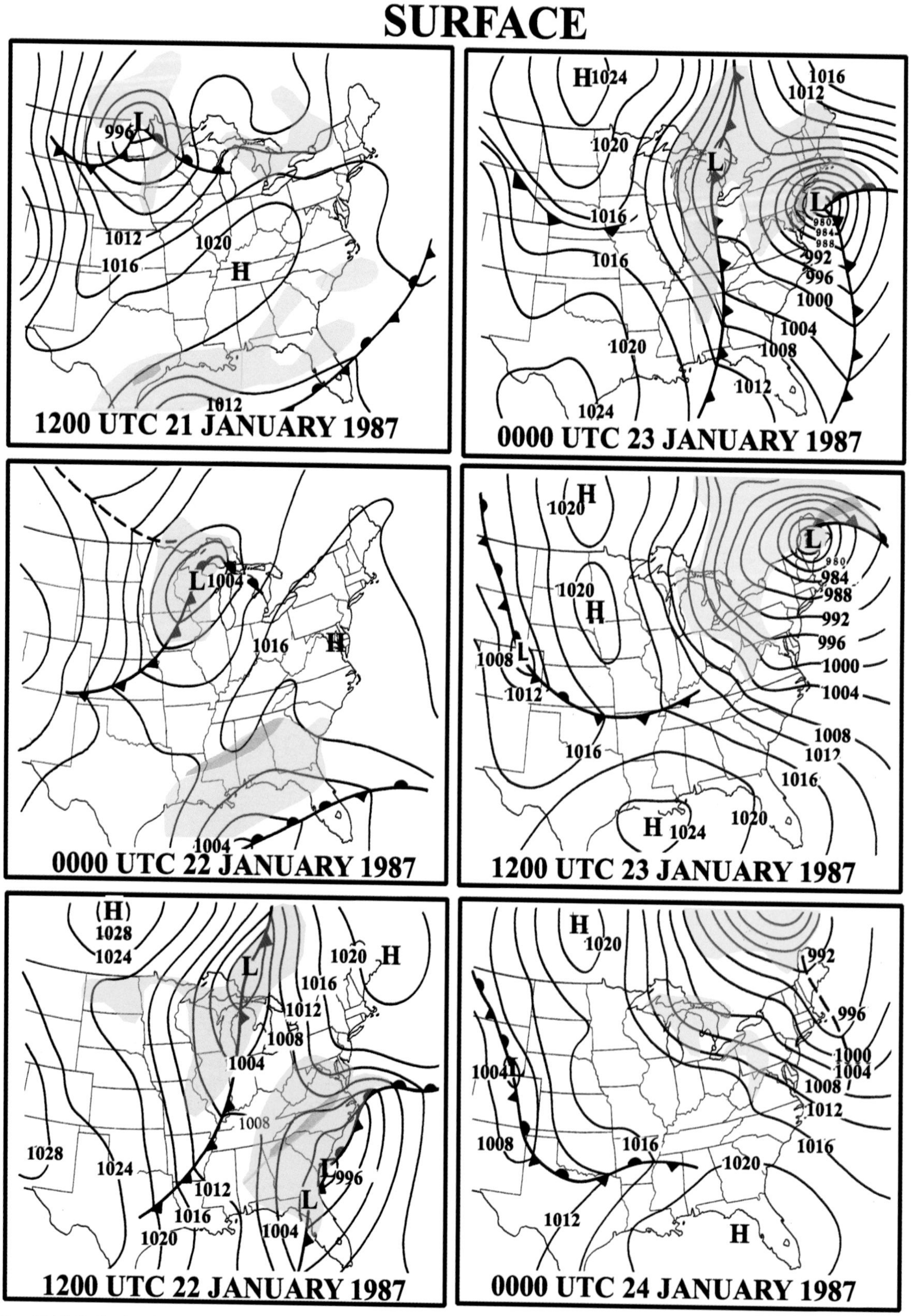

SURFACE

1200 UTC 21 JANUARY 1987

0000 UTC 23 JANUARY 1987

0000 UTC 22 JANUARY 1987

1200 UTC 23 JANUARY 1987

1200 UTC 22 JANUARY 1987

0000 UTC 24 JANUARY 1987

FIG. 10.21-2. Twelve-hourly surface weather analyses for 1200 UTC 21 Jan–0000 UTC 24 Jan 1987. See Fig. 10.1-2 for details.

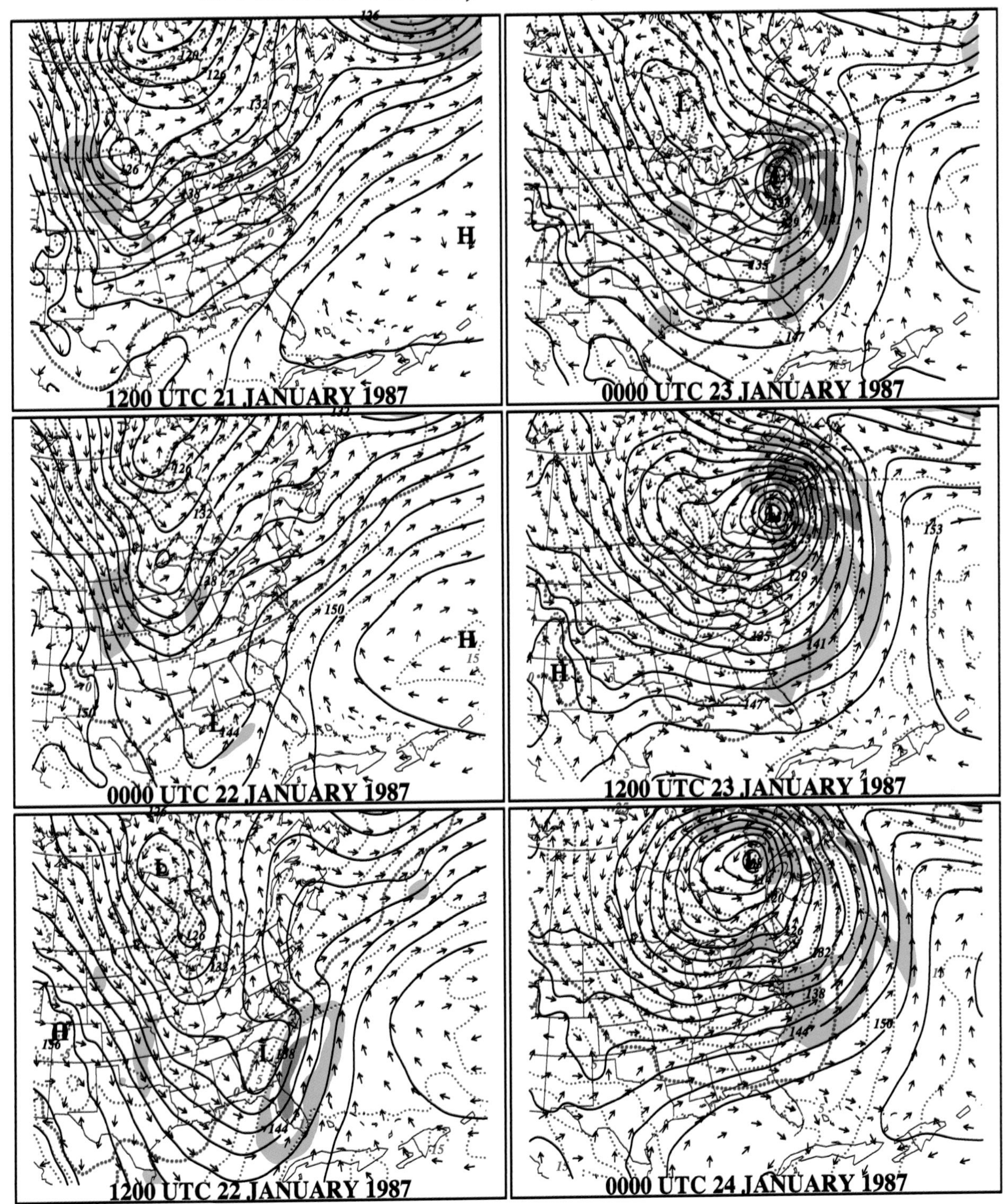

FIG. 10.21-3. Twelve-hourly 850-hPa analyses for 1200 UTC 21 Jan–0000 UTC 24 Jan 1987. See Fig. 10.1-3 for details.

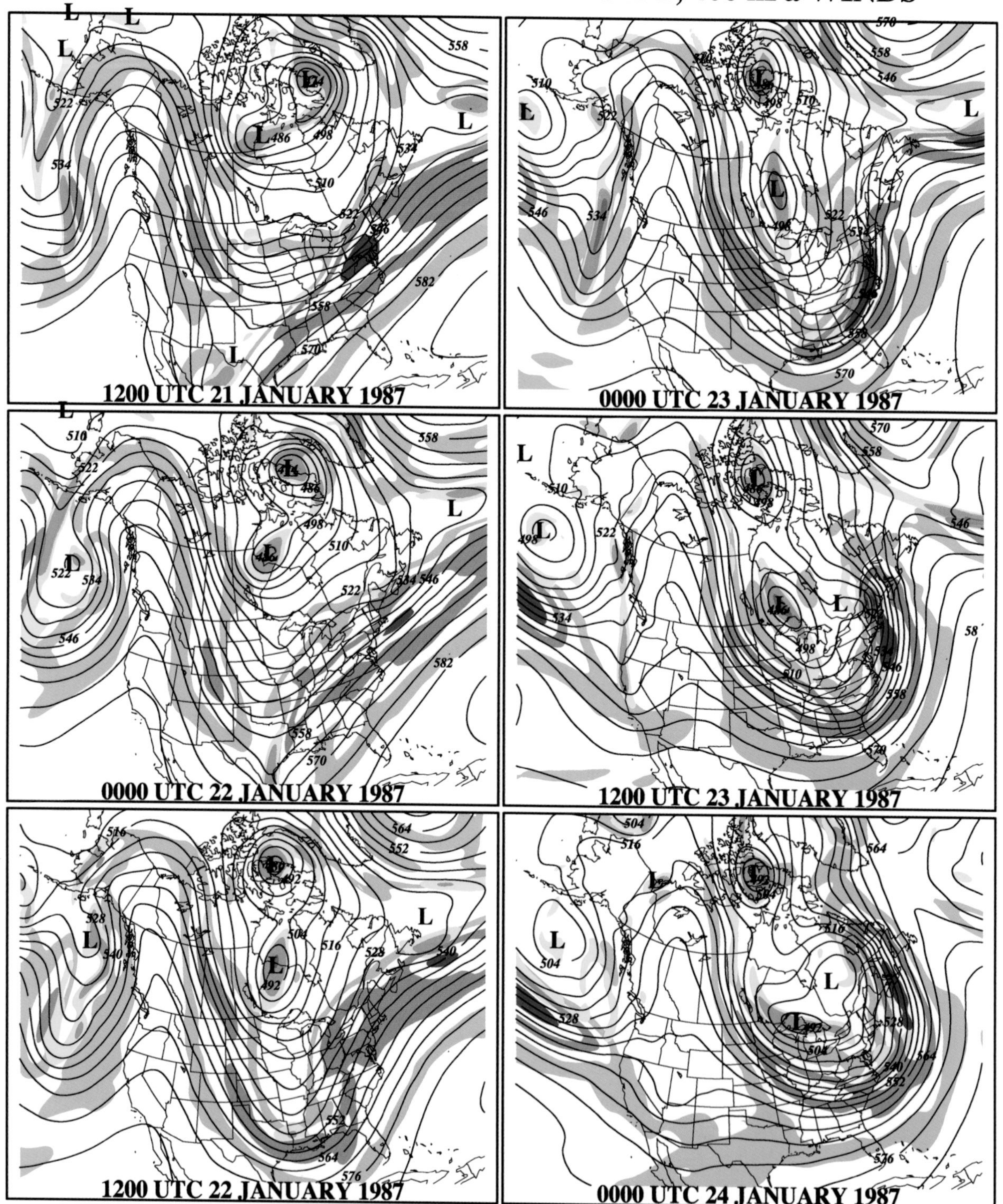

FIG. 10.21-4. Twelve-hourly analyses of 500-hPa geopotential heights and upper-level wind fields for 1200 UTC 21 Jan–0000 UTC 24 Jan 1987. See Fig. 10.1-4 for details.

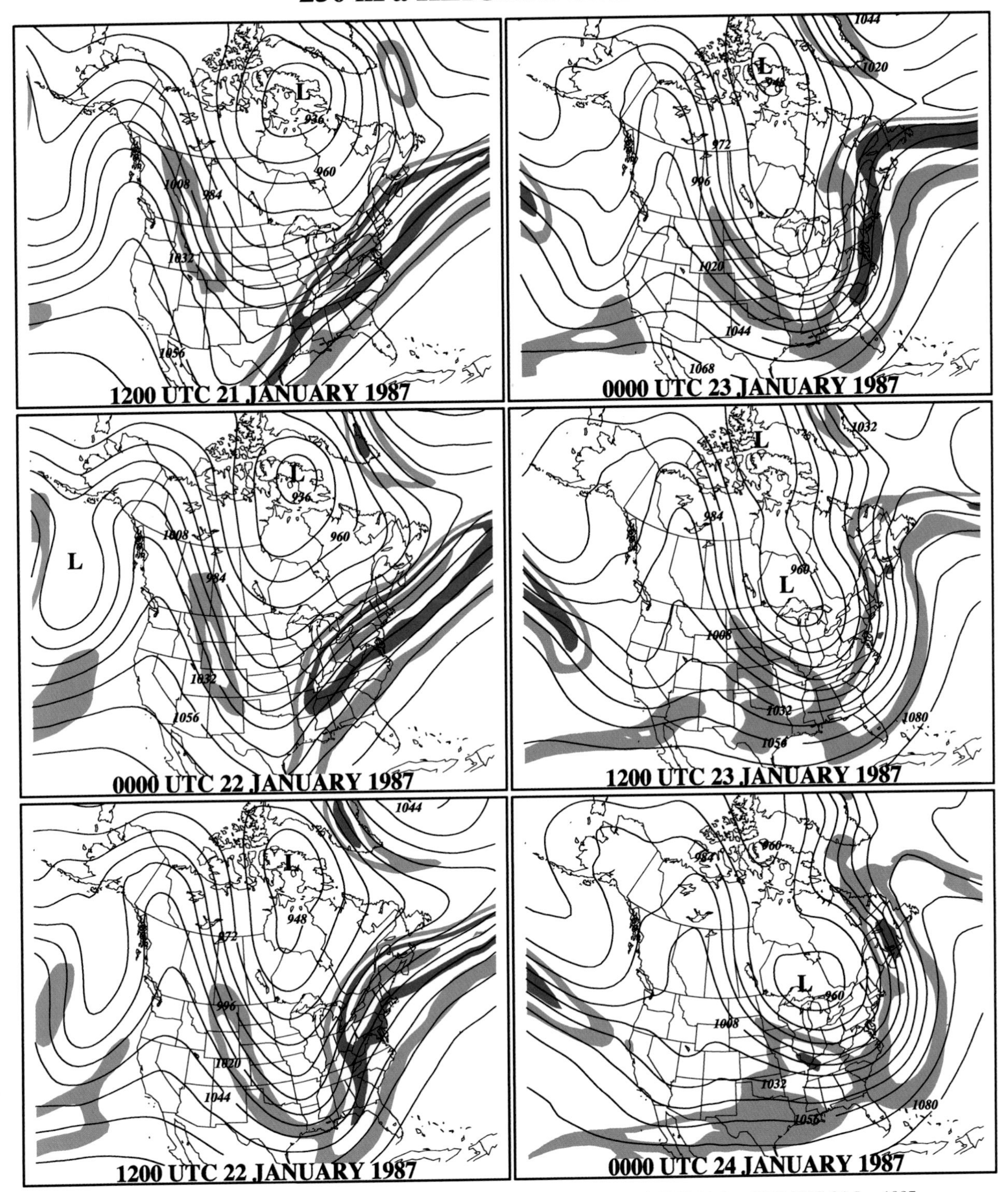

FIG. 10.21-5. Twelve-hourly analyses of 250-hPa geopotential height for 1200 UTC 21 Jan–0000 UTC 24 Jan 1987. See Fig. 10.1-5 for details.

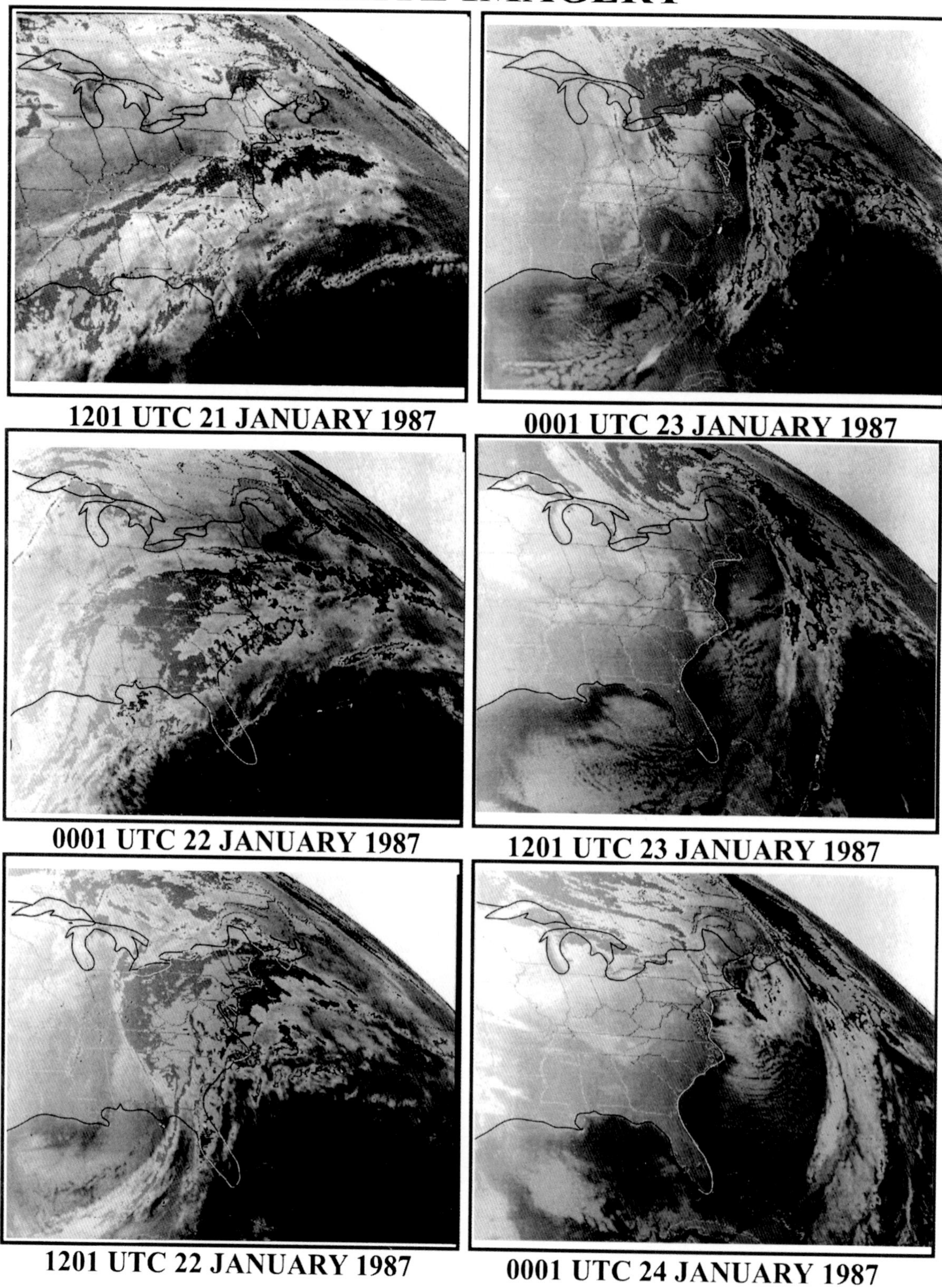

FIG. 10.21-6. Twelve-hourly infrared satellite images for 1201 UTC 21 Jan–0001 UTC 24 Jan 1987. See Fig. 10.16-6 for more details.

rate as it progressed northeastward from the Gulf of Mexico to Maine from 21 through 23 January. The deepening rate increased to −90 m per 12 h in the period ending at 1200 UTC 22 January, and to −180 m per 12 h in the following 12-h period, as the cyclone moved northeastward along the Atlantic coast. The increased deepening rate was reflected in the amplifying local height falls observed around the low center.

- A south-to-southwesterly LLJ began developing along the Gulf coast at 0000 UTC 22 January, amplified, and shifted to the Southeast coast by 1200 UTC 22 January, immediately upwind of the heaviest precipitation. As the storm intensified rapidly, easterly winds to the north of the 850-hPa low center increased to 30–35 m s^{-1} by 0000 and 1200 UTC 23 January.

d. 500-hPa geopotential height and 400-hPa wind analyses

- This storm and the following two snowstorms occurred during a winter marked by the strongest El Niño since 1982–83 (Fig. 2–19, volume I) and was a particularly stormy winter. While 1986 ended with 3 months dominated by the positive phase of the North Atlantic Oscillation (NAO), the NAO became negative around 1 January and remained mostly negative through mid-March. Daily values of the NAO were becoming increasingly negative during the period in which the storm developed and moved up along the East Coast, reaching its nadir on 25 January. Analyses of 500-hPa data show an upper trough over Greenland on 21 January that yields to a building ridge centered east of Iceland by 0000 UTC 23 January, suggestive of the negative NAO.
- The cyclone developed as a deepening large-scale 500-hPa trough was becoming established over the central and eastern United States on 21–22 January downwind of a high-amplitude ridge over western North America. A major 500-hPa anticyclone became established over the North Atlantic Ocean east of Greenland on 21–22 January.
- Prior to cyclogenesis along the East Coast on 22 January, a region of confluence was observed behind a 500-hPa trough moving rapidly along the northeastern United States–Canada border. Rising sea level pressures and a weak surface anticyclone extended northeastward within this confluence zone and supplied cold air for this snow event.
- The surface cyclone developed in the Gulf of Mexico as a trough initially over Mexico began to move eastward and merged with the larger-scale trough developing over the central United States on 22 January with a distinct vorticity maximum located over Louisiana by 1200 UTC 22 January 1987.
- By 0000 and 1200 UTC 23 January, a distinct 500-hPa height minimum and associated cyclonic vorticity maximum moved northeastward along the East Coast, as the surface cyclone reached its lowest central pressure over northern New England.
- A diffluent geopotential height pattern was not evident downwind of the trough axis until 0000 UTC 23 January, after the onset of rapid sea level development and the entire system moved toward New England.
- During the 24-h period between 1200 UTC 22 January and 1200 UTC 23 January, the 500-hPa trough rotated from a north–south line (from Wisconsin to Louisiana), to a northwest–southeast axis (a negative tilt), then to a west–east orientation (from southern Quebec to Maine). This tendency was accentuated by the interaction of the short-wave trough with a major trough moving south from Canada to the Great Lakes and deepening on 23 January. Concurrently, the surface low moved from Florida to Maine and deepened nearly 30 hPa.
- The amplitude of the trough and downstream ridge increased through 0000 UTC 23 January, with a concurrent decrease of the half-wavelength between the axes.
- The jet streak aspects of this case were not like many others, in that a southwest–northeast-oriented jet extending from Texas to Maine dominated the precyclogenetic period, rather than a confluent jet oriented west–east across the northeast United States. As the trough deepened over the southeast United States, a distinct southern jet system developed during and after surface cyclogenesis.
- A jet streak with wind speeds exceeding 70 m s^{-1} was located near the base of the trough moving rapidly along the United States–Canada border on 21 January. The surface anticyclone built northeastward from the south-central United States within the confluent entrance region of this intense jet streak.
- Two separate jet streaks were observed over the eastern United States by 1200 UTC 22 January, as rapid cyclogenesis commenced along the southeast coast. The rapidly expanding area of heavy snowfall from Georgia to Maryland appeared to lie between the entrance region of the jet over the Northeast and the exit region of the jet that extended from the eastern Gulf of Mexico to southeastern Virginia. Unfortunately, the detailed structure of the jet streak along the East Coast may be difficult to define beyond this reanalysis, since many upper-level wind reports were missing. Maximum wind speeds within this jet exceeded 70 m s^{-1} by 0000 UTC as the surface cyclone deepened rapidly off the middle Atlantic coast on 23 January.
- A jet streak with wind speeds of at least 50–60 m s^{-1} was located to the west of the deepening large-scale trough at all times. A separate jet streak located over central Canada through 1200 UTC 22 January was associated with a southward-propagating vortex that maintained very cold conditions over the northeastern United States during the following week.

e. 250-hPa geopotential height and wind analyses

- The upper-tropospheric geopotential analyses clearly show the ridging along the West Coast, the deepening of the downstream trough into the northern Gulf of Mexico as the surface cyclogenesis commenced in that region by 0000 UTC 22 January, and the merger with a second deepening trough over the Great Lakes by 0000 UTC 24 January.
- Perhaps the most important feature gleaned from this analysis is the southwest–northeast-oriented 90–100 m s^{-1} jet streak extending from the south-central United States northeastward toward New England and southeastern Canada on 22 January, within a noticeably confluent region.
- The expanding precipitation shield on 22 January appears to extend northeastward between 1200 UTC 21 January and 1200 UTC 22 January, along with the elongated surface low pressure system developing along the East Coast, within the right-entrance region of this intense jet streak.
- As the trough deepened over the United States by 1200 UTC 22 January and a cutoff low developed over the Great Lakes, a separate, more polar, jet system propagated into the central United States. This jet increased to greater than 70 m s^{-1} by 0000 UTC 24 January, well after the snowstorm had moved into eastern Canada and cold air had settled over the eastern half of the nation. This jet and a short-wave trough over the Rocky Mountains at 0000 UTC 24 January played a significant role in the development of the next snowstorm in the middle Atlantic states on 25–27 January.

f. Infrared satellite imagery sequence

- The infrared satellite images show a cloud pattern similar to that exhibited by the storm of 19–20 January 1978, with the rapid expansion of high clouds well to the north of the developing surface cyclone. The rapid expansion of the cloud mass occurred within the entrance region of the intense jet streak over the Ohio Valley. The northern and western edges of this cloud mass were aligned with the axis of the 90–100 m s^{-1} upper-level jet extending from the Ohio Valley to New England at 0000 and 1200 UTC 22 January.
- At 1200 UTC 22 January, much of the heavy precipitation was located from beneath the highest cloud tops southward to the edge of the highest cloud tops over Georgia. By 0000 UTC 23 January, however, the heavy precipitation was located primarily under the warmer cloud-top region over the middle Atlantic states and New England as an easterly 30–35 m s^{-1} LLJ emerged north of the 850-hPa low. A distinct comma shape defined the cloud mass at that time.
- A comma shape to the cloud mass began to appear as dry air began to race northeast over the northeastern Gulf of Mexico at 1200 UTC 22 January, southwest of the developing surface low. This dry-air wedge expanded and raced northward in the following 12 h, probably in association with the second, more southern, upper-level jet that developed on the eastern flank of the upper trough after 1200 UTC 22 January.

Silver Spring, Maryland, 26 Jan 1987 (photo courtesy of Paul Kocin).

22. 25–27 January 1987

a. General remarks

- The second major snowstorm in 4 days severely disrupted the middle Atlantic region but missed most of New England, except southeastern Massachusetts. Although the heavy snow was not as widespread as in the previous storm, it combined with the deep snow cover already in place and some of the coldest temperatures of the season to produce severe winter weather conditions from Virginia to New Jersey. Washington and its suburbs were at a standstill following these two storms, with schools closed for a week or more. In a continuing tradition pitting snowstorms against urban mayors (i.e., Mayor Lindsay in New York City following the 9–10 February 1969 snowstorm, and Mayor Jane Byrne, following a blizzard in Chicago during January 1979), this snowstorm turned into a political hot potato for Washington Mayor Marion Barry who was out of town during this and the previous snowstorm.
- The following regions reported snow accumulations exceeding 10 in. (25 cm): southeastern West Virginia, central and northern Virginia, central and southern Maryland, Delaware (except the extreme north), southern New Jersey, eastern Long Island, and southeastern Massachusetts.
- The following regions reported snow accumulations exceeding 20 in. (50 cm): none.

TABLE 10.22-1. Snowfall amounts for urban and selected locations for 25–27 Jan 1987; see Table 10.1-1 for details.

NESIS = 1.70 (category 1)	
Urban center snowfall amounts	
Washington, D.C.–Dulles Airport	10.2 in. (26 cm)
Baltimore, MD	9.6 in. (24 cm)
Philadelphia, PA	4.1 in. (10 cm)
New York, NY–Avenue V, Brooklyn	4.1 in. (10 cm)
Boston, MA	3.3 in. (8 cm)
Other selected snowfall amounts	
Patuxent River, MD	18.0 in. (46 cm)
Salisbury, MD	17.1 in. (43 cm)
Atlantic City, NJ	16.3 in. (41 cm)
Edgartown, MA	16.0 in. (41 cm)
Roanoke, VA	13.9 in. (35 cm)
Dover, DE	12.0 in. (30 cm)
Bridgehampton, NY	11.0 in. (28 cm)

- A band of light to moderate snowfall developed across the middle Atlantic states on 25 January, north of a stationary front anchored over Tennessee and North Carolina. As the cyclone neared the coast, heavy snowfall developed from Virginia into southern Maryland, Delaware, and central and southern New Jersey. The heavy snow area expanded east-northeastward across eastern Long Island and extreme southeastern New England by 1200 UTC 26 January, as the storm moved out over the Atlantic Ocean.
- The cyclone deepened very slowly as it crossed the southeastern United States. The central sea level pressure fell from 1006 to 1003 hPa between 0000 UTC 25 January and 0000 UTC 26 January. As the surface low turned northeastward along the Carolina coast, a period of more significant deepening began. The central sea level pressure fell from 1003 hPa at 0000 UTC 26 January to 990 hPa east of Virginia at 1200 UTC 26 January. This period of relatively rapid deepening coincided with the heavy snowfall across Virginia, Maryland, and Delaware.
- The heaviest snow fell with a well-defined 150-km-wide band centered parallel to and about 300 km to the north of the surface low track.

b. Surface analyses

- Cold air for this snowfall was established over the north-central and northeastern United States following the storm of 21–23 January. A broad area of high pressure (1025–1030 hPa) extended from the northern plains to the middle Atlantic states and New England on 25–26 January. Another high pressure cell emerged and drifted from the upper Midwest southward to the Gulf coast on 26–27 January.
- Cold-air damming, marked by an inverted sea level pressure ridge, developed over the middle Atlantic states by 0000 UTC 26 January. Coastal frontogenesis also occurred by this time along the coast of the Carolinas.
- The surface low advanced from eastern Texas to eastern Georgia in the 24-h period ending at 0000 UTC 26 January, averaging 12 m s^{-1}. A center "jump" occurred at 0000 UTC 26 January as a new low developed along the coastal front over eastern South Carolina. This low moved to a position near Cape Hatteras, North Carolina, by 0600 UTC, then east-northeastward over the Atlantic Ocean, passing approximately 200 km southeast of southern New England. The center propagated at a rate of about 15 m s^{-1} as it moved from South Carolina to a position about 340 km east of the Virginia coast in the 12-h period ending at 1200 UTC 26 January.

c. 850-hPa analyses

- West-to-northwesterly flow and strong cold-air advection brought an extensive area of −10° to −25°C 850-hPa temperatures over the middle Atlantic states and New England by 25 January. A large temperature gradient had developed north of the 0°C isotherm, from the lower Ohio Valley across the middle Atlantic states, by 0000 UTC 25 January.
- The 0°C isotherm generally extended from the 850-hPa low center in the south-central United States eastward to the Carolina coast at 0000 and 1200 UTC 25 January. An S-shaped isotherm pattern became more evident by 0000 UTC 26 January, as the warm-air advection east of the 850-hPa low and cold-air ad-

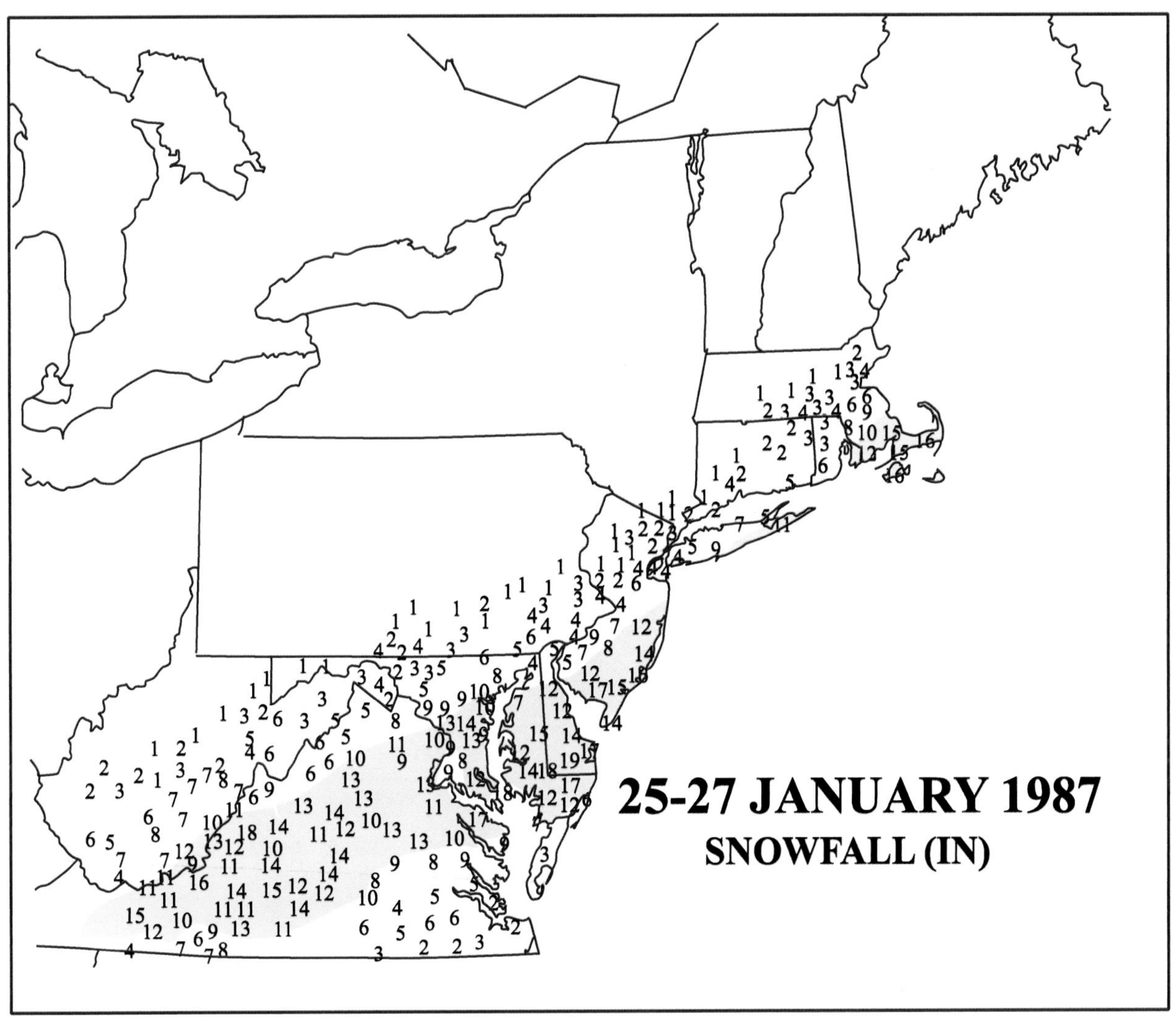

FIG. 10.22-1. Snowfall (in.) for 25–27 Jan 1987. See Fig. 10.1-1 for details.

vection to the west both strengthened. The increasing warm-air advection was accompanied by an increase of the temperature gradient along the Southeast coast, south of the 0°C isotherm. The intensifying temperature gradient and warm-air advection were associated with the development of heavy snowfall over the middle Atlantic states by 0000 UTC 26 January.

- The 850-hPa low developed slowly over the south-central United States on 25 January, similar to its surface counterpart. Very little deepening was observed through 0000 UTC 26 January, with moderate deepening of -60 to -90 m (per 12 h) thereafter.
- A 20–25 m s^{-1} southerly low-level jet (LLJ) was observed over the Gulf coast region through 1200 UTC 25 January, directed toward the expanding precipitation area east and northeast of the surface low. The LLJ shifted to the Carolina coast and was noticeably ageostrophic by 0000 UTC 26 January as the surface low pressure center "jumped," or redeveloped farther northeastward along the coast during this period. The combination of the ageostrophic LLJ, falling pressure along the coastal front, and the expanding precipitation shield is similar to the precyclogenetic period of the February 1979 Presidents' Day snowstorm.
- As the 850-hPa low deepened off the Virginia coast by 1200 UTC 26 January and off the New England coast by 0000 UTC 27 January, increasing wind speeds of 20 m s^{-1} and greater appear to the north of the low. These wind speeds spiraling around the 850-hPa low center are slightly less than those that developed within many other cases.

d. 500-hPa geopotential height and 400-hPa wind analyses

- This was another storm that occurred during the El Niño of 1986–87. It also occurred during a period in

SURFACE

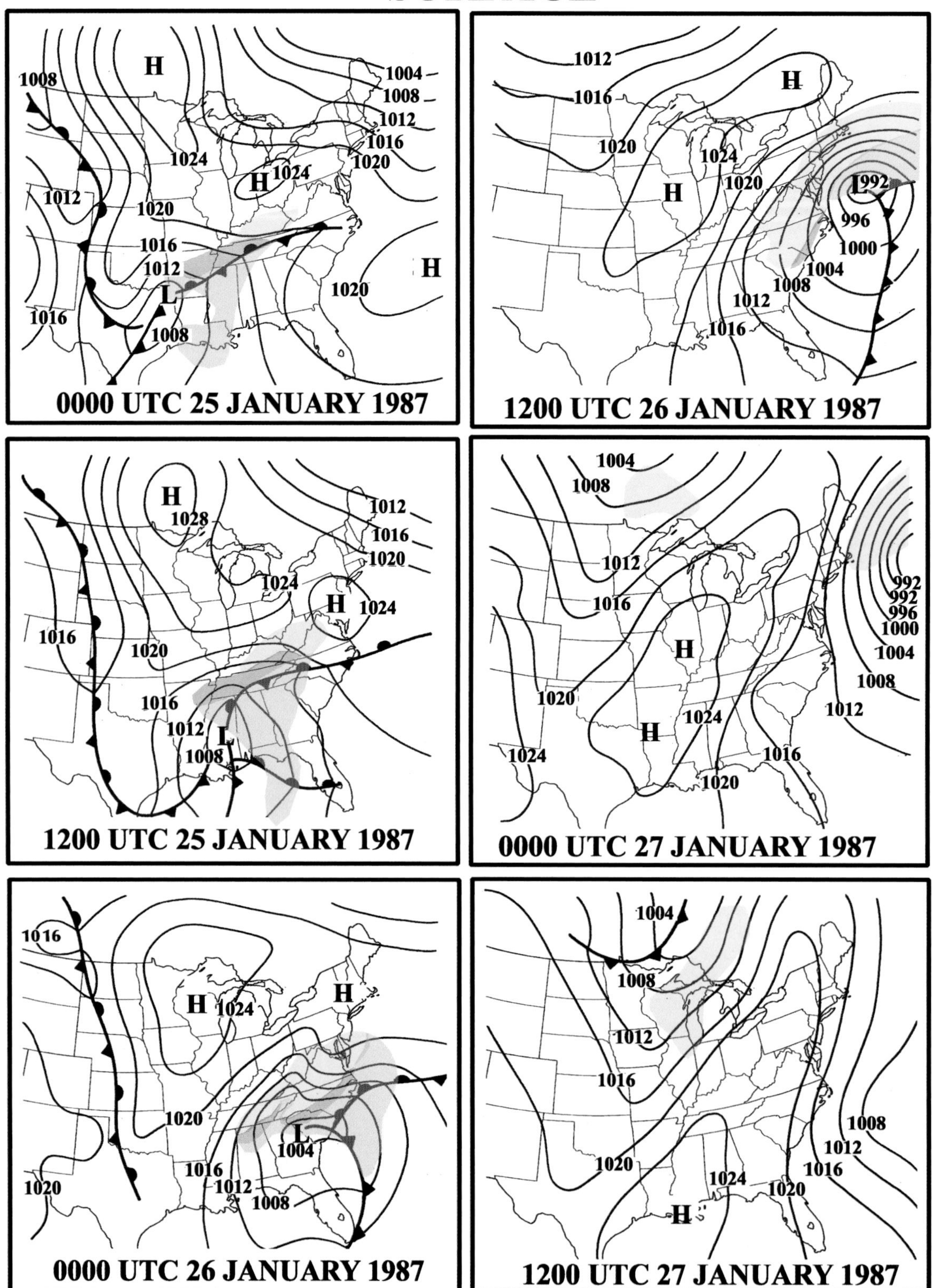

FIG. 10.21-2. Twelve-hourly surface weather analyses for 0000 UTC 25 Jan–1200 UTC 27 Jan 1987. See Fig. 1.1-2 for details.

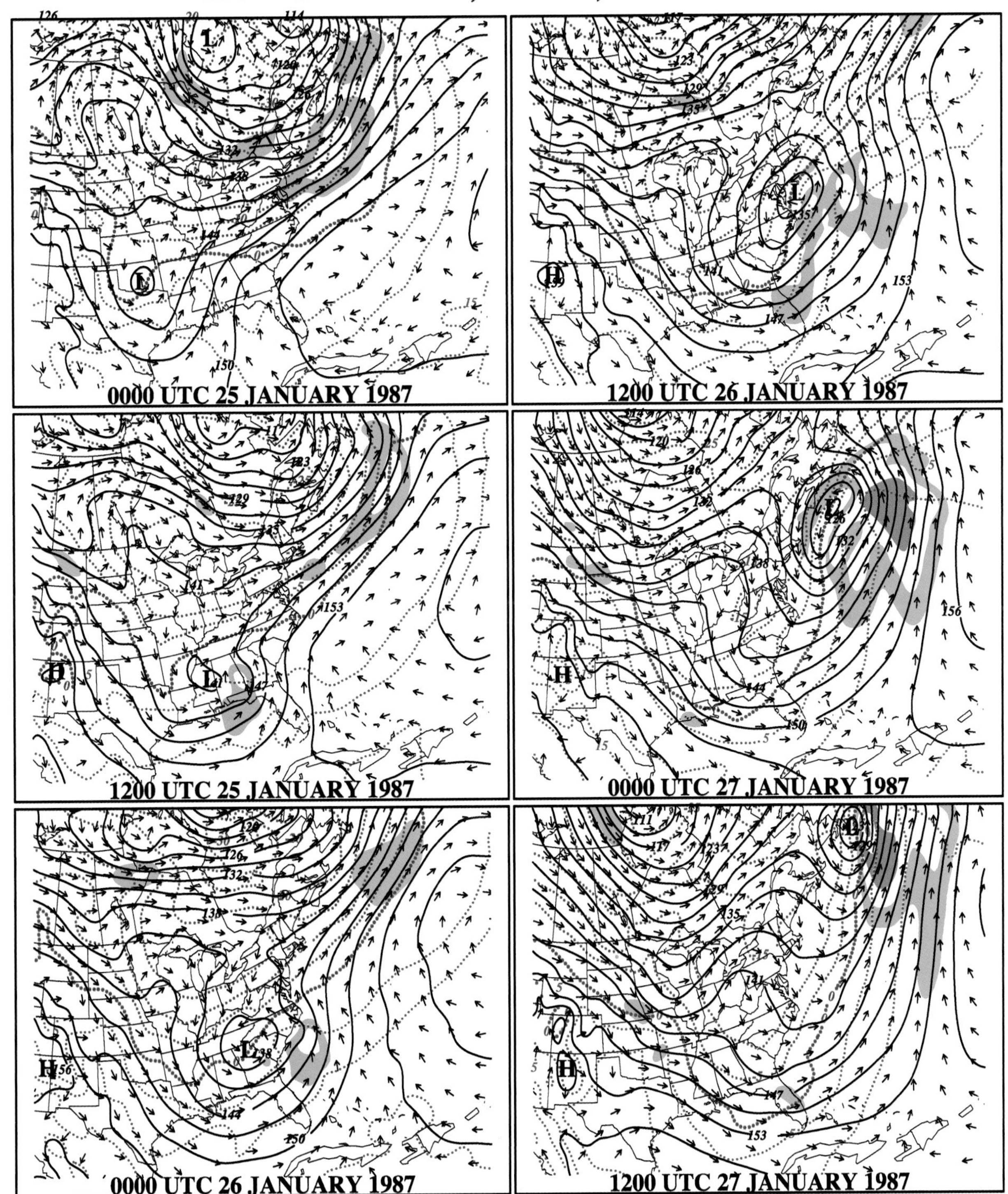

FIG. 10.22-3. Twelve-hourly 850-hPa analyses for 0000 UTC 25 Jan–1200 UTC 27 Jan. See Fig. 10.1-3 for details.

500 hPa HEIGHTS AND VORTICITY, 400 hPa WINDS

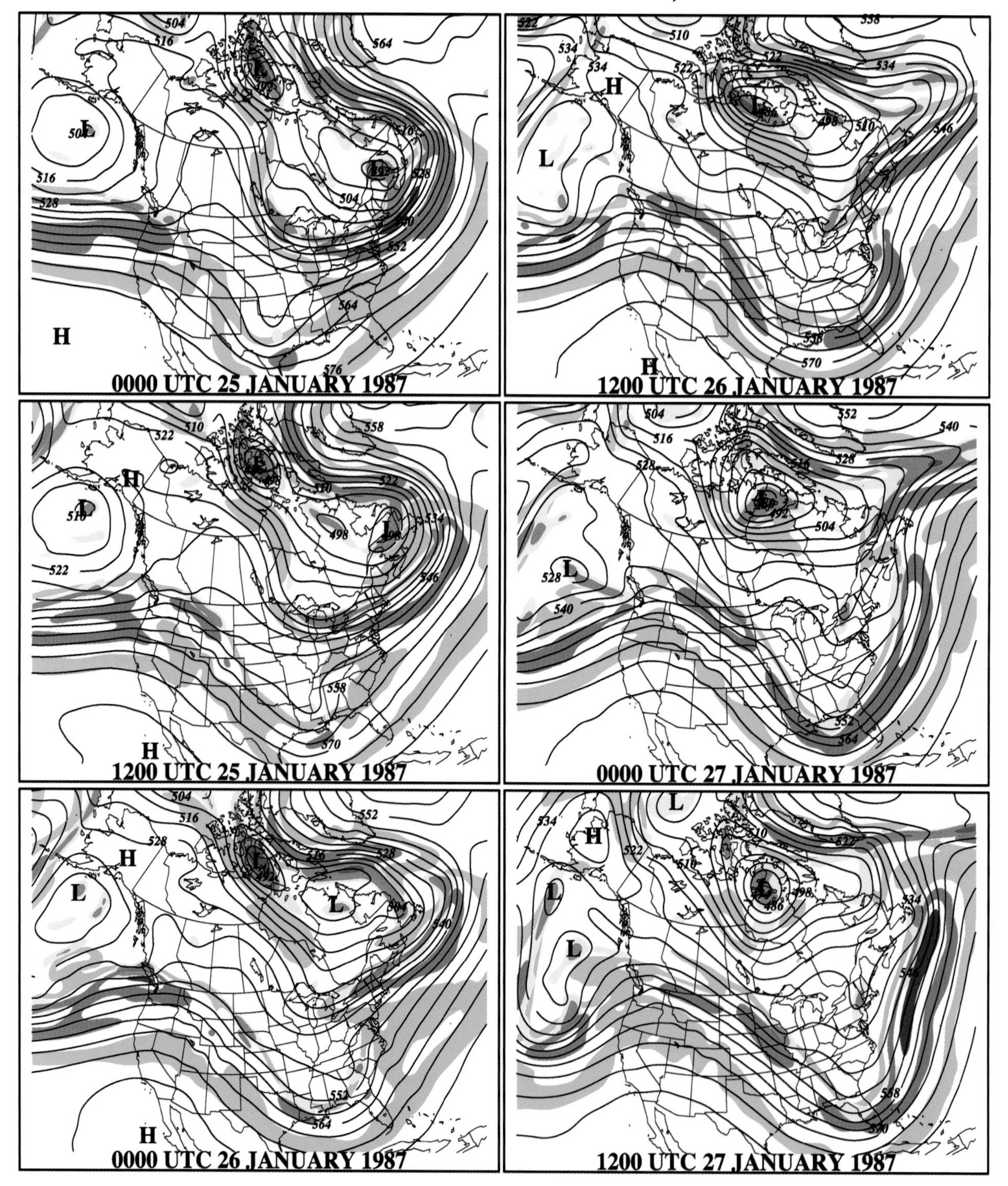

FIG. 10.22-4. Twelve-hourly analyses of 500-hPa geopotential heights and upper-level wind fields for 0000 UTC 25 Jan–1200 UTC 27 Jan 1987. See Fig. 10.1-4 for details.

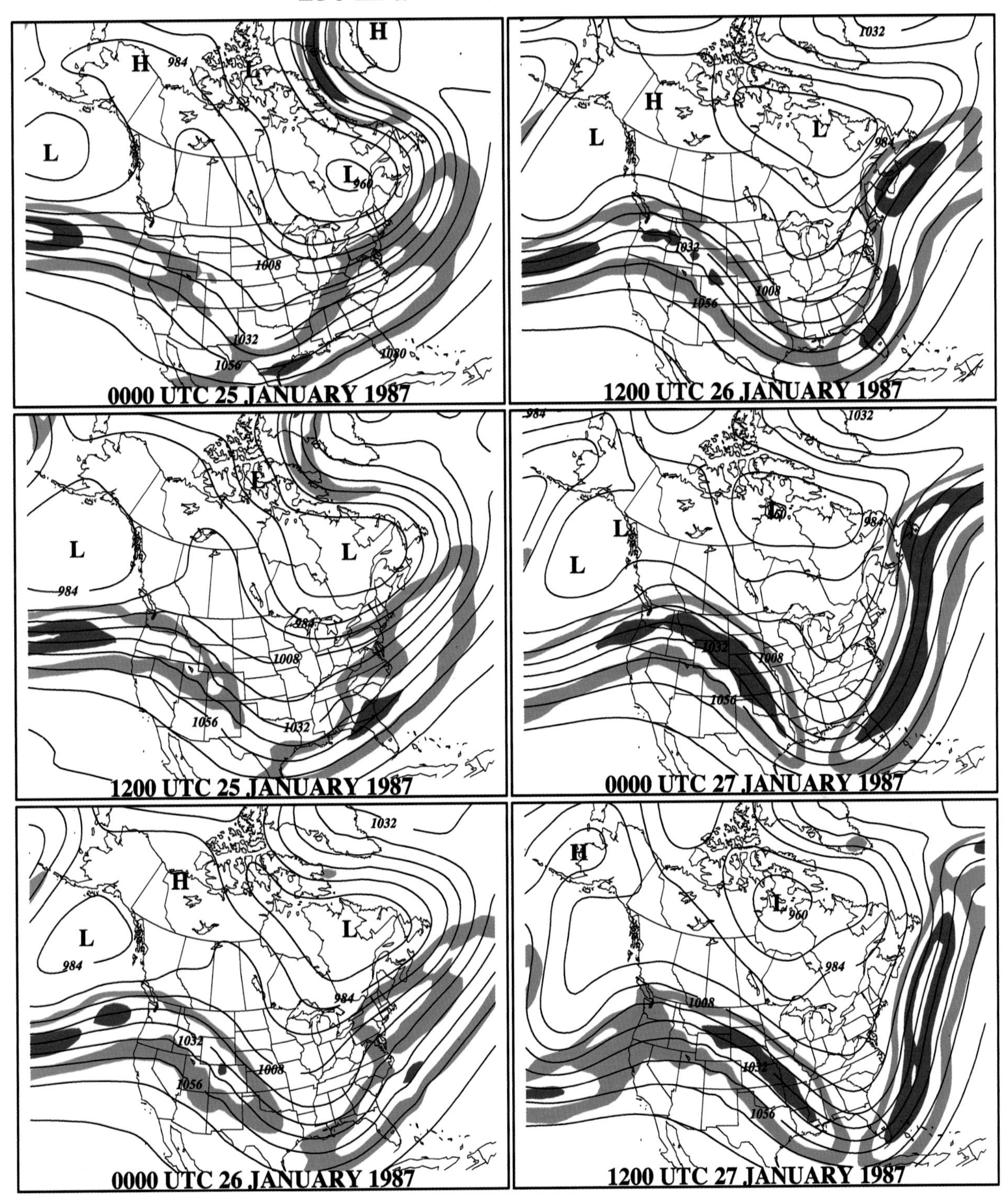

FIG. 10.22-5. Twelve-hourly analyses of 250-hPa geopotential height for 0000 UTC 25 Jan–1200 UTC 27 Jan 1987. See Fig. 10.1-5 for details.

SATELLITE IMAGERY

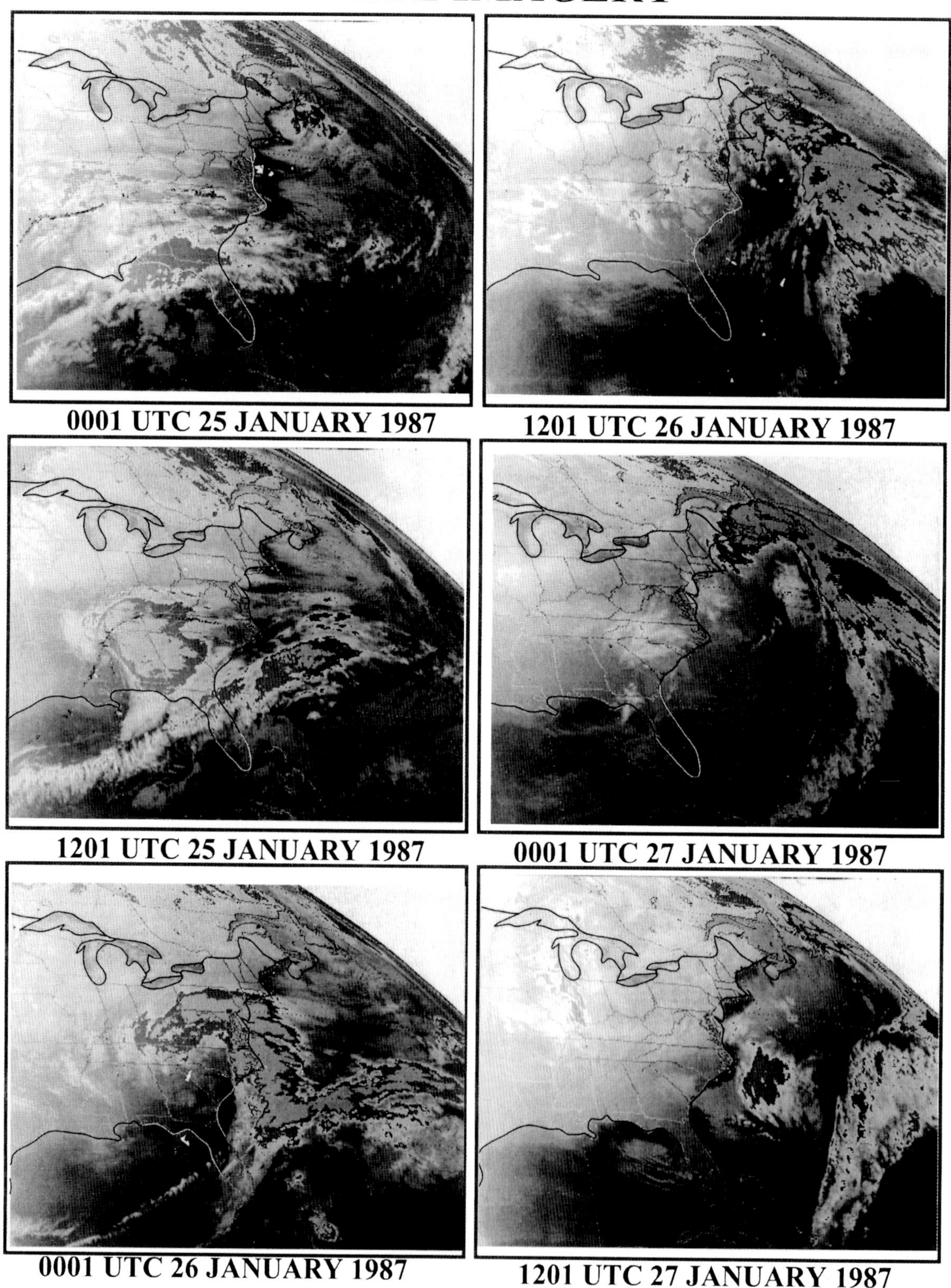

FIG. 10.22-6. Twelve-hourly infrared satellite images for 0001 UTC 25 Jan–1201 UTC 27 Jan 1987. See Fig. 10.16-6 for more details.

which the North Atlantic Oscillation (NAO) was negative but trending less negative, and was marked by a distinct cutoff upper ridge at 500 hPa east of Greenland.

- Prior to East Coast cyclogenesis, the 500-hPa circulation was dominated by an intense vortex over eastern Canada that was associated with the East Coast storm a few days earlier and the cold-air outbreak that followed it. A nearly stationary 500-hPa anticyclone just east of Greenland slowed the movement of this vortex, maintaining cold air in the north-central and northeastern United States. The west coast of North America was characterized by a deamplifying ridge across Canada and a diffluent jet system crossing the coastline. The trough that spawned this storm crossed the southwestern and south-central United States between 23 and 25 January in a westerly flow regime.
- The geopotential height field showed pronounced confluence across the northeastern United States through 26 January. This confluence zone was located above the sea level high pressure area and the west-northwesterly flow and cold-air advection at 850 hPa. The presence of a large 500-hPa vortex over eastern Canada, a 500-hPa anticyclone near Greenland, confluence across the northeastern United States, and a trough moving eastward through the southern United States produced a pattern similar to many other snowstorms examined in this chapter.
- The trough that was associated with the Northeast snowstorm was marked by pronounced cyclonic curvature at the base as it propagated from Texas to the southeastern United States on 25 January. A distinct cyclonic vorticity maximum and its associated cyclonic vorticity advection pattern were observed with this trough as it moved into the south-central United States and then east-northeastward along the East Coast on 26 January. The 500-hPa trough remained an open wave, with little increase in amplitude during cyclogenesis.
- The amplitude between the trough and its downstream ridge increased between 1200 UTC 25 January and 1200 UTC 26 January, a period marked by small, but increasing, surface deepening rates.
- Diffluence was evident downwind of the trough axis at all times. A northwest-to-southeast trough orientation (negative tilt) evolved on 25 January and was most prominent at and after 0000 UTC 26 January. The surface low and precipitation developed within the diffluent flow downwind of the trough axis.
- This case is marked by the dual-jet pattern with a northern and southern jet streak interacting during and after the onset of cyclogenesis and the rapid expansion of heavy snowfall across the middle Atlantic states.
- A polar jet was located within the highly confluent flow over the northeastern United States on 25–26 January. Maximum wind speeds within this jet streak exceeded 60 m s^{-1} on 25–26 January, with wind speeds greater than 70 m s^{-1} analyzed at 0000 UTC 25 January. A band of snowfall developed across the middle Atlantic states within the anticyclonic entrance region of the jet on 25 January.
- Separate jet features were evident upstream and downstream of the trough axis in the south-central United States during most of the observing period. A 50+ m s^{-1} wind maximum remained near the base of the trough as it propagated toward the East Coast, with some hint of a separate wind maximum developing east of this trough axis over northern Florida at 0000 UTC 26 January. By 0000 UTC 26 January, this complex jet streak structure exhibited a defined diffluent exit region over the southeastern United States, above the area marking the surface low pressure center jump and associated surface cyclogenesis. This jet–trough system, in combination with the well-defined jet streak entrance region over the northeastern United States indicated the presence of the "laterally coupled" transverse circulations noted for other cases and described in chapter 4 of volume I.

e. 250-hPa geopotential height and wind analyses

- The 250-hPa geopotential height field illustrates the confluent zone in the Northeast that preceded cyclogenesis and maintained cold air across the middle Atlantic states that resulted in snow, rather than rain. The 250-hPa analyses also illustrate the amplifying trough over the south-central United States by 0000 UTC 26 January.
- A jet streak exceeding 70 m s^{-1} developed east of the trough axis moving from Texas to northern Florida by 1200 UTC 25 January. This wind maximum represents the lower extension of the subtropical jet extending from the northern Gulf of Mexico to the northeast United States and overlaying a polar jet located east of the 500-hPa trough axis at 0000 UTC 27 January.
- The laterally coupled jets are evident at this level at 1200 UTC 26 January with the 992-hPa surface cyclone, LLJ east of the 850-hPa low center, and extensive precipitation area located in the right-entrance region of the northern polar jet and the exit region of the southern subtropical jet.

f. Infrared satellite imagery sequence

- By 1201 UTC 25 January, a cloud mass with cold cloud-top temperatures developed across Kentucky, Tennessee, Alabama, and Georgia, to the north and east of the surface low and was associated with an expanding area of light to moderate rain. A band of high cloud tops, though not as cold as the main cloud mass, streamed eastward from Kentucky across Virginia and New Jersey. This cloud area was associated with a band of light snow to the north of a stationary surface front and within the confluent entrance region of the upper-level jet streak evident at 400 hPa extending from Pennsylvania across New England.

- A separate region of high clouds streaming eastward from the southeastern states at 0000 UTC 25 January moved well offshore in advance of the main comma cloud by 0001 UTC 26 January, in association with the subtropical jet observed at 250 hPa.
- By 0000 UTC 26 January, the cloud mass from the Tennessee Valley had expanded northeastward and attained a comma shape. This cloud mass was located within the diffluent exit region of the upper-level jet streak near and to the east of the base of the trough over the southeastern United States. The heaviest snow was falling over Virginia, near the southern edge of the coldest cloud tops that formed the comma head. The comma cloud continued to drift northeastward without any marked change in its shape through 1200 UTC 27 January.
- A clear tongue was observed in the southwestern quadrant of the developing cyclone after 0000 UTC 26 January, coincident with the combination of polar–subtropical jets located to the south of the surface low pressure center.

Silver Spring, Maryland, 23 Feb 1987 (photo courtesy of Paul Kocin).

23. 22–24 February 1987

a. General remarks

- This rapidly developing cyclone produced an 8–12-h period of very heavy, wet snowfall across portions of the middle Atlantic states. It was later responsible for the deaths of 48 at sea over the western Atlantic (Reed et al. 1992). The heavy loading of snow produced great damage to tree limbs and power lines, resulting in the loss of electricity to hundreds of thousands of people in the middle Atlantic states.
- The snowstorm occurred primarily during the night of 22/23 February, and came in stark contrast to the pleasant, mild, 10°C conditions of the preceding afternoon. An important aspect of this case is the widely varying operational numerical model forecasts of the location and intensity of cyclogenesis. Despite the inconsistency of the model forecasts and the unusually warm temperatures, NWS forecasters (e.g., David Caldwell) made accurate predictions and related warnings of the timing and amount of the snowfall. The intense cyclogenesis associated with this snowstorm was investigated in a study by Reed et al. (1992).
- The following regions reported snow accumulations exceeding 10 in. (25 cm): eastern West Virginia, northern Virginia, central and northern Maryland, northern Delaware, southeastern Pennsylvania, central and southern New Jersey, and part of Long Island.
- The following region reported snow accumulations exceeding 20 in. (50 cm): portions of southeastern Pennsylvania.

TABLE 10.23-1. Snowfall amounts for urban and selected locations for 22–24 Feb 1987; see Table 10.1-1 for details.

NESIS = 1.46 (category 1)	
Urban center snowfall amounts	
Washington, DC–Dulles Airport	12.0 in. (30 cm)
Baltimore, MD	10.1 in. (26 cm)
Philadelphia, PA	6.8 in. (17 cm)
New York, NY–Kennedy Airport	6.1 in. (15 cm)
Other selected snowfall amounts	
Coatesville, PA	23.5 in. (60 cm)
Clarksville, MD	18.2 in. (46 cm)
Wilmington, DE	14.4 in. (37 cm)
Martinsburg, WV	13.3 in. (34 cm)
Dover, DE	10.5 in. (27 cm)

b. Surface analyses

- This snow event was preceded and followed by near to slightly above normal temperatures. Prior to the heavy snow, daytime temperatures on 22 February were generally between 8° and 10°C, hardly typical for a major snowstorm. As the snow began, temperatures quickly fell to near freezing, but very few locations recorded readings lower than 0° or −1°C. At Washington's National Airport, 10 in. (25 cm) fell in spite of above-freezing air temperatures before and during the storm. The warm temperatures, snowfall rates of up to 5 in. (12 cm) h^{-1}, and unsteady wind speeds at the height of the storm enabled the snow to accumulate on a host of objects unable to withstand its weight. Tree limbs and power lines suffered extensive damage.
- The relatively mild temperatures preceding the storm were associated with a modest anticyclone (1020–1025 hPa) over the northeastern United States that weakened as the cyclone developed late on 22 February. Although the surface air mass was atypical for a heavy snow event, the high pressure ridge extending southward east of the Appalachian Mountains at 1200 UTC 22 February and 0000 UTC 23 February, indicated that cold-air damming occurred prior to cyclogenesis.
- The cyclone that produced the heavy snow evolved from a complex system of low pressure centers that consolidated into one explosively developing storm along the Southeast coast. By 0000 UTC 22 February, a low pressure system had developed in the western Gulf of Mexico, while another low pressure center and a cold front were located over the northern plains. In the following 24 h, the cold front from the plains states drifted to the Midwest as the low center over the Gulf of Mexico moved east-northeastward to Georgia, accompanied by an expanding shield of light to moderate rain.
- At 0000 UTC 23 February, rapid cyclogenesis and heavy snowfall were about to commence over the middle Atlantic states. The Gulf low was crossing southern Georgia while two new lows formed. One center appeared over eastern Tennessee, immediately ahead of the upper-level vorticity maximum, as another center emerged off Charleston, South Carolina, along a developing coastal front. In the following 3–6 h, the South Carolina low became the dominant cyclone.
- The low along the Carolina coast intensified explosively after 0000 UTC 23 February, as it tracked northeastward along the coast from eastern North Carolina to southeastern Virginia. The central pressure fell from 1004 hPa at 0000 UTC to 990 hPa by 0900 UTC on 23 February. The cyclone then followed a more easterly path off the Maryland coast and deepened at even greater rates, reaching 981 hPa by 1200 UTC and 964 hPa by 1800 UTC 23 February. The heavy snowfall from West Virginia and Virginia through Pennsylvania and New Jersey occurred between 0000 and 1800 UTC, during this period of rapid intensification. Lightning and thunder with strong, gusty winds were also observed at many locations in the middle Atlantic states early on 23 February.
- Winds increased along the middle Atlantic coast as the cyclone deepened rapidly early on 23 February, but the wind speeds were not excessive, probably due

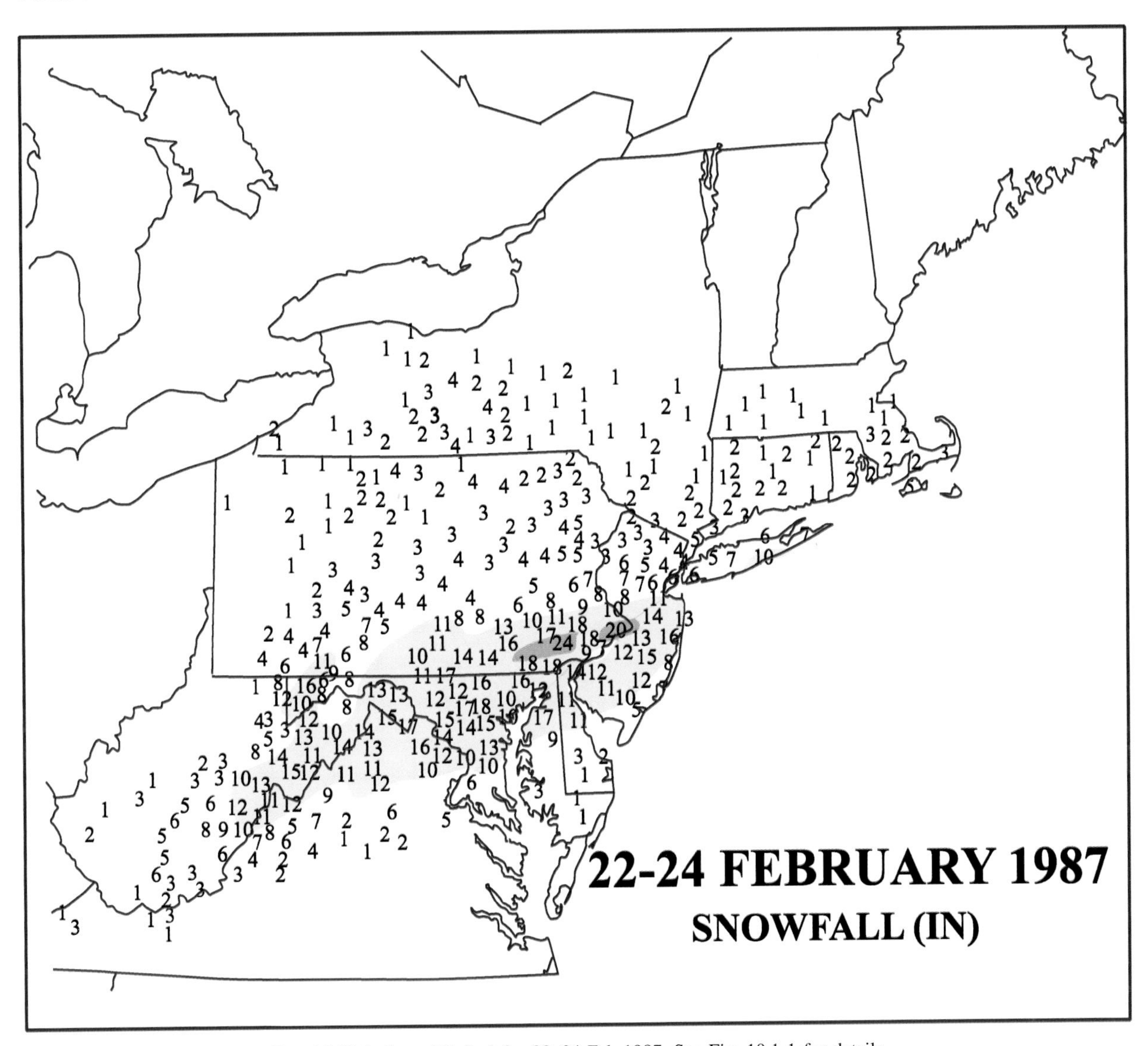

FIG. 10.23-1. Snowfall (in.) for 22–24 Feb 1987. See Fig. 10.1-1 for details.

to the relatively small preexisting sea level pressure gradients related to the weak anticyclone that preceded this storm. Increasing winds were more likely related to the isallobaric effects associated with the onset of rapid development.

- The snow melted quickly, as temperatures rose to 5°–10°C on the afternoon of 23 February.

c. *850-hPa analyses*

- A weak northwesterly flow regime and weak cold-air advection at 850 hPa characterized the precyclogenetic period over the northeastern United States on 21–22 February. The 850-hPa temperatures north of the Virginia–North Carolina border ranged between 0° and −5°C prior to the onset of the storm.
- Two 850-hPa low centers were observed at 0000 and 1200 UTC 22 February: one over Minnesota and another over Texas–Louisiana. The southern low center, which was associated with the surface low over the Gulf of Mexico, deepened significantly as it propagated northeastward. The northern low did not show any change in intensity until it disappeared after 0000 UTC 23 February.
- As the southern low moved northeastward, a pronounced S-shaped isotherm pattern developed over the southeastern United States between 0000 and 1200 UTC 23 February, as warm-air advection increased east and southeast of the center. This development accompanied the more rapid deepening of the low beginning at 0000 UTC 23 February and the outbreak of heavy snowfall in Virginia, West Virginia, and Maryland.

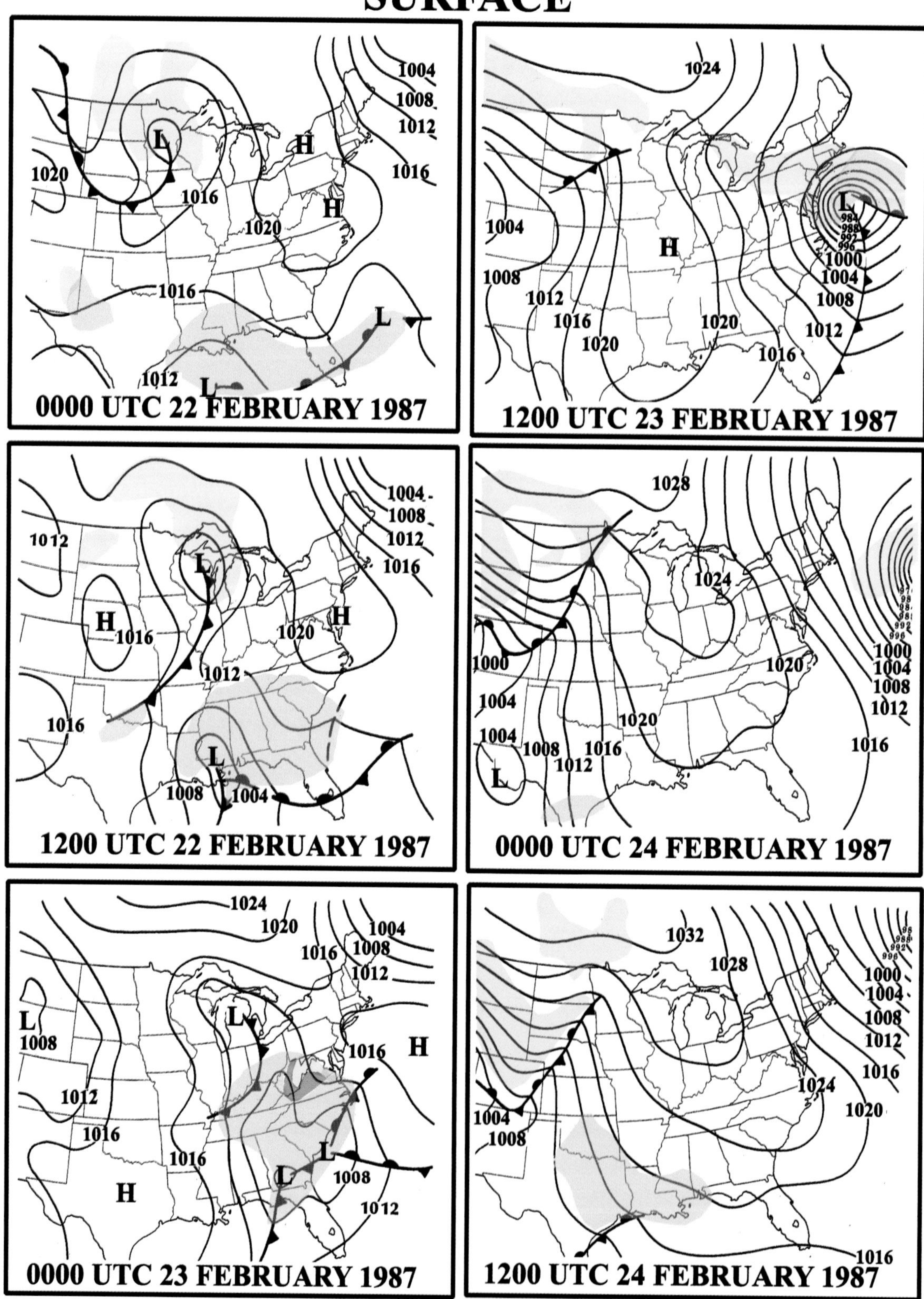

FIG. 10.23-2. Twelve-hourly surface weather analyses for 0000 UTC 22 Feb–1200 UTC 24 Feb 1987. See Fig. 10.1-2 for details.

850 hPa HEIGHTS, WINDS, TEMPERATURE

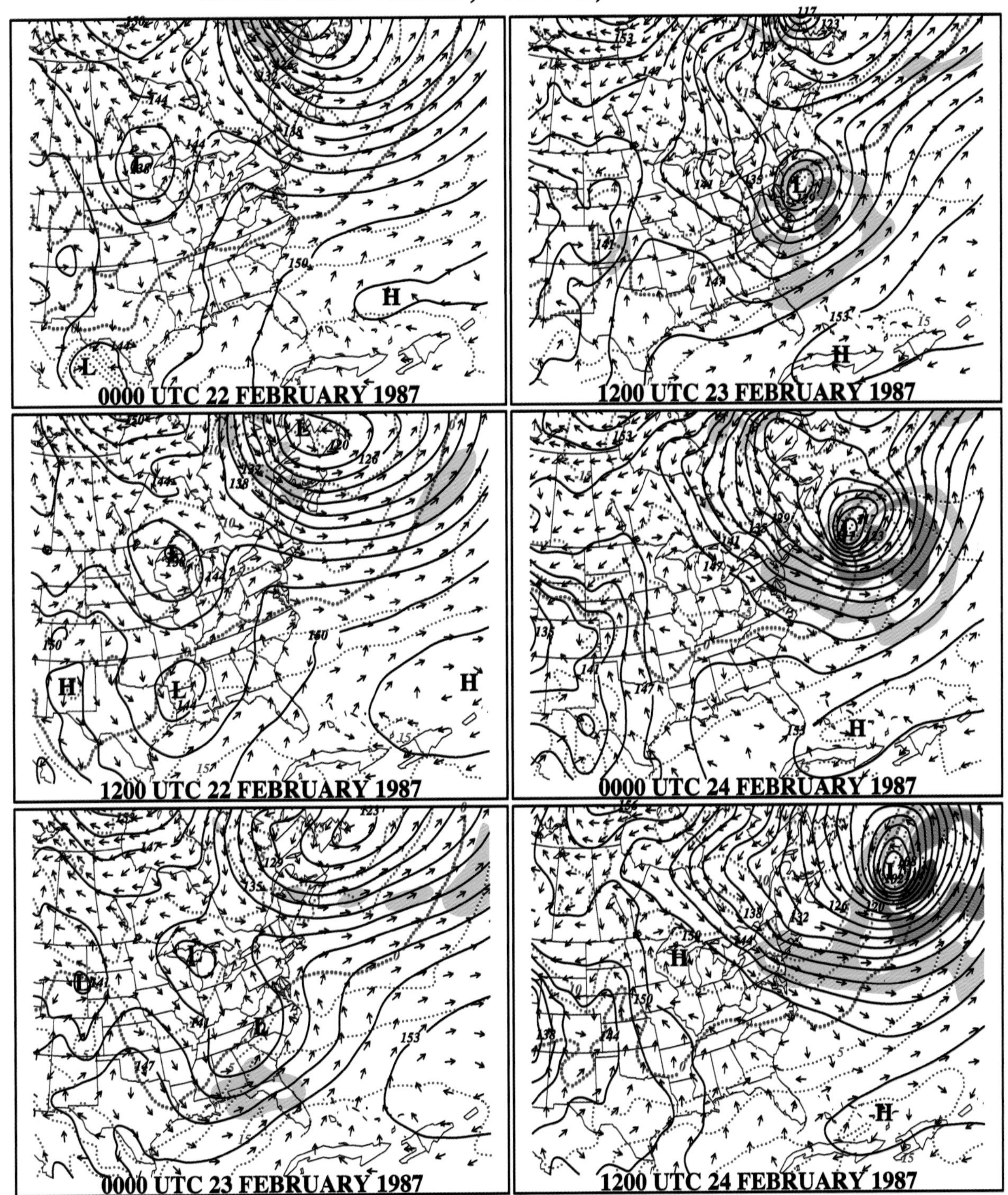

FIG. 10.23-3. Twelve-hourly 850-hPa analyses for 0000 UTC 22 Feb–1200 the 24 Feb 1987. See Fig. 10.1-3 for details.

500 hPa HEIGHTS AND VORTICITY, 400 hPa WINDS

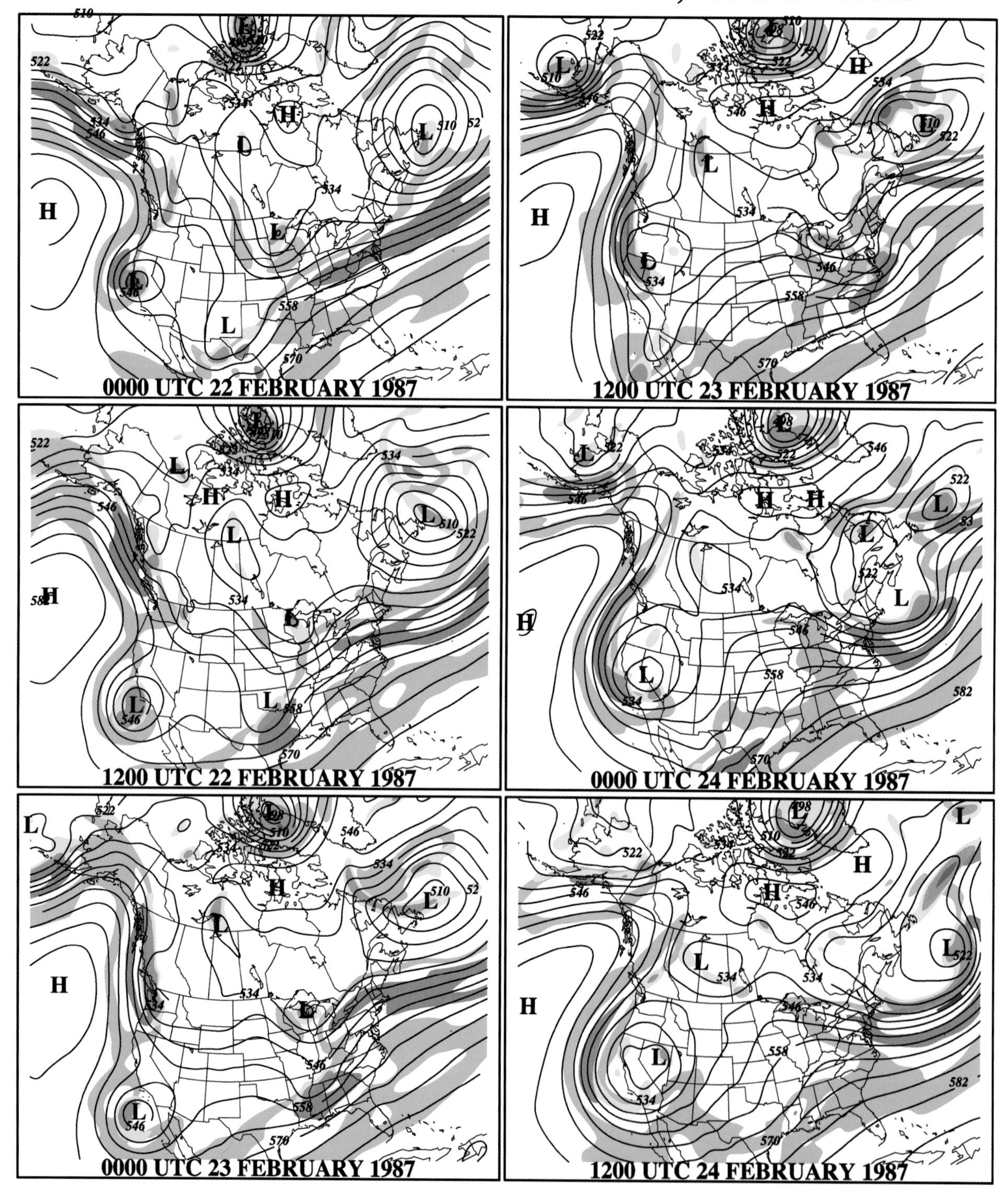

FIG. 10.23-4. Twelve-hourly analyses of 500-hPa geopotential heights and upper-level wind fields for 0000 UTC 22 Feb–1200 UTC 24 Feb 1987. See Fig. 10.1-4 for details.

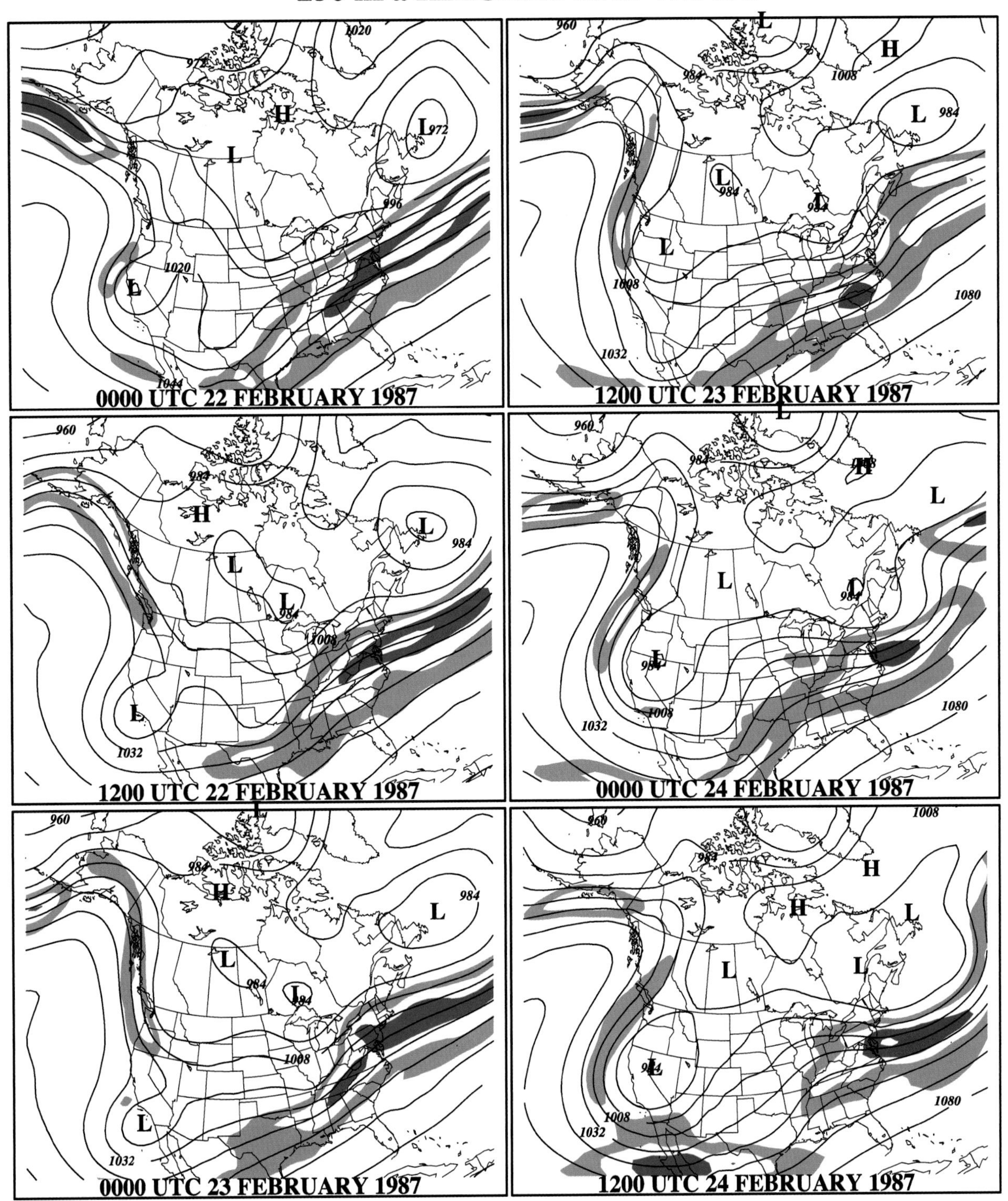

FIG. 10.23-5. Twelve-hourly analyses of 250-hPa geopotential height for 0000 UTC 22 Feb–1200 UTC 24 Feb 1987. See Fig. 10.1-5 for details.

SATELLITE IMAGERY

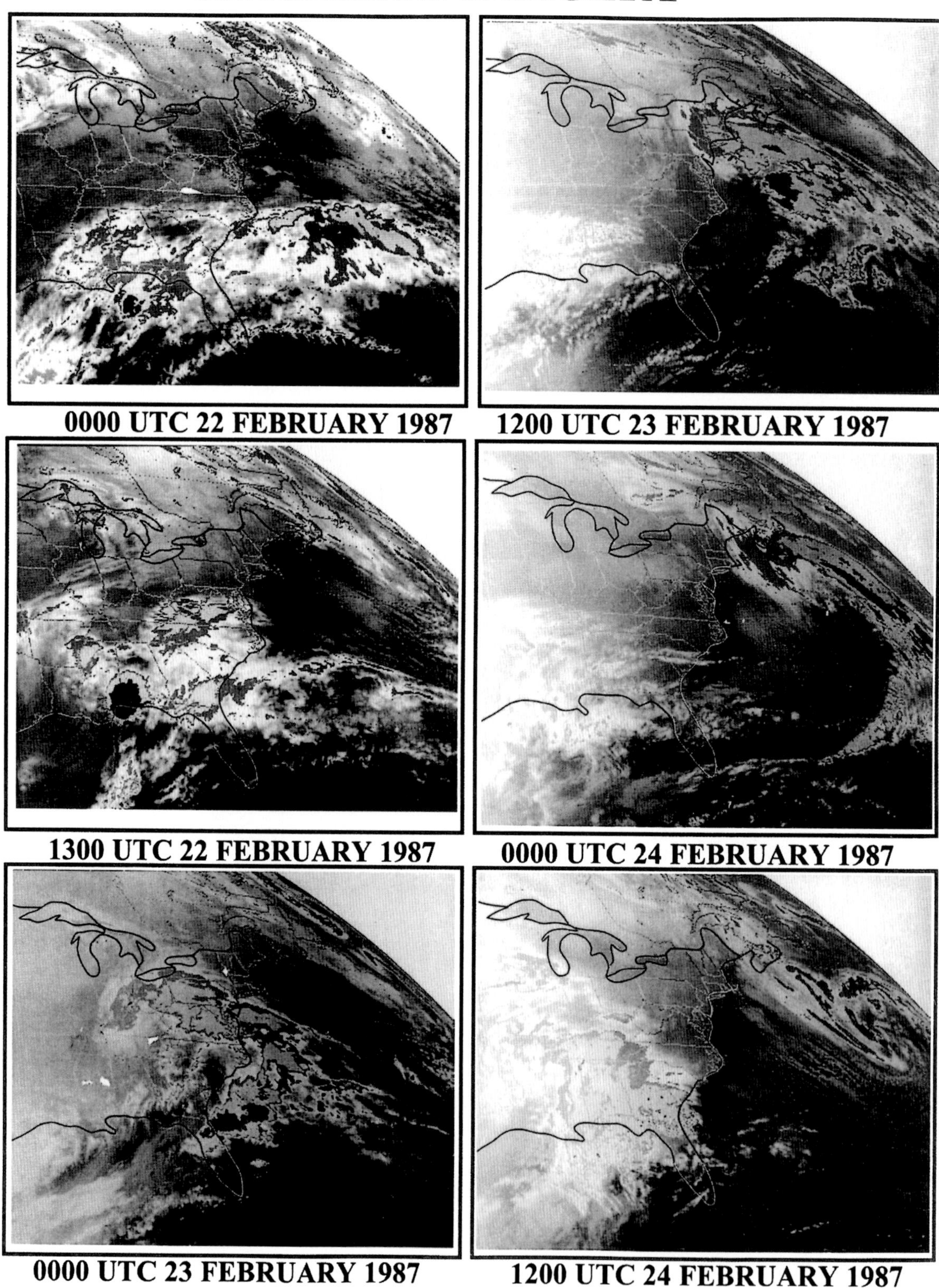

FIG. 10.23-6. Twelve-hourly infrared satellite images for 0000 UTC 22 Feb–1200 UTC 24 Feb 1987. See Fig. 10.16-6 for more details.

- Vigorous intensification of the 850-hPa low commenced as the complex surface low pressure area consolidated into one rapidly developing center along the Carolina coast. Deepening rates increased from −60 m (per 12 h) by 0000 UTC 23 February to −180 m (12 h)$^{-1}$ by 1200 UTC 23 February. The accelerating deepening rates were reflected in the local height changes, which amplified from −80 m (per 12 h) over the southern Appalachians at 0000 UTC 23 February to −210 m (per 12 h) over eastern Virginia by 1200 UTC 23 February.
- A southerly low-level jet (LLJ) first appeared over Louisiana and Mississippi after 1200 UTC 22 February, as rain overspread the Gulf states. Wind speeds continued to increase over the southeastern United States to speeds greater than 25 m s^{-1} by 0000 UTC 23 February, at the beginning of the cyclone's rapid deepening stage.
- An easterly to southeasterly, noticeably ageostrophic, wind component was observed across North Carolina and Virginia by 0000 UTC 23 February, directed toward the expanding area of heavy snow spreading across the middle Atlantic states. The wind speeds increased rapidly and were generally greater than 30 m s^{-1} south and east of the developing 850-hPa circulation by 1200 UTC 23 February.
- The continued intensification of this system as it moved over the Atlantic Ocean is illustrated by the deep vortex and greater than 40 m s^{-1} wind speeds at 1200 UTC 24 February.

d. 500-hPa geopotential height and 400-hPa wind analyses

- This storm also occurred during the continuing El Niño that marked the 1986/87 winter season and during a significant period of negative North Atlantic Oscillation (NAO) conditions. Daily values of the NAO show a pronounced dip in the NAO beginning around 11 February, reaching its low point near the time that this storm was occurring, with values rising rapidly thereafter. The 500-hPa analyses indicate an upper cutoff ridge east of Greenland several days prior to the storm that weakened as the storm developed, with a new cutoff ridge developing near Baffin Bay. The presence of the ridge and its weakening are consistent with the negative NAO and its transition around 24 February.
- The precyclogenetic environment at 500 hPa was dominated by relatively stationary features over eastern Canada, while conditions over the western and southern United States were changing rapidly. A large vortex was located near Newfoundland, where it had remained nearly stationary for the previous week. A 500-hPa anticyclone was also relatively stationary from Hudson Bay to Greenland.
- A confluent flow was reinforced over the Great Lakes and northeast United States as a distinct trough moved over the Great Lakes on 23 February. The confluent flow sustained a cool northwesterly flow across the northeast United States on 22 February. The separate trough over the Great Lakes supported the distinct 850-hPa circulation observed over Lake Superior in the precyclogenetic period on 22 February.
- Farther west, the rapidly changing 500-hPa height field featured a trough moving eastward from the southwestern United States and a separate trough propagating southward along the Pacific coast. Each of these troughs was marked by a separate cyclonic vorticity maximum.
- The East Coast storm developed when the trough over northern Mexico and southwestern Texas on 22 February accelerated eastward, as the trough along the Pacific coast moved southward through California. The trough exiting the southwestern United States initiated surface low pressure development over the Gulf of Mexico, and later along the South Carolina coast.
- The trough that spawned the cyclone passed to the south and then east of a slower-moving trough over the northern United States on 22–23 February. These troughs phased by 0000 UTC 23 February and then developed a northwest-to-southeast orientation (a negative tilt). The rapidly deepening East Coast storm was located beneath the diffluent height contours east of the trough axis by 1200 UTC 23 February. The heavy precipitation and cyclogenesis occurred in advance of a well-defined cyclonic vorticity center, in the region of strong cyclonic vorticity advection associated with the southern trough.
- During the week preceding the storm, upper-level troughs progressing eastward across the United States encountered the large vortex centered over eastern Canada and were suppressed and weakened. Operational numerical weather prediction models initialized prior to 1200 UTC 22 February forecast a continuation of this trend, as the trough over the upper Midwest was predicted to move at a faster rate than the southwestern trough. The larger phase speed forecast for the northern trough resulted in the complete deamplification of the southwestern trough as it reached the East Coast, with the subsequent cyclogenesis predicted to be very weak and suppressed to the south of its actual location.
- The operational model simulations initialized at 1200 UTC 22 February captured the merging more accurately and alerted forecasters to the likelihood of rapid surface development, who then issued warnings 6 h before the onset of heavy snowfall in the Washington–Baltimore metropolitan area. It therefore appears that the interactions of the Great Lakes and more southern trough systems (and their associated jet streaks) were crucial to the evolution of this storm.
- As the trough from the southwestern United States moved eastward, the amplitude between the upstream ridge and trough decreased sharply by 0000 UTC 23

February. The amplitudes of the trough and downstream ridge, however, remained the same or decreased only slightly. Although the amplitudes of the trough and downstream ridge did not increase during rapid cyclogenesis between 0000 and 1200 UTC 23 February, the half-wavelength between the trough and the downstream ridge decreased dramatically.

- This case, like many others, is marked by a complex evolution of jet maxima that evolve toward a dual-jet pattern: one to the north of the developing cyclone and a separate jet associated with the trough to the south.
- The northern jet was located within the confluent zone extending from the Ohio Valley off the New England coast during the precyclogenetic period (0000 UTC 23 February).
- A southern jet evolves into a well-defined 60 m s^{-1} jet streak east of the trough axis by 0000 UTC 23 February extending from Texas to the Carolinas.
- The rapid cyclogenesis occurred within the diffluent exit region of the southern jet and right-rear entrance region of the northern jet after 0000 UTC 23 February.
- This dual-jet pattern was sustained as rapid cyclogenesis was occurring off the East Coast after 1200 UTC 23 February, and as wind speeds near the base of the upper-level trough increased to greater than 70 m s^{-1} by 1200 UTC 24 February with the surface cyclone continuing to deepen in the exit region of this jet.
- The explosive development of the surface low, the formation of the LLJ, and the expanding area of precipitation within the right-rear entrance region of the northern jet and the exit region of the southern jet between 0000 and 1200 UTC 23 February are consistent with the "lateral coupling" concept discussed by Uccellini and Kocin (1987) and in chapter 4 of volume I.

e. 250-hPa geopotential height and wind analyses

- The deep trough over southeastern Canada and the complex series of troughs in the precyclogenetic period are quite evident for the upper troposphere, with the confluent flow accentuated over the middle Atlantic states on 22 February.
- A subtropical jet is evident extending from Texas into the confluent flow with wind speeds exceeding 70 to near 80 m s^{-1} through 1200 UTC 22 February.
- The jet streaks evolve into a dual-jet pattern as the rapid cyclogenesis commences at 0000 UTC 23 February, especially as the southern jet maximum becomes more distinct over Tennessee.
- The surface cyclone appears to be located within the entrance region of the northern jet and the exit region of the southern jet through 24 February as the surface low continues to deepen into a major storm over the Atlantic Ocean.

f. Infrared satellite imagery sequence

- The satellite images initially showed a broad expanse of clouds across the southeastern United States, to the south of the upper-level jet system centered over the middle Atlantic states at 0000 UTC 22 February. By 1300 UTC 22 February, patches of enhanced cold cloud tops were observed from Alabama to Missouri. The black-enhanced region near the Gulf coast depicts an outbreak of convection near the surface low center. The expanding cloud region from Mississippi to southern Missouri was associated with the upper-level trough and related jet streaks propagating eastward across Texas at this time.
- By 0000 UTC 23 February, a comma-shaped cloud mass had developed over the eastern United States. The arc-shaped comma head was located within the exit region of the jet streak at the base of the trough over Mississippi and Alabama and the entrance region of the more northern jet. The comma head resembles the cold conveyor belt described in chapter 4.4 of volume I, but there was no well-defined cloud signature to identify the warm conveyor belt.
- Rapid cyclogenesis was occurring within relatively shallow clouds over South Carolina at 0000 UTC 23 February. Although the highest cloud tops were located well to the north, an area of convective cloud tops was observed just south of the surface low center. Heavy snow developed from the Virginias to southern Pennsylvania, along the southern edge of the highest cloud tops associated with the comma head that were extending into the entrance region of the northern jet streak.
- By 1200 UTC 23 February, the comma-shaped cloud had become better organized, with the heaviest snow located near the southwestern edge of the comma head over New Jersey and Long Island where the 850-hPa LLJ intensified to the north of the low-level circulation. In the following 12 h, the comma cloud became even more pronounced as the storm continued to deepen explosively over the ocean.

Morning commuters cross over one of many walls of ice in Manhattan, 15 Mar 1993 (photo courtesy of Kevin Ambrose, *Great Blizzards of New York City*, Historical Enterprises, 1994).

24. 12–14 March 1993

a. General remarks

- The March 1993 "Superstorm" was one of the most paralyzing extratropical cyclones of the 20th century (see Gilhousen 1994; Kocin et al. 1995; Uccellini et al. 1995; Caplan 1995; Bosart et al. 1996; Dickinson et al. 1997; Huo et al. 1995, 1998, 1999a,b). This storm produced snow over much of the eastern United States, affecting over 100 million people and likely produced the greatest distribution of significant snow of any storm during the 20th century (Kocin et al. 1995).
- Heaviest snows fell along the Appalachians, where amounts exceeded 40 in. (100 cm) in several states while much of the Northeast urban corridor received between 6 and 15 in. of snow (15–38 cm), with lesser amounts related to snow changing to ice pellets and rain and then back to snow. The storm was responsible not only for record snows, but also produced a tornado outbreak across Florida as well as storm surge flooding in northwestern Florida. Damage from the storm was estimated in the billions of dollars, produced the largest air traffic disruption in history to date, and resulted in numerous fatalities. The storm is also notable because it was, with some exceptions, very well forecast and publicized up to 5 days in advance, resulting in an enormous response from the public and state officials to prepare for its arrival. While the storm was well forecast, the storm's intensification over the Gulf of Mexico was underforecast while the East Coast development was overforecast [see Kocin et al. (1995), Uccellini et al. (1995), and Caplan (1995) for a review of the cyclone and forecast challenges].
- The following regions reported snow accumulations exceeding 10 in. (25 cm): northern and western Virginia, central and northern Maryland, northern Delaware, central and northern New Jersey, Pennsylvania, New York, Connecticut, northern Rhode Island, Massachusetts (except the southeast), Vermont, New Hampshire, and Maine
- The following regions reported snow accumulations exceeding 20 in. (50 cm): northwestern Virginia, southwestern Virginia, western Maryland, about one-half of Pennsylvania, much of New York, northwestern Connecticut, western Massachusetts, portions of Vermont, New Hampshire, and Maine

b. Surface analyses

- The storm developed on 11–12 March in the western Gulf of Mexico, east of Brownsville, Texas, while high pressure covered much of the northern half of the United States. As the storm intensified on the morning of 12 March, significant rainfall developed along the Texas and Gulf coasts and spread northward into Arkansas and northern Mississippi, where snow mixed with the rain. By 1200 UTC, the storm's estimated pressure was around 1000 hPa.
- One of the most significant aspects of this storm is its rapid development in the Gulf of Mexico on 12 March. This development was accompanied by the onset of very strong winds in the Gulf (as reported by oil rigs), approaching hurricane force, especially north and northwest of the surface low. By evening, the storm had moved northeastward to a position south of New Orleans with a pressure of 984 hPa, a fall of approximately 16 hPa in 12 h, while the center of the storm made "landfall" in northwestern Florida around 0600 UTC 13 March. At this time, the central pressure was 975 hPa, similar to that observed in some significant hurricanes. It is unlikely that pressure values this low and this far south have occurred during any other recent extratropical cyclone in the Gulf of Mexico (Kocin et al. 1995).
- As the storm deepened, precipitation spread north and eastward across Mississippi and Alabama into Tennessee and the western Appalachians, with mostly light to moderate wet snow in northern portions and rain near the Gulf. The explosive development of heavy snowfall occurred later that night as the storm exited the Gulf of Mexico and moved northeastward along the Atlantic coast.
- One of the more dramatic surface charts in this monograph is from 1200 UTC 13 March as the storm center moved into east-central Georgia, with the central pres-

TABLE 10.24-1. Snowfall amounts for urban and selected locations for 12–14 Mar 1993; see Table 10.1-1 for details.

NESIS = 12.52 (category 5)	
Urban center snowfall amounts	
Washington, DC–Dulles Airport	13.9 in. (35 cm)
Baltimore, MD	11.9 in. (30 cm)
Philadelphia, PA	12.3 in. (31 cm)
New York, NY–La Guardia Airport	12.3 in. (31 cm)
Boston, MA	12.8 in. (32 cm)
Other selected snowfall amounts	
Syracuse, New York	42.9 in. (109 cm)
Beckley, WV	30.9 in. (78 cm)
Albany, NY	26.6 in. (68 cm)
Pittsburgh, PA	25.3 in. (64 cm)
Scranton, PA	21.4 in. (54 cm)
Harrisburg, PA	20.4 in. (52 cm)
Worcester, MA	20.1 in. (51 cm)
Charleston, WV	18.9 in. (48 cm)
Elkins, WV	18.8 in. (48 cm)
Portland, ME	18.6 in. (47 cm)
Allentown, PA	17.6 in. (46 cm)
Concord, NH	17.0 in. (44 cm)
Roanoke, VA	16.0 in. (41 cm)
Hartford, CT	14.8 in. (38 cm)
Wilmington, DE	13.7 in. (35 cm)
Birmingham, AL	13.0 in. (33 cm)
Newark, NJ	12.7 in. (32 cm)
Bridgeport, CT	10.8 in. (27 cm)
Providence, RI	10.2 in. (26 cm)

(a)

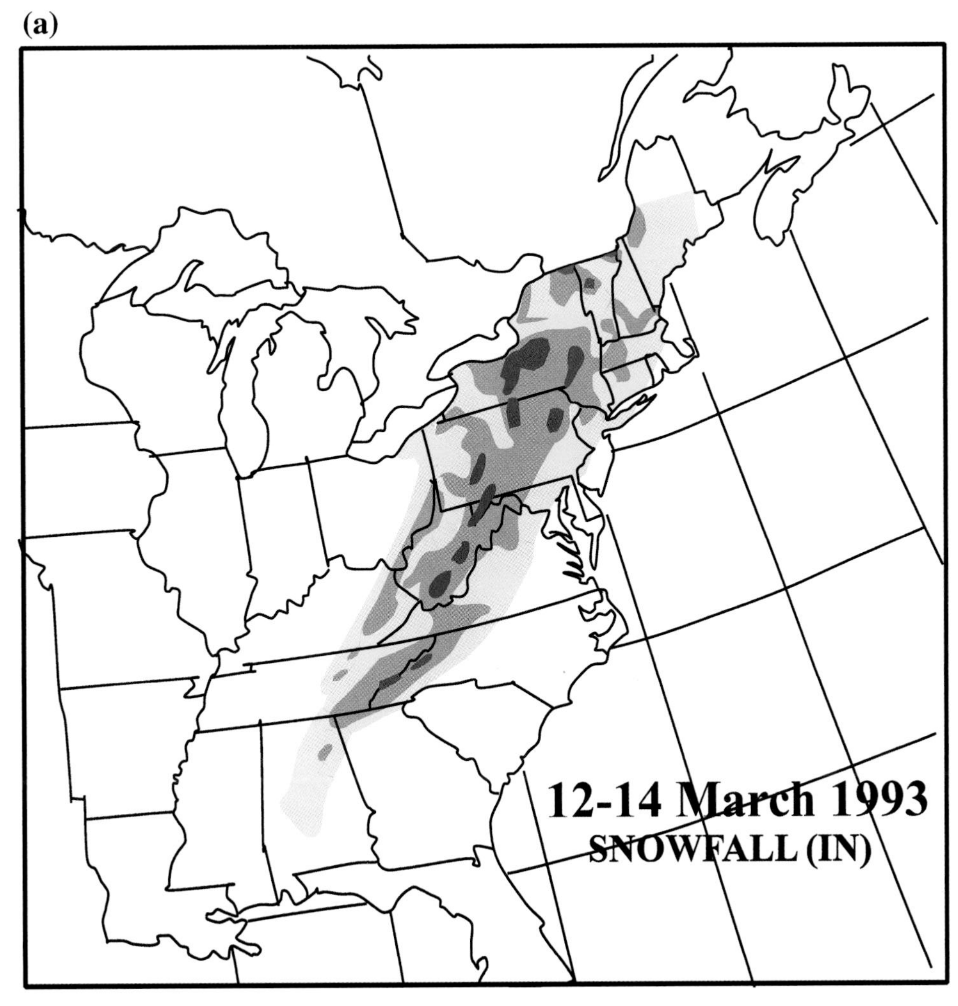

FIG. 10.24-1a. Snowfall (in.) in the eastern United States for 12–14 Mar 1993. See Fig. 10.1-1 for details.

sure down to 971 hPa. Heavy snow and high winds enveloped portions of Alabama and Georgia, and snow spread rapidly northeastward overnight into the western Carolinas, Virginia, West Virginia, Maryland, Delaware, New Jersey, and New York. A strong coastal front was located along the Atlantic coast, along which the low pressure center tracked. Snow changed to rain over eastern North Carolina, Virginia, and Maryland early on 13 March.

- High pressure drifted slowly eastward overnight from New England toward the Canadian Maritime Provinces, a position favoring a changeover from snow to rain in the Northeast urban corridor (see volume I, chapter 5). Cold-air damming was less evident in the pressure field than in some cases due to the overwhelming presence of the deepening low pressure system.
- The storm moved northeastward along the East Coast during 13 March, passing on a track slightly inland, reaching the Chesapeake Bay by evening and deepening to 960 hPa, the lowest pressure it would reach. Deepening rates were already diminishing, with the central pressure dropping 11 hPa in 12 h by 0000 UTC 14 March.
- With the deepening low along the East Coast, near–hurricane force winds were experienced along much of the East Coast before the storm center neared. The strong winds lasted for 12 h or less, sparing more major destruction.
- Heavy snow paralyzed all of the Appalachians, where snow continued through 13 March, and heavy snows and high winds in the Northeast urban corridor left an average of 10–12-in. accumulations (25–30 cm), before mixing with and changing over to ice pellets and briefly to rain from Philadelphia to southern New England. The changeover spared the Northeast cities an additional foot (30 cm) of snowfall, although the heavy coating of ice and snow and falling temperatures left a thick layer of ice on top of the snow as the storm moved by.
- The surface low moved at a fairly constant speed as

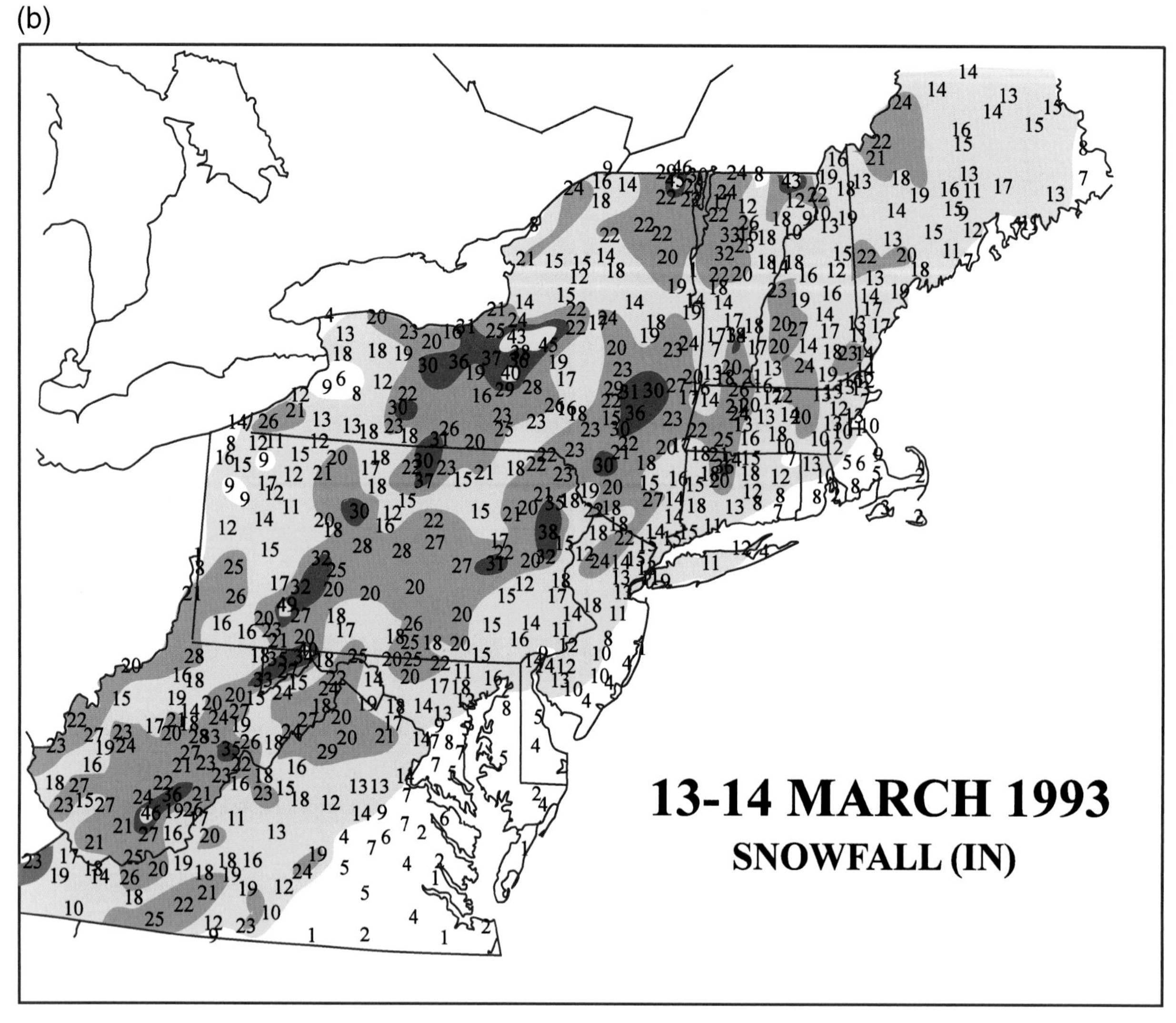

FIG. 10.24-1b. Snowfall (in.) for 12–14 Mar 1993. See Fig. 10.1-1 for details.

it moved northeastward along the East Coast, averaging 15 m s^{-1} on 13–14 March.

- Following 0000 UTC 14 March, the storm slowly diminished in intensity, reaching eastern New England by 1200 UTC 14 March with a pressure around 965 hPa. Overnight, rain or ice pellets changed briefly back to snow in the Northeast urban corridor while snows continued across the Appalachians and began to taper off from southwest to northeast. The snows gradually ended across New England late on 14 March.

c. 850-hPa analyses

- Much of the eastern half of the United States north of the Gulf states remained below 0°C at 0000 UTC 12 March before the surface low began to develop in the Gulf of Mexico.
- The developing 850-hPa low center over the western Gulf of Mexico at 1200 UTC 12 March was located along the warm side of the lower-tropospheric baroclinic zone.
- Between 0000 and 1200 UTC 12 March, the 850-hPa 0°C isotherm drifted southward into Alabama and points west while remaining nearly stationary in the Carolinas. The rain–snow boundary appeared to be collocated with the 0°C isotherm, with very heavy snow stretching from the southwest to the northeast just west of the 850-hPa freezing isotherm.
- As the storm was developing rapidly over the Gulf of Mexico on 12 March, the 850-hPa low intensified commensurately, still on the warm side of the 0°C isotherm. However, the lower-tropospheric temperature gradient surrounding the 0°C isotherm was increasing, indicating frontogenesis was occurring as the low developed and moved northeastward toward the Florida coast.

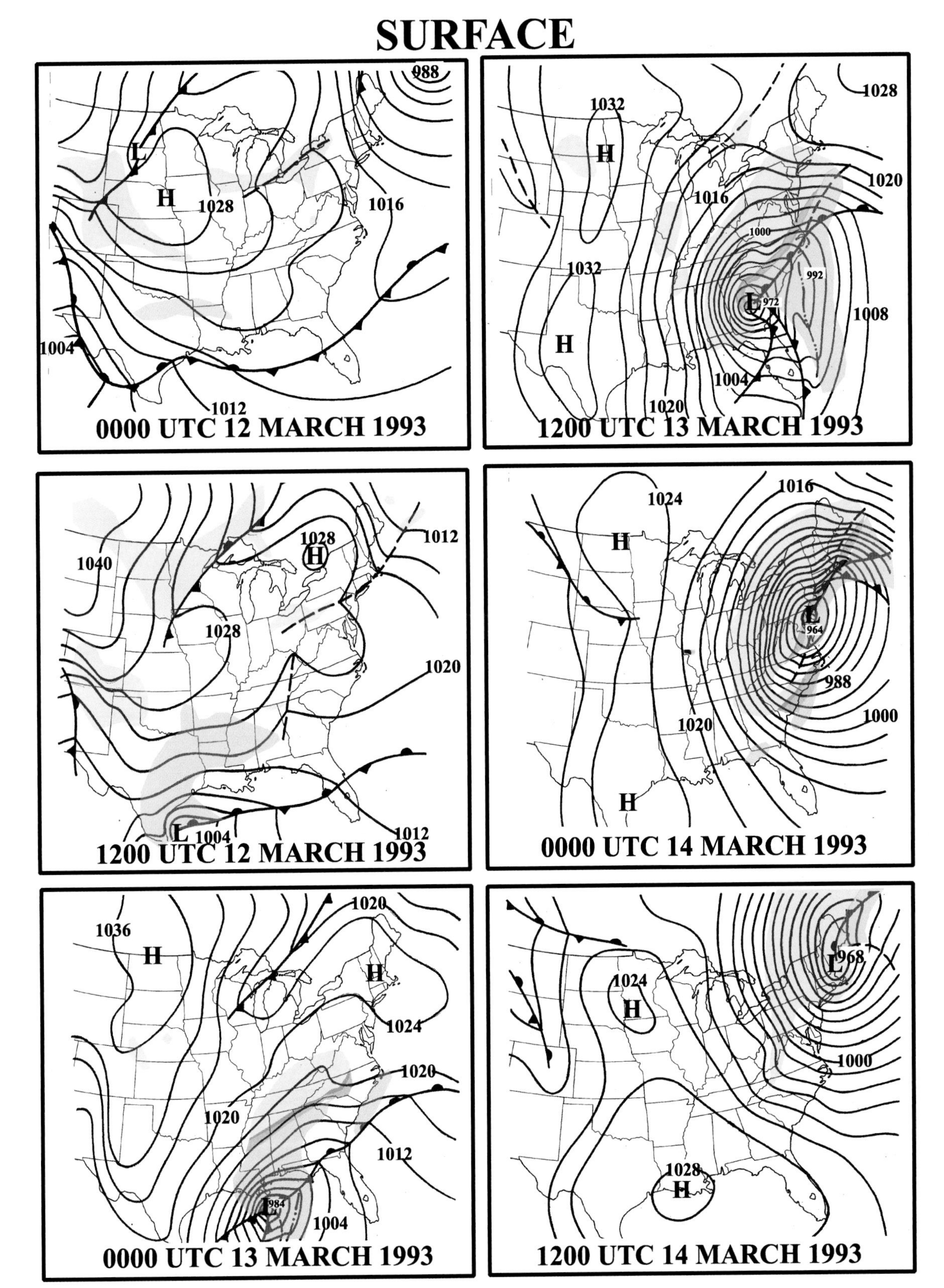

FIG. 10.24-2. Twelve-hourly surface weather analyses for 0000 UTC 12 Mar–1200 UTC 14 Mar 1993. See Fig. 10.1-2 for details.

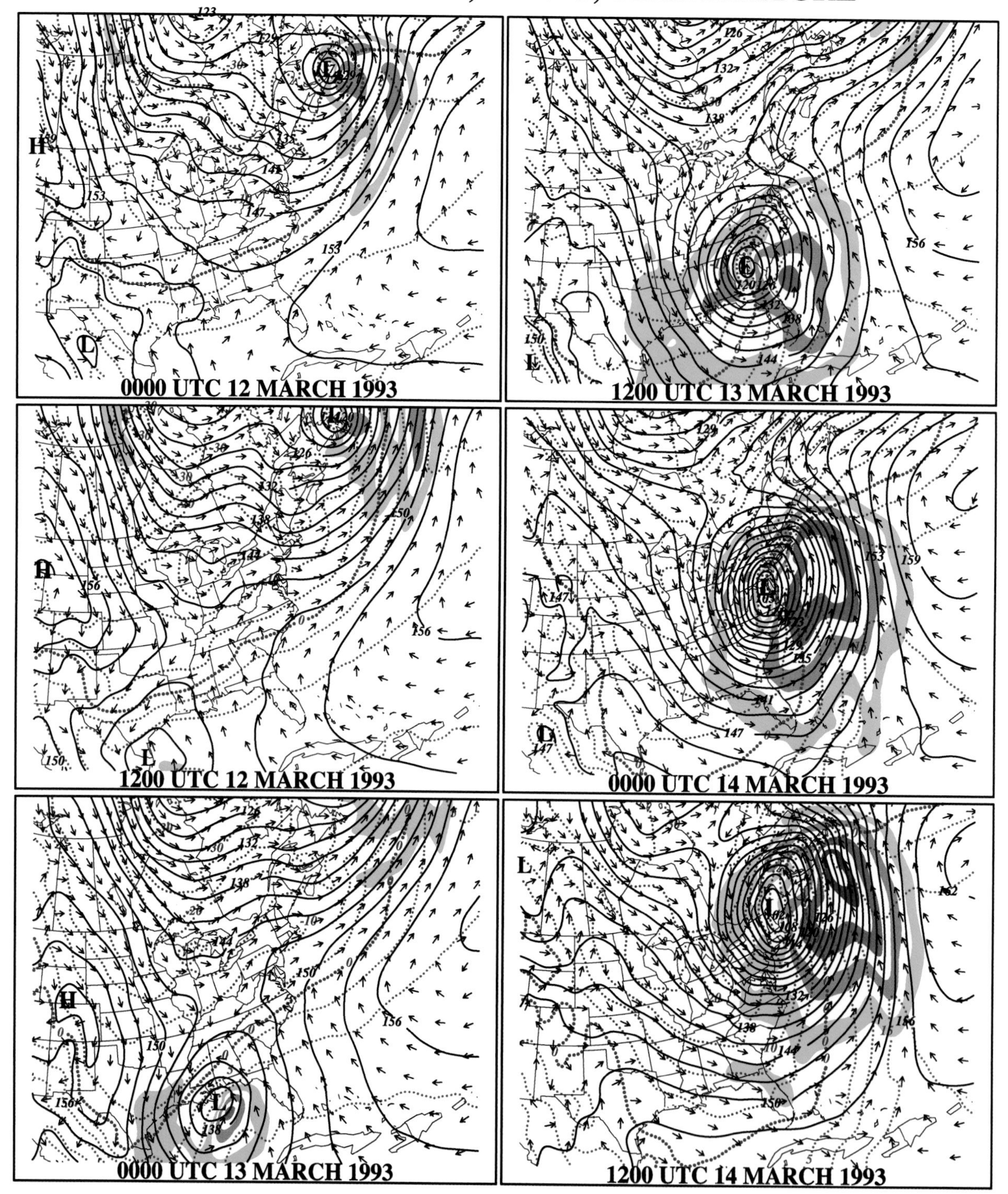

FIG. 10.24-3. Twelve-hourly 850-hPa analyses for 0000 UTC 12 Mar–1200 14 Mar 1993. See Fig.10.1-3 for details.

500 hPa HEIGHTS AND VORTICITY, 400 hPa WINDS

FIG. 10.24-4. Twelve-hourly analyses of 500-hPa geopotential heights and upper-level wind fields for 0000 UTC 12 Mar–1200 UTC 14 Mar 1993. See Fig. 10.1-4 for details.

250 hPa HEIGHTS AND WINDS

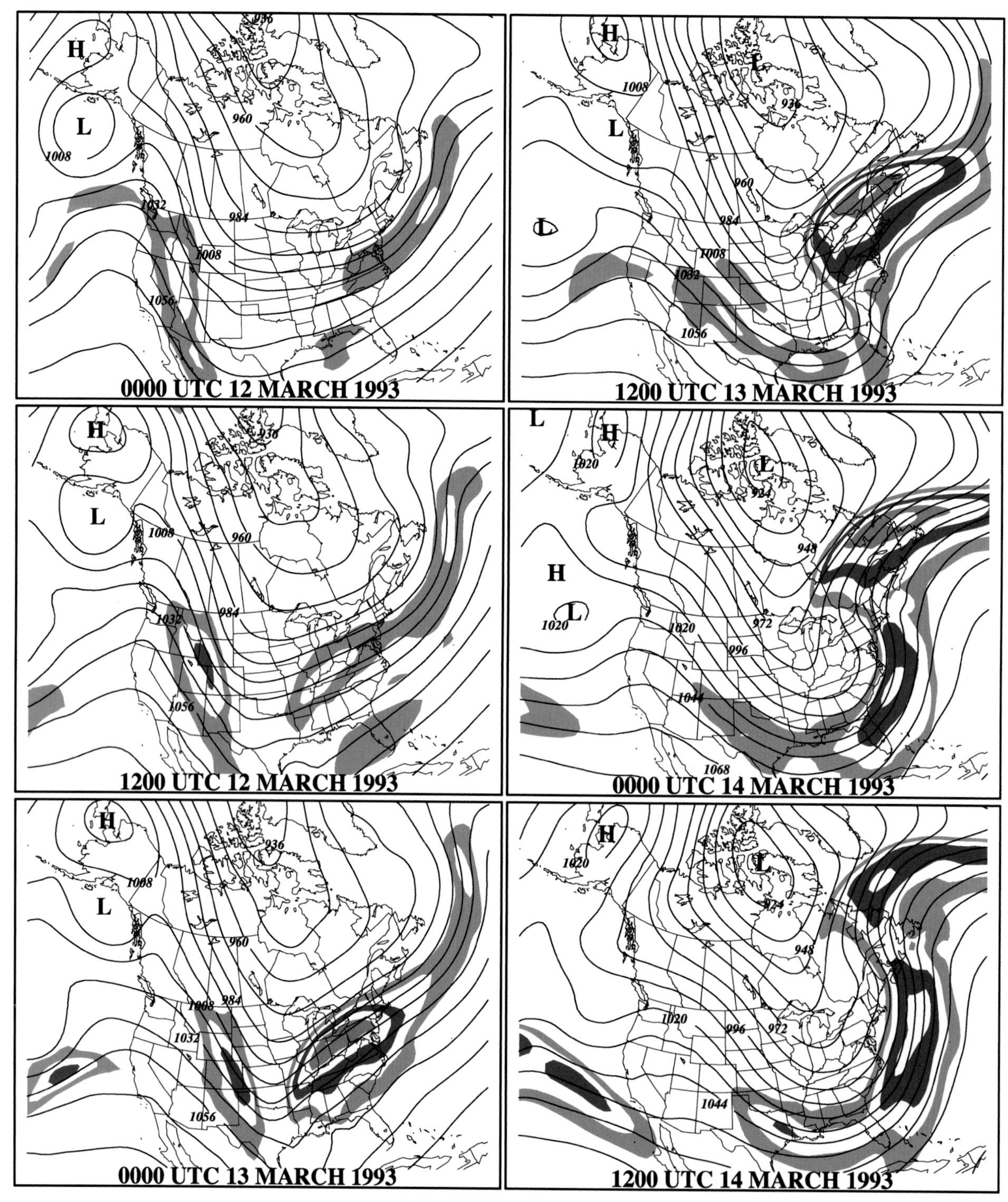

FIG. 10.24-5. Twelve-hourly analyses of 250-hPa geopotential height for 0000 UTC 12 Mar–1200 UTC 14 Mar 1993. See Fig. 10.1-5 for details.

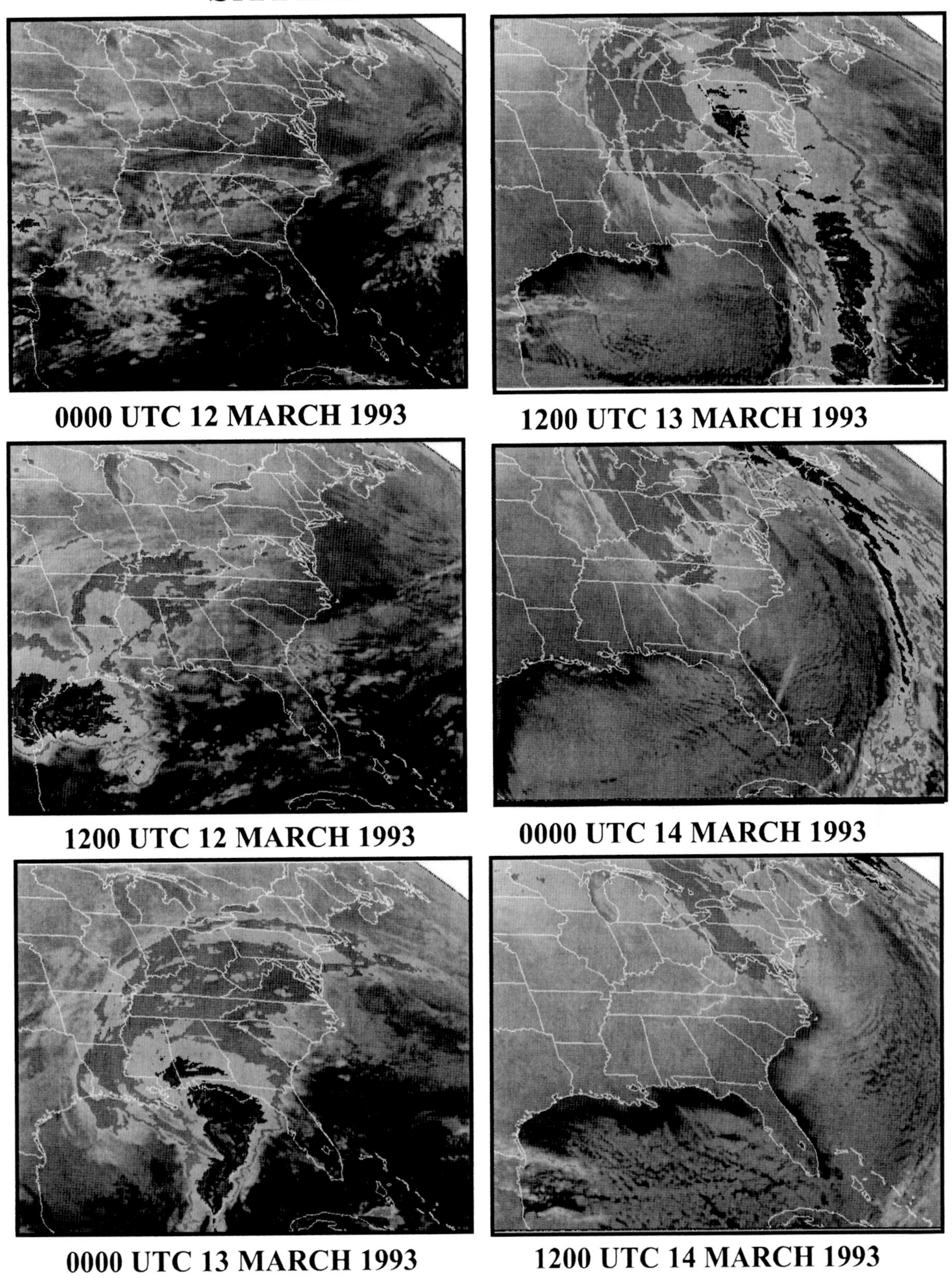

FIG. 10.24-6. Twelve-hourly infrared satellite images for 0000 UTC 12 Mar–1200 UTC 14 Mar 1993. See Fig. 10.16-6 for more details.

- By 1200 UTC 13 March, a massive lower-tropospheric circulation accompanied by increasing temperature gradients had developed along the baroclinic zone from Florida northeastward to the Carolina. By 0000 UTC 14 March, a huge cyclonic vortex covered the eastern third of the country with very strong temperature gradients found near and to the northeast of the 850-hPa low.
- The isotherm pattern had taken on the classic "S" shape on 13 March, as warm-air advection was concentrated north and east of the surface low and cold-air advections occurred to the south and west.
- During the period of rapid intensification between 1200 UTC 12 March and 1200 UTC 13 March, wind speeds surrounding the 850-hPa low increased dramatically and represent the largest areal coverage of high wind speeds of *any* of the cases.
- A 40+ m s^{-1} low-level jet (LLJ) developed in the western Gulf of Mexico as the storm system underwent the initial explosive development phase between 1200 UTC 12 March and 0000 UTC 13 March.
- The LLJ shifted to a position off the South Carolina coast by 1200 UTC and continued to intensify, attaining wind speeds greater than 55 m s^{-1} east of the Canadian coast by 1200 UTC 14 March. Easterly winds to the north of the deepening vortex also increased during this 24-h period, exceeding 30–35 m s^{-1} and directed toward the area of heaviest snowfall.

d. 500-hPa geopotential height and 400-hPa wind analyses

- A long period of El Niño conditions existed from 1990 through early 1995 (Fig. 2–19, volume I). Although the El Niño peaked in early 1992, a secondary peak occurred during spring 1993. The winter of 1992/93 was marked by the North Atlantic Oscillation (NAO) being primarily positive from November 1992 through mid-February 1993. Daily values tended to be negative from 15 February until 13 March, the date the storm formed in the Gulf of Mexico and the start of a distinct positive phase of the NAO that lasted the rest of March. Therefore, the storm occurs during a transition from a negative to a positive NAO.
- The development of the March 1993 "Superstorm" was accompanied by a massive amplification of the upper trough–ridge pattern over the eastern United States and a tremendous increase in upper-level wind speeds (discussed below).
- The precyclogenetic environment at 500 hPa was dominated by a broad cyclonic flow across much of the United States and southern Canada, with a developing upper trough across the southwestern United States and a digging upper trough over western Canada on 10–11 March.
- A broad region of confluence extended over much of the eastern United States by 0000 UTC 12 March. The confluence was located upwind of a mobile upper trough, but no closed low, over eastern Canada.
- At 0000 and 1200 UTC 12 March, two separate short-wave troughs, each associated with separate vorticity maxima, are observed over the western United States with one trough over southern Texas and the other broad trough and upper jet streak centered over the eastern Rockies and plains at 1200 UTC. The initial development of the storm over the western Gulf of Mexico occurred as the southern trough moved eastward over Texas early on 12 March. Its continued development occurred as the second trough dove southeastward toward the Gulf coast between 1200 UTC 12 March and 0000 UTC 13 March.
- The two troughs began to merge by 0000 UTC 13 March (see volume I, chapter 4) over the northern Gulf of Mexico and Louisiana. By 1200 UTC 13 March, an exceptionally amplified, negatively tilted upper trough dominated the southeastern United States, characterized by very large geopotential height falls (−290 m per 12 h). The explosive development of the surface low occurred within the diffluent region just downwind of the negatively tilted trough axis, as the systems merged.
- The wavelength between the trough and the amplifying downstream ridge also decreased dramatically. The deepening trough so far south, and the negative tilt occurring over Alabama and Georgia by 1200 UTC 13 March, are consistent with other cases in which snow changes to mixed precipitation or rain in the urban corridor sometime during the event. The merger of these two systems was also demonstrated from a potential vorticity perspective [see Kocin et al. (1995); also see volume I, chapter 7).
- As the two troughs merged and heights rose rapidly across the eastern United States on 12–13 March, the confluent zone shifted northeastward toward the U.S.–Canadian border by 1200 UTC 13 March and then lifted rapidly northward toward eastern Canada in the subsequent 12 h.
- A large upper closed low at 500 hPa developed over western Virginia by 0000 UTC 14 March as the surface low ceased developing over the eastern United States. In the following 12 h, the 500-hPa low continued to deepen and lift rapidly northeastward as the surface low weakened and moved toward New England.
- This case was marked by intense jet streak systems that evolved into the classic dual-jet pattern during the period when the storm system was moving northeastward along the East Coast.
- One jet with wind speeds near 50 m s^{-1} at 400 hPa was associated with the confluent flow over the northeastern United States on 12 March. This jet amplified to speeds greater than 80 m s^{-1} as the storm system developed and moved up the East Coast by 1200 UTC 13 March.
- Another jet streak was associated with the dual-trough

system over the western United States with maximum wind speeds exceeding 75 m s^{-1} that propagated over the western United States toward Texas as the low began to develop in the western Gulf of Mexico on 12 March. The initial development of the surface low in the Gulf of Mexico occurred in the entrance region of the northern jet but was far removed from the exit region of the western jet during its initial development.

- As cyclogenesis and thunderstorms blossomed over the Gulf of Mexico early on 12 March, the precipitation developed within the entrance region of the rapidly strengthening upper jet extending over the north-central to northeastern United States. As the two upper troughs merged, the western jet streak continued to dive southward toward the Gulf of Mexico, rounding the base of the trough at 1200 UTC 13 March with wind speeds of 50–60 m s^{-1} at 400 hPa. The expansion of the heavy precipitation northeastward across the eastern United States occurred within the exit region of the southern jet and the entrance region of the northern jet.
- As the trough deepened toward the southeastern United States and developed a negative tilt by 1200 UTC 13 March, a separate jet maximum formed east of the trough axis and amplified rapidly to speeds greater than 90 m s^{-1} at 400 hPa by 1200 UTC 14 March. As the surface low deepened to near 960 hPa over Delaware on 13 March, the storm became located within the diffluent exit region of the jet and within the entrance region of the northern jet.

1) 250-hPa GEOPOTENTIAL HEIGHT AND WIND ANALYSES

- The geopotential height analyses illustrate the strong upper-level ridging in the West, the amplifying trough into the center of the country, and the distinct confluent zone extending from the north-central United States to southeastern Canada during the evolution of this storm.
- Perhaps the most remarkable upper-level feature is the rapid enhancement of the jet streak within the confluent zone and downstream ridge between 0000 UTC 12 March and 0000 UTC 13 March, when wind speeds over the Ohio Valley exceeded 90 and 100 m s^{-1} over southeastern Canada by 1200 UTC 13 March.
- This increase to over 90 m s^{-1} at 0000 UTC 13 March, an increase of 30 m s^{-1} in only 12 h, occurred at the same time as the surface low was developing rapidly over the Gulf of Mexico and at the time the two upper troughs appear to merge. This increase occurred at the crest of an amplifying ridge over the southeastern United States, coinciding with the large values of latent release (implied with the rapid development of convection over the Gulf of Mexico). This hypothesis has been documented for other cases of rapid cyclogenesis [see, e.g., Chang et al. (1982) and Keyser and Johnson (1984); see also volume I, chapter 7].
- Between 0000 and 1200 UTC 13 March, the upper confluent zone shifted to the north-northeast over the Northeast and southeastern Canada with wind speeds exceeding 90 m s^{-1}. In the following 24 h, the jet streak continued to retreat northeastward across eastern Canada and then south of Greenland by 1200 UTC 14 March 1993, where it weakened. Meanwhile, the jet streak developing east of the trough axis after 1200 UTC 12 March becomes more evident, with the laterally coupled jet pattern well established off the Northeast coast by 0000 UTC 14 March.

e. *Infrared satellite imagery sequence*

- Satellite imagery indicates that a diffuse area of cloudiness with two separate areas of thunderstorms at 0000 UTC 12 March developed near the Texas–Mexico coast and in the Gulf of Mexico and merged into a rapidly expanding region of cold cloud tops over the western Gulf of Mexico by 1200 UTC 12 March. This expansion occurs during the initial stages of cyclogenesis in the western Gulf of Mexico.
- The rapidly expanding region of cold cloud tops is indicative of the explosive development of precipitation during the initial development of the storm and shows a well-defined comma-shaped cloud region over the Gulf of Mexico by 0000 UTC 13 March. As the surface low began to develop over the Gulf of Mexico, drier air in the satellite image works its way to the south and west of the developing surface low and the overall cloud mass expands to the north and east, an illustration of the asymmetric development of the cloud mass associated with the storm. The dry air is related to descending stratospheric air and large potential vorticity values (Uccellini et al. 1995; Kocin et al. 1995).
- A separate region of colder upper-level clouds, not nearly as intense as the cloud development over the Gulf of Mexico, began to stream northeastward toward the Tennessee Valley by 1200 UTC 12 March. This separate area spread rapidly northeastward by 0000 UTC 13 March and its northern edge corresponded with the entrance region of the rapidly intensifying jet streak over the Ohio Valley characterized by 90 m s^{-1} winds.
- The growing comma-shaped cloud covers much of the eastern United States by 1200 UTC 13 March with the heavily convective comma tail having passed east of the Florida coast. This comma tail was associated with the tornado outbreak and was followed by the storm surge along Florida's northwest coast. Compared with other storms described in this chapter, the comma cloud can be described as enormous.

- By 0000 UTC 14 March, the comma tail had weakened over the Atlantic Ocean indicative of diminishing convection while the head of the broad comma-shaped cloud was located over the middle Atlantic states. Highest cloud tops have moved rapidly northeastward away from the surface low pressure center and coincide with the cessation of cyclogenesis. In the following 12 h, the comma cloud continued to drift northeastward accompanied by a smaller area of cold cloud tops.

The Flatiron Building, Feb 1994 (photo courtesy of Kevin Ambrose, *Great Blizzards of New York City*, Historical Enterprises, 1994).

25. 8–11 February 1994

a. General remarks

- This snowstorm was a combination of two separate storm systems that produced an extensive and deep snow cover over much of southern New England, New York City, and northern portions of the middle Atlantic states. These storms occurred during one of the most severe winters of the 1990s when the Northeast urban corridor was plagued by numerous storms that produced snow, freezing rain, and ice pellets.
- The first storm occurred on 8–9 February and brought heavy snows from Pennsylvania and central New Jersey northward into southern New England with the heaviest snows falling in eastern New England, where 15–20 in. (37–50 cm) fell. From central New Jersey southward to Maryland and Delaware, freezing rain made travel hazardous.
- The second storm occurred on 11 February and was associated with one of the most damaging ice storms on record, affecting much of the southeast United States northward into Virginia and southern Maryland. This was primarily an ice pellet storm in Washington and Baltimore. While ice pellets are often not as serious as either freezing rain or snow, accumulations of 2 to as much as 6 in. (5–15 cm), with a liquid equivalent of greater than an inch (2.5 cm), brought these two cities to a standstill, closing airports and impeding auto travel. To the north, heaviest snows fell across the New York City metropolitan area into southern New England, leaving a widespread mantle of snow nearly 2 feet (60 cm) deep from New York City to Boston.
- The following regions reported snow accumulations exceeding 10 in. (25 cm): eastern and southeastern Pennsylvania, central and northern New Jersey, southeastern New York and Long Island, Connecticut, Rhode Island, Massachusetts, and scattered areas of northern and western Maryland, western Pennsylvania and New York, and southern New Hampshire.
- The following regions reported snow accumulations exceeding 20 in. (50 cm): portions of extreme eastern Pennsylvania, northern New Jersey, New York City and Long Island, Connecticut, and Massachusetts.

b. Surface analyses

- The first heavy snow-producing system was associated with several weak low pressure systems that traveled along a nearly stationary front extending from Kentucky eastward to the Virginia coast on 8–9 February. These low pressure systems were not strong but formed along a frontal boundary separating very cold air from very mild air over the southeast United States. Early on the morning of 8 February, temperatures ranged from below 0° (°F; below −18°C) across northern New England and New York to the 10°s (°F; −12°C) in Boston, the upper 10°s (°F; −6°C) in New York City, the mid- to high 20°s (°F; −4°C) in Philadelphia, to near freezing (0°C) near Baltimore, and to the mid-30°s (°F; 1°–2°C) in Washington. South of the front in Virginia, temperatures were primarily in the 50°s (°F; 10°–15°C).
- Snow spread rapidly from west to east from Pennsylvania and New York eastward into southern New England as low pressure traveling along the front developed early on 8 February over the Ohio Valley, and moved toward West Virginia by 0000 UTC 9 February (central pressure of 1007 hPa), with a new weak low pressure center developing east of the Maryland coast. This low pressure center developed slowly and continued moving east, passing southeast of southern New England by 1200 UTC 9 February 1994, with its central pressure falling to 1002 hPa.
- Snow fell across northern Pennsylvania, northern New Jersey, southeastern New York, and southern New England. Snow mixed with sleet, and freezing rain occurred in central and southern Pennsylvania and New Jersey and changed to freezing rain over Maryland and extreme northern Virginia.

TABLE 10.25-1. Snowfall amounts for urban and selected locations for 8–11 Feb 1994; see Table 10.1-1 for details.

NESIS = 4.81 (category 3)			
	Combined	8–9 Feb	11 Feb
Urban center snowfall amounts			
Washington, DC–Dulles Airport	3.5 in. (9 cm)	0.4 in. (1 cm)	3.1 in. (8 cm)
Baltimore, MD	4.6 in. (12 cm)	0.5 in. (1 cm)	4.1 in. (10 cm)
Philadelphia, PA	10.3 in. (26 cm)	5.7 in. (14 cm)	4.6 in. (12 cm)
New York, NY–Central Park	21.7 in. (55 cm)	9.0 in. (23 cm)	12.7 in. (32 cm)
Boston, MA	26.9 in. (68 cm)	18.7 in. (47 cm)	8.2 in. (21 cm)
Other selected snowfall amounts			
Newark, NJ	30.1 in. (76 cm)	12.1 in. (31 cm)	18.0 in. (45 cm)
Bridgeport, CT	23.4 in. (59 cm)	10.8 in. (27 cm)	12.6 in. (32 cm)
Hartford, CT–Bradley International Airport	19.5 in. (50 cm)	10.0 in. (25 cm)	9.5 in. (24 cm)
Allentown, PA	19.0 in. (48 cm)	9.2 in. (23 cm)	9.8 in. (25 cm)
Providence, RI	16.8 in. (43 cm)	7.0 in. (18 cm)	9.8 in. (25 cm)
Harrisburg, PA	15.0 in. (38 cm)	6.2 in. (16 cm)	8.8 in. (22 cm)

(a)

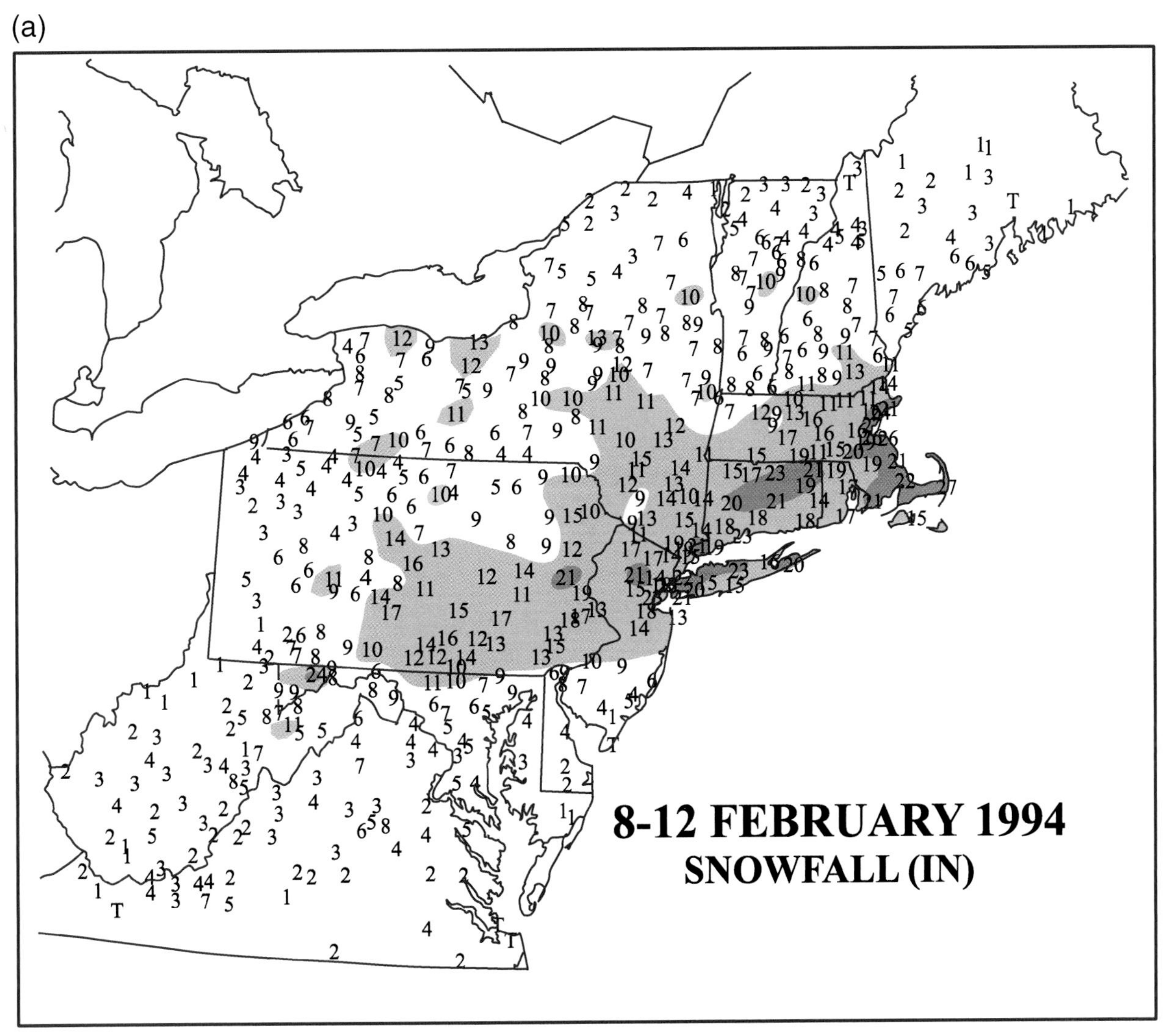

FIG. 10.25-1a. Snowfall (in.) for combined storms of 8–9 Feb and 10–11 Feb 1994. See Fig. 10.1-1 for details.

- Generally, 8–12 in. of snow (20–30 cm) fell from eastern Pennsylvania, northern New Jersey, and southeastern New York, to southern New England (Fig. 10.25-1b). Heaviest snows fell in eastern Massachusetts where more than 18 in. (45 cm) fell in Boston.
- Cold air for the snow event was supplied by a large high pressure system extending across the northern half of the United States. One center of high pressure (1035 hPa) was centered north of New York at 1200 UTC 8 February. This air mass remained locked over New England, with its corresponding surface ridge extending southward to Virginia, where the cold air did not progress any farther south.
- As the first low pressure system began to intensify off southern New England, by 1200 UTC 9 February, the nearly stationary front began to move slowly south as a cold front. Across the south-central and southeastern United States, the very mild conditions were suddenly replaced by much colder temperatures near the surface.
- As the cold front moved southward to the Gulf of Mexico, high pressure over the Midwest and Northeast intensified to1040 hPa over northern New York by 0000 UTC 11 February. Cold air continued to spill southward along the eastern slopes of the Appalachians, with an impressive wedge of cold air moving southward from the Carolinas to southern Georgia, indicative of a significant cold-air damming event.
- As high pressure moved eastward toward the Northeast, low pressure developed along the Gulf coast frontal boundary early on 10 February (1008 hPa) and began moving northeastward over the following 24–36 h. With very shallow cold air and milder air aloft, heavy freezing rain developed across Louisiana, Mississippi, Arkansas, and Tennessee late on 10 February. This was a devastating ice storm.
- As this low pressure moved northeastward from the Gulf of Mexico to Alabama (while not intensifying) and cold-air damming occurred east of the Appala-

(b)

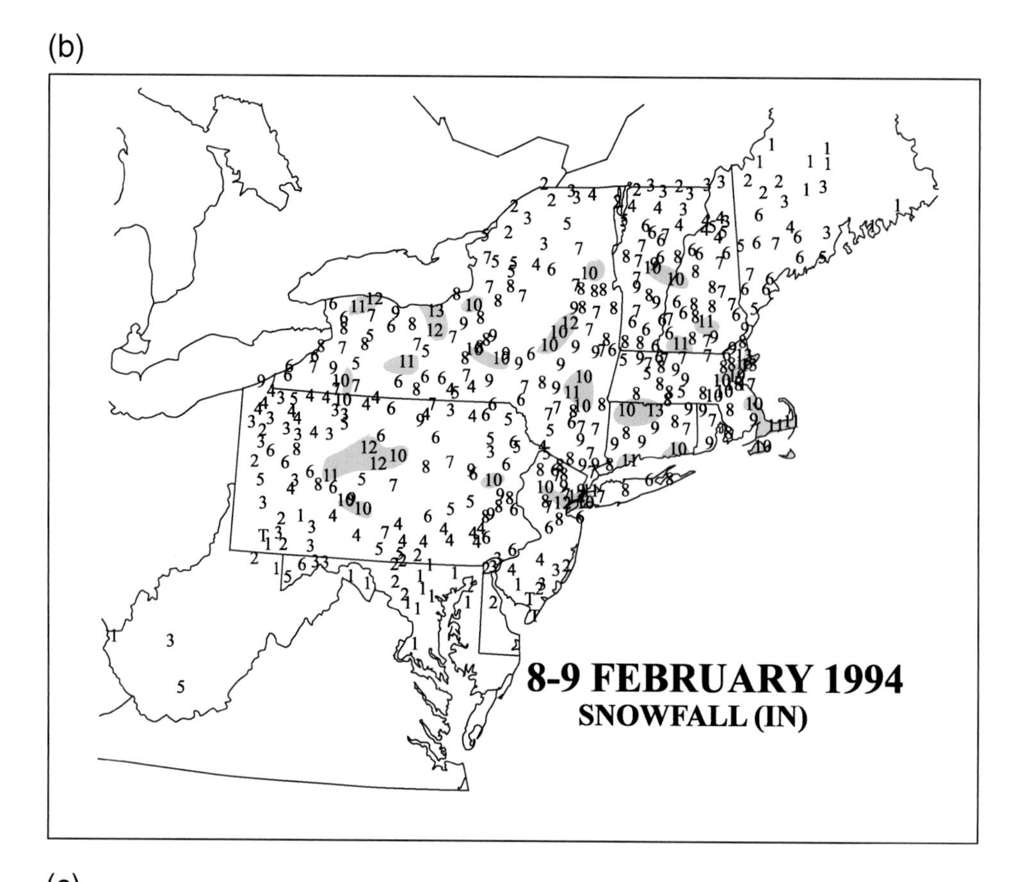

(c)

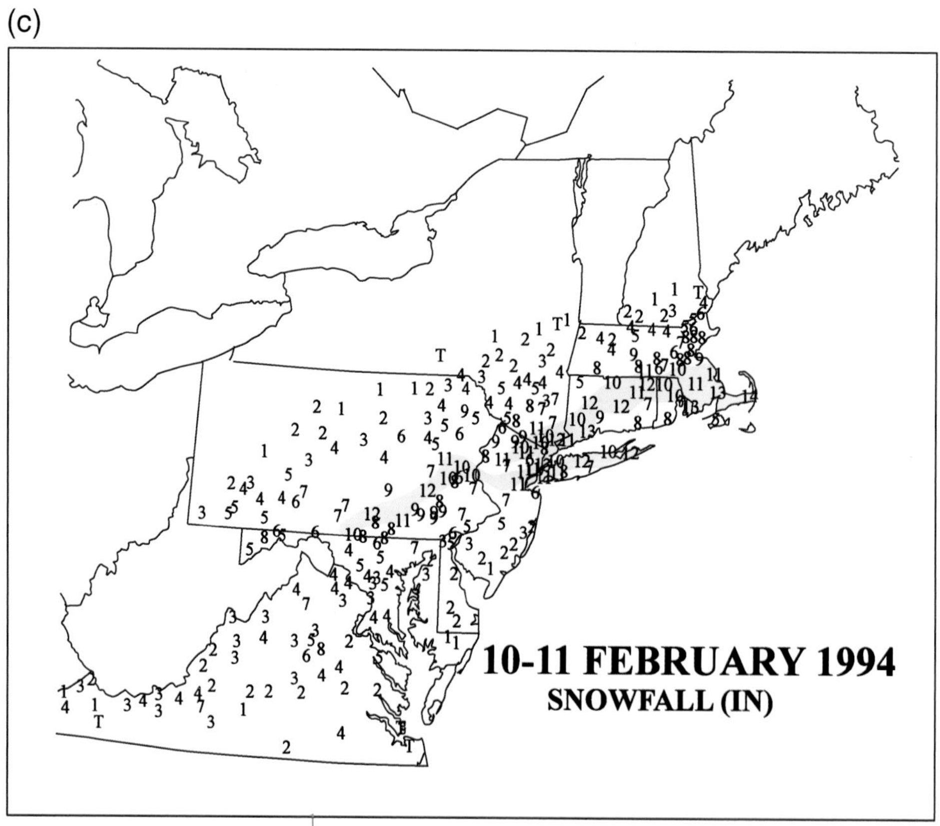

FIG. 10.25-1b, c. Snowfall (in.) for storms of (top) 8–9 Feb and (bottom) 10–11 Feb 1994. See Fig. 10.1-1 for details.

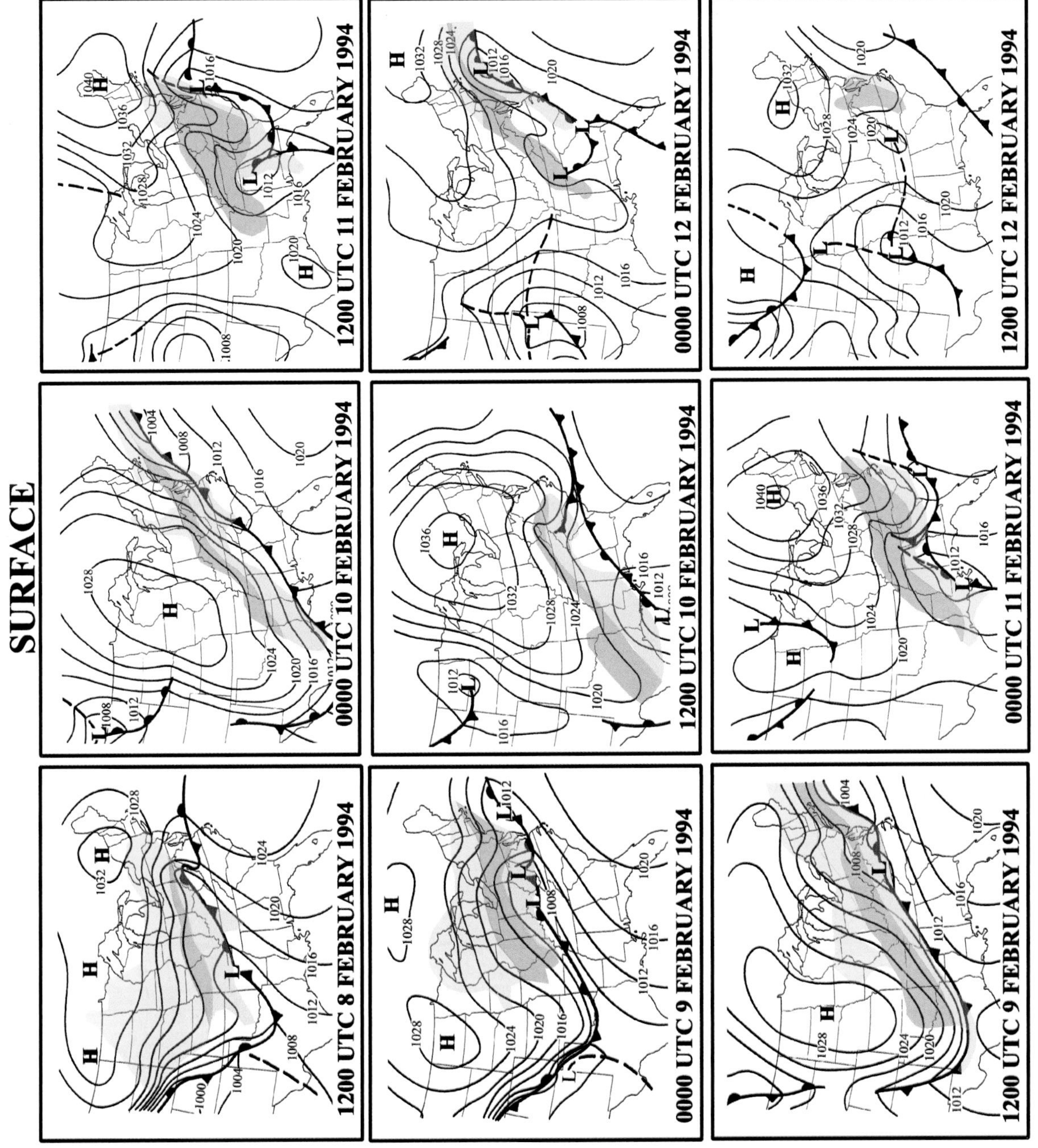

FIG. 10.25-2. Twelve-hourly surface weather analyses for 1200 UTC 8 Feb–1200 UTC 12 Feb 1994. See Fig. 10.1-2 for details.

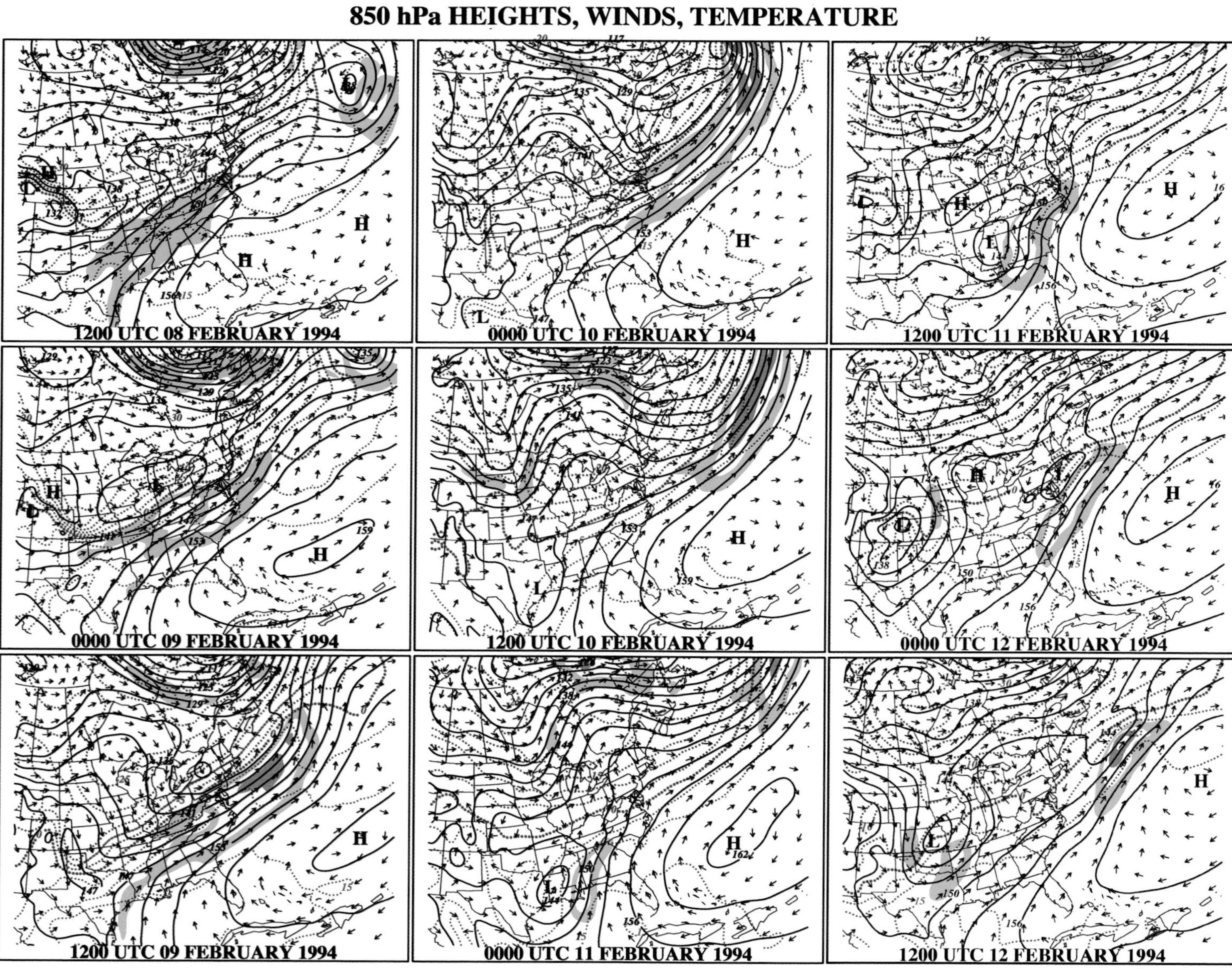

FIG. 10.25-3. Twelve-hourly 850-hPa analyses for 1200 UTC 8 Feb–1200 UTC 12 Feb 1994. See Fig. 10.1-3 for details.

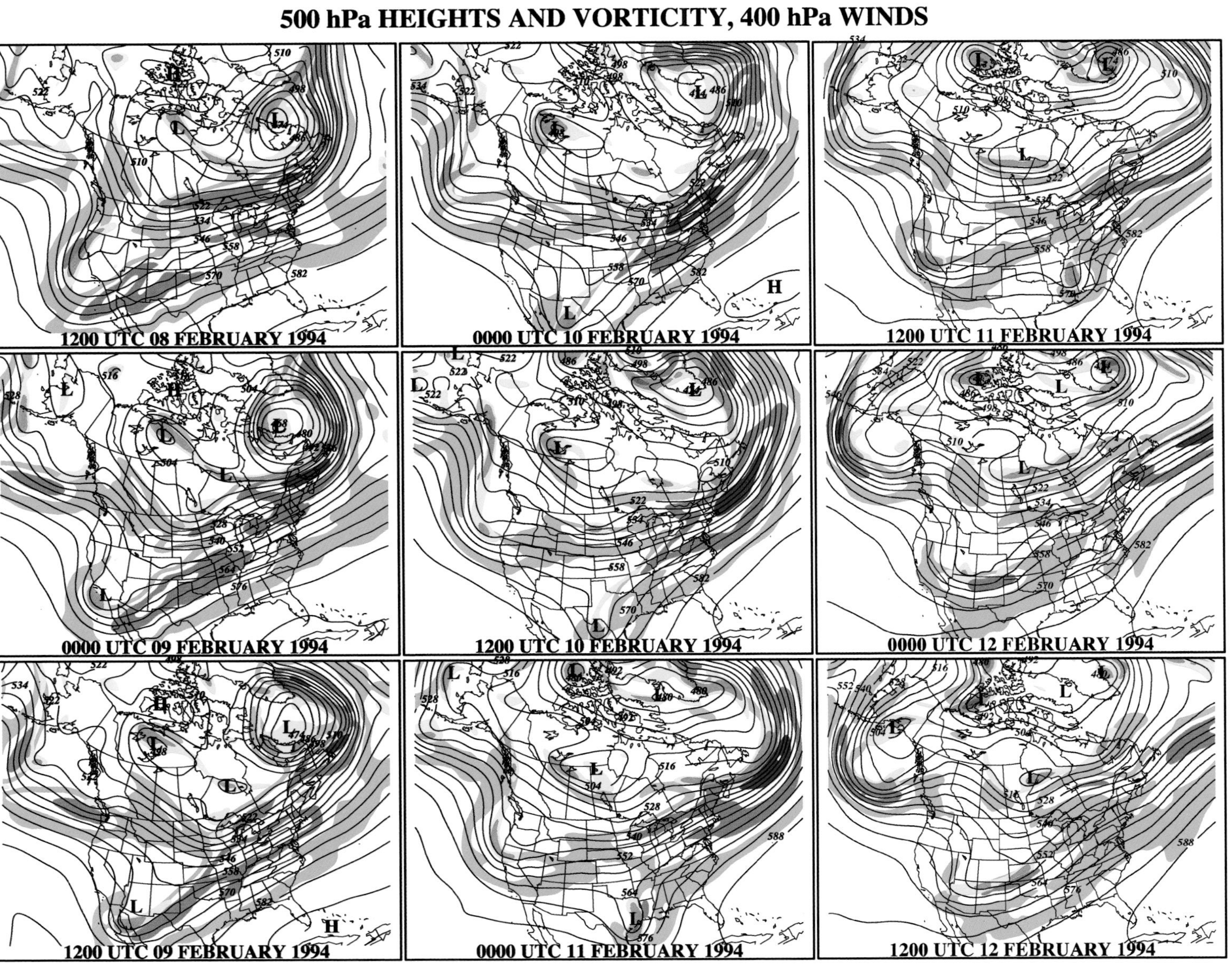

FIG. 10.25-4. Twelve-hourly analyses of 500-hPa geopotential heights and upper-level wind fields for 1200 UTC 8 Feb–1200 UTC 12 Feb 1994. See Fig. 10.1-4 for details.

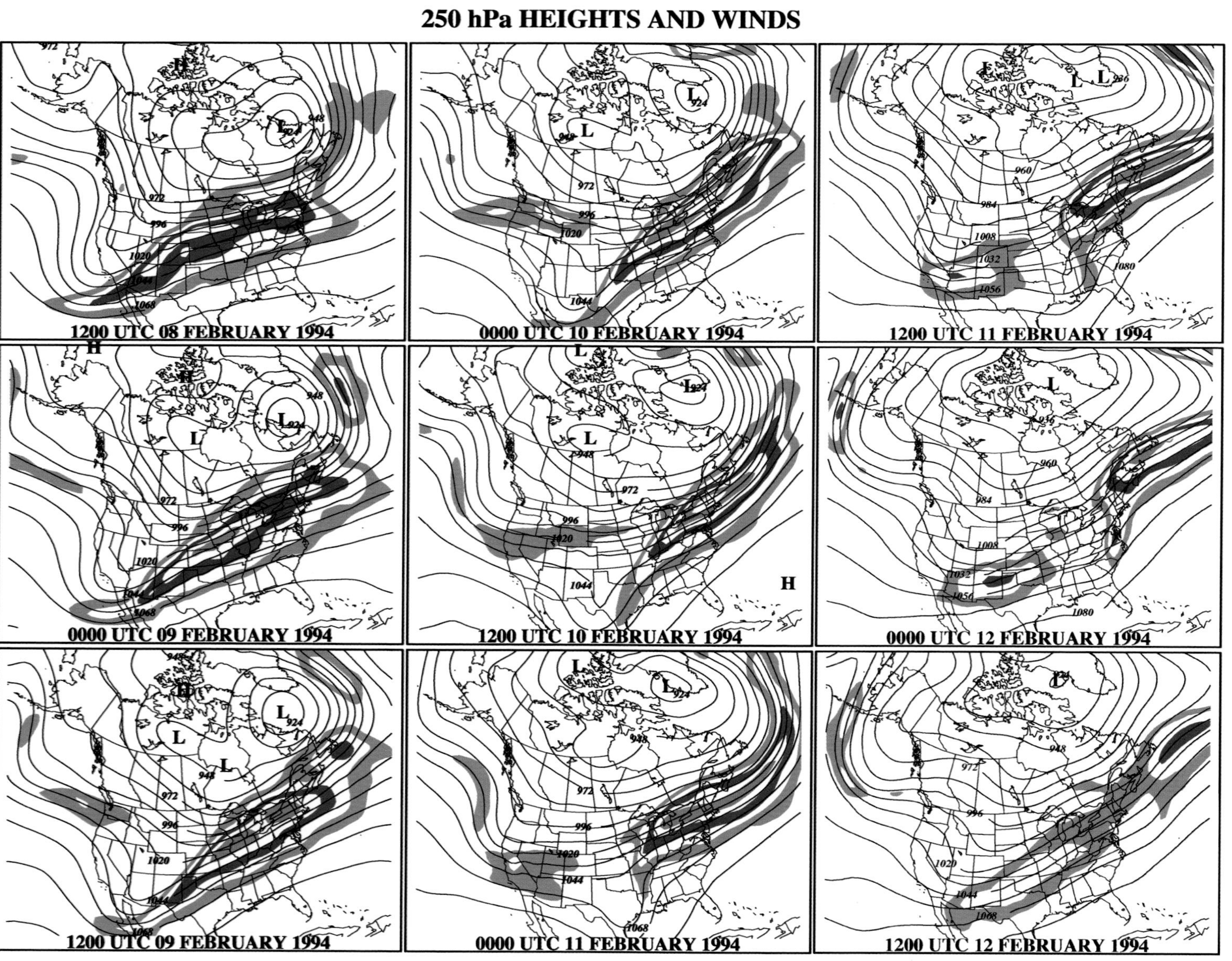

FIG. 10.25-5. Twelve-hourly analyses of 250-hPa geopotential height for 1200 UTC 8 Feb–1200 UTC 12 Feb 1994. See Fig. 10.1-5 for details.

(a)
SATELLITE IMAGERY

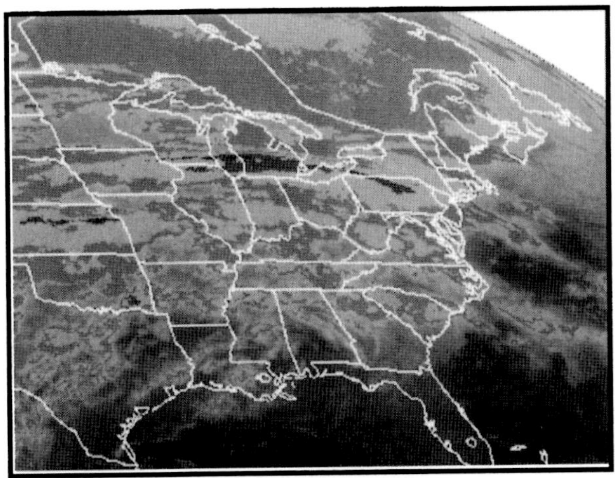

1200 UTC 8 FEBRUARY 1994

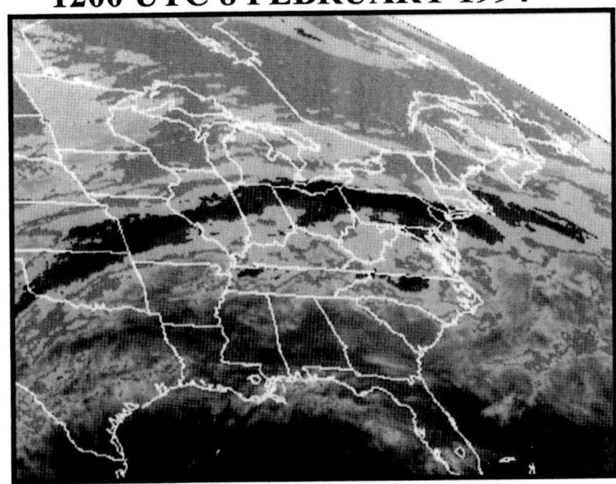

0000 UTC 9 FEBRUARY 1994

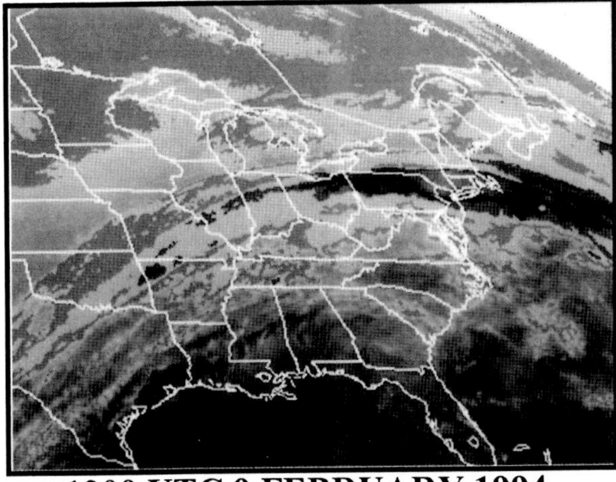

1200 UTC 9 FEBRUARY 1994

FIG. 10.25-6a. Twelve-hourly infrared satellite images for 1200 UTC 8 Feb–1200 UTC 9 Feb 1994. See Fig. 10.16-6 for more details.

chians, a coastal front developed just east of the Southeast coast late on 10 February and a second low pressure system developed along this front (1018 hPa at 0000 UTC 11 February). While this low pressure system was not a rapidly developing cyclone, it was still marked by significant pressure falls along the mid-Atlantic coast and deepened to 1012 hPa by 1200 UTC 11 February. The surface cyclone moved northeastward along the East Coast before veering east-northeastward over the Atlantic south of New England late on 11 February (1010 hPa at 0000 UTC 12 February), and early on 12 February this cyclone was located well east of southern New England.

- Freezing rain spread northeastward from the lower Tennessee Valley to Virginia and southern Maryland late on 10 February while mostly ice pellets occurred across Maryland and southern New Jersey on 11 February, where temperatures remained in the 10°s (°F; <-7 °C). To the north, precipitation was mostly in the form of moderate to heavy snow from eastern Pennsylvania and central New Jersey northward into southern New England. Most of the snow ended late on 11 February. More than 10 in. (25 cm) of snow fell during this event from northern New Jersey, New York City, and into southern New England (Fig. 10.25-1c).

c. 850-hPa analyses

- As the first period of significant snow was developing on 8–9 February, the main features at 850 hPa were a weak low or trough moving eastward from the Midwest to the Northeast along a very strong zone of temperature contrast. The 0°C isotherm generally extended from north of the Ohio River to the Maryland–Pennsylvania border and moved slowly northeastward to southern Pennsylvania and New Jersey by 9 February. Snow fell north of this isotherm while freezing rain occurred just to the south over Maryland and northern Virginia, where above-freezing temperatures at 850 hPa overlay subfreezing surface temperatures north of the surface stationary front.
- As the 850-hPa trough moved eastward off New England on 10 February, colder temperatures moved southward in its wake with the 0°C isotherm reaching southern Kentucky and Virginia by 0000 UTC 10 February. After this time, warmer air began moving northward in the 600–800-hPa layer (not shown) as cold air at the surface continued to drift southward toward southern Georgia through 0000 UTC 11 February. With milder air moving north and cold air remaining at the surface, freezing rain and sleet occurred over a large region from Arkansas eastward to Virginia, Maryland, and New Jersey.
- As low pressure developed in the western Gulf of Mexico early on 10 February, a corresponding 850-hPa low moved from Texas northeastward to northern

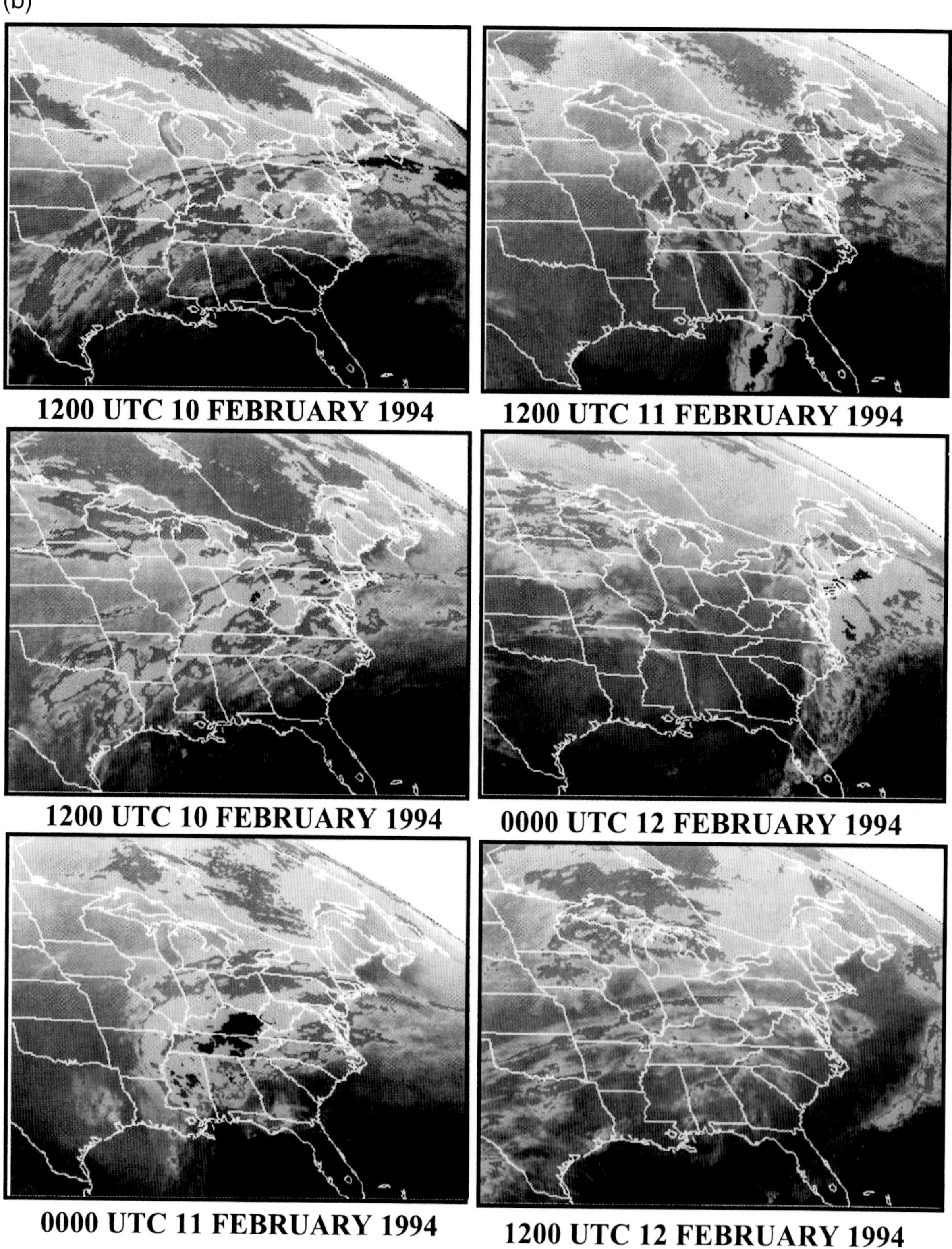

FIG. 10.25-6b. Twelve-hourly infrared satellite images for 1200 UTC 10 Feb–1200 UTC 12 Feb 1994. See Fig. 10.16-6 for more details.

Mississippi at 1200 UTC 11 February. Meanwhile, as a secondary surface low developed off the Southeast coast to the middle Atlantic coast by 1200 UTC 11 February, it corresponded with an enhanced temperature gradient over the middle Atlantic states and a weak 850-hPa trough. A weak low developed within this trough by 0000 UTC 12 February, with the surface low moving south of the New England coast by 1200 UTC 12 February.

- The 0°C isotherm continued to move toward the Washington–Baltimore region, being located over the region by 0000 UTC 12 February. With surface temperatures near 20°F (−7°C) in this area, ice pellets were the predominant precipitation mode. Farther south, where milder temperatures and a deeper layer of milder air overlay the cold surface air, precipitation fell mainly as freezing rain across North Carolina northward into eastern Virginia and southern Maryland.
- The second area of heavy snow was associated with an enhanced and narrow region of warm-air advection at 850 hPa with only a weak low center. Once this region of enhanced advections passed east of New England, it was replaced by a much weaker region of temperature difference by 0000 UTC 12 February.
- Low-level wind speeds greater than 20 m s^{-1} were directed from the southwest to northeast over a broad area from the south-central United States to New England and marked the first precipitation event on 8–9 February.
- The second event was marked by a low-level jet (LLJ), which exceeded 25 m s^{-1} as it developed along the Gulf coast by 0000 UTC 11 February and extended northeastward during the next 24 h. This jet feature appears to be transporting moisture from the Gulf region into the area of heavy freezing rain, sleet, and snow as it moved northeastward during the second event on 11–12 February.

d. 500-hPa geopotential height and 400-hPa wind analyses

- The winter of 1993/94 was cold and stormy across the Northeast and occurred during weak El Niño conditions that reverted toward a more neutral position around February 1994 (Fig. 2-19, volume 1). Beginning in November 1993, daily values of the North Atlantic Oscillation (NAO) became distinctly positive and continued in that mode through mid-February. Therefore, the two snowstorms that occurred from 8 February through 12 February occurred during the end of a long positive phase of the NAO.
- As the first area of heavy snow began to affect portions of the Northeast urban corridor, the main features governing the 500-hPa circulation were a deep vortex over eastern Canada, a strong upper ridge south of Florida, and a distinct, low-amplitude short-wave trough moving through the Great Lakes. Very strong southwesterly flow in between the trough–ridge pattern extended from the southwest United States to the midwest and northeast. The presence of a deep vortex in eastern Canada and a strong ridge over the southeast United States is consistent with the large north–south temperature contrasts observed at the surface (as indicated by the frontal boundary) and at 850 hPa.
- At 1200 UTC 8 February, large-scale confluence was found over much of the central and eastern United States and southern Canada. This region of confluence was located over the extended region of high pressure and cold surface temperatures found from the plains states to the Northeast.
- One short-wave trough over southern Canada at 1200 UTC 8 February moved eastward across southern Canada and the western Great Lakes through the end of 9 February and appears to be associated with the eastward movement of weak low pressure along the surface frontal boundary. As the trough continued eastward after that time, the surface stationary front moved south as a cold front and cold air moved southward across the south-central and southeast United States.
- As this first trough lifted to the northeast, 500-hPa heights rose across much of the eastern United States and a second short-wave trough began to lift northeastward from the southwestern United States on 9–10 February. Upper-level confluence was reinforced across the eastern United States on 10–11 February as the southeast ridge began to build northward during this period and was associated with building high pressure over the eastern region during this time.
- As the southwestern trough lifted northeastward into the confluent 500-hPa environment on 10–11 February, the southwestern trough lost amplitude and damped considerably. At 0000 UTC 11 February, a significant short-wave trough is found over Texas. By 1200 UTC 11 February, the trough is still a significant feature but has begun to deamplify. By 0000 UTC 12 February, the trough is barely discernible over the Northeast as it continues to lift northeast. This is an example of heavy snow in the Northeast occurring despite a weakening upper-level trough.
- Both of these heavy snow events were dominated by a confluent jet streak in the Ohio Valley that moved off the New England coast.
- At 1200 UTC 9 February, a 60+ m s^{-1} jet streak at 400 hPa propagated between the eastern Canadian vortex and the Southeast ridge. As the 500-hPa trough moved eastward toward eastern Canada by 0000 UTC 10 February, the upper-level jet consolidated and amplified to greater than 80 m s^{-1}. Maximum wind speeds passed just to the north of the regions experiencing the heavy snows associated with the first snow-producing system. The first snow event appears to occur within the exit region of this jet as it propagated northeastward on 9–10 February.

- As the southwestern upper trough moved northeastward toward the East Coast by 1200 UTC 11 February, there was a weak jet associated with this system east of the trough axis that was directed toward the entrance region of the jet system described above. At this time, the southerly LLJ had formed and moved northeastward and heavy frozen precipitation was occurring to the east of the lifting upper trough and in the right-entrance region of the more northern jet. Therefore, this second area of snow was associated with a strong jet entrance region of the northern jet and weak exit region of the more diffuse jet associated with the weakening, lifting southern trough.
- The lack of a strong southern jet and weakening short-wave trough is consistent with the relatively weak and slowly deepening surface low. It would appear that the heavy snow associated with this case is related almost entirely to the ascent within the entrance region of a strong polar jet streak over New England.

e. 250-hPa geopotential height and wind analyses

- The height field for the entire period from 8 through 12 February was dominated by confluent flow extending from the central United States through New England to southeastern Canada, with a deep vortex moving very slowly from eastern Canada toward Greenland.
- Within this confluent flow field and above the surface frontal zone, a series of jet streaks on 8 and 9 February with wind speeds increasing to 90 m s^{-1} consolidated into a jet streak with wind speeds exceeding 100 m s^{-1} over northern New York and New England by 10 February.
- The first band of heavy snow extending into New York City and southern New England on 0000 UTC 9 February developed within the elongated exit region of the intensifying jet streak. The more significant snow and ice storm on 10–12 February evolved within the entrance region of this jet as it propagated eastward through New England. The southern trough that propagated northeastward from Texas on 10–11 February appears to diminish and deamplify as it interacts with this intense jet system.

f. Infrared satellite imagery sequence

- The strong frontal boundary and first snowstorm can be observed as an enhancing stream of high anticyclonically curved cloud tops associated with an intensifying upper-level jet with maximum wind speeds increasing to over 90 m s^{-1} by 0000 UTC 9 February.
- The first area of enhanced cold cloud tops moves eastward over New England and over the western Atlantic by 1200 UTC 9 February as the first low pressure system moves east of New England. However, cold cloud tops still extend far southwestward toward Texas as the next short-wave trough approaches Texas. This region of clouds remains disorganized through 1200 UTC 10 February.
- As the Texas short-wave lifted northeastward, a well-defined anticyclonically curved cloud organized over the Tennessee and Ohio Valleys as the surface low moved northeastward from the Gulf of Mexico into Louisiana. Note the enhancing darker cloud tops from Mississippi northeastward into Kentucky above the areas being affected by freezing rain.
- A comma-shaped cloud mass is evident over the eastern United States by 1200 UTC 11 February with cyclogenesis occurring along the East Coast. The darker, higher cloud tops have disappeared, coinciding with the lack of amplification of the upper-level system associated with this cyclone. The cloud drifted eastward without significant enhancement as the surface low deepened marginally and the upper-level trough deamplified.

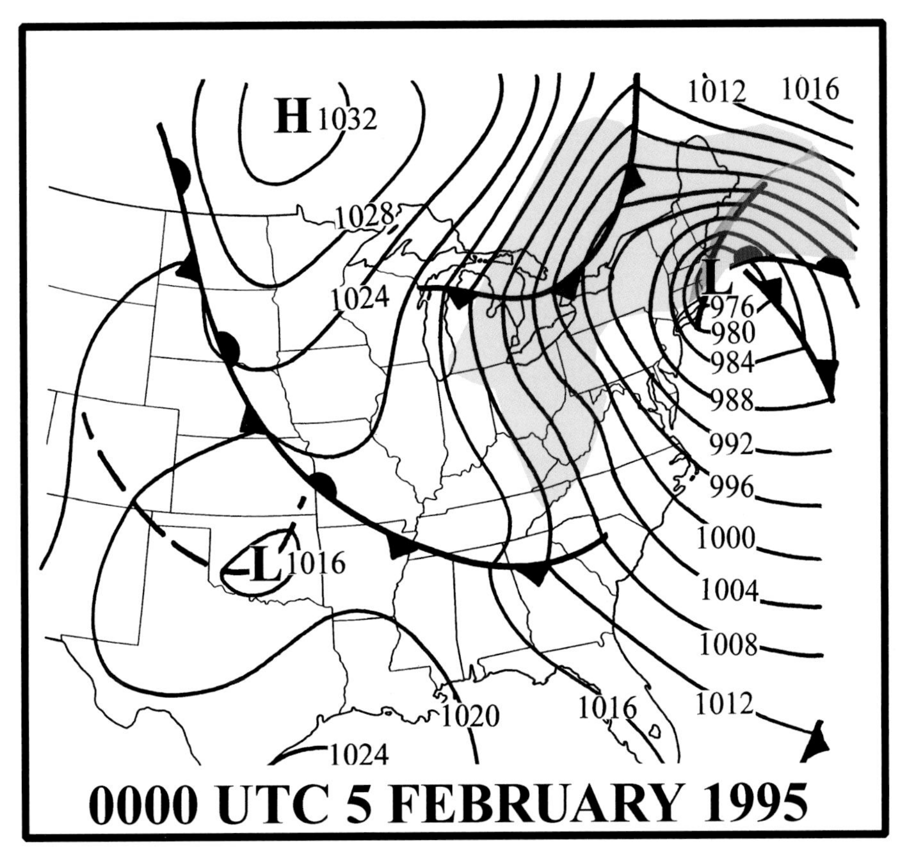

Surface weather analysis for 0000 UTC 5 Feb 1995.

26. 3–5 February 1995

a. General remarks

- This storm was the only significant Northeast snowstorm to occur during one of the least snowiest winters of the last half of the 20th century. This storm leans toward being classified as an "interior snowstorm" (see volume I, chapter 5) for which there was a tendency for snow to change to rain within the immediate coastal areas. This storm is related to a rapidly developing cyclone that hugs the Northeast coast, with heaviest snowfall amounts occurring immediately inland. However, even in the urban regions, significant accumulations of 6–10 in. (15–25 cm) fell. Particularly hard hit were the northern and western suburbs of Philadelphia, where up to 16 in. (41 cm) fell, and New York City, where up to 18 in. (46 cm) fell. In New York City and Boston, the close path of the low pressure center along the coastline allowed snow to change to rain, but not before significant snow accumulations occurred.
- The following regions reported snow accumulations exceeding 10 in. (25 cm): eastern Pennsylvania, central and northern New Jersey, eastern New York, western Connecticut, Massachusetts (except the coast and southeast), Vermont, New Hampshire, and Maine.
- The following regions reported snow accumulations exceeding 20 in. (50 cm): scattered portions of north-central New York and Vermont.

TABLE 10.26-1. Snowfall amounts for urban and selected locations for 3–5 Feb 1995; see Table 10.1-1 for details.

NESIS = 3.51 (category 2)	
Urban center snowfall amounts	
Washington, DC–Dulles Airport	6.2 in. (16 cm)
Baltimore, MD	5.3 in. (13 cm)
Philadelphia, PA	8.8 in. (22 cm)
New York, NY–LaGuardia Airport	10.4 in. (26 cm)
Boston, MA	6.6 in. (17 cm)
Other selected snowfall amounts	
Concord, NH	14.1 in. (36 cm)
Albany, NY	13.3 in. (34 cm)
Worcester, MA	12.6 in. (32 cm)
Hartford, CT–Bradley International Airport	9.8 in. (25 cm)

b. Surface analyses

- The surface low tracked from the southern plains states on 2 February eastward toward the Appalachians late on 3 February and intensified slowly as it reached eastern Kentucky at 0000 UTC 4 February, deepening to 1002 hPa.
- As the low reached West Virginia, continuing to deepen slowly, a new secondary low began to develop in northeastern South Carolina, deepening rapidly on the night of 3/4 February. As the secondary low deepened rapidly, the primary low over West Virginia was quickly absorbed into the expanding circulation as the coastal low deepened to 992 hPa over eastern Maryland by 1200 UTC 4 February.
- The location of a surface low over the Delmarva Peninsula is consistent with snowstorms characterized by a changeover from snow to rain immediately along the Northeast coast.
- The surface low deepened explosively as it moved northeastward from eastern Maryland to southeastern New England on 4 February. The low deepened 22 hPa from the onset of secondary cyclogenesis at 0600 UTC to 1800 UTC 4 February, when the central pressure fell to 980 hPa near Long Island. The storm system continued to deepen as it moved northeastward, reaching eastern Maine by the morning of 5 February with a central pressure of 962 hPa.
- The storm was preceded by a modest cell of high pressure that drifted southeastward from north of the Great Lakes at 0000 UTC 3 February to near Boston in 24 h with a central pressure of only between 1020 and 1023 hPa. The high pressure was accompanied by moderately cold air that was sufficient to produce snow along the entire breadth of the Northeast urban corridor.
- Light to moderate precipitation fell as all snow in the Washington and Baltimore metropolitan areas. As the storm gained intensity farther north, snow became heavier but also tended to change over to rain near the coastline as temperatures rose above freezing. With the storm track remaining close to the mid-Atlantic coast, snow changed to rain in Philadelphia, New York City, and Boston, but not before 8–10 or more inches (20–25 cm) fell over much of the Philadelphia and New York regions. As the storm crossed southeastern New England, snow changed to rain more rapidly in the Boston metropolitan area, sparing them a more significant snowfall.
- Wind speeds were relatively light along the coast despite the storm's rapid development. Lack of a strong surface anticyclone diminished the impact of high winds from this storm until the snow ended.

c. 850-hPa analyses

- This is a case characterized by an 850-hPa trough that evolves into a closed circulation only after the storm system moved up along the East Coast. The lack of a preexisting 850-hPa low may help explain low snowfall totals in Baltimore and Washington since a strong low-level jet and warm-air advections were not present until the low-level circulation developed after the storm moved by, after 0000 UTC 4 February.
- Prior to cyclogenesis at 1200 UTC 3 February, cold air marked by temperatures of −5° to −15°C extended across the northeastern United States within a weak

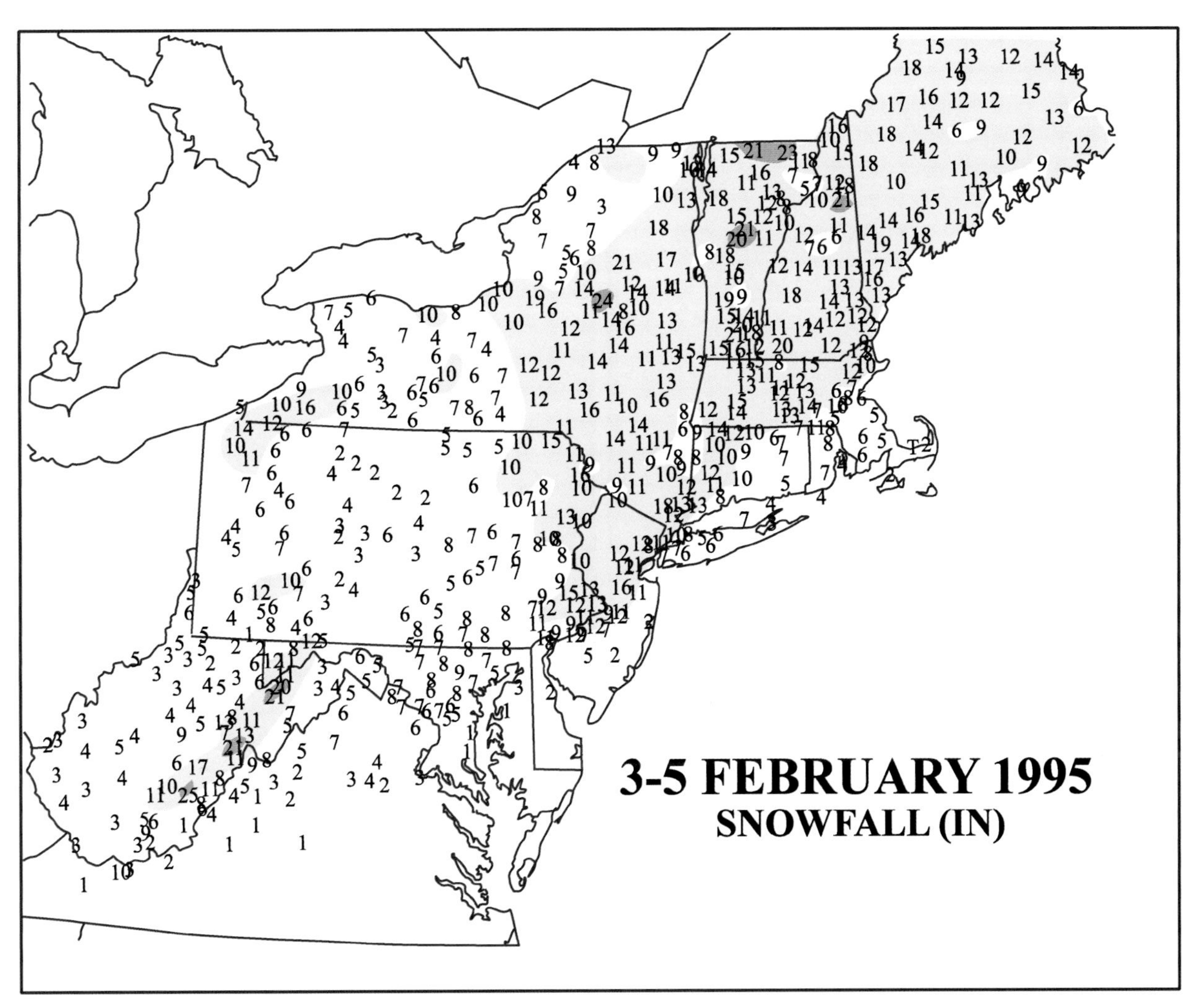

FIG. 10.26-1. Snowfall (in.) for 3–5 Feb 1995. See Fig. 10.1-1 for details.

ridge. This ridge moved eastward by 0000 UTC 4 February and was replaced by significant south-to-southwesterly flow, characterized by warm-air advection along the entire coast and toward northern New York and southeastern Canada.

- Consistent with explosive surface cyclogenesis, an 850-hPa trough deepened over the eastern United States, developing a closed center by 1200 UTC 4 February and intensified rapidly over the next 24 h off the Long Island coast and over New England.
- During the initial intensification of the 850-hPa center between 0000 and 1200 UTC 4 February, southerly winds in advance of the trough increased from 20 m s^{-1} over South Carolina to greater than 30 m s^{-1} but off the North Carolina coast by 1200 UTC 4 February. This low-level jet (LLJ) increased to greater than 45 m s^{-1} by 1200 UTC 5 February, reflecting the rapid intensification as the system moved into New England.
- For this case, most of the Northeast urban corridor was affected by a southerly 850-hPa wind component out ahead of this system. The easterly component to the intensifying low-level jet (as seen in many other heavy snow cases) developed only farther north over interior New England as the 850-hPa low deepened rapidly by 0000 UTC 5 February as the cyclone developed explosively.
- The 0°C isotherm advanced northward along the Northeast coast between 1200 UTC 4 February and 0000 UTC 5 February as the 850-hPa low deepened rapidly and was associated with the changeover from snow to rain from Philadelphia to Boston. An S-shaped pattern to the isotherms also developed by 0000 UTC 5 February 1995.

d. 500-hPa geopotential height and 400-hPa wind analyses

- The storm occurred at the end of a very long period of El Niño, dating to 1990. It also occurred during a winter that was overwhelmingly dominated by a positive North Atlantic Oscillation (NAO). However, dai-

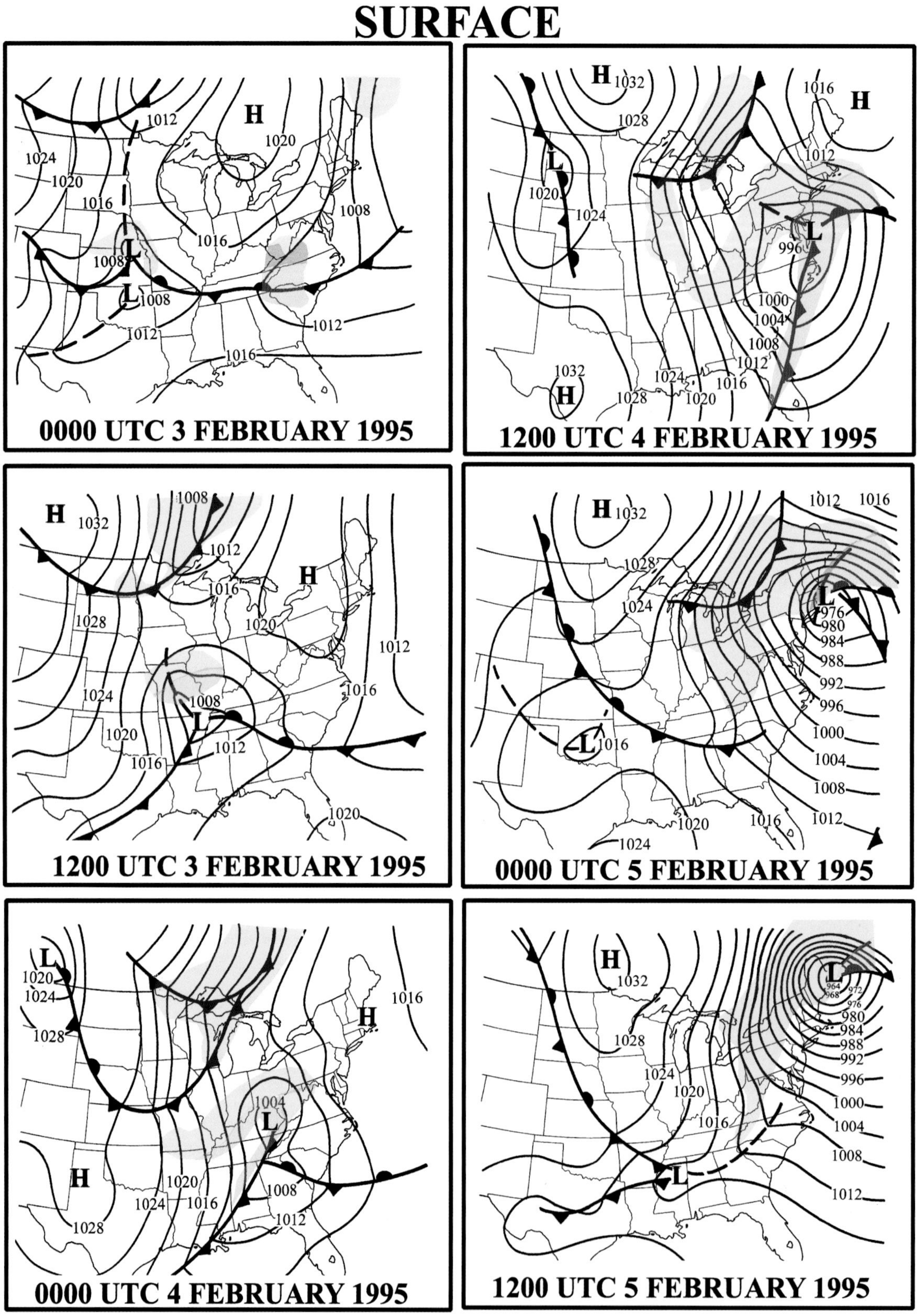

SURFACE

0000 UTC 3 FEBRUARY 1995

1200 UTC 4 FEBRUARY 1995

1200 UTC 3 FEBRUARY 1995

0000 UTC 5 FEBRUARY 1995

0000 UTC 4 FEBRUARY 1995

1200 UTC 5 FEBRUARY 1995

FIG. 10.26-2. Twelve-hourly surface weather analyses for 0000 UTC 3 Feb–1200 UTC 5 Feb 1995. See Fig. 10.1-2 for details.

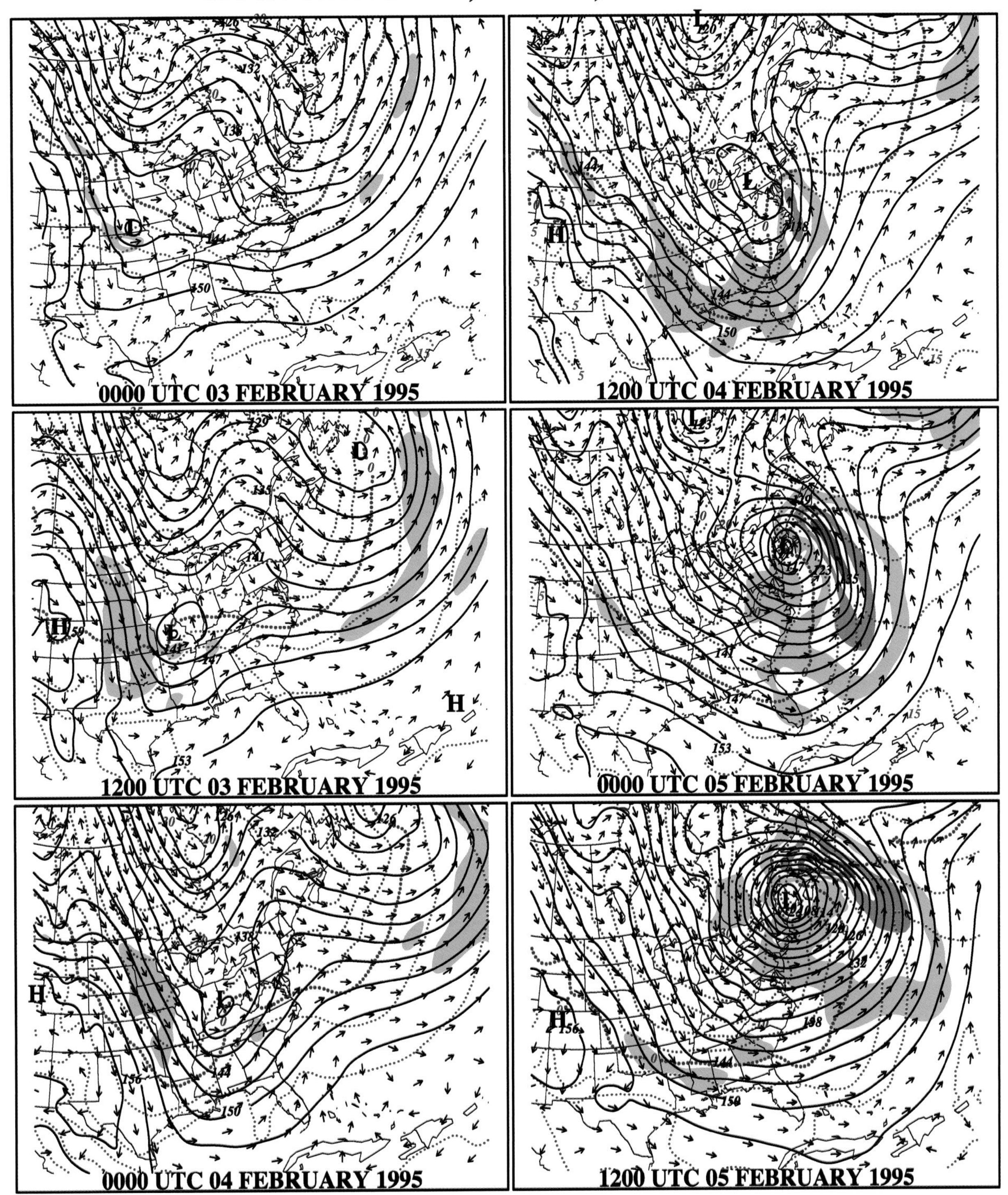

FIG. 10.26-3. Twelve-hourly 850-hPa analyses for 0000 UTC 3 Feb–1200 UTC 5 Feb 1995. See Fig.10.1-3 for details.

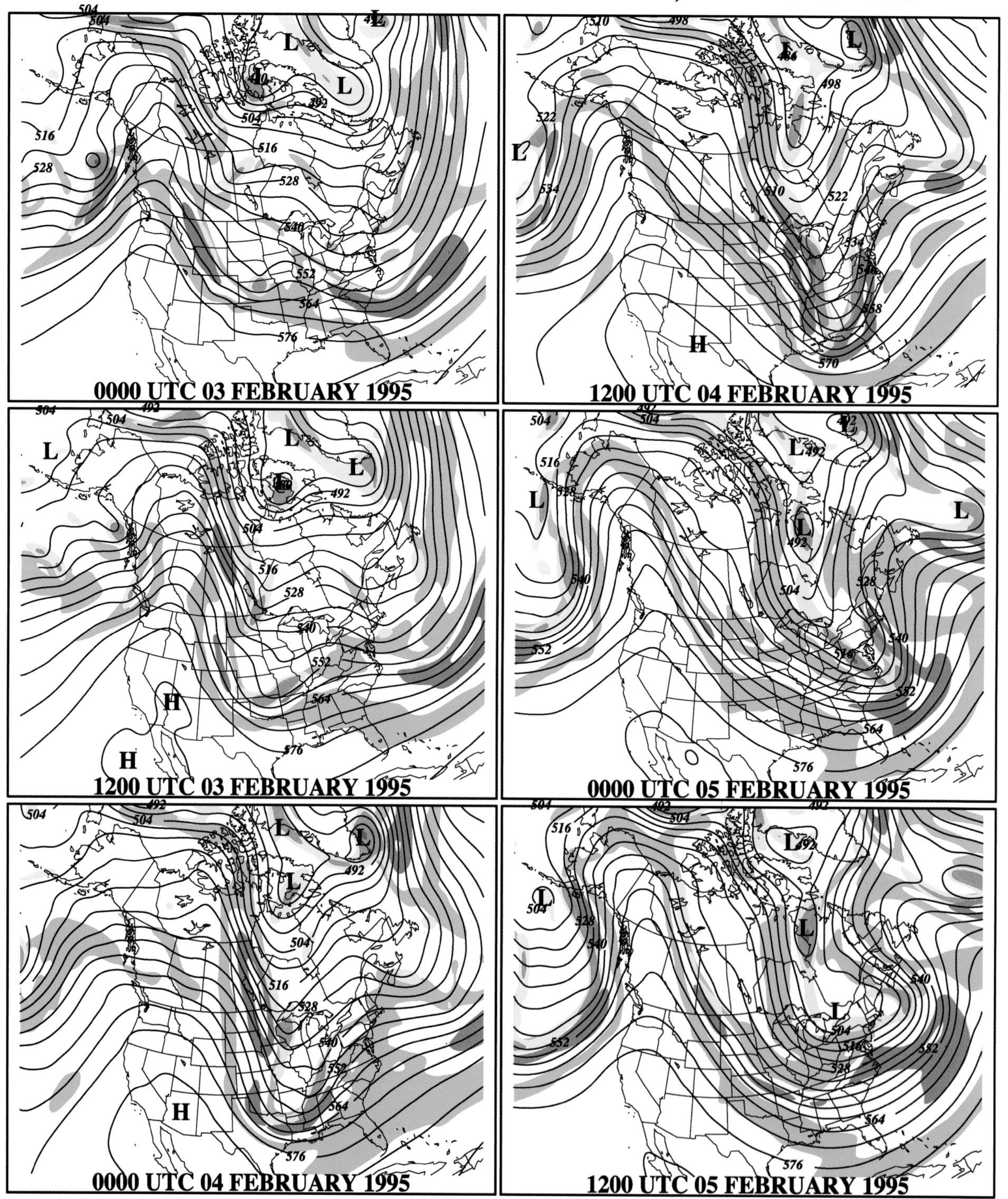

FIG. 10.26-4. Twelve-hourly analyses of 500-hPa geopotential heights and upper-level wind fields for 0000 UTC 3 Feb–1200 UTC 5 Feb 1995. See Fig. 10.1-4 for details.

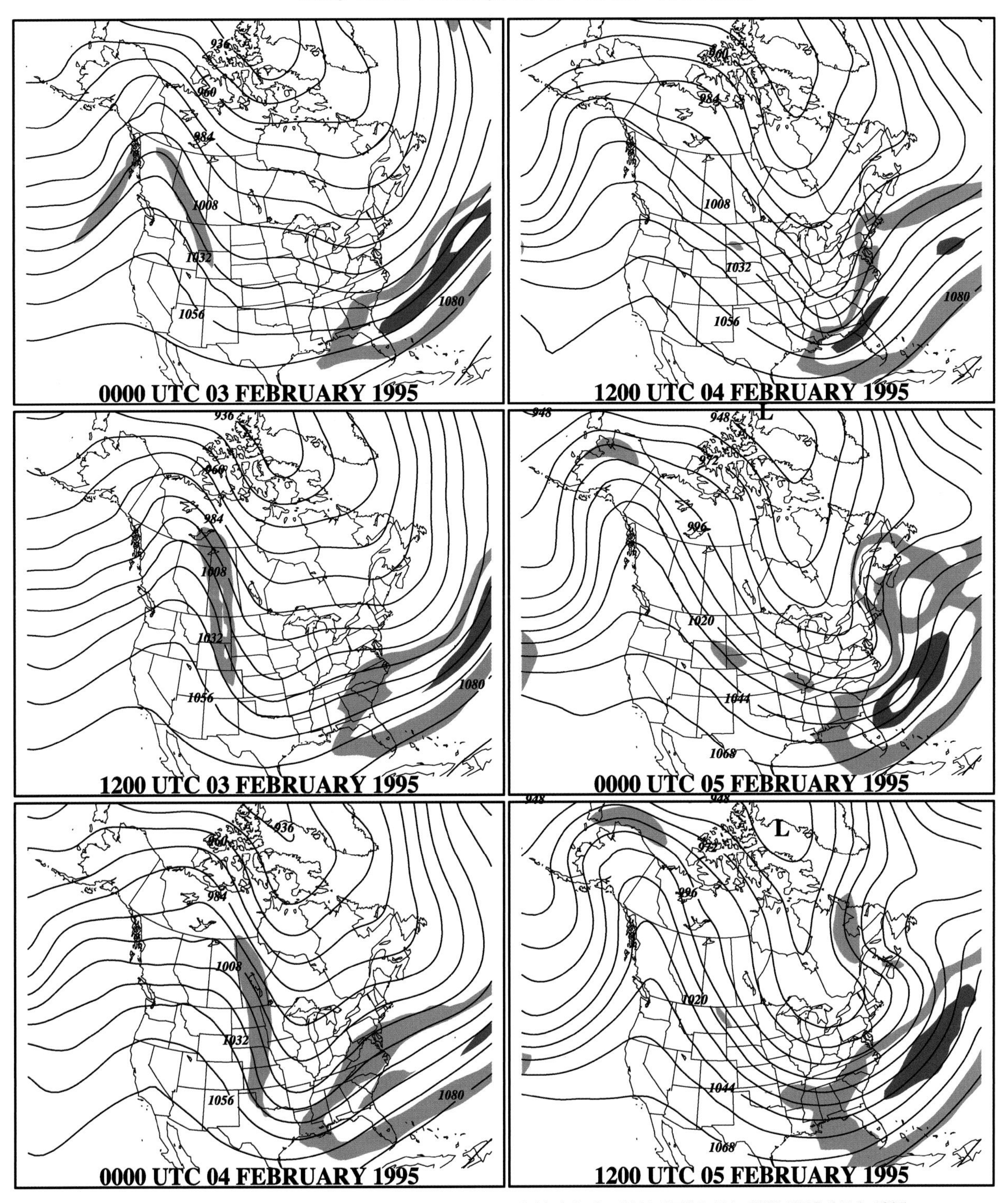

FIG. 10.26-5. Twelve-hourly analyses of 250-hPa geopotential height for 0000 UTC 3 Feb–1200 UTC 5 Feb 1995. See Fig. 10.1-5 for details.

SATELLITE IMAGERY

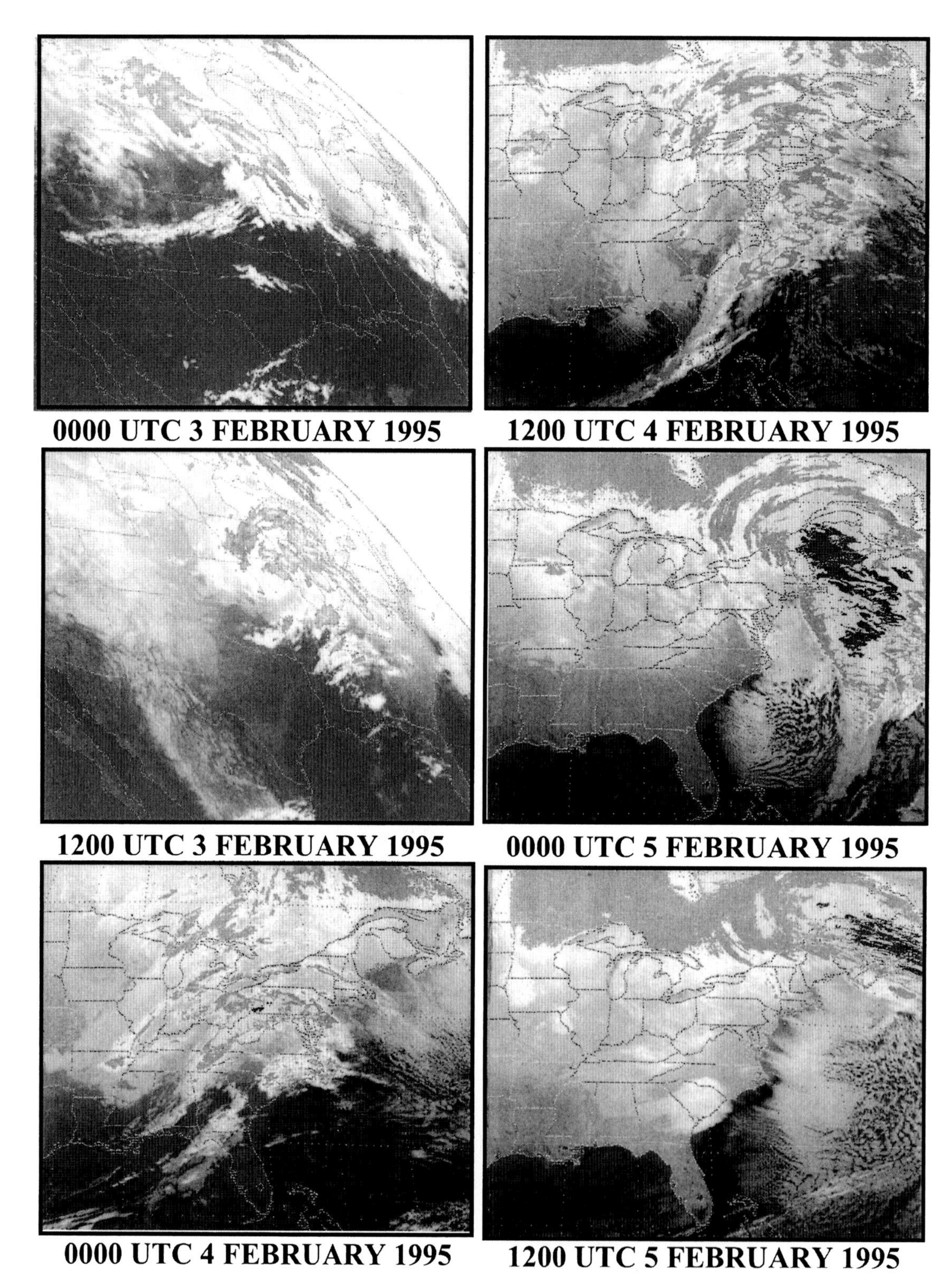

FIG. 10.26-6. Twelve-hourly infrared satellite images for 0000 UTC 3 Feb–1200 UTC 5 Feb 1995. See Fig. 10.16-6 for more details.

ly values of the NAO fluctuated in late January and early February and this storm developed during a week in which the NAO was negative.

- This is a case marked by a uniform height gradient across southeastern Canada and the northern United States, and a lack of confluence over New England and eastern Canada. This is consistent with the lack of a strong surface anticyclone for the Northeast during the precyclogenetic period and related changeover from snow to rain immediately along the coast.
- The precyclogenetic environment at 500 hPa was characterized at 0000 UTC 3 February by an amplifying ridge along the west coast of North America and a succession of digging troughs over the central United States.
- The upper troughs over central North America continued to amplify into a long-wave trough extending from Hudson Bay southward to Louisiana by 0000 UTC 4 February with distinct vorticity maxima over Wisconsin and another extending northeastward from northern Mississippi. This amplification continued through 1200 UTC 4 February, as the southern system moved toward the East Coast. The trough became negatively tilted by 0000 5 February and more diffluent downwind of this trough axis as the surface low developed rapidly. By 1200 UTC 5 February, a highly negatively tilted trough moved northeastward into New England as the surface low deepened to 962 hPa.
- As the trough amplified and became negatively tilted between 0000 UTC 4 February and 1200 UTC 5 February, the wavelength between the trough axis and the downwind ridge axis decreased significantly, as in other cases marked by rapidly developing surface cyclones.
- This case is marked by the lack of a "dual jet" pattern that marked other cases. The lack of a strong trough characterized by a cutoff low over eastern Canada or a downstream blocking ridge contributes to the lack of a strong confluent pattern and upper-level jet streak over the northeastern United States or eastern Canada with this case.
- As the upper troughs amplified over central North America, a 60–65 m s^{-1} northwesterly jet streak propagated southeastward at 400 hPa between 1200 UTC 3 February and 0000 UTC 4 February. As the upper trough amplified, a separate jet streak developed near and to the east of the trough axis, extending northeastward from Louisiana to northern Alabama by 0000 UTC 4 February. The surface low amplified within the increasingly diffluent exit region of this jet after 1200 UTC 4 February.

e. 250-hPa geopotential height and wind analyses

- The upper-tropospheric height field illustrates the uniform height gradients in the Northeast in the precyclogenetic environment, the amplifying ridge along the west coast, and the deepening trough digging into the central United States on 3–4 February.
- A subtropical jet extending eastward from the East Coast is the most prominent wind feature for this case. There is a tendency for the wind maximum to consolidate along the Southeast coast by 1200 4 February, but with only a weakly defined exit region. The lack of a northern jet across the Great Lakes and New England is consistent with the lack of confluent flow in that region.
- After the cyclone underwent a period of rapid development, there is evidence of a separate 60+ m s^{-1} wind maximum over Maine at 0000 UTC 5 February, extending northeastward from the region of rapid cyclogeneisis and heavy snow toward the downstream ridge crest.

f. Infrared satellite imagery sequence

- A comma-shaped cloud mass is associated with the short-wave trough that moved southeastward between 0000 and 1200 UTC 3 February 1995. As the trough moved toward the Ohio Valley and the surface low deepened slowly, the cloud mass appears relatively disorganized with many different areas of higher cloud tops.
- This area ultimately organized into the distinct comma-shaped cloud mass over New England and the western Atlantic by 0000 UTC 5 February, the period of the storm's greatest intensification. The cloud mass extending northeastward over New England and over southeastern Canada coincides with the 60 m s^{-1} wind maximum analyzed northeast of the storm center at the 250-hPa level.

R and 17th Street, Washington D.C., 7 Jan 1996 (photo by Hiram Ruiz).

27. 6–8 January 1996

a. General remarks

- The "Blizzard of '96" will rank as one of the most significant snowstorms of the 20th century since many of the largest population centers within the Northeast urban corridor were buried under more than 20 in. (50 cm) of snow. This was the most significant storm during the snowiest winter of the 20th century for much of the area from Virginia through southern New England. The large snow depths were due to the long duration of the storm, lasting as long as 36 h, as opposed to very large snowfall rates that typify other storms such as the 1979 Presidents' Day Storm. The storm is estimated to have cost more than $2 billion in damage and paralyzed the Northeast urban corridor for an entire week.
- The storm was followed by a couple of smaller snowstorms that led to one of the snowiest weeks on record from Washington northward to New York. Heavy rain and very mild conditions melted much of the snow on 19 January 1996, unleashing severe flooding across New York, Pennsylvania, and Maryland (Leathers et al. 1998).
- Medium-range model forecasts of this storm initially kept the region from New York City to Boston free of heavy snow. Forecasters, however, alerted the East Coast to the potential of significant snow several days in advance. Within a day to two days prior to the onset of snow, many numerical models accurately predicted the strength and slow movement and related heavy precipitation of the storm. As the low began to develop, it was clear that this storm would produce a tremendous amount of snow over the heavily populated Northeast corridor, with forecasters indicating that snow would be measured in "feet rather than inches."
- The following regions reported snow accumulations exceeding 10 in. (25 cm): Virginia, West Virginia, Maryland, Delaware, southern and eastern Pennsylvania, New Jersey, southeastern New York, Connecticut, Rhode Island, Massachusetts, and southeastern New Hampshire.
- The following regions reported snow accumulations exceeding 20 in. (50 cm): western, central, and northern Virginia; central and western Maryland; eastern and southeastern Pennsylvania; northern New Jersey; portions of southern New Jersey; southeastern New York; Long Island; western Connecticut; portions of western and southeastern Massachusetts and Rhode Island.
- The following regions reported snow accumulations exceeding 30 in. (75 cm): northern Virginia, central and western Maryland, portions of south-central and southeastern Pennsylvania, portions of northern New Jersey, portions of northwestern Connecticut, and southwestern Massachusetts.

TABLE 10.27-1. Snowfall amounts for urban and selected locations for 6–8 Jan 1996; see Table 10.1-1 for details.

NESIS = 11.54 (category 5)	
Urban center snowfall amounts	
Washington, DC–Dulles Airport	24.5 in. (62 cm)
Baltimore, MD	22.0 in. (56 cm)
Philadelphia, PA	30.7 in. (78 cm)
New York, NY–LaGuardia Airport	23.8 in. (60 cm)
New York, NY–Central Park	20.2 in. (51 cm)
Boston, MA	18.2 in. (46 cm)
Other selected snowfall amounts	
Big Meadows, VA	47.0 in. (119 cm)
Newark, NJ	27.8 in. (71 cm)
Allentown, PA	25.6 in. (65 cm)
Roanoke, VA	24.9 in. (63 cm)
Providence, RI	24.0 in. (61 cm)
Elkins, WV	23.4 in. (59 cm)
Harrisburg, PA	22.2 in. (56 cm)
Wilmington, DE	22.0 in. (56 cm)
Lynchburg, VA	21.4 in. (54 cm)
Scranton, PA	21.0 in. (53 cm)
New York, NY–John F. Kennedy Airport	20.7 in. (52 cm)
Charleston, WV	20.5 in. (51 cm)
Washington, D.C.–National Airport	17.1 in. (43 cm)
Hartford, CT–Bradley International Airport	15.8 in. (40 cm)
Bridgeport, CT	15.0 in. (38 cm)
Portland, ME	10.2 in. (26 cm)

b. Surface analyses

- This storm was a classic Northeast snowstorm with cold high pressure entrenched over the eastern half of the United States and a developing storm in the western Gulf of Mexico that evolved into a major cyclone along the East Coast.
- On 6 January, a large dome of high pressure covered the central and eastern United States as low pressure began developing in the western Gulf of Mexico. Two centers of high pressure were found: one over the north-central United States (central pressure 1042 hPa at 1200 UTC 6 January) and a second center over northern New York (central pressure 1038 hPa). High pressure was accompanied by very cold air with subfreezing temperatures across much of the eastern half of the United States. The morning of 6 January was bitterly cold across the Northeast with subzero (°F) temperatures covering much of the area inland from the coast.
- Low pressure in the western Gulf of Mexico was associated with a broad inverted pressure trough extending northward into the Ohio Valley at 1200 UTC 6 January. In the following 12 h, the trough became more pronounced across the eastern United States and heavy snows developed across much of the southern Appalachians into western Virginia by 0000 UTC 7 January.
- With cold air anchored over the East, cold-air damming is clearly identified by the inverted pressure ridge along the East Coast at 0000 UTC 7 January with the development of a pronounced coastal front

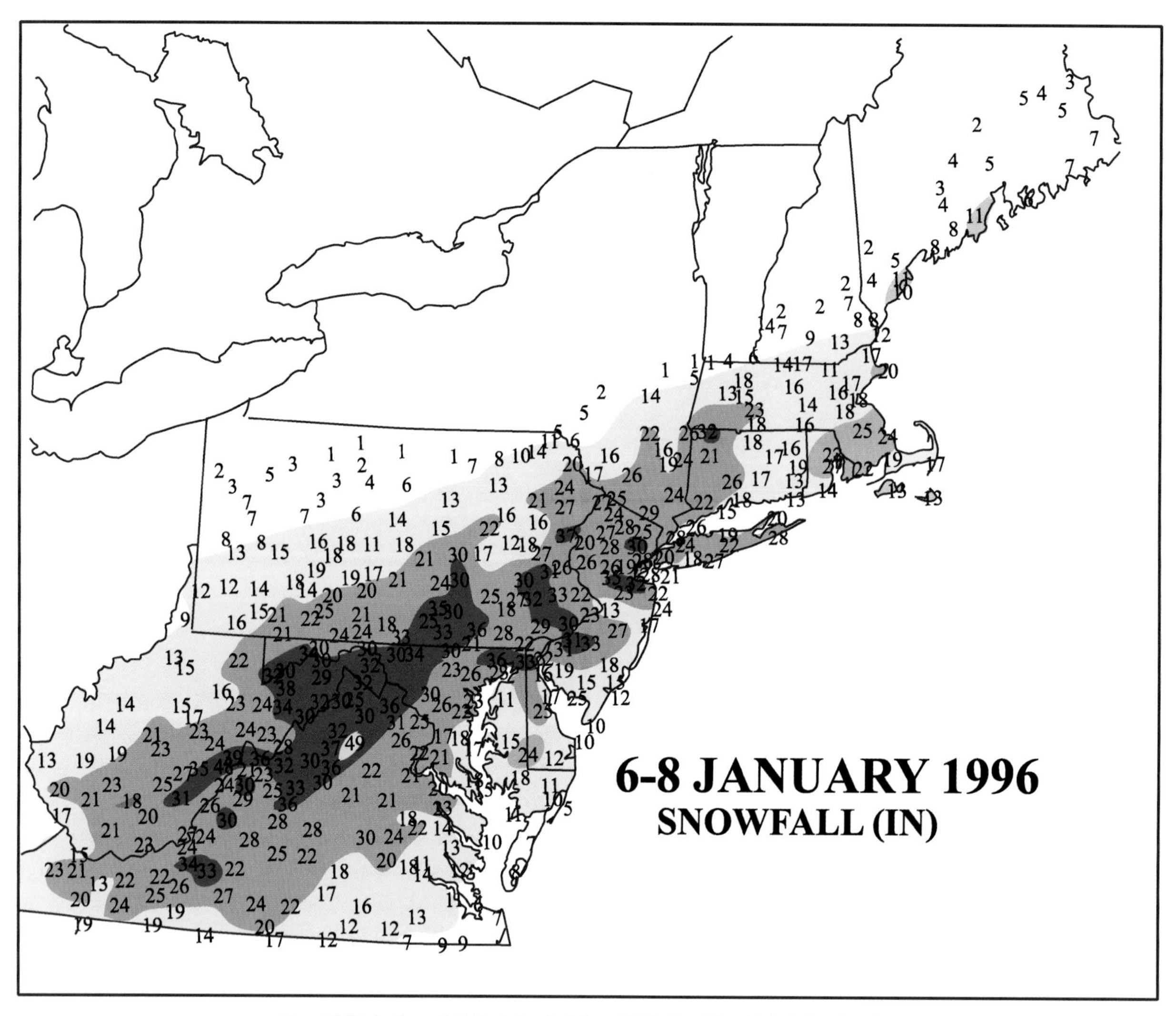

FIG. 10.27-1. Snowfall (in.) for 6–8 Jan 1996. See Fig. 10.1-1 for details.

extending from northern Florida northeastward just off the Southeast coast.

- By 1200 UTC 7 January, low pressure within the inverted pressure trough was deepening slowly with a 1004-hPa low pressure system over eastern Georgia. A new low developed farther northeastward along the coastal front with a separate center near Cape Hatteras. The development of the low along the coastal front accompanied a widespread outbreak of heavy snow from the Carolinas northeastward through the mid-Atlantic states overnight, with snows beginning in New York City near 1200 UTC 7 January.
- During 7 January, the Georgia low dissipated at the expense of the intensifying and slow-moving low pressure system just east of the Virginia coast. There was a brief period of rapid intensification as pressures fell 12 hPa in about 6 h but in general, the low pressure system was not the explosive deepener that characterized other significant snowstorms.
- Nevertheless, the surface low deepened to 990 hPa by 0000 UTC 8 January with heavy snow occurring throughout the Northeast, through much of the northern Appalachians, and also in the Ohio Valley. The surface low deepened to 980 hPa by 1200 UTC 8 January as heavy snow continued to affect the Northeast urban corridor but ended across the Ohio Valley. Temperatures through much of the Northeast were generally less than 20°F (-7°C) with north-northeasterly winds gusting to 40 mi h^{-1} (20 m s^{-1}).
- As a narrow layer of warm air moved in aloft between 600 and 700 hPa, snow mixed with and changed to ice pellets in the Washington–Baltimore metropolitan area after 1800 UTC 7 January and remained as mixed precipitation through much of the afternoon.
- While precipitation rates diminished over portions of the middle Atlantic states during the afternoon of 7 January, moderate to heavy snow redeveloped in Virginia during the evening, spread northeastward, and persisted through the morning of 8 January. This extended period of snowfall is sometimes referred to as

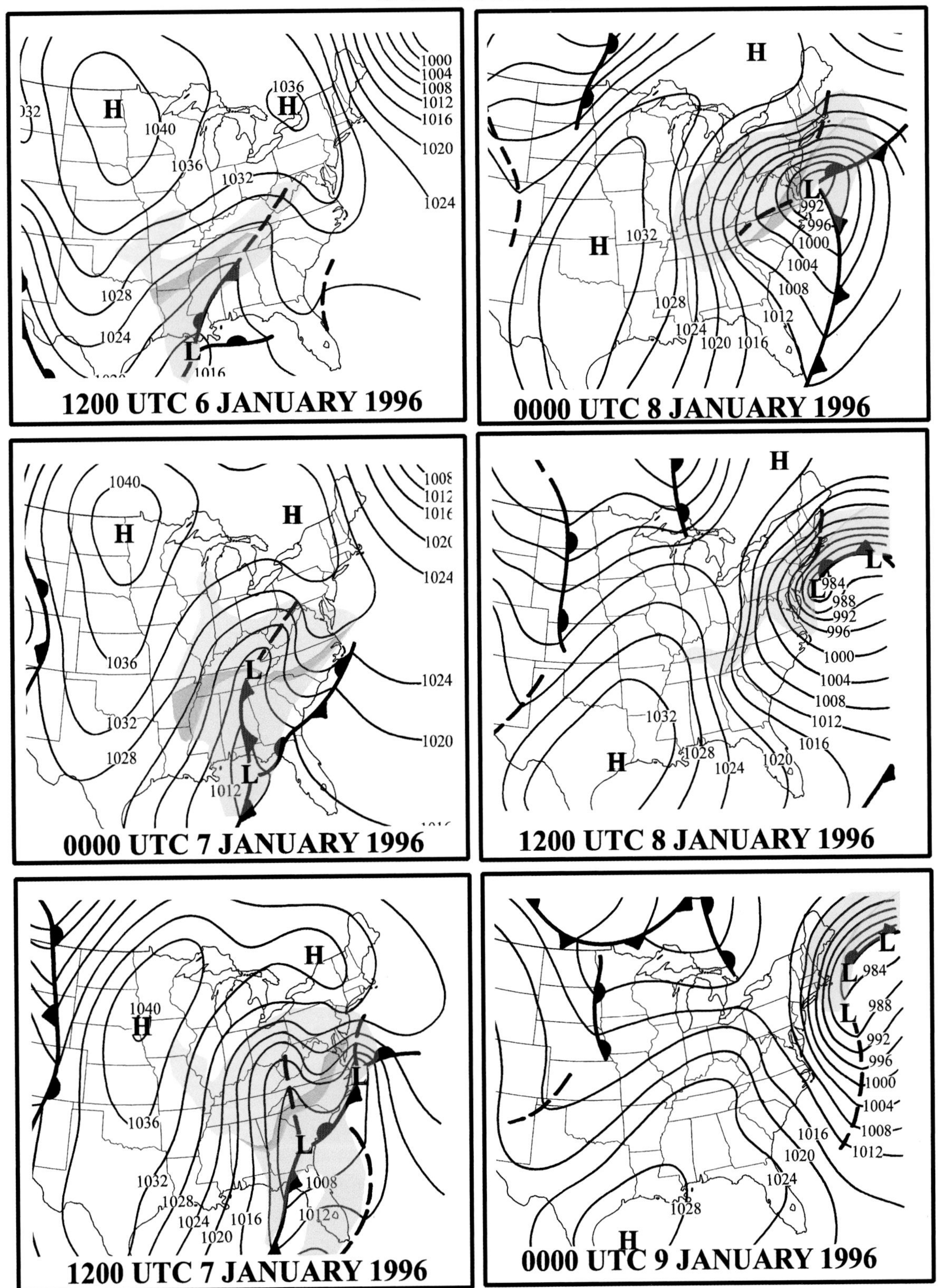

FIG. 10.27-2. Twelve-hourly surface weather analyses for 1200 UTC 6 Jan–0000 9 Jan 1996. See Fig. 10.1-2 for details.

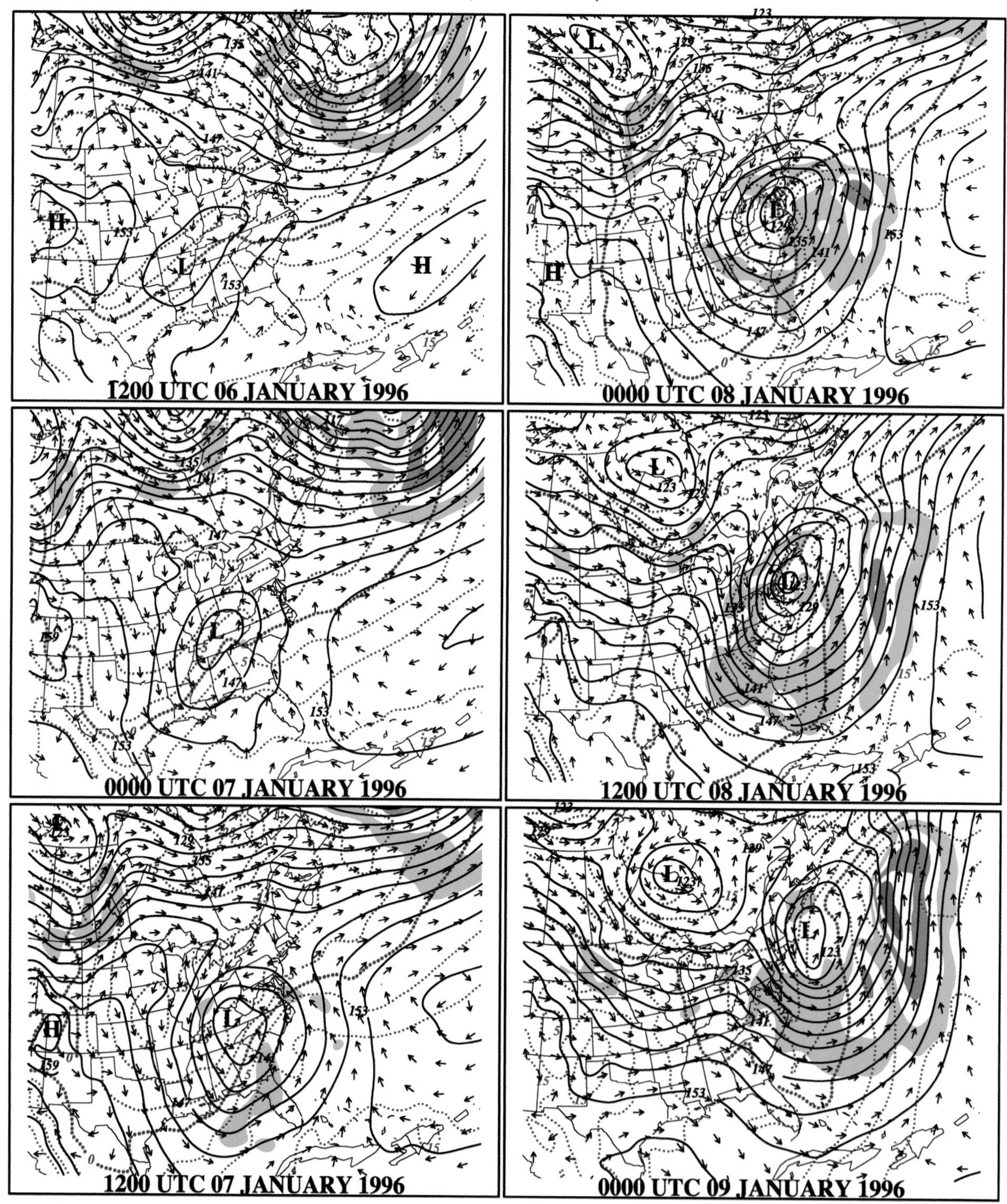

FIG. 10.27-3. Twelve-hourly 850-hPa analyses for 1200 UTC 6 Jan–0000 9 Jan 1996. See Fig.10.1-3 for details.

500 hPa HEIGHTS AND VORTICITY, 400 hPa WINDS

FIG. 10.27-4. Twelve-hourly analyses of 500-hPa geopotential heights and upper-level wind fields for 1200 UTC 6 Jan–0000 UTC 9 Jan 1996. See Fig. 10.1-4 for details.

250 hPa HEIGHTS AND WINDS

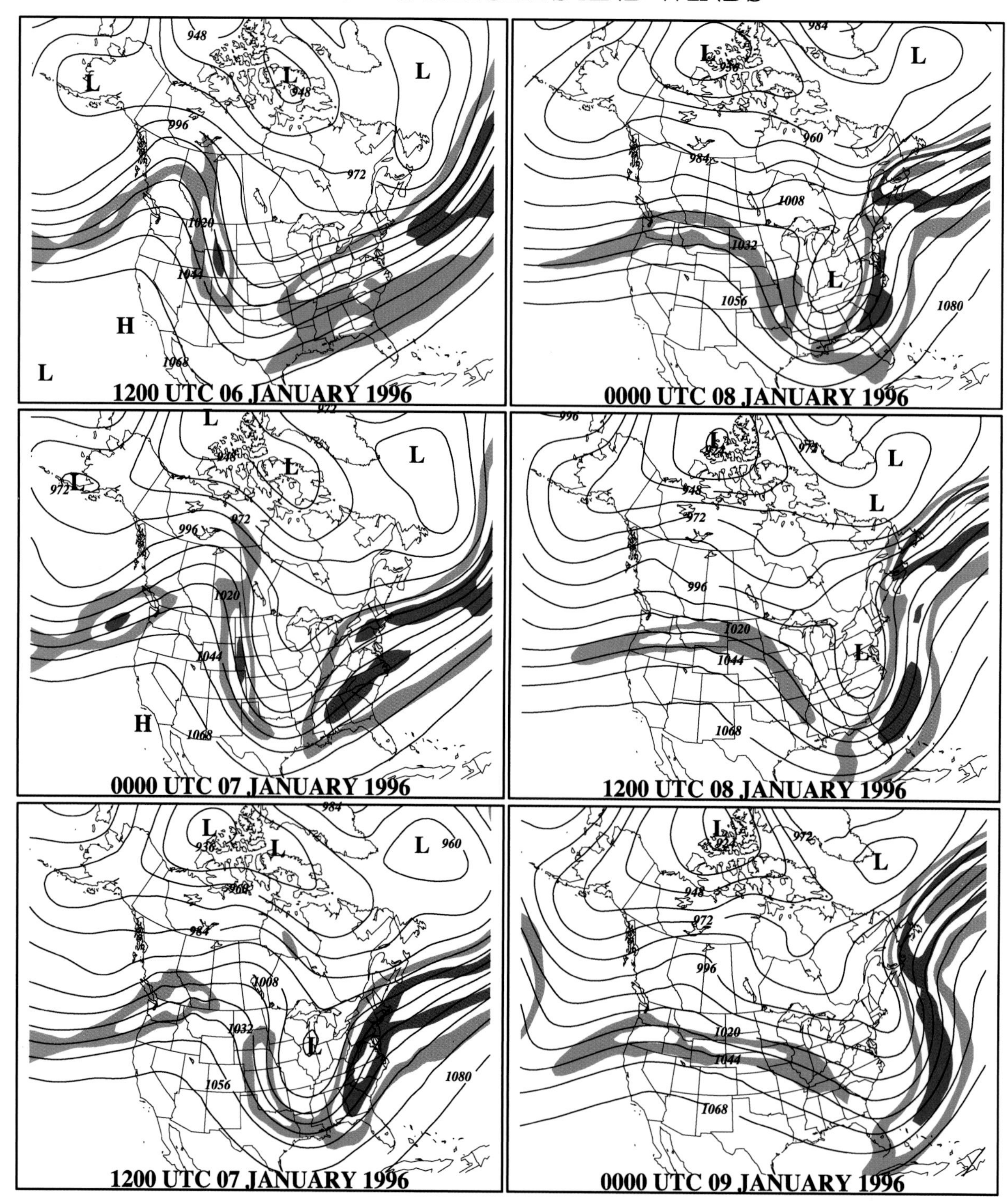

FIG. 10.27-5. Twelve-hourly analyses of 250-hPa geopotential height for 1200 UTC 6 Jan–0000 UTC 9 Jan 1996. See Fig. 10.1-5 for details.

SATELLITE IMAGERY

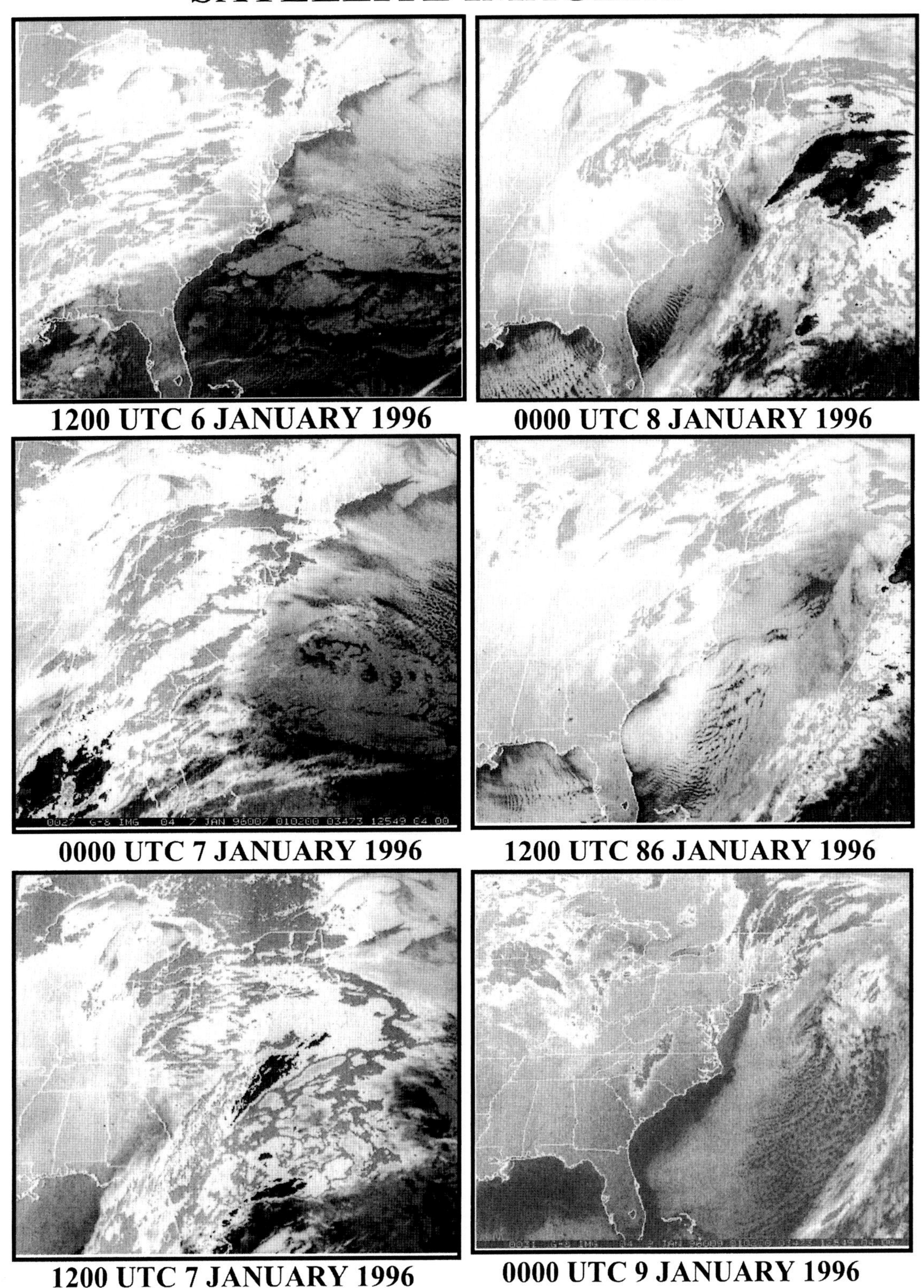

FIG. 10.27-6. Twelve-hourly infrared satellite images for 1200 UTC 6 Jan–0000 UTC 9 Jan 1996. See Fig. 10.16-6 for details.

the "backlash" of the storm since the initial precipitation had already diminished. As the surface low reached its lowest pressure early on 8 January, the region of heavy snow had diminished in size but was still affecting the heavily populated Northeast urban corridor.

- Snows ended gradually from southwest to northeast, from the middle Atlantic states to New England, during the day on 8 January as the surface low began to fill and move east-northeastward over the Atlantic.

c. 850-hPa analyses

- The precyclogenetic period was characterized by a northwesterly flow regime across the Northeast with a core of −20°C temperatures just northwest of New York, 850-hPa temperatures generally between −5° and −15°C across the Northeast urban corridor, and a developing 850-hPa low over the Tennessee Valley. The northwesterly flow regime maintained cold lower-tropospheric temperatures across the Northeast.
- The 850-hPa low over Tennessee and Kentucky slowly intensified as it drifted northeastward between 1200 UTC 6 and 7 January. By 1200 UTC 7 January, southerly flow increased to the east of the center with 20 m s^{-1} winds east of the Carolina and Virginia coastline, and the southeasterly–easterly flow north of the low center also began to strengthen. This flow regime acted to increase the moisture transport toward an expanding area of moderate to heavy snowfall from Virginia to just south of New York City.
- By 0000 UTC 8 January easterly winds exceeding 20–25 m s^{-1} increased over a large region from Maryland north to New Jersey, coincident with the development of the low-level jet and the redevelopment of heavy snowfall over that region. Note that the easterly jet did not appear to attain speeds greater than 30 m s^{-1}, which may help explain why snowfall rates rarely exceeded 1 or 2 in. (2.5–5 cm) per hour.
- The 0°C isotherm only progressed northward to southeastern Virginia by 1200 UTC 7 January while the 850-hPa temperature gradient strengthened and developed into a classic S-shaped pattern with the 0°C isotherm extending northward along the immediate mid-Atlantic coast to just south of Long Island and New England.
- As the storm reached its peak intensity by 1200 UTC 8 January, the 850-hPa low began to elongate on a southwest–northeast axis accompanied by an extensive region of northeasterly winds exceeding 25 m s^{-1} over much of the Northeast into Pennsylvania. Between 1200 UTC 8 January and 0000 UTC 9 January, the low lifted northeastward east of New England and the strongest winds, now northwesterly, shifted to the mid-Atlantic states, while southerly low-level winds increased to greater than 40 m s^{-1}, well to the east of the storm center.

d. 500-hPa geopotential height and 400-hPa wind analyses

- The Blizzard of '96 developed during a weak La Niña (Fig. 2-19, volume I). It also occurred during a period in which the North Atlantic Oscillation (NAO) was primarily negative, the first real extended period of negative NAO since November and December 1989, which was characterized by cold weather in the eastern United States. A strong negative NAO developed in early December and continued into January 1996, when it entered a period of positive and negative fluctuations beginning after the Blizzard of '96. Therefore, the storm occurred as negative values of the NAO were retreating from their lowest values during the last half of December 1995. Analyses at 500 hPa of a persistent upper cutoff ridge near Greenland are consistent with the negative phase of the NAO.
- At 1200 UTC 6 January, the precyclogenetic environment at 500 hPa was characterized by a "classic" presnowstorm pattern. An amplifying trough over the central United States was located downwind of a ridge along the west coast and upwind of a strongly confluent flow over the northeastern United States and southeastern Canada to the rear of an intense upper low just south of Newfoundland. The confluent flow pattern in the Northeast overlaid the large area of high pressure, cold temperatures, and 850-hPa northwesterly flow and cold-air advection across the Northeast.
- Between 1200 UTC 6 January and 1200 UTC 7 January, the eastern Canadian trough moved rapidly eastward but the strong confluent 500-hPa flow persisted over the Northeast and southeastern Canada.
- The upper trough over the central United States amplified between 1200 UTC 6 and 1200 UTC 7 January with the development of a large cutoff low over the Ohio Valley by 1200 UTC 7 January. This cutoff continued to deepen through 1200 UTC 8 January as it shifted eastward, covering much of the eastern United States by 0000 UTC 8 January, and continued to move eastward, reaching the Virginia–North Carolina coast by 1200 UTC 8 January. The wavelength between the amplifying trough and downstream ridge decreased significantly after 1200 UTC 7 January, with an enhanced region of diffluence developing east of the trough axis.
- The cutoff nature of this trough and its slow movement allowed warmer air within the 600–700-hPa layer to move in from the south over the middle Atlantic states, changing the heavy snow to sleet in the Washington–Baltimore area for a 4–6-h period. The strong circulation at 0000 and 1200 UTC 8 January also contributed to the redeveloping heavy snows throughout the region from northern Virginia to New York City after 0000 UTC 8 January.
- This case is marked by the classic "dual jet" pattern with a northern jet extending across the middle Atlantic and New England states in the precyclogenetic

period and a southern jet associated with the amplifying trough over the central and eastern United States.

- The northern jet was located within the confluent flow field, with the entrance region shifting only gradually to New York and New England through 0000 UTC 8 January.
- The amplifying trough over the central United States is associated with a 50–60 m s^{-1} jet streak diving southeastward toward the base of the trough between 1200 UTC 6 January and 1200 UTC 7 January. After 0000 UTC 7 January, a separate wind maximum appears to develop east of the trough axis, and amplifies to greater than 70 m s^{-1} by 0000 UTC 8 January. The surface low and heavy snowfall appear to develop in the increasingly diffluent exit region of this jet and the entrance region of the northern jet on 7–8 January.

e. 250-hPa geopotential height and wind analyses

- The geopotential height analyses depict that the precyclogenetic confluence over the Northeast became even better defined after 1200 UTC 7 January as the ridge built into New England downwind of the amplifying trough in the middle part of the country.
- The dual-jet pattern was distinct by 0000 and 1200 UTC 7 January, with the surface low and heavy snows developing within the exit region of the southern jet (located east of the trough axis) and right-entrance region of the northern jet in the confluent region near the downstream ridge crest.
- Between 0000 UTC 7 January and 0000 UTC 8 January, maximum winds within the northern jet increased from greater than 70 to greater than 90 m s^{-1} as the confluent zone strengthened and heavy snowfall spread rapidly northeastward within the entrance region of the jet.

f. Infrared satellite imagery sequence

- High clouds streamed northeastward toward the East Coast at 1200 UTC 6 January into the entrance region of the jet extending from the Ohio Valley toward the Atlantic Ocean, and as the trough amplified over the central United States. This area of cloudiness continued to expand over the next 24 h as the upper trough amplified and the surface low developed over the southeastern United States.
- As one main mass of enhanced upper clouds moved eastward toward the Atlantic coast by 1200 UTC 7 January, a second area of enhanced cloudiness developed to the west of the original cloud mass, complete with a banded structure of higher and lower clouds, extending northward from North Carolina to New York. These banded clouds are associated with the strong vertical motions associated with cyclogenesis near Cape Hatteras and the rapid expansion of the snowfall into Pennsylvania and New Jersey.
- These two areas of enhanced clouds split by 0000 UTC 8 January as the first mass extended northeastward over the Atlantic Ocean, appearing to be collocated with the entrance region of the northern jet moving slowly off the southeastern Canadian coast during this time. The second cloud mass takes the shape of a comma "head" from Kentucky northeastward into New York and New England. The heaviest snows were falling along the southern edge of the enhanced cloud tops over the Ohio Valley into the northern middle Atlantic states and southern New England.
- The comma head splits from the comma tail as the storm completes its intensification by 1200 UTC 8 January and the comma head becomes more disorganized. This area of enhanced cloudiness drifted east as the surface low moved east of the coast by 0000 UTC 9 January 1996.

Back Bay/South End in Boston, Massachusetts, 1 Apr 1997 (Source is Gormie's G-Spot Gallery, photo courtesy of Donald Sutherland's digital photography library, www.geocities.com/donsutherland1/snow1.html).

28. 31 March–1 April 1997

a. General remarks

- This late season storm was the greatest spring snowstorm on record for portions of eastern New York, northwestern New Jersey and portions of New England, especially Massachusetts. Heavy snow occurred as a storm developed rapidly beneath an amplifying short-wave upper trough that evolved into a strong 500-hPa vortex.
- Only 36 h before the height of the storm, the surface weather map provided few clues of the upcoming storm's severity. However, computer models available 2–3 days prior to the storm correctly indicated that a storm, with the potential for deep snows, would affect the Northeast. Once the storm formed east of New Jersey, it deepened rapidly and moved slowly just south of Long Island before drifting southeastward out over the Atlantic on 1 April.
- Northeastern Pennsylvania and New York's Catskill Mountains received from 20 to over 30 in. of snow (25 to over 50 cm). Suburban Philadelphia was hit particularly hard as 6 in. (15 cm) to as much as 15 in. (38 cm) of snow and high winds felled trees and toppled power lines. Across south-central New Jersey, up to 10 in. (25 cm) fell. New York City escaped the brunt of the storm but 1–2 feet (30–60 cm) of snow was common across northwestern New Jersey. Heavy snows spread from western to eastern New England on 31 March, where it continued to fall through the night into the early afternoon before ending on 1 April. While snow was initially slow to accumulate in Boston after it had begun as rain [with only 3 in. (8 cm) by midnight], the snow quickly piled up overnight into the next day, closing most highways, the airport, and the subway system. More than 22 in. (56 cm) fell in Boston on 1 April, the greatest April snow on record, and 25.4 in. (64 cm) total occurred on 31 March–1 April, the third greatest snow total in Boston history and the most in 24 h. In Worcester, Massachusetts, 33 in. (84 cm) fell, the biggest snowstorm in that city's snowy history and up to 3 feet (90 cm) fell in portions of central and eastern Massachusetts. An analysis of this storm by Martin (1999) showed the importance of a trough of warm air aloft ("trowal") in the development of the heavy snow band in the northwest sector of the cyclone.
- The following regions reported snow accumulations exceeding 10 in. (25 cm): portions of southeastern Pennsylvania, northeastern Pennsylvania, southeastern New York (except Long Island), central and northern Connecticut and Rhode Island, Massachusetts, southern Vermont, and New Hampshire
- The following regions reported snow accumulations exceeding 20 in. (50 cm): northwestern New Jersey, portions of northeastern Pennsylvania, southeastern New York, northern Connecticut, northern Rhode Island, portions of western Massachusetts, much of central and eastern Massachusetts, and portions of southern New Hampshire.

TABLE 10.28-1. Snowfall amounts for urban and selected locations for 30 Mar–1 Apr 1997; see Table 10.1-1 for details.

NESIS = 2.37 (category 1)	
Urban center snowfall amounts	
Boston, MA	25.4 in. (64 cm)
Baltimore, MD	1.4 in. (3 cm)
Philadelphia, PA	3.9 in. (10 cm)
New York, NY–Kennedy Airport	1.7 in. (4 cm)
Other selected snowfall amounts	
East Jewett, NY	37.0 in. (94 cm)
Milton, MA	36.0 in. (91 cm)
Worcester, MA	33.0 in. (84 cm)
Milton, MA–Blue Hill Observatory	30.0 in. (76 cm)
Albany, NY	15.3 in. (38 cm)
Bradley Field, Windsor Locks, CT	14.7 in. (36 cm)
Martinsburg, WV	13.3 in. (34 cm)
Concord, NH	9.9 in. (25 cm)

b. Surface analyses

- The development of this storm was unusual in that the surface weather charts prior to the storm had little indication of an impending intense East Coast storm (although numerical prediction models did quite well). The evolution of the cyclone is more complex than most of the other cases.
- Unlike most of its wintertime counterparts, no cold surface high pressure center was affecting the Northeast corner of the nation prior to this snowstorm. Instead, an area of cold high pressure lingered over eastern Canada as the storm was beginning to take shape late on 30 March. The colder air appeared to be drawn into the northeastern United States only after the storm began to develop on 31 March.
- An expanding area of mostly light to moderate rains accompanied a developing frontal system over the south-central United States early on 30 March. As this system moved east over the next day, a cold front and weak low pressure consolidated into a well-defined low pressure area over southwestern Virginia by 0000 UTC 31 March 1997.
- Between 0000 UTC 31 March and 0000 UTC 1 April, a surface low developed over Virginia and moved east-northeastward as a broader center of low pressure consolidated off the mid-Atlantic and New York coasts early on 31 March.
- Precipitation began as rain across much of eastern Pennsylvania, eastern New York, and New England during this stage of the storm's development. Following 1200 UTC 31 March, the surface low began to deepen, colder air began to wrap into the cyclone's circulation from the high pressure system building southward over Canada, and stronger vertical motions

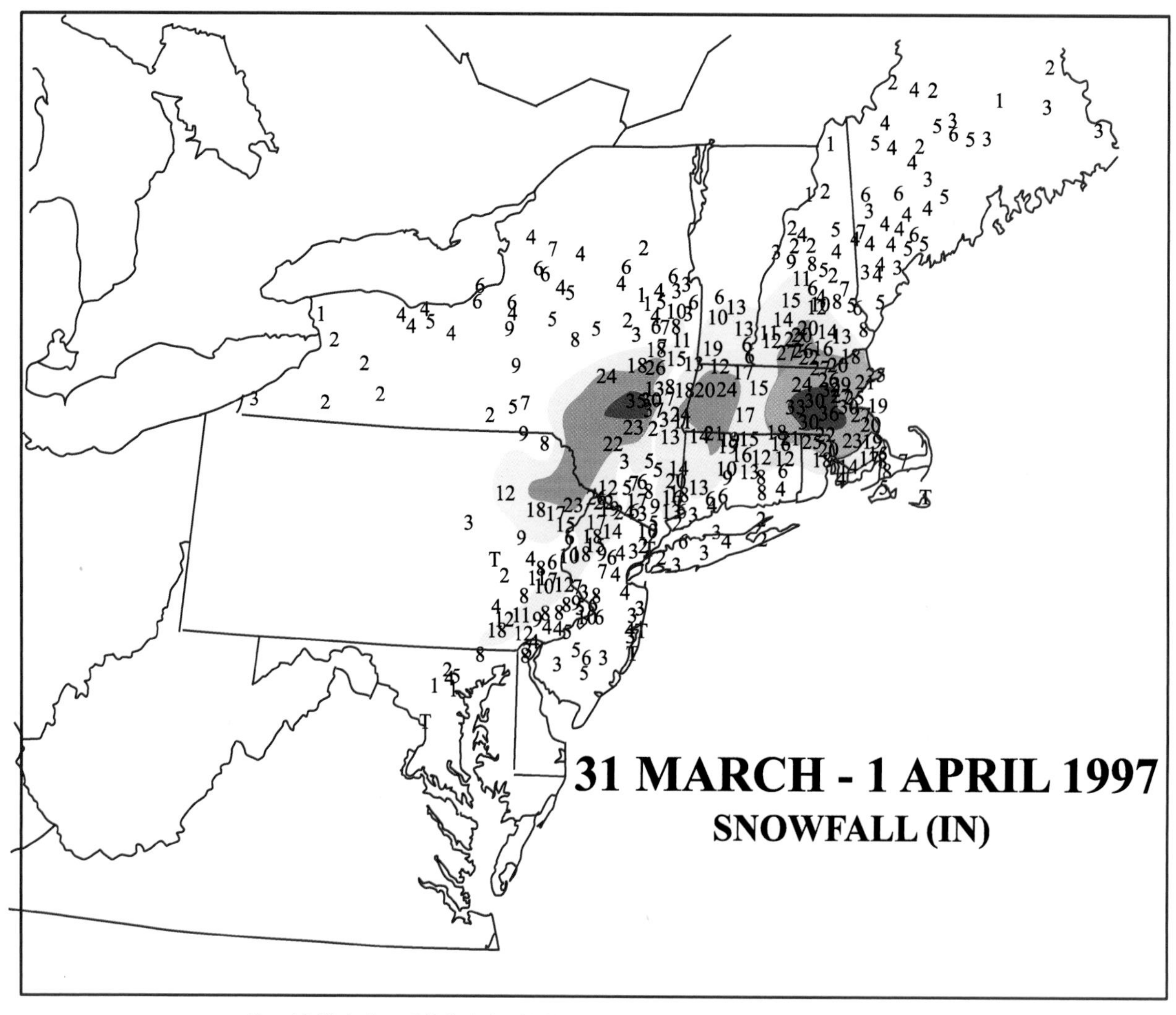

FIG. 10.28-1. Snowfall (in.) for 31 Mar–1 Apr 1997. See Fig. 10.1-1 for details.

helped change the rain to snow, first across northeastern Pennsylvania and southeast New York, then over southeastern Pennsylvania, New Jersey, Long Island, and New England.

- The surface low deepened rapidly east of New Jersey on 31 March, and by the evening, the central sea level pressure was near or slightly below 980 hPa with high winds and heavy precipitation surrounding the cyclone. As cold air and moisture wrapped around the intense circulation, snow became heavy at times across New Jersey, southeastern Pennsylvania, and into northeastern Maryland with snowy conditions spreading southward into the Washington area.
- New York City avoided heavy snowfall accumulations as total precipitation amounts were not heavy and temperatures remained above freezing.
- Heavy snows developed during the day on 31 March in eastern New York and western New England and spread toward eastern New England, where precipitation began as rain or mixed rain and snow. As the storm intensified during the evening, precipitation changed to all snow throughout much of eastern New England, including the Boston metropolitan area.
- On the night of 31 March/1 April, the storm stalled just south of the New England coast and then began to head to the southeast away from New England. The heaviest snowband remained over eastern Massachusetts through the night and into the morning before finally drifting east-southeastward off the coast during the afternoon of 1 April.

c. 850-hPa analyses

- During the precyclogenetic period, a reservoir of colder air was located well north of the Northeast urban corridor in south-central Canada. There was a developing lower-tropospheric baroclinic zone over the

FIG. 10.28-2. Twelve-hourly surface weather analyses for 0000 UTC 30 Mar–1200 UTC 1 Apr 1997 See Fig. 10.1-2 for details.

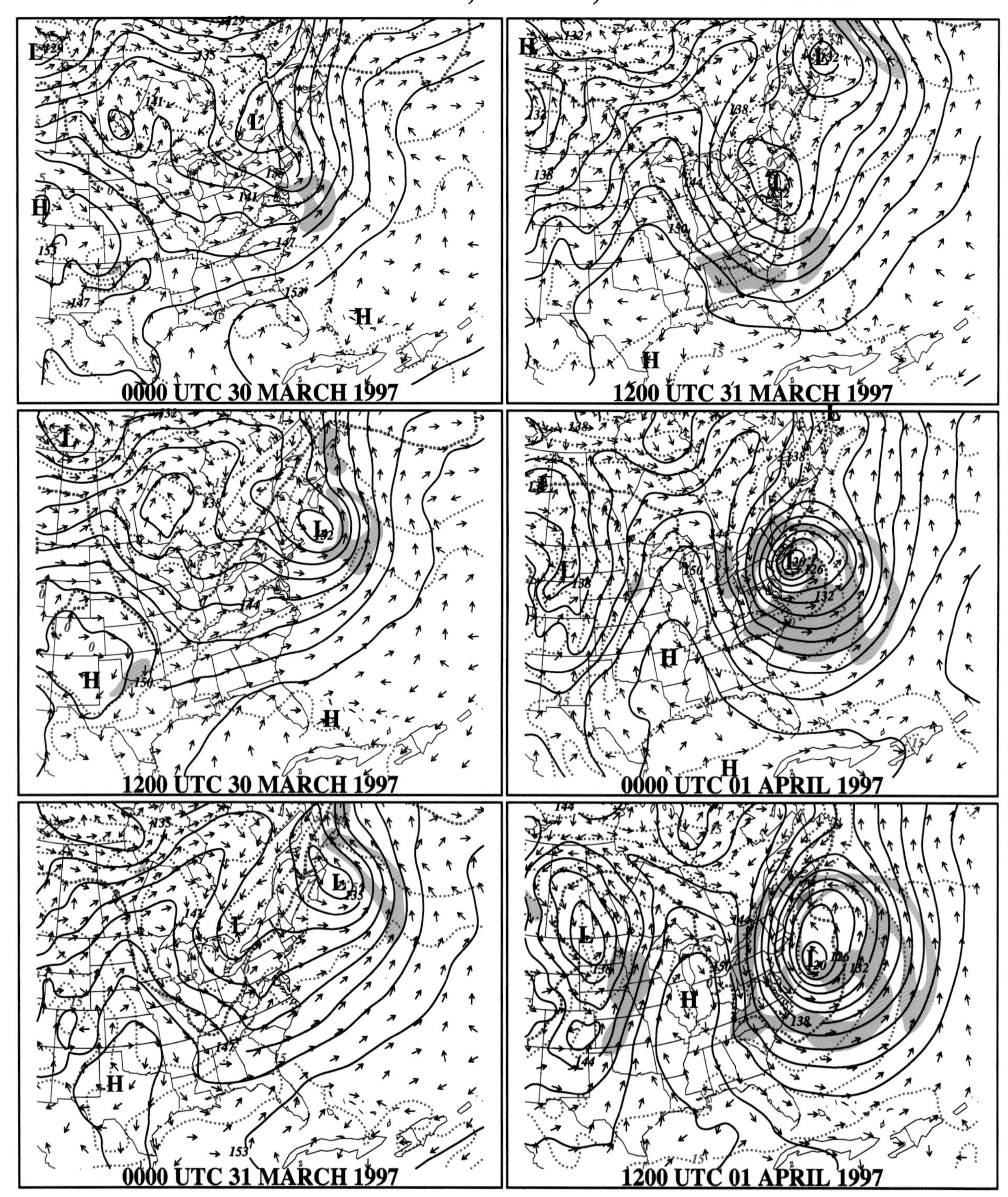

FIG. 10.28-3. Twelve-hourly 850-hPa analyses for 0000 UTC 30 Mar–1200 UTC 1 Apr 1997. See Fig. 10.1-3 for details.

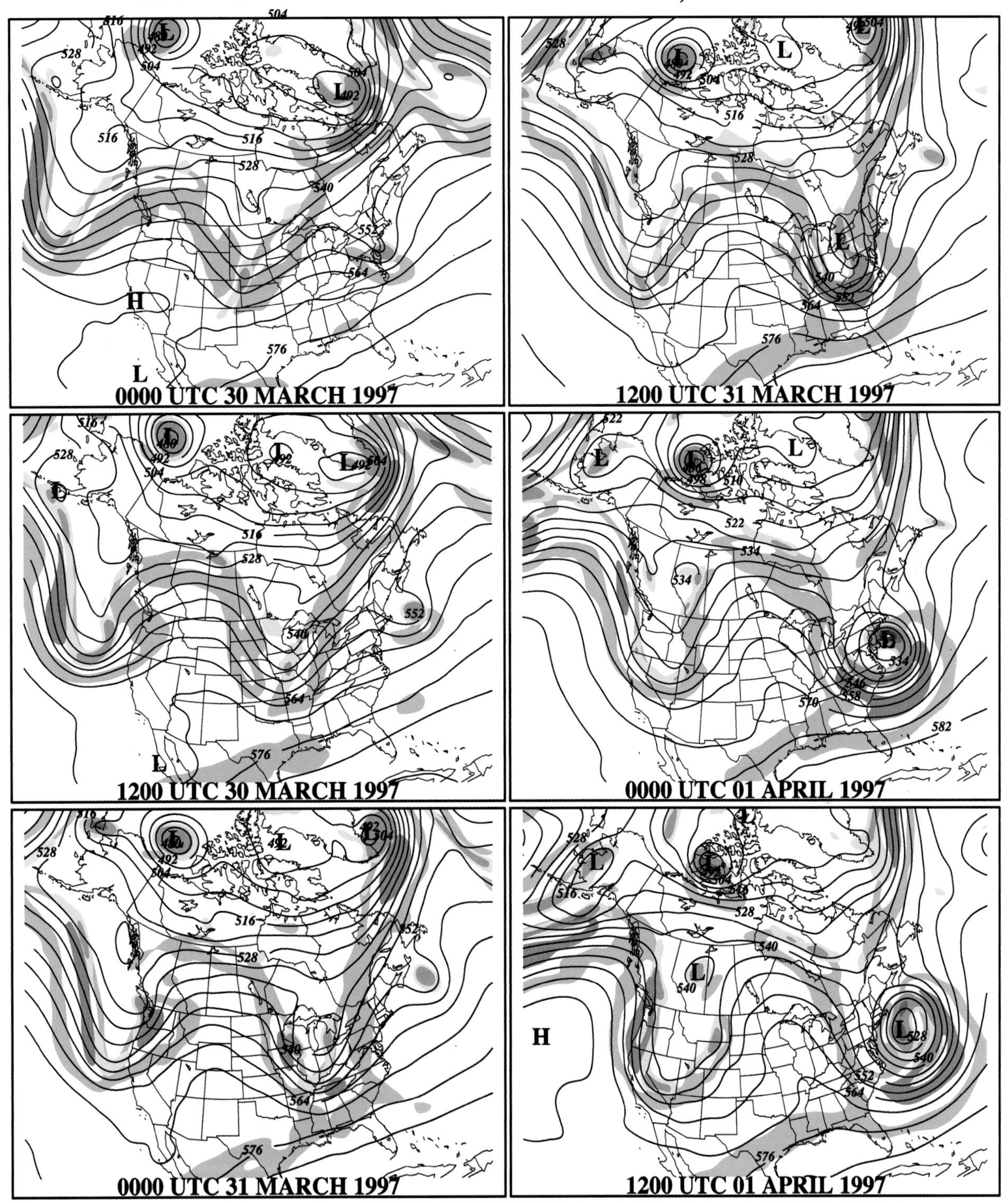

FIG. 10.28-4. Twelve-hourly analyses of 500-hPa geopotential heights and upper-level wind fields for 0000 UTC 30 Mar–1200 UTC 1 Apr 1997. See Fig. 10.1-4 for details.

250 hPa HEIGHTS AND WINDS

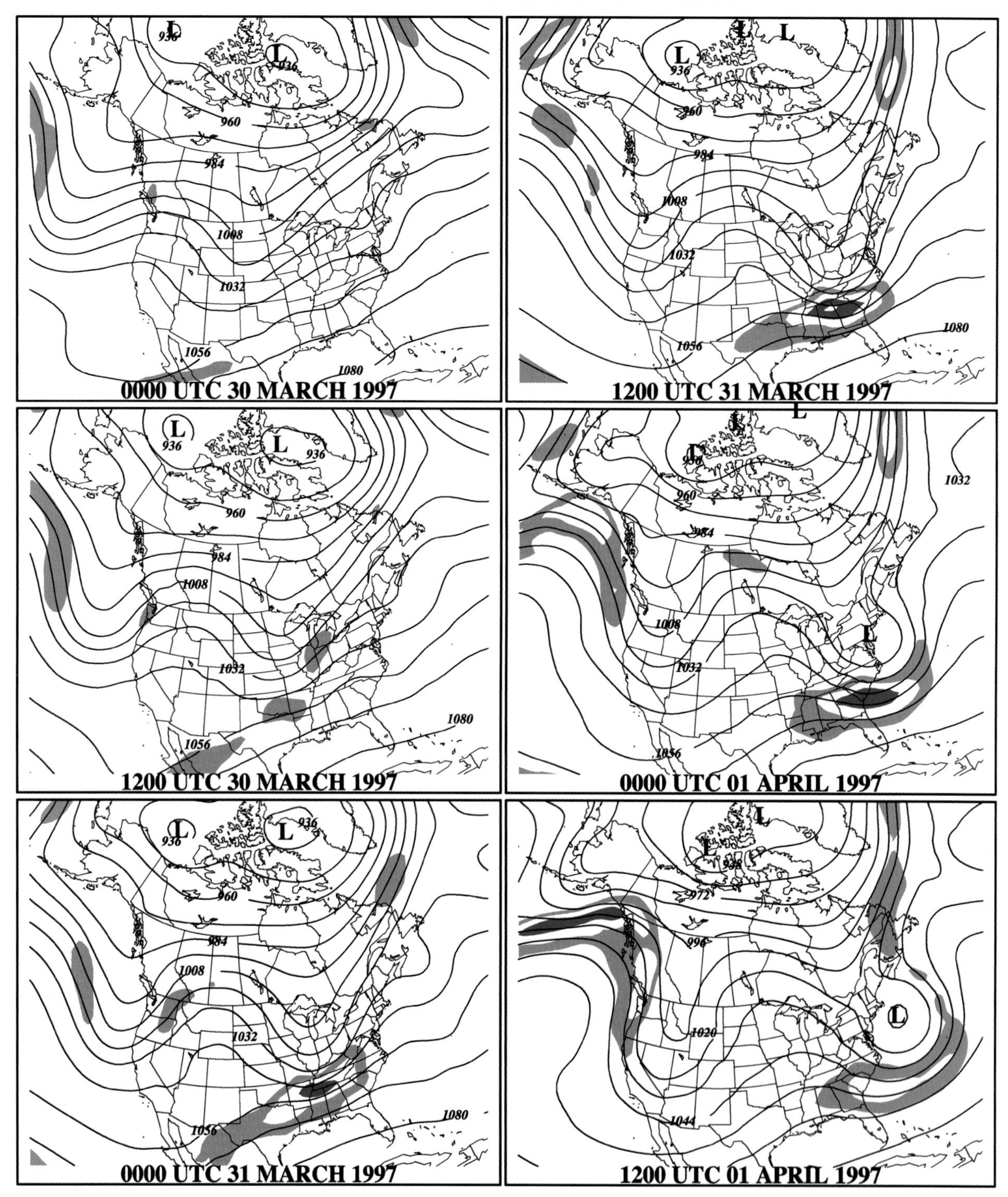

FIG. 10.28-5. Twelve-hourly analyses of 250-hPa geopotential height for 0000 UTC 30 Mar–1200 UTC 1 Apr 1997. See Fig. 10.1-5 for details.

SATELLITE IMAGERY

FIG. 10.28-6. Twelve-hourly infrared satellite images for 0000 UTC 30 Mar–1200 UTC 1 Apr 1997. See Fig. 10.16-6 for details.

south-central United States at 0000 UTC 30 March that became better defined over the Appalachians by 0000 UTC 31 March with little evidence of cyclogenesis. By 0000 UTC 31 March, an 850-hPa low had developed east of Lake Huron and then appeared to drop southward and intensify along the developing baroclinic zone, with the center of the low deepening rapidly off the New Jersey coast by 1200 UTC 31 March.

- At 0000 UTC 30 March, the 0°C isotherm moved from near the northeast U.S.–Canadian border and drifted slowly southward in the following 24 h into northern New England but the coldest 850-hPa temperatures were still found over Michigan behind the developing 850-hPa low over Lake Huron. This cold air push was located to the north of the primary baroclinic zone at 0000 UTC 31 March.
- Between 0000 and 1200 UTC 31 March, a broad cyclonic center extended from western New York to eastern Virginia with the rapidly deepening center over New Jersey as cold lower-tropospheric temperatures dropped southward from the upper Midwest into New York, Pennsylvania, and Ohio. In the subsequent 12 h, the 850-hPa low consolidated into an intense circulation just south of Long Island at 0000 UTC 1 April and then moved slowly southeastward over the Atlantic.
- With cyclogenesis under way at 1200 UTC 31 March, the colder air over the Midwest was now surging rapidly southeastward toward the middle Atlantic coast with temperatures of $-5°$ to $-10°C$ extending into the Ohio Valley, New York, Pennsylvania, and the northwestern portions of the mid-Atlantic states. Note that the 0°C isotherm only progressed southward into central New England at this time, where most of the precipitation was still falling as rain (except over elevated terrain).
- The intensification of the 850-hPa low from 1200 UTC 31 March to 0000 UTC 1 April was accompanied by falling temperatures across all of the Northeast except southeastern Massachusetts as much of the rain in New England changed to snow. In the following 12 h, when the heaviest snows fell across eastern New England, 850-hPa temperatures continued to drop slowly below $-5°C$. The coldest 850-hPa temperatures of $-10°C$ or lower were found from central New York and Pennsylvania into northern Virginia.
- The development of an intense 850-hPa low between 1200 UTC 31 March and 0000 UTC 1 April was accompanied by the development of a southerly low-level jet exceeding 30 m s^{-1} well east of the low center. A more significant development was the easterly and northerly jet exceeding 25 m s^{-1} north and west of the 850-hPa low center. At 0000 UTC 1 April, the development of heavy snowfall across eastern New England was accompanied by easterly 850-hPa winds of 25–30 m s^{-1}, a clear indication that the strengthening flow north of the low center was acting to transport moisture from the North Atlantic toward the expanding area of heavier snowfall.

d. 500-hPa geopotential height and 400-hPa wind analyses

- This storm developed during the beginning of a very strong El Niño that persisted from spring 1997 through mid-1998 (Fig. 2–19, volume I). The storm also developed at the end of a brief period of positive values of the North Atlantic Oscillation (NAO). There was a considerable amount of fluctuation between positive and negative phases of the NAO from mid-March through early April with no clear preference.
- There are two notable aspects of the 500-hPa and upper-level jet evolutions of the storm. The first is the amplification of the trough in the central United States by 0000 UTC 31 March and the closing off of the 500-hPa low center along the East Coast by 0000 UTC 1 April. The second is that there was neither a preceding confluent zone nor a northern jet system, which is consistent with the lack of a cold surface high pressure system in the Northeast prior to the storm.
- Surface high pressure over east-central Canada was found beneath confluent upper flow displaced farther north than was found in other cases (eastern Canada) to the rear of an intense 500-hPa vortex west of Greenland at 0000 UTC 30 March. The confluent flow does not appear to play a significant role in bringing colder temperatures to the northeastern United States since it was located too far north of the region.
- The upper trough over the central United States amplified between 1200 UTC 30 March and 0000 UTC 1 April, in conjunction with the rapid cyclogenesis, leading to the development of an intense 500-hPa closed low and a consolidated pattern near the base of the trough at 0000 UTC 1 April. As the trough amplified over the eastern United States, a marked area of diffluence developed east of the trough axis by 1200 UTC 31 March over the rapidly developing surface cyclone.
- The 400-hPa wind analyses show that the most significant wind features for this case were the jet streaks associated with the amplifying trough with no indication of a northern, confluent jet system (identified for other cases) over New England or the middle Atlantic states. Confluent flow and a separate jet were located well to the north off the northeastern Canadian coast.
- As the upper trough over the northern plains began to deepen, a 40–50 m s^{-1} jet streak propagated into the central United States by 0000 UTC 31 March while a separate 50 m s^{-1} jet maximum began developing just east of the trough axis over Tennessee. The amplification and apparent merger of these jets was accompanied by a developing diffluent jet exit region over the middle Atlantic states between 0000

and 1200 UTC 31 March as the surface low began to consolidate and intensify off the East Coast. It appears that this development of the surface low and evolution of the southeasterly low-level jet were both coupled with an amplifying upper-level jet and the distinct along-stream wind variation in the jet's diffluent exit region.

e. 250-hPa geopotential heights and wind analyses

- This case is marked by a rather bland westerly flow regime in the precyclogenetic period with generally uniform height gradients from central Canada to the Gulf coast at 0000 UTC 30 March.
- The main feature at this level is the amplifying trough over the eastern United States by 0000 UTC 31 March, which becomes negatively tilted off the east coast by 0000 UTC 1 April as the surface cyclone is rapidly intensifying.
- The most noteworthy jet feature at this level is a weak subtropical jet system moving northeastward out of Mexico and Texas at 1200 UTC 30 March, which becomes better defined by 31 March. Wind speeds in the jet increase to over 70 m s^{-1} by 0000 UTC 31 March and 80 m s^{-1} by 1200 UTC 31 March. The diffluent exit region of this jet overlies the exit region of the polar jet system at both times, coincident with the rapid development of the surface low immediately off the East Coast on 31 March.
- There is no distinct northern jet system factoring into the case until 2 April as the wind speeds in the confluent region and ridge crest downwind of the trough increase and appear to maintain the entrance region jet flow downwind of the precipitation area after the surface low has developed rapidly off the New England coast.

f. Infrared satellite imagery sequence

- The cloud mass associated with this storm underwent a dramatic amplification from a fairly unimpressive satellite cloud signature. The cloud mass associated with this storm was a loose, comma-shaped system without much upper cloud definition through 0000 UTC 31 March. Once the upper-level trough deepened dramatically at 1200 UTC 31 March, the cloud mass underwent a rapid transformation into a distinct area of enhanced colder cloud tops, with a sharp back edge over eastern Pennsylvania and New York and several east–west banded cloud features embedded within this area, often indicative of a developing storm system.
- This cloud mass evolved rapidly into a circular, vortex structure by 0000 UTC 1 April with a distinct center near Long Island and a distinct area of high cloud elements forming the comma head from New England west-southwestward into New York and Pennsylvania. Heaviest snow was falling along the southern edge of the enhanced cloud tops from eastern Pennsylvania into interior New York and southern New England.
- As the upper-level and the surface lows drifted southeastward overnight, the axis of the main cloud band to the north-northwest of the low drifted southeastward and extended from south of Long Island northeastward into eastern New England at 1200 UTC 1 April. This cloud band was associated with the record snowfall that fell across southeastern New England.

Dupont Circle in Washington D.C., 25 Jan 2000 (copyright Keith Stanley).

29. 24–26 January 2000

a. General remarks

- During the warmest winter (December—February) ever recorded in the United States, the northeast was one of the few regions to experience a brief period of winter weather conditions, which occurred during the last half of January. During this period, the most significant winter storm developed off the Southeast coast on 24 January and then intensified as it moved toward the middle Atlantic states on 25 January. The storm moderated as it moved toward New England, sparing a more significant snowfall there. Near-blizzard conditions affected portions of North Carolina, Virginia, Maryland, and Washington, as strong north-northwesterly winds combined with up to 20 in. (50 cm) of snow to bring these areas to a standstill. This was one of the heaviest snowstorms on record in central North Carolina.
- Perhaps the most noteworthy aspect of this snowstorm was the consistent failure of many numerical weather prediction models to adequately forecast the evolution of this storm (Zhang et al. 2002; Langland et al. 2002). During the 2–4-day period prior to the storm, medium-range forecast models did suggest a major East Coast snowstorm but incorrectly predicted it for 26 January and attributed it to an amplifying upper trough moving southeastward from the plains states toward the East Coast.
- The upper-level trough that was actually associated with the developing storm was forecast to be much weaker and move farther east than actually occurred. The short-range models began to forecast a significant storm system only once the storm had already developed on 24 January, thereby providing no model guidance for the record snowfall in North Carolina. Subsequent short-range forecast models kept the heaviest snow too far east, leading to underestimates of the storm's impact from the Carolinas to the mid-Atlantic states to New England. The lack of a snow forecast even 12 h before fed a media-driven perception that the entire weather community totally missed this storm, even though warnings were issued 6–8 h before heavy snow enveloped the area from Washington northward.
- The following regions reported snow accumulations exceeding 10 in. (25 cm): central North Carolina, eastern Virginia, central Maryland, northern Delaware, eastern Pennsylvania, and portions of New Jersey, eastern New York, western Massachusetts, Vermont, and New Hampshire.

TABLE 10.29-1. Snowfall amounts for urban and selected locations for 24–26 Jan 2000; see Table 10.1-1 for details.

NESIS = 3.14 (category 2)	
Urban center snowfall amounts	
Washington, DC–Dulles Airport	10.3 in. (26 cm)
Baltimore, MD	14.9 in. (38 cm)
Philadelphia, PA	9.2 in. (23 cm)
New York, NY–LaGuardia Airport	6.5 in. (17 cm)
Boston, MA	3.5 in. (9 cm)
Other selected snowfall amounts	
Raleigh, NC	20.2 in. (51 cm)
Annapolis, MD	17.0 in. (43 cm)
Richmond, VA	11.0 in. (28 cm)

b. Surface analyses

- The storm developed as a wave of low pressure along a surface front that had previously moved southward to the Gulf coast. The surface low then moved northeastward from the northern Gulf along the frontal boundary, which edged back toward the East Coast on 24–25 January.
- A series of weak low pressure systems consolidated into one rapidly developing center in the northeastern Gulf of Mexico and southern Georgia–northern Florida just before 1200 UTC 24 January. An expanding area of thunderstorms over Florida and southern Georgia and an expanding area of rainfall across the Southeast heralded the initial development of the storm.
- The surface low deepened rapidly during 24 January with heavy snows developing in northwestern South Carolina that later spread northeastward into central North Carolina that evening. The surface low moved from southern Georgia northeastward to a position east of Charleston, South Carolina, by 0000 UTC 25 January as the pressure fell about 14 hPa to 990 hPa in the 12-h period ending at 0000 UTC.
- The low pressure system continued to deepen to 978 hPa as it moved north-northeastward to just east of Cape Hatteras at 1200 UTC 25 January. The central pressure of the surface low then remained steady as the storm drifted slowly northward east of the Virginia coast and then began to fill as it moved toward eastern Long Island after 0000 UTC 26 January.
- A large band of heavy snow spread rapidly northward across central North Carolina late on 24 January and early on 25 January. This band then expanded northeastward toward the middle Atlantic states early on 25 January. The eastern portion of the heavy precipitation band located initially over the western Atlantic Ocean pivoted toward the north-northwest, resulting in the rapid development of heavy snow from eastern Virginia north-northeastward through the Northeast urban corridor at 1200 UTC 25 January, affecting all the major cities from Richmond to Boston. As the day progressed, the band continued to pivot toward the west with the heaviest snow remaining on an axis from eastern Virginia across central Maryland and eastern Pennsylvania. From New Jersey through New England, the band moved west and weakened as snowfall diminished and changed to light freezing rain and ice

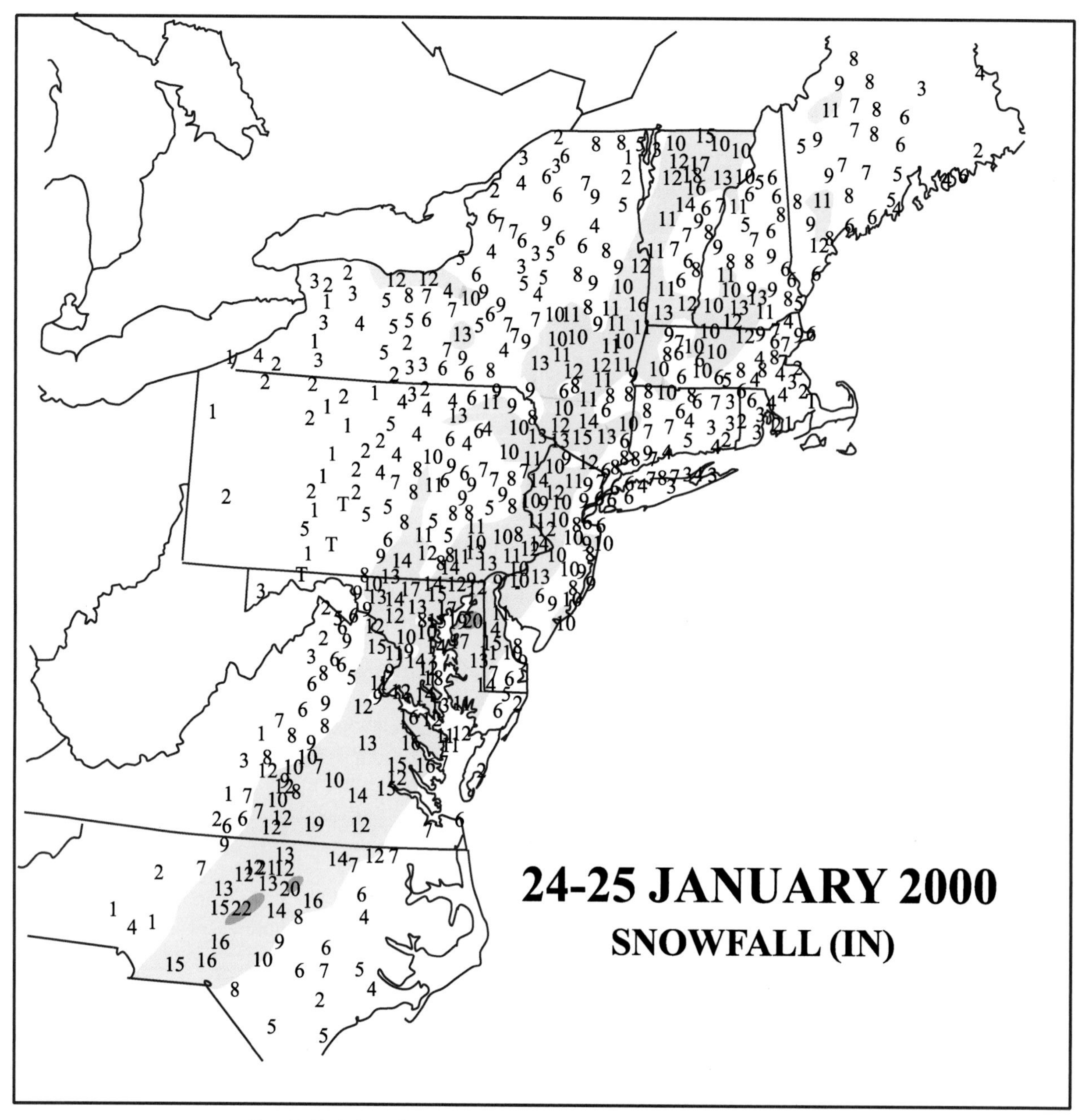

FIG. 10.29-1. Snowfall (in.) for 24–26 Jan 2000. See Fig. 10.1-1 for details.

pellets, curtailing more significant accumulations in these coastal areas.

- The intense snow band shrank over the middle Atlantic states during the afternoon and evening of 25 January. This development occurred as the surface low ceased to intensify and began to fill as it moved northeastward toward New England. By the morning of 26 January, the central pressure of the low had risen to 990 hPa as it reached eastern Maine.
- Cold air was in place before the storm affected the East although the surface high pressure pattern did not resemble many of the other cases. One high pressure cell moved east across eastern Canada while a second area of high pressure over the central United States built northeastward across the Ohio Valley and Appalachians as the storm developed on 24–25 January.

c. 850-hPa analyses

- A storm system that passed east of the New England coast on 22–23 January had the effect of advecting cold 850-hPa temperatures south and eastward over southeastern Virginia and northern North Carolina by

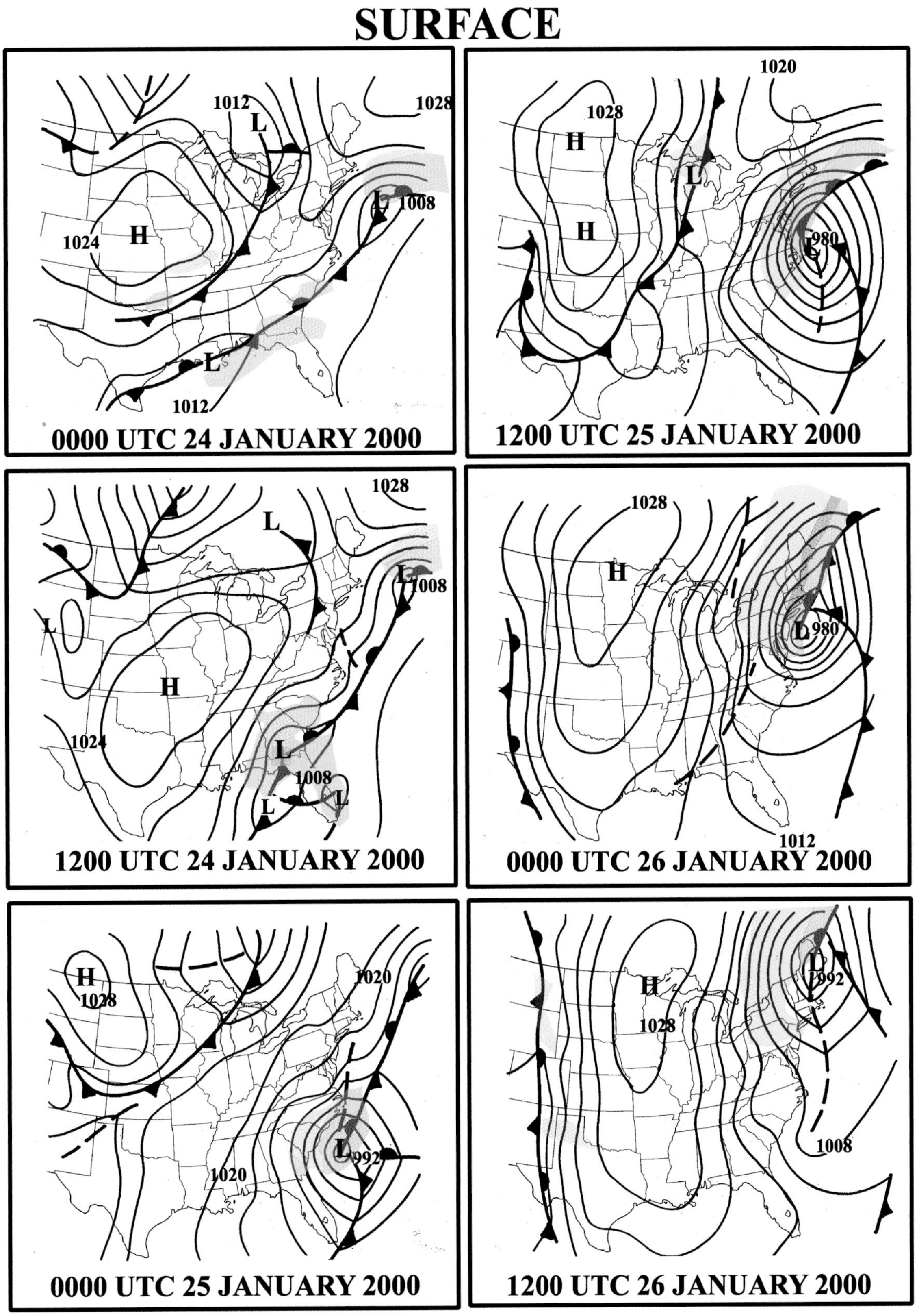

SURFACE

FIG. 10.29-2. Twelve-hourly surface weather analyses for 0000 UTC 24 Jan–1200 UTC 26 Jan 2000. See Fig. 10.1-2 for details.

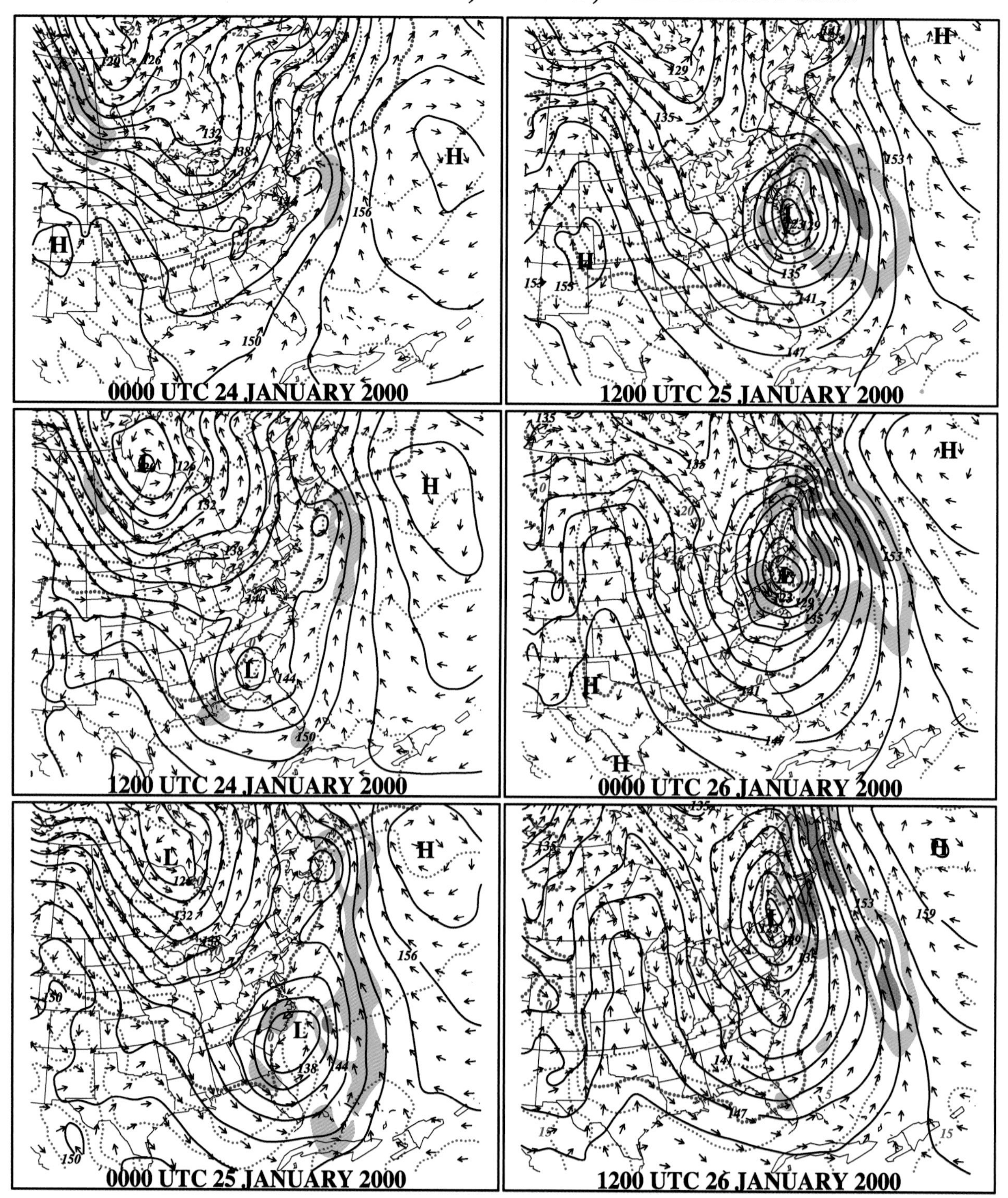

FIG. 10.29-3. Twelve-hourly 850-hPa analyses for 0000 UTC 24 Jan–1200 UTC 26 Jan 2000. See Fig. 10.1-3 for details.

500 hPa HEIGHTS AND VORTICITY, 400 hPa WINDS

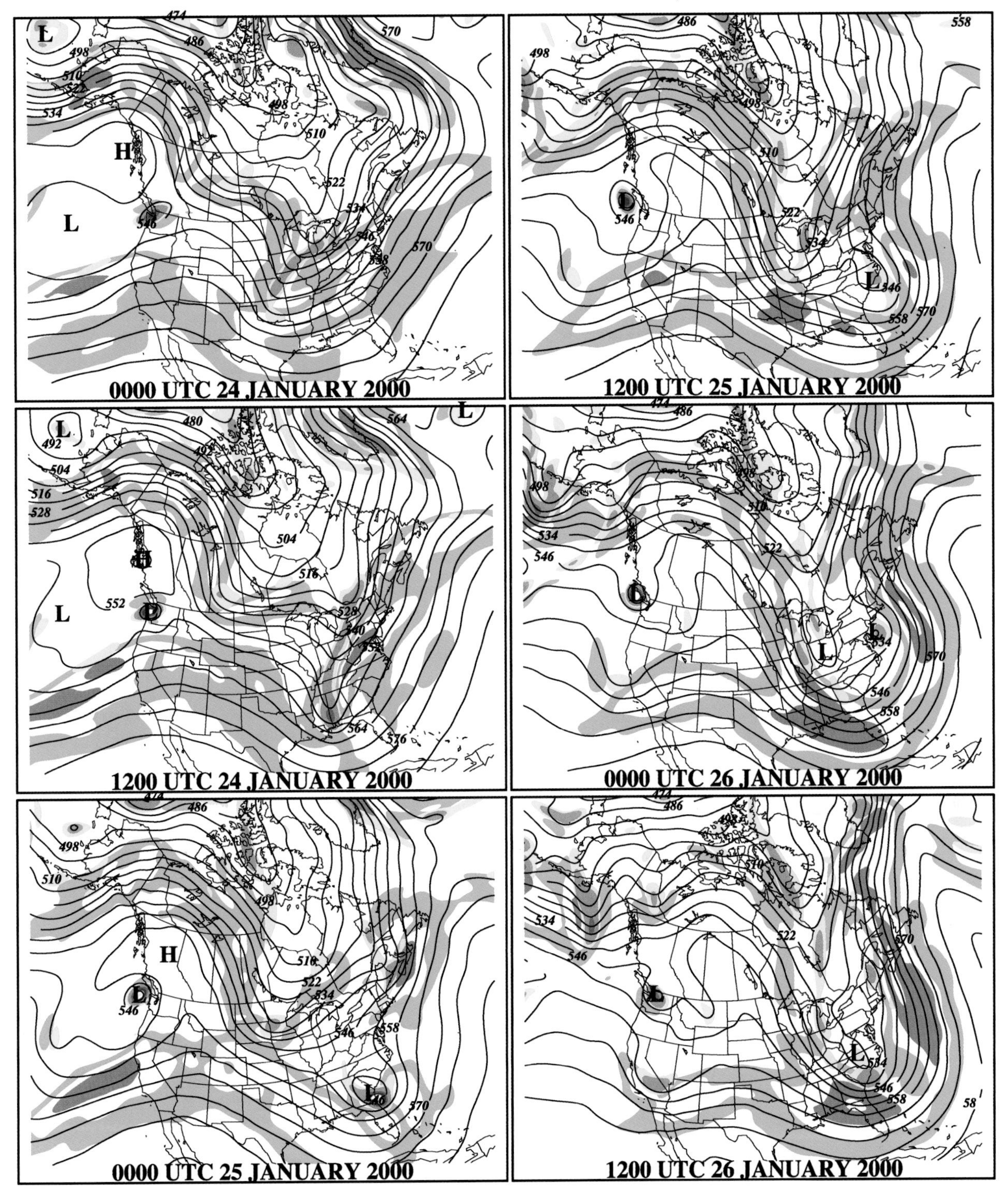

FIG. 10.29-4. Twelve-hourly analyses of 500-hPa geopotential heights and upper-level wind fields for 0000 UTC 24 Jan–1200 UTC 26 Jan 2000. See Fig. 10.1-4 for details.

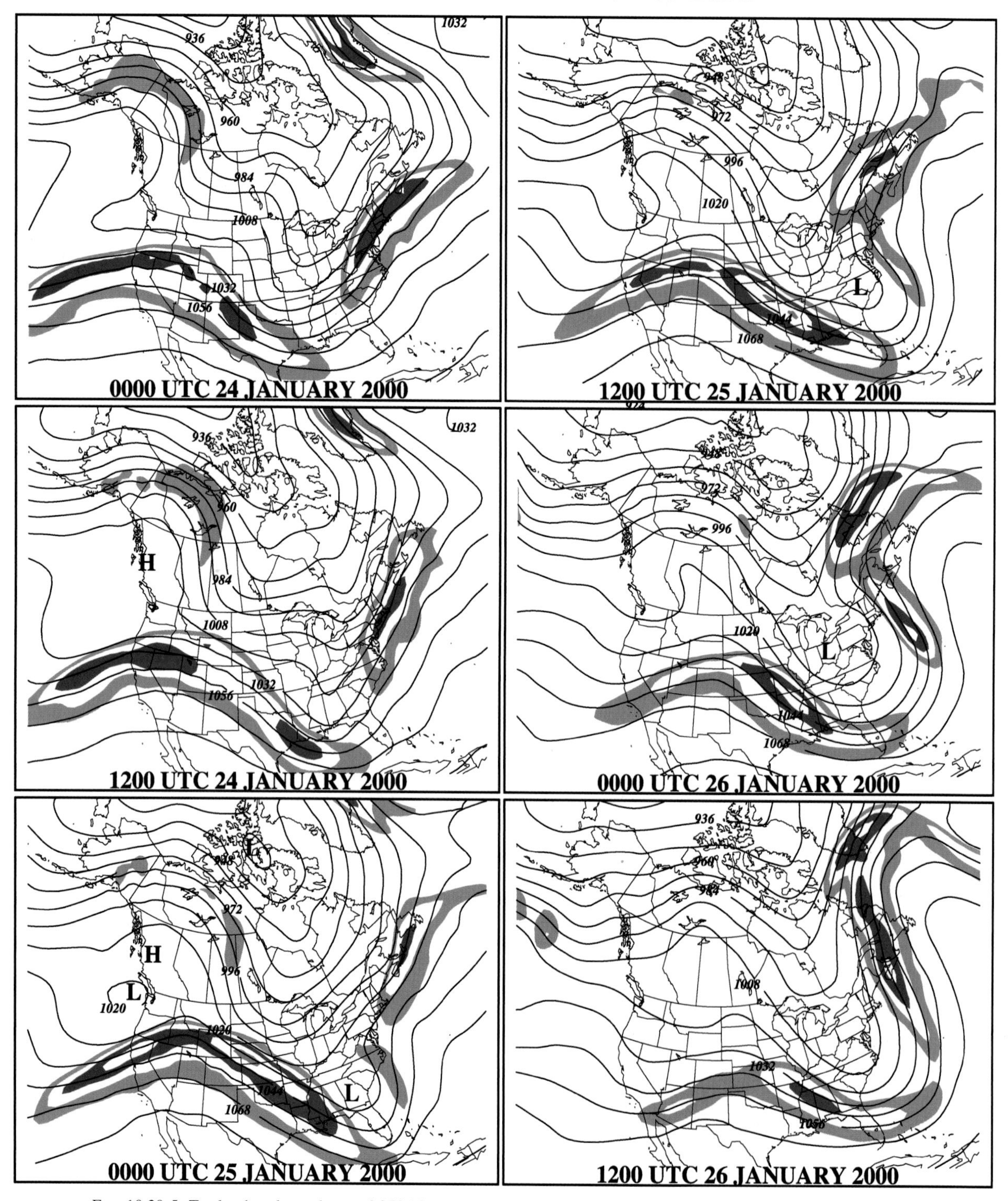

FIG. 10.29-5. Twelve-hourly analyses of 250-hPa geopotential height for 0000 UTC 24 Jan–1200 UTC 26 Jan 2000. See Fig. 10.1-5 for details.

SATELLITE IMAGERY

FIG. 10.29-6. Twelve-hourly infrared satellite images for 0000 UTC 24 Jan–1200 UTC 26 Jan 2000. See Fig. 10.16-6 for details.

0000 UTC 24 January. A northeast–southwest temperature gradient was established behind this first system. As a new short-wave trough associated with the developing surface low plunged southward to the southeast United States by 1200 UTC 24 January and 0000 UTC 25 January, the baroclinic zone appeared to intensify along the Carolina coast.

- A rather weak 850-hPa trough over Mississippi and Louisiana at 0000 UTC 24 January amplified into a closed low over southeastern Georgia at 1200 UTC 24 January as the surface low began developing. The 850-hPa low developed on the warm side of the 0°C isotherm.
- In the 12 h ending at 0000 UTC 25 January, the 850-hPa low deepened rapidly and moved off the South Carolina coast within the strengthening baroclinic zone. An easterly low-level jet developed to the north of the low over North Carolina and Virginia with maximum winds exceeding 20 m s^{-1}. This jet appeared to be a critical factor in enhancing the moisture transport into the area of heavy snows in the Carolinas during this period.
- The 850-hPa low moved northeastward and deepened rapidly between 0000 and 1200 UTC 25 January. It also changed orientation from a nearly symmetrical low to one tilted on a southwest–northeast axis as warmer air was advected westward to the north of the low. Heaviest snow across Virginia, Maryland, and eastern Pennsylvania occurred with 850-hPa temperatures near −5°C, while the 0°C isotherm became oriented along the coastline from the low center east of North Carolina northward through New England. The rotation of the thermal gradient, and 0°C isotherm, toward the west-northwest was consistent with the movement of the distinct bands of heavy snow from east to west across the mid-Atlantic states during this period.

d. 500-hPa geopotential height and 400-hPa wind analyses

- This storm developed during a winter characterized by La Niña conditions and North Atlantic Oscillation (NAO) values that were primarily positive. However, the NAO became negative on 12 January and remained negative through 27 January. During this 2-week period, cold weather covered much of the eastern United States, with this storm developing at the end of the negative NAO period (see Fig. 2–22, volume I). Analyses of 500-hPa data showed a persistent cutoff upper ridge during this entire 2-week period, and some evidence still remains of the ridge, located to the east of Greenland at 0000 UTC 24 January. During the ensuing 3 days, the ridge retreated eastward, indicating an end to the negative phase of the NAO.
- The precyclogenetic environment at 500 hPa was dominated by a nearly stationary cutoff ridge near Iceland, a feature observed with a significant number of other snowstorms.
- However, there was no closed low over eastern Canada, another prominent feature associated with many snowstorms. Upper-level confluence was not very evident over the eastern United States and Canada by 0000 UTC 24 January. However, as the upper trough responsible for cyclogenesis deepened and a second trough over the Great Lakes lifted northeastward, a confluent pattern developed over the northeastern United States, supporting a gradual increase in the wind speeds in that region as the surface cyclone intensified on 25 January.
- The surface low developed as an upper trough located over the central United States at 0000 UTC 24 January deepened dramatically over the southeastern United States between 1200 UTC 24 January and 0000 UTC 25 January. The deepening trough developed a closed center at 500 hPa and attained a negative tilt by 0000 UTC 25 January.
- As the trough deepened and developed a negative tilt, the distance between the trough axis and downstream ridge axis (half-wavelength) decreased dramatically to less than 750 km by 0000 UTC 25 January, with an noticeable diffluence developing off the Carolina and Virginia coasts above the developing surface cyclone.
- After the initial deepening, the upper trough lifted northeastward toward the middle Atlantic states between 0000 and 1200 UTC 25 January, as heavy snow spread throughout the area from North Carolina to New England and the surface low deepened to 978 hPa.
- As the trough continued to lift northward and the amplitude decreased, a second upper-level trough deepened over the Ohio Valley at 0000 UTC 26 January and reached the Carolina coast with a significant vorticity maximum by 1200 UTC 26 January. The period of decreasing amplitude of the first trough moving northeastward along the East Coast is consistent with the period during which the surface low weakened and snowfall rates diminished over New England. The 3–7-day model forecasts that pointed to an East Coast storm were predicting a storm with this second trough on 26 January, completely missing the storm system and related negatively tilted trough along the East Coast on 25 January.
- The jet patterns at 400 hPa are not as easily described with this case as in other cases. The most important features evolve before, during, and just after 1200 UTC 24 January as the trough amplified in the southeastern United States. One jet system extends from the Pacific coast to the greater than 40 m s^{-1} jet core just west of the trough axis over Mississippi. A diffluent exit region of this jet is located over the northeast Gulf of Mexico and northern Florida where the surface low has begun to develop at 1200 UTC 24 January.

- A second "outflow" jet exceeding 50 m s^{-1} extends north-northeastward from northern Georgia toward New Jersey and into a more confluent region over New England at 1200 UTC 24 January. An outbreak of convection (see the corresponding satellite imagery) and the area of more general precipitation that developed over the Southeast appears to explain the development of the entrance region of the jet at this time (see the section on latent heat release in chapter 4 of volume I) and as the precipitation continued to spread northward into North Carolina and Virginia during the day.
- Between 0000 and 1200 UTC 25 January, a separate jet maximum exceeding 50 m s^{-1} developed east of the trough axis. By 1200 UTC, the surface low and easterly low-level jet (LLJ) were located in the left-exit region of this new jet streak and the entrance region of a broadly defined area of higher wind speeds over New England and southeastern Canada.

e. 250-hPa geopotential height and wind analyses

- The 250-hPa height analyses provide a clear picture of the trough deepening into the south-central United States on 24 January, becoming negatively tilted over the southeast coast by 0000 UTC 25 January, and merging with the next amplifying trough as it propagated northeastward along the east coast and deamplified by 0000 UTC 26 January.
- The jet analyses at 250 hPa, unlike those at the 400-hPa level, are more straightforward in describing the evolution of this case. During the precyclogenetic period on 24 January, a northern jet exceeding 70 m s^{-1} extends from the middle Atlantic coast to New England, while a well-defined jet stream with embedded jet streaks extends from the California coast to the northeastern Gulf of Mexico, where it amplifies to greater than 90 m s^{-1} at 0000 UTC 25 January as the trough amplifies over the Southeast and the surface low intensifies rapidly off the southeast coast.
- The initial cyclogenesis occurs in the entrance region of the northern jet and in the well-defined exit region of the southern jet as the trough becomes negatively tilted and more diffluent between 1200 UTC 24 January and 0000 UTC 25 January.
- The separate jet streak evident at 400 hPa east of the trough axis at 0000 UTC 25 January also is evident at this level. This jet streak amplifies to greater than 80 m s^{-1} by 0000 UTC 26 January, with the surface low occurring in its left-exit region and the right-entrance region of the northern jet.
- The northern jet also intensifies to speeds greater than 80 m s^{-1} as the downstream ridge builds into southeastern Canada and the precipitation area expands northeastward into its right-entrance region on 25 and 26 January.

f. Infrared satellite imagery sequence

- Satellite images show two main cloud features at 0000 UTC 24 January, prior to the development of a strong surface low off the Southeast coast. The primary area of clouds was the long enhanced line of clouds extending from Florida to the middle Atlantic coast. The second area of clouds is the small enhanced mass extending from Oklahoma into Illinois south to Tennessee. This small region is evidence of the short-wave trough that later amplified and produced the intense East Coast storm.
- The amplification of the upper-level trough moving southeastward across the Gulf coast on 24 January is reflected by the rapid expansion of a comma-shaped cloud over Georgia and northwestern Florida, associated with an outbreak of thunderstorms and heavy rains at 1200 UTC 24 January as the surface low developed over the southeastern United States.
- This small comma-shaped cloud grew rapidly into a classic comma-cloud signature within the next 24 h as the upper-level trough became negatively tilted, the wavelength between the trough axis and downstream ridge shortened dramatically, and the surface low developed rapidly as it moved just off the Atlantic coast. The heaviest snow developed on the eastern side of the comma "head" and extended to the south-southwest of the main comma tail at 1200 UTC 25 January from Virginia north-northeastward toward New Jersey.
- The comma cloud became less organized as the surface low began to weaken and as the upper-level trough began to deamplify by 0000 UTC 26 January 2000. A second digging trough moving southeastward from the Midwest to the North Carolina coast between 1200 UTC 25 and 26 January is not marked by a distinct cloud signature in the satellite image because much of the moisture associated with the initial East Coast storm has been drawn northward and eastward away from the deepening system.

New York City, 30 Dec 2000 (source unknown)

30. 30–31 December 2000

a. General remarks

- This storm occurred at the end of the coldest November–December period on record nationwide, an astonishing fact given that the previous winter was the warmest on record for the nation. The snowstorm began as a low pressure system moving southeastward across the plains and Midwest that later weakened over the Ohio Valley, then redeveloped off the North Carolina coastline. The storm then developed rapidly early on 30 December, accelerated northeastward along the coast, and affected the Northeast over a very short period of time, approximately 18 h.
- Numerical prediction models indicated the potential for a major East Coast storm as much as 7 days in advance. Short-range forecast models, however, had difficulty predicting the sharp western boundary of the heavy snow–no-snow gradient, sometimes predicting heavy snow for the Washington–Baltimore region through central Pennsylvania, which did not occur.
- Heavy snow developed with a sharp western edge, sparing Washington and Baltimore. The snow gradient bisected the Philadelphia metropolitan area, with little to no snow in the western suburbs and up to a foot (30 cm) of snow in the northern and eastern suburbs. New York City was hard hit with 10–14 in. (25–35 cm) of snow. The Boston metropolitan area was spared significant accumulations as snow changed to rain, while Boston's western suburbs received up to a foot (30 cm) of snow. Heaviest snows fell in northern New Jersey, where up to 30 in. (75 cm) were measured and in eastern New York, especially the Catskills, where totals of more than 20 in. (50 cm) of snow were common.
- The following regions reported snow accumulations exceeding 10 in. (25 cm): central and northern New Jersey, western Long Island, extreme eastern Pennsylvania, eastern New York, western Connecticut, western and central Massachusetts, and portions of Vermont, New Hampshire, and Maine.
- The following regions reported snow accumulations exceeding 20 in. (50 cm): northern New Jersey and southeastern New York.

TABLE 10.30-1. Snowfall amounts for urban and selected locations for 30–31 Dec 2000; see Table 10.1-1 for details.

NESIS = 2.48 (category 1)	
Urban center snowfall amounts	
Philadelphia, PA	9.0 in. (23 cm)
New York, NY–La Guardia Airport	13.0 in. (33 cm)
Boston, MA	0.5 in. (9 cm)
Other selected snowfall amounts	
Randolph, NJ	26.0 in. (66 cm)
Elizabeth, NJ	16.0 in. (41 cm)
White Plains, NY	16.0 in. (41 cm)
Oceanside, NY	15.6 in. (40 cm)
Newark, NJ	13.9 in. (35 cm)
Albany, NY	12.6 in. (31 cm)
Burlington, VT	12.6 in. (31 cm)
Portland, ME	10.8 in. (27 cm)
Hartford, CT–Bradley International Airport	10.0 in. (25 cm)

b. Surface analyses

- Cold air was in place prior to the storm across the northeast as low pressure over eastern Canada circulated cold air southward across the northeastern United States for several days preceding the storm. Temperatures in the Northeast urban corridor the morning before the storm were primarily in the teens (°F; −8° to −10°C). The primary high pressure system was located over central Canada but a wedge of high pressure extended southeastward into the Northeast.
- The storm initially developed as an "Alberta clipper"-type system (see volume I, chapter 3) with a low pressure system moving southeastward from the northern plains to the Midwest on 28–29 December. Heavy snows accompanied this storm across North Dakota, Minnesota, and Iowa, where up to 10 in. (25 cm) fell. The low weakened as it approached the Midwest and Ohio Valley late on 29 December, accompanied by primarily light snows.
- A storm system along the Gulf and southeast coasts that produced a damaging ice storm across Oklahoma and Arkansas on 25–26 December moved across the southeastern United States on 28 December. Early numerical model forecasts attempted to merge this weather system with the one diving southeastward from the plains states, creating the potential for a tremendous East Coast storm. As it turned out, the Southeast system moved quickly offshore and did not merge with the approaching system from the northwest. Despite the lack of merging, a significant Northeast cyclone developed nonetheless.
- As low pressure weakened over the Ohio Valley late on 29 December, a secondary low developed just east of the North Carolina coastline between 0000 and 0600 UTC 30 December. Precipitation east of the North Carolina coast developed rapidly and moved northward toward extreme southeastern Virginia, eastern Maryland, and Delaware as light freezing rain, ice pellets, and snow.
- The secondary low deepened rapidly from approximately 1008 hPa at 0200 UTC 30 December off the North Carolina coast to 995 hPa by 1200 UTC 30 December east of the Delaware coast, a drop of 13 hPa in 10 h. The storm continued to deepen to 990 hPa by 1500 UTC and then ceased deepening. The cyclone's entire deepening occurred for only 13–15 h.
- Precipitation expanded rapidly northward after developing east of North Carolina around 0000 UTC 30

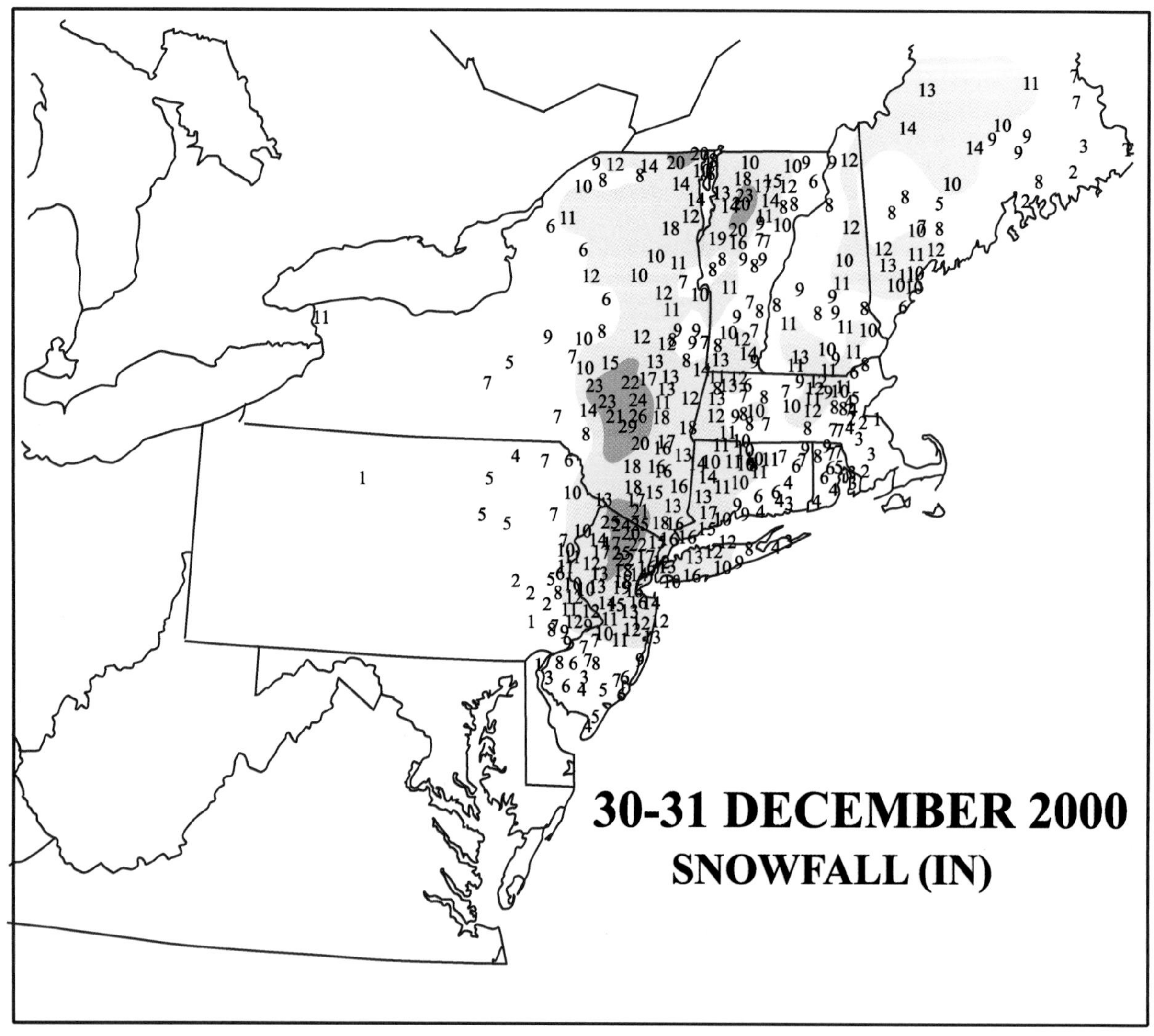

FIG. 10.30-1. Snowfall (in.) for 30–31 Dec 2000. See Fig. 10.1-1 for details.

December, reaching southern New Jersey between 0600 and 0800 UTC, and reaching New York City by 1000 UTC 30 December. Heavy snow developed shortly after the precipitation began, and by 1200 UTC heavy snow was falling in New York City while up to 4–6 in. of snow (10–15 cm) had already fallen across a large portion of southern and central New Jersey. Thunder and lightning accompanied the development of heavy snow in New York City and other locations. Much of the New York City metropolitan area received from 10 to 16 in. (25 to 40 cm) of snow with heavier amounts over New Jersey.

- The intense snow band was characterized by a very sharp western edge, with no precipitation and clearing skies in Washington and Baltimore. The Philadelphia metropolitan area represented the western gradient of snowfall with no snow to a couple inches (5 cm) in the western suburbs, 10 in. (24 cm) in the city, and more than a foot (30 cm) of snow to the north and east.
- Heaviest snow fell across northern New Jersey, just west of New York City and then across southeastern New York. Twenty to 30 inches of snow (50–75 cm) fell in north-central New Jersey, primarily from the early morning of the 30th through the late afternoon and early evening, accompanied by snowfall rates approaching 4 in. (10 cm) h^{-1}.
- As the surface low ceased deepening off the New Jersey coast, it drifted northward, reaching central Long Island by 2100 UTC 30 December. Snow mixed with ice pellets and rain and changed to rain over eastern Long Island and then did the same across eastern Connecticut, Rhode Island, and eastern Massachusetts as the surface low drifted northeastward into eastern Connecticut late on 30 December. In eastern New England, up to 12 in. (30 cm) of snow fell before

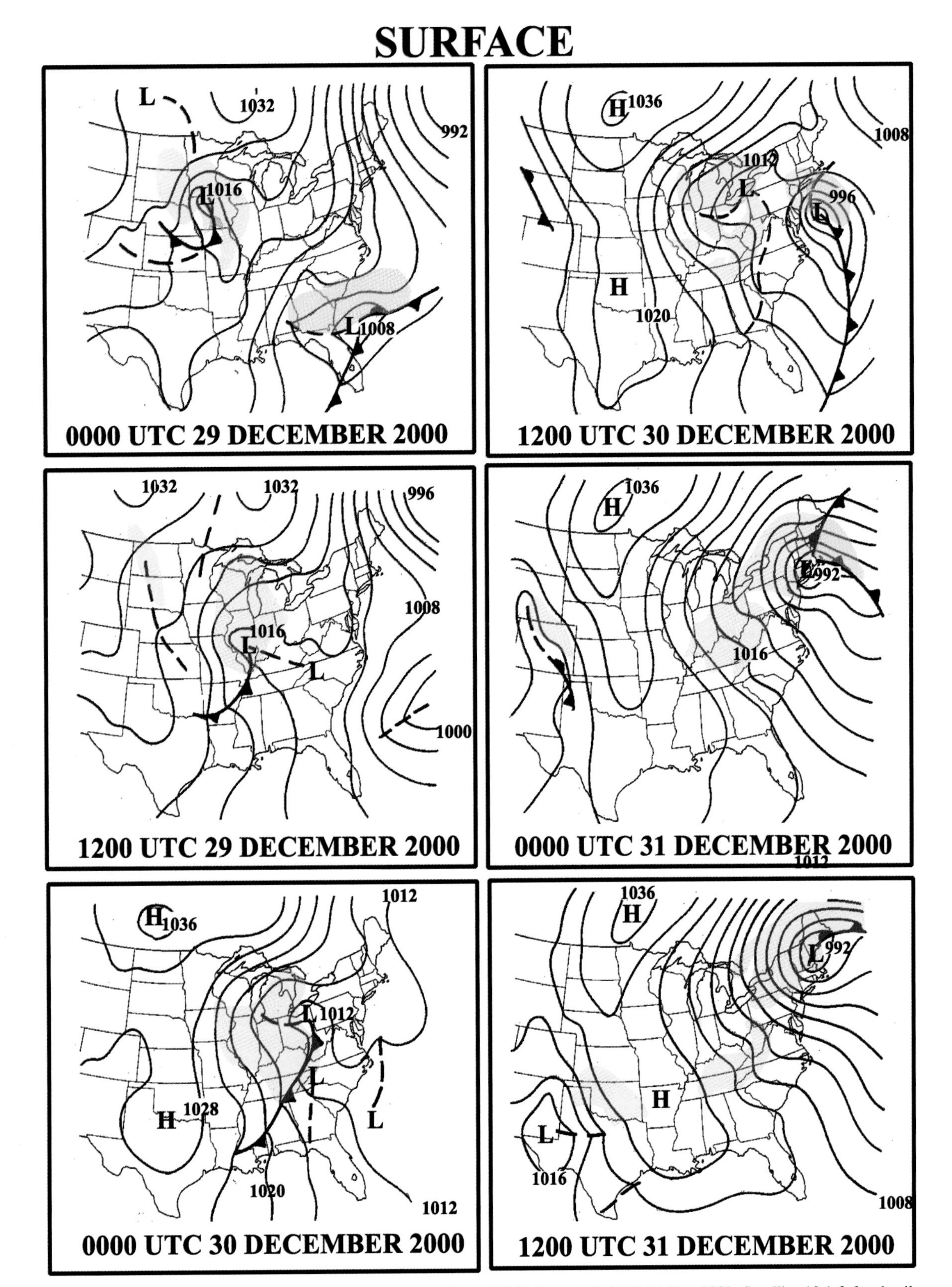

FIG. 10.30-2. Twelve-hourly surface weather analyses for 0000 UTC 29 Dec–1200 UTC 31 Dec 2000. See Fig. 10.1-2 for details.

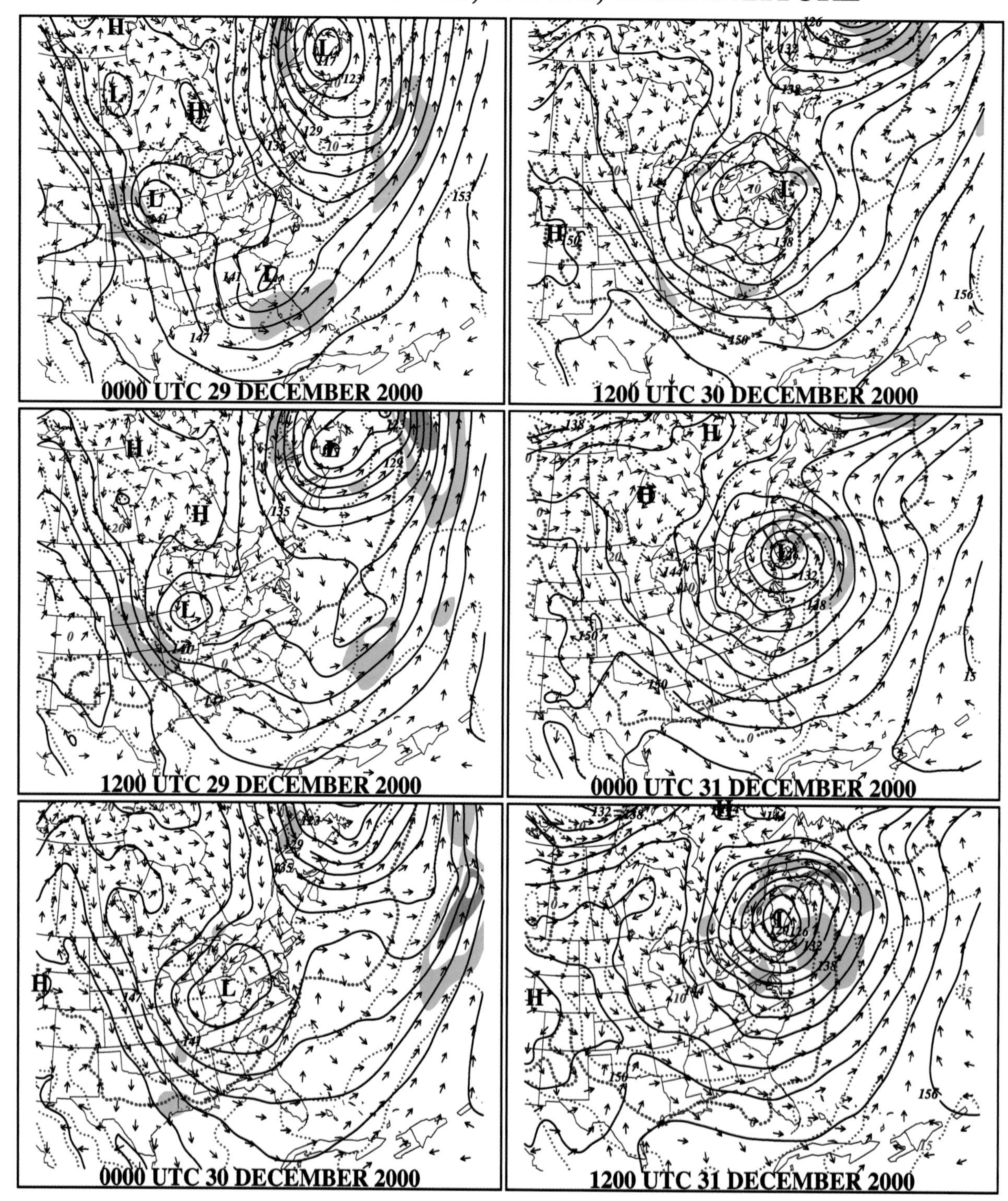

FIG. 10.30-3. Twelve-hourly 850-hPa analyses for 0000 UTC 29 Dec–1200 UTC 31 Dec 2000. See Fig. 10.1-3 for details.

FIG. 10.30-4. Twelve-hourly analyses of 500-hPa geopotential heights and upper-level wind fields for 0000 UTC 29 Dec–1200 UTC 31 Dec 2000. See Fig. 10.1-4 for details.

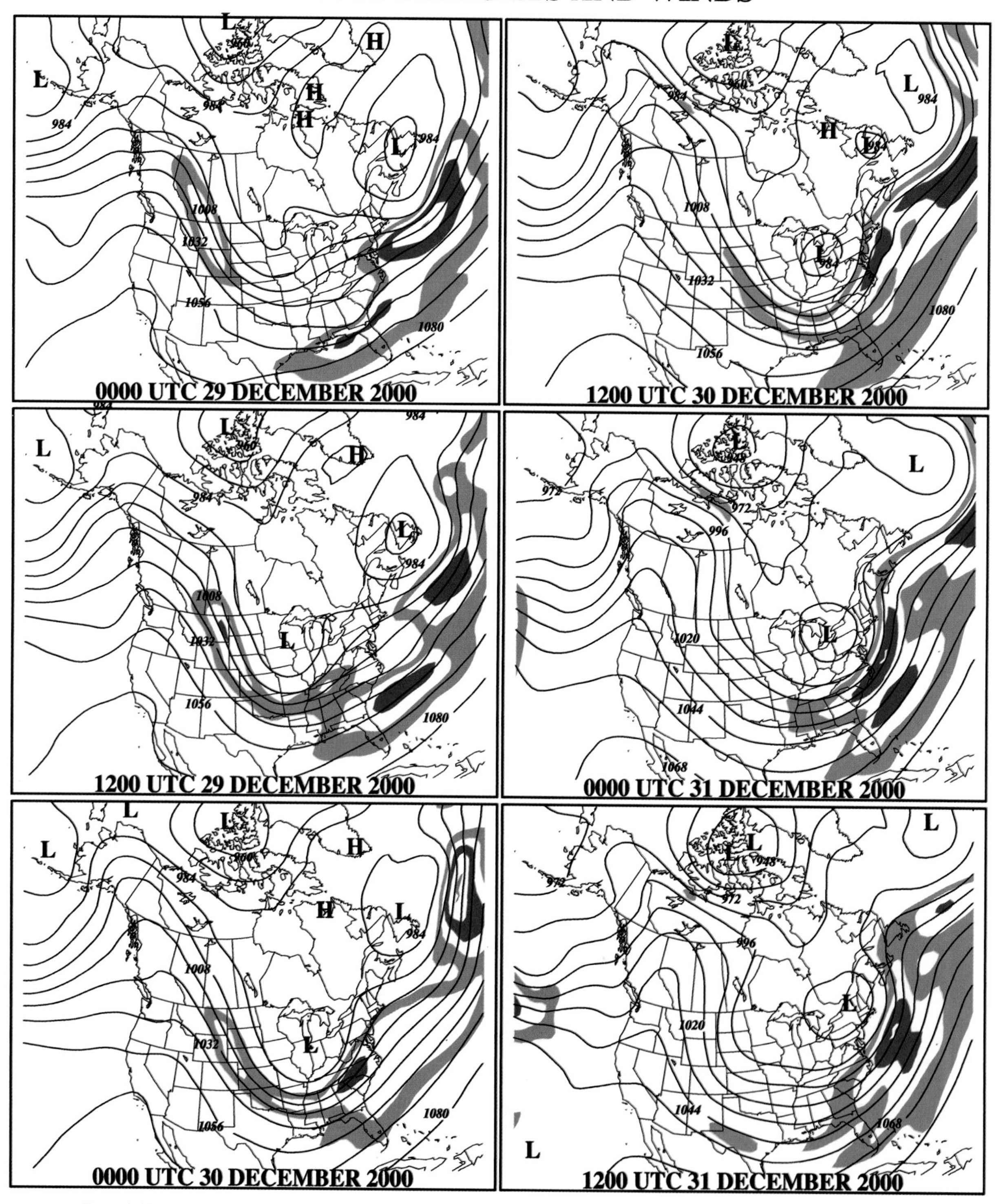

FIG. 10.30-5. Twelve-hourly analyses of 250-hPa geopotential height for 0000 UTC 29 Dec–1200 UTC 31 Dec 2000. See Fig. 10.1-5 for details.

SATELLITE IMAGERY

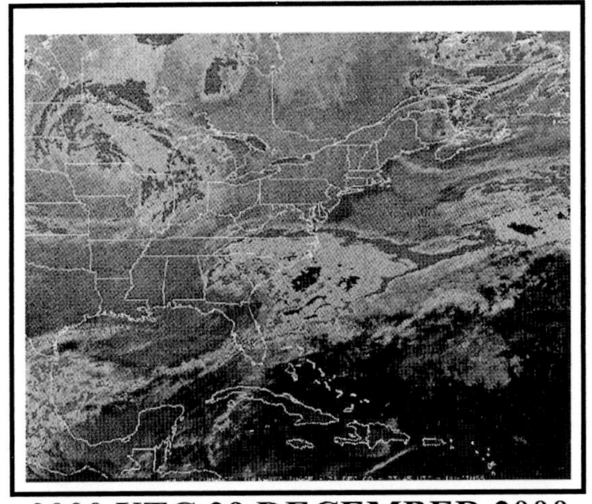

0000 UTC 29 DECEMBER 2000

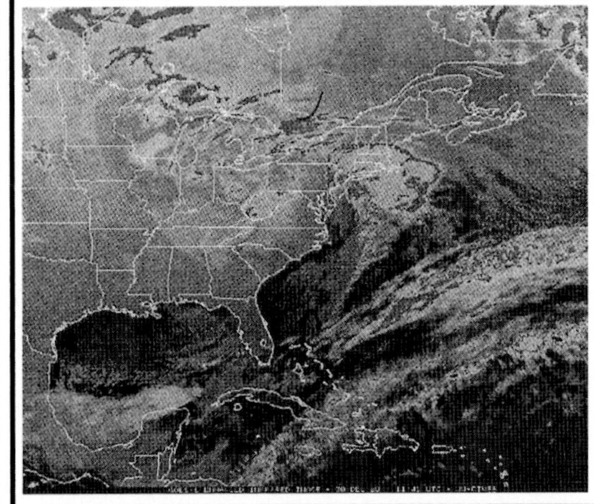

1200 UTC 30 DECEMBER 2000

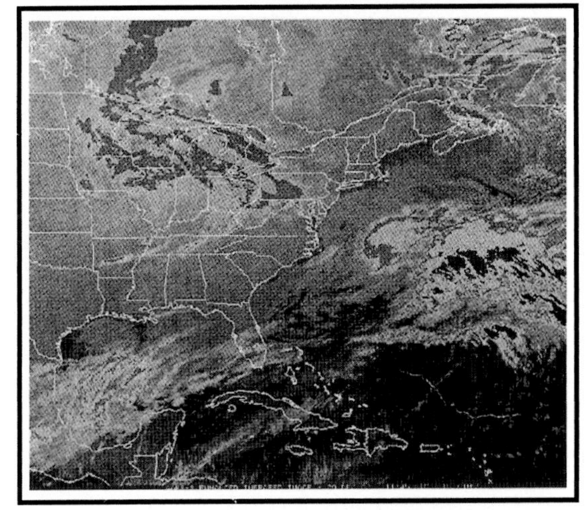

1200 UTC 29 DECEMBER 2000

0000 UTC 31 DECEMBER 2000

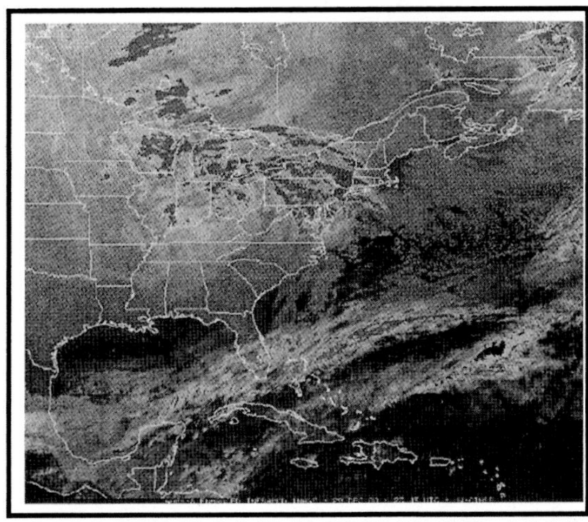

0000 UTC 30 DECEMBER 2000

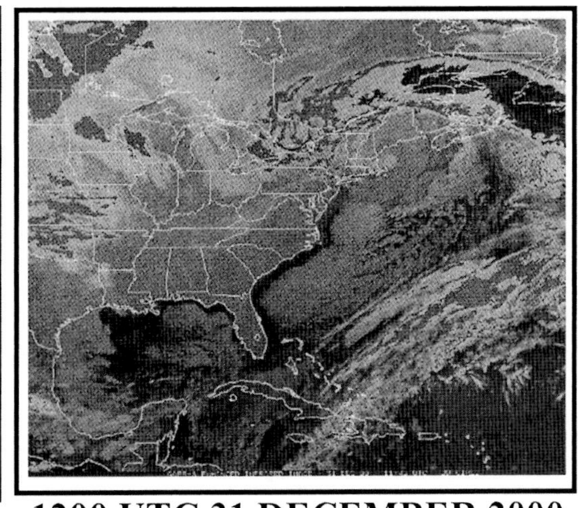

1200 UTC 31 DECEMBER 2000

FIG. 10.30-6. Twelve-hourly infrared satellite images for 0000 UTC 29 Dec–1200 UTC 31 Dec 2000. See Fig. 10.16-6 for details.

dry air from the south reduced the duration of the heavy snows.

- As the surface low drifted northeastward across eastern Connecticut, a new center developed near Boston and moved slowly northeastward toward coastal Maine and also moved onshore early on 31 December.

c. 850-hPa analyses

- Northwesterly flow around a storm system over Newfoundland advected low-level cold air through much of the Northeast prior to 30 December. The 850-hPa temperatures averaged $-5°$ to $-10°$ C over the region on 29 December.
- An 850-hPa low dropped southeastward across the Midwest and reached the Ohio Valley by 0000 UTC 30 December. As the secondary low formed shortly thereafter off the Carolina coast, the 850-hPa low also redeveloped along the mid-Atlantic coastline by 1200 UTC with the 0°C isotherm displaced east of the center. The 850-hPa low deepened rapidly by 0000 UTC 31 December and reached southern Connecticut.
- The development of a southeasterly 20–25 m s^{-1} to easterly low-level jet is indicated in the analyses from 1200 UTC 30 December directed toward New York City and northern New Jersey. This jet feature continued to expand and wrap around the rapidly developing vortex by 0000 UTC 31 December. This easterly jet increased in speed to greater than 30 m s^{-1} by 1200 UTC 31 December.
- Heaviest snow developed in New Jersey and moved into southeastern New York within the exit region of the low-level jet and just west of the 850-hPa low as it tracked from the mid-Atlantic coast at 1200 UTC 30 December to southern Connecticut at 0000 UTC 31 December. Heaviest snows developed in the 12-h period accompanying the rapid surface pressure and 850-hPa height falls just before and after 1200 UTC 30 December. The 850-hPa low deepened marginally between 0000 and 1200 UTC 31 December as it moved from southern Connecticut to southern Maine.
- The 850-hPa secondary low developed in a region of more pronounced isotherm gradients than its parent low over the Ohio Valley. The development of an "S shape" to the isotherms accompanied the intensification of the 850-hPa low along the mid-Atlantic coast. By 0000 UTC, 850-hPa temperatures were uniformly well below zero across all of New England, despite the occurrence of mostly rain in southeastern New England. Despite the relatively cold middle and upper troposphere, warm air at the lowest levels below 850 hPa changed the snow to rain in eastern Massachusetts and may have contributed to increased instability with numerous reports of thunder and lightning.
- The complete lack of snow in the Washington–Baltimore metropolitan areas and the sharp western boundary of the snowfall in eastern Pennsylvania can be attributed to the strong westerly flow that followed the 850-hPa circulation off the Delaware coast between 0000 and 1200 UTC 30 December. An east-to-southeasterly flow, accompanied by a pattern of warm-air advections and moisture transports, was confined to a narrow strip from southern New Jersey at 0000 UTC 30 December toward New York City at 1200 UTC 30 December.

d. 500-hPa geopotential height and 400-hPa wind analyses

- This storm developed during a weak La Niña and also during a distinct period of negative North Atlantic Oscillation (NAO) conditions. The NAO entered a period in which it tended toward negative starting at the end of October and became very negative during the second half of December. Like the storm in January 2000, this storm occurred after the NAO reached a minimum and started to become less negative (also see Fig. 2-22, volume I). The 500-hPa analyses of an upper ridge over Greenland that moves westward across northeastern Canada are suggestive of a negative NAO that is weakening.
- The precyclogenetic environment at 500 hPa was dominated by a pronounced cutoff ridge over Greenland that had forced the southeasterly track of weather systems over Canada into the eastern United States for much of December, resulting in persistent cold weather. Coincident with the cutoff ridge was a cutoff 500-hPa trough over Newfoundland, a typical location prior to many major Northeast snowstorms.
- Upper-level confluence was evident over the northeastern United States and southeastern Canada throughout the entire precyclogenetic period (prior to 0000 UTC 30 December). The confluence could have suppressed the approaching storm system to the south but lifted northeastward over the western Atlantic as the cyclone-inducing trough reached the eastern United States. The confluence was not associated with a pronounced surface high, but a surface ridge extending from central Canada southeastward toward the Northeast.
- A distinct short-wave trough that produced a severe ice storm over the southern United States on 25–26 December moved off the east coast and weakened by 1200 UTC 29 December. This system appeared to disrupt the low-level southerly flow regime off the southeast coast, and likely influenced the early cyclogenetic flow into the primary system amplifying into the central United States on 29–30 December.
- A strongly "digging" or southeasterly propagating upper trough amplified as it crossed the northern plains and Midwest on 28–29 December. As this system moved eastward, the Midwest surface low weakened and secondary cyclogenesis began off the East Coast by 0600 UTC 30 December.

- As the trough amplified on 29–30 December, it attained a negative tilt and became noticeably diffluent. Rapid cyclogenesis ensued as these features reached the East Coast after 0000 UTC 30 December.
- As the trough deepened and cyclogenesis ensued, the distance between the trough axis and downstream ridge axis (half-wavelength) decreased significantly between 0000 UTC 30 December and 0000 UTC 31 December.
- Another complicated jet structure at the 400-hPa level marked the precyclogenetic period. Although there was a northern jet, the entrance region was well offshore before the cyclone developed off the New Jersey coast on 30 December. The dominant jet feature was the polar jet associated with the southeast-propagating (digging) trough into the central United States.

e. 250-hPa geopotential height and wind analyses

- The digging trough over the northern plains and the Midwest was accompanied by a strengthening 50 m s^{-1} upper-level jet streak diving southeastward to the rear of the trough. As this jet approached the base of the trough at 1200 UTC 29 December, a separate jet appeared to develop and rapidly extend eastward, amplifying to greater than 55 m s^{-1} by 0000 UTC 30 December.
- The rapid cyclogenesis and low-level jet developed within the diffluent exit region of this amplifying jet streak as it approached the East Coast after 0000 UTC 30 December. The heavy snowfall also developed within the exit region of this strong jet nearing the East Coast.
- The 500-hPa trough lifted northeastward as the upper jet rounded the base of the trough and lifted northeastward after 1200 UTC 30 December.

f. Infrared satellite imagery sequence

- One comma-shaped cloud mass moved off the Southeast coast at 0000 UTC 29 December, the remnants of the ice storm that affected the south-central United States on 25–26 December. A second comma-shaped cloud is observed over the upper Midwest, associated with a narrow band of heavy snow. This second cloud feature became more disorganized as it moved east-southeastward toward the East Coast on 29 December.
- Once the upper-level trough, its dominant jet streak, and diffluent exit region reached the East Coast at 0000 UTC 30 December, a new comma-shaped cloud mass developed rapidly along the North Carolina–Virginia coast between 0000 and 1200 UTC 30 December. This rapidly expanding comma cloud was associated with the rapid development of heavy snows that developed across New Jersey and the New York City metropolitan area.
- The comma-shaped cloud continued to expand as it moved northward through New England into southeastern Canada by 1200 UTC 31 December.

Sunset after the President's Day snowstorm of 2003 (photo courtesy of Kevin Ambrose).

31. 15–18 February 2003

a. General remarks

- This storm is known as the Presidents' Day Snowstorm II since it occurred during the Presidents' Day weekend, as did its predecessor in February 1979 (chapter 10.18). The storm was a long-duration and very cold snowstorm that was accompanied by a large, arctic anticyclone but only a weak surface cyclone. Over 3 days, Baltimore recorded its heaviest snow event on record with 28.2 in. (73 cm); Boston received 27.5 in. (70 cm), the heaviest 24-h snowfall on record; and New York City's John F. Kennedy Airport recorded 25.6 in. (64 cm), the heaviest fall on record at that location. Officially, New York City's Central Park measured its fourth greatest snowfall and heaviest February snowstorm with 19.8 in. (51 cm). Washington's Reagan National and Dulles Airports recorded 16.7 (42) and 22.4 in. (57 cm), respectively, while Philadelphia recorded 20.8 in. (53 cm), including 2 in. (5 cm) on the morning of 15 February and 18.7 in. (48 cm) from the second round of snow on 16–17 February. Portions of the metropolitan areas from Washington through Boston recorded more than 20 in. (50 cm) of snow. Blizzard conditions were not widespread as winds were only of marginal strength immediately along the coastline.
- The following regions reported snow accumulations exceeding 10 in. (25 cm): northern Virginia, northern West Virginia, Maryland, Delaware, southern and eastern Pennsylvania, New Jersey, eastern New York, Connecticut, Rhode Island, Massachusetts, southern Vermont and New Hampshire, and extreme southern Maine.
- The following regions reported snow accumulations exceeding 20 in. (50 cm): northern Virginia, northern West Virginia, Maryland, Delaware, southern and southeastern Pennsylvania, much of New Jersey, southeastern New York, and portions of Connecticut, Rhode Island, Massachusetts, and southern New Hampshire.
- The following regions reported snow accumulations exceeding 30 in. (75 cm): scattered portions of northeastern West Virginia, northern Virginia, Maryland, and southern Pennsylvania.
- The following editorial appeared in the *New York Times* (18 February 2003) expressing the impact of the blizzard during a turbulent period in American history.

TABLE 10.31-1. Snowfall amounts for urban and selected locations for 15–18 Feb 2003; see Table 10.1-1 for details.

NESIS = 8.91 (category 4)	
Urban center snowfall amounts	
Washington, DC	16.7 in. (42 cm)
Baltimore, MD	28.2 in. (73 cm)
Philadelphia, PA	20.8 in. (53 cm)
New York, NY–Central Park	19.8 in. (51 cm)
Boston, MA	27.6 in. (70 cm)
Other selected snowfall amounts	
Keysers Ridge, WV	44.0 in. (112 cm)
Burke, VA	35.0 in. (89 cm)
Martinsburg, WV	28.0 in. (73 cm)
New York, NY–John F. Kennedy Airport	25.6 in. (64 cm)
Newark, NJ	23.1 in. (59 cm)
Washington Dulles, VA	22.4 in. (57 cm)
Wilmington, DE	22.2 in. (57 cm)
Atlantic City, NJ	21.6 in. (55 cm)
Harrisburg, PA	20.3 in. (52 cm)
Allentown, PA	20.1 in. (51 cm)
New York, NY–LaGuardia Airport	17.5 in. (44 cm)
Bridgeport, CT	17.3 in. (44 cm)
Pittsburgh, PA	15.3 in. (39 cm)
Hartford, CT–Bradley International Airport	15.1 in. (38 cm)
Providence, RI	15.0 in. (38 cm)
Albany, NY	12.9 in. (32 cm)
Concord, NH	11.0 in. (28 cm)

The Blizzard of 2003

There's one thing to be said for the powerful snowstorm that brought the East Coast to a near-standstill over the past two days. It restored us all to the immediacy of the moment. This was an event that had nothing to do with human will. All the official resolutions and declarations in the world would not have abated a single flake of falling snow. Humans have always been disposed to read something symbolic into the grand cataclysms that nature brings, and in the hush that fell over New York, Washington and the other cities on the Eastern Seaboard you could hear an extraordinary peace, as if the storm had momentarily overshadowed the war against terrorism and the escalating crisis with Iraq. The run on duct tape and plastic sheeting stopped temporarily, replaced by a run on snow blowers and ice salt. That free-floating dread was replaced by free-floating flurries.

The explanation of this storm has a classic regionalism to it. A stream of warm, wet Southern air ran up against a cold dome of Arctic air stationed over the Northeast. The storm cut a broad swath through the mid-Atlantic states, creating one disaster area after another, threatening to break some of the East's most venerable records. Washington was especially hard hit.

In New York, a city that often seems impervious to weather, the snow gained the upper hand yesterday. Airports shut down, trains were delayed and side streets that went unplowed were all but impassable to cars. Pedestrians, finding snow-clogged sidewalks hard to navigate, walked boldly down the middle of main thoroughfares like Broadway and Fifth Avenue, only occasionally scampering to the side as a lone bus or snowplow approached. Cross-country skiers, liberated by the national holiday, flocked to Central Park, where they found an eerily peaceful Siberian landscape. Absent a

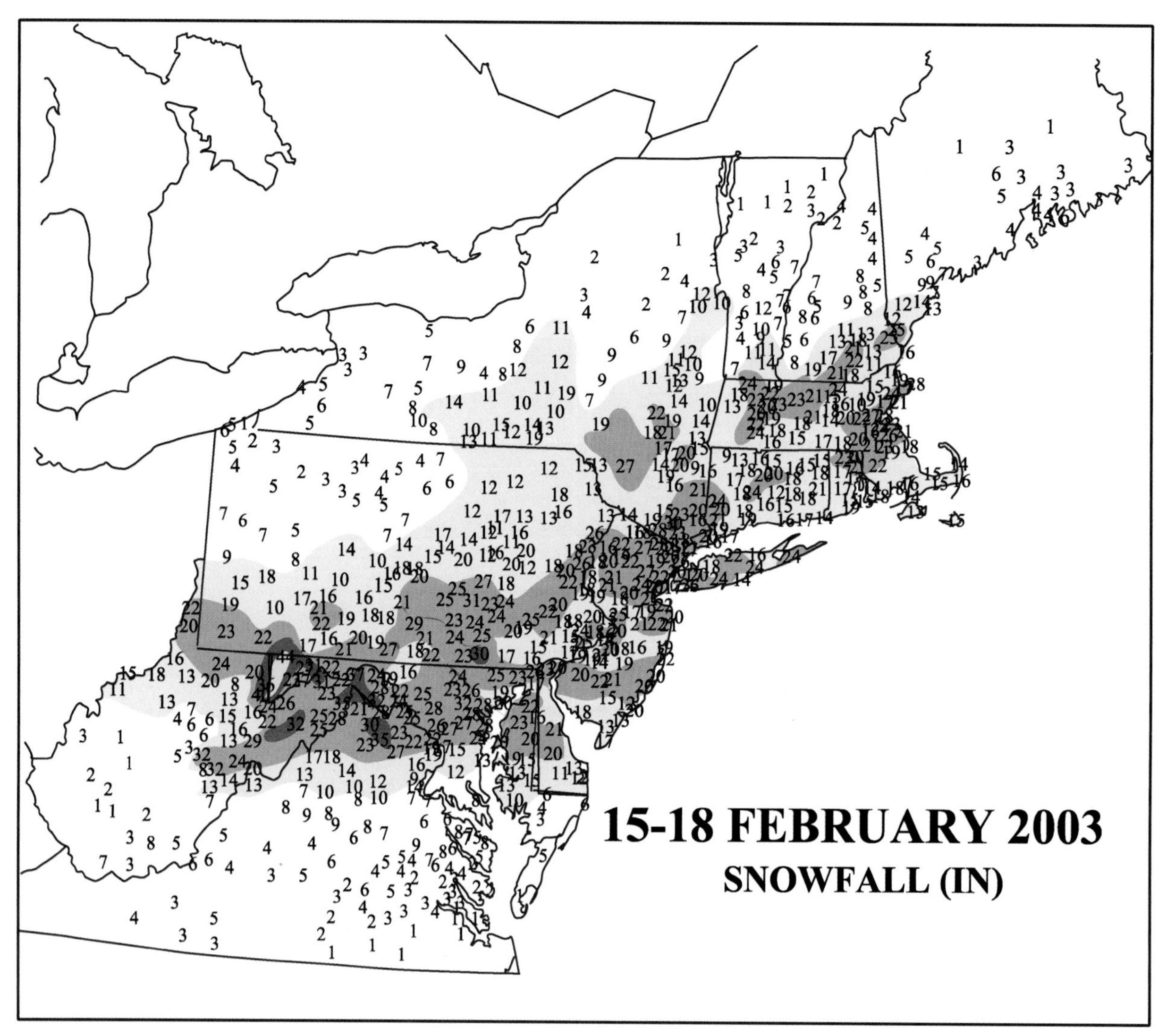

FIG. 10.31-1. Snowfall (in.) for 15–18 Feb 2003. See Fig. 10.1-1 for details.

snow-removal miracle, the wintry scene, including chest-high drifts, promises to give way to huge traffic snarls and other transportation nightmares as the work-week begins today. New Yorkers will be watching closely to see how quickly Mayor Michael Bloomberg can get the city back up and running–including the outer boroughs, where snow clearing has not always been a high priority for City Hall.

Along with all the other things that a storm like this delivers–inconvenience to most, severe risk to some, death to a few–it also delivers a strangely uplifting sense that life might just be lived at a different pace. Sweeping up from the south, shutting down malls and monuments and entire cities, this storm came upon us like a vestige of the past, as if the rising winds were somehow blowing us back in time. Compared with the need to stay warm and dry, the other necessities of life seem less pressing. Suddenly there is nowhere to get to and no getting there fast. All the urgencies that crowd the calendar look entirely postpone-able. The storm brought an almost meditative calm, time to slow down and remember, once again, that human life can be lived only within the frame of nature.

It was natural to watch this storm with mixed emotions. The plowing bill will wreak havoc on already wrecked city budgets. Undoing the damage in the worst-hit areas will take days. And yet there's something oddly gratifying and pleasantly distracting about the snow. After all, it would be nice to think that the historic times we live in were made historic by nothing more than a major blizzard.

b. Surface analyses

- The storm developed in a series of stages that began on 14 February and lasted into 18 February. An initial

FIG. 10.31-2. Twelve-hourly surface weather analyses for 1200 UTC 15 Feb–0000 UTC 18 Feb 2003. See Fig. 10.1-2 for details.

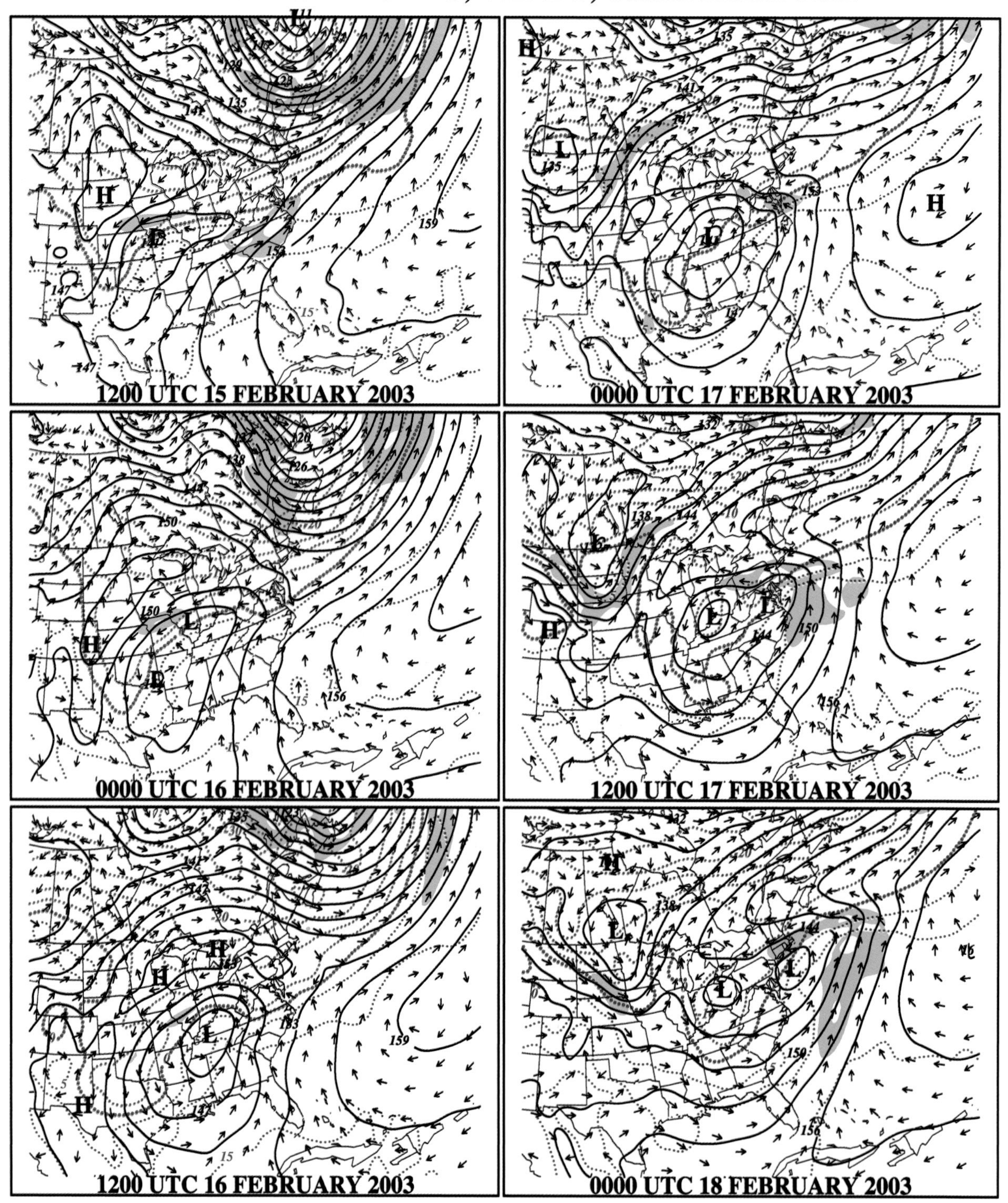

FIG. 10.31-3. Twelve-hourly 850-hPa analyses for 1200 UTC 15 Feb–0000 UTC 18 Feb 2003. See Fig. 10.1-3 for details.

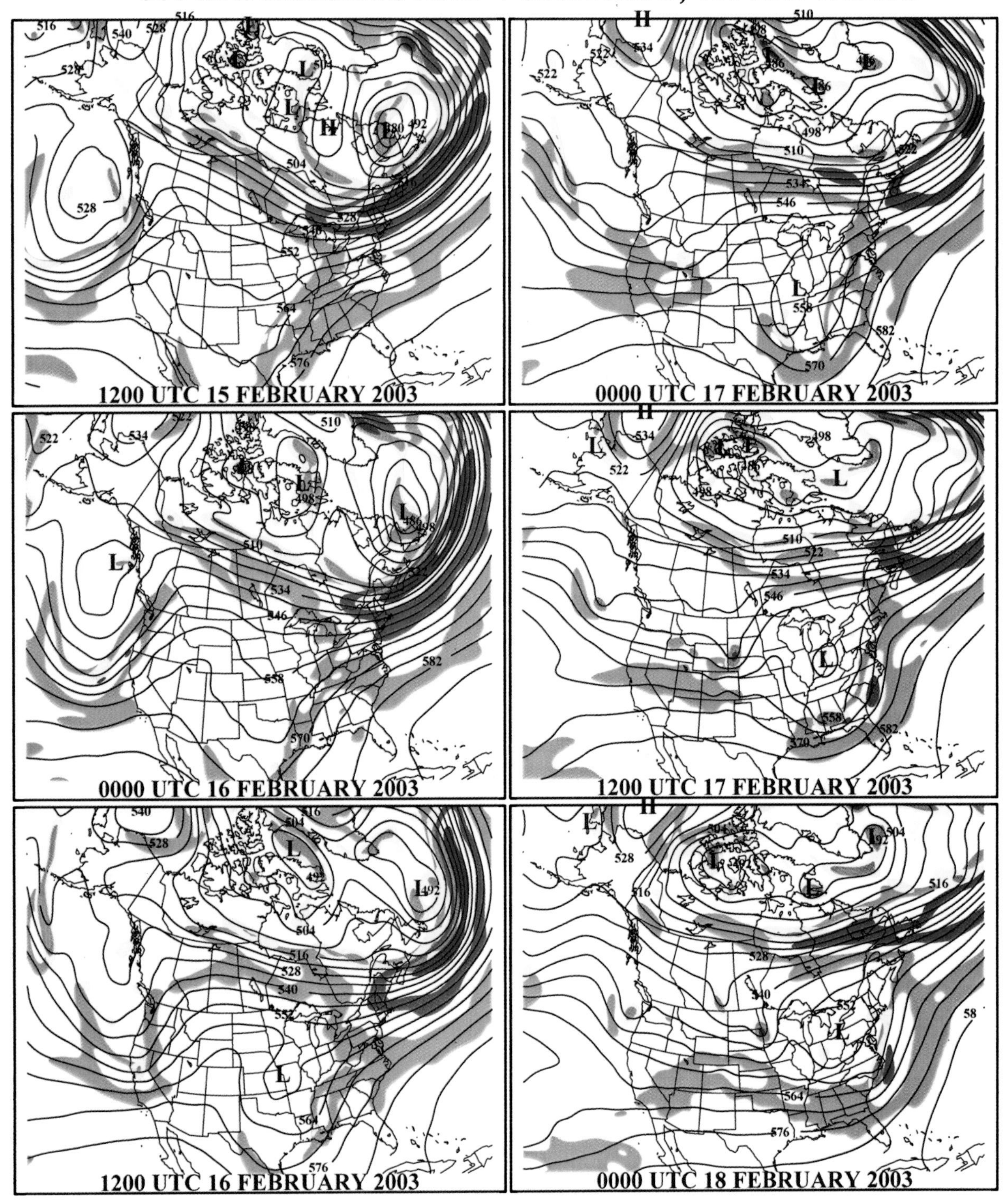

FIG. 10.31-4. Twelve-hourly analyses of 500-hPa geopotential heights and upper-level wind fields for 1200 UTC 15 Feb–0000 UTC 18 Feb 2003. See Fig. 10.1-4 for details.

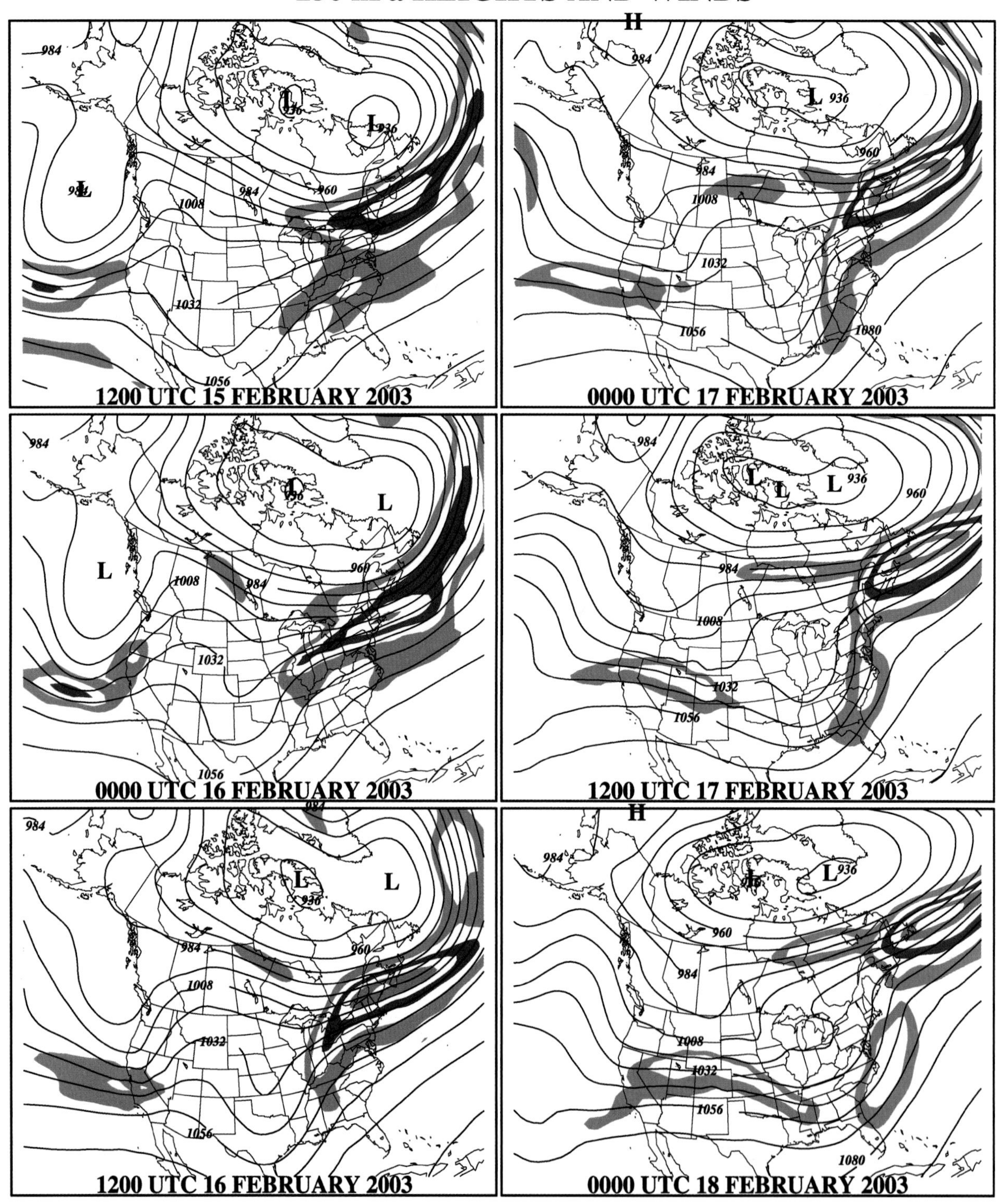

FIG. 10.31-5. Twelve-hourly analyses of 250-hPa geopotential height for 1200 UTC 15 Feb–0000 UTC 18 Feb 2003. See Fig. 10.1-5 for details.

SATELLITE IMAGERY

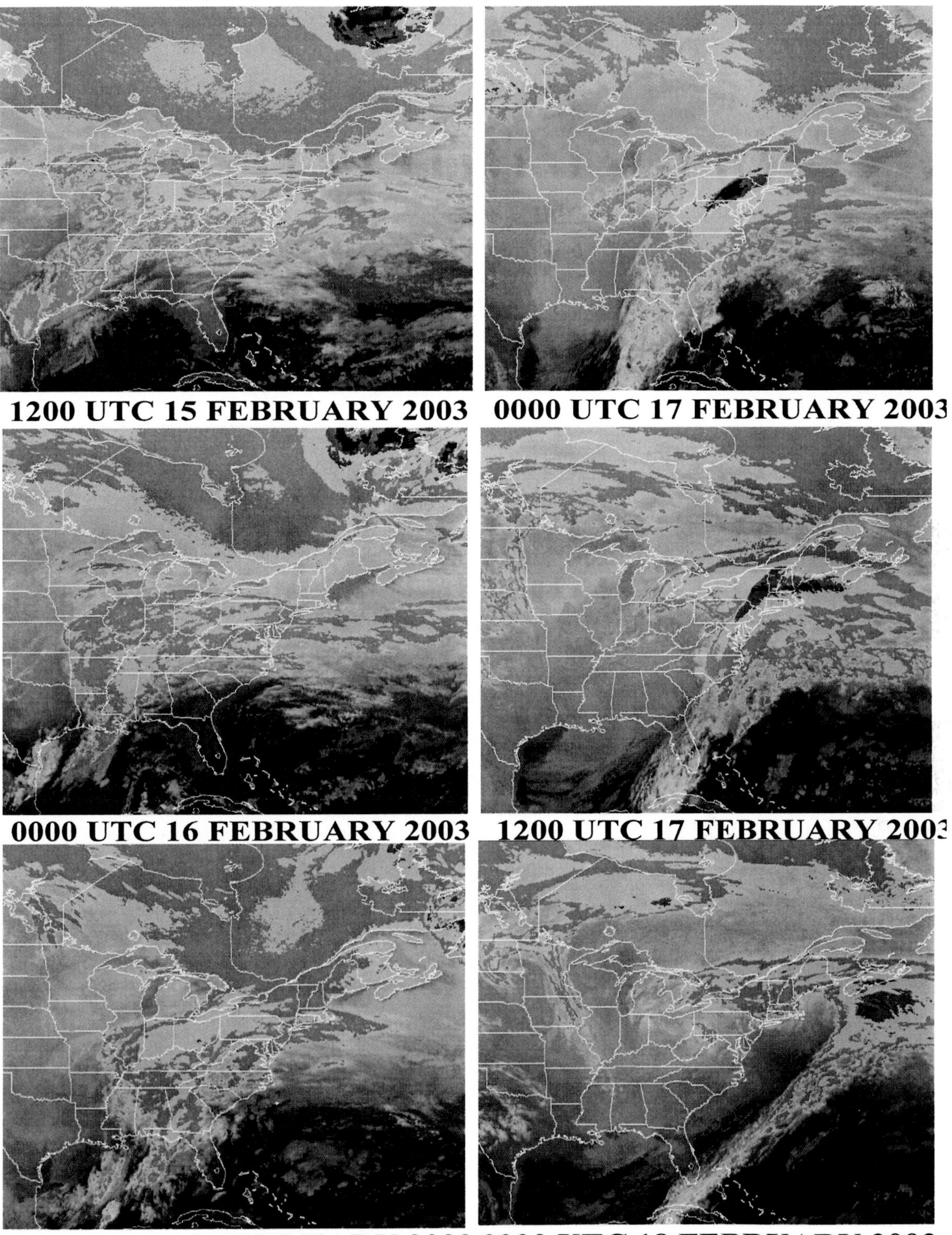

FIG. 10.31-6. Twelve-hourly infrared satellite images for 1200 UTC 15 Feb–0000 UTC 18 Feb 2003. See Fig. 10.16-6 for details.

burst of snow early on 15 February produced a narrow band of snowfall accumulations between Washington and Philadelphia, with as much as 4–5 in. (10–12 cm) of snow in suburban Maryland. During this phase, only 0.2 in. (0.5 cm) fell at Washington's Reagan–National Airport while 2 in. (5 cm) or more fell at Washington's Dulles Airport, and Baltimore and Philadelphia.
- A large anticyclone strengthened over eastern Canada and central pressures increased to 1046 hPa by 1200 UTC 16 February. As the anticyclone strengthened, cold air surged southward along the East Coast following the initial snowfall. By 0000 UTC 16 February, the leading edge of the cold air had drifted southward to the North Carolina–South Carolina border with freezing temperatures progressing southward to the Virginia–North Carolina border. By 1200 UTC 16 February, the cold air continued to surge southward along the east slopes of the Appalachians into Georgia and later reached as far west as northeastern Alabama.
- The cold anticyclone provided bitterly cold temperatures during the second phase of the storm, with temperatures averaging mostly in the teens (°F) in the major metropolitan areas. During the snow, temperatures fell to as low as 10°F (−12°C) at Philadelphia.
- The second phase of the storm developed as precipitation expanded eastward from the Ohio Valley within the cold wedge east of the Appalachians later on 15 February and then began to spread slowly northward through Maryland, Delaware, and southern Pennsylvania on 16 February. The development of heavy snow from northern Virginia to Pennsylvania occurred with a fairly weak area of low pressure over Tennessee and no significant low pressure system along the Atlantic coast. Once snow started falling, it fell heavily at rates approaching 2 in. (5 cm) h^{-1}, or greater.
- Coastal frontogenesis and weak East Coast cyclogenesis developed later on 16 February, after heavy snow had already brought the Washington–Baltimore metropolitan area to a standstill and as snowfall spread slowly northeastward into eastern Pennsylvania, New Jersey, and into New York City by evening. By 0000 UTC 17 February, a 1010-hPa low was analyzed just east of northeastern South Carolina.
- The low pressure system deepened slowly between 0000 and 1200 UTC 17 February as it moved from the Carolina coastline to a position east of the Delmarva Peninsula, having deepened only to 1005 hPa. Heavy snow spread into New York City and southern New England overnight.
- The low pressure system then moved slowly east-northeastward and did not deepen any further. However, the heavy snow continued to spread into eastern New England, where Boston received their record-setting snow late in the day and at night. It was during this phase of the storm that a strong pressure gradient developed north of the low as it moved toward the weakening, but still strong anticyclone centered over Maine by 1200 UTC 17 February. The strong winds created near-blizzard conditions from New York City to Boston on 17 February.
- A final area of light to moderate snow moved across the Appalachians late on 17 February and brought generally light accumulations from northern Virginia to New York late on 17 February and early on 18 February before the snow finally ended.

c. 850-hPa analyses

- A deep low over Labrador was associated with strong northwesterly flow and advection of very cold air into the Northeast as the surface anticyclone was building eastward across southeastern Canada. A core of −35° to −40°C air was located over Quebec at 1200 UTC 15 February with temperatures below −30°C in northern Maine.
- The 0°C isotherm was located over northern Virginia at 1200 UTC 15 February as a weak short-wave trough crossed the Maryland–Virginia coast associated with the initial burst of snow. Once this system passed eastward and snow ended, the 0° isotherm sagged southward into southern Virginia by 0000 UTC 16 February and near the North Carolina border by 1200 UTC 16 February where freezing rain, sleet, and snow continued into Virginia. The 850-hPa temperature at Blacksburg, Virginia, dropped from 2° to −10°C by 1200 UTC 16 February with the development of a very strong temperature gradient over North Carolina and Virginia while the corresponding 850-hPa temperature at Greensboro was 7°C. Heavy snow and sleet developed over Virginia during this time.
- An 850-hPa low was located over western Missouri at 1200 UTC 15 February and drifted very slowly eastward into the Ohio Valley by 1200 UTC 17 February while barely deepening.
- By 0000 UTC 17 February, the very intense temperature gradient along the mid-Atlantic coast was attended by greater than 20 m s^{-1} east-southeasterly winds at both Washington's Dulles Airport and Wallops Island, Virginia, indicative of a developing low-level jet (LLJ), directed from the Atlantic toward northern Virginia and the Appalachian Mountains. At this time, heavy snow had already blanketed the mid-Atlantic region and was spreading slowly northward into northern New Jersey and New York City.
- A secondary 850-hPa low formed in the region of an intense temperature gradient over Virginia by 1200 UTC 17 February. This secondary 850-hPa low corresponds to the period of heavy snow developing from New Jersey and New York City into southern New England. This low drifted slowly northeastward to off the New Jersey coast by 0000 UTC 18 February, when heavy snow was affecting much of southern New England.
- As the separate low developed along the coast, the

easterly LLJ amplified to greater than 25 m s^{-1} by 1200 UTC 17 February, directed toward the area of heaviest snows in northeastern Pennsylvania through New York City. This low and its attending easterly LLJ drifted slowly eastward to off the southeastern New England coastline by 1200 UTC 18 February (not shown) and did not deepen.

- The lack of intensification of the secondary 850-hPa low is consistent with the lack of intensification of the surface low pressure center. The slow evolution and presence of an intense baroclinic zone at 850 hPa and an LLJ transporting copious moisture into the area of heaviest snow contributed to the heavy snowfall occurring over an extended period, especially from Virginia to southeastern Pennsylvania.

d. 500-hPa geopotential height and 400-hPa wind analyses

- This storm developed during a winter during which a moderate El Niño was present and the North Atlantic Oscillation (NAO) tended toward a negative value. That said, the storm actually developed in a period of the winter where the El Niño was weakening and when the sign of the NAO was actually positive.
- The precyclogenetic environment at 500 hPa was not dominated by a pronounced cutoff ridge over Greenland, hence, in a pattern where the NAO is not expected to be negative. However, a pronounced cutoff 500-hPa trough was entrenched over Newfoundland at 1200 UTC 15 February 2003, a typical location prior to many major Northeast snowstorms.
- This is a case in which upper-level confluence is evident over the eastern half of North America through 1200 UTC 17 February, even as the upper-level trough over eastern Canada recedes northeastward to near Greenland. The confluence is associated with the strengthening surface high pressure center over eastern Canada that supplied the very cold air in the low levels just prior to and during this snowstorm.
- An upper trough that originally supplied heavy rainfall over California on 12–13 February moved slowly eastward and weakened. By 1200 UTC 15 February, this trough was a rather broad feature in the middle of the country and contained several shorter wavelength features. This trough remained broad with separate short-wave features, and amplified slowly between 1200 UTC 16 February and 1200 UTC 17 February as it moved slowly eastward toward the East Coast. The trough then weakened after 0000 UTC 18 February as it crossed the East Coast, coinciding with the slow movement of a surface low that did not deepen appreciably with time.
- While the amplitude of the trough does not appear to deepen, the half-wavelength between the trough and downstream ridge appears to shorten as the trough approaches the East Coast on 16–17 February. More importantly, from 1200 UTC 15 February to 0000 UTC 17 February, the eastern and southern portions of the trough extend well to the south, with south-to-southwesterly flow extending from Mexico and the Gulf of Mexico toward the developing snowstorm. As is shown in the satellite imagery, this flow appears to support a connection from the Tropics toward higher latitudes, from well south of Texas and the Gulf of Mexico toward the developing snowstorm on 16–17 February.
- This case is defined by the overwhelming presence of a strong upper-level jet streak within the confluent zone over New England upstream of the upper trough over eastern Canada. This jet is marked by greater than 70 m s^{-1} winds along its axis and a well-defined entrance region that remains over the northeast quarter of the United States through 1200 UTC 17 February, after which it drifts slowly eastward over the Atlantic. This jet streak and its entrance region are located above the strengthening surface anticyclone over southeastern Canada.
- A separate southern jet with wind speeds greater than 40 m s^{-1} developed east of the deepening trough over the southeast United States by 1200 UTC 16 February. This jet slowly amplified to greater than 50 m s^{-1} along the Carolina coast by 1200 UTC 17 February. The surface low that slowly developed off the Virginia coast on 17 February formed within the exit region of the southern jet as precipitation streamed northeastward toward the right-entrance region of the northern jet streak.

e. 250-hPa geopotential height and wind analyses

- This case has a well-defined confluence region maintained over the northeast United States through 1200 UTC 17 February. The deepening trough over the central United States is well defined by 15–16 February at 250 hPa. The southern extension of the trough extends over the western Gulf of Mexico as it amplified across the central United States by 1200 UTC 16 February.
- The northern jet streak within the confluent zone over the Northeast intensified from greater than 70 m s^{-1} to greater than 100 m s^{-1} between 1200 UTC 15 February and 0000 UTC 17 February as the entrance region of the northern jet extended from the Ohio Valley to the middle Atlantic states to New England.
- A slow increase in wind speeds east of the southern trough axis is also analyzed with a greater than 60 m s^{-1} jet streak extending from the Gulf region northeastward along the East Coast by 17 February.
- The heavy snow on 16 February developed within the right-entrance region of the northern jet. The cyclone that slowly developed east of the Virginia coast and its expanding region of precipitation on 17 February

developed within the exit region of the southern jet streak and right-entrance region of the northern jet.

f. Infrared satellite imagery sequence

- A broad west–east band of clouds extends from the Ohio Valley eastward to off the middle Atlantic coast at 1200 UTC 15 February and 0000 UTC 16 February in association with the precipitation streaming eastward in the cold air north of the stationary boundary and within the entrance region of the northern jet, especially evident in the 250-hPa analyses.
- While a comma "tail" is present during this entire case as moisture streams northward in an anticyclonic arc from the Tropics toward the eastern half of the United States, the flow of moisture becomes more pronounced by 1200 UTC 16 February and 0000 UTC 17 February from the Gulf of Mexico northeastward into the expanding cloud mass over the middle Atlantic states and Northeast. This tropical–extratropical interaction between the moisture flowing northward from the Tropics and the cold air parked over the Northeast is related to the upper jets on the east side of the trough nearing the East Coast interacting with the pronounced entrance region of the intensifying northern jet, especially visible at 250 hPa.
- An area of colder cloud-top temperatures develops from the panhandle of West Virginia northeastward into southeastern New York by 0000 UTC 17 February as heavy snows develop from the Washington–Baltimore area northeastward toward New York City. This area of cold cloud tops drifts northeastward into New England by 1200 UTC 17 February as heavy snow expands into southern New England. The classic "comma shape" to the cloud mass is fully formed over New England by 0000 UTC 18 February with the heaviest snow in the Boston area occurring within the southern area of the comma "head" over New England.

Rockefeller Center after the 5–7 Dec 2003 snowstorm (photo by R. S. Guskind).

32. 5–7 December 2003

a. General remarks

- This storm occurred more than 2 weeks before the official start of winter and was one of the heaviest early December snowstorms on record for New York City [14.0 in. (36 cm)] and Boston [16.9 in. (43 cm)]. This was a complex storm system that occurred in distinct stages in portions of the middle Atlantic states, while in New England, it was perceived to be just one long-duration snowstorm.
- The distribution of snowfall from this storm was remarkably nonuniform or mesoscale in character. Even while the surface low was deepening south of New England on 6 December, snow fell heavily in some places, such as eastern Massachusetts and extreme eastern New York, while it was snowing barely at all in other locations, such as eastern Connecticut and extreme southeastern New Hampshire. The net result was a widely varying snowfall distribution with nearly 3 feet (90 cm) of snow falling in some of Boston's northern suburbs, while only 8 in. (20 cm) of snow fell in Nashua, New Hampshire, only 50 km to the northwest.
- The following regions reported snow accumulations exceeding 10 in. (25 cm): extreme northern Virginia, northern Maryland, south-central and southeastern Pennsylvania, northern New Jersey, southeastern and eastern New York, Connecticut (except the southeast), Rhode Island, Massachusetts, Vermont, New Hampshire, and Maine.
- The following regions reported snow accumulations exceeding 20 in. (50 cm): eastern New York, Vermont, central and northern New Hampshire and Maine, eastern Massachusetts, and northwestern Rhode Island.
- The following regions reported snow accumulations exceeding 30 in. (75 cm): scattered portions of eastern New York, eastern Massachusetts and northern Vermont, northeastern New Hampshire, and northwestern Maine.

TABLE 10.32-1. Snowfall amounts for urban and selected locations for 5–7 Dec 2003; see Table 10.1-1 for details.

NESIS = 4.63 (category 3)	
Urban center snowfall amounts	
Washington, DC–Reagan National Airport	2.6 in. (6 cm)
Washington, DC–Dulles Airport	7.8 in. (20 cm)
Baltimore, MD–Baltimore–Washington Airport	6.8 in. (17 cm)
Philadelphia, PA	4.8 in. (12 cm)
New York, NY–Central Park	14.0 in. (35 cm)
Boston, MA	16.9 in. (42 cm)
Other selected snowfall amounts	
Pinkham Notch, NH	52.0 in. (132 cm)
Rangeley, ME	41.0 in. (104 cm)
Peabody, MA	35.6 in. (90 cm)
Rowley, MA	34.0 in. (86 cm)
Norwood, MA	26.0 in. (66 cm)
Taunton, MA	25.9 in. (66 cm)
Concord, NH	22.5 in. (57 cm)
East Boston, MA	22.0 in. (56 cm)
Centereach, NY	20.5 in. (52 cm)
Upton, NY	19.8 in. (50 cm)
Farmingdale, NY	19.0 in. (48 cm)
Hartford, CT–Bradley International Airport	19.0 in. (48 cm)
Albany, NY	18.0 in. (46 cm)
Burlington, VT	18.0 in. (46 cm)
Bryant Park, NY	17.8 in. (45 cm)
Providence, RI	17.0 in. (43 cm)
Newark, NJ	16.4 in. (42 cm)
Worcester MA	14.5 in. (37 cm)
Baltimore, MD	12.0 in. (30 cm)
Bridgeport, CT	12.0 in. (30 cm)
Northeast Philadelphia, PA	10.5 in. (27 cm)

b. Surface analyses

- This was a complex storm that developed in a series of stages, similar to that of the "Presidents' Day II" snowstorm earlier in the year (chapter 10.31), although in a much different manner. Snow fell in two distinct stages for a number of places throughout the middle Atlantic states, but blended into one prolonged event over New England.
- An initial burst of snow occurred in the middle Atlantic states on 5 December as low pressure developed along a coastal front off the Southeast coast and moved slowly northeastward to a position east of the Maryland coastline by 0000 UTC 6 December 2003. This low pressure system developed east of South Carolina late on 4 December and deepened from 1016 to 1005 hPa in 24 h ending at 0000 UTC 6 December.
- Snow changed to rain in the southern and eastern suburbs of Washington and Baltimore but remained as all snow in the northern and western suburbs of Washington and from the city of Baltimore north and west. Up to 10 in. (25 cm) of snow fell in northern Virginia in a narrow band while 4–6 in. (10–15 cm) were common in the northern suburbs of Washington and in Baltimore on 5 December. Snow changed to sleet and rain in Philadelphia after only a couple of inches fell during the day on 5 December, while precipitation remained as all snow in New York City, where 8 in. (20 cm) accumulated by late evening on 5 December.
- The surface low drifted eastward from the Maryland coast and weakened after 0000 UTC 6 December as a new low pressure center formed close to the Maryland coast by 1200 UTC 6 December. This represents secondary cyclogenesis as a primary low pressure center moved across the Ohio Valley on 4–5 December and dissipated over the Appalachian Mountains. The coastal low pressure system then moved northeastward and deepened more rapidly than the first coastal low, deepening from 1002 hPa at 1200 UTC 6 December, to 992 hPa by 0000 UTC 7 December, and to 988 hPa by 1200 UTC 7 December.
- The development of this new low pressure center was accompanied by the redevelopment of snow in the Washington–Baltimore area late on 5 December and early on 6 December. An additional 4–6 in. (10–15

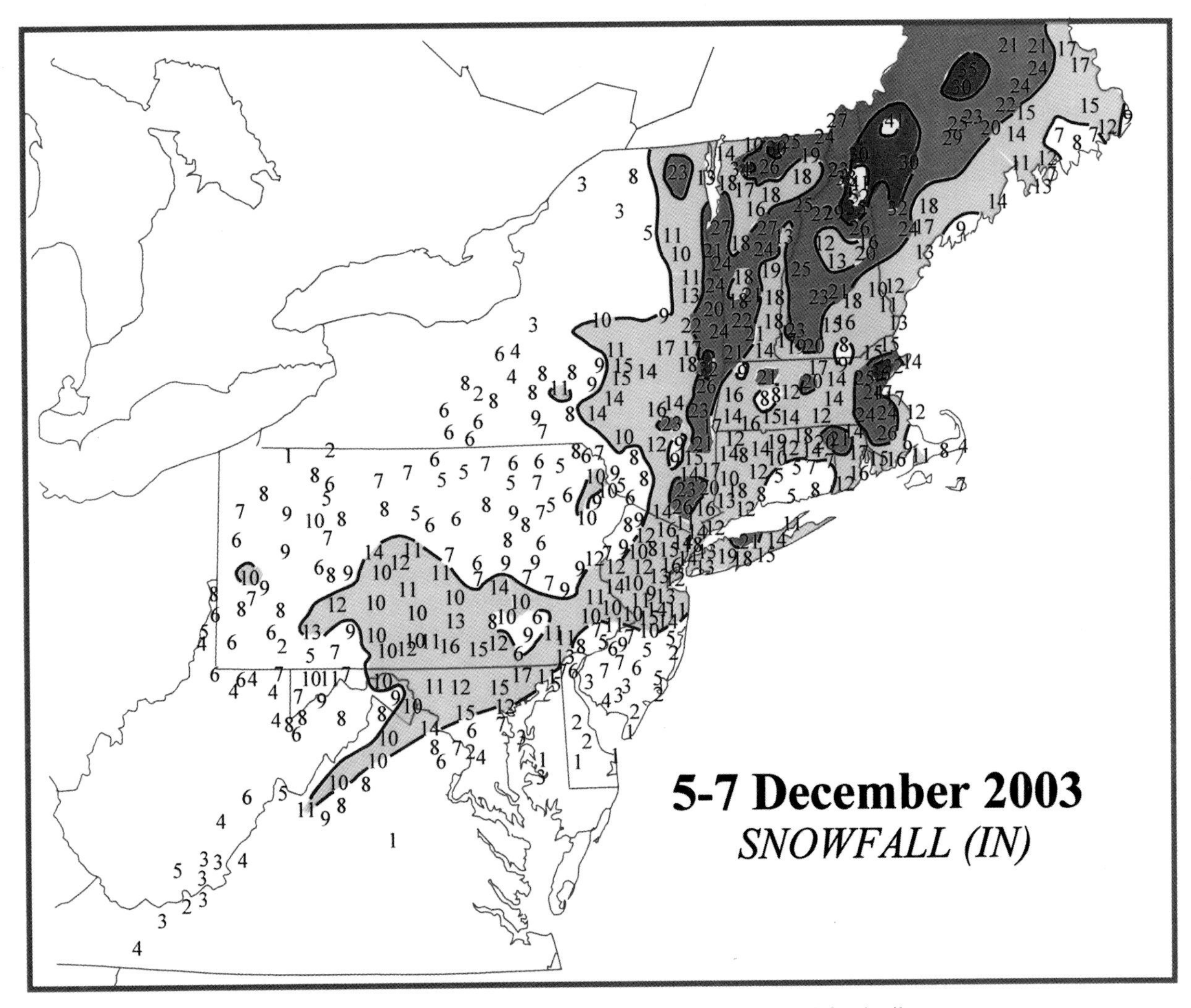

FIG. 10.32-1. Snowfall (in.) for 5–7 Dec 2003. See Fig. 10.1-1 for details.

cm) of snow fell in the northern suburbs of Washington and Baltimore. An area of moderate to heavy snow developed across eastern Maryland, southeastern Pennsylvania, and New Jersey on 6 December and then moved back into the New York City area during the afternoon. This band of heavy snow extended northward into eastern New York and western New England, where up to 30 in. (75 cm) of snow fell.

- Snow began over southern New England late on 5 December and spread slowly northward early on 6 December. As the second surface low moved northeastward and deepened on 6 December, snow continued to fall across much of New England, becoming very heavy across eastern Massachusetts and across much of northern New England, where some of the heaviest snow fell into the early morning of 7 December. In some areas of New England, snow fell for 36–48 h, contributing to the excessive snowfall accumulations.
- With the surface low intensifying east of the New Jersey coast on 6 December, a strengthening surface anticyclone north of New York and New England provided cold air for snowfall in the Northeast. This high pressure system strengthened from 1032 hPa at 0000 UTC 6 December to 1041 hPa by 0000 UTC 7 December. The combination of a strengthening high pressure system and a deepening coastal low produced wind gusts exceeding 40 mi h^{-1} from New York City to Boston, with winds greater than 50 mi h^{-1} recorded along the coast of Maine, resulting in blowing and drifting snow from New Jersey northward into Maine.

c. 850-hPa analyses

- The precyclogenetic period was marked by a trough over eastern Canada, with northwesterly flow advecting cold air into the Northeast as the surface anticyclone was building eastward across southeastern

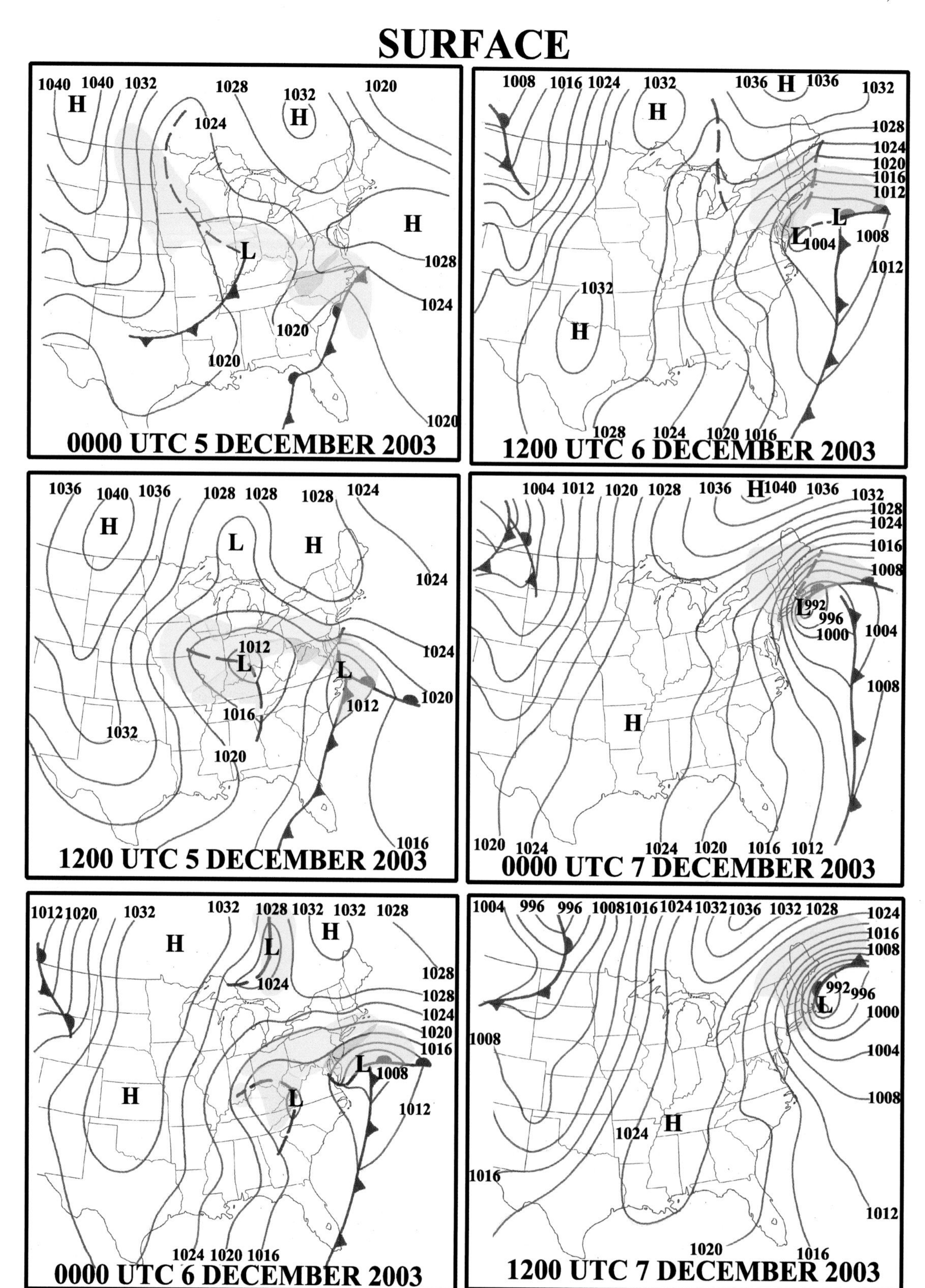

FIG. 10.32-2. Twelve-hourly surface weather analyses for 0000 UTC 5 Dec–1200 UTC 7 Dec 2003. See Fig. 10.1-2 for details.

850 hPa HEIGHTS, WINDS, TEMPERATURE

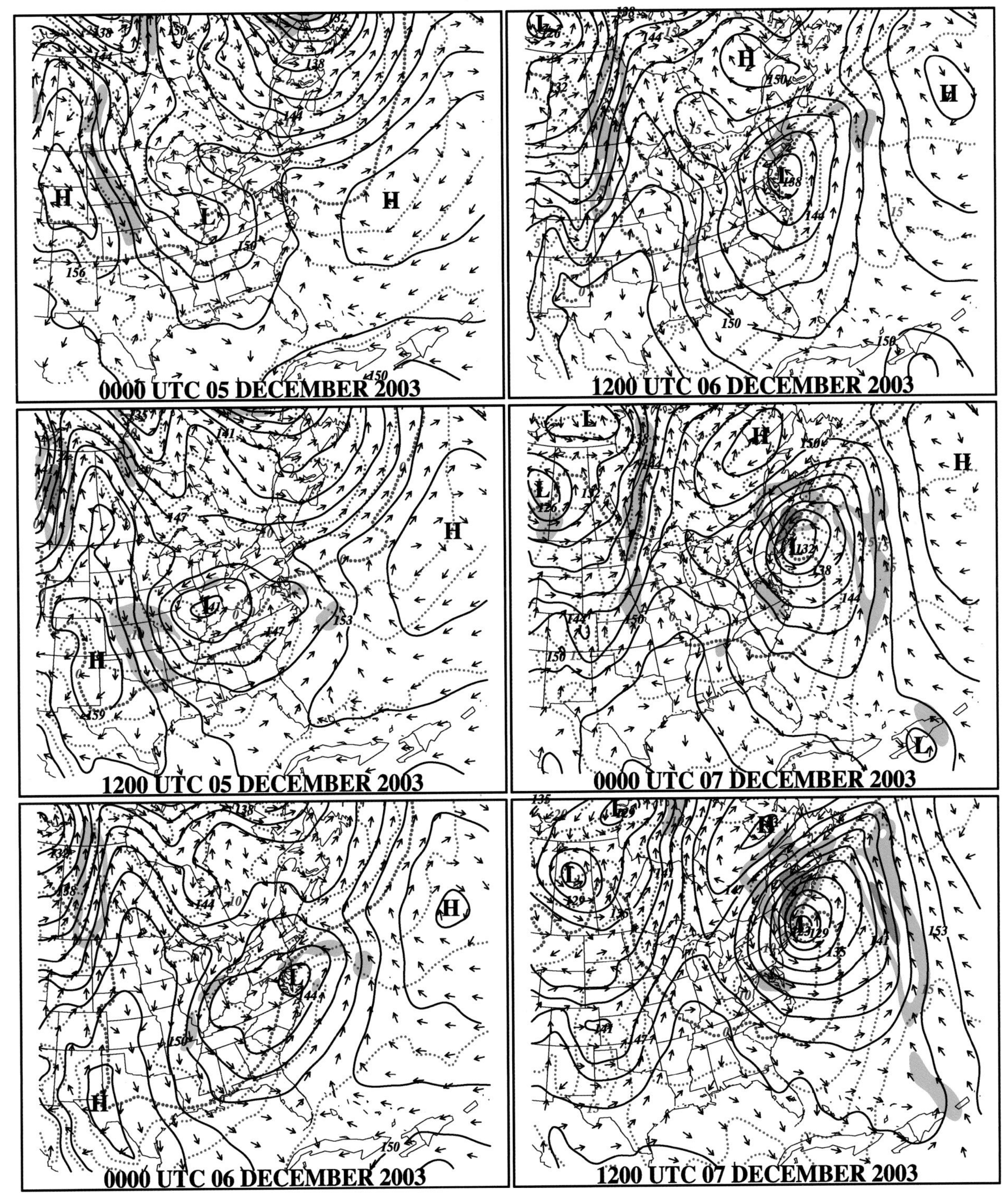

FIG. 10.32-3. Twelve-hourly 850-hPa analyses for 0000 UTC 5 Dec–1200 UTC 7 Dec 2003. See Fig. 10.1-3 for details.

500 hPa HEIGHTS AND VORTICITY, 400 hPa WINDS

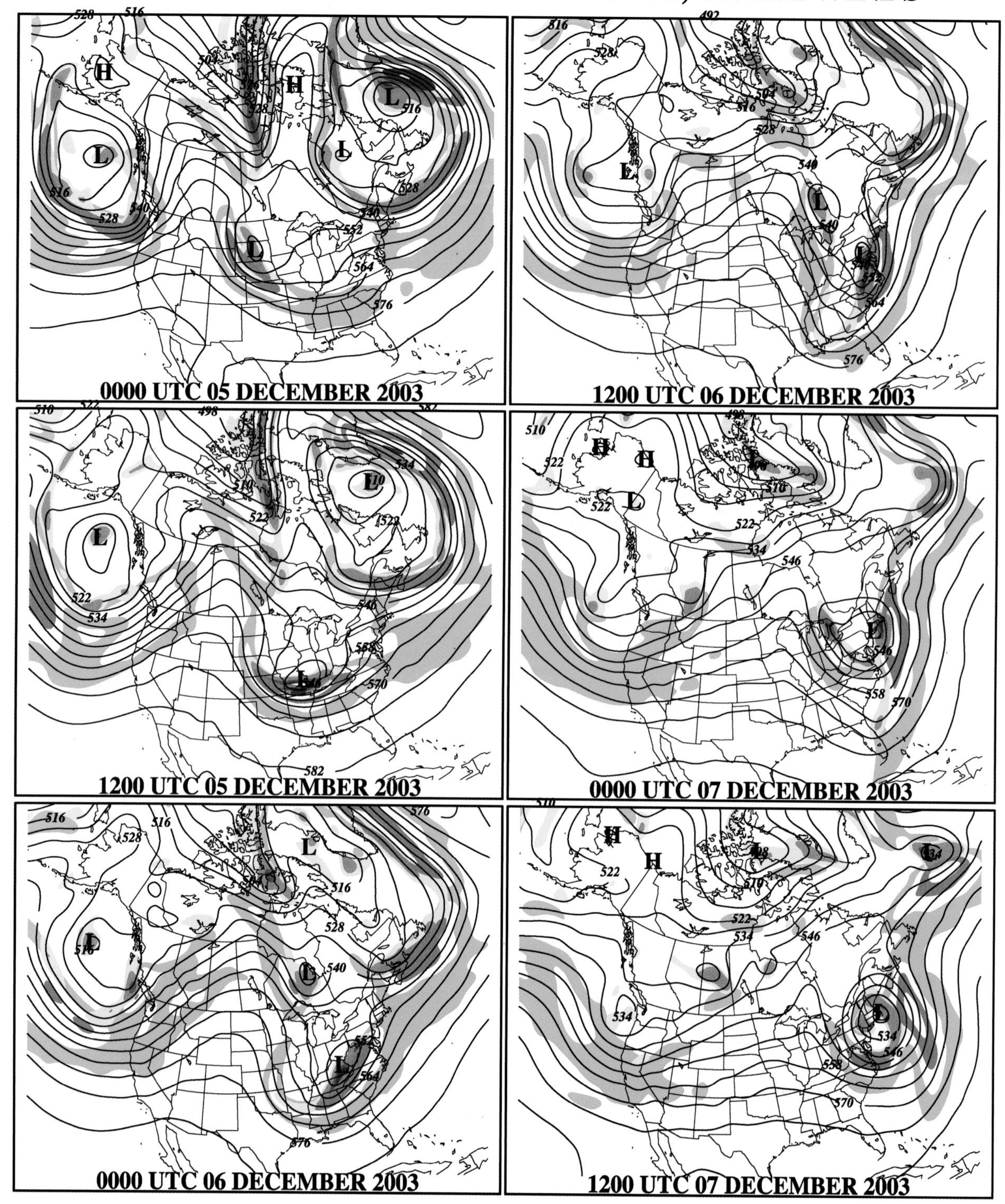

FIG. 10.32-4. Twelve-hourly analyses of 500-hPa geopotential heights and upper-level wind fields for 0000 UTC 5 Dec–1200 UTC 7 Dec 2003. See Fig. 10.1-4 for details.

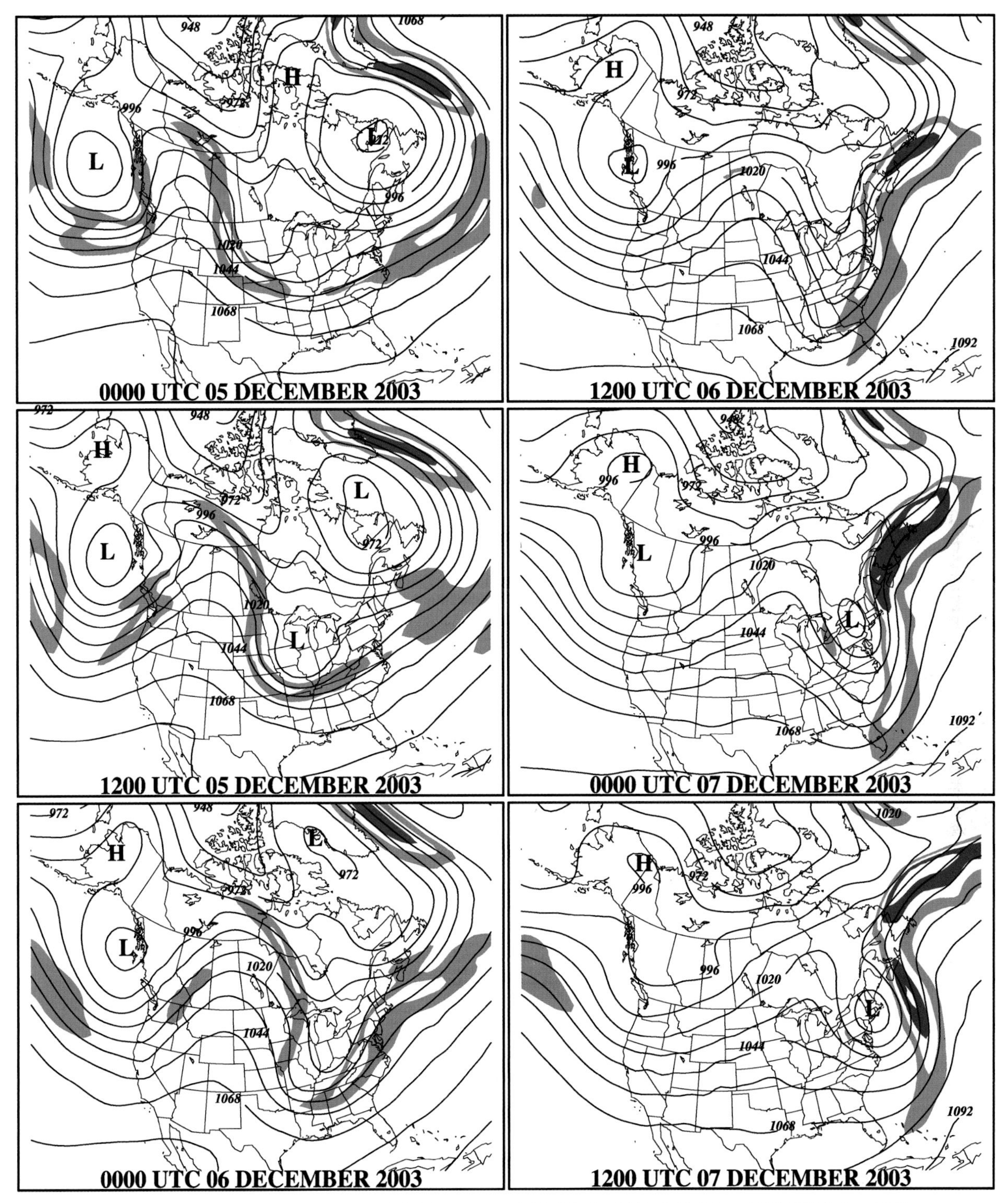

FIG. 10.32-5. Twelve-hourly analyses of 250-hPa geopotential height for 0000 UTC 5 Dec–1200 UTC 7 Dec 2003. See Fig. 10.1-5 for details.

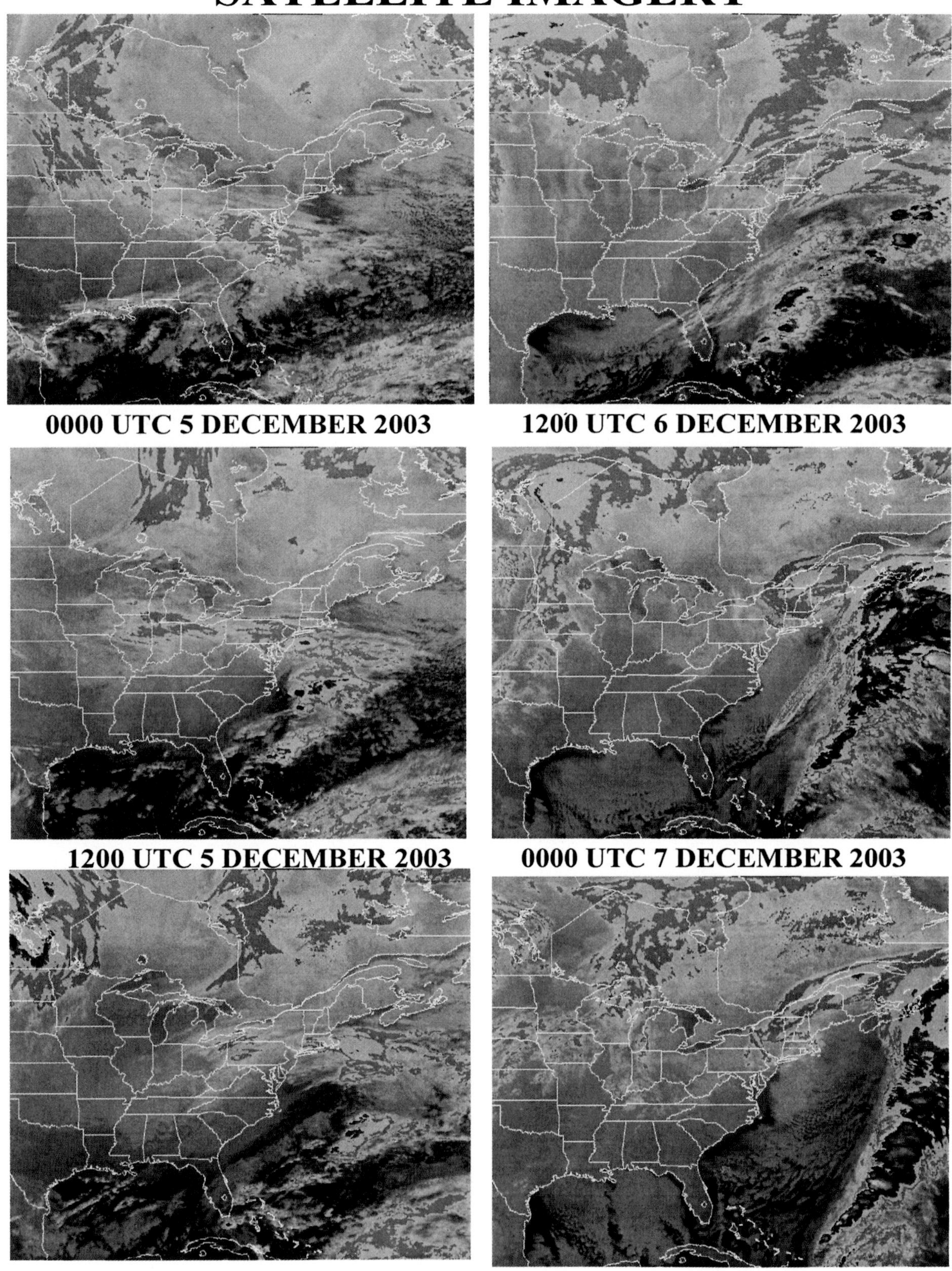

FIG. 10.32-6. Twelve-hourly infrared satellite images for 0000 UTC 5 Dec–1200 UTC 7 Dec 2003. See Fig. 10.16-6 for details.

Canada. The 0°C isotherm progressed as far south as central Virginia by 0000 UTC 5 December.

- The precyclogenetic period was also marked by a significant 850-hPa low over the Ohio Valley that developed into a closed circulation by 1200 UTC 5 December. A secondary 850-hPa low developed along the middle Atlantic coast after 1200 UTC 5 December within a zone of an increasing temperature gradient along the immediate middle Atlantic coast and related area of warm-air advection as snow developed across northern Virginia, Maryland, and southern Pennsylvania. During this same period of time, the temperature gradient surrounding the 850-hPa low attained an S shape as the second surface low began a period of more rapid intensification.
- A southeast-to-easterly low-level jet with winds exceeding 20 m s^{-1} developed, first to the east, and then to the north of the developing 850-hPa circulation along the middle Atlantic coast. The low-level jet (LLJ) was directed toward the area of moderate to heavy precipitation in the middle Atlantic states at 1200 UTC 5 December and 0000 UTC 6 December, and toward the Northeast states on 6–7 December.
- As the original 850-hPa low drifted slowly southeastward across the Ohio Valley and weakened by 0000 6 December, it seemed to eventually combine with the coastal circulation by 1200 UTC 6 December. Therefore, the 850-hPa low appeared to remain virtually stationary just off the Maryland coast between 0000 and 1200 UTC 6 December. This was the same period during which one coastal surface low drifted east and weakened while a new surface low developed along the Maryland coast.
- Following 1200 UTC 6 December, the 850-hPa low moved northeastward and deepened for the next 24 h, resulting in an intense circulation nearly collocated with the surface low by 1200 UTC 7 December. The LLJ to the north of the low intensified from greater than 25 m s^{-1} at 1200 UTC 6 December to greater than 30 m s^{-1} over the New England coast by 0000 and 1200 UTC 7 December, directed at each time toward the areas of heaviest snowfall.

d. 500-hPa geopotential height and 400-hPa wind analyses

- This storm and related circulation pattern represented the early start of the 2003/04 winter, during a period in which the El Niño–Southern Oscillation (ENSO) was relatively weak and the sign of the NAO was trending from positive to negative.
- The precyclogenetic environment at 500 hPa was dominated by a large cutoff trough over eastern Canada, with several short-wave features rotating about the vortex, maintaining the trough over eastern Canada.
- Pronounced upper-level confluence was found over the northeast United States on 4–5 December during the precyclogenetic period and over southeastern Canada during the cyclogenetic period on 5–6 December. The confluence is associated with the relatively stationary surface high pressure center over eastern Canada that strengthened on 6 December and supplied the low-level cold air for the snowstorm. This confluence was associated with the entrance region of an upper-level jet streak that appears relatively stationary east of Maine before drifting slowly northward to Newfoundland on 6 December.
- The evolution of the upper troughs associated with the complex low pressure system with this case is complicated and contributed to the prolonged period of snowfall that marked this case. A precursor upper-level trough over the Tennessee Valley at 0000 UTC 5 December was associated with the development of the initial coastal low pressure system on 5 December. The development of the second, intensifying surface low was associated with a separate and more significant trough that initially propagated southeastward across the northern Midwest on 4–5 December and then drifted east-northeastward toward the East Coast on 5–6 December. This trough was associated with a greater than 60 m s^{-1} jet streak at 400 hPa that rotated from the western side of the trough at 0000 UTC 5 December, amplified to 70 m s^{-1} at the base of the trough at 1200 UTC 5 December and then propagated to the eastern side of the trough at 0000 and 1200 UTC 6 December. This occurred as the second surface low developed off the Maryland coast, within the exit region of this jet streak.
- A third trough dropping southeastward from south-central Canada into the northeastern United States on 5–7 December also played a role in maintaining the upper trough over the northeast United States through 7 December as it propagated southward over the eastern Great Lakes on 6 December and rotated about the main trough moving slowly northeastward toward the New England coast on 6–7 December.
- At 1200 UTC 5 December and 0000 UTC 6 December, the initial area of developing snowfall across the middle Atlantic states and southern New England occurred within the transverse circulation patterns associated with the classic "dual-jet pattern," within the right-entrance region of the northern jet streak and the left-exit region of the jet streak associated with the approaching trough.

e. 250-hPa geopotential height and wind analyses

- The 250-hPa level reflects the trough pattern over southeastern Canada and confluence across the northeast United States on 5–6 December needed to sustain cold air for this snowstorm. The trough and related jet streak that amplified across the central United

States by 5 December, contributing to this slow-moving coastal system, are also clearly evident.

- The Canadian trough drifted northeastward at and after 0000 UTC 6 December as a trough over the center of the country drifted east and attained a negative tilt by 0000 UTC 7 December. The rapid development of the surface cyclone coincided with the amplification and development of the negatively tilted trough.
- The second area of snowfall on 6–7 December appears to develop primarily in the exit region of the 400-hPa jet rotating about the east side of the upper trough but also appears to develop in the entrance region of an intensifying jet streak at 250 hPa drifting northward from the northeast United States to southeastern Canada between 0000 UTC 6 December and 0000 UTC 7 December.

f. Infrared satellite imagery sequence

- An extended area of clouds from Minnesota southeastward to the middle Atlantic states at 0000 UTC 5 December represents the multiple troughs–jet streaks that influenced the complex evolution of this storm system. One comma-shaped cloud mass is observed at this time over North Carolina and Virginia that represents the first coastal low pressure system and area of snowfall that drifts northeastward over the next 12–24 h. The area of highest clouds shrinks as it moves northward toward southern New England by 0000 UTC 6 December.
- A second area of colder cloud-top temperatures drifts from the Midwest at 1200 UTC 5 December to Pennsylvania at 0000 UTC 6 December and begins to expand as a separate area of colder cloud-top temperatures drops southeastward across southeastern Canada between 0000 and 1200 UTC 6 December. This occurs at the same time as a region of colder cloud-top temperatures remains anchored over extreme eastern New England as snow begins to envelop much of New England.
- Between 1200 UTC 6 December and 0000 UTC 7 December, the separate cloud masses organize into the classic comma shape that is maintained through 1200 UTC 7 December, the period in which the surface low intensified south of New England and the heaviest snows fell from Long Island to Maine.

Chapter 11

DESCRIPTIONS OF "NEAR MISS" EVENTS

In this chapter, examples of three categories of "near miss" storms (see volume I, chapter 5) are presented. The first category is called interior snowstorms, since these are events that posed a threat of heavy snow accumulations along the coast, but the threat did not materialize as heaviest snowfall amounts fell inland and snow changed to rain or mixed precipitation within the more heavily populated urban corridor. The second category is termed moderate snowstorms and describes cases that posed a potential heavy snow threat to the Northeast urban corridor but were marked by more moderate snowfall amounts, generally 4–10 in. (10–25 cm). In these cases, heavy snow amounts exceeding 10 in. (25 cm) may have fallen, but over smaller domains than described for the 32 cases discussed in chapter 10. The third category is termed ice storms and describes significant storms involving other frozen precipitation types such as freezing rain or ice pellets.

Synoptic overviews of interior snowstorms, moderate snowstorms, and ice storms from 1950 to 2000 are presented to contrast these winter weather events with the major snowstorms described in chapter 10. Fifteen cases each of interior and moderate snowstorms and seven ice storm cases are described for the period from 1950 through the end of the 20th century. Snowfall figures are provided (except for a few ice storm cases in which snowfall did not play an important role), as well as a 12-hourly sequence of surface charts and 850-, 500-, 400-, and 250-hPa analyses derived from the National Centers for Environmental Prediction–National Center for Atmospheric Research (NCEP–NCAR) reanalysis datasets. Some of the following information was derived from summaries appearing in the publications *Climatological Data* and *Storm Data.*

1. Interior snowstorms

Fifteen interior snowstorms are described in this section, with emphasis placed on the evolution of the surface cyclone and anticyclones and their associated upper-level troughs, confluence, diffluence patterns, and upper-level jet streaks. As described in chapter 5 of volume I the surface low pressure systems associated with these storms tend to move closer to the coastline than the snowstorms described in chapter 10, with the surface low often passing near or to the west of the coasts of the Carolinas and the Delmarva Peninsula. Surface anticyclones are located farther north and east than those shown in chapter 10, favoring a flow of maritime air near the coastline, resulting in surface warming that would favor rain or a change from snow to rain.

These snowstorms are marked by upper-level troughs that tend to have more amplitude and become negatively tilted farther south and west of the upper-level troughs associated with the heavy snow cases presented in chapter 10. Across the western Atlantic, the downstream upper-level ridge is more amplified in these cases than the cases in chapter 10. As the upper ridge builds over the Atlantic, upper-level confluence over the northeast United States and southeastern Canada is forced northward and eastward, or diminishes. The cold surface anticyclone in these cases drift eastward, allowing warmer air to reach the coastline, favoring rain over snow near the coast and snow farther inland.

The 15 cases described in the following sections include several notable storms from the last half of the 20th century and include one of the most damaging storms on record, the Ash Wednesday storm of March 1962, a slow-moving storm system that produced record coastal flooding along the mid-Atlantic coast and a significant snowfall in portions of the middle Atlantic states. Many of the other storm systems produced widespread snowfall inland with snow and rain along the coast.

While many of these storms produced similar areas of heavy snowfall as in the 30 cases shown in the previous chapter, they typically affected smaller populations since the heaviest snow fell outside the urban corridor. As a result, Northeast Snowfall Impact Scale values (NESIS; see volume I, chapter 8; also see Kocin and Uccellini 2004) generally ranged between 2 and 4 (Table 8-2, volume I), and averaged near 3.1, significantly less than the average of 4.8 for the first 30 snowstorm cases in chapter 10. The smaller NESIS values for the interior snowstorms illustrates how storms with similar areas of snowfall have a smaller impact, related to the smaller population affected by the inland storms.

a. 17–18 February 1952: NESIS = 2.17

During the relatively quiescent early 1950s, this was one of a few storms that produced heavy snow, with

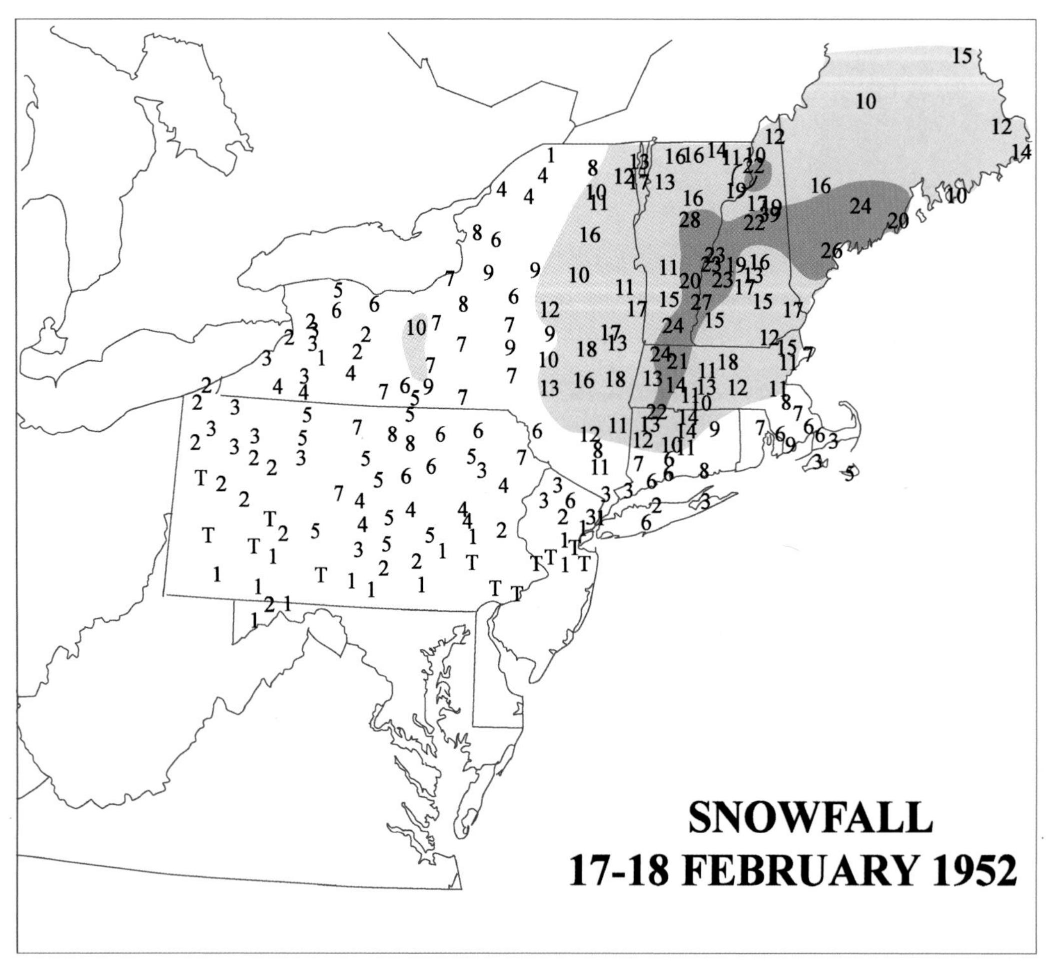

FIG. 11.1-1. Snowfall (in.) for 17–18 Feb 1952. Shading contours are for 10 (25), 20 (50), and 30 in. (75 cm).

12–30 in. (30–75 cm) of snow across much of central and northern New England (Fig. 11.1-1). This storm spared much of the urban corridor from the heaviest snows, with the exception of the Boston metropolitan area, where 10–15 in. (25–38 cm) of snow was common.

The storm developed as a diffuse area of low pressure consolidated off the southern New England coast into an intense cyclone, with central pressures falling rapidly near southeastern Massachusetts between 1830 UTC 17 February and 0630 UTC 18 February 1952 (Fig. 11.1-2). This period of rapid deepening resulted in heavy snow late on 17 February and early on 18 February, with high winds across much of central New England northeastward into Maine. The loss of two ocean tankers resulted in 15 fatalities and 28 others are estimated to have lost their lives during the storm. The snowstorm was especially severe in Maine, where more than a thousand people were stranded on the roads.

The storm was associated with an 1) intensifying 850-hPa circulation immediately along the coast and related low-level jets (LLJs; see Fig. 11.1-2; also note the amplification of wind speeds to greater than 40 m s^{-1} in the easterly LLJ between 1500 UTC 17 February and 0300 UTC 18 February 1952) and 2) an amplifying upper-level trough that attained a negative tilt by 1830 UTC 17 February (Fig. 11.1-3), coincident with the onset of rapid deepening. With the negative tilt, 500-hPa heights rose downstream of the trough along the western Atlantic, with a significant jet streak at 400 hPa extending northeastward up the East Coast on 17 February (Fig. 11.1-3). The rapid development phase of the surface and 850-hPa lows occurred within the exit region of that jet and entrance region of a jet streak at the 250-hPa level farther north and east. The building ridge along the East Coast eliminated the confluence over southeastern Canada as high pressure retreated eastward across eastern Canada. The eastward movement of the surface high allowed warmer air to advect inland along the northeastern coastline, where rain fell rather than snow.

An upper-level ridge appears to be anchored across the North Atlantic to the east of Greenland and Iceland (Fig. 11.1-3), consistent with the negative phase of the North Atlantic Oscillation (NAO; Table 2-5, volume I). An upper low east of Canada drifts eastward over the North Atlantic and is located farther north and east than

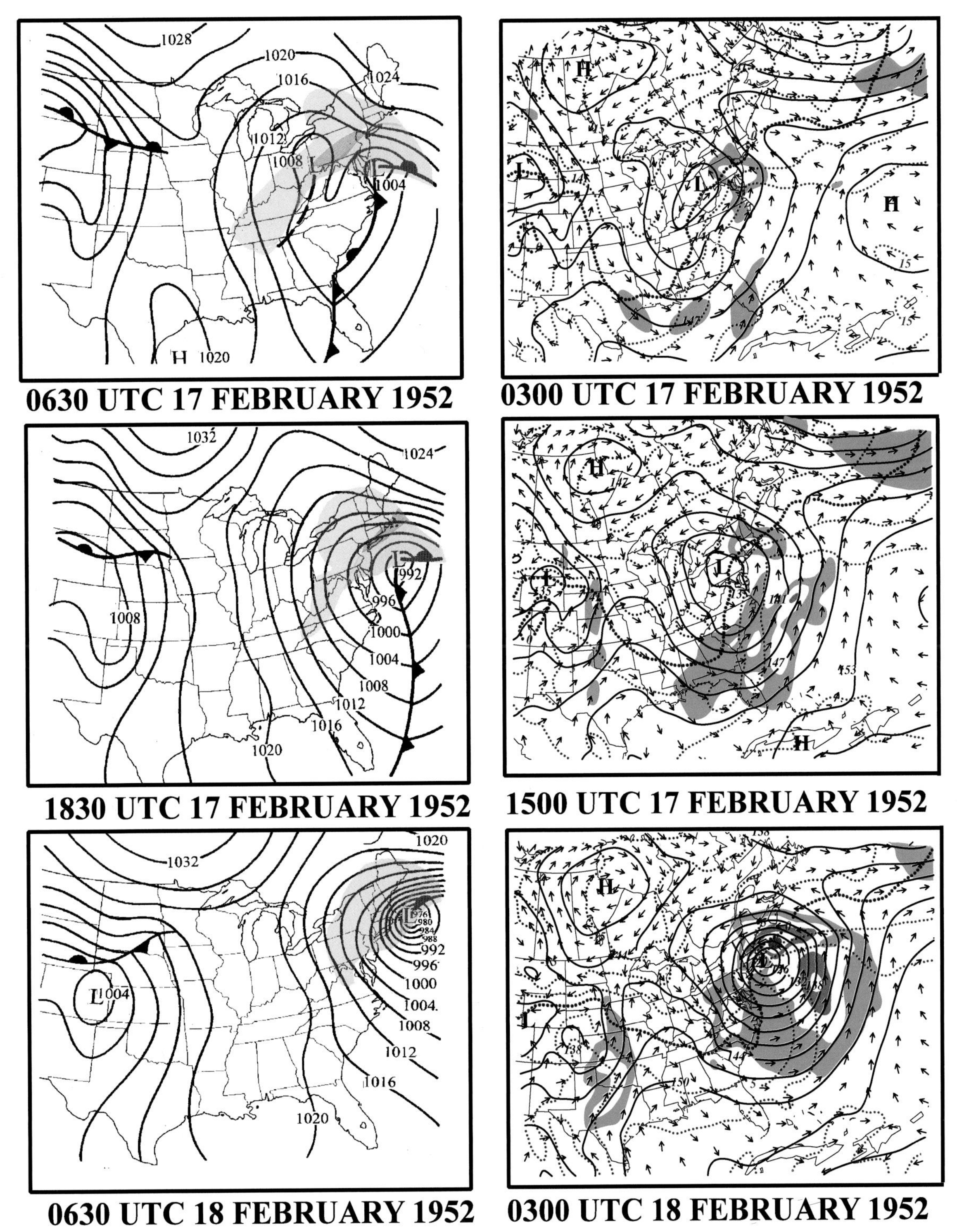

FIG. 11.1-2. Sequence of (left) surface and (right) 850-hPa charts for the 24-h period between 0630 UTC 17–18 Feb 1952 (surface), and 0300 UTC 17 and 18 Feb 1952 (850 hPa). Surface maps include surface high and low pressure centers and fronts. Color shading indicates blue, snow; violet, mixed precipitation; and green, rain. Solid lines are isobars (4-hPa intervals), and dotted lines represent axes of surface troughs not considered to be fronts. The 850-hPa analyses include contours of geopotential height (solid, at 30-m intervals; 156 = 1560 m), isotherms (dashed, °C, 5°C intervals; blue, 0°C and less; red, 5°C and greater), and intervals of wind speed greater than 20 m s^{-1} (at 5 m s^{-1} intervals; alternating blue/white shading; red shading, 40 m s^{-1}).

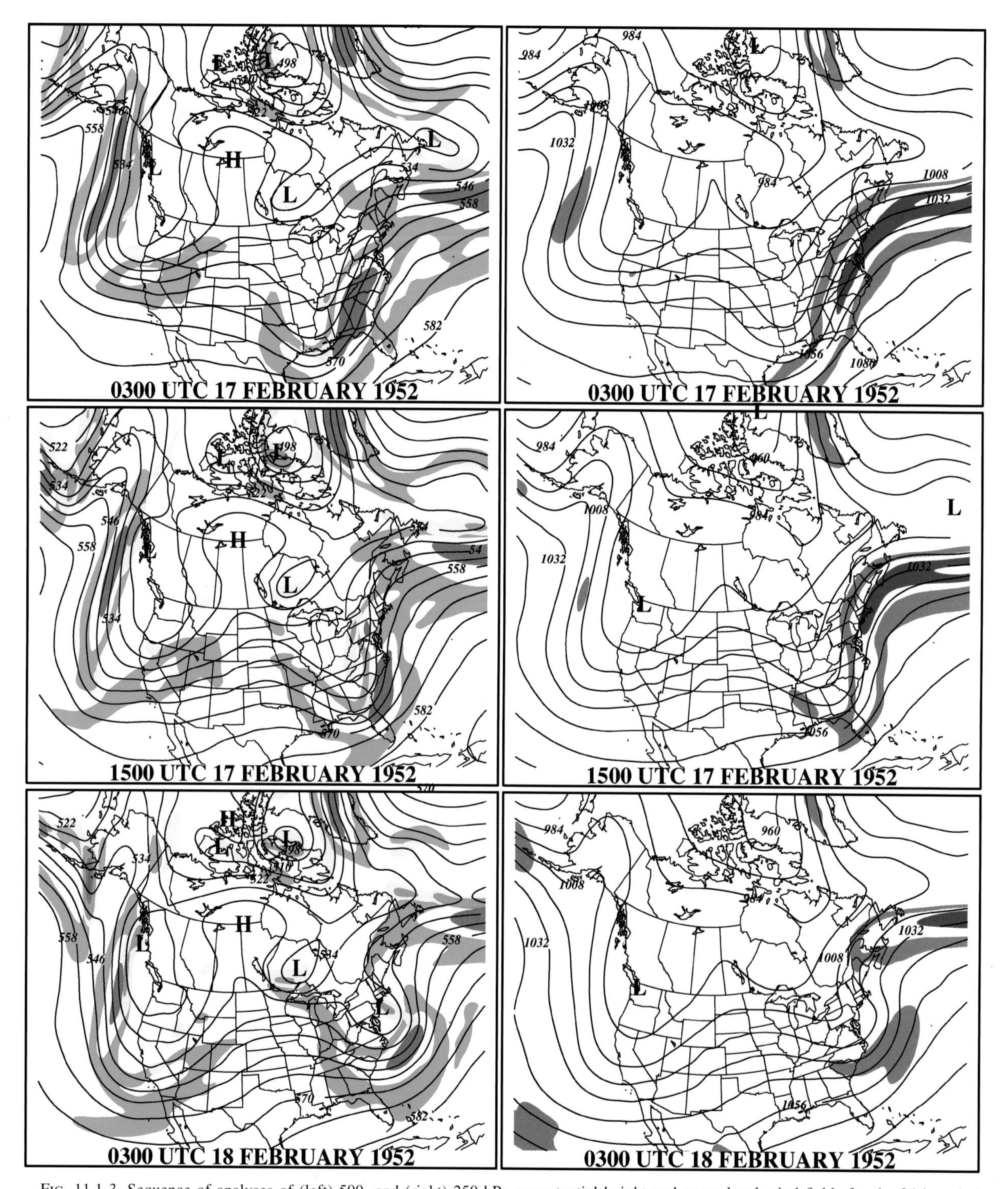

FIG. 11.1-3. Sequence of analyses of (left) 500- and (right) 250-hPa geopotential height and upper-level wind fields for the 24-h period between 0300 UTC 17 and 18 Oct 1952. Twelve-hourly analyses of 500-hPa geopotential height and upper-level wind include locations of geopotential height maxima (H) or minima (L), contours of geopotential height (solid, at 60-m intervals; 522 = 5220 m), locations of 500-hPa absolute vorticity maxima (yellow/orange/brown areas beginning at 16×10^{-5} s^{-1}; intervals of 4×10^{-5} s^{-1}), and 400-hPa wind speeds exceeding 30 m s^{-1} (at 10 m s^{-1} intervals; alternate blue/white shading). Analyses of 250-hPa geopotential height and winds (rhs) include heights (solid at 120 m intervals; 1032=10 320 m), and wind speeds exceeding 50 m s^{-1} (at 10 m s^{-1} intervals; alternate blue/white shading).

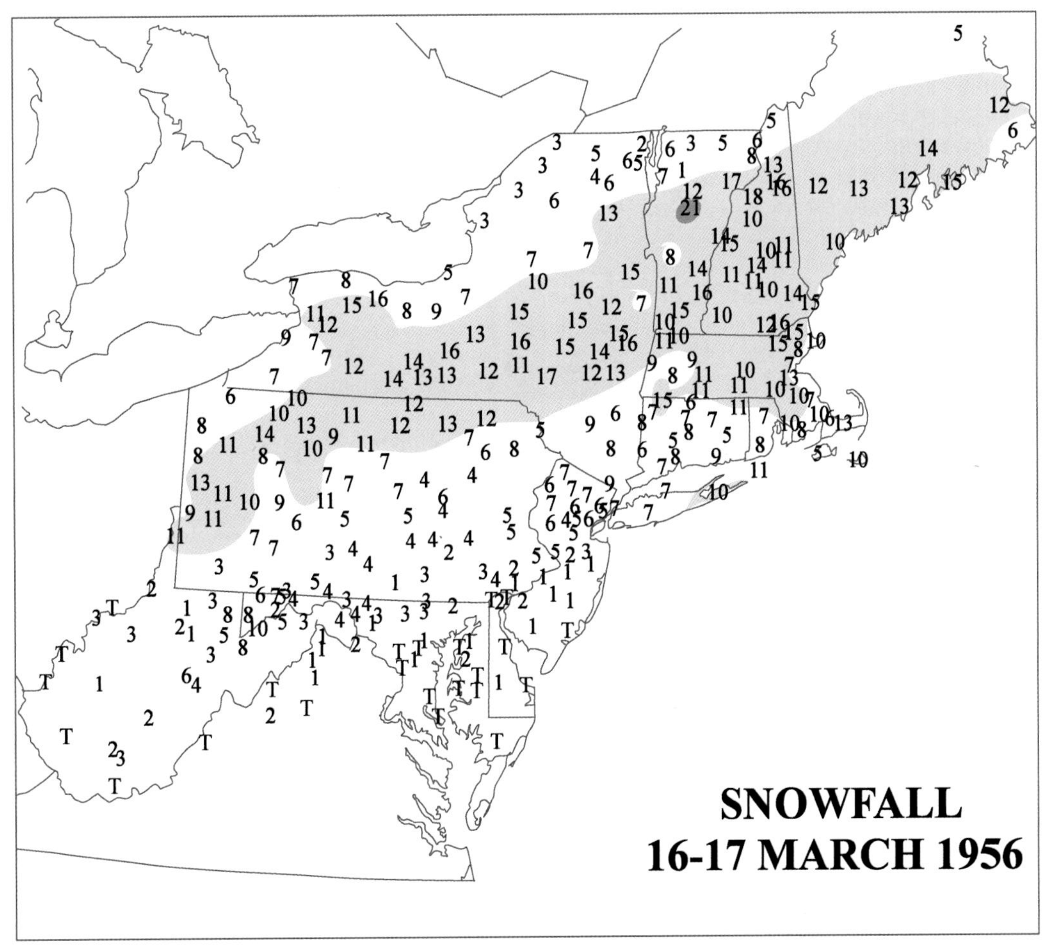

FIG. 11.1-4. Snowfall (in.) for 16–17 Mar 1956. See Fig. 11.1-1 for details.

many of the heavy snow cases. This location may have helped anchor the surface anticyclone in southeastern Canada, supporting the major snowstorm that occurred over much of central and northern New England as the upper trough became negatively tilted along the East Coast. Nevertheless, the large easterly fetch from the Atlantic Ocean along the southern edge of this anticyclone and the more inland track of the surface and 850-hPa low ensured that precipitation fell mostly as rain within the northeast urban corridor.

b. 16–17 March 1956: NESIS = 2.93

Even though this snowstorm occurred more than halfway through the 1950s it was the first powerful storm system in that decade to affect much of the Northeast urban corridor, producing severe blizzard conditions in interior and eastern New England. It was followed by a less intense snowstorm that had a greater impact on the Northeast urban corridor on 18–19 March, described in depth in section 10.1.

In Washington, Baltimore, and Philadelphia, most of the precipitation fell as rain that briefly changed to snow (Fig. 11.1-4). However, late on 16 March, rain and snow in New York City changed to all snow and accumulated up to 8 in. (20 cm) in parts of the city, snarling traffic.

In the Boston area, up to 10 in. (25 cm) fell in the city and up to 15 in. (38 cm) accumulated northwest of the city, accompanied by rapidly falling temperatures and high winds. In New England, the rapid development of the storm resulted in an estimated 48 deaths from exposure, overexertion, or traffic accidents. In addition, many ships were driven ashore or torn from their moorings at the height of the storm late on 16 March and early on 17 March.

The evolution of this storm is very similar in structure to many of the major snow events shown in chapter 10. Upper-level confluence is found to the rear of a shortwave trough exiting the eastern Canadian coastline in the precyclogentic period with a pronounced 80–90 m s^{-1} jet streak extending from the northeastern United States to just off the Canadian coast (Fig. 11.1-6). The rapid development phase of the surface cyclone occurred within the right entrance region of the northern jet and exit region of the southern jet streak (as depicted at the 400- and 250-hPa levels), like many of the snowstorm described in chapter 10. The explosive deepening of a secondary low pressure system off the middle Atlantic coast (Fig. 11.1-5) occurred within a distinct diffluent exit region of a well-defined jet streak, as a negatively tilted trough moved toward the East Coast.

Several important differences between this storm and many of the major urban corridor snowstorms include the following: 1) the surface low in this case tracked 50 km inland across the Carolinas (Fig. 11.1-5) and 2) a closed upper low developed south of Greenland (Fig. 11.1-6), rather than across eastern Canada, allowing the upper-level confluence in the northeast United States to retreat northward during the evolution of the storm. As a result, the surface high moved from near Maine northeastward to eastern Canada, a position that permitted low-level northeasterly flow along the Northeast urban corridor to turn more easterly and pass over the ocean. This airflow warmed the lower troposphere immediately along the coast and within the urban corridor, out ahead of the developing storm system. This storm track and easterly fetch from over the ocean restricted the heaviest snow to the region north of New York City, as rain fell along the immediate coastline. As the storm circulation intensified off the southeast New England coastline, colder air wrapped around the storm center and rain changed back to snow from New York City to northward to Boston.

c. 12–13 March 1959: NESIS = 3.64

This is a representative example of an interior snowstorm with inland snow and rain falling along the coast. The storm was also the most significant storm of the winter of 1958/59, occurring on the 71st anniversary of the March blizzard of 1888 (see chapter 9). In New York and New England, 44 deaths were attributed to the storm as much of eastern New York received more than a foot (30 cm) of snow (Fig. 11.1-7). Heavy snow changed to rain in the urban corridor from New York City to Boston, where 4–8 in. (10–20 cm) of snow fell before the changeover to rain. Much of New England received 1–2 ft (30–60 cm) of snow, with lesser amounts in northern Maine and along the southern coast of New England. Farther south, no measurable snow fell in Washington, while an inch (2.5 cm) of snow fell in Baltimore and Philadelphia before changing to rain. Hurricane to near-hurricane force winds occurred along the southeastern New England coast as the storm developed explosively, with 70 mi h^{-1} (35 m s^{-1}) wind speeds and gusts near 100 mi h^{-1} (50 m s^{-1}) reported at Provincetown, Massachusetts. In Connecticut, Massachusetts, Rhode Island, New Hampshire, and Maine, thunderstorms (some with hail) were reported with the snow and a man was reported to have been knocked off his snowplow by lightning.

The precyclogenetic period was not marked by confluent flow and a northern jet system, as a weak high pressure system was analyzed over New England late on 11 March (Fig. 11.1-8). The surface low was a typical example of "secondary development" and was accompanied by the formation of a low-level jet with wind speeds greater than 20–30 m s^{-1} east and north of the 850-hPa low center (Fig. 11.1-8).

The development of the storm system is related to a relatively complex upper trough at 500 hPa and a distinct diffluent exit region of a jet streak analyzed at greater than 60 m s^{-1} at 400 hPa and greater than 80 m s^{-1} at 250 hPa (Fig. 11.1-9). As the trough neared the coast, the northern portion cut off, developing a closed low, and became negatively tilted while the surface low deepened along the New England coastline. As the upper trough associated with the developing cyclone approached the East Coast, the upper-level ridge surged northward along the East Coast into New England. The storm took a more inland track passing over southeastern Virginia before undergoing rapid deepening off the New Jersey coast on 12 March. Increasing easterly to southeasterly flow advected warmer air toward the coastline from the surface to 850 hPa north and east of the developing storm, yielding rain rather than snow in the major cities (Fig. 11.1-8).

d. 13–15 February 1960: NESIS = 4.17

Only 3 weeks before the major snowstorm of 2–4 March 1960 affected much of the Northeast urban corridor (see section 10.4), this storm posed an equivalent threat for heavy snowfall in the Northeast. The precursor setup for a major snowfall included a cold surface high poised north of the Great Lakes and a low pressure system developing in the Gulf of Mexico (see Fig. 11.1-11), a classic nor'easter.

However, the storm moved more on a north-northeastward, rather than northeastward, track, with the storm center hugging the coastline as it moved from the Carolinas toward New York City on 14 February, causing the heaviest snows to fall west of the urban corridor across western portions of Virginia, West Virginia, Pennsylvania, and New York. Near-blizzard conditions were reported across western Pennsylvania and New York, with up to 2 ft (60 cm) of snow reported (Fig. 11.1-10). In New York, 17 deaths were attributed to the storm. In New Jersey, freezing rain caused many traffic accidents. In Connecticut, only 1 or 2 in. (2.5–5 cm) of snow fell before a changeover to rain, although up to a foot (30 cm) of snow fell in northwestern Connecticut. The main impact in New England was limited to high winds and high tides that resulted in major flooding along the southern New England shore.

This is a case that initially had many of the precursors of a major urban corridor snowstorm, with upper-level confluence, a jet streak entrance region over New England, and a blocking upper-level ridge over Greenland (Fig. 11.1-12). The cutoff upper-level ridge located near Greenland is consistent with the highly negative NAO that was diagnosed during this period (Table 2-5, volume I). However, the amplification of an upper trough over the south-central United States on 13–14 February and a building ridge along the coastline toward southeastern Canada by 1200 UTC 14 February quickly eliminated the confluence zone and related jet streak entrance region as the storm moved northward along the coast.

The deep upper trough extending southward into the Gulf of Mexico resulted in a surface low developing unusually far south over the Gulf of Mexico (Fig. 11.1-11). As the upper trough amplified and became negatively tilted south and west of the Northeast urban corridor, an upper ridge developed along the East Coast and western Atlantic Ocean as the surface low intensified rapidly along the immediate coastline. A southerly to southeasterly low-level jet expanded and amplified as the 850-hPa low moved northeastward along the coast (Fig. 11.1-11). The negative tilt and amplifying ridge weakened the confluent flow originally located across southeastern Canada and New England. Therefore, the surface high pressure ridge and cold air could not be sustained and the surface low took a northerly track that allowed warmer air to be drawn into the coastal regions of the middle Atlantic states and much of New England, shifting the heaviest snows well to the west of the urban corridor.

e. 6–7 March 1962: NESIS = 2.76

This famous and infamous storm is often referred to as either the Ash Wednesday storm or the "Great Atlantic Storm." This was one of the most damaging coastal storms on record in the eastern United States. Coastal flooding affected much of the Atlantic coast for several days, with flooding occurring over as many as five consecutive high tide cycles, permanently reshaping the Virginia coastline. Damage was estimated at well over $100 million in 1960 dollars (this would have been a multibillion dollar storm in 2000 dollars).

The surface cyclone was a slowly deepening storm system that covered a large portion of the western Atlantic by 7 March (Fig. 11.1-14). This storm is a "near miss" snowstorm since heavy snows occurred across the interior of Virginia, West Virginia, Maryland, and Pennsylvania, producing record-breaking snows, particularly in interior Virginia and northeastern West Virginia. Many of these inland areas received 30 in. (75 cm) or greater, including 42 in. (107 cm) at Big Meadows, Virginia (Fig. 11.1-13). The heavy snowfall was supported, in part, by a well-defined 30–35 m s^{-1} easterly 850-hPa low-level jet analyzed from the Atlantic toward the Appalachian Mountains (Fig. 11.1-14), as the storm system developed on 6–7 March. In eastern Virginia, heavy wet snow downed numerous power and telephone lines and crippled transportation, with some areas losing power and heat for 4 days. Thunder and lightning were reported with the snow in Virginia. Heavy snow accumulations also occurred within portions of the Washington, Baltimore, and Philadelphia metropolitan areas, although heaviest snows fell west of these cities. Little precipitation fell from New York to Boston.

The dominance of large-scale atmospheric blocking over much of North America contributed to the development of this storm, marked by the presence of a large upper cutoff ridge across Quebec, a very anomalous pattern for early March (Fig. 11.1-15) and consistent with very strong negative daily values of the NAO (see Fig. 2-23 and chapter 2 of volume I). With the upper ridge anchored in place, an upper trough and a related jet streak system over the central United States drifted to the southeast of the blocking ridge, and then cut off, a process that supported a very large, slow-moving, and slowly intensifying low pressure system covering much of the western Atlantic by 6 March (Fig. 11.1-14). The blocking pattern prevented the storm from turning northward toward New England, as was predicted to occur. High pressure anchored over northeastern Canada provided the cold air for heavy snow in the inland areas of the middle Atlantic states. However, the long fetch of easterly winds along the immediate mid-Atlantic coastline for an extended period of time caused the snow to change to rain, sparing the coastal region a major snowstorm.

f. 18–19 February 1964: NESIS = 2.39

This was a classic interior snowstorm marked by snow changing to rain and then back to snow along the coast. This storm occurred during the snowy and stormy El Niño winter of 1963/64 (see Fig. 2-19, volume I) and was the last of three snow-producing storms in the northeast United States (the others occurred on 10–11 and 16–17 February). The surface low evolved from a cyclonic circulation that nearly covered the eastern half of the United States on 18 February (Fig. 11.1-17), with the primary center taking a track from southern Georgia on 1200 UTC 18 February to southern New Jersey by 19 February. Very heavy snows fell in the interior of Pennsylvania (Fig. 11.1-16), including greater than 20 in. (50 cm) at Harrisburg, Pennsylvania, their heaviest snowstorm of record at the time. Wet snow changed to rain in Washington, Baltimore, and Philadelphia with up to 6 in. (15 cm) in Baltimore, 4–8 in. (10 to 20 cm) of snow fell in the New York City area, and more than a foot of snow fell over northwestern New Jersey and southeastern New York. Up to a foot (30 cm) of snow fell in the Boston metropolitan area while much of southern New England reported 5–10 in. (13–25 cm) of snow. The heaviest snow fell between 0000 and 1200 UTC 19 February as the surface system intensified in southeastern Virginia and the 850-hPa closed low and associated 20–30 m s^{-1} low-level jet became focused along the East Coast (Fig. 11.1-17).

This storm was marked by the presence of an upper cutoff anticyclone near Iceland, consistent with the negative phase of the NAO (see Table 2-5, volume I). While

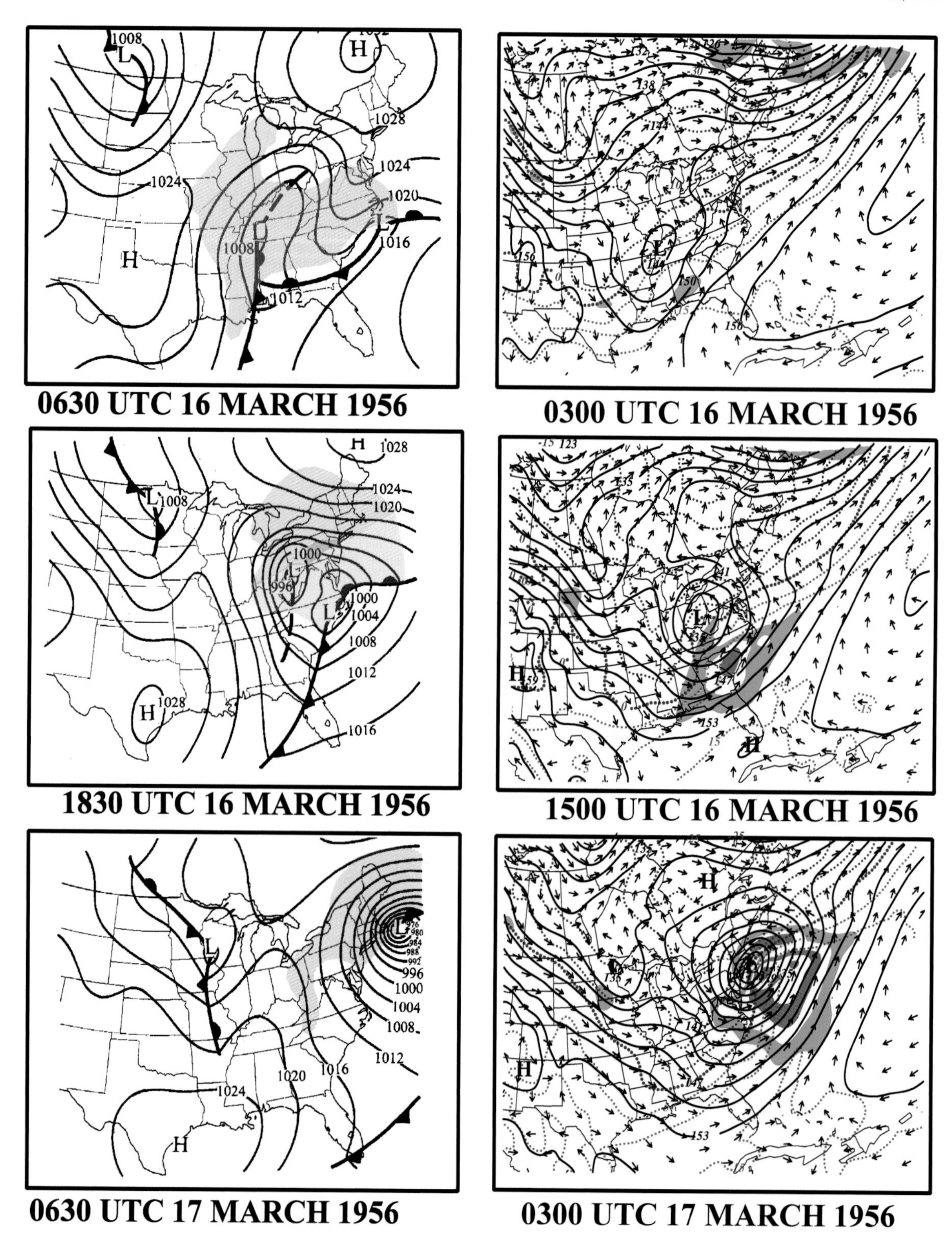

FIG. 11.1-5. Sequence of (left) surface and (right) 850-hPa charts for the 24-h period between 0630 UTC 16 and 17 Mar 1956 (surface) and 0300 UTC 16 and 17 Mar 1956 (850 hPa). See Fig. 11.1-2 for details.

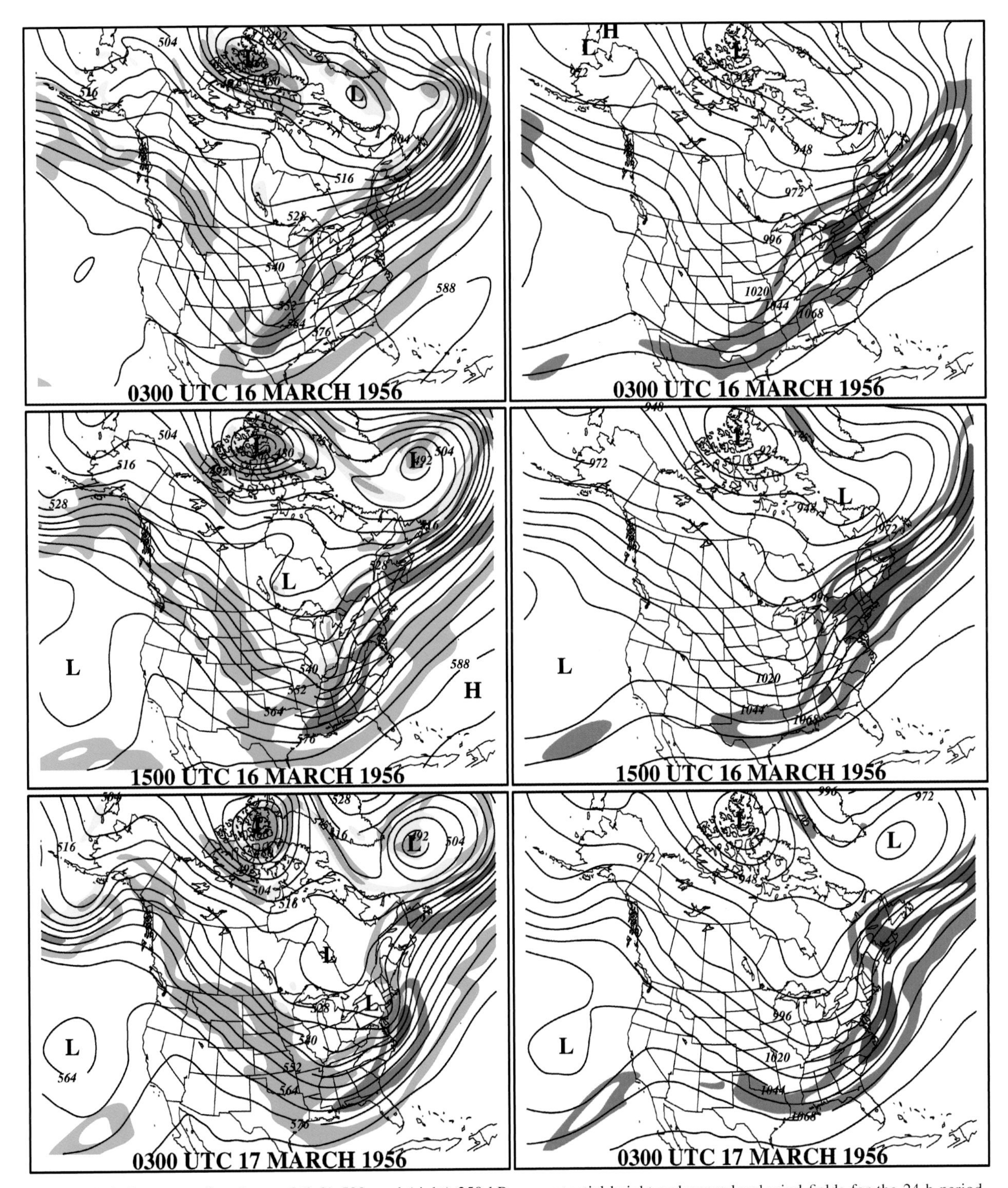

FIG. 11.1-6. Sequence of analyses of (left) 500- and (right) 250-hPa geopotential height and upper-level wind fields for the 24-h period between 0300 UTC 16 and 17 Mar 1956. See Fig. 11.1-3 for details.

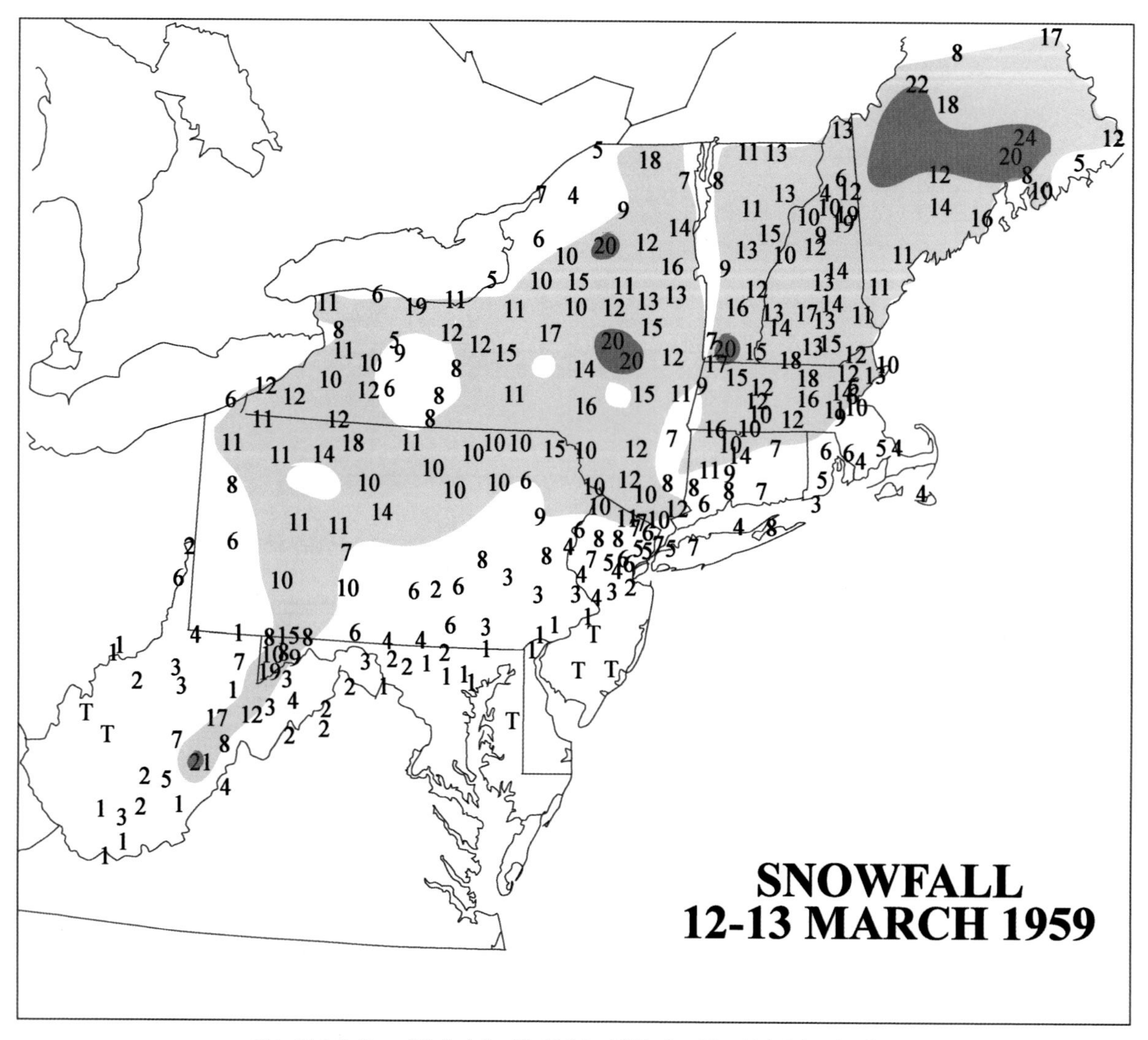

FIG. 11.1-7. Snowfall (in.) for 12–13 Mar 1959. See Fig. 11.1-1 for details.

the negative phase of the NAO may be present in many of the snowstorms in the urban corridor, this case shows that the negative NAO does not guarantee that snow will fall in the major metropolitan areas of the Northeast.

The 500-, 400-, and 250-hPa analyses (Fig. 11.1-18) show that the precyclogenetic period was marked by a confluent flow in the Northeast, a deep trough in the central United States, and a double-jet structure common to many of the major snow events immediately along the coast (volume I, chapter 4). However, in this case, the confluent flow was not maintained as the upper trough and southern jet stream lifted northeastward, west of the coastal plain. As the upper-level trough system lifted northeastward across the Tennessee Valley prior to reaching the East Coast, a distinct ridge built northeastward along the East Coast, while the coastal cyclone developed on 18 February. Although a significant closed upper low was positioned off the southeastern coast of Canada, the negative tilt of the storm-producing upper-level trough and building ridge along the East Coast weakened the upper-level confluence initially located across eastern Canada. Weak high pressure located near New England in the precyclogenetic period drifted eastward (Fig. 11.1-17) as the upper-level ridge developed along the East Coast. All these features combined to sustain a mild east to southeasterly low-level flow ahead of the surface low pressure center and kept precipitation primarily rain in the Northeast urban corridor from Washington to New York City as the snow fell farther inland.

g. 22–24 January 1966: NESIS = 4.45

This storm occurred during a very active El Niño period (see Fig. 2-19, volume I) dominated by several major East Coast snowstorms and also during a period in which the NAO was in a negative phase (see Table 2-5 and chapter 2 of volume I). During late January

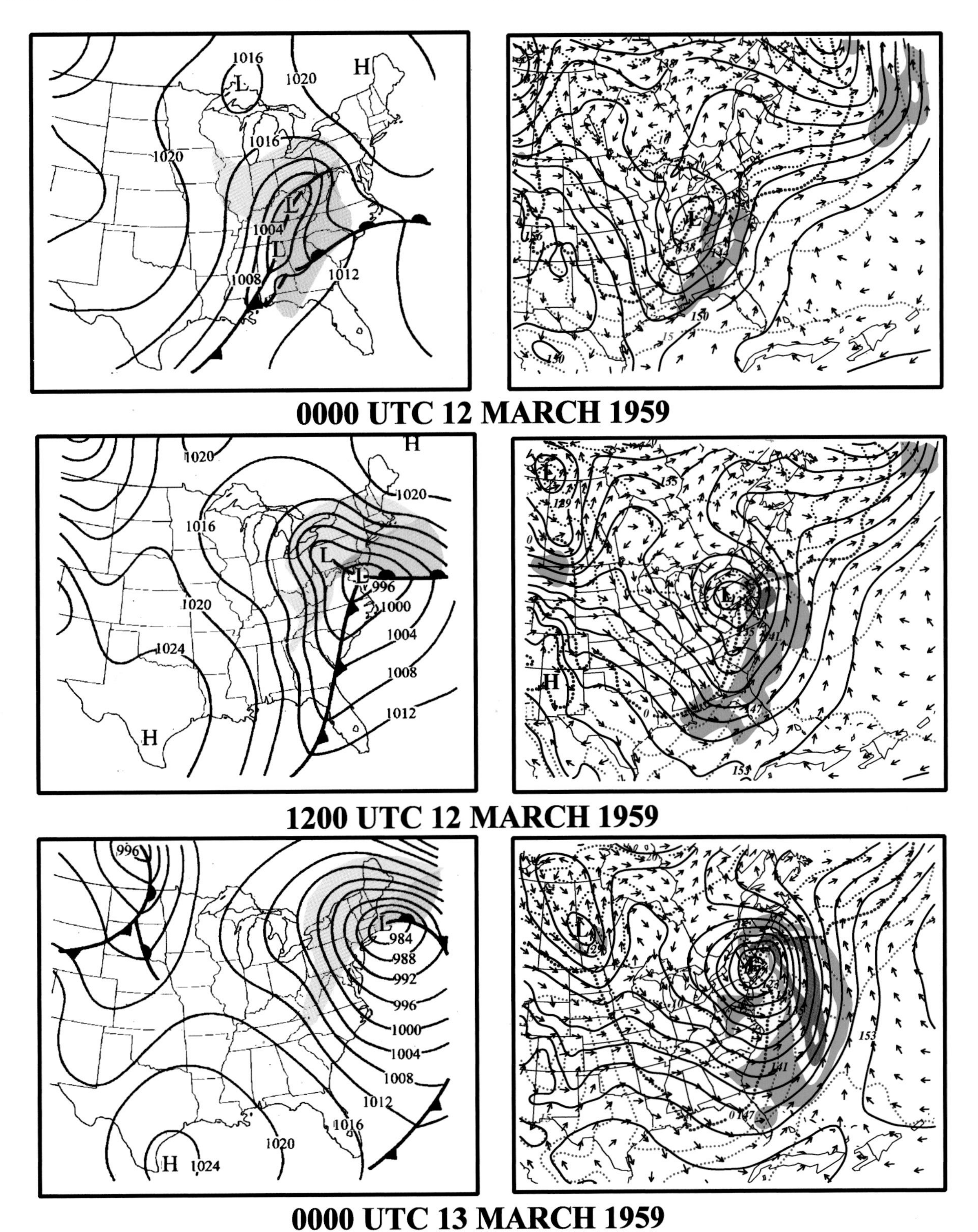

FIG. 11.1-8. Sequence of (left) surface and (right) 850-hPa charts for the 24-h period between 0000 UTC 12 and 13 Mar 1959. See Fig. 11.1-2 for details.

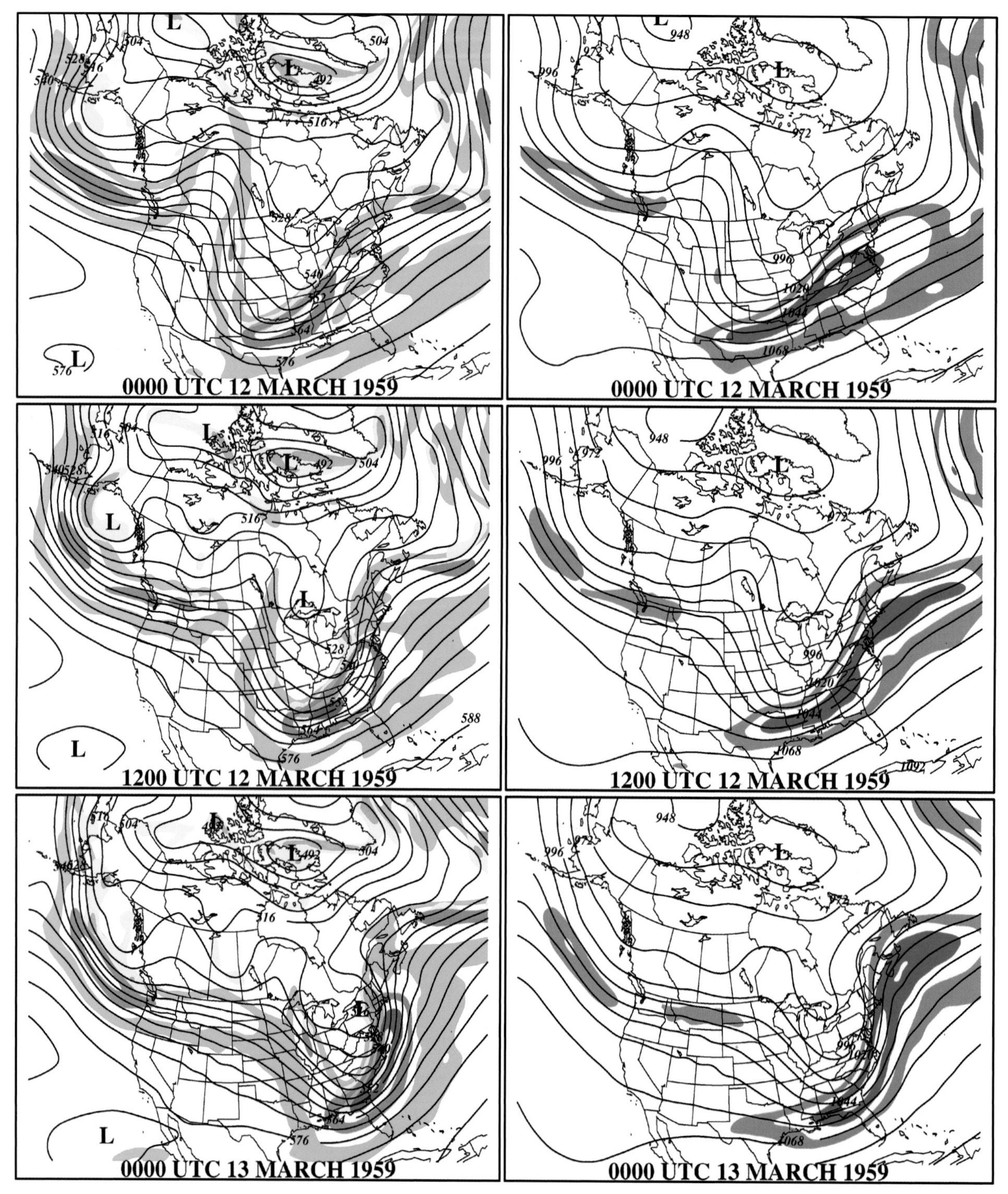

FIG. 11.1-9. Sequence of analyses of (left) 500- and (right) 250-hPa geopotential height and upper-level wind fields for the 24-h period between 0000 UTC 12 and 13 Mar 1959. See Fig. 11.1-3 for details.

1966, a series of three snowstorms occurred, dominating the stormy winter of 1965/66. This was the first of these storms, a major interior snowstorm with snow, mixed precipitation, and rain in the urban corridor. The second snowstorm occurred a few days later, producing heavy snows in the Southeast. The third and final storm occurred on 29–31 January 1966, the "blizzard of '66" (see section 10.9).

The storm evolved from a distinct pair of inverted troughs: one in the Tennessee Valley and one along the

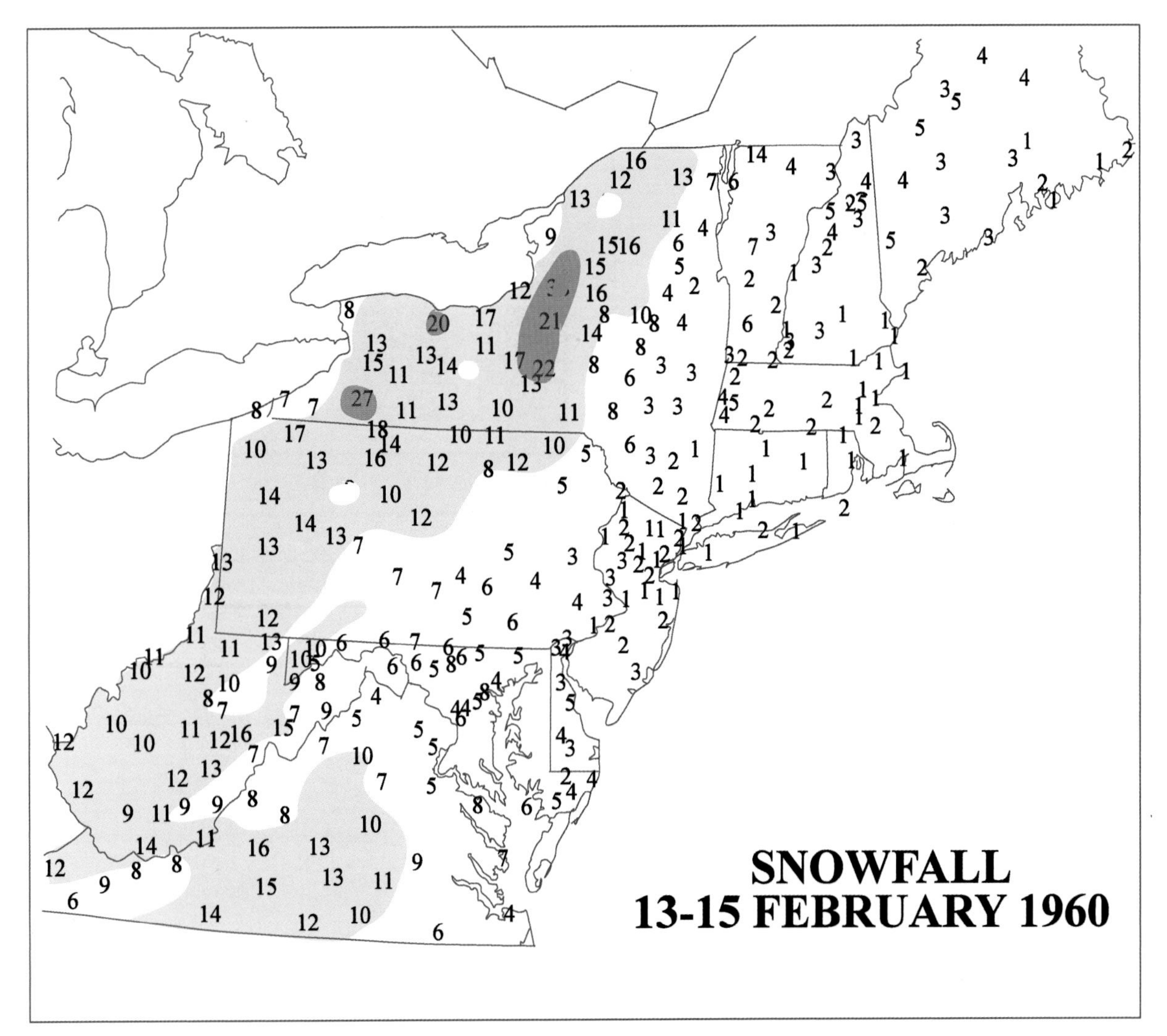

FIG. 11.1-10. Snowfall (in.) for 13–15 Feb 1960. See Fig. 11.1-1 for details.

East Coast on 1200 UTC 22 January (Fig. 11.1-20). By 0000 UTC 23 January, a deepening low pressure system moved along and slightly west of the Carolina coastline and was located over the Delmarva Peninsula by 1200 UTC 23 January. The 850-hPa circulation also intensified during this period, with a greater than 30 m s^{-1} low-level jet directed from the Atlantic Ocean toward the developing precipitation shield by 0000 UTC 23 January. As the surface low intensified over the Delmarva Peninsula, a 35+ m s^{-1} easterly low-level jet wrapped around the northern part of the deep 850-hPa circulation.

With the track of the surface low slightly inland, rain soaked the Northeast urban corridor between Washington and New York City. Coastal flooding was significant in New York, Connecticut, and Massachusetts with tides of 3–4 ft (90–120 cm) above normal. The storm was considered the worst storm of the winter to date in Massachusetts with near paralysis of all transportation and falling trees due to the heavy wet snow. More than 20 in. (50 cm) of wet snow fell across a large section of western New York (Fig. 11.1-19), but modest winds only produced minor drifting. Nevertheless, more than a thousand people were marooned on the New York Thruway in New York's Mohawk Valley. Six fatalities related to exposure occurred in New York as people in stalled vehicles waited to be rescued or were seeking shelter.

A large cutoff ridge dominated this case over Labrador, with a surface high pressure cell extending across eastern Canada. But only a weak upper trough and related upper- level confluence are noted east of the Canadian east coast (Fig. 11.1-21) in the precyclogenetic environment. This is another typical interior snowstorm marked by an upper trough that 1) deepens over the central United States with a well-defined jet streak, near the base of the trough as it develops a negative tilt well to the south and west of the coast, and 2) begins to lift

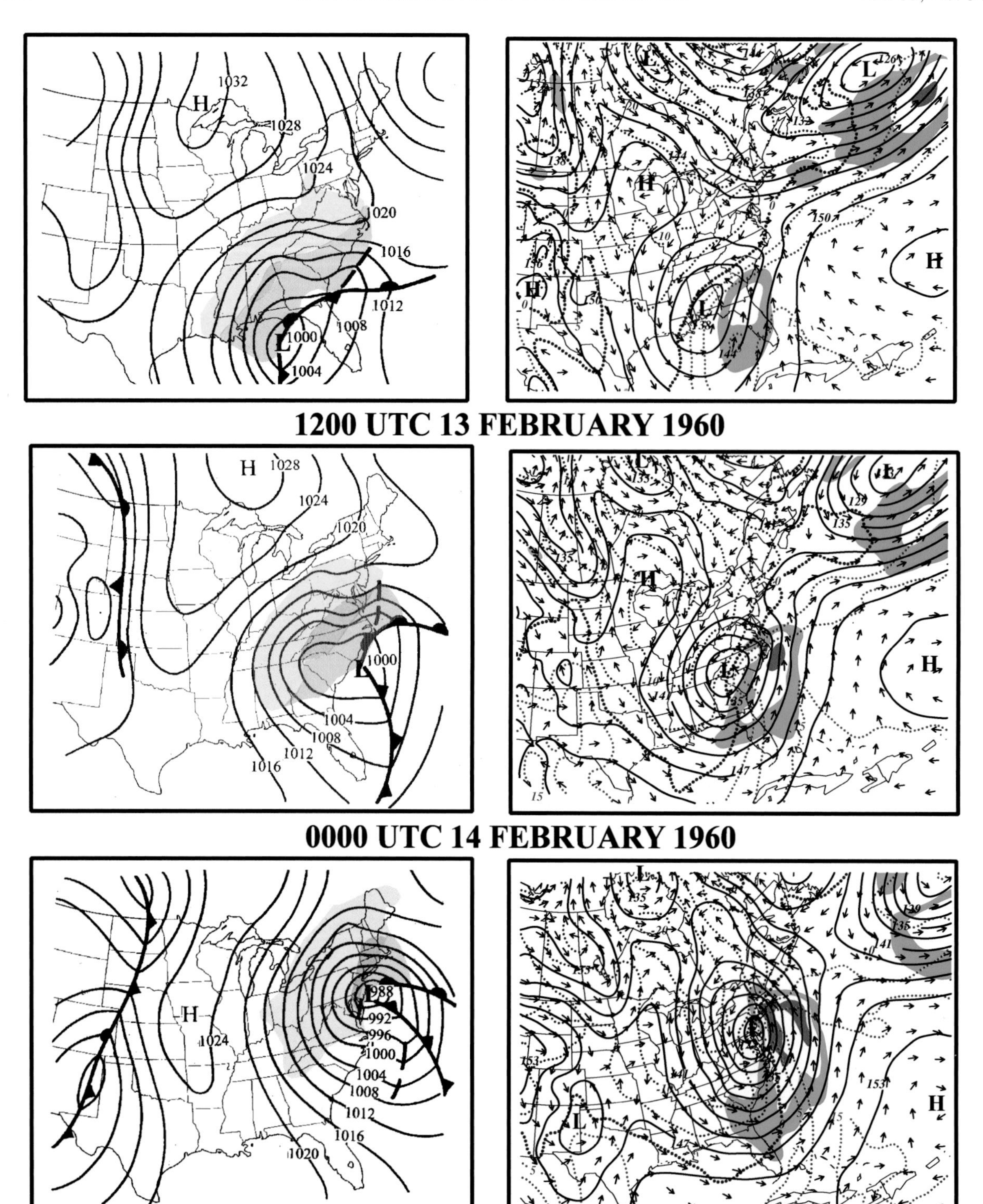

FIG. 11.1-11. Sequence of (left) surface and (right) 850-hPa charts for the 24-h period between 1200 UTC 13 and 14 Feb 1960. See Fig. 11.1-2 for details.

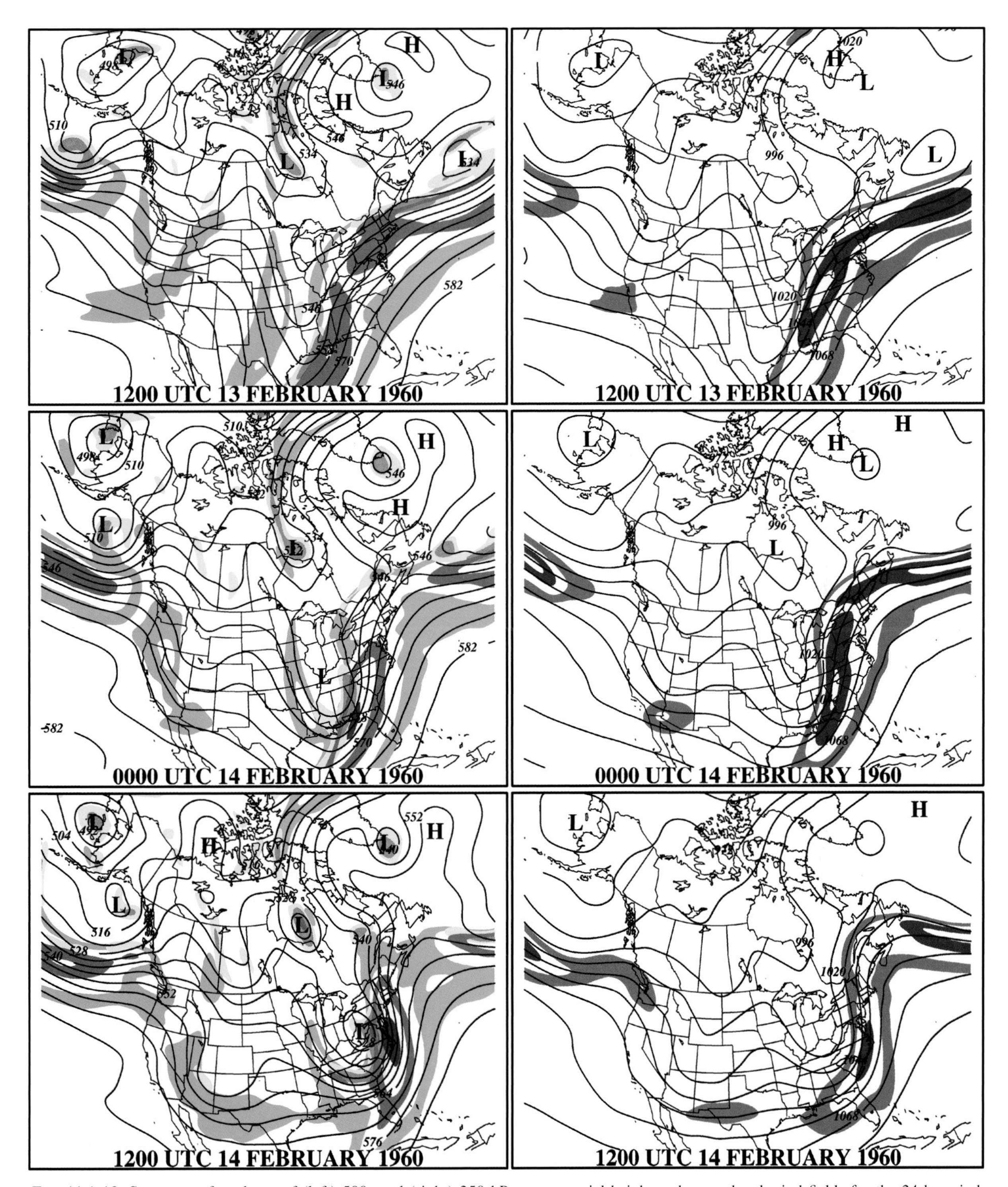

FIG. 11.1-12. Sequence of analyses of (left) 500- and (right) 250-hPa geopotential height and upper-level wind fields for the 24-h period between 1200 UTC 13 and 14 Feb 1960. See Fig. 11.1-3 for details.

northeastward across the Tennessee Valley with a rapidly building ridge along the Atlantic coast. As a result, the surface low developed within the diffluent exit region of the jet streak (as depicted in the 400- and 250-hPa analyses), located to the west of the sample of cases that produced heavy snow along the coast. The net result is that the surface low took a track along and to the west of the East Coast, rather than just offshore. The surface winds turned more easterly rather than nouth-easterly, allowing mild air to reach the coastline, re-

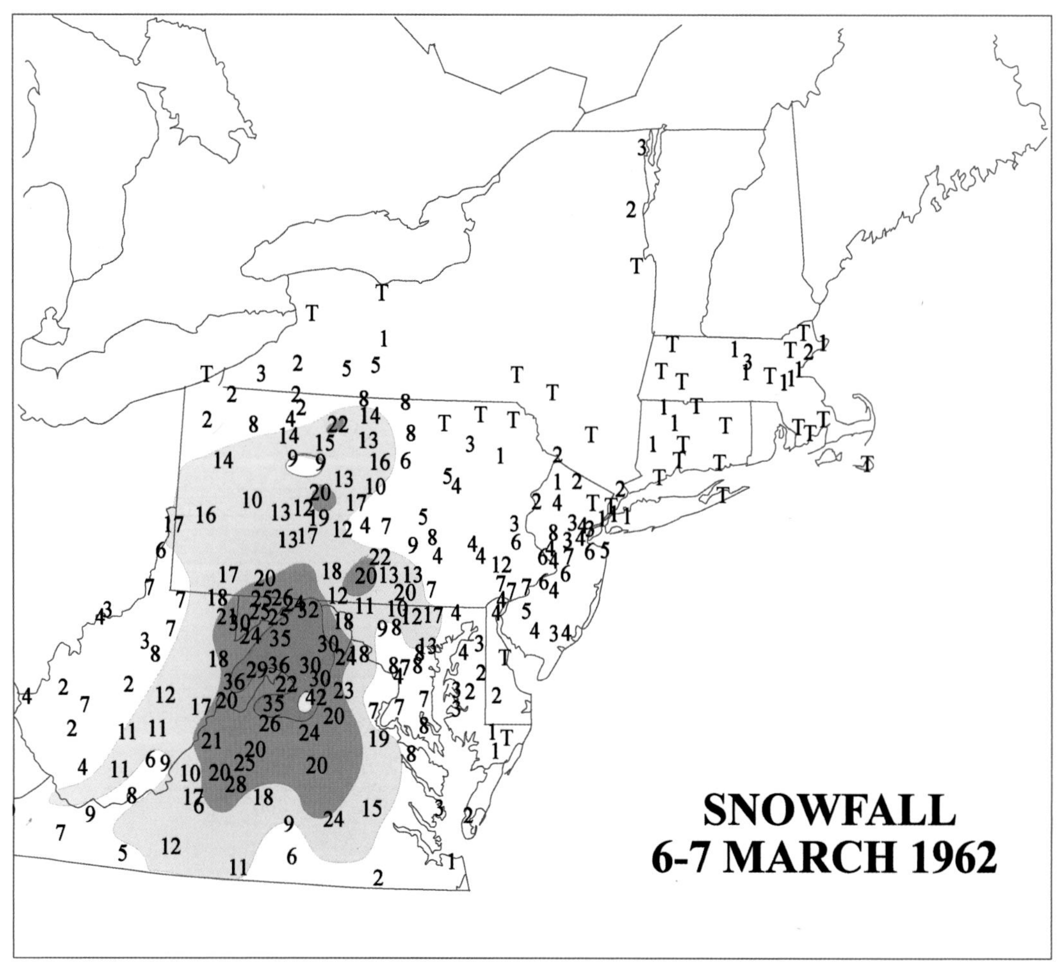

FIG. 11.1-13. Snowfall (in.) for 6–7 Mar 1962. See Fig. 11.1-1 for details.

sulting in snow changing to rain from Washington to southern New England. The coastal track, the building ridge along the Atlantic coastline, combined with a lack of strong confluent upper-level flow, allowed warmer air to advect into the lower troposphere along the immediate coastline, producing the coastal rains and restricting the heavy snowfall to the interior locations.

h. 3–5 March 1971: NESIS = 3.73

This storm capped off a particularly snowy winter in northern New York and New England, but with little snowfall recorded in the Northeast urban corridor. The storm produced record snowfall accumulations over a 2-day period in interior New York and New England, but only produced a brief period of snow changing to heavy rain and then briefly back to snow in the coastal areas. In New York, the storm was one of the heaviest March snowstorms of record. Western, central, and northern New York were paralyzed by 20–40 in. (50–100 cm) of snow (Fig. 11.1-22), accompanied by winds gusting to 40–50 mi h^{-1} (20–25 m s^{-1}). The New York Thruway was closed from eastern New York to Buffalo, as thousands of motorists were stranded with a snow emergency declared in over 50 counties.

The interior snowstorm was related to a complex surface low that moved from Georgia to eastern North Carolina before exploding into a major cyclone just off the New Jersey coast before moving toward eastern New England on 4 March (Fig. 11.1-23). The central sea level pressure dropped to 961 hPa in eastern New England as an intense 850-hPa circulation with winds exceeding 20 m s^{-1} (Fig. 11.1-23) affected the eastern third of the United States (See also Boyle and Bosart 1986).

This is yet another example of a deep upper-level trough that becomes negatively tilted well to the south and west of the Northeast urban corridor, with the upper ridge building rapidly along the East Coast and western Atlantic (Fig. 11.1-24). The confluent flow over New England early in the storm period quickly gave way to rising heights as an upper ridge developed over the western Atlantic. This is also another example of a surface low that becomes focused in the diffluent exit of a jet streak on 3–4 March, the axis of which is farther west extending from Georgia up the Appalachian Mountains

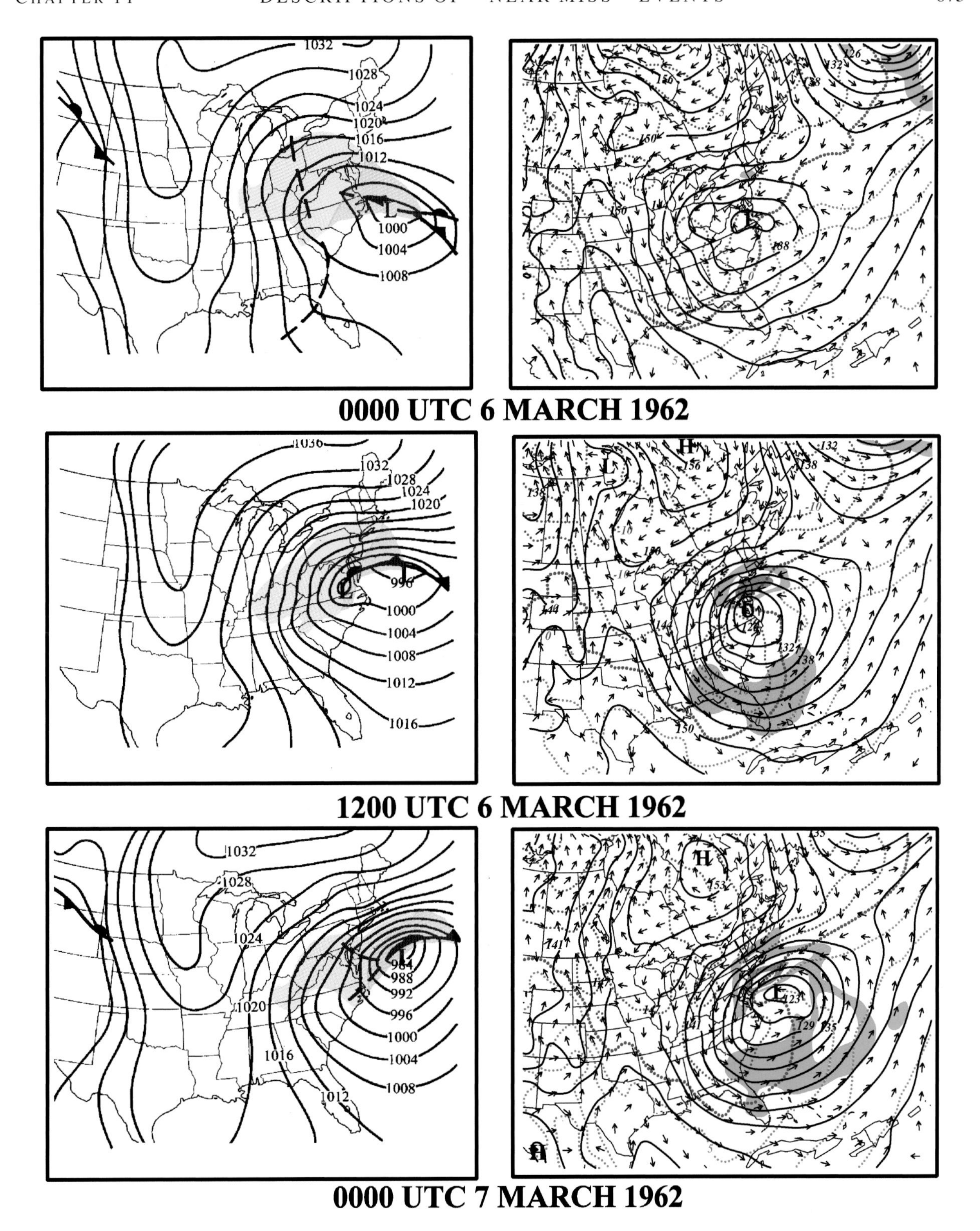

FIG. 11.1-14. Sequence of (left) surface and (right) 850-hPa charts for the 24-h period between 0000 UTC 6 and 7 Mar 1962. See Fig. 11.1-2 for details.

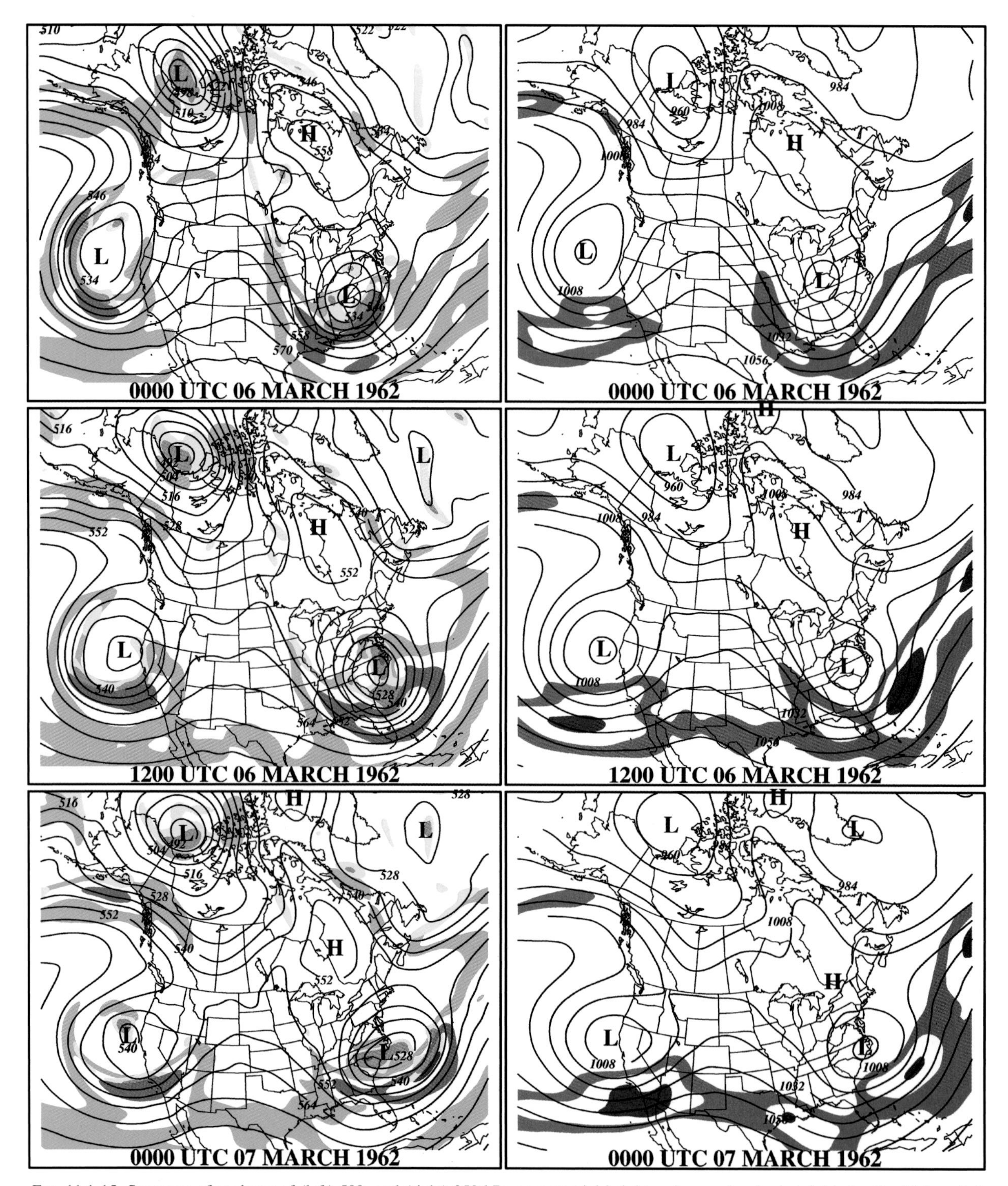

FIG. 11.1-15. Sequence of analyses of (left) 500- and (right) 250-hPa geopotential height and upper-level wind fields for the 24-h period between 0000 UTC 6 and 7 Mar 1962. See Fig. 11.1-3 for details.

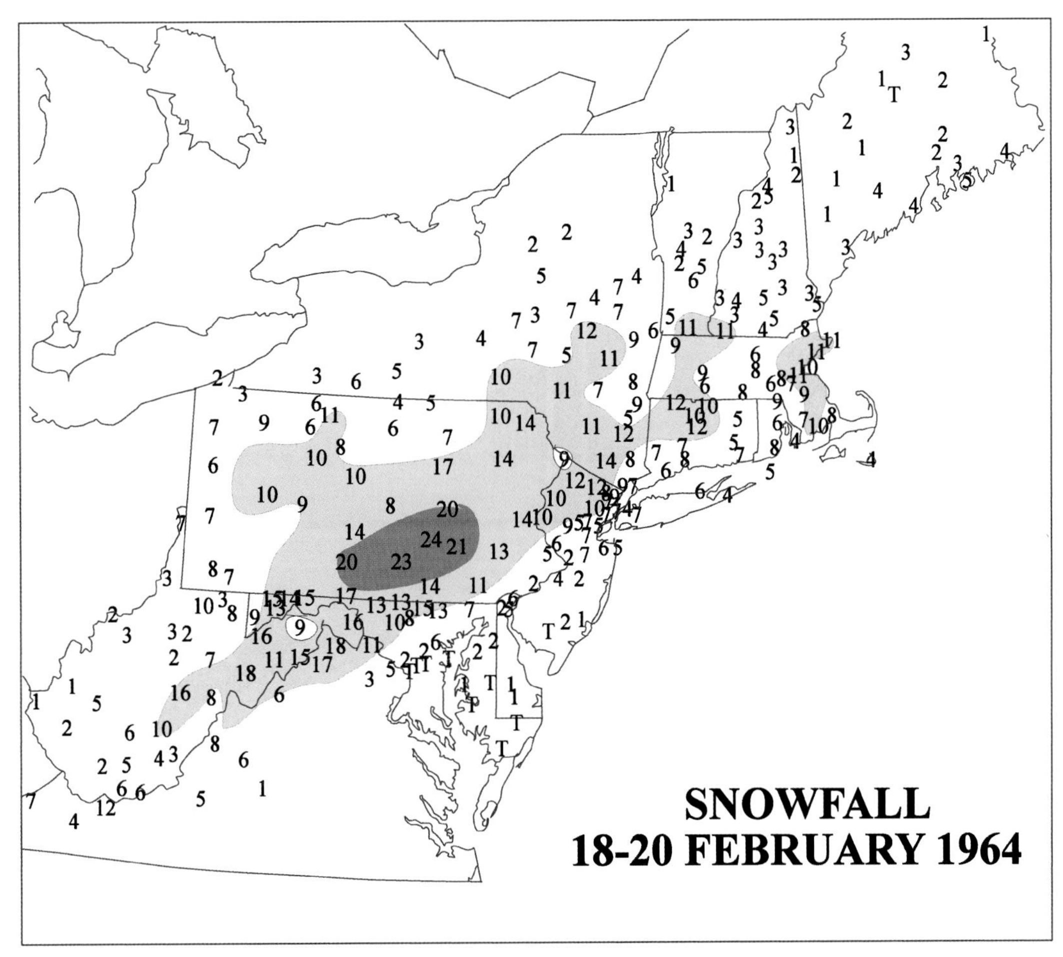

FIG. 11.1-16. Snowfall (in.) for 18–20 Feb 1964. See Fig. 11.1-1 for details.

at 0000 UTC 4 March (Fig. 11.1-24). The surface low responds by hugging the coastline, heading on a more northerly track, as the upper trough attains a negative tilt and the low-level flow of warmer air above 0°C is directed from the south and east well inland of the coastline. Rain fell throughout the Northeast urban corridor, with massive amounts of snow falling inland as the upper trough evolved into a negatively tilted cutoff low over the northeast United States.

i. 25–27 November 1971: NESIS = 2.33

Record November snowfall occurred throughout northeastern Pennsylvania, the Catskill Mountains, and the upper Hudson Valley of New York into central and northern New England on Thanksgiving Day 1971, crippling surface transportation and stranding holiday travelers throughout the inland portions of the Northeast. More than 20 in. (50 cm) of snow fell in a large area of northeastern Pennsylvania and eastern New York (Fig. 11.1-25), although winds were not strong enough in the heavy snow area to cause significant drifting. One report in *Storm Data* noted the extremely narrow transitional zone between snow and rain over southwestern Maine, southern Vermont, eastern Massachusetts, and northern Rhode Island (also see volume I, chapter 6). Rain fell from Washington to Boston with the heavy snowfall area located only 30–50 mi (50–90 km) inland.

As in other inland cases, this storm evolved from several distinct inverted troughs west and east of the Appalachian Mountains into one center by 0000 UTC 25 November, located just west of the Virginia and North Carolina coastline (Fig. 11.1-26). The surface low continued its more inland track through the morning of 25 November, emerging over the Atlantic near the Delaware and New Jersey coastlines. During this period the surface low underwent an explosive development phase with the tight circulation and rapid development of a 25–30 m s^{-1} easterly low-level jet streak evident from the surface up to the 850-hPa level by 1200 UTC 25 November, centered near the New Jersey coast, during the period of heaviest snowfall in northern Pennsylvania.

While the upper-level jet streak located in the central United States on 24 November was not as intense as some of the other interior snowstorms, the trough am-

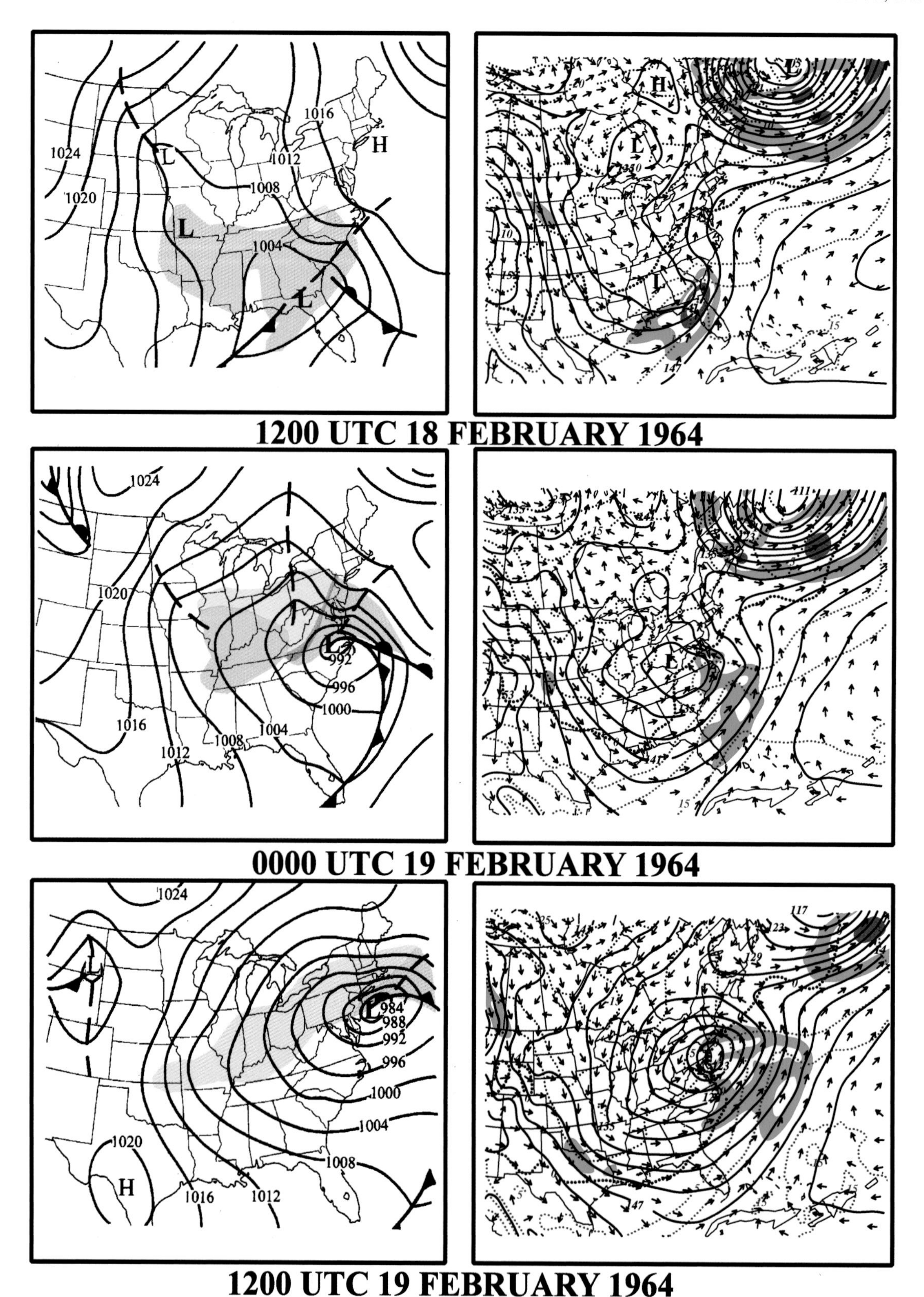

FIG. 11.1-17. Sequence of (left) surface and (right) 850-hPa charts for the 24-h period between 1200 UTC 18 and 19 Feb 1964. See Fig. 11.1-2 for details.

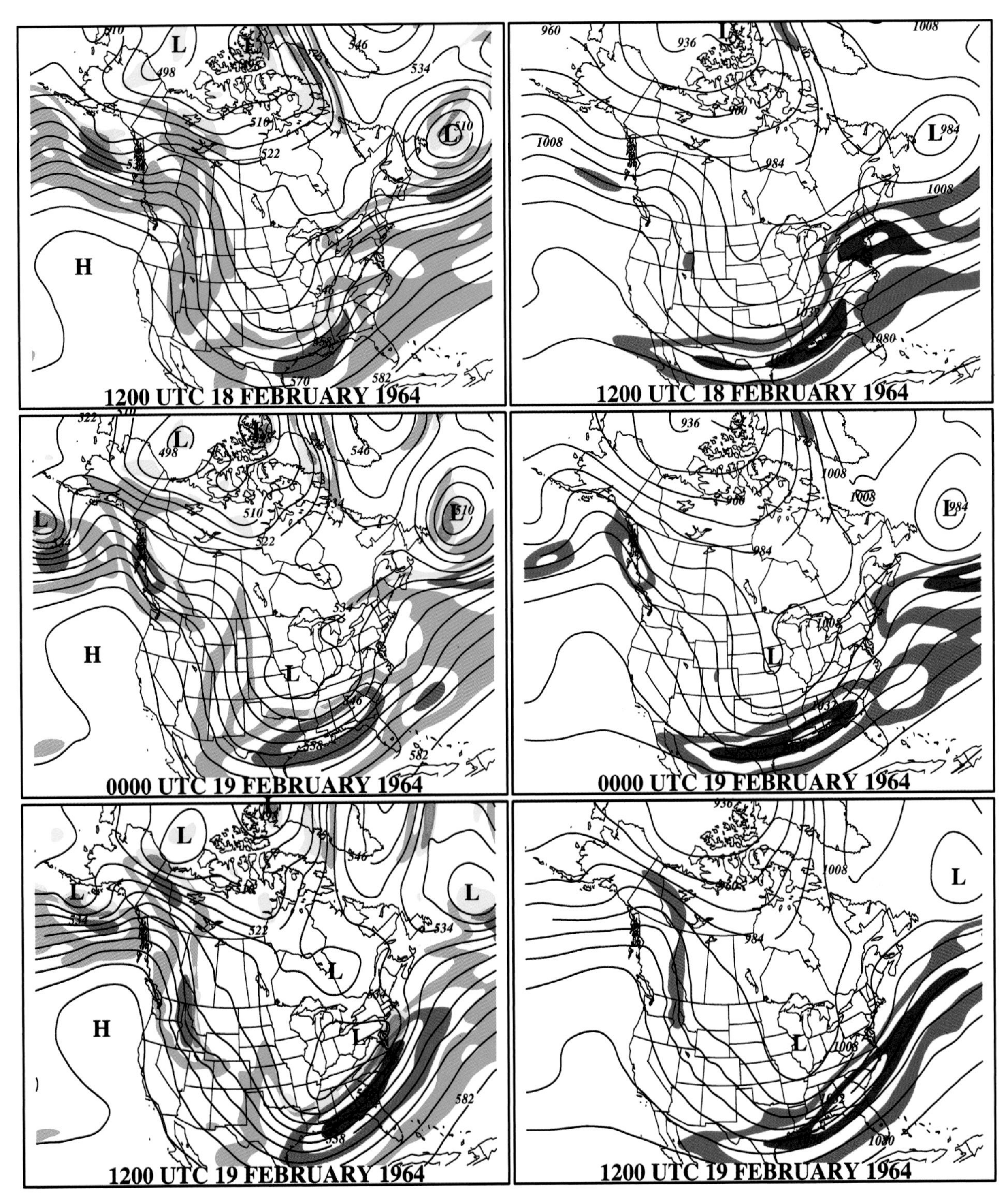

FIG. 11.1-18. Sequence of analyses of (left) 500- and (right) 250-hPa geopotential height and upper-level wind fields for the 24-h period between 1200 UTC 18 and 19 Feb 1964. See Fig. 11.1-3 for details.

plified in the 12-h period ending at 1200 UTC 25 November with increasing vorticity advections (Fig. 11.1-27). The upper trough over eastern Canada and the confluent flow located immediately upstream, which is associated with the surface high pressure center over Maine early in the storm period, propagated rapidly eastward. The strong surface high over northern New England also moved eastward out over the western Atlantic Ocean (Fig. 11.1-26). Even though the trough nearing the East Coast was not highly amplified, it became neg-

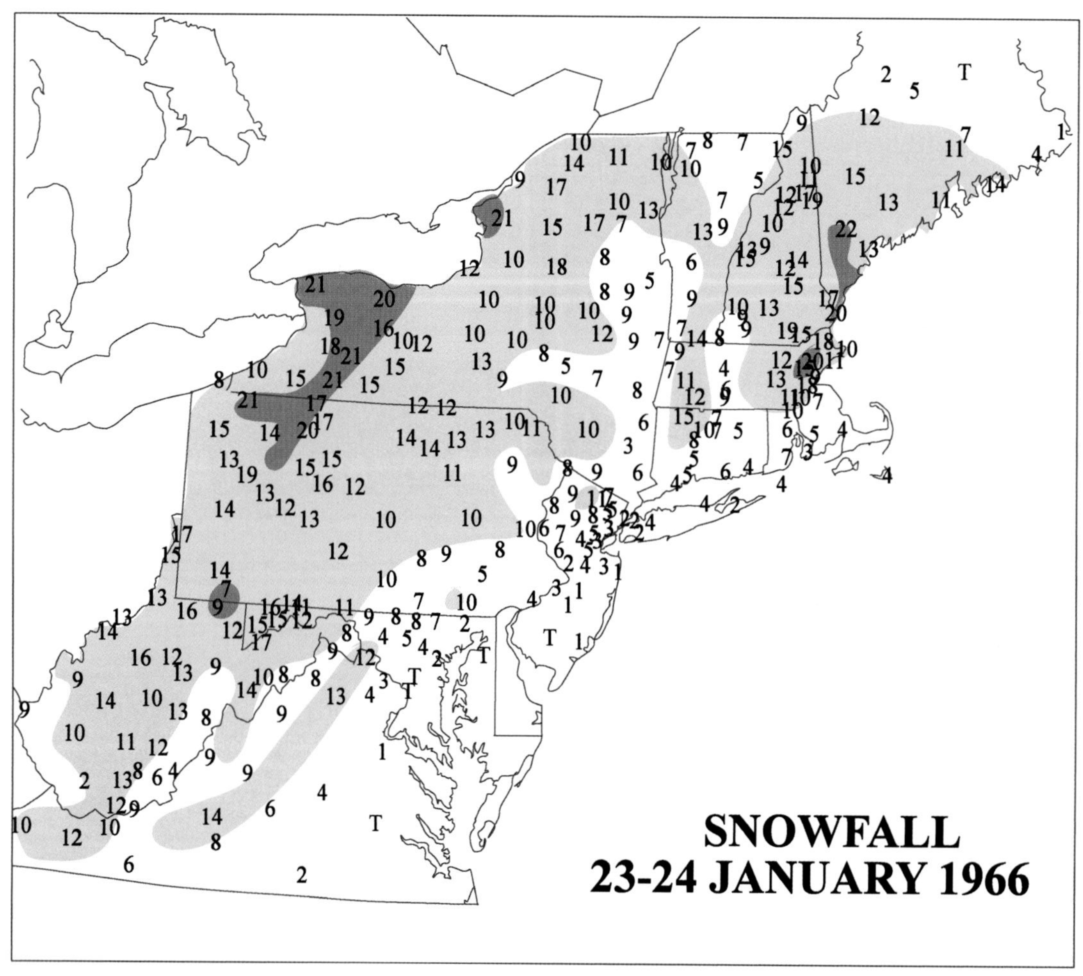

FIG. 11.1-19. Snowfall (in.) for 23–24 Jan 1966. See Fig. 11.1-1 for details.

atively tilted over the Tennessee Valley well to the south and west of the urban corridor, resulting in the surface low tracking close and to the west of the coastline. As a result, easterly low-level flow and strong warm-air advection resulted in rain along the immediate coastline, with heavy snows falling farther west, just west and north of the major metropolitan centers.

j. 16–18 January 1978: NESIS = 4.10

A series of three back-to-back-to-back Northeast storms brought record snowfall accumulations during January 1978 from Pennsylvania to New England. These storms occurred during a winter characterized by moderate El Niño conditions (Fig. 2-19, volume I) and a generally negative phase of the NAO (Fig. 2-20), although this storm occurred during a period when daily values of the NAO were positive (Table 2-5, volume I).

The first storm on 11–15 January produced locally heavy snow in Pennsylvania and New York and a severe ice storm in coastal New York and New England (see section 11.3). The 16–18 January case was the second event and forecasters had feared that this storm would produce heavy snow in the urban corridor. But, snow changed to rain and the heaviest snows fell west of the coastal plain in interior Pennsylvania, New York, and New England (Fig. 11.1-28). Following this storm, the roof of the Hartford, Connecticut, Civic Center Coliseum collapsed under the weight of snow, ice, and water only 6 h after a crowd of 5000 was inside. The third of the three events occurred on 19–20 January and was a major urban corridor snowstorm (see section 10.16).

The surface low developed in the Gulf coastal region on 17 January with two distinct inverted troughs extending northeastward west and east of the Appalachian Mountains. The storm then evolved into a broad low pressure system moving slowly northeastward just west of the coastline, located off of New Jersey and Long Island by 1200 UTC 18 January. The more inland track of the surface low and broad 850-hPa cyclonic circulation (Fig. 11.1-29) ensured that snow changed to rain from Washington to southern New England, despite the existence of a strong and cold anticyclone extending into New England in the precyclogenetic period on 17

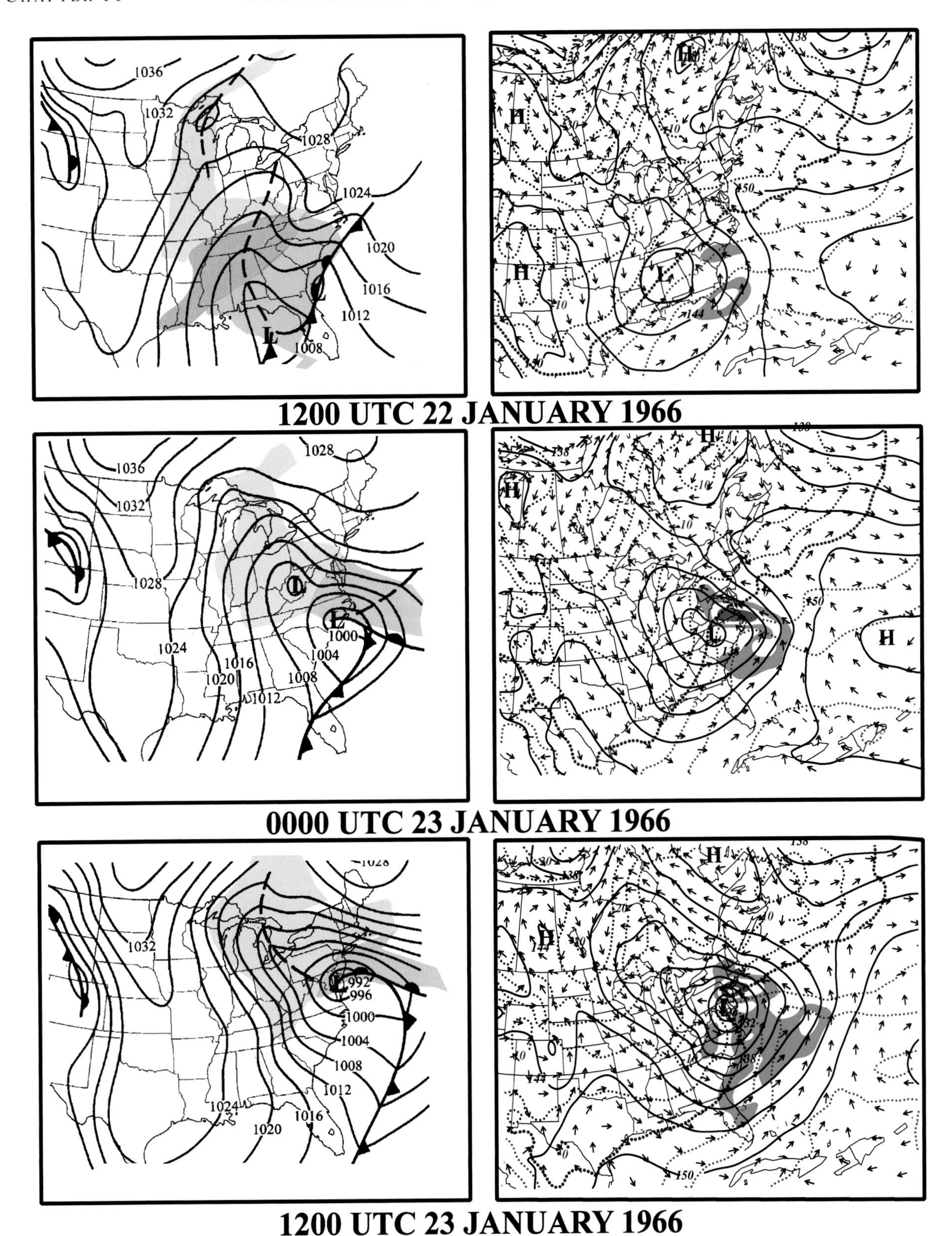

FIG. 11.1-20. Sequence of (left) surface and (right) 850-hPa charts for the 24-h period between 1200 UTC 22 and 23 Jan 1966. See Fig. 11.1-2 for details.

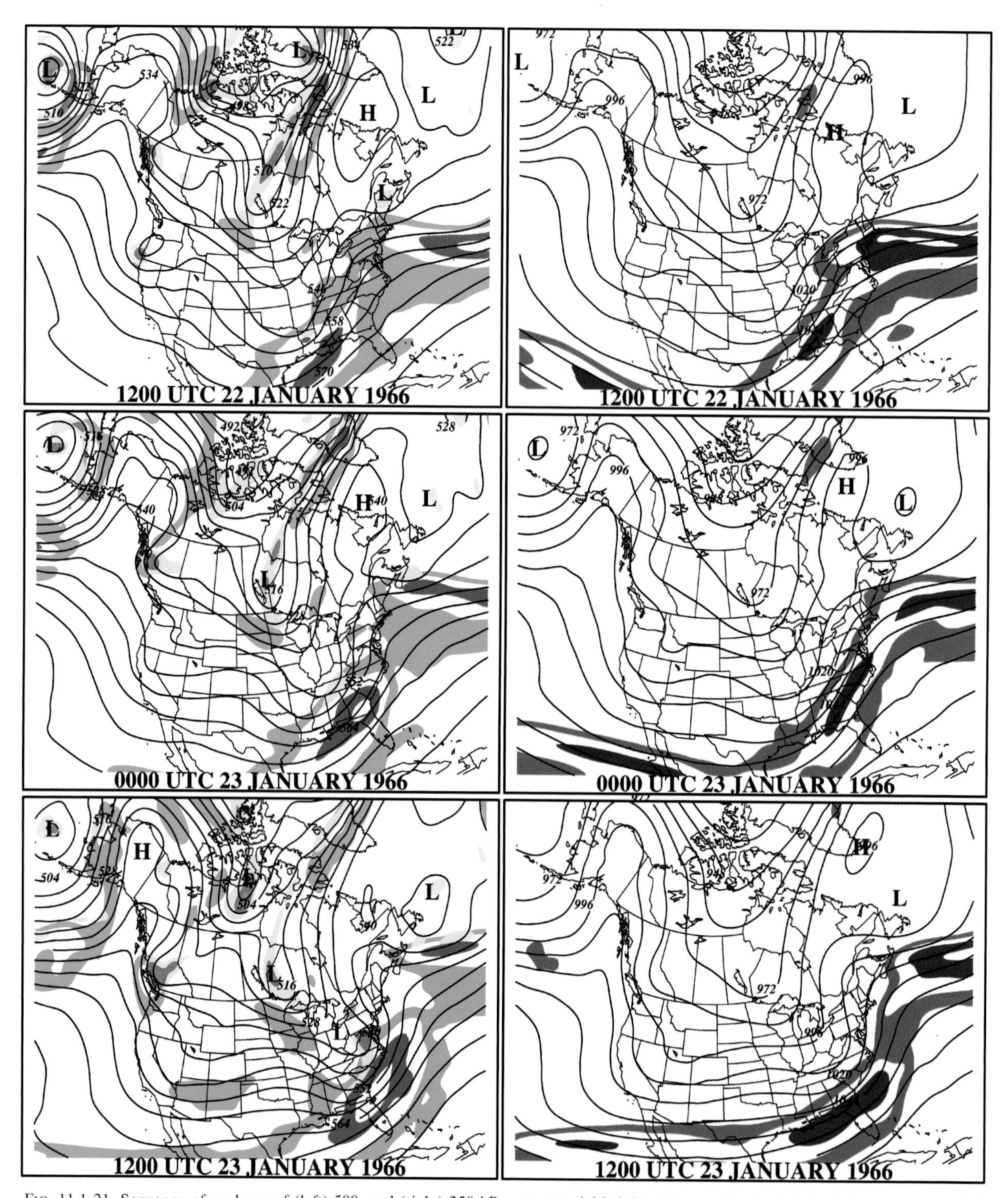

FIG. 11.1-21. Sequence of analyses of (left) 500- and (right) 250-hPa geopotential height and upper-level wind fields for the 24-h period between 1200 UTC 22 and 23 Jan 1966. See Fig. 11.1-3 for details.

January, although the anticyclone drifted east of New England by 0000 UTC 18 January.

At 500, 400, and 250 hPa (Fig. 11.1-30), the precyclogenetic period was marked by strong confluent flow and a distinct jet streak entrance region extending eastward from New York to New England on 1200 UTC 17 January. The southern trough–jet streak is rather modest compared to other cases presented in this monograph. Nevertheless, this trough developed a negative tilt over the southeast United States by 0000 UTC 18

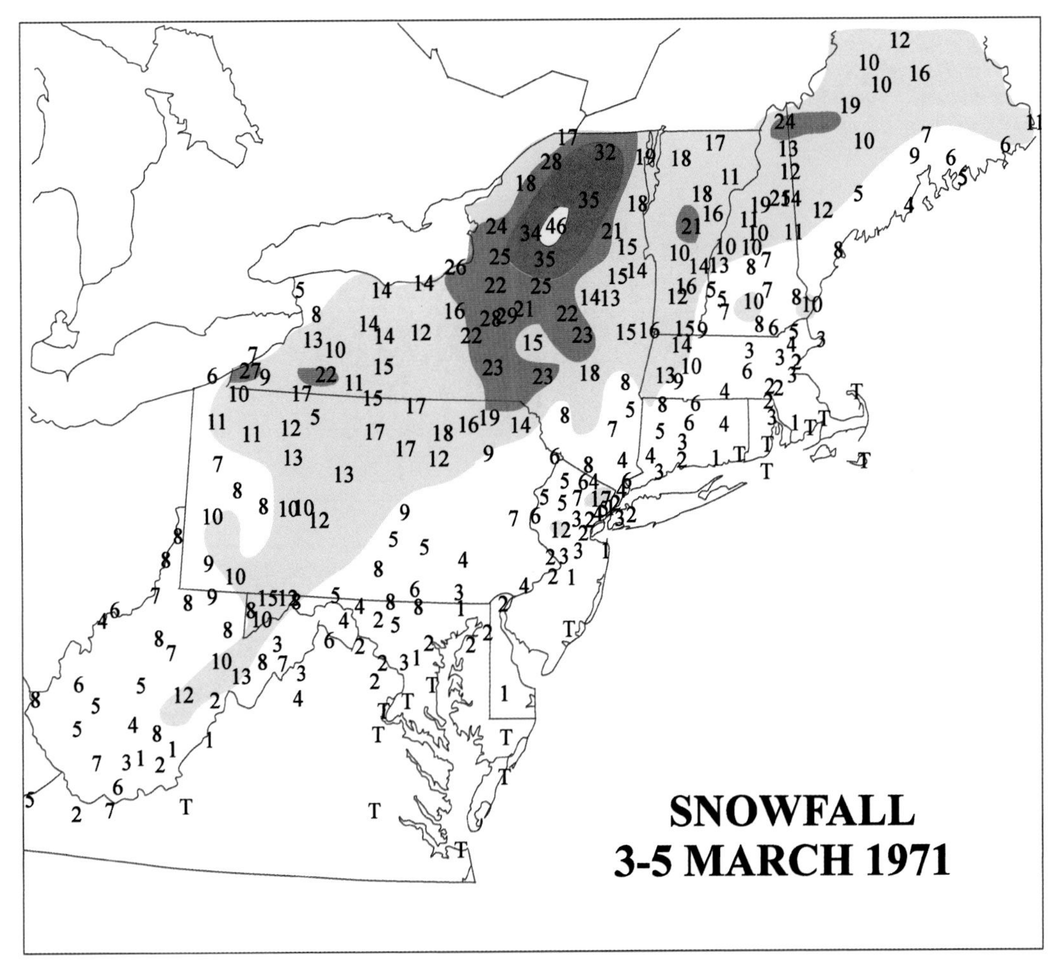

FIG. 11.1-22. Snowfall (in.) for 3–5 Mar 1971. See Fig. 11.1-1 for details.

January. As the trough assumed a negative tilt, a ridge developed rapidly along the East Coast. The upper-level confluence and an embedded jet streak (and related strong surface high pressure system to the north of New England early in the period) shifted rapidly east-northeastward since the flow over eastern Canada was not blocked and no cutoff or strong upper-level trough existed. The net result was rapid eastward progression of the surface high pressure system, allowing warmer air to move into the coastal areas. In addition, the surface low took a more inland track before emerging at the coast over New Jersey, with a classic snow-to-rain situation developing within the Northeast urban corridor.

k. 28–29 March 1984: NESIS = 1.86

This early spring snowstorm was associated with a historic intense outbreak of tornadoes across the southeast United States in association with an intense mesoscale cyclone (documented by Gyakum and Barker 1988; Kuo et al. 1995; Gyakum et al. 1995; Mesinger et al. 1993; Kocin et al. 1986). The larger-scale circulation and related synoptic-scale surface low pressure system produced a snowstorm across interior portions of Pennsylvania, New York, and New England (Fig. 11.1-31); a microburst in Massachusetts; high winds and coastal flooding throughout the Northeast; and record low pressures along the middle Atlantic coast.

This was a massive storm system that involved a cyclonic circulation over the entire eastern half of the United States by 1200 UTC 28 March (Fig. 11.1-32). Embedded within this slow-moving system was a primary low moving up into the Tennessee Valley on 28 March and the explosive development of a separate mesoscale surface low in Mississippi and Alabama that traveled northeastward to South Carolina by 0000 UTC 29 March. This separate low was associated with the convective outbreak across the Southeast late on 28 March, which then continued to deepen rapidly along the East Coast on 29 March. The broad 850-hPa circulation intensified along the East Coast with a greater than 45 m s^{-1} low-level jet developing north of the low centered by 1200 UTC 29 March (Fig. 11.1-32).

This storm system combined elements typical of both

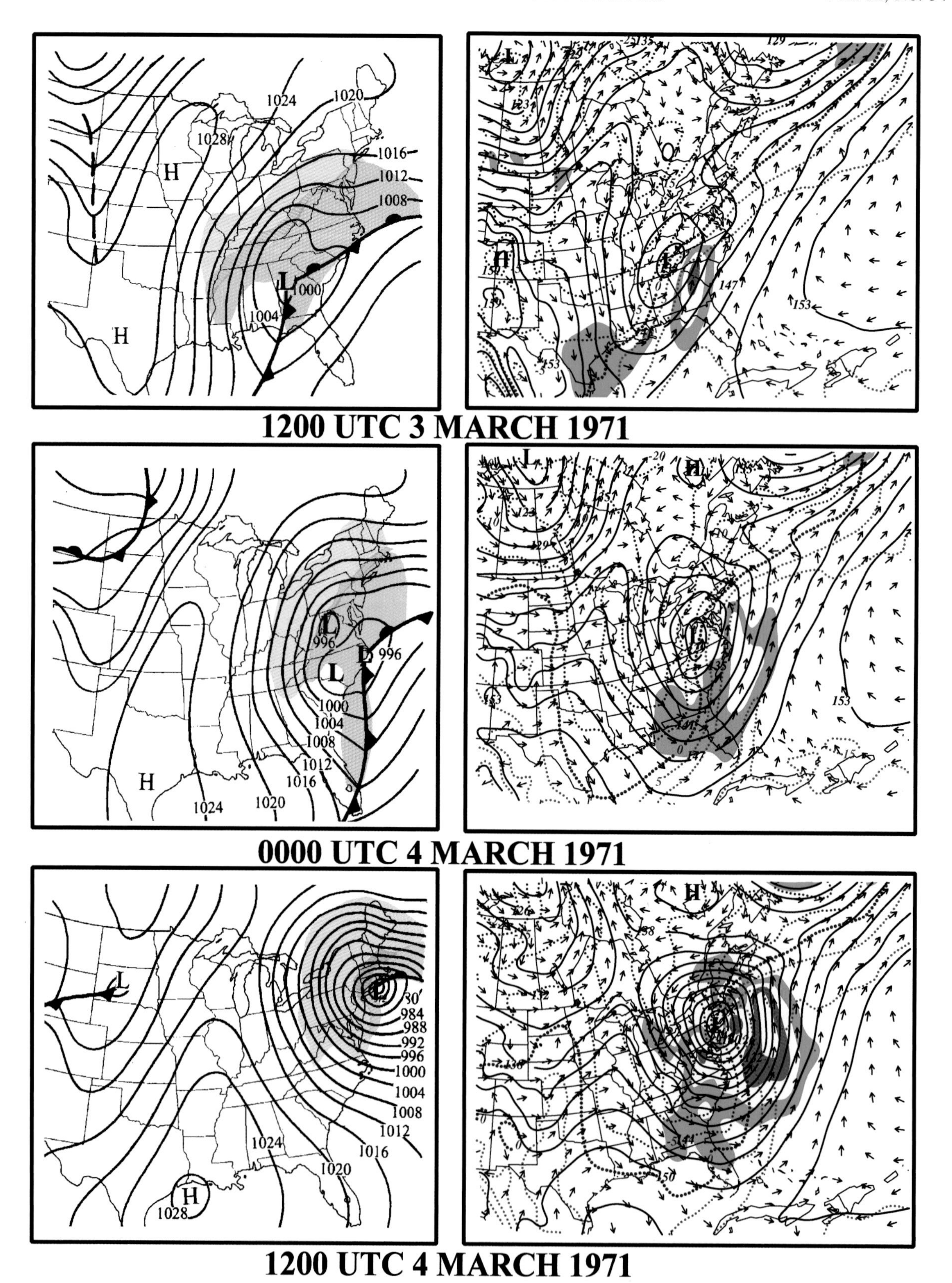

FIG. 11.1-23. Sequence of (left) surface and (right) 850-hPa charts for the 24-h period between 1200 UTC 3 and 4 Mar 1971. See Fig. 11.1-2 for details.

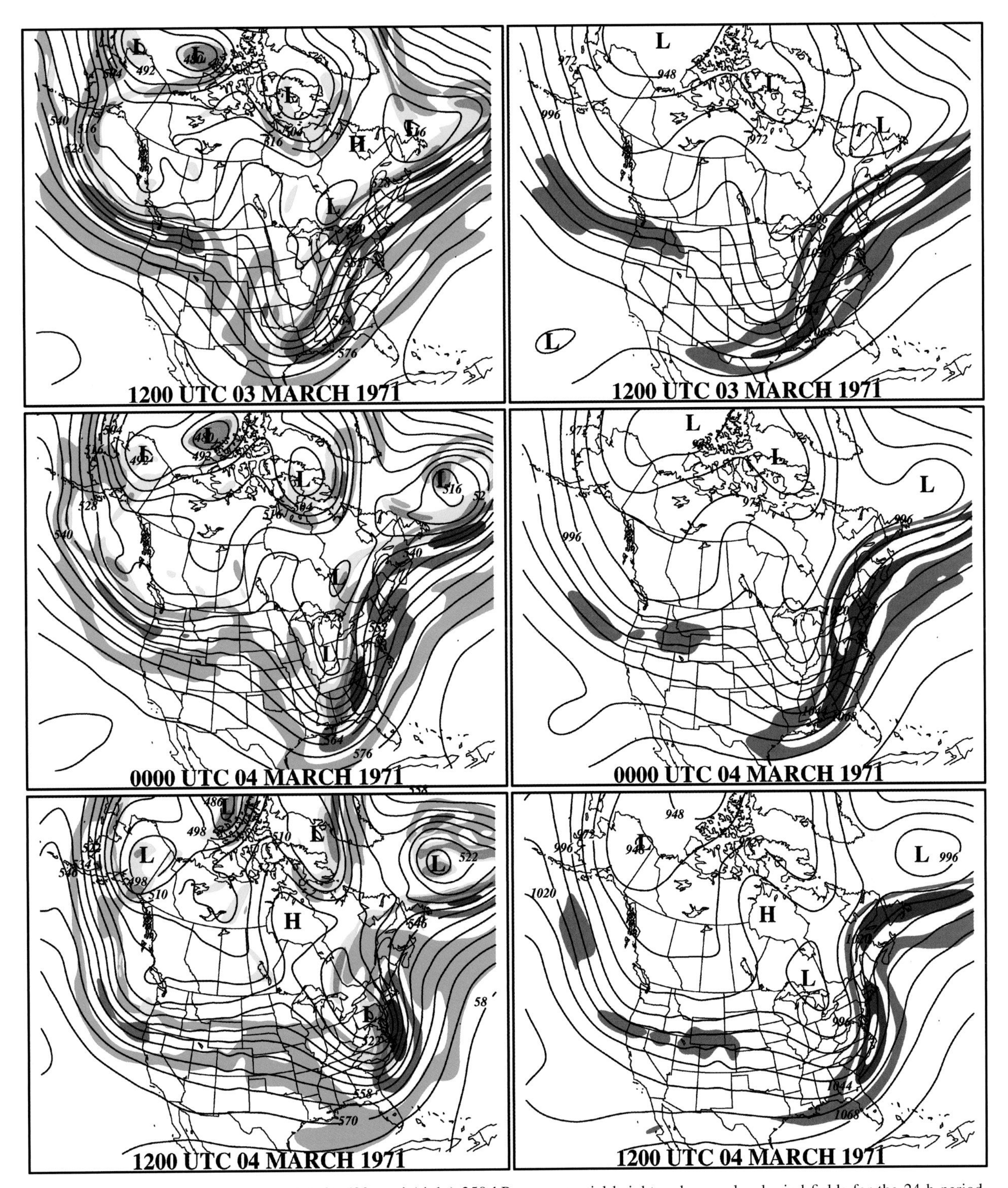

FIG. 11.1-24. Sequence of analyses of (left) 500- and (right) 250-hPa geopotential height and upper-level wind fields for the 24-h period between 1200 UTC 3 and 4 Mar 1971. See Fig. 11.1-3 for details.

spring and winter. The development of a slow, cutoff weather system extending over the eastern half of the country is more typical of spring. The large-scale pattern also resembles one conducive to major snowstorms. A slow-moving cutoff low was located over eastern Canada during a negative phase of the NAO and a strong confluent jet flow was evident in the precyclogenetic period over the Great Lakes (Fig. 11.1-33), providing

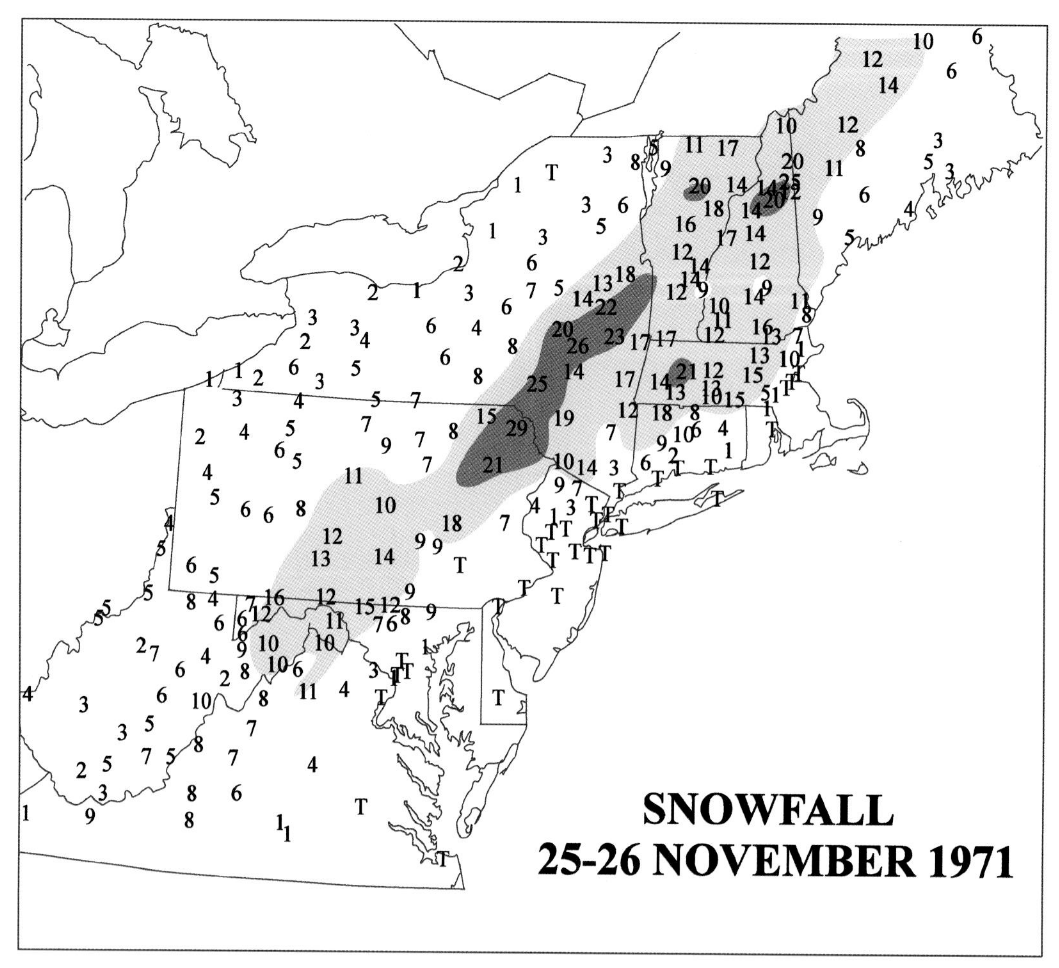

FIG. 11.1-25. Snowfall (in.) for 25–26 Nov 1971. See Fig. 11.1-1 for details.

support for cold high pressure to the north of New England. A vigorous upper trough located over the central United States marked by a well-defined jet streak and diffluent exit region was located over the southern United States and provided ample support for the developing low pressure system over the southeast. The baroclinic support for the storm was reinforced by very warm air across the southern United States [the temperature peaked at 106°F (41°C) at Brownsville, Texas, on 27 March—an early season heat record], a factor that supported the outbreak of severe weather in the southeast.

As the southern trough amplified and developed a neutral to negative tilt at 0000 UTC 29 March, and the exit region of the southern jet approached the East Coast, the surface low took an inland path from the eastern Carolinas to the Delmarva Peninsula, allowing a tremendous fetch of easterly flow into the Northeast urban corridor. This track, combined with a surface high pressure system receding well into Canada, and the long easterly fetch north of the storm center, produced more of a rain situation for the coastal domain with the heavier snow falling farther inland. The surface low then tracked eastward over the Atlantic, rather than up along the coastline, allowing snow to fall across southern New England and changing the rain back to snow in the middle Atlantic states as the storm passed out to sea.

l. 1–2 January 1987: NESIS = 2.26

During the stormy winter of 1986/87, a winter marked by moderate El Niño conditions (Fig. 2-19, volume I), a negative period of the North Atlantic Oscillation (Table 2-5, volume I), and several major urban corridor snowstorms (see chapters 10.21, 10.22, and 10.23), the New Year was greeted by a major coastal storm that left rain along the immediate coast from Virginia to Long Island and a mix of rain and snow for much of the urban corridor. Significant snowfall occurred across interior northern New Jersey to southeastern New England (Fig. 11.1-34). Farther inland, heavy snows of 10–20 in. (25–50 cm) fell from the highlands of West Virginia into central Pennsylvania, much of interior New York, and eastern New England.

This storm was associated with an intense surface cyclone that tracked immediately along the coastline from Georgia to North Carolina on 1 January before

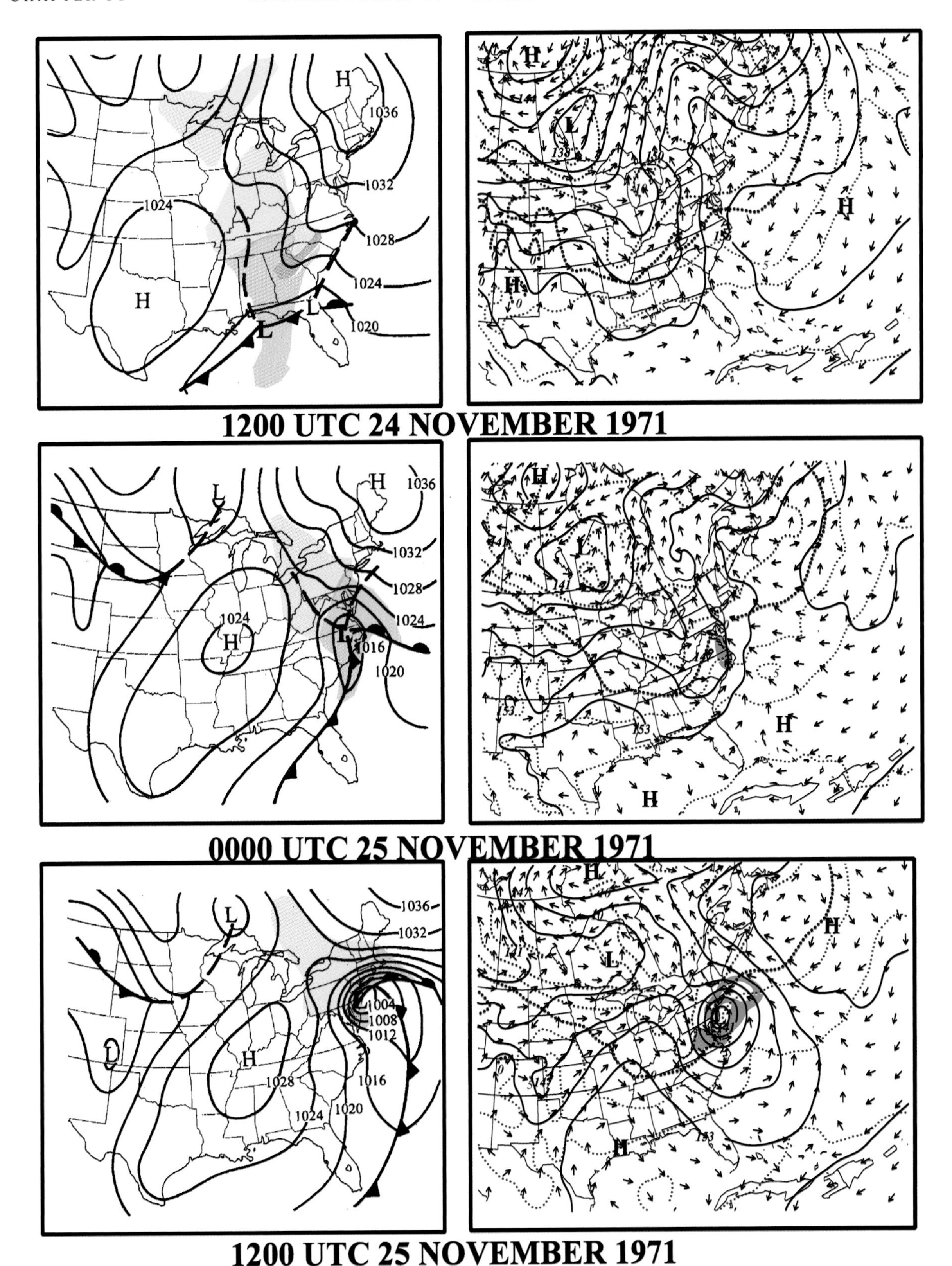

FIG. 11.1-26. Sequence of (left) surface and (right) 850-hPa charts for the 24-h period between 1200 UTC 24 and 25 Nov 1971. See Fig. 11.1-2 for details.

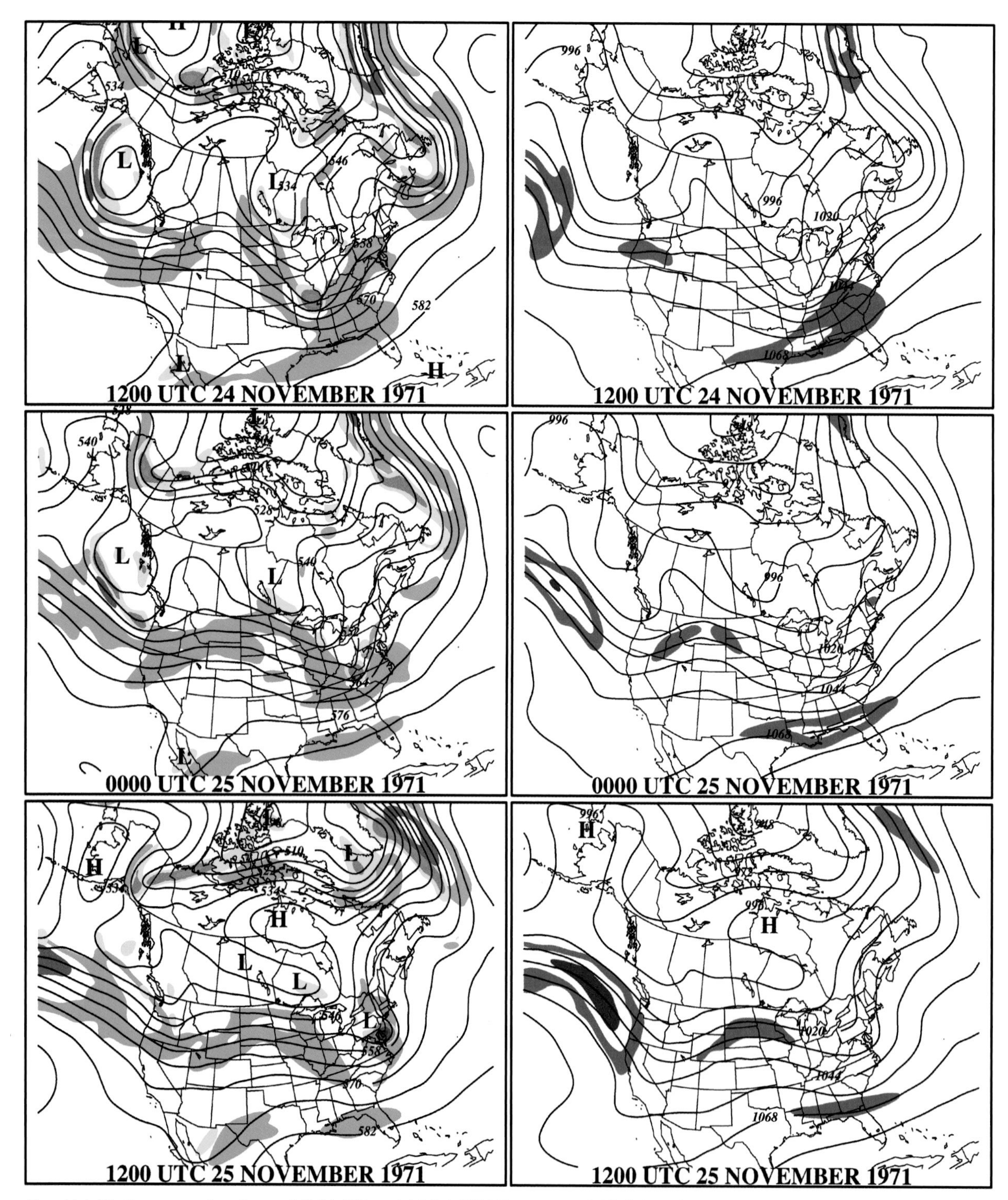

FIG. 11.1-27. Sequence of analyses of (left) 500- and (right) 250-hPa geopotential height and upper-level wind fields for the 24-h period between 1200 UTC 24 and 25 Nov 1971. See Fig. 11.1-3 for details.

moving east-northeastward on 2 January (Fig. 11.1-35). A cold surface high pressure system located over Maine at 1200 UTC 1 January quickly moved east as the intensifying surface low moved up the coast. The intensifying easterly flow extended up to 850 hPa (Fig. 11.1-35) as the storm intensified along the Carolina coast, providing enough warm air in the lower troposphere to ensure a changeover to rain from Washington to southern New England.

The upper levels (Fig. 11.1-36) were marked by a

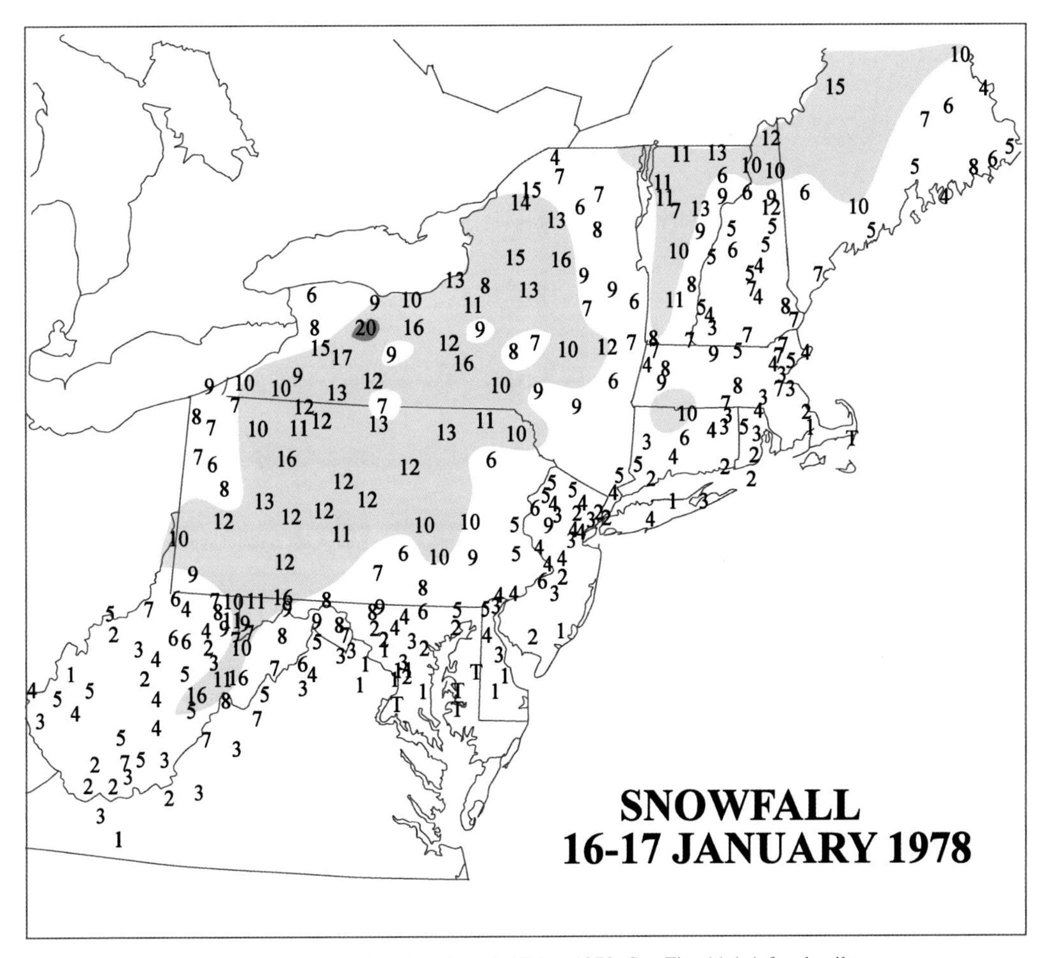

FIG. 11.1-28. Snowfall (in.) for 16–17 Jan 1978. See Fig. 11.1-1 for details.

confluent flow pattern over the Great Lakes in the precyclogenetic period that quickly receded northeastward to southeastern Canada as the storm developed along the East Coast. The lack of upper-level confluence over the northeastern United States and southeastern Canada during the evolution of the storm allowed the high pressure originally over Miami to drift east-northeastward and weaken. The dominant feature at 500 hPa in this case was the amplifying trough over the central United States that closed off well west of the Appalachian Mountains by 0000 UTC 2 January. The combination of a jet streak near the base of the southern trough on 2 January, with a well-defined, diffluent exit region, provide support for the rapid development of the surface low immediately along the coastline. The strong ridging developing along the Northeast coast ahead of the trough and the northeastward displacement of the confluence over southeastern Canada supported a strong easterly fetch out ahead of the storm system allowing warmer air to penetrate the immediate coastline. The net result was rain, rather than snow, in the urban corridor as the storm moved northeastward along the coastline.

m. 11–12 December 1992: NESIS = 3.10

Prior to the 12–14 March 1993 "Superstorm" (see chapter 10.24), this early season storm made headlines as the "Storm of the Century" in the local media since it brought severe coastal flooding and hurricane force wind gusts from Massachusetts to New Jersey. In Massachusetts, coastal flooding was not as severe as during the Blizzard of 1978 (chapter 10.17) or during the "Halloween Gale" or "Perfect Storm" of 1991, but coastal flooding in New York City and New Jersey led to presidential disaster declarations. The storm also produced some exceptionally heavy snowfall amounts of up to 4 ft (120 cm) across elevated portions of interior New York and the Berkshires in Massachusetts (Fig. 11.1-37), with the heavy, wet snow leading to power outages for thousands of people in those areas. In addition, heavy wet snow fell across western Pennsylvania, Maryland, northeastern West Virginia, and the Blue Ridge Mountains of Virginia, with accumulations in excess of 40 in. (100 cm) in the highlands of southwestern Pennsylvania and western Maryland, setting records that would be broken in some areas by the March 1993

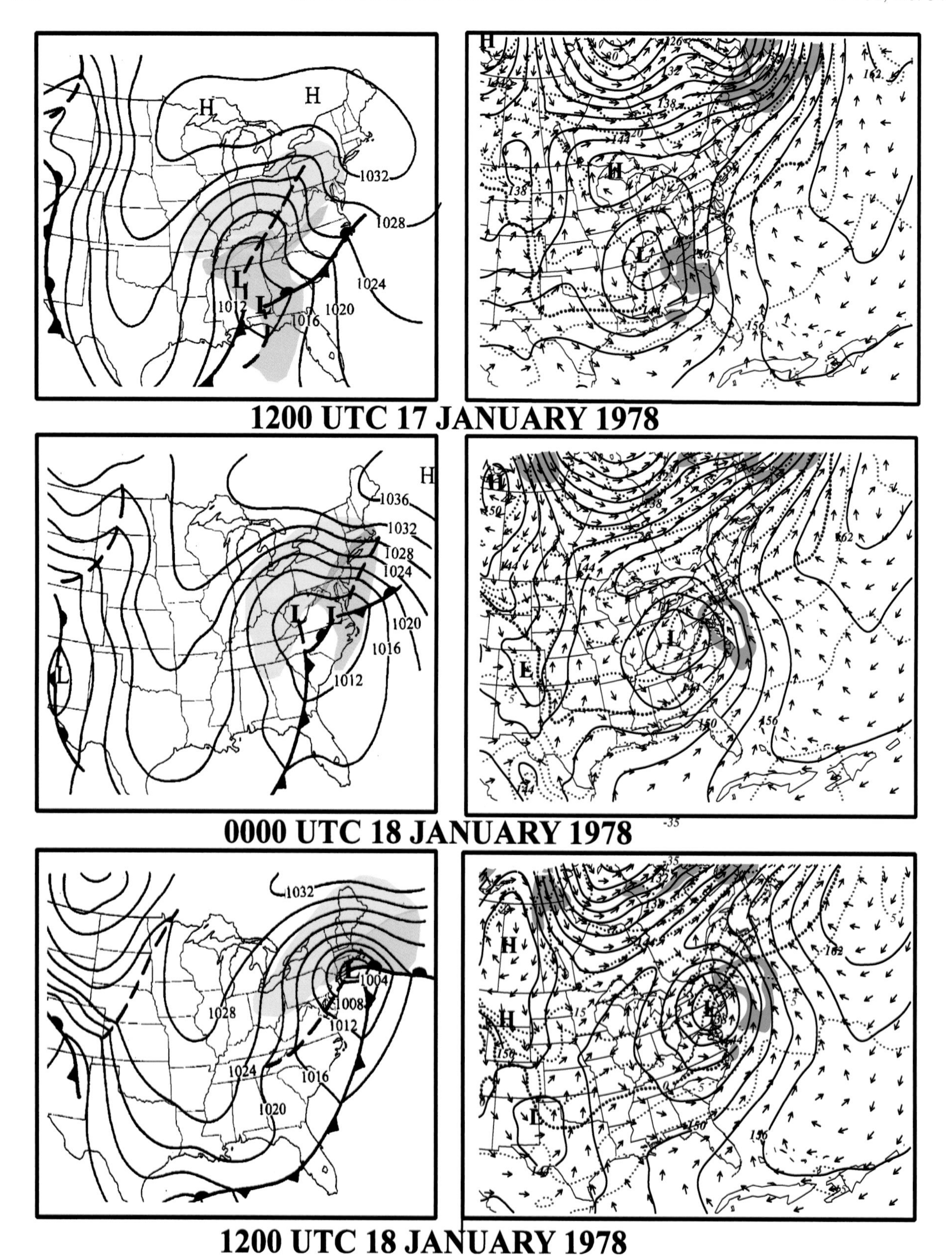

FIG. 11.1-29. Sequence of (left) surface and (right) 850-hPa charts for the 24-h period between 1200 UTC 17 and 18 Jan 1978. See Fig. 11.1-2 for details.

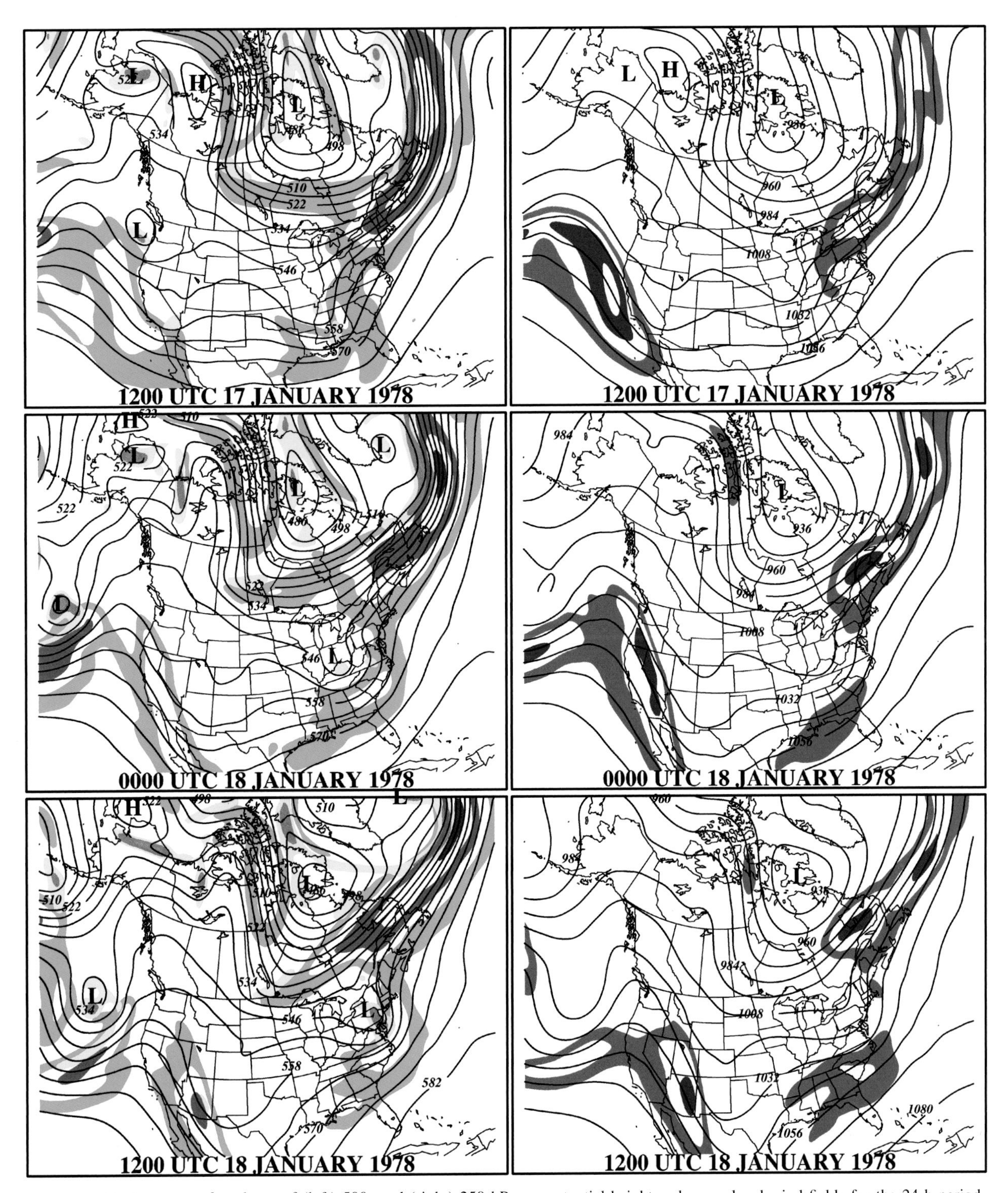

FIG. 11.1-30. Sequence of analyses of (left) 500- and (right) 250-hPa geopotential height and upper-level wind fields for the 24-h period between 1200 UTC 17 and 18 Jan 1978. See Fig. 11.1-3 for details.

Superstorm (see chapter 10.24). In the heavily populated urban corridor, precipitation began as snow in the Washington–Baltimore corridor, which quickly changed to rain, and began as rain in the major cities farther northeastward. However, as the storm moved slowly northeastward over the western Atlantic, colder air began to funnel into southern New England and New York, changing the rain back to snow in Boston. Nearly 10

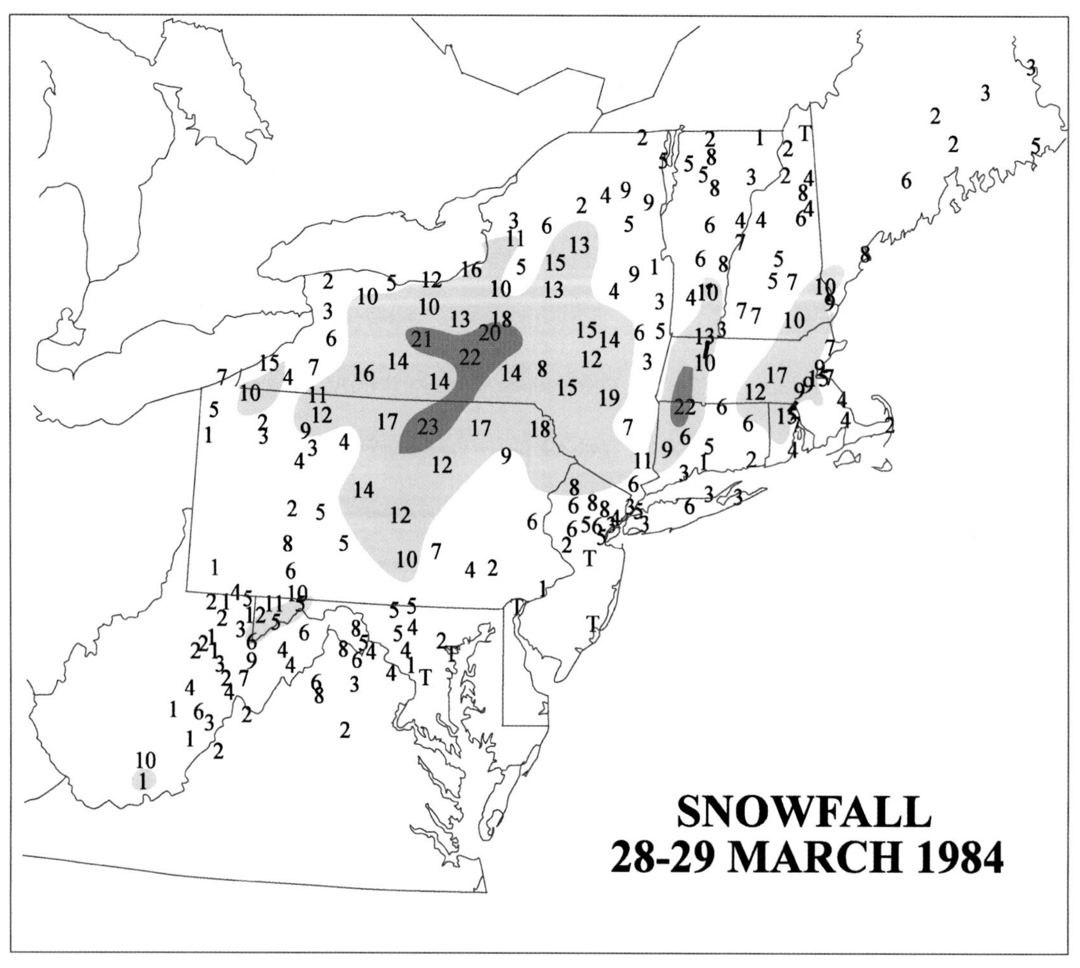

FIG. 11.1-31. Snowfall (in.) for 28–29 Mar 1984. See Fig. 11.1-1 for details.

in. (25 cm) of snow fell in the Boston metropolitan area, but much more snow fell in the western suburbs, including more than 30 in. (75 cm) at Worcester, Massachusetts. Up to 10 in. (25 cm) of snow fell in the northern suburbs of New York City.

The precyclogenetic environment was marked by a strong cold high pressure system centered over Maine at 1200 UTC 10 December (Fig. 11.1-38). The storm evolved from a primary low pressure system moving southeastward from the northern Midwest and a slow-moving secondary low pressure system that tracked northward along an inverted trough, just inland of the coastline. By 1200 UTC 11 December, an intense cyclone was located over the Delmarva Peninsula, with a strong easterly fetch north of the storm center marked by hurricane force surface winds. The strong and expanding cyclonic circulation extended up beyond the 850-hPa level, where an easterly low-level jet north of the intensifying low exceeded 40+ m s^{-1} by 1200 UTC 12 December (Fig. 11.1-38).

This case was marked by a highly amplified trough (and well-defined jet streak) that became negatively tilted in the Southeast, without the presence of an upper confluent pattern and 500-hPa cutoff low over eastern Canada (Fig. 11.1-39). The amplification of the southern trough and ridging extending northward along the Atlantic coast to central New England was part of a well-defined blocking pattern that slowed the forward movement of the storm system and maintained the strong cell of high pressure north of New England. These factors generated the persistent, strong easterly flow along the Northeast coast that not only produced the excessive flooding but also favored rainfall rather than snow in the Northeast urban corridor, especially as the center of the surface high drifted to the northeast of Maine. As the cutoff low passed south of New England, temperatures fell aloft and rain changed back to snow throughout southern New England and southeastern New York.

n. 3–4 January 1994: NESIS = 2.87

This multifaceted storm contained several interesting mesoscale phenomena (see volume I, chapter 6), including 1) a precursor surface low that produced moderate snow amounts over coastal New England; 2) some

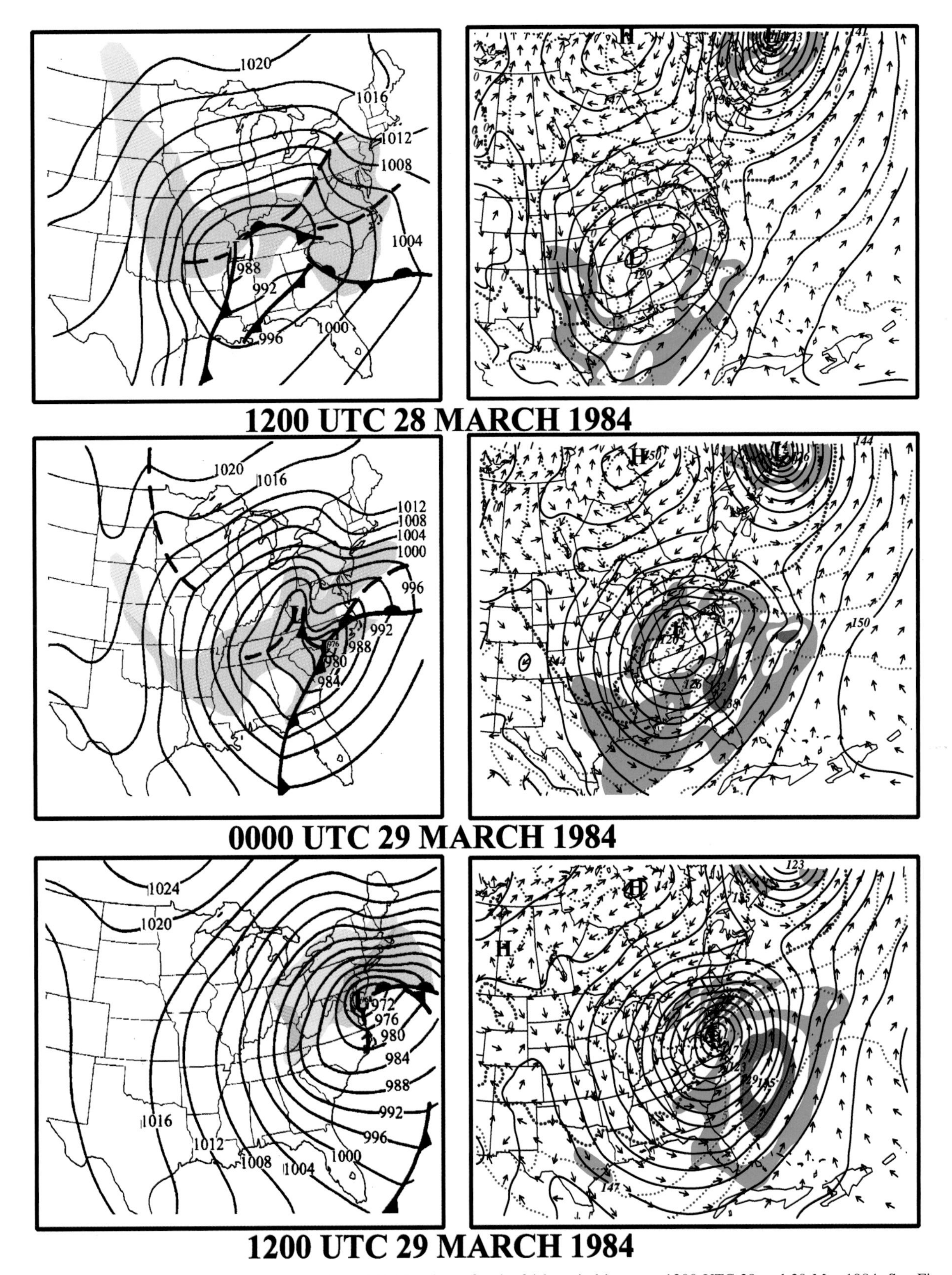

FIG. 11.1-32. Sequence of (left) surface and (right) 850-hPa charts for the 24-h period between 1200 UTC 28 and 29 Mar 1984. See Fig. 11.1-2 for details.

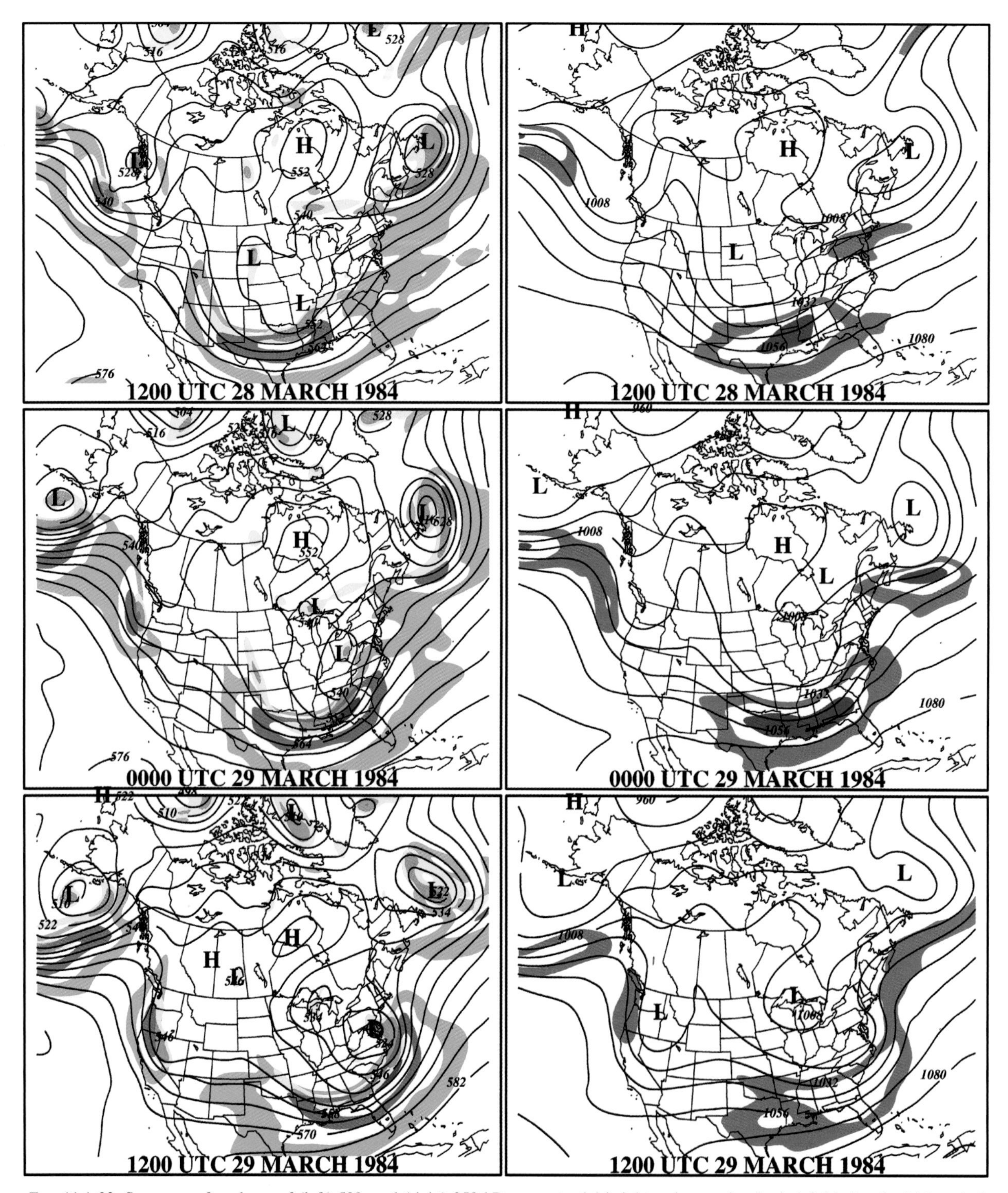

FIG. 11.1-33. Sequence of analyses of (left) 500- and (right) 250-hPa geopotential height and upper-level wind fields for the 24-h period between 1200 UTC 28 and 29 Mar 1984. See Fig. 11.1-3 for details.

of the greatest 1-hourly snowfall rates observed in recent years, with reports of 6-in. (15 cm) hourly snowfall rates at locations in West Virginia, western Pennsylvania, and western New York; and 3) a significant gravity wave event that affected the precipitation distribution and rates [see volume I, chapter 6; Bosart et al. (1998)].

The band of heaviest snows fell from West Virginia into western Pennsylvania and western New York with

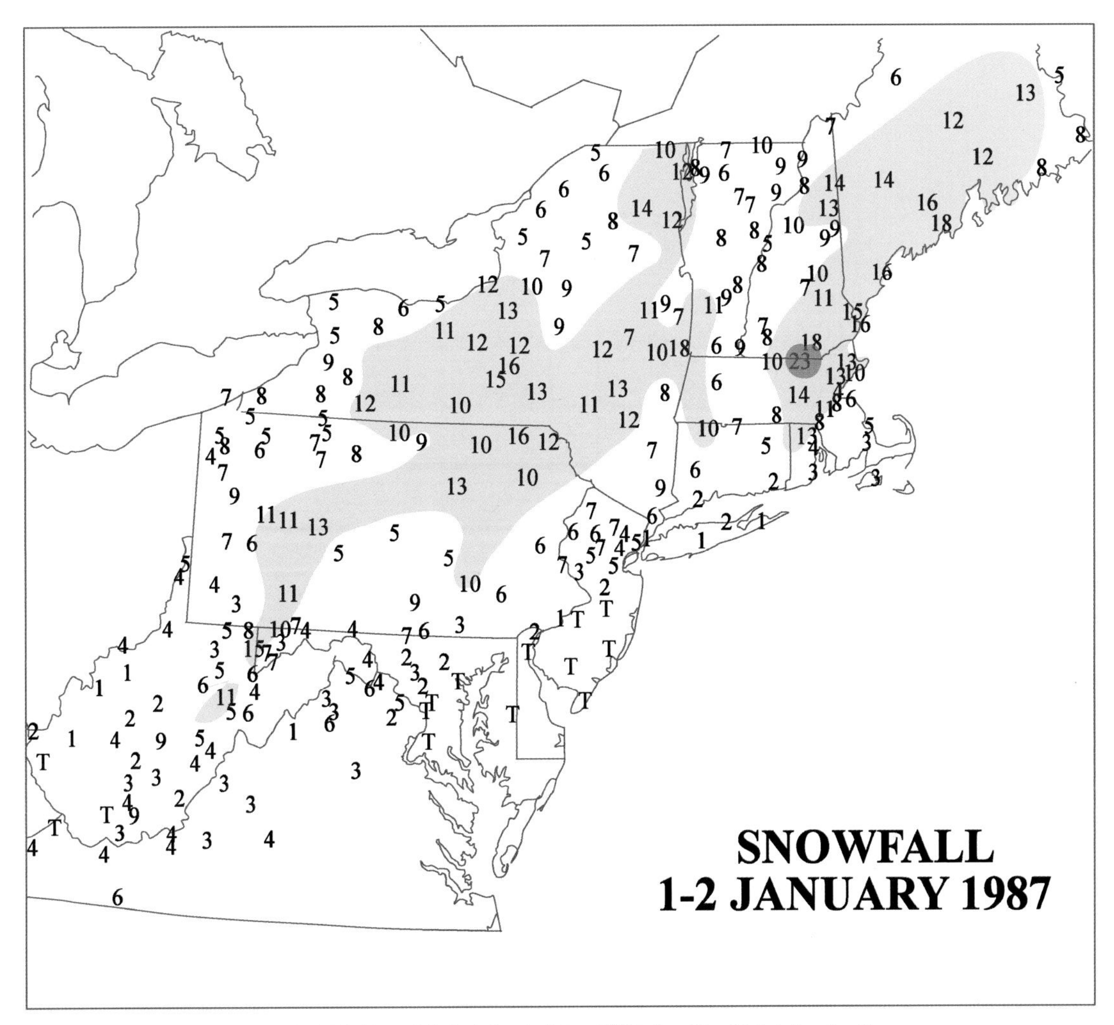

FIG. 11.1-34. Snowfall (in.) for 1–2 Jan 1987. See Fig. 11.1-1 for details.

a large area of 2-ft (60 cm) snows in southwestern Pennsylvania and 33 in. (83 cm) at Waynesboro, Pennsylvania (Fig. 11.1-40). As an example of the intense snowfall rates associated with the storm, Syracuse, New York, reported 10 in. (25 cm) of snow in only 2 h. The storm closed the Pennsylvania Turnpike, numerous roofs collapsed, and a state of emergency was declared throughout southwestern Pennsylvania. In West Virginia, the large snowfall rates contributed to the uprooting of thousands of trees because the heavy snows followed a significant rainfall. Up to 10 in. (25 cm) of snow fell in the Boston metropolitan area, but mostly rain fell from New York City to Washington.

While the storm was anything but typical, it did have some signatures typical of interior snowstorms, including the extension of two inverted troughs west and east of the Appalachian Mountains in the precyclogenetic period, as a strong, cold anticyclone over New England receded north and east (Fig. 11.1-41). The surface low developed over Virginia on 4 January, well west of the coast within a cyclonic circulation pattern that extended over the eastern third of the country. A deep upper trough extended to the Gulf coast and became negatively tilted south and west of the Northeast urban corridor, allowing strong upper ridging along the western Atlantic and East Coast (Fig. 11.1-42). As a result, a strong upper confluent flow over New England and southeastern Canada moved northeastward, as did a strong surface high pressure system initially located just north of New England, allowing milder oceanic-modified air to penetrate the immediate coastline. Therefore, the Northeast urban corridor experienced mostly rain in spite of the existence of a distinct confluent entrance region of an upper-level jet streak in northern New England and southeastern Canada at 1200 UTC 4 January (Fig. 11.1-42) and a notable dual-jet pattern that have marked many of the snowstorm cases in chapter 10.

o. 2–4 March 1994: NESIS = 3.46

During one of the stormiest winters since the late 1970s, one characterized by a weak El Niño (Fig. 2-19,

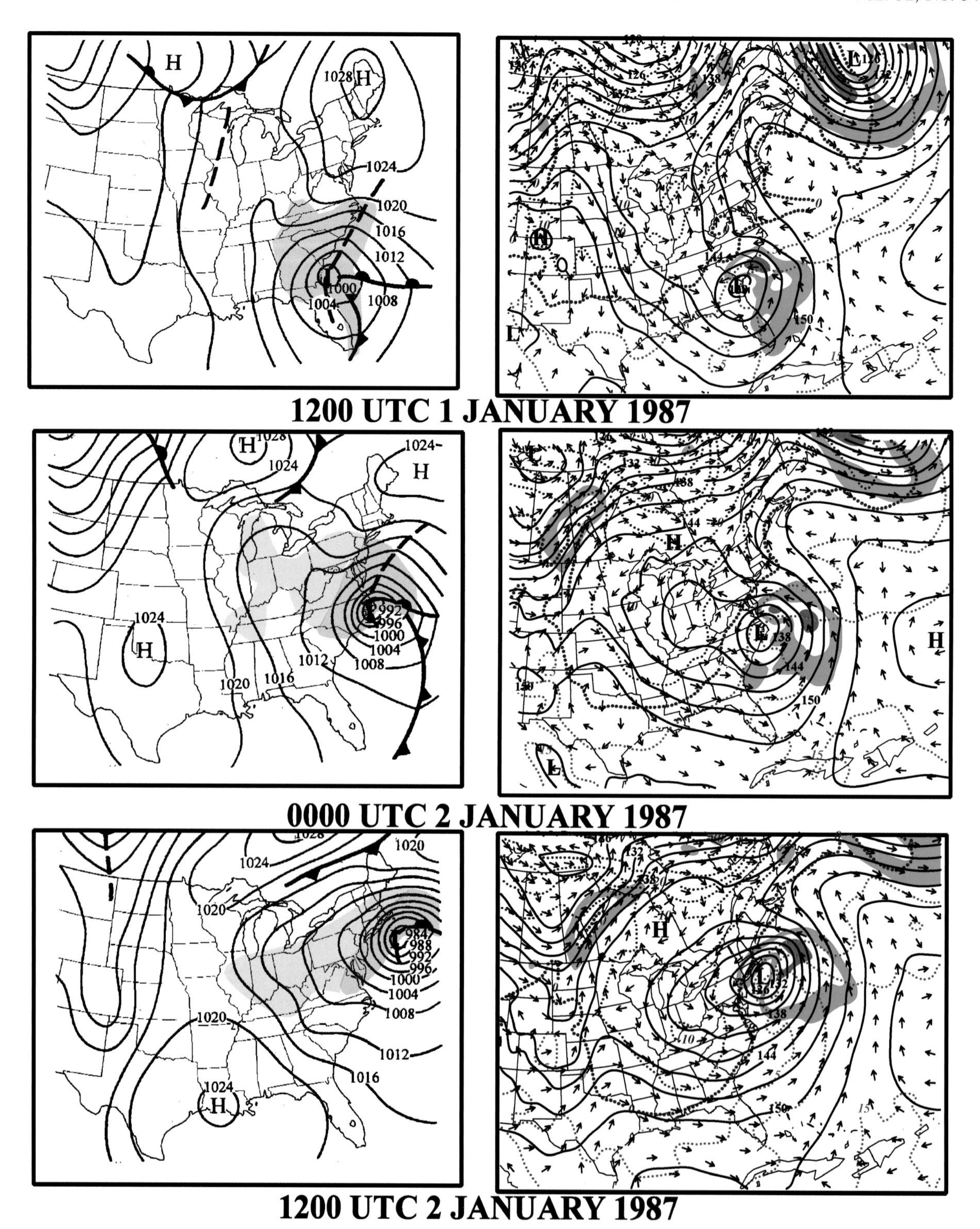

FIG. 11.1-35. Sequence of (left) surface and (right) 850-hPa charts for the 24-h period between 1200 UTC 1 and 2 Jan 1987. See Fig. 11.1-2 for details.

volume I) and positive phase of the NAO (see Table 2-5, volume I), this storm produced snow changing to rain or ice pellets across much of the Northeast urban corridor and a band of greater than 25-in. (60-cm) snow accumulations from Pennsylvania into central New York (Fig. 11.1-43). State College, Pennsylvania, recorded a new 24-h snowfall record with 26.1 in. (63 cm) and a storm total of 28 in. (68 cm). Many roads were closed

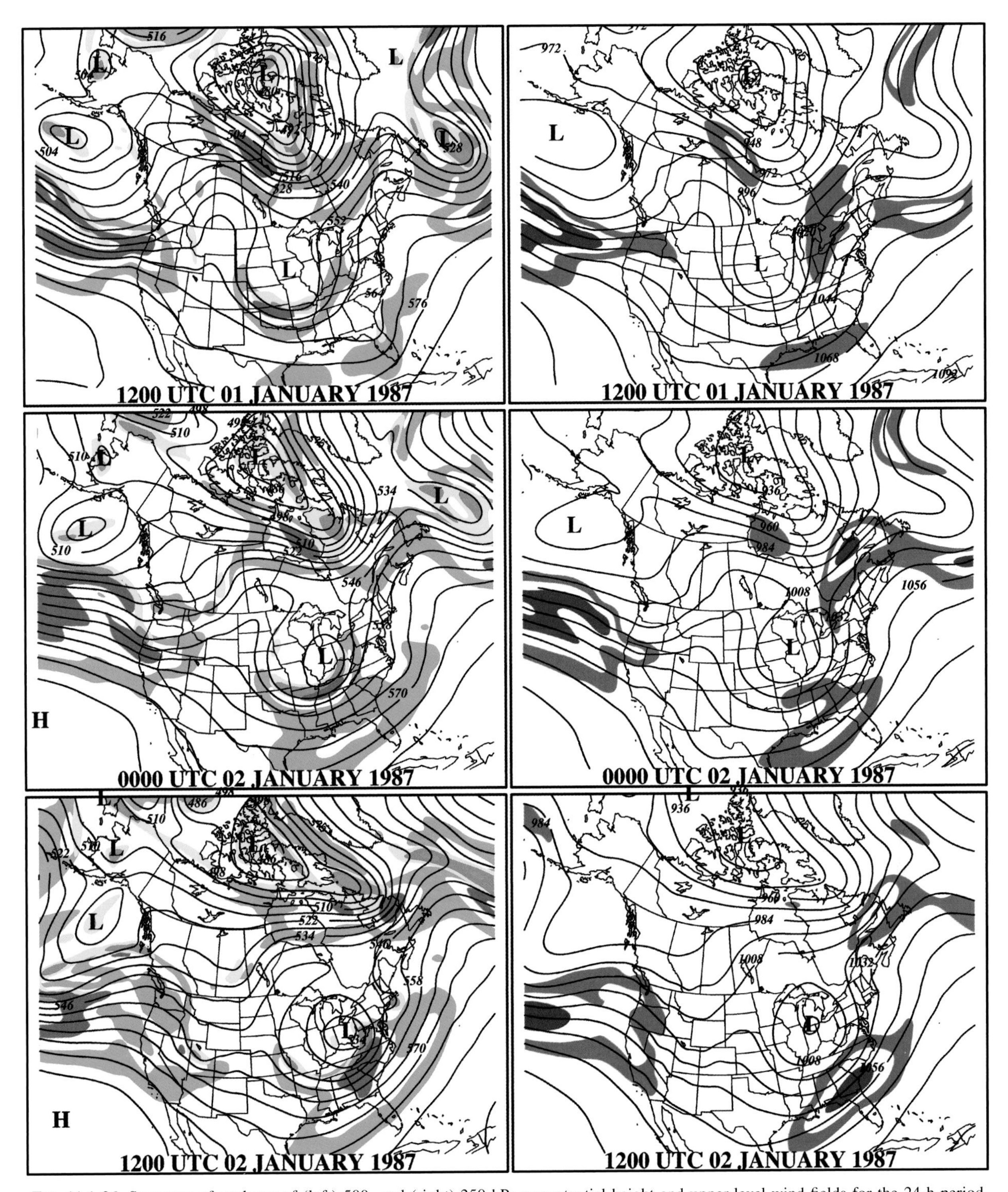

FIG. 11.1-36. Sequence of analyses of (left) 500- and (right) 250-hPa geopotential height and upper-level wind fields for the 24-h period between 1200 UTC 1 and 2 Jan 1987. See Fig. 11.1-3 for details.

in central Pennsylvania and snowdrifts stranded numerous motorists. Before snow changed to rain and/or sleet in the urban corridor, 3–6 in. (8–16 cm) of snow fell from Washington to New York City with up to 8 in. (20 cm) recorded in Boston. Gale force winds produced coastal flooding from Virginia northward to southern New England.

This storm was another typical interior snowstorm, with the surface low moving from the northern Gulf of Mexico and following an inland path across eastern

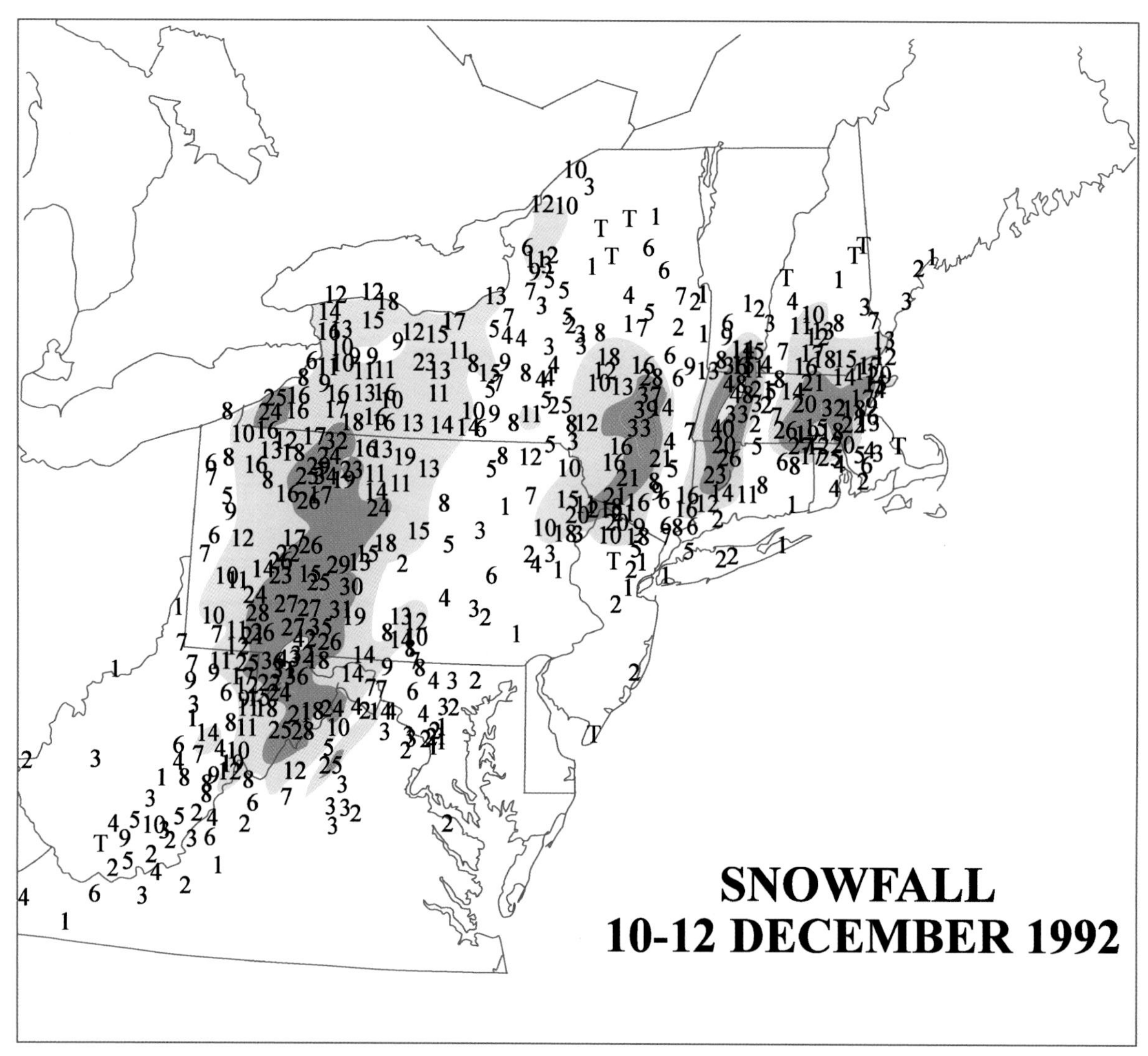

FIG. 11.1-37. Snowfall (in.) for 10–12 Dec 1992. See Fig. 11.1-1 for details.

North Carolina toward the Delmarva Peninsula (Fig. 11.1-44). On 3 March, rapid cyclogenesis commenced along the Virginia coastline as a cold anticyclone (originally located over New England) shifted rapidly eastward. The intense pressure gradient between this high and the rapidly developing cyclone produced a long easterly fetch directed toward the New Jersey and New England coastlines, with strong winds producing coastal flooding. The strong circulation is also reflected by the rapid deepening 850-hPa low, with greater than 30 m s^{-1} easterly winds developing north of the circulation center at 0000 UTC 2 March, directed toward the area of heavy rain along the coast and heavy snowfall in the inland areas of Pennsylvania and New York.

This storm system, like other interior storms, was associated with a deep trough, first amplifying into the south-central United States on 2 March and then lifting northeastward as it neared the East Coast, while the upper-level ridge amplified along the Atlantic coast toward New England (Fig. 11.1-45). The confluent 500-hPa trough initially located over southeastern Canada moved rapidly eastward as the ridge built northward toward New England allowing the surface high pressure and upper-level confluence to drift northeastward over the Canadian Maritime Provinces during the evolution of the storm. Thus, despite a well-defined double-jet pattern, the shifting of the confluence zone toward the north and east depleted the support for a cold surface high to lock in over the northeast. This, combined with the southern trough digging too far south and too far to the west of the coast, forced the surface low to track to the west of and immediately along the East Coast. This track plus the strong east and southeasterly flow warmed the lower troposphere throughout the urban corridor, changing the snow over to heavy rain.

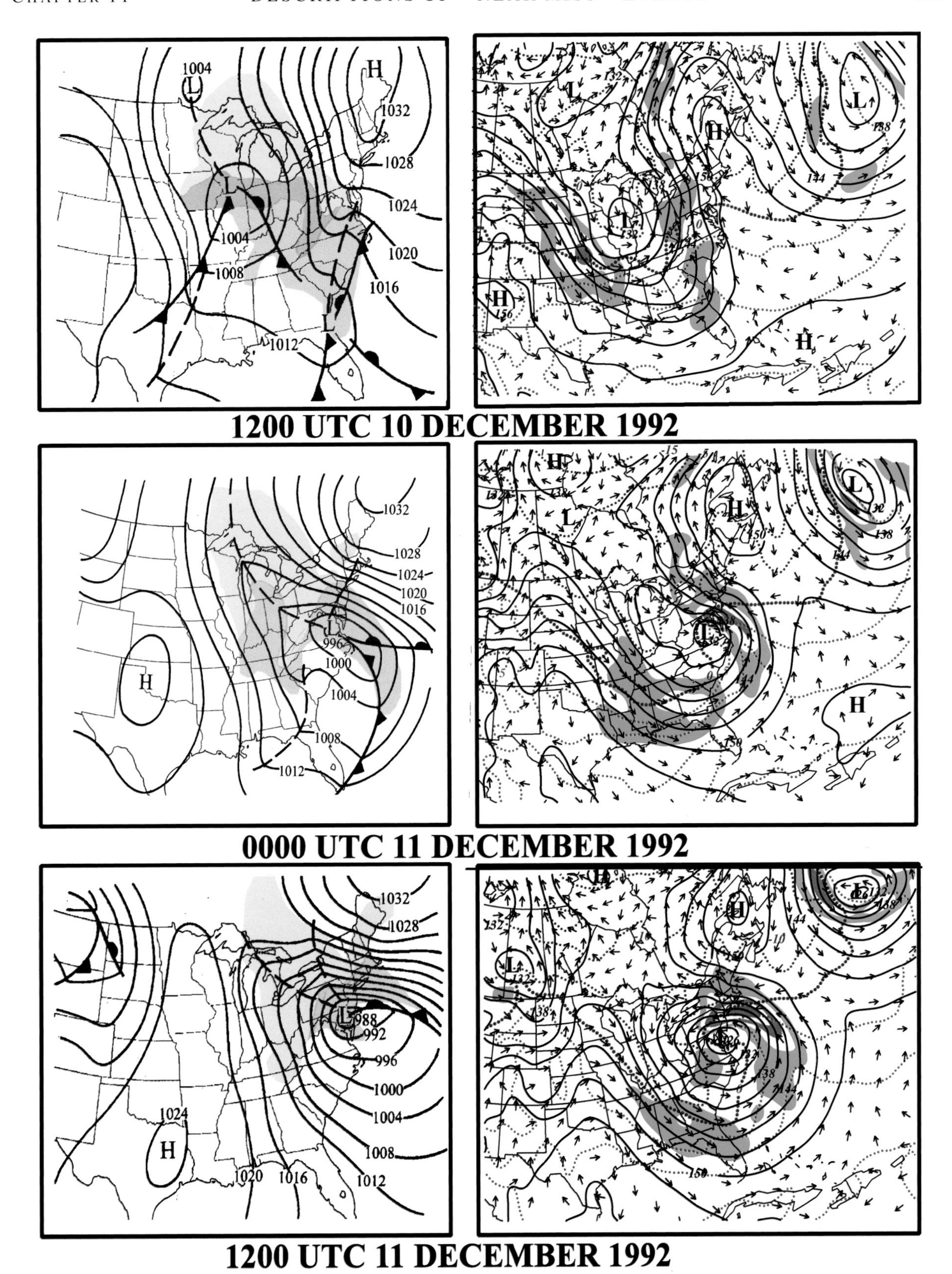

FIG. 11.1-38. Sequence of (left) surface and (right) 850-hPa charts for the 24-h period between 1200 UTC 10 and 11 Dec 1992. See Fig. 11.1-2 for details.

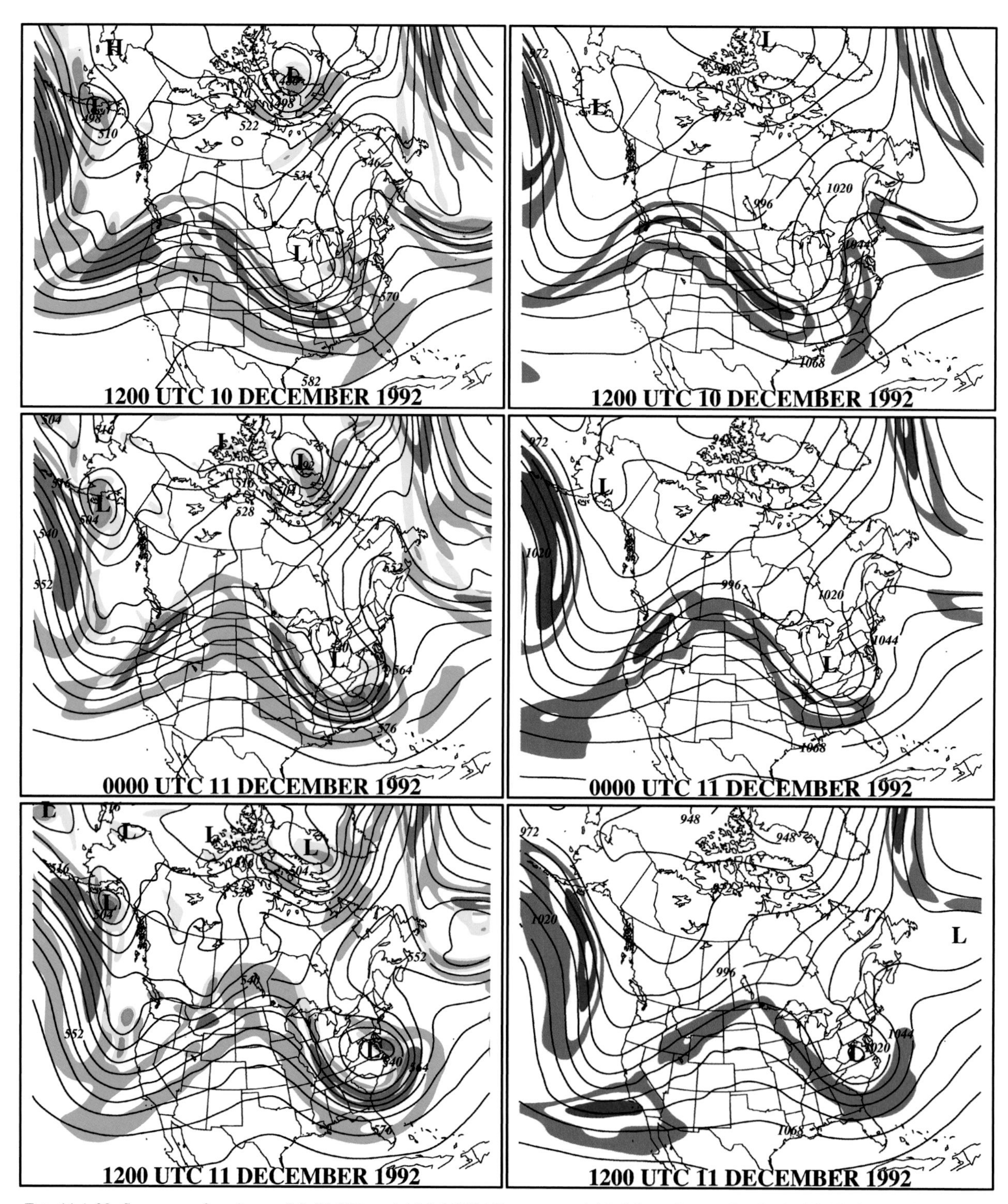

FIG. 11.1-39. Sequence of analyses of (left) 500- and (right) 250-hPa geopotential height and upper-level wind fields for the 24-h period between 1200 UTC 10 and 11 Dec 1992. See Fig. 11.1-3 for details.

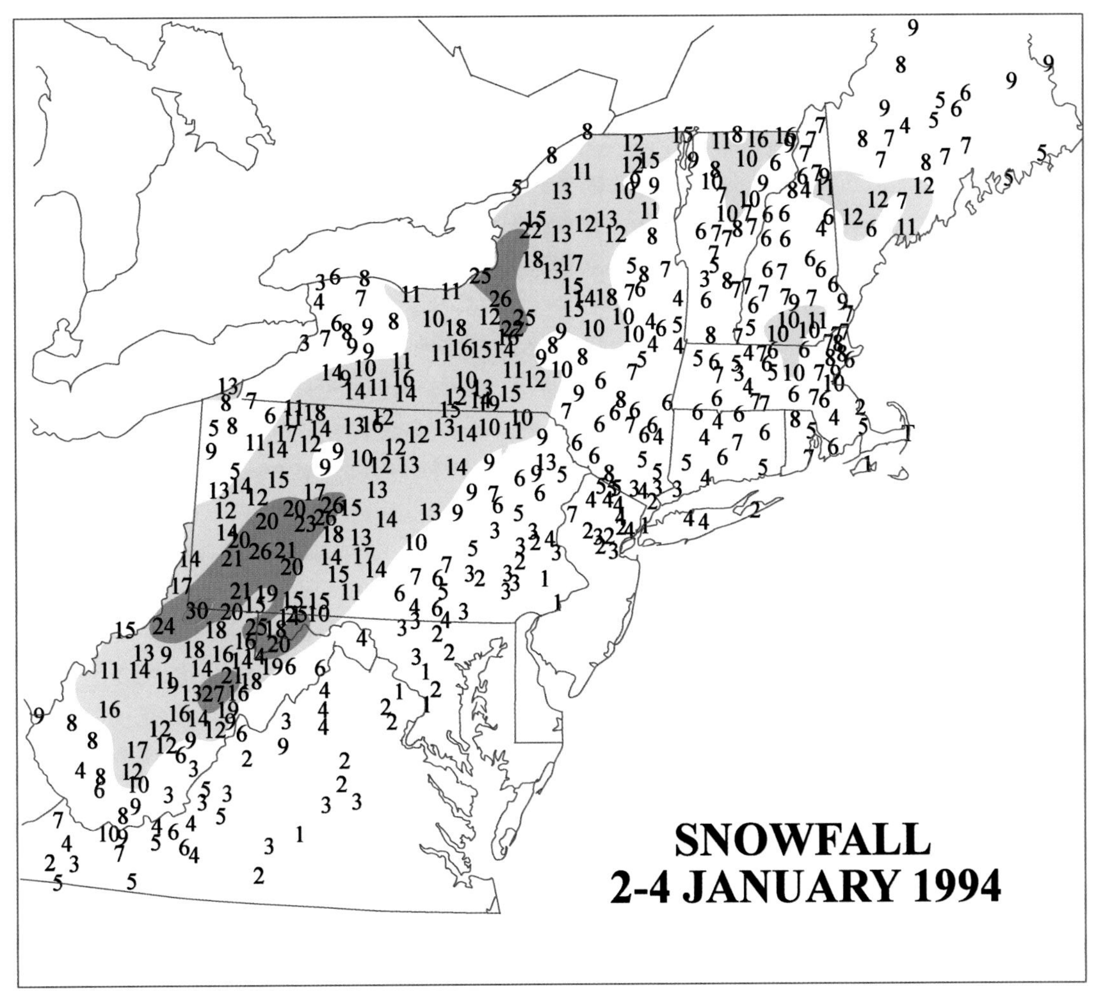

FIG. 11.1-40. Snowfall (in.) for 2–4 Jan 1994. See Fig. 11.1-1 for details.

2. Moderate snowstorms

Fifteen "moderate" snowstorms are described in this section. These are snowstorms marked by snowfall amounts that average 4–10 in. (10–25 cm) and contain relatively small areas of heavier snowfall [10–20 in. (25–50 cm)] when compared to the heavy snow cases in chapter 10. The NESIS values [see volume I, chapter 8.4; also Kocin and Uccellini (2004c)] for these storms range between 1.2 and 4.85, with an average around 2.2. The majority of storms score between 1 and 2 with the larger values typically including snowfall outside the Northeast. These values are much smaller than those of the major snowstorms in chapter 10 and smaller than most interior snowstorms as well, reflecting both smaller areas of heavier snowfall and the smaller populations affected by only moderate to heavy snowfall.

The surface cyclones in this sample tend to be much weaker than their heavy snow counterparts in chapter 10 and are associated with upper-level troughs with smaller amplitudes (see Fig. 5-19, volume I). However, the consistent appearance of upper-level confluence, with its associated upper-level jet streaks and the surface anticyclone in the northern United States and southeastern Canada, is similar to the 32 heavy snow cases (chapter 10), although some of the surface anticyclones have lower pressures than their heavy snow counterparts. In any event, the direct circulation in the entrance region of these confluent jet systems seems to provide the support for these moderate snow events. The lack of an amplifying upper-level trough in the south-central United States appears to be an important factor in limiting the potential of several of these snowstorms to grow into major snow producers over large areas. In a number of cases, there may be multiple upper troughs that fail to merge or phase together, resulting in either separate moderate snow events or one event that fails to develop heavy snowfall rates. In some cases, the faster speeds of these weather systems were another reason for limited snowfall amounts, while in other cases a mixture of freezing rain and sleet acted to limit the final snowfall accumulations. The net effect is to either limit the areal extent of these snowstorms, the maximum snow that is produced, or both.

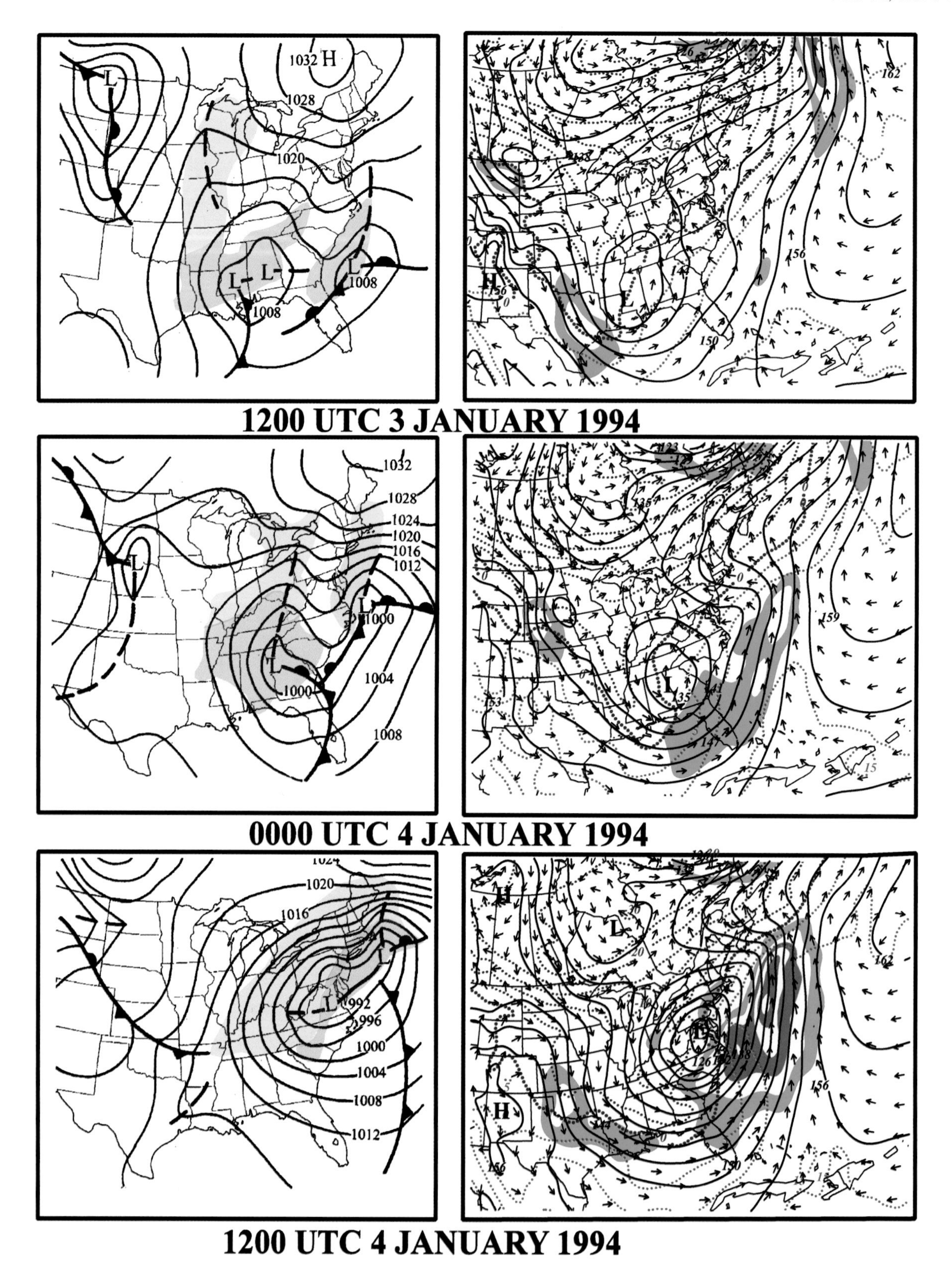

FIG. 11.1-41. Sequence of (left) surface and (right) 850-hPa charts for the 24-h period between 1200 UTC 3 and 4 Jan 1994. See Fig. 11.1-2 for details.

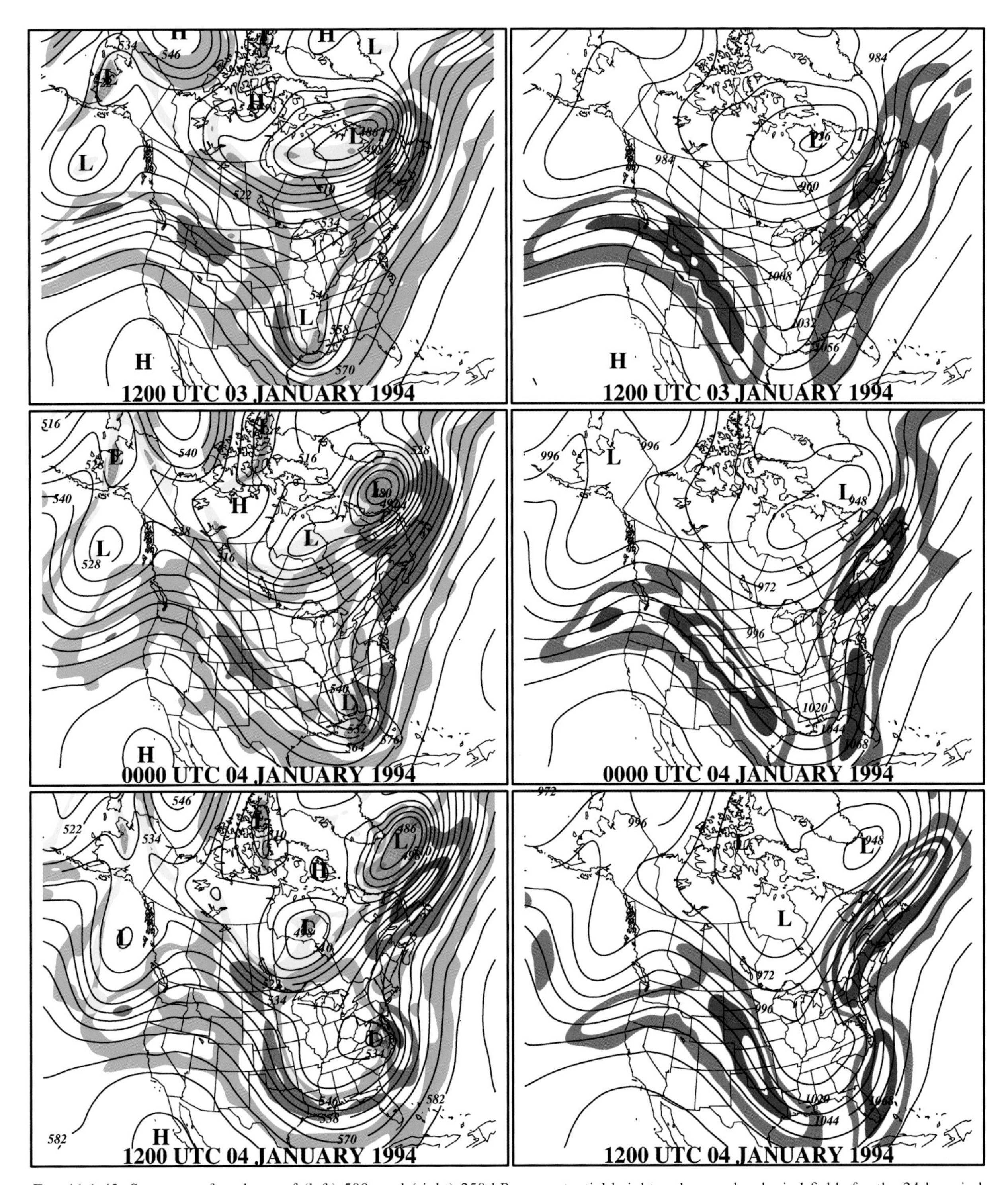

FIG. 11.1-42. Sequence of analyses of (left) 500- and (right) 250-hPa geopotential height and upper-level wind fields for the 24-h period between 1200 UTC 3 and 4 Jan 1994. See Fig. 11.1-3 for details.

Another interesting aspect of many of these cases is that the velocity of some of the wind maxima of the jet streaks involved with these storms appear to be less than those of the heavy snow cases. The along-stream wind variation in the exit region of the southern jet and/or the entrance region of the northern jet is also less distinct. The lack of well-defined exit and entrance regions acts to diminish the vertical ascent patterns associated

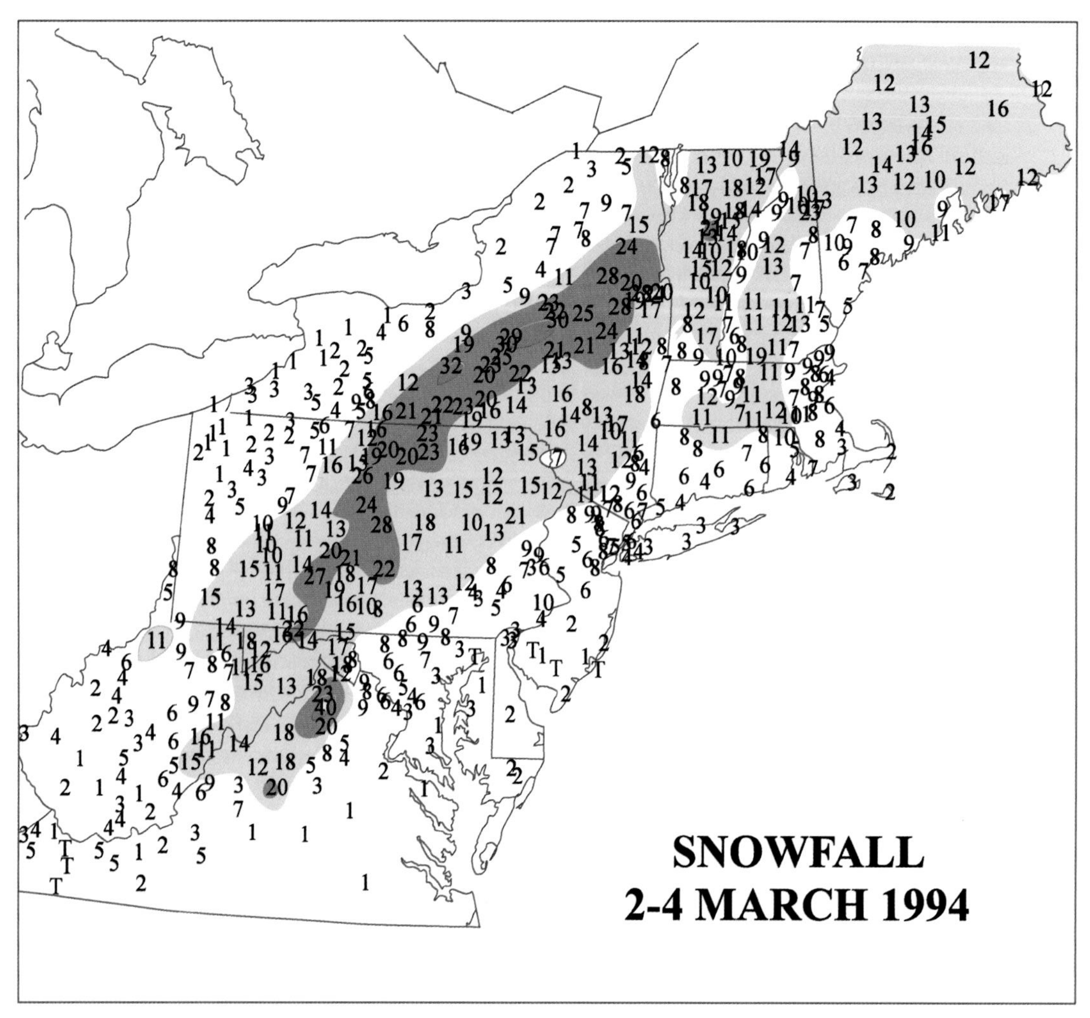

FIG. 11.1-43. Snowfall (in.) for 2–4 Jan 1994. See Fig. 11.1-1 for details.

with the jet streaks and is also a likely factor in preventing these storm systems from not fully developing until they pass well east of the populated Northeast urban corridor.

a. 4–5 December 1957: NESIS = 1.32

The first significant snowfall of the snowy winter of 1957/58, characterized by a strong El Niño and a negative phase of the North Atlantic Oscillation (NAO) was an early season snowstorm that produced mostly moderate snows of 6–10 in. (15–25 cm) (Fig. 11.2-1). It also produced a narrow band of snowfall accumulations of 10–15 in. (25–38 cm) across the Northeast urban corridor from northern Virginia and Washington, to the New York City metropolitan area. Up to 14 in. (35 cm) of snow fell in Loudoun County in northern Virginia. In Pennsylvania, snow drifted to 6 feet (180 cm) in places and a dozen deaths were attributed to overexertion. In north-central Maryland, the snow was described as the heaviest in more than a decade and thousands of motorists were stranded on blocked highways. In Baltimore, telephone lines were reported to be the busiest on record as hotels filled with stranded travelers.

The snowstorm of 4–5 December 1957 may be considered a smaller version of the major snowstorms highlighted in chapters 3 and 4 in volume I, and chapter 10. The precyclogenetic environment was marked by a cold high pressure ridge extending eastward from the Great Lakes (Fig. 11.2-2). A primary low moved eastward into Kentucky on 4 December, with a secondary cyclone developing off the Virginia coast, supported by the concurrent rapid deepening and expansion of the 850-hPa low off the Delmarva Peninsula between 1200 UTC 4 December and 0000 UTC 5 December (Fig.11.2-2). The rapid speed of the system as it redeveloped along the East Coast likely spared the region from heavier total snowfall. The subsequent eastward propagation of the surface low off the Virginia coast kept the heaviest snows south of New England.

Upper-level conditions are very similar to many of the heavy-snow-producing storms shown in chapter 4

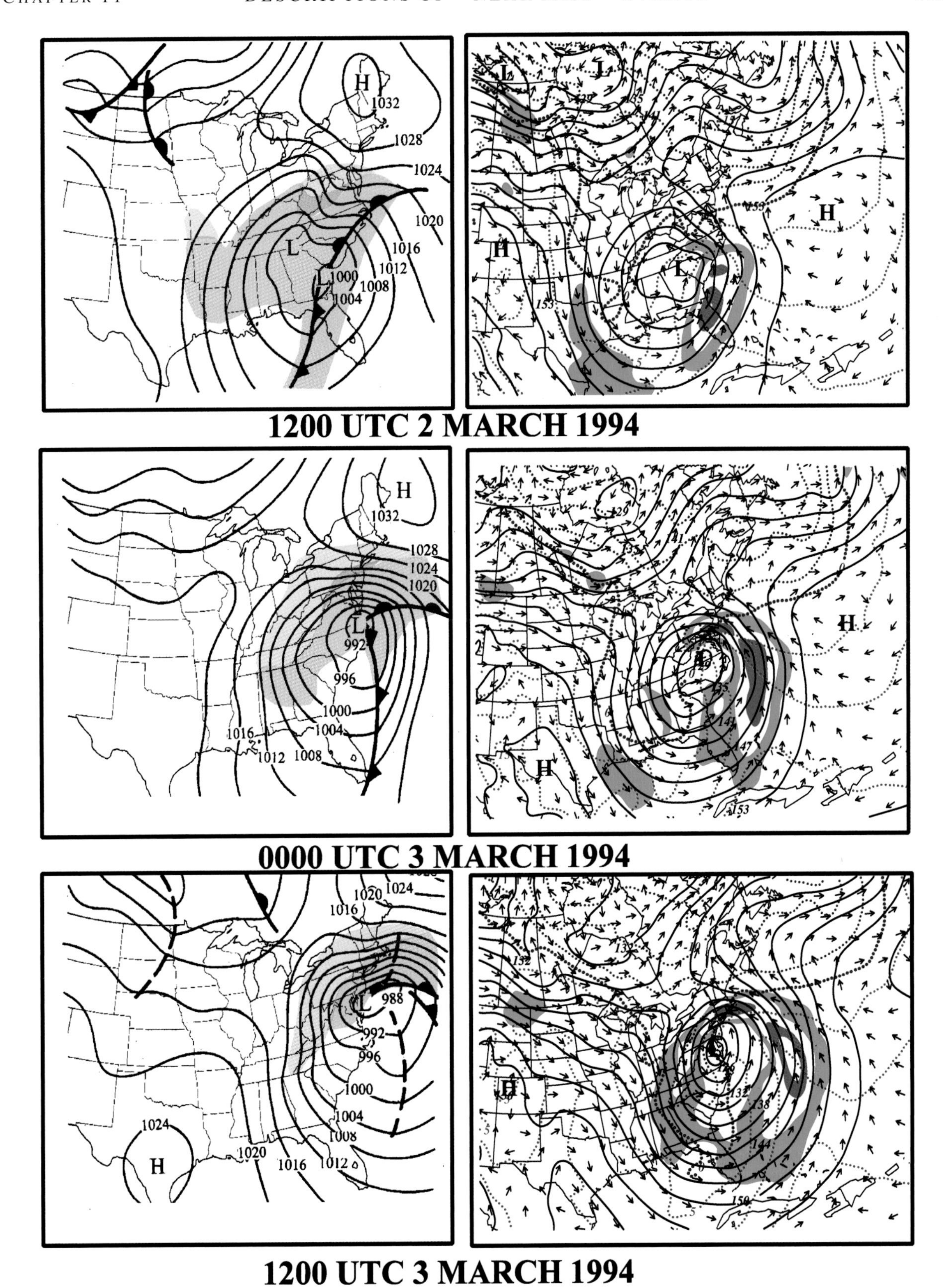

FIG. 11.1-44. Sequence of (left) surface and (right) 850-hPa charts for the 24-h period between 1200 UTC 2 and 3 Mar 1994. See Fig. 11.1-2 for details.

FIG. 11.1-45. Sequence of analyses of (left) 500- and (right) 250-hPa geopotential height and upper-level wind fields for the 24-h period between 1200 UTC 2 and 3 Mar 1994. See Fig. 11.1-3 for details.

of volume I (Fig. 11.2-3). A slow-moving upper closed low off Labrador in eastern Canada at 500 hPa is located upwind of a broad area of confluence covering much of eastern Canada. While this pattern is similar to many other cases of heavy snowfall, only weak surface high pressure is found beneath the confluent upper-level jet streak extending east off the New England coast at 0000 UTC 4 December (Fig. 11.2-2). Meanwhile, a moderate-amplitude upper-level short-wave trough associated with the surface cyclogenesis moved eastward from the

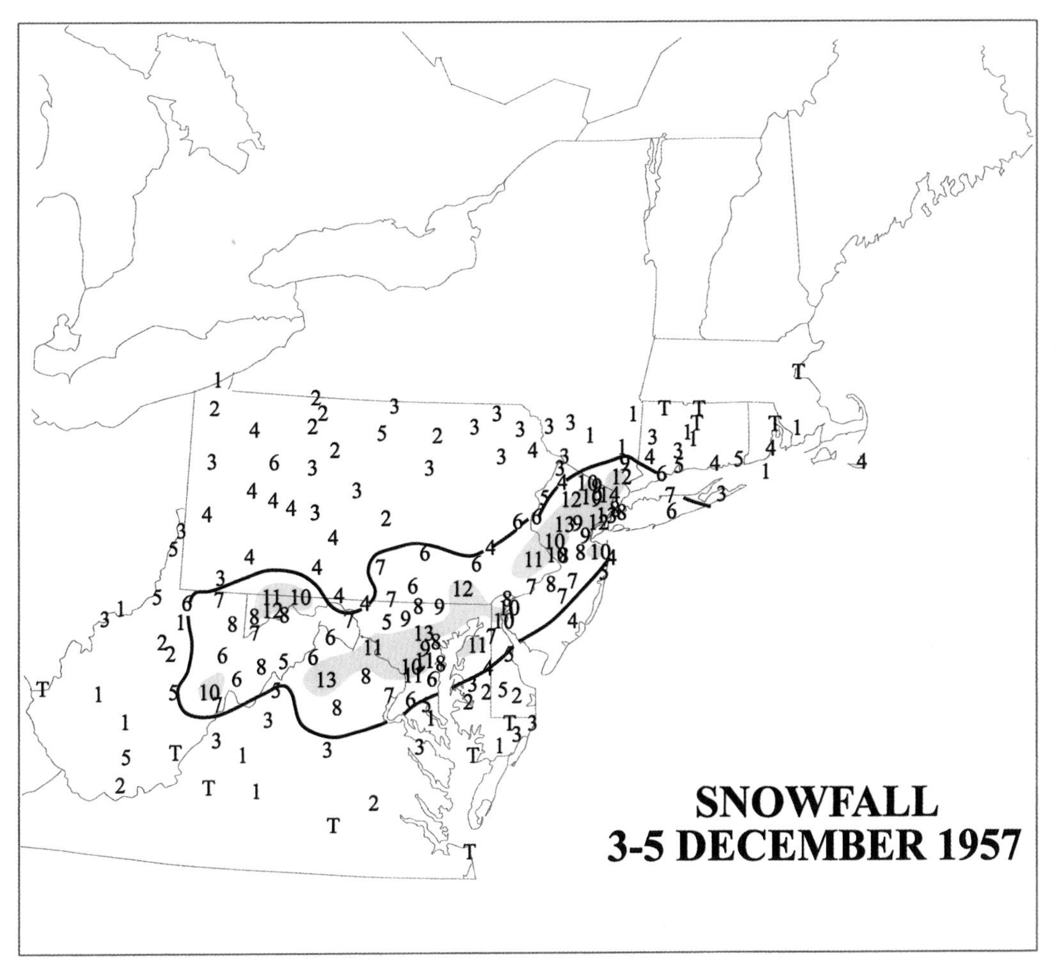

FIG. 11.2-1. Snowfall (in.) for 3–5 Dec 1957. See Fig. 11.1-1 for details.

Ohio Valley and amplified along the East Coast and became negatively tilted off the coast. This evolution supported the rapid eastward propagation of the surface low from the Ohio Valley to off the Virginia coast by 1200 UTC 4 December. Once offshore, the surface low deepened rapidly, while a classic dual-jet pattern became most evident at 400 and 250 hPa by 0000 UTC 5 December (Fig. 11.2-3). All of these factors came together too far east to support an evolution of this moderate snow event to an even more significant storm for the Northeast urban corridor.

b. 23–25 December 1961: NESIS = 1.37

This storm was responsible for a widespread area of 6–10 in. (15–25 cm) of snow from eastern Pennsylvania to coastal New England (Fig. 11.2-4). Snowfall was particularly heavy in eastern Massachusetts, where reports of 10–20 in. (25–50 cm) of snow were common and the heaviest amounts were found in the northeastern sections of Massachusetts and southeastern New Hampshire. Hundreds of cars were stranded between Boston and Worcester with police rescues a common occurrence. Over much of the Northeast urban corridor, snow combined with freezing rain, lessening total snowfall accumulations. Nevertheless, the combination produced a very treacherous Christmas for the Northeast.

This snowstorm developed in a pattern conducive to Northeast snowfall marked by a negative phase of the NAO, with the presence of a large 500-hPa cutoff low over eastern Canada to the south of a cutoff upper-level high over southern Greenland (Fig. 11.2-6), and during a period when the El Niño–Southern Oscillation (ENSO) was near neutral (neither El Niño nor La Niña). A surface low consolidated off the mid-Atlantic and intensified from 1000 to 985 hPa on 24 December (Fig. 11.2-5), moving more slowly toward the New England coast than many other significant Northeast snowstorms. The sea level pressure gradients north of the storm center were relatively weak for this case since high pressure was anchored north of Minnesota and pressures remained low over the northeast United States. It was not until the storm deepened more rapidly after 1200 UTC 24 December that pressure gradients increased over New England, combining with heavier snowfall rates to create a more serious snow event for that region.

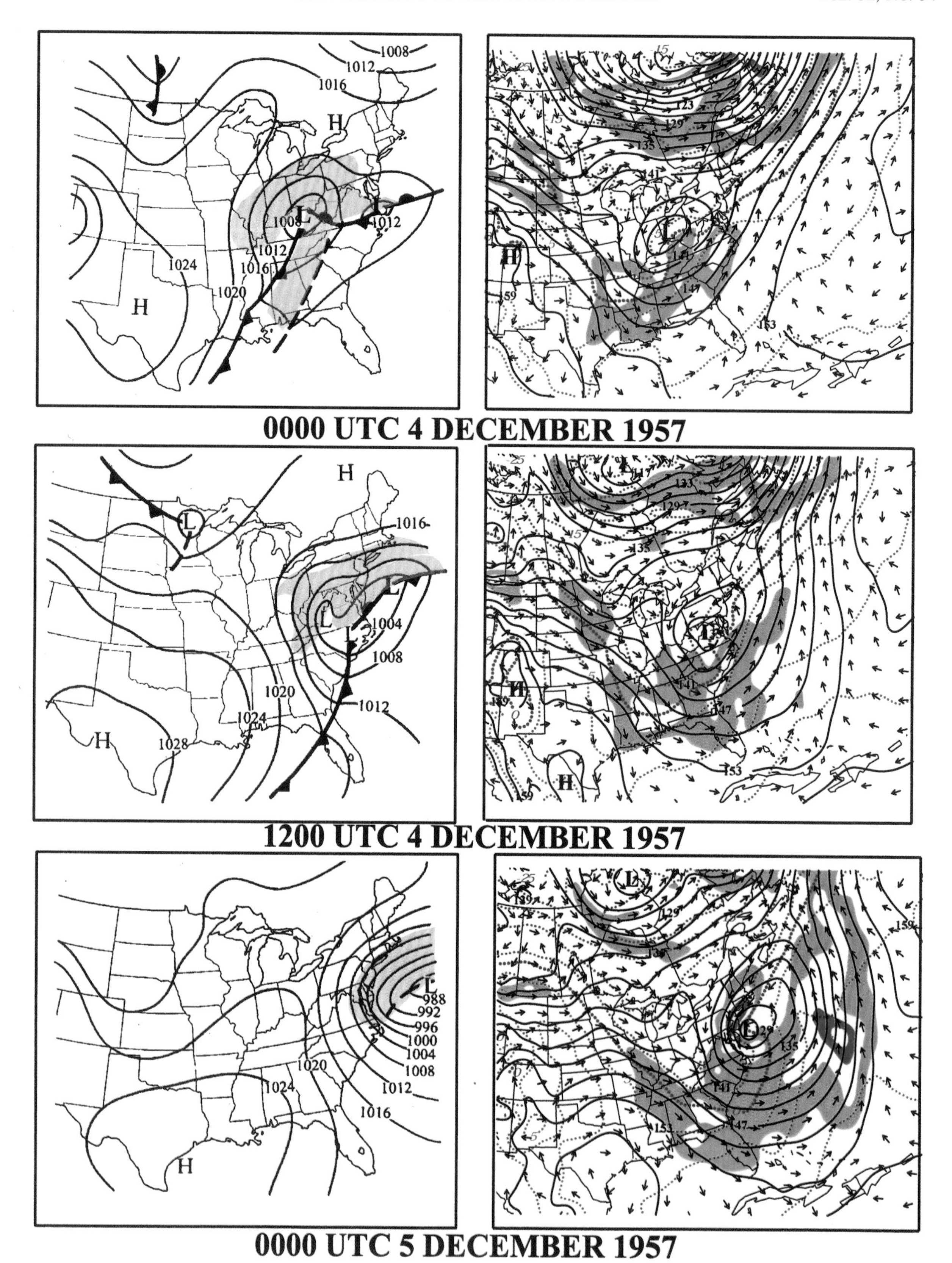

FIG. 11.2-2. Sequence of (left) surface and (right) 850-hPa charts for the 24-h period between 0000 UTC 4 and 5 Dec 1957. See Fig. 11.1-2 for details.

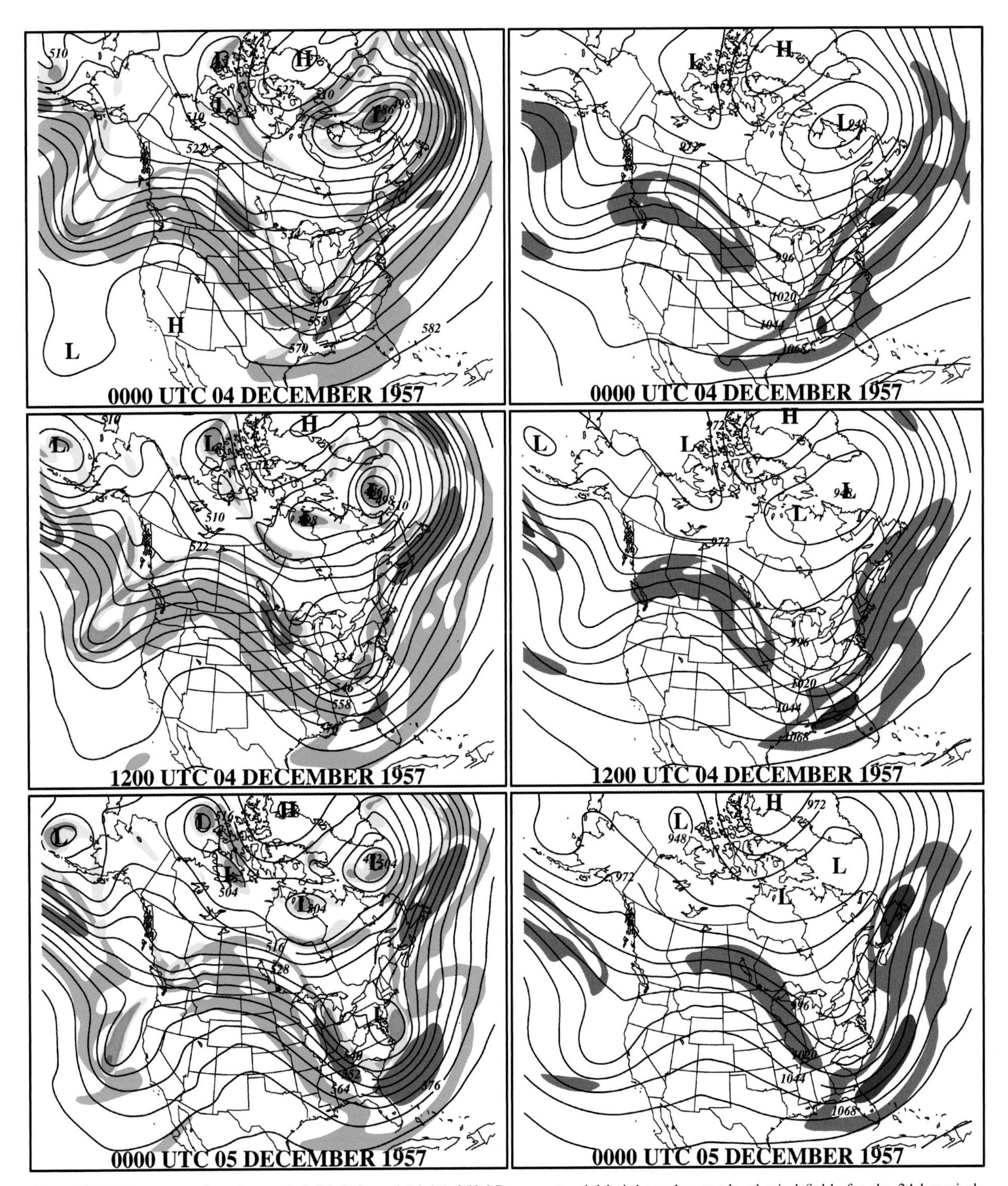

FIG. 11.2-3. Sequence of analyses of (left) 500- and (right) 250-hPa geopotential height and upper-level wind fields for the 24-h period between 0000 UTC 4 and 5 Dec 1957. See Fig. 11.1-3 for details.

This storm was associated with an upper-air pattern conducive to heavy snowfall. This pattern includes a confluent, jet streak entrance region at 400 hPa over New England during the precyclogenetic period and a distinct closed low and associated jet streak moving from the Ohio Valley to the New Jersey coast by 0000 UTC 25 December (Fig. 11.2-6). Nevertheless, the snow amounts remained in the moderate range throughout the mid-Atlantic region, related perhaps to 1) the slow evolution the 850-hPa circulation with rel-

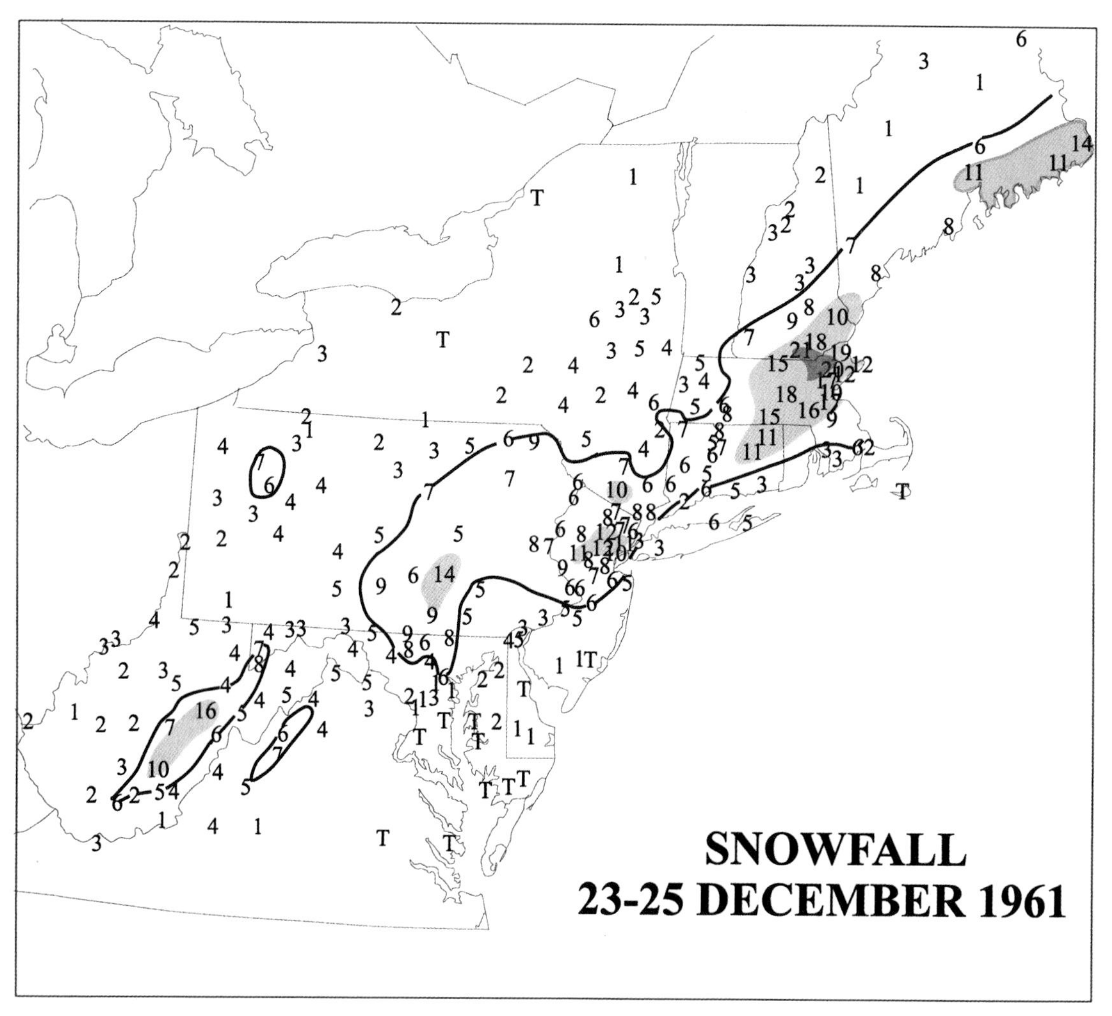

FIG. 11.2-4. Snowfall (in.) for 23–25 Dec 1961. See Fig. 11.1-1 for details.

ativity weaker temperature gradients and associated temperature advection patterns up to 1200 UTC 24 December (Fig. 11.2-5), and 2) the relatively weak along-stream wind gradients at the jet levels along the East Coast on 24 December. Both of these factors became more distinct by 0000 UTC 25 December, which likely accounted for the heavier snowfall in New England. Nevertheless, the delay in all of these factors coming together until after the storm moved through the Washington to New York City area kept this as a more moderate snowstorm for the Northeast urban corridor.

c. 14–15 February 1962: NESIS = 1.59

The Valentine's Day Snowstorm of 1962, although more limited in areal extent than most of the Northeast snowstorms covered in this monograph, was particularly severe across southern and central New England and was one of the greatest snowstorms on record for Rhode Island. Heaviest snows of greater than 20 in. (50 cm) affected several counties in Massachusetts (Fig. 11.2-7), where drifts of 3–5 feet (90–150 cm) were generated by gale force winds. In Boston, Logan International Airport was closed for 24 h and seven deaths were blamed on overexertion. Up to 2 feet (60 cm) of snow fell in the Catskill Mountains of New York and 15–25 in. (40–60 cm) fell across Rhode Island, resulting in either the second or third greatest February snowfalls of the century in Rhode Island to that point.

This New England snowstorm developed consistently with secondary development, as described in volume I, chapter 3. However, the paths of both the primary and secondary lows are different than the paths shown in Fig. 3-5 (volume I). The primary low moved along a northeastward track into the Great Lakes region, farther north than any case shown in Fig. 3-5. The secondary surface low developed south of New England and drifted slowly eastward (Fig. 11.2-8). The more northern location of the storm track for the primary low and the area in which the secondary low developed acted to constrict the area of heavy snowfall to eastern New York

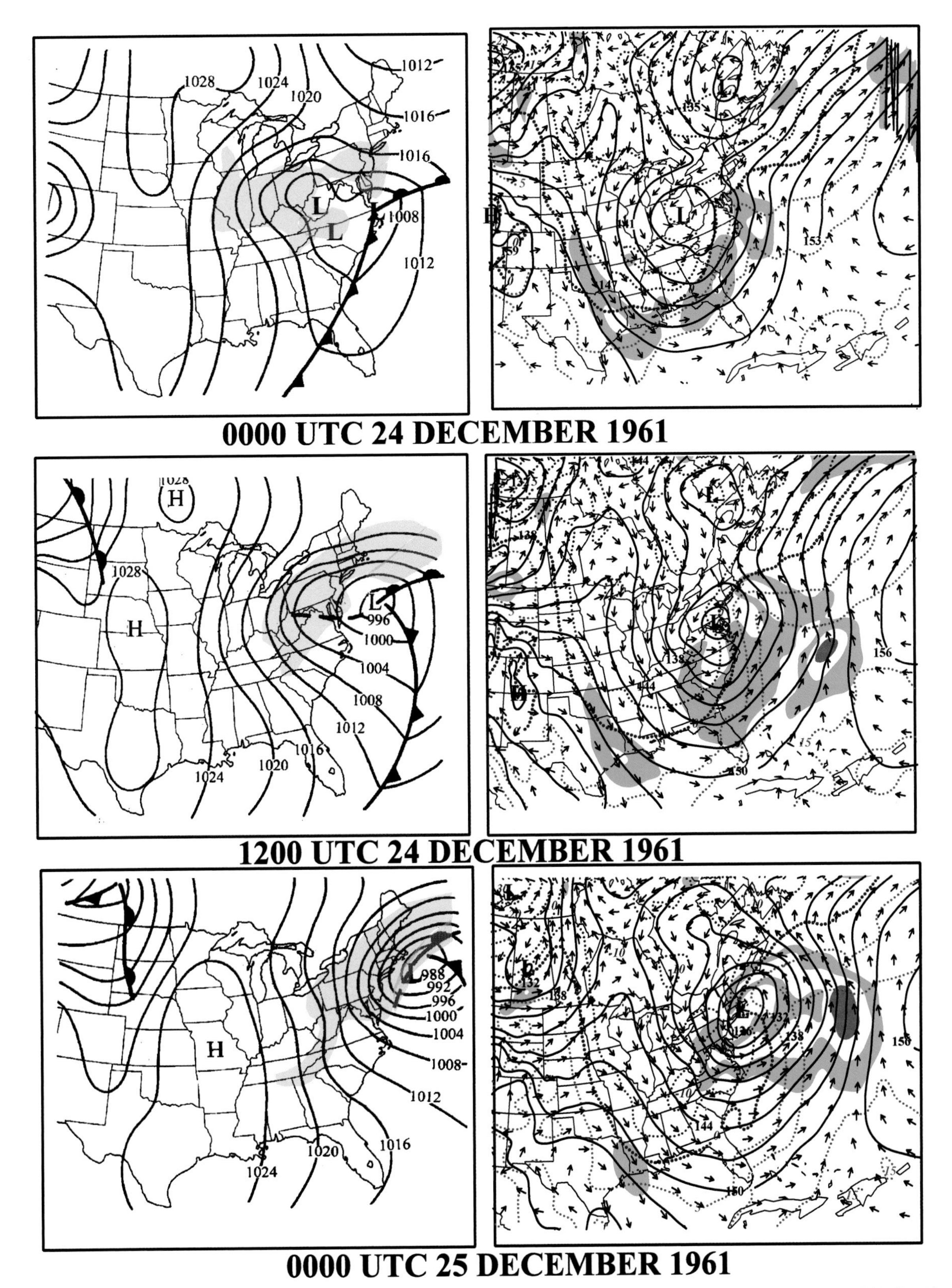

FIG. 11.2-5. Sequence of (left) surface and (right) 850-hPa charts for the 24-h period between 0000 UTC 24 and 25 Dec 1961. See Fig. 11.1-2 for details.

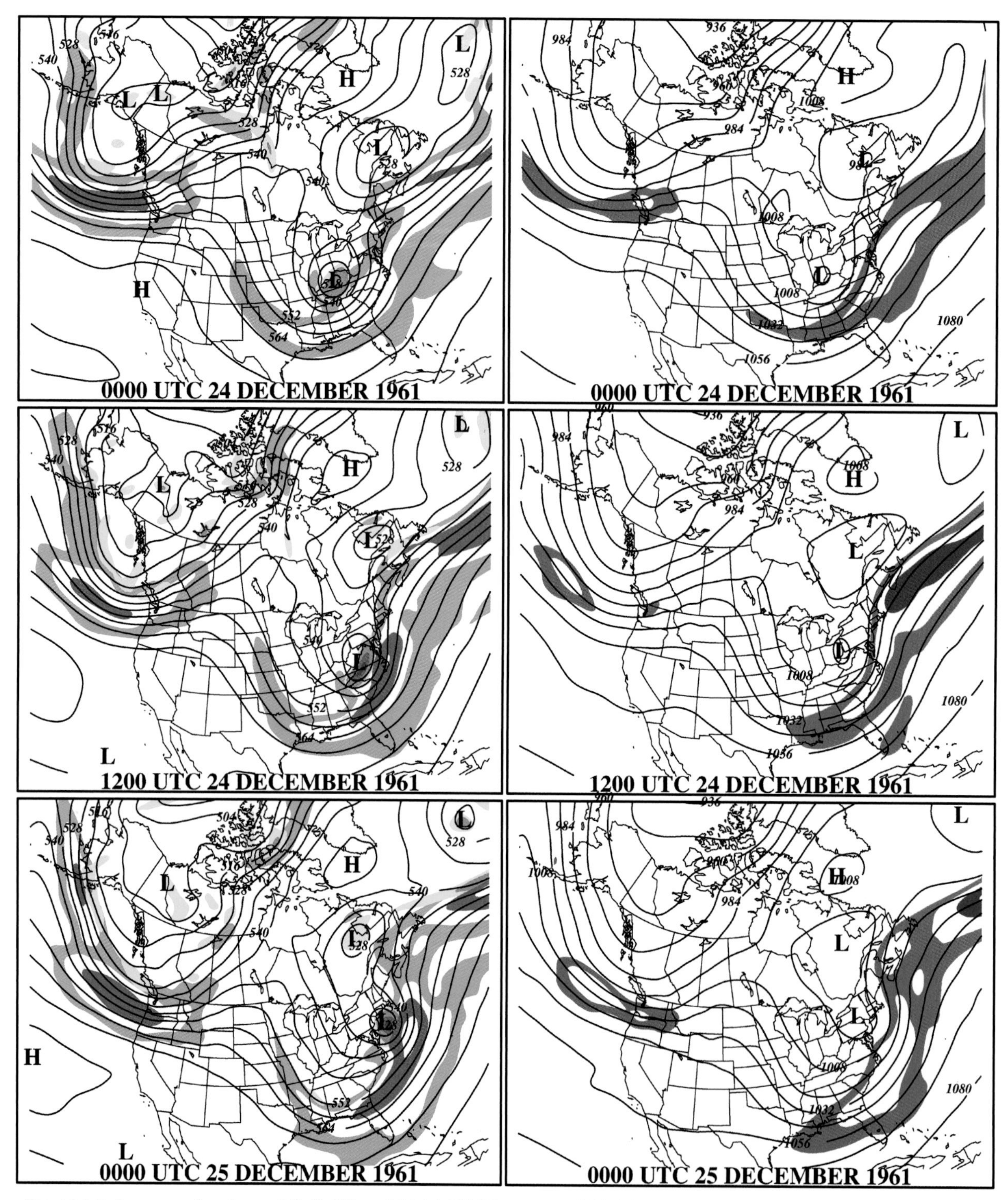

FIG. 11.2-6. Sequence of analyses of (left) 500- and (right) 250-hPa geopotential height and upper-level wind fields for the 24-h period between 0000 UTC 24 and 25 Dec 1961. See Fig. 11.1-3 for details.

and central and southern New England, keeping the mid-Atlantic clear of any snowfall for this case. The initial rapid development of the secondary cyclone occurred well east of the New Jersey coast early on 15 February, with the central pressure of the low dropping 12 hPa in 12 h, but deepened at a slow pace later on 15 February. This rapid intensification before 1200 UTC 15 February is also reflected by the development of the intense 850-hPa vortex southeast of Long Island and the strengthening easterly flow (exceeding 20 m s^{-1}) north of the

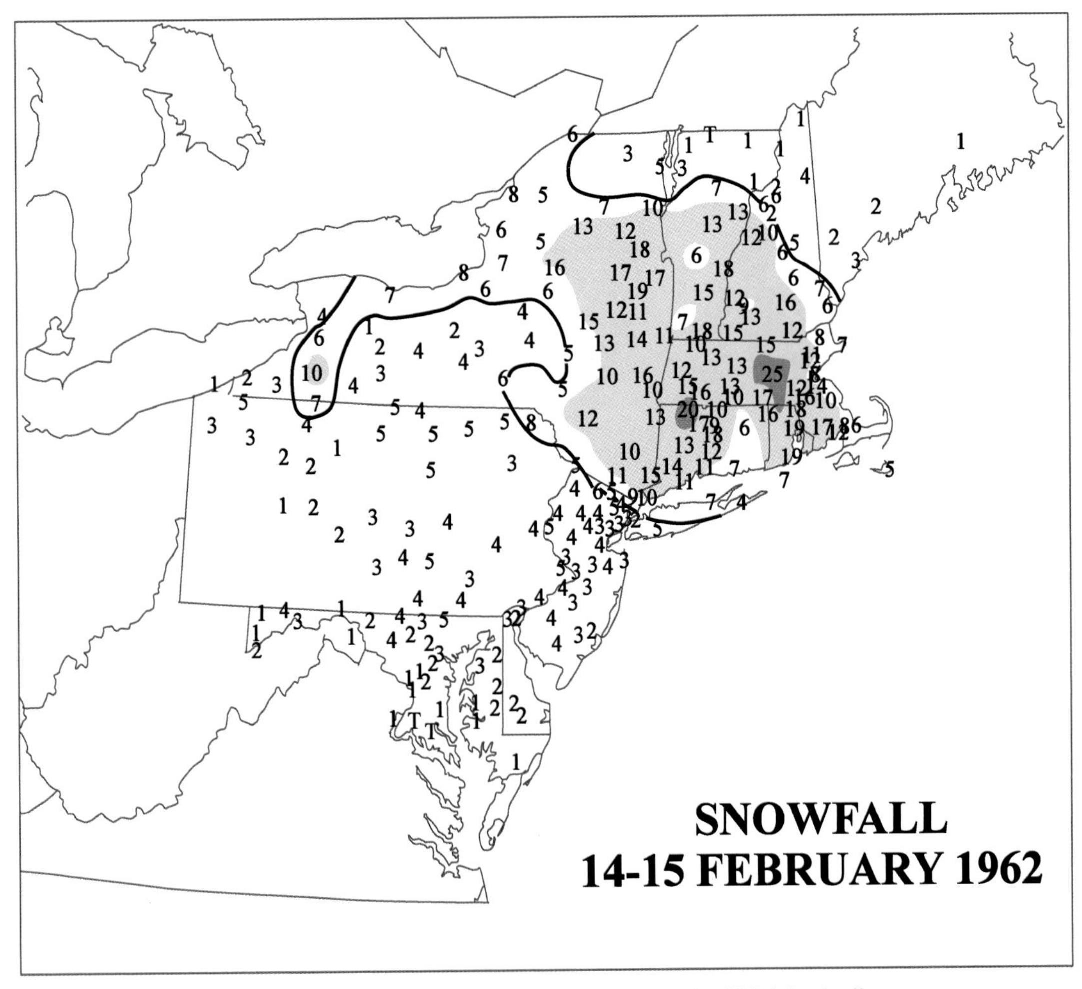

FIG. 11.2-7. Snowfall (in.) for 14–15 Feb 1962. See Fig. 11.1-1 for details.

center that was directed toward the area of heaviest snowfall in southern New England at 1200 UTC 15 February.

The upper levels show an unusual 500-hPa pattern (Fig. 11.1-9) with a low-amplitude, slow-moving negatively tilted trough crossing New England on 15 February, while a very large upper ridge covered the North Atlantic south of Greenland, impeding the trough's eastward movement. This contributed to the long duration of snowfall in New England, resulting in the heavy snow amounts. Another interesting factor is the unusual orientation of a northern jet system and related confluent flow over Duebec and Labrador, especially at 0000 UTC 15 February. The surface low appeared to redevelop rapidly within the diffluent exit region of a jet streak east of New Jersey between 0000 and 1200 UTC 15 February, with the surface low then becoming vertically aligned with the deepening trough by 1200 UTC 15 February.

d. 22–24 December 1963: NESIS = 3.17

This storm developed on 22 December near the Gulf coast states with an inverted trough extending northward into the southern Ohio Valley, producing record-setting snow in Arkansas, Mississippi, and Tennessee. Up to 18 in. (46 cm) of snow fell in northern Mississippi, and 14 in. (35 cm) fell at Memphis, Tennessee, followed by −13°F (−24°C) temperatures. As the storm redeveloped along the East Coast on 23 December, heavy snows fell across western Virginia, where 10–16 in. (25–40 cm) fell. But snowfall amounts diminished farther northeastward where only 6–10 in. (15–25 cm) of snow (or less) fell from Washington to southern New England (Fig. 11.2-10). Nonetheless, the storm disrupted traffic and canceled many airline flights from Washington to New York City. The relatively high NESIS value of 3.17 reflects the inclusion of the heavy snowfall in the southern states.

This is an example where a major snow producer over the lower Mississippi Valley weakened as the system moved toward the East Coast. The surface low developed in the northeastern Gulf of Mexico with a pool of very cold air located to its north (Fig. 11.2-11). As the storm moved northeastward along the Atlantic coast, the surface low deepened only slightly on 23 December

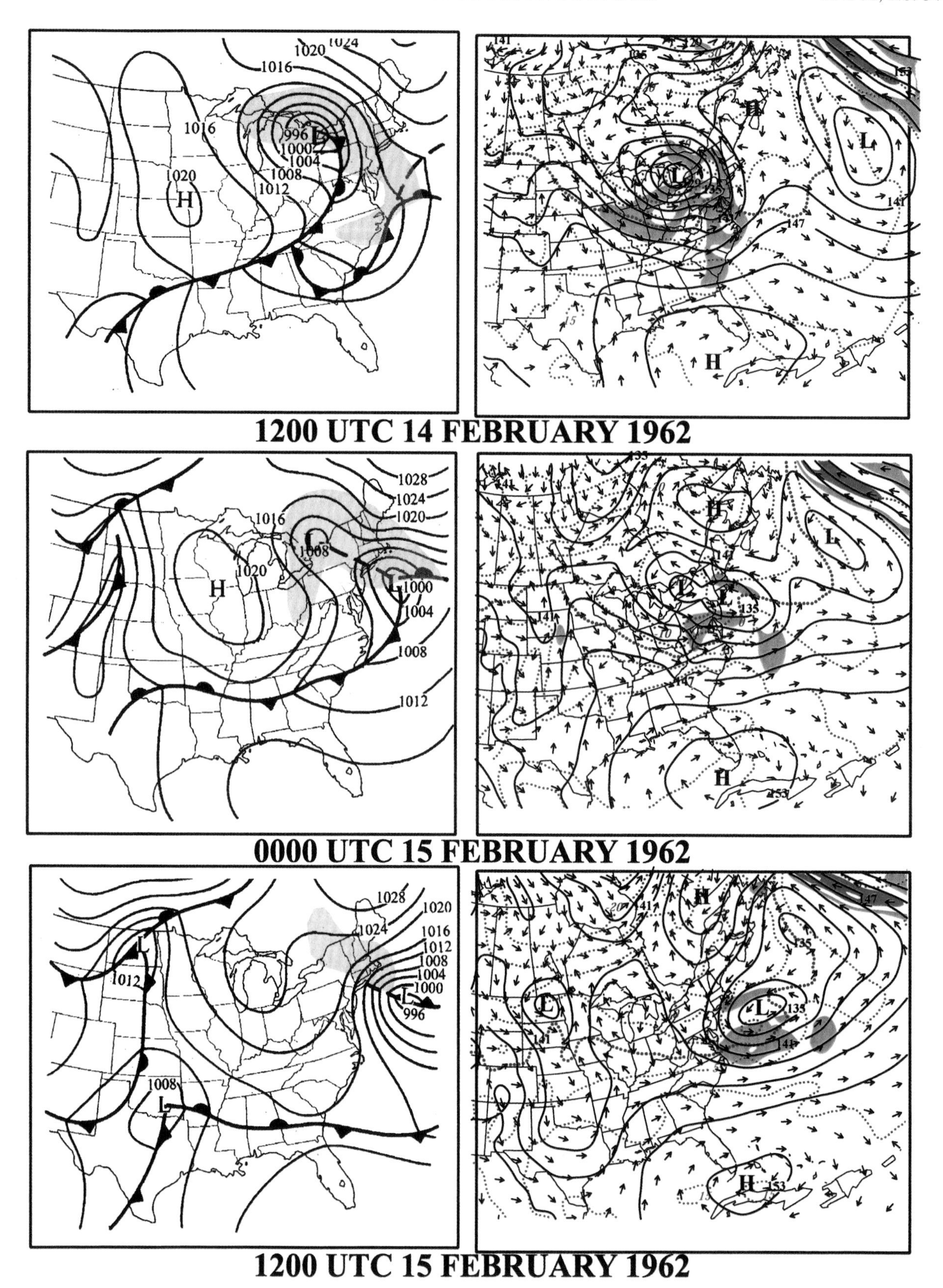

FIG. 11.2-8. Sequence of (left) surface and (right) 850-hPa charts for the 24-h period between 1200 UTC 14 and 15 Feb 1962. See Fig. 11.1-2 for details.

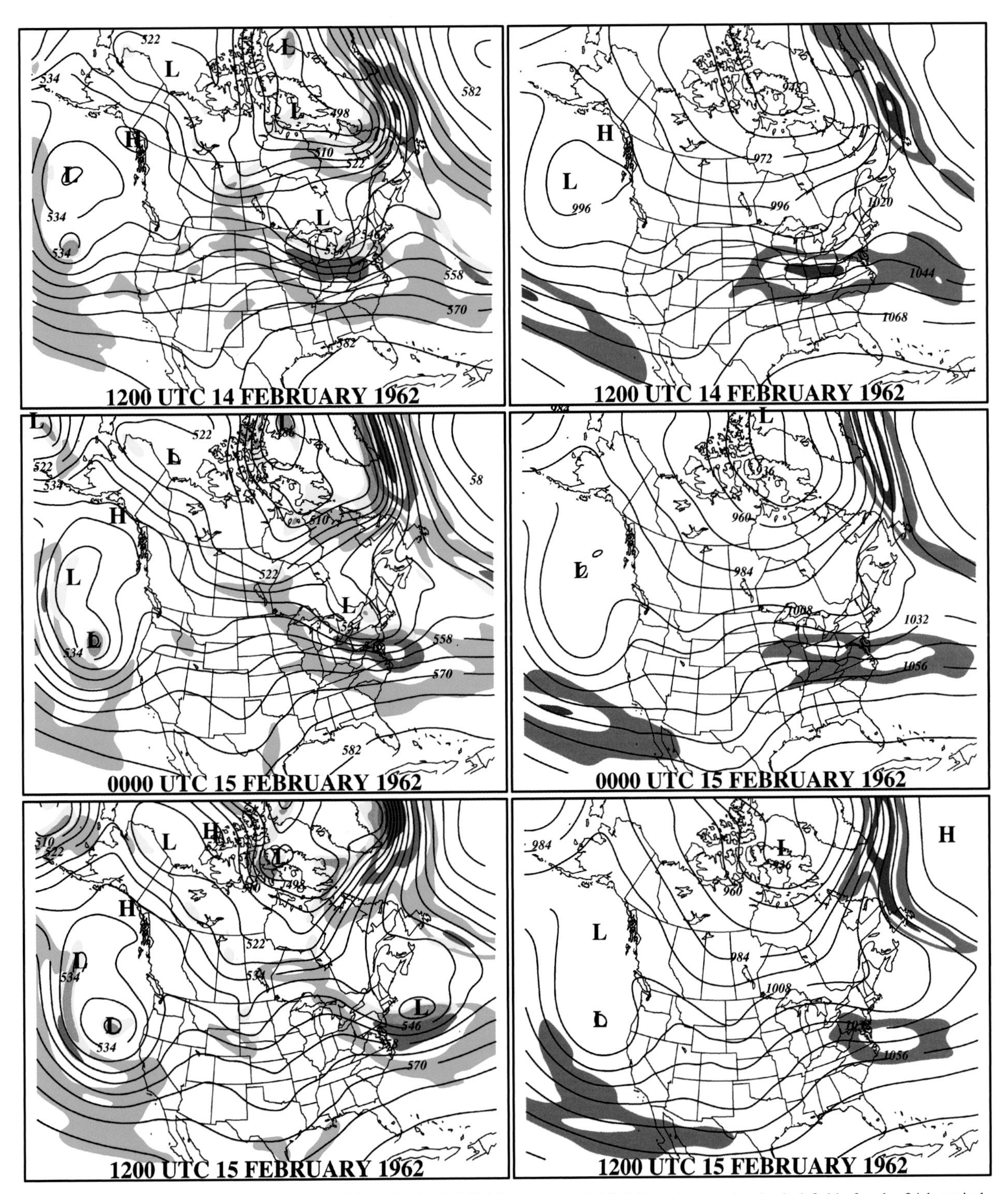

FIG. 11.2-9. Sequence of analyses of (left) 500- and (right) 250-hPa geopotential height and upper-level wind fields for the 24-h period between 1200 UTC 14 and 15 Feb 1962. See Fig. 11.1-3 for details.

and took on the appearance of an inverted trough along the East Coast (rather than a distinct low pressure center) as the surface high over New England drifted eastward. This same weakening track is reflected in the 850-hPa analyses, where a distinct closed circulation centered in Arkansas at 0000 UTC 23 December and a temperature advection pattern that would seem to support continued development devolved into an elongated and weakening system along the East Coast on 24 December.

The weakening of this storm occurred as a deep up-

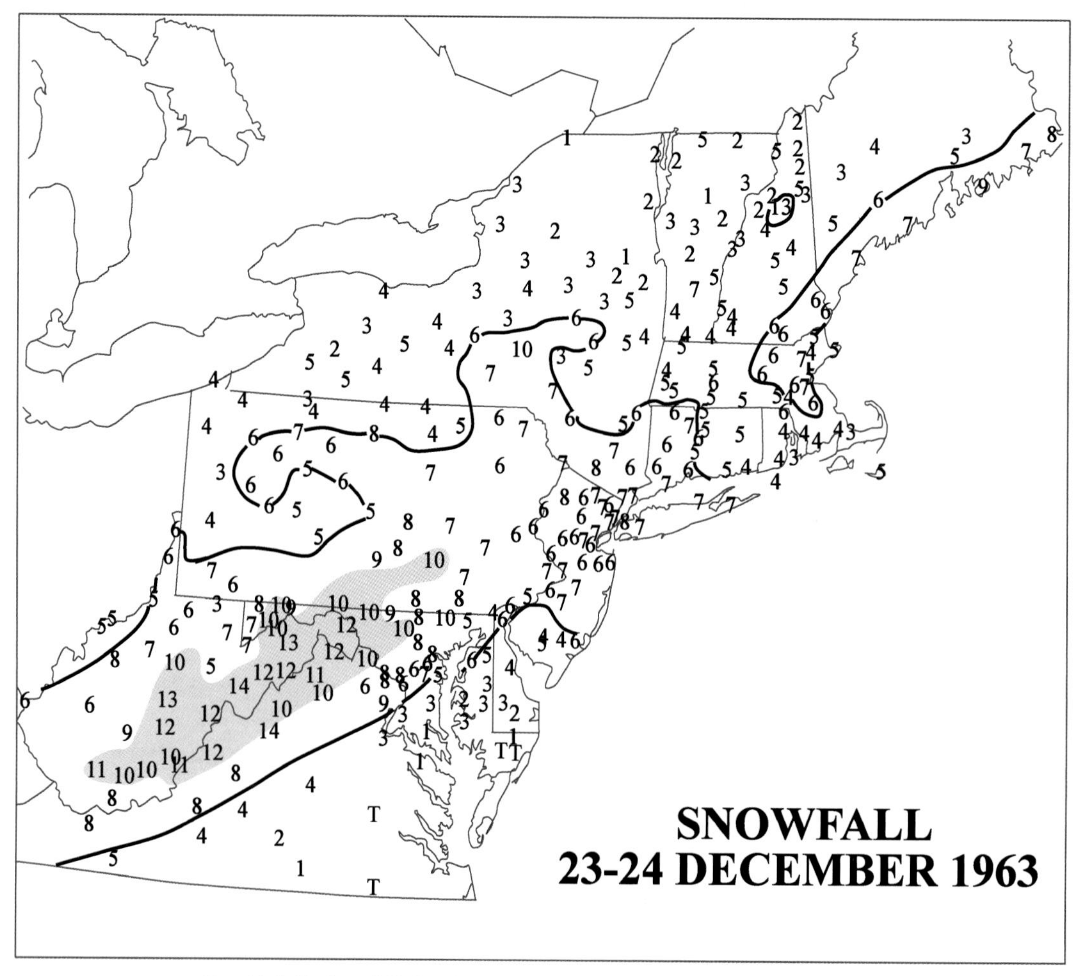

FIG. 11.2-10. Snowfall (in.) for 23–24 Dec 1963. See Fig. 11.1-1 for details.

per-level trough over the central United States on 23 December appeared to "shear" northeastward as the trough took on a "positive" tilt by 0000 UTC 24 December (Fig. 11.2-12). In addition, while plenty of cold air was entrenched across much of the eastern United States, large-scale confluence is largely lacking over eastern Canada without the presence of a strong upper trough over eastern Canada. As a result, the surface high over the Northeast weakened as it moved to the east and temperatures rose to near freezing along the immediate coastline with heavy snow in the New York City area and Long Island turning to freezing drizzle before ending.

Another factor in the storm's weakening along the East Coast on 24 December is the nature of the upper-level jet streaks. A well-defined and diffluent exit region of the southern jet is analyzed at the 400-hPa level as the storm developed along the Gulf coast at 0000 and 1200 UTC 23 December (Fig. 11.2-12). But, as the upper-level trough sheared and developed a more positive tilt by 0000 UTC 24 December, the along-stream wind variation became more diffuse within the complex jet streak pattern, evident at both 400 and 250 hPa. As a result of the weakening of upper-level support mechanisms, the surface low and associated snowfall rates diminished, preventing it from becoming a major snowfall producer in the Northeast.

e. 16–17 January 1965: NESIS = 1.95

During the snowy, cold decade of the 1960s, one of several heavy snow threats occurred during January 1965. A large cold anticyclone located over central Canada provided bitterly cold air for much of the northeastern United States, and upper-level troughs amplified into the central and southern United States. But several short-wave disturbances did not consolidate into a major storm, although moderate snow, high winds, and temperatures in the teens (°F) created blizzard conditions across much of the Northeast on 16 January. Although 12–18 in. (30 46 cm) of snow fell across Cape Cod and Nantucket, Massachusetts, with drifts up to 5 feet (150 cm) in places, most snowfall amounts across the North-

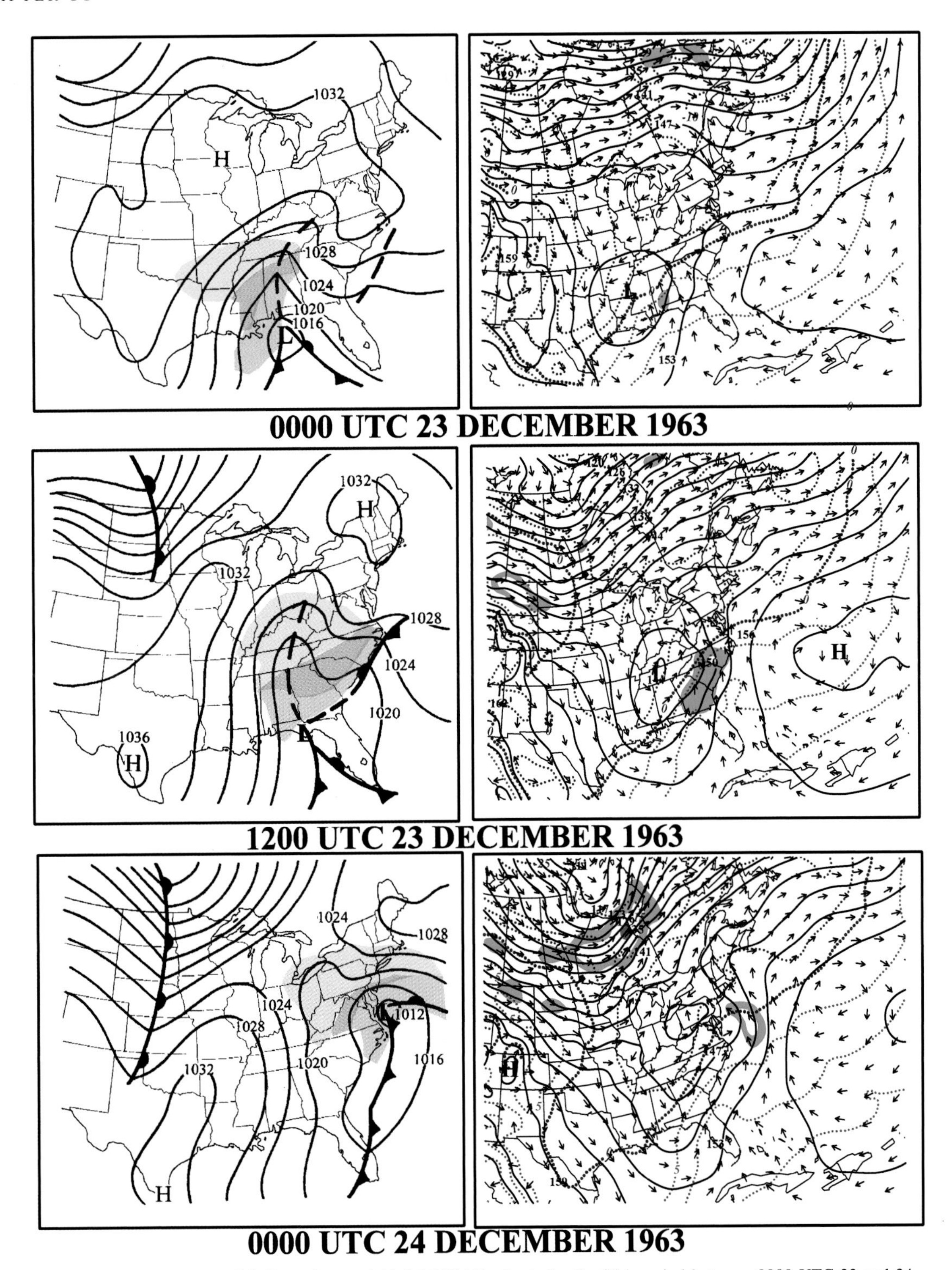

FIG. 11.2-11. Sequence of (left) surface and (right) 850-hPa charts for the 24-h period between 0000 UTC 23 and 24 Dec 1963. See Fig. 11.1-2 for details.

east urban corridor were 6–10 in. (15–25 cm) or less (Fig. 11.2-13).

The 16–17 January 1965 event is selected because its initial development was associated with typical precursors to a major snowstorm, with an amplifying upper trough over the central United States (Fig. 11.2-15) and the appearance of a large surface anticyclone associated with a confluent upper-level jet over the northeastern United States. However, a major snowstorm did not occur as both the surface low and the 850-hPa circulation

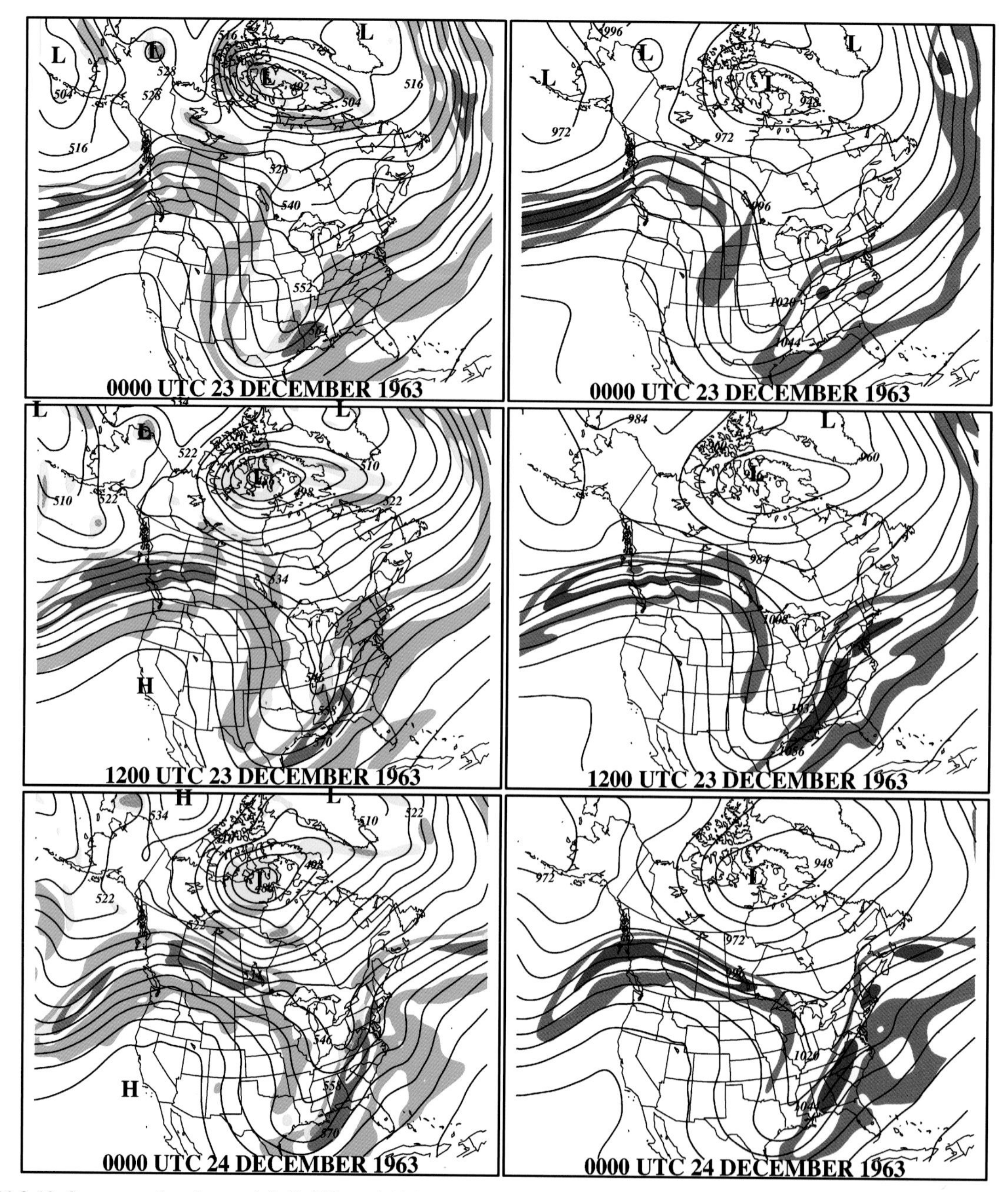

FIG. 11.2-12. Sequence of analyses of (left) 500- and (right) 250-hPa geopotential height and upper-level wind fields for the 24-h period between 0000 UTC 23 and 24 Dec 1963. See Fig. 11.1-3 for details.

became elongated along the East Coast on 16 January rather than consolidating into one major storm system (Fig. 11.2-14). The presence of the strong surface anticyclone located to the north combined with the elongated surface low pressure system to sustain a strong pressure gradient and high winds, creating blizzard conditions. A surface low developed off the Southeast coast late on 16 January and early on 17 January that then passed out to sea and also did not result in a rapidly deepening cyclone or significant snowfall along the East Coast.

The elongated nature of the surface lows that marked this case is related to the evolution of the upper-level height and wind features that initially looked promising, but failed to evolve toward a classic pattern that could produce a major snowstorm. Although the confluent entrance region of the northern jet over New York to New England remained intact on 16 January (Fig. 11.2-15), the trough that was digging into the central and southeast United States maintained an elongated structure and positive tilt through 17 January. Furthermore, a dual-jet pattern at both 400 and 250 hPa at 1200 UTC 16

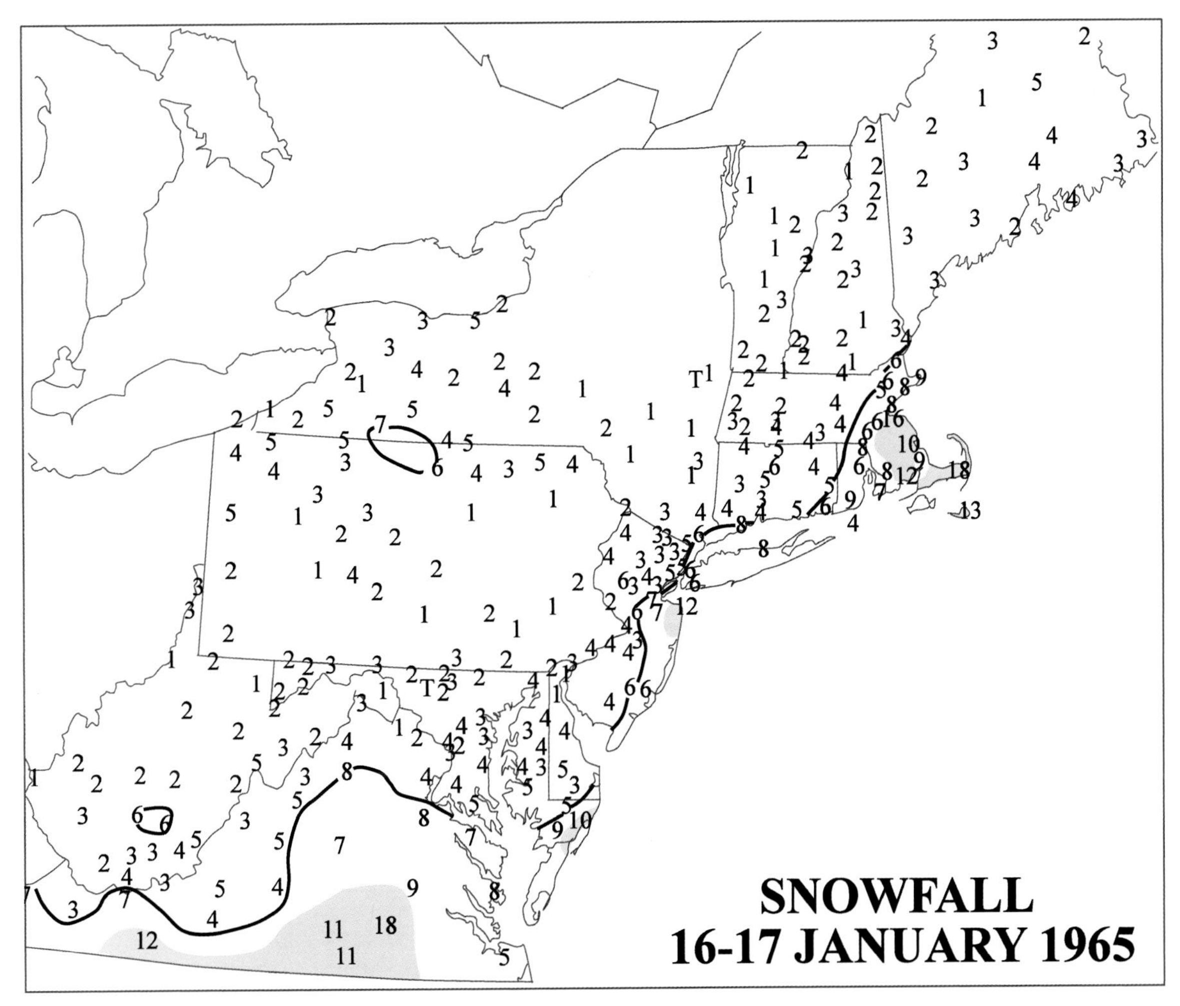

FIG. 11.2-13. Snowfall (in.) for 16–17 Jan 1965. See Fig. 11.1-1 for details.

January was not sustained as the entrance region wind pattern elongated by 0000 UTC 17 January, resulting in a diminished along-stream wind variation in both the exit region of the southern jet and the entrance region of the northern jet. Thus, an initially favorable situation to deliver a heavy snowfall event along the East Coast evolved into mostly a prolonged light to moderate snow event for much of the coast.

f. 21–22 March 1967: NESIS = 1.20

An unexpected heavy snowfall in March 1967 affected New York City, Long Island, eastern Connecticut, southeastern Massachusetts, and Rhode Island with up to 18 in. (46 cm) of snow, closing schools, catching city officials unprepared, and crippling most forms of transportation, particularly on the morning of 22 March. An area of 6–10 in. (15–25 cm) of snow fell from northern New Jersey and northeastern Pennsylvania through New York City and Long Island and much of southern New England (Fig 11.2-16).

The surface and upper-level features of this storm bear some resemblance to that of the storm of 14–15 February 1962 (see earlier in this chapter). A primary low remained relatively far north and weakened over the upper Midwest as a secondary surface low developed south of New England. Unlike the February 1962 storm, the surface low did not intensify appreciably (Fig. 11.2-17) and remained relatively stationary off Long Island. At 850 hPa, a closed cyclonic circulation was located over the Great Lakes at 0000 UTC 22 March, usually not a good indication for snowfall in the New Jersey and New York area since that location often favors warming to the east of the low, often resulting in a change to rain. However, by 1200 UTC 22 March, a separate and distinct circulation intensified at 850 hPa south of Long Island and remained stationary for the next 12 h supporting the snowfall from New Jersey to southern New England.

At upper levels (Fig. 11.2-18), a relatively low-amplitude trough moved slowly eastward across the northeast United States between a nearly stationary cutoff low south of Newfoundland and south of a small upper-level ridge over eastern Quebec. The ridge–cutoff low

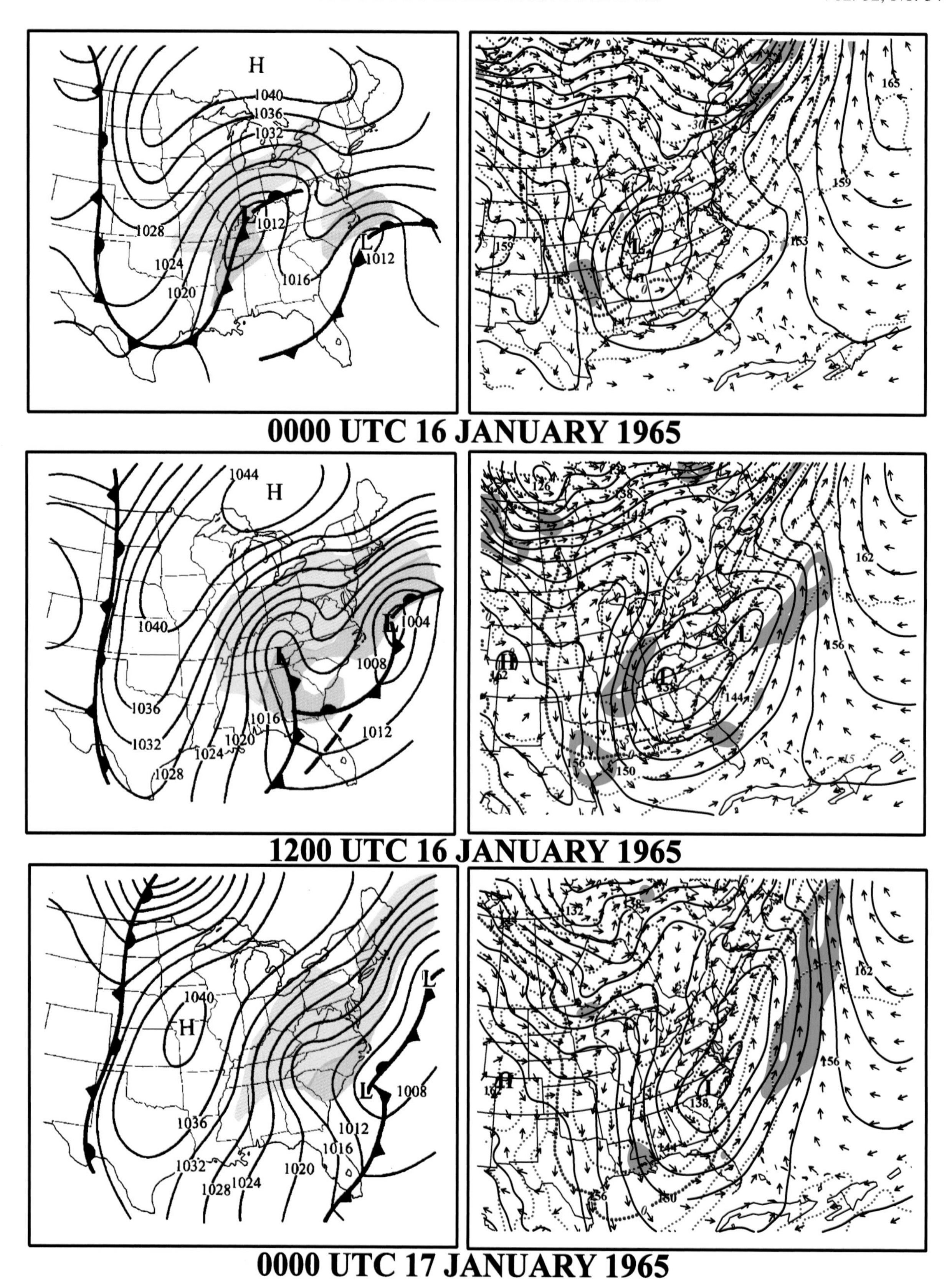

FIG. 11.2-14. Sequence of (left) surface and (right) 850-hPa charts for the 24-h period between 0000 UTC 16 and 17 Jan 1965. See Fig. 11.1-2 for details.

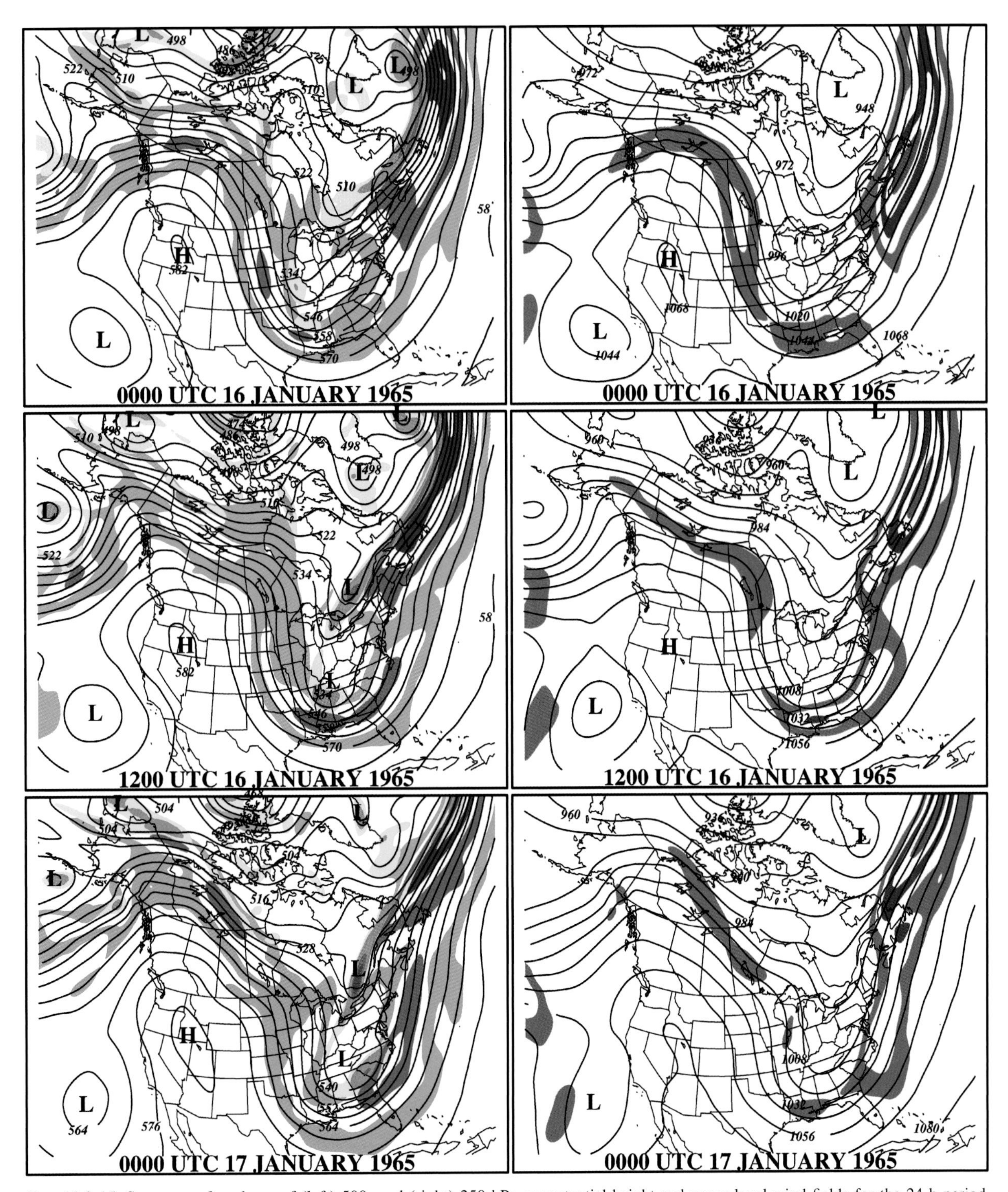

FIG. 11.2-15. Sequence of analyses of (left) 500- and (right) 250-hPa geopotential height and upper-level wind fields for the 24-h period between 0000 UTC 16 and 17 Jan 1965. See Fig. 11.1-3 for details.

combination across eastern Canada favored a nearly stationary, surface high pressure system north of Maine that provided enough cold air to support a significant snowfall. Furthermore, the system seemed to take on the characteristics of a synoptic scale "omega" blocking pattern, which slowed the movement of the surface low and prolonged the heavy snowfall in the New York City metropolitan area and southern New England on 22

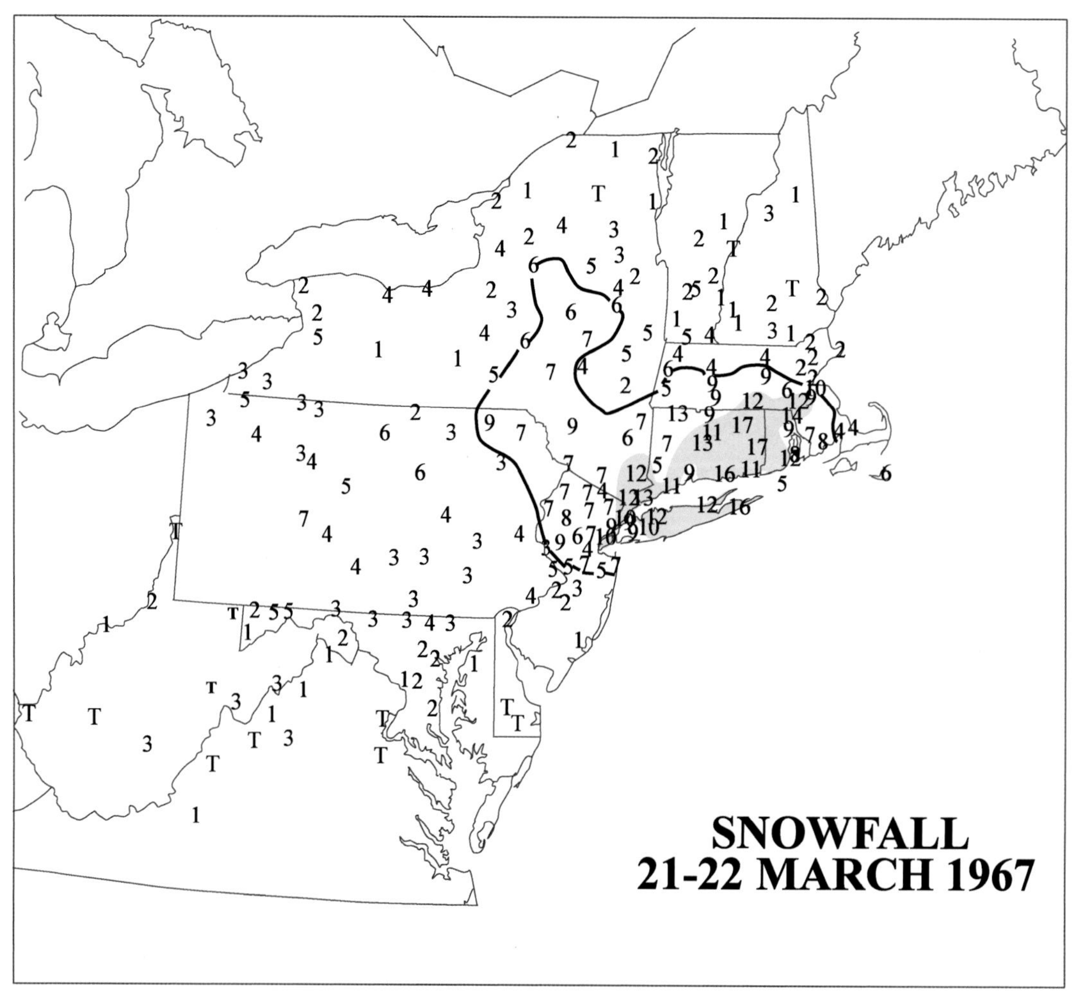

FIG. 11.2-16. Snowfall (in.) for 21–22 Mar 1967. See Fig. 11.1-1 for details.

March. The short wavelength of the upper trough, the lack of surface deepening once the low formed south of Long Island, the small amplitude, and the relativity small scale of this system are all factors that are different from the larger, more significant snowstorms described in chapters 3 and 4 of volume I, and chapter10. The net effect was to confine the moderate to heavy snowfall to a small area from New Jersey through New York City and Long Island to southern New England.

g. 31 December 1970–1 January 1971: NESIS = 2.10

This New Year's Day snowstorm represents an event that had the potential to develop into a major Northeast snowstorm, but never quite lived up to that potential. A band of heavy snow, 10–20 in. (25–50 cm), was located primarily west of the Northeast urban corridor with a more moderate snowfall over the metropolitan areas. Snowfalls of greater than 20 in. (50 cm) were reported in portions of eastern West Virginia, western Maryland, and south-central Pennsylvania (Fig. 11.2-19).

This is a case that appears to have surface and upper-level characteristics that would support a major snowstorm event for the Northeast urban corridor, resembling many of the cases in chapter 10. The track of the surface low, the location of the surface anticyclone during the precyclongenetic period, and the general upper-level trough–ridge pattern and jet streak evolution are all similar to those of the classic snowstorms (Figs. 11.2-20 and 11.2-21). Indeed, the diffluent exit region of the southern jet at 400 hPa and the entrance region of the northern jet at 250 hPa seem ideally located at 0000 UTC 1 January to support a more significant snowfall along the coast than actually occurred. High pressure remained relatively stationary north of New York, while a surface low moved from the southeastern United States to off the middle Atlantic coast. Strong confluent flow remained in place over northern New England, while a negatively tilted trough supported the surface low formation off the Carolina coast. Yet, the heaviest

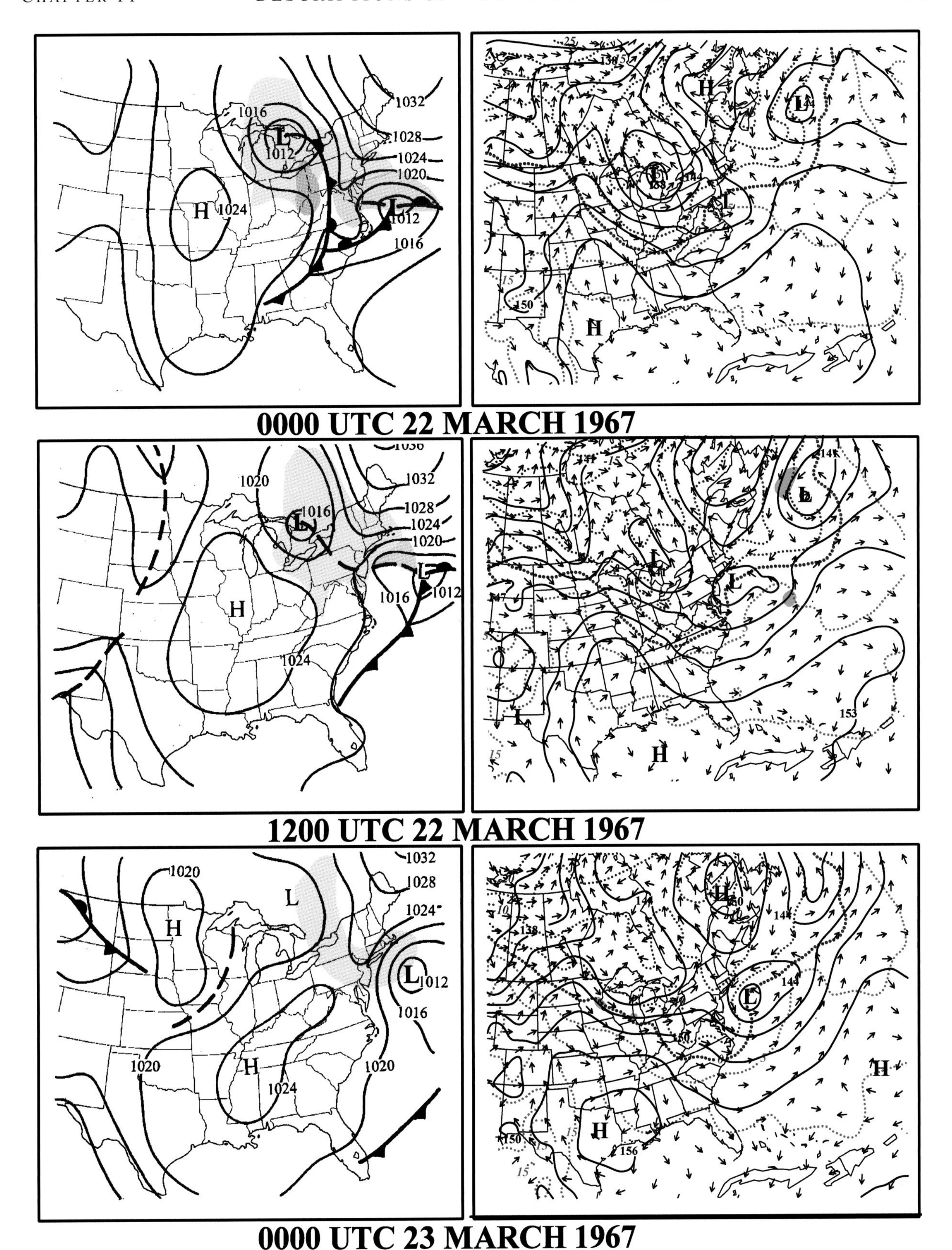

FIG. 11.2-17. Sequence of (left) surface and (right) 850-hPa charts for the 24-h period between 0000 UTC 22 and 23 Mar 1967. See Fig. 11.1-2 for details.

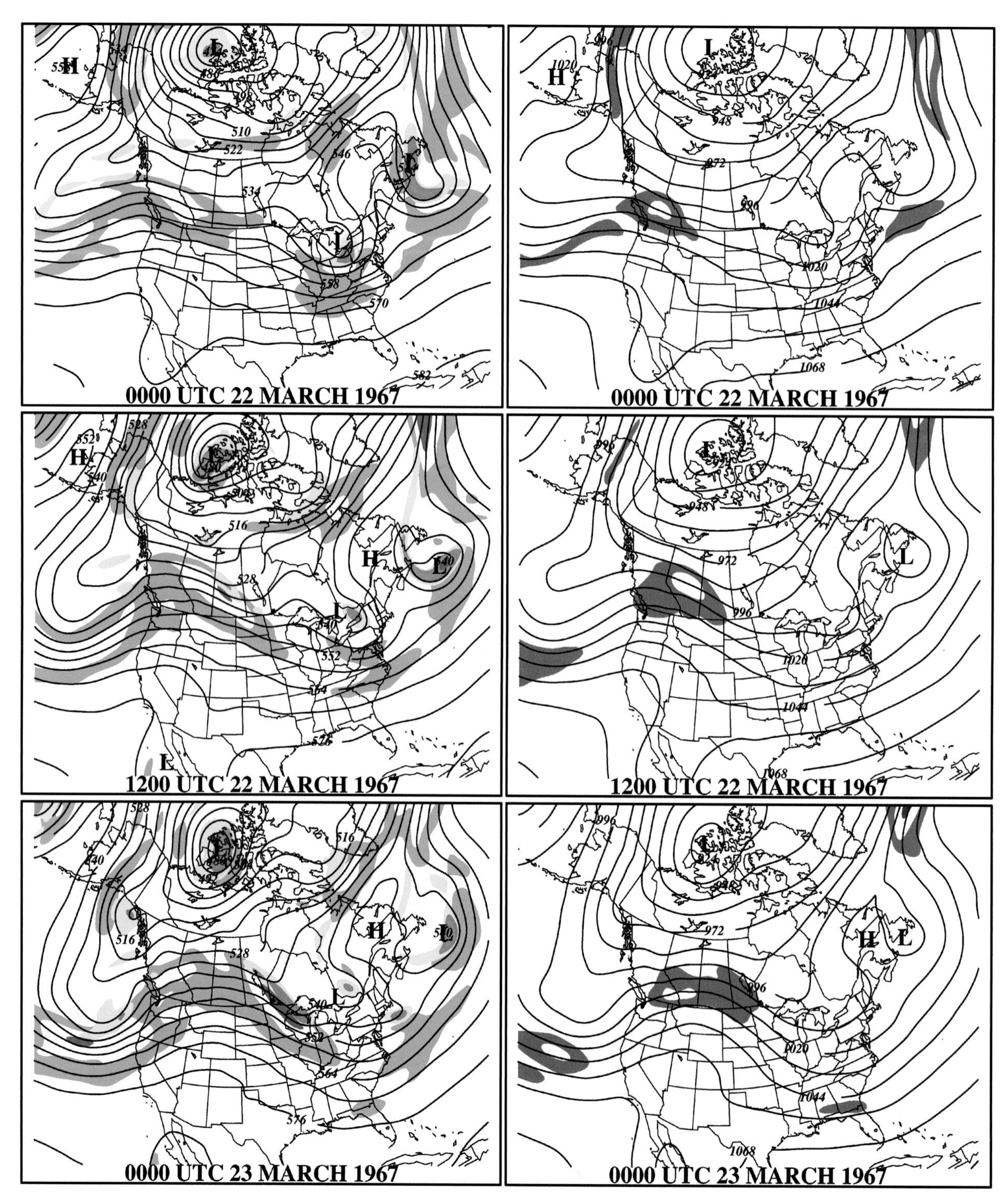

FIG. 11.2-18. Sequence of analyses of (left) 500- and (right) 250-hPa geopotential height and upper-level wind fields for the 24-h period between 0000 UTC 22 and 23 Mar 1967. See Fig. 11.1-3 for details.

snows of 10–20 in. (25–50 cm) of snow or greater were confined to West Virginia, western Virginia, western Maryland, and south-central Pennsylvania, while totals of no more than 6–10 in. (15–25 cm) were common in the Northeast urban corridor with primarily light to moderate snowfall rates during this event.

Some general differences exist between this storm and other more significant Northeast snowstorms, possibly

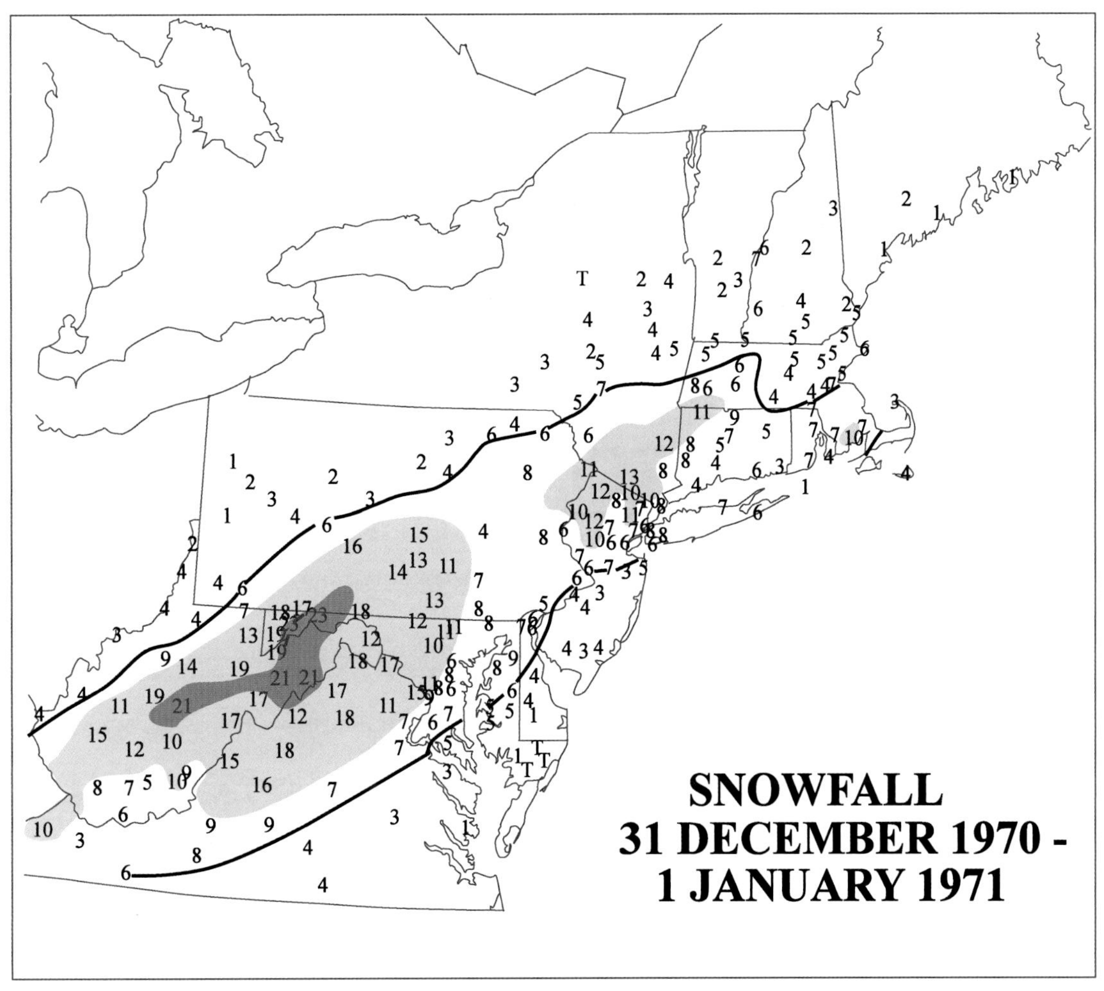

FIG. 11.2-19. Snowfall (in.) for 31 Dec 1970–1 Jan 1971. See Fig. 11.1-1 for details.

contributing to the storm's lesser snowfall in the Northeast urban corridor. First, central pressures within the surface anticyclone north of New York (1022 hPa) were somewhat lower relative to many major snow cases. Second, while the surface low deepened about 8 hPa between 1200 UTC 31 December to 0000 UTC 1 January as it moved northeastward along the Southeast coast, the surface low did not deepen at all early on 1 January as it moved east of the Virginia coast (Fig. 11.2-20). Furthermore, although the 850-hPa circulation appears to intensify between 1200 UTC 31 December and 1200 UTC 1 January, 850-hPa temperature gradients remain relatively weaker than in other cases, reducing the thermal advection pattern and related ascent. In addition, the easterly low-level jet analyzed north of the 850-hPa center remains relatively weak, barely reaching 20 m s^{-1} by 1200 UTC 1 January, with a more southerly low-level jet displaced farther east over the Atlantic Ocean. During this period, the surface low became more elongated, and light to moderate snowfall rates produced less snowfall in New York City to New England than was expected.

h. 13–15 January 1982: NESIS = 3.08

This snowstorm was the result of two separate shortwave troughs that produced two distinct periods of snow across the middle Atlantic states and New England. The first storm system brought moderate snowfall from Virginia to southern New England on 13 January and contributed to a disastrous commercial jet crash into the Potomac River at Washington. During this storm, snow changed to rain or freezing rain along the immediate coast from southeastern Virginia to southern New Jersey. The second storm moved rapidly northeastward on 14 January and produced a brief but heavy period of snow immediately along the coastal plain, with new snowfall accumulations of up to 10 in. (25 cm) in coastal Virginia and Maryland (Fig. 11.2-22).

The storm occurred during a month when some of the coldest air masses observed during the second half of the 20th century penetrated southward across all of the central and eastern United States, with daily NAO values (Table 2-5, volume I) remaining negative for much of the month of January. ENSO was near neutral

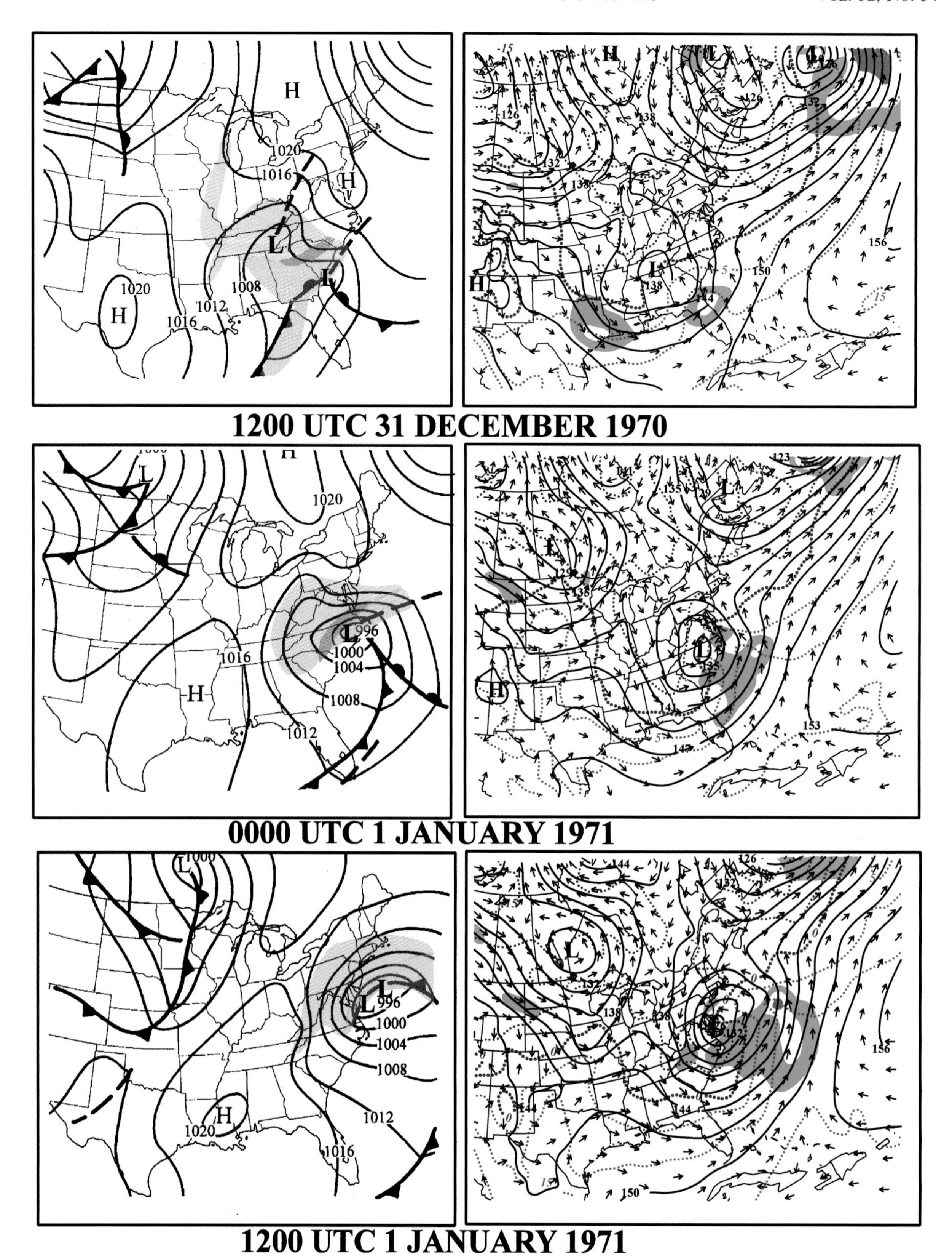

FIG. 11.2-20. Sequence of (left) surface and (right) 850-hPa charts for the 24-h period between 1200 UTC 31 Dec 1970 and 1200 UTC 1 Jan 1971. See Fig. 11.1-2 for details.

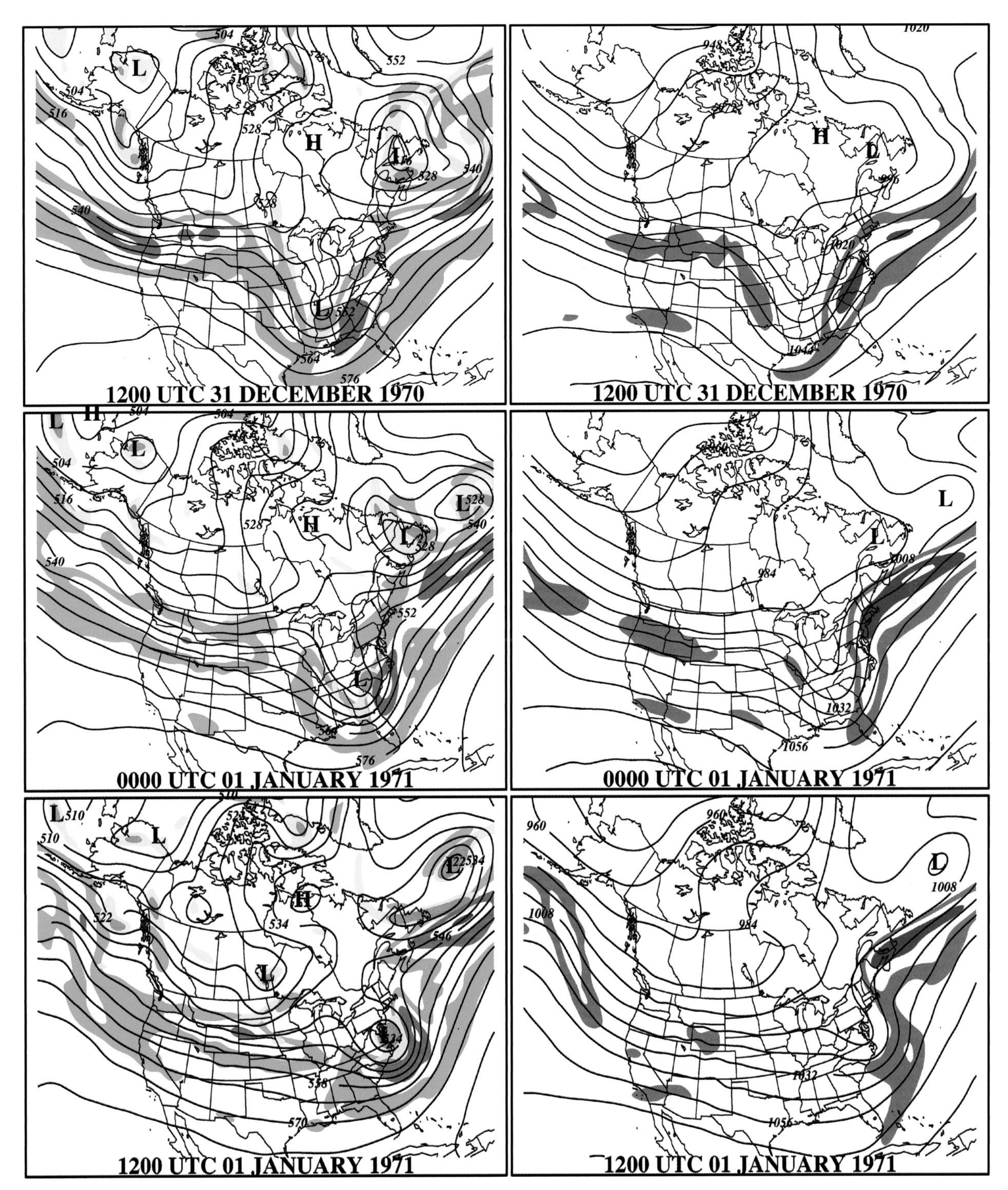

FIG. 11.2-21. Sequence of analyses of (left) 500- and (right) 250-hPa geopotential height and upper-level wind fields for the 24-h period between 1200 UTC 31 Dec 1970 and 1200 UTC 1 Jan 1971. See Fig. 11.1-3 for details.

at this time (Fig. 2-19, volume I), prior to the beginning of a major El Niño event in 1983.

The moderate snowfall events occurred in association with a series of upper-level trough–jet systems. The first short-wave trough and relatively weak cyclone moved northeastward from the Gulf of Mexico along the Atlantic coast (Fig. 11.2-23) and produced a general 4–8-in. snowfall (10–20 cm) from Virginia northward to

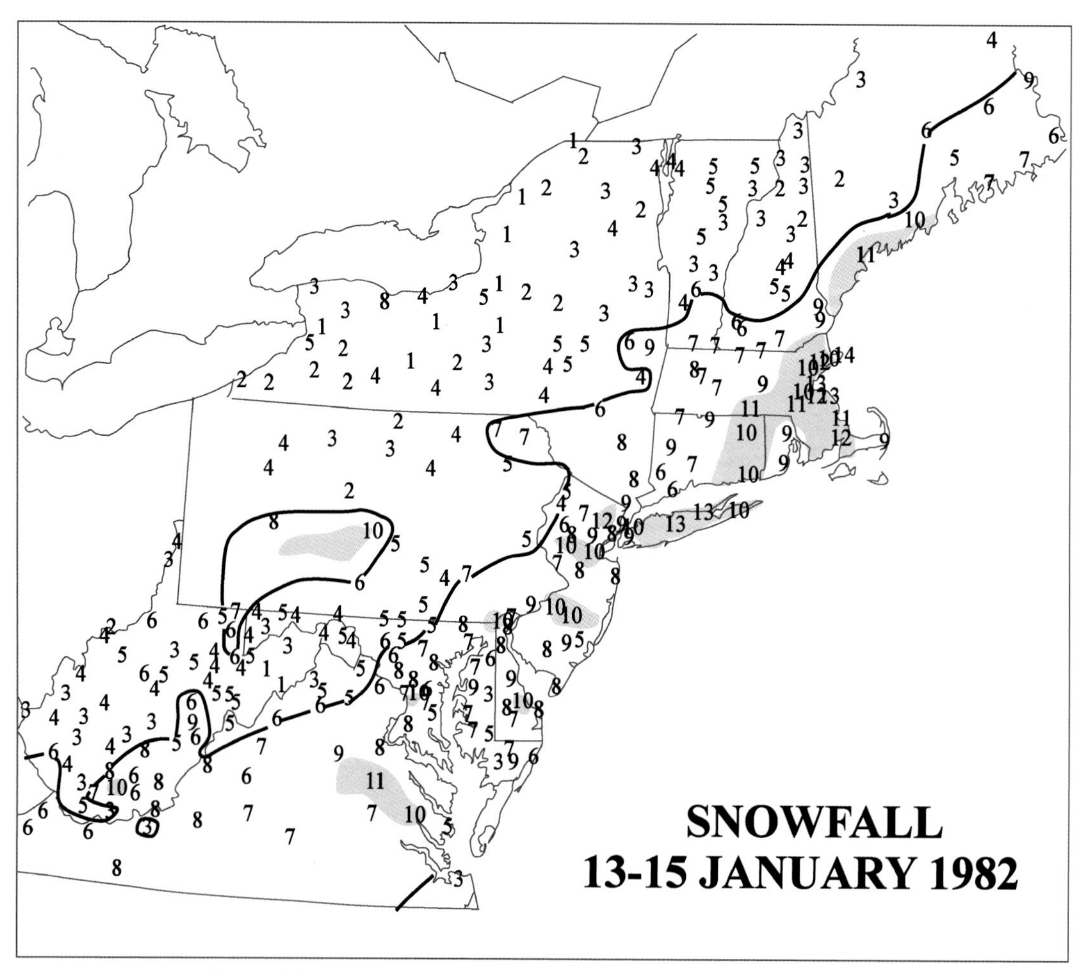

FIG. 11.2-22. Snowfall (in.) for 13–15 Jan 1982. See Fig. 11.1-1 for details.

New England on 13 January. A second storm system moved rapidly northeastward from the Gulf of Mexico the next day (14 January) and was off the North Carolina coast by 0000 UTC 15 January (Fig. 11.2-23). A relatively narrow band of briefly moderate to heavy snowfall with up to 4–8-in. snowfall accumulations occurred along the coastal sections of the middle Atlantic states. The fast movement of this system and the sharp boundary between snow and no snow on the western edge of the precipitation shield restricted the areal coverage of the snowfall associated with this system. The combination of these two systems did result in a significant snowfall, but total snowfall amounts were less than most of the major snowstorms described earlier, especially in the urban corridor.

This case is an example of two upper short-wave and jet streak systems that did not merge nor phase soon enough to produce a major snowfall, but remained separate entities that only began to merge together off the East Coast by 0000 UTC 15 January. The first system appeared as a relatively weak upper-level short-wave trough with a distinct jet streak at 400 hPa at 0000 UTC 14 January (Fig. 11.2-24), with a strong warm-air advection pattern and inverted trough along the East Coast (Fig. 11.2-23) that contributed to the moderate snowfall. The second system was a more significant upper-level trough but acted on a baroclinic field significantly modified by the first storm system (Fig. 11.2-23) that shifted the temperature gradients farther east. The second short-wave disturbance also deamplified by 0000 UTC 15 January as it became negatively tilted along the Carolina coast (Fig. 11.2-24), but with little diffluence evident downstream of the trough axis extending from the Great Lakes to Georgia. The lack of diffluence with this trough and the elongated advection pattern along the East Coast restricted the moderate to heavy snow along a narrow area immediately along the coast on 15 January, as the surface low tracked 100–200 km east of the North Carolina coast.

i. 8–9 March 1984: NESIS = 1.29

This rapidly moving storm produced a swath of 6–10 in. (15–25 cm) of snow from the suburbs of Wash-

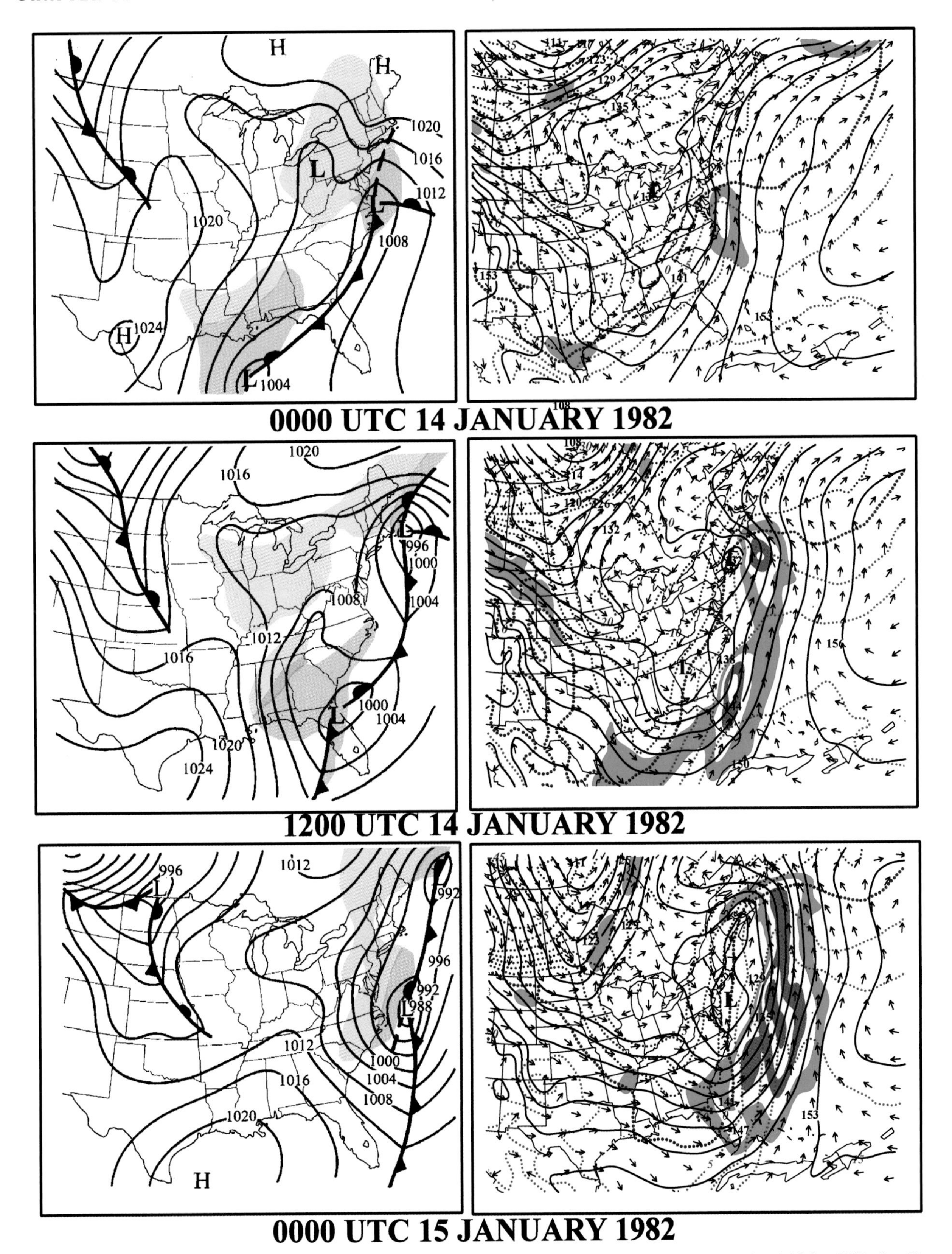

FIG. 11.2-23. Sequence of (left) surface and (right) 850-hPa charts for the 24-h period between 0000 UTC 14 and 15 Jan 1982. See Fig. 11.1-2 for details.

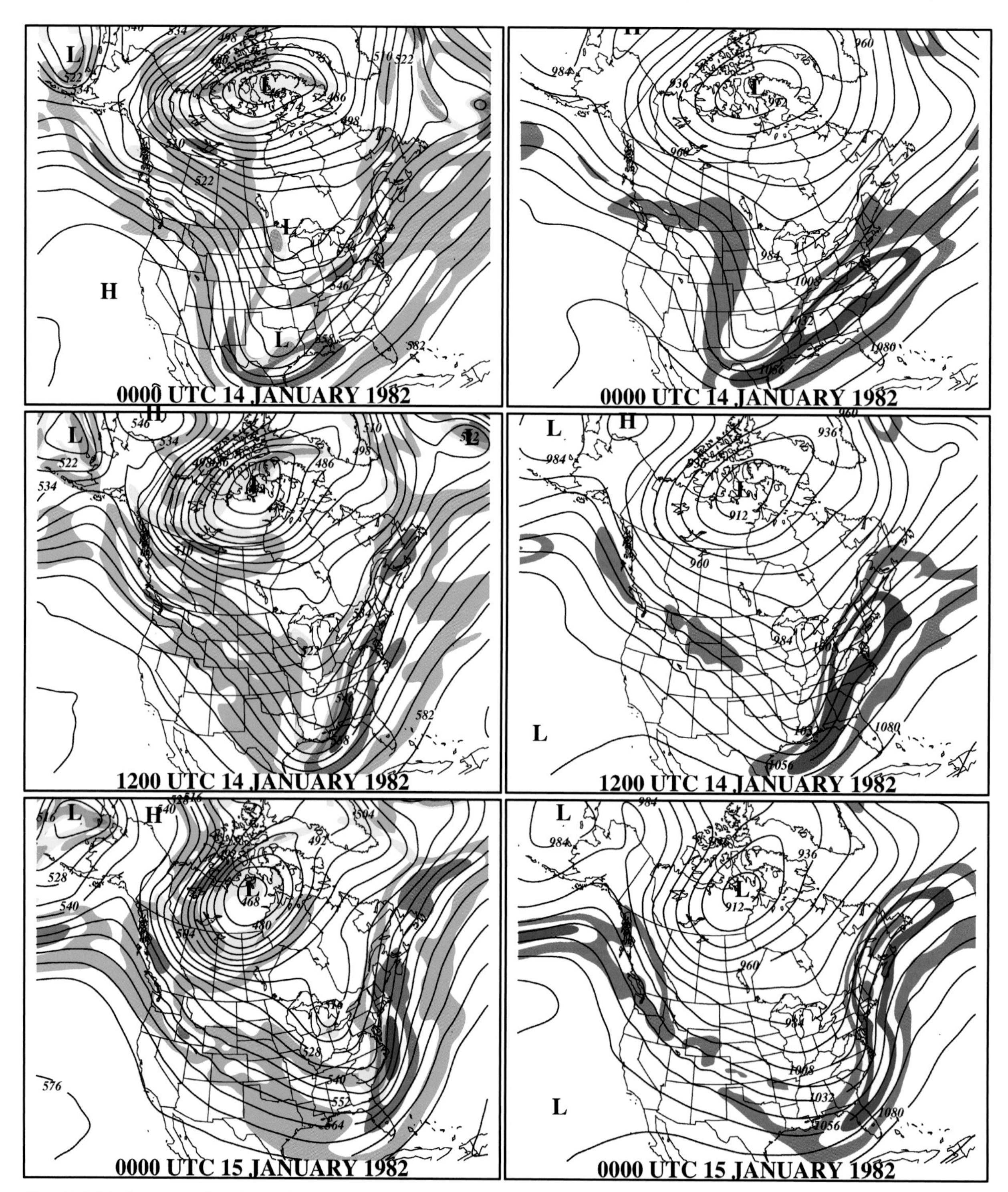

FIG. 11.2-24. Sequence of analyses of (left) 500- and (right) 250-hPa geopotential height and upper-level wind fields for the 24-h period between 0000 UTC 14 and 15 Jan 1982. See Fig. 11.1-3 for details.

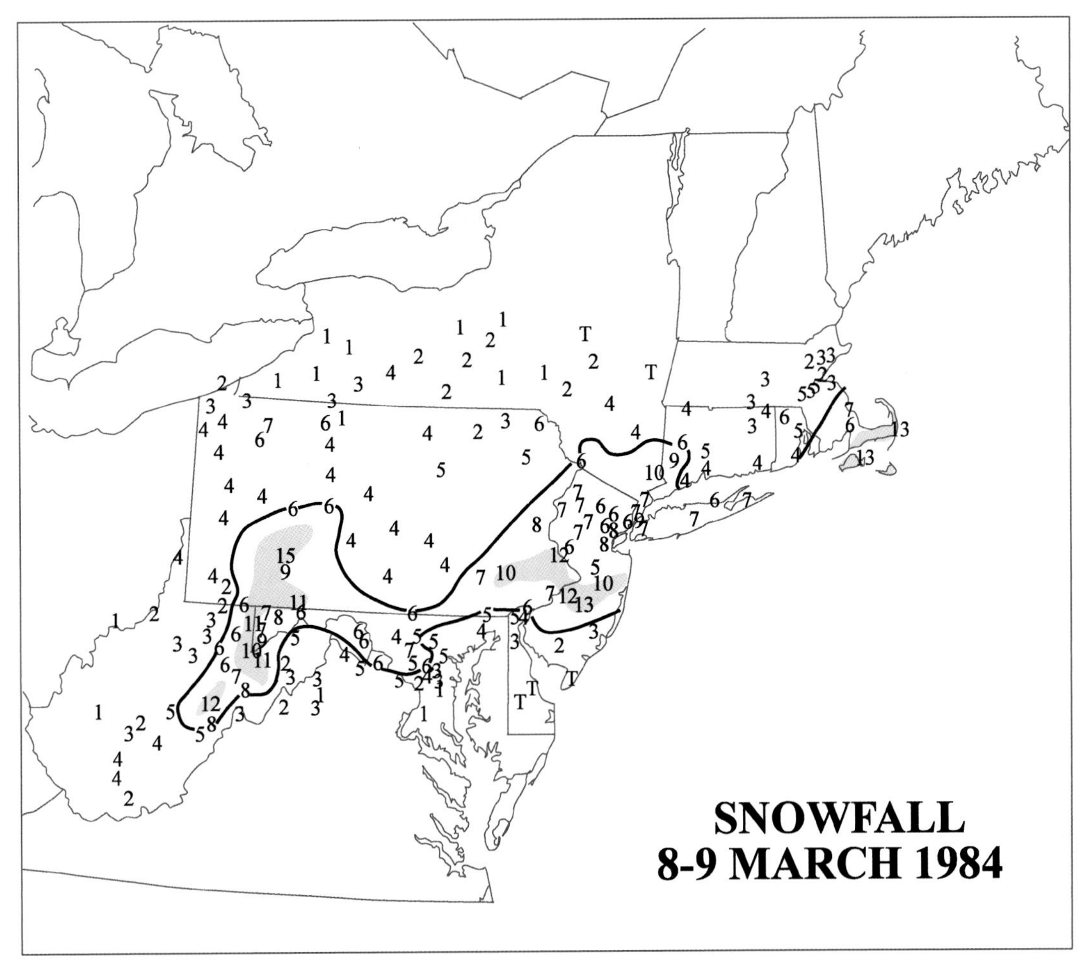

FIG. 11.2-25. Snowfall (in.) for 8–9 Mar 1984. See Fig. 11.1-1 for details.

ington to southern New England, with isolated reports of up to 13 in. (33 cm) over central New Jersey (Fig. 11.2-25). The storm is noted for an outbreak of convectively driven snow that occurred during the evening rush hour in the Washington–Baltimore metropolitan area that dropped nearly 6 in. (15 cm) of snow in 1.5 h, creating gridlock conditions that lasted well into the night. The precipitation then evolved into a more general snowfall pattern that spread northeastward toward New York and New England. Snow fell for periods of generally 12 h or less, with rates briefly approaching 1–2 in. (2.5–5 cm) h^{-1} (See Kocin et al. 1985).

The surface low had the characteristics of a typical "Alberta clipper" that progresses southeastward from the northern plains into the Ohio Valley. As the surface low moved toward the Atlantic coast and weakened, a pronounced secondary low developed over Virginia and intensified 12 hPa in 12 h as it moved northeastward off the southern New England coast (Fig. 11.2-26).

An amplifying upper-level short-wave trough and associated jet streak with a very distinct exit region at 400 hPa (Fig. 11.2-27) supported the development of the primary low at 1200 UTC 8 March and the rapid redevelopment of the secondary surface low over Virginia by 0000 UTC 9 March. Cold air was provided by the confluent upper-level flow pattern across southeastern Canada, which supported the extension of the cold surface high over the Great Lakes eastward to New England and southward to the mid-Atlantic states. The 850-hPa analysis depicts a closed circulation pattern moving rapidly from west to east on 8–9 March, with warm-air advection ahead of the vortex center and a cold-air advection pattern behind. The warm-air advection was enhanced by a 25 + m s^{-1} low-level jet in the Ohio Valley at 1200 UTC 8 March and in the mid-Atlantic states at 0000 UTC 9 March, located beneath the diffluent exit region of the polar jet (at 400 hPa) and directed toward the region of heaviest snow.

Although this case was marked by distinct upper- and lower-troposphere features conducive to heavy snowfall and the outbreak of convection during the secondary

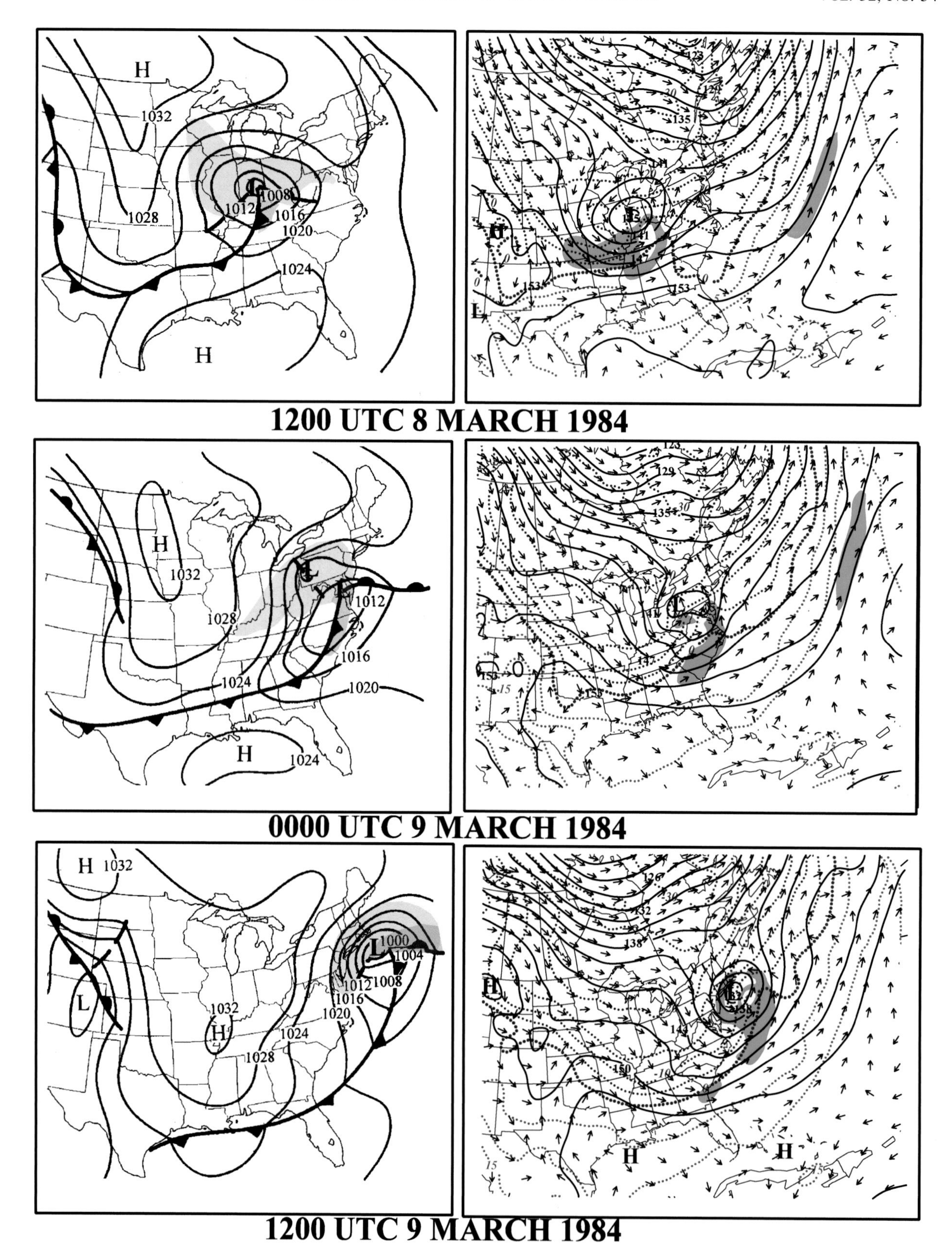

FIG. 11.2-26. Sequence of (left) surface and (right) 850-hPa charts for the 24-h period between 1200 UTC 8 and 9 Mar 1984. See Fig. 11.1-2 for details.

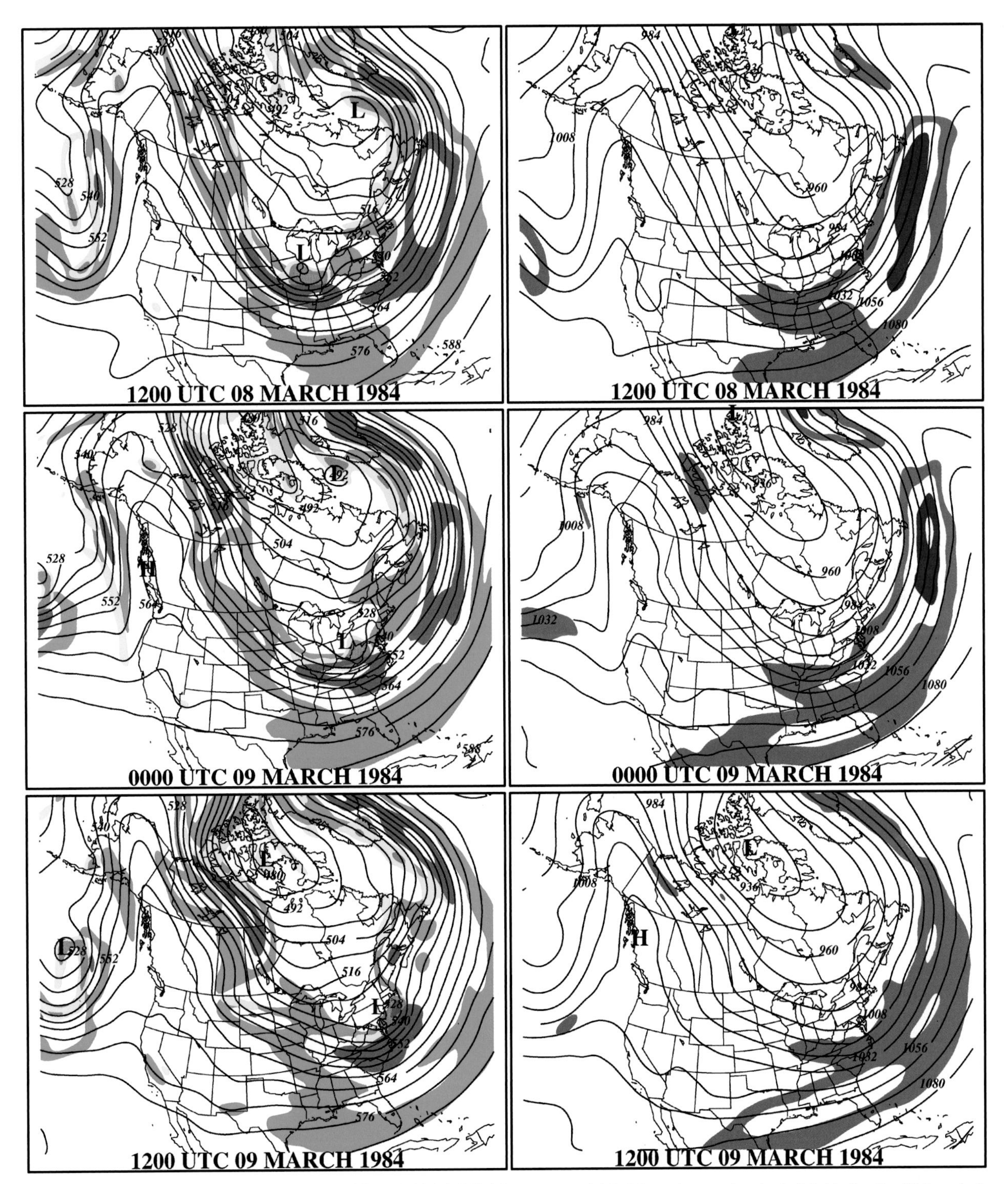

FIG. 11.2-27. Sequence of analyses of (left) 500- and (right) 250-hPa geopotential height and upper-level wind fields for the 24-h period between 1200 UTC 8 and 9 Mar 1984. See Fig. 11.1-3 for details.

redevelopment of the surface low, snowfall remained in the moderate category due to the short duration of the snowfall (less than 12 h) related to the rapid speed of this system.

j. 7–8 January 1988: NESIS = 4.85

This case is a typical moderate Northeast urban corridor snowfall that had the potential to produce greater

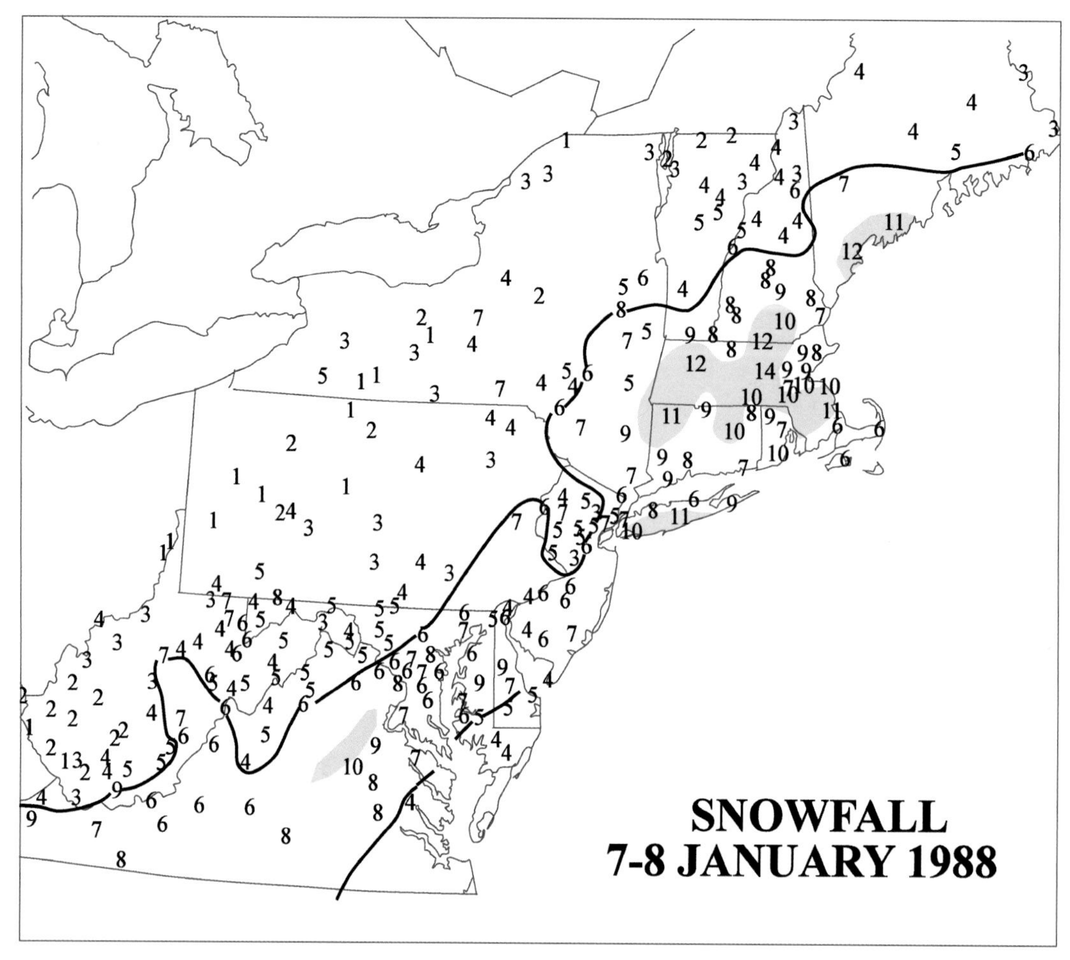

FIG. 11.2-28. Snowfall (in.) for 7–8 Jan 1988. See Fig. 11.1-1 for details.

accumulations since many of the ingredients for heavy snowfall were present. The storm originated in the Gulf of Mexico on 7 January, with two inverted troughs extending northward: one west of the Appalachians and one up the southeast Atlantic coastline. During this stage, the storm produced record snowfall amounts over Arkansas and Tennessee (Mote et al. 1997) and elsewhere, accounting for the high NESIS value of 4.85. As the surface low moved from the northern Gulf of Mexico to the North Carolina coast at 1200 UTC 8 January and then northeastward along the East Coast (Fig. 11.2-29), only moderate snowfall accumulations of 4–10 in. (10–25 cm) materialized in the Northeast urban corridor (Fig. 11.2-28), without the significant snowfall rates observed in other cases.

An examination of the upper-level charts (Fig. 11.2-30) reveals that as this storm began moving toward the East Coast on 8 January, the upper confluence within the entrance region of a well-defined jet streak in the Northeast and a related cold anticyclone were clearly in place. But the upper-level short-wave trough in the south-central United States was deamplifying with time. The associated upper-level jet streaks were poorly defined as the system approached the East Coast, especially at the 400-hPa level. As the upper trough neared the East Coast and weakened, the short-wave system was barely discernible in the 500-hPa height field at 1200 UTC 8 January, during the period when the surface low was developing along the East Coast. Furthermore, the surface low seemed to accelerate northeastward as the upper-level trough damped within the strong west-to-southwest flow field. The net result was a modest surface cyclone and associated weak 850-hPa circulation and modest temperature advection pattern (Fig. 11.2-29) yielding snowfall rates that remained at less than an inch (2.5 cm) per hour as the storm accelerated up the coast.

k. 27–28 December 1990: NESIS = 1.56

This is a case marked by a surface low pressure system that slowly developed along the East Coast within a highly confluent upper-level flow regime. Despite the presence of a larger-scale upper ridge across the eastern

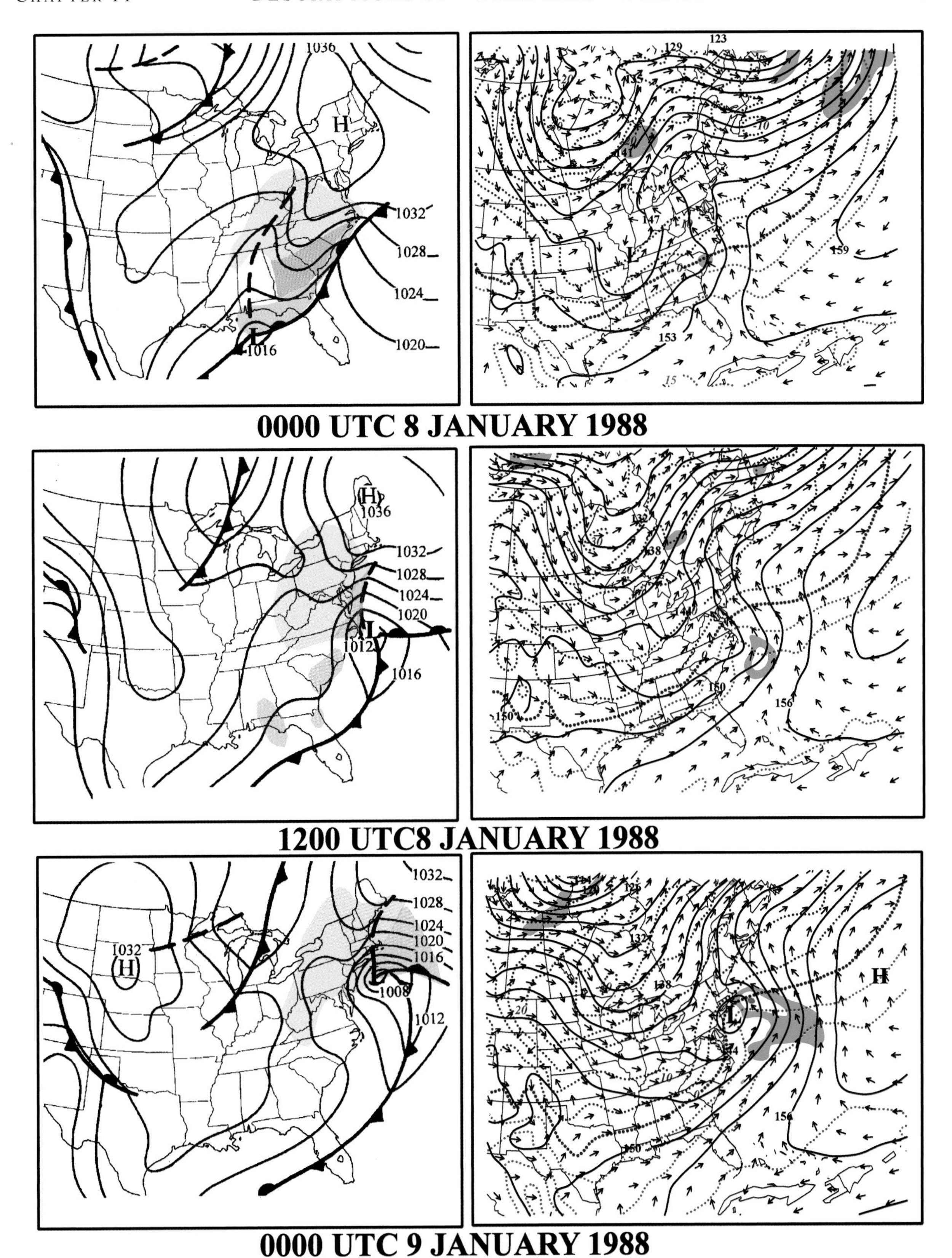

FIG. 11.2-29. Sequence of (left) surface and (right) 850-hPa charts for the 24-h period between 0000 UTC 8 and 9 Jan 1988. See Fig. 11.1-2 for details.

FIG. 11.2-30. Sequence of analyses of (left) 500- and (right) 250-hPa geopotential height and upper-level wind fields for the 24-h period between 0000 UTC 8 and 9 Jan 1988. See Fig. 11.1-3 for details.

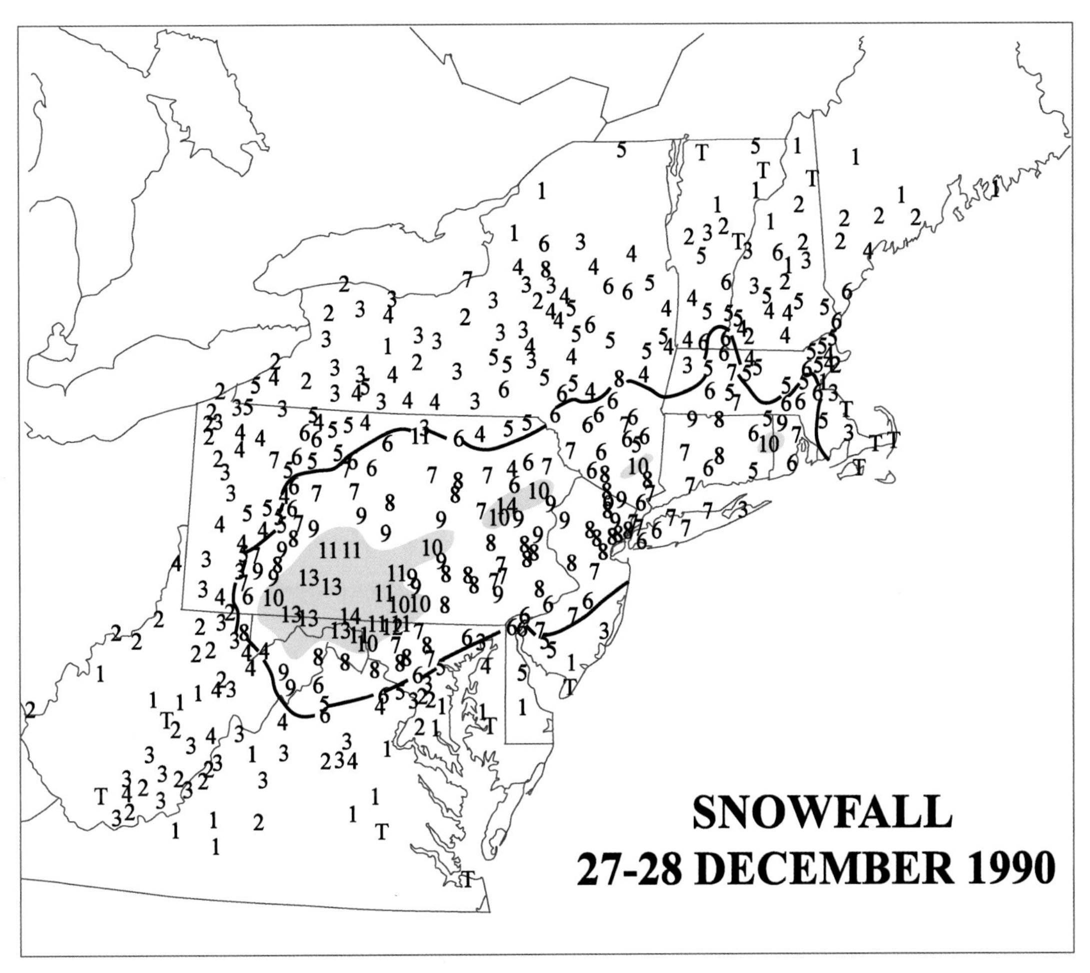

FIG. 11.2-31. Snowfall (in.) for 27–28 Dec 1990. See Fig. 11.1-1 for details.

United States, strong confluence over the northeast United States acted to establish a high pressure cell and related cold surface temperatures along much of the east coast during the prestorm period. Two inverted troughs, one to the west of the Appalachian Mountain chain and one along the East Coast, were dominant features, with a surface low developing along the coast by 1200 UTC 28 December (Fig. 11.2-32). Moderate snowfall developed in the Ohio Valley and moved into the middle Atlantic states as these inverted troughs extended northeastward toward the cold anticyclone anchored over the Northeast, all within the confluent entrance region of the upper-level jet streak extending from the Great Lakes to southeastern Canada on 27–28 December at both 400 and 250 hPa (Fig. 11.2-33). An area of 6–10 in. (15–25 cm) of snow occurred from northern Maryland and West Virginia northeastward into southern New England (Fig. 11.2-31). Heaviest snows of 10–15 in. (25–38 cm) were found in southern Pennsylvania and the panhandles of Maryland and West Virginia.

A short-wave trough at 500 hPa over the south-central United States at 1200 UTC 27 December (Fig. 11.2-33) also played a role in the development of the inverted trough and related snowfall. The short wave at first amplified, then deamplified, and was absorbed within the strong confluent flow marking the entrance region of the jet streak over New England as moderate to heavy snows developed from West Virginia to Massachusetts and cold, shallow air remained entrenched east of the Appalachian Mountains.

This was a difficult forecast situation since the operational forecast models [Limited Fine Mesh (LFM) and Nested-Grid Model (NGM)] predicted little surface low development, rising 500-hPa geopotential heights and 1000–500-hPa thicknesses, and a weak upper short-wave trough approaching from the southwest. Forecasters anticipated a rapid warming trend as the large-scale upper trough amplified over the western United States and an upper-level ridge amplified across the eastern United States. The presence of the weak short-wave disturbance within the highly confluent flow was not expected to produce the outbreak of widespread snow and weak cyclogenesis that resulted in a moderate snowfall.

The case of 27–28 December 1990 provides an interesting contrast with the preceding January 1988 mod-

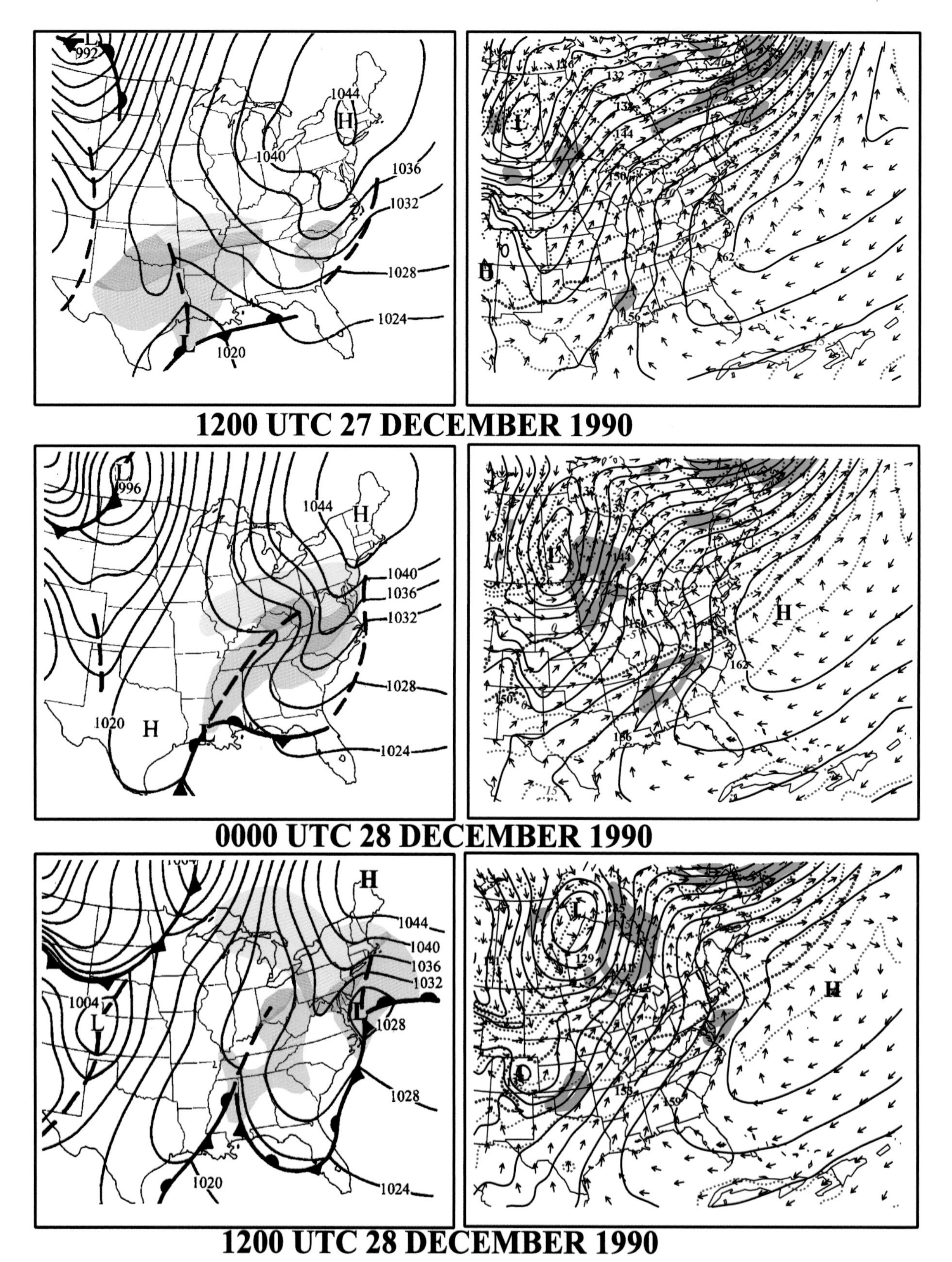

FIG. 11.2-32. Sequence of (left) surface and (right) 850-hPa charts for the 24-h period between 1200 UTC 27 and 28 Dec 1990. See Fig. 11.1-2 for details.

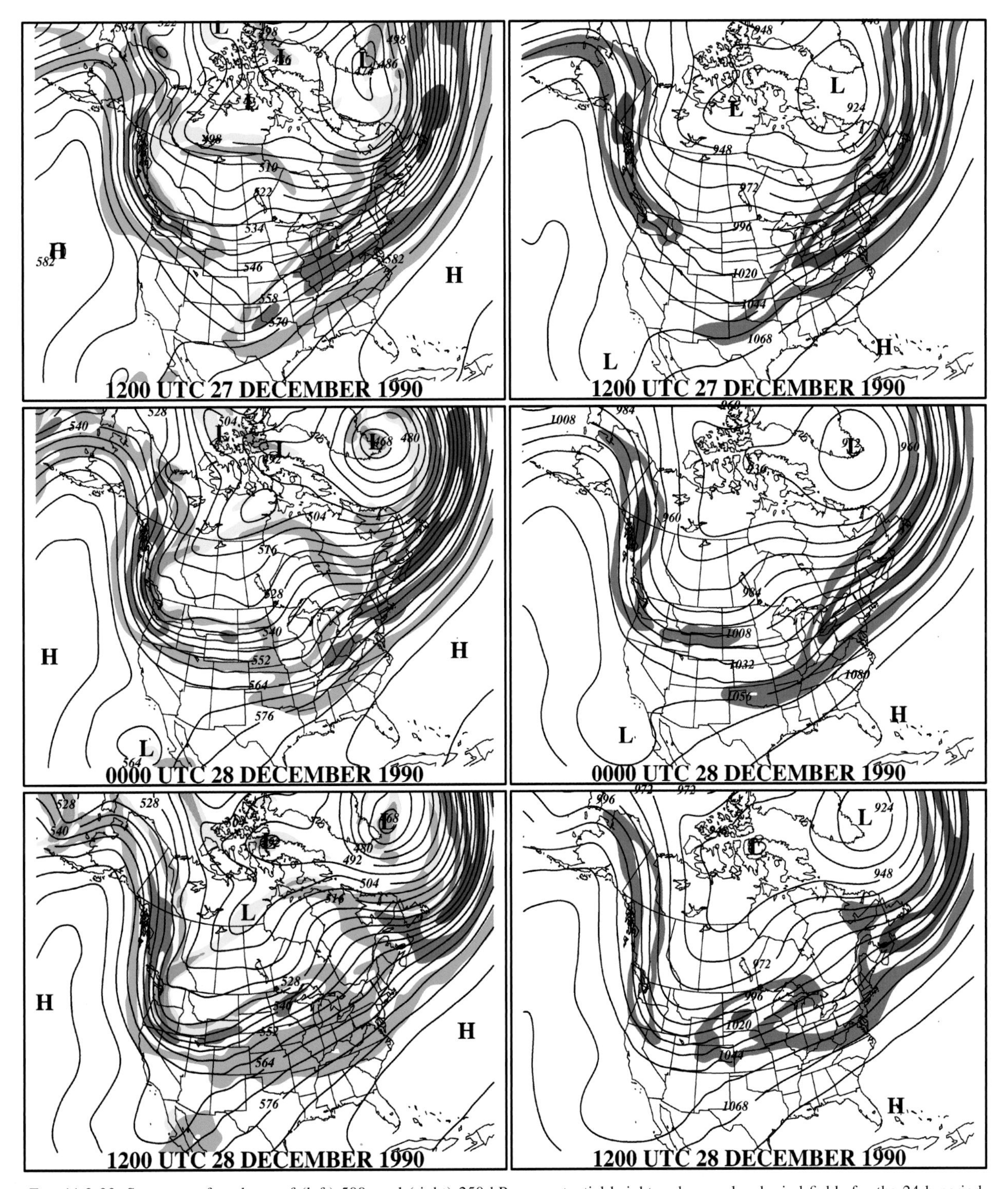

FIG. 11.2-33. Sequence of analyses of (left) 500- and (right) 250-hPa geopotential height and upper-level wind fields for the 24-h period between 1200 UTC 27 and 28 Dec 1990. See Fig. 11.1-3 for details.

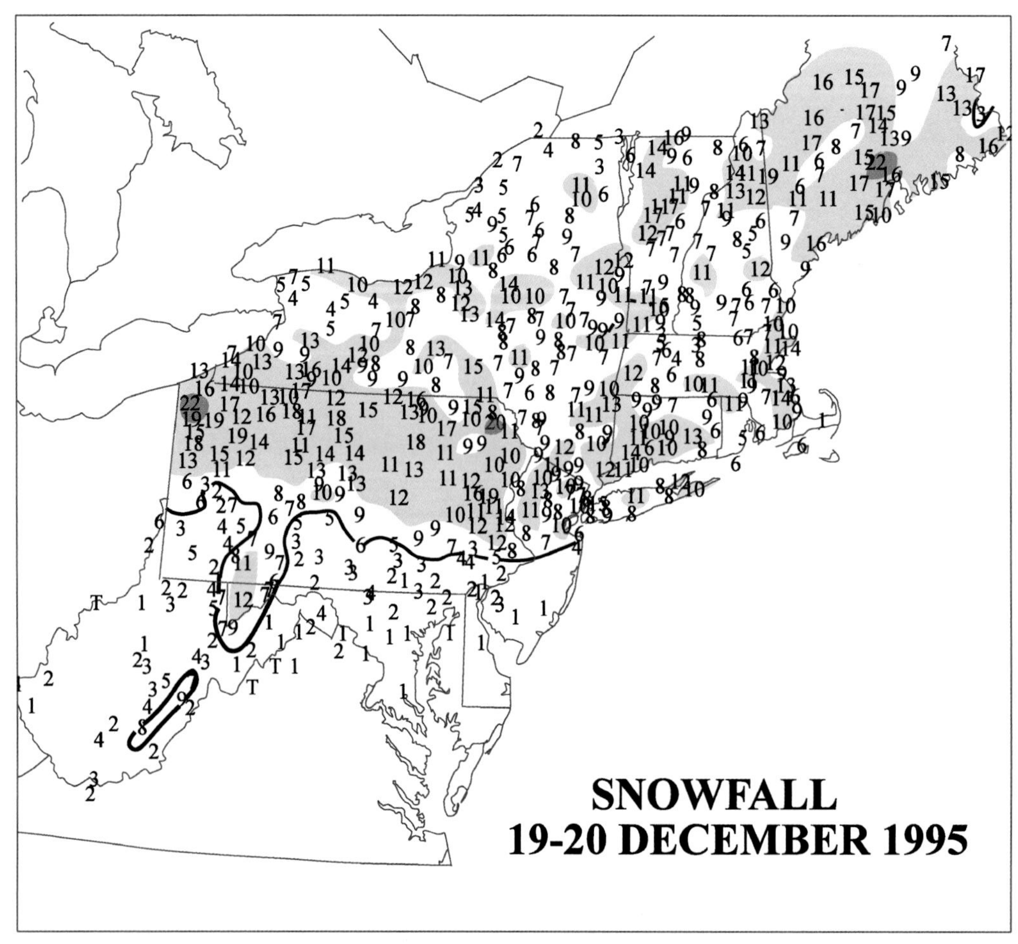

FIG. 11.2-34. Snowfall (in.) for 19–20 Dec 1995. See Fig. 11.1-1 for details.

erate snowfall case in a number of ways. Prior to its occurrence, the January 1988 case appeared to present the threat of a major snowstorm to the Northeast urban corridor because of the presence of a significant cold anticyclone, a significant upper-level trough, and the associated dual-jet-streak pattern. However, that threat later diminished because the upper-level trough weakened and the surface low did not strengthen appreciably. The December 1990 case, by contrast, initially did not appear to present a threat for producing moderate or heavy snowfall because of the absence of a significant upper-level trough. In this case, a large, cold surface anticyclone was located beneath the entrance region of a significant jet streak embedded within highly confluent flow across eastern Canada and the northeastern United States (Fig. 11.2-33). Both cases are examples of moderate snowstorms that occur with relatively weak upper-level trough disturbances that are marked by decreasing amplitude as the trough becomes absorbed within the confluent southwesterly flow over New England. The cases are also marked by cold anticyclones (and weak surface low pressure systems) that appear to be sustained by the general confluent flow regime in the Northeast. Finally, the outbreak of snowfall is consistent with the rising motion expected within the entrance region of the jet streaks located within the confluent zone. As such, this factor seems to be a critical element for the snowfall noted in both of these cases.

l. 19–21 December 1995: NESIS = 3.32

During the snowiest winter of the 20th century for the Northeast urban corridor, one dominated by the negative phase of the NAO (Fig. 2-20, volume I; also see Table 2-5), this case could be considered a major urban snowstorm because there is a significant band of snow greater than 10 in. (25 cm) (Fig. 11.2-34). However, much of the snow fell over a fairly long period of time without excessive snowfall rates. This case is considered a "moderate" event because it did not deliver the big snows even though there are notable similarities with other major snowstorms, especially for the precyclogenetic period.

By 1200 UTC 19 December, cold high pressure was poised north of the U.S.–Canadian border and a significant low pressure system was moving into the Ohio Valley, marked by a well-defined 850-hPa circulation and upper-level trough at 500 and 250 hPa (Figs. 11.2-

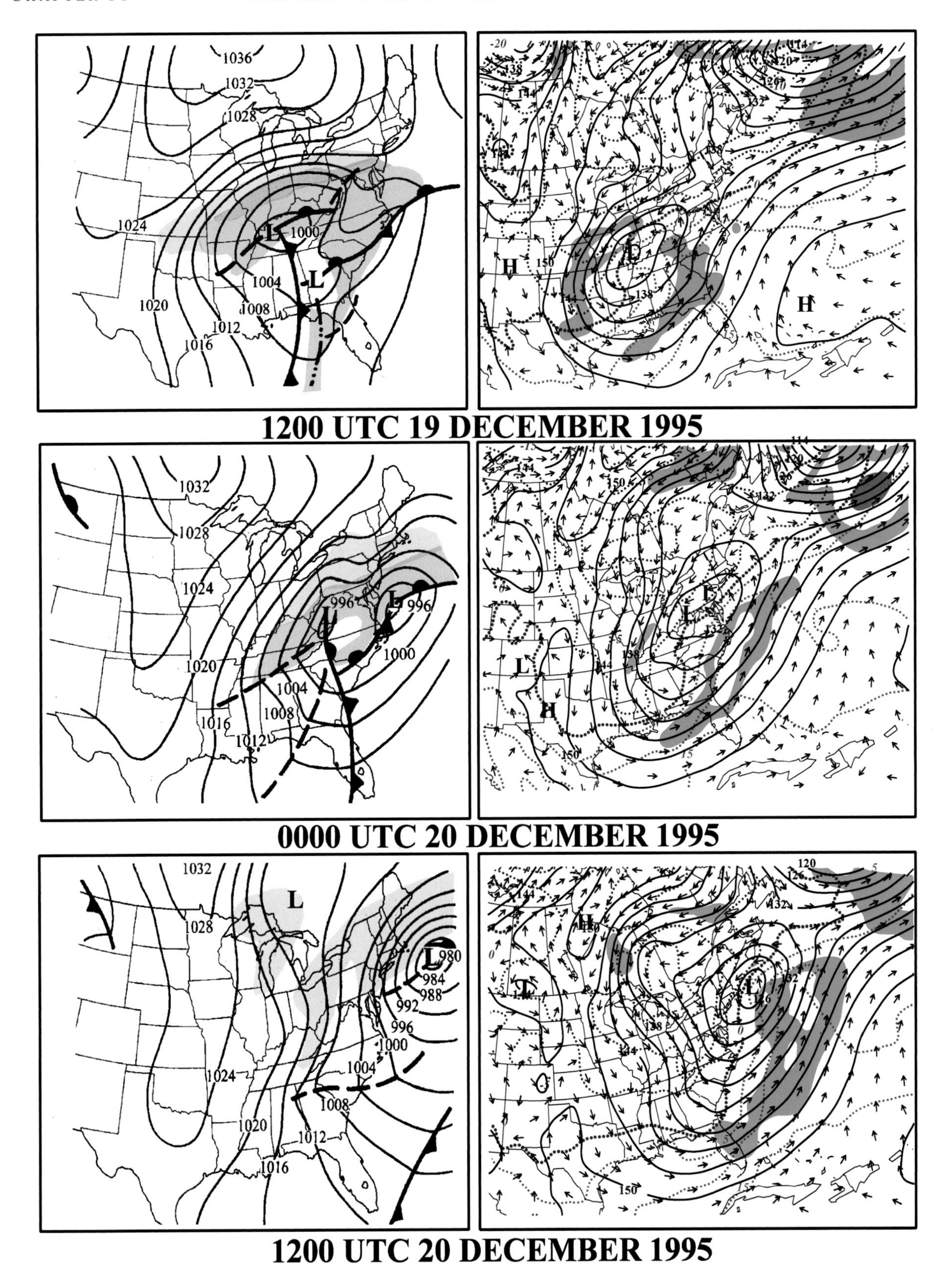

FIG. 11.2-35. Sequence of (left) surface and (right) 850-hPa charts for the 24-h period between 1200 UTC 19 and 20 Dec 1995. See Fig. 11.1-2 for details.

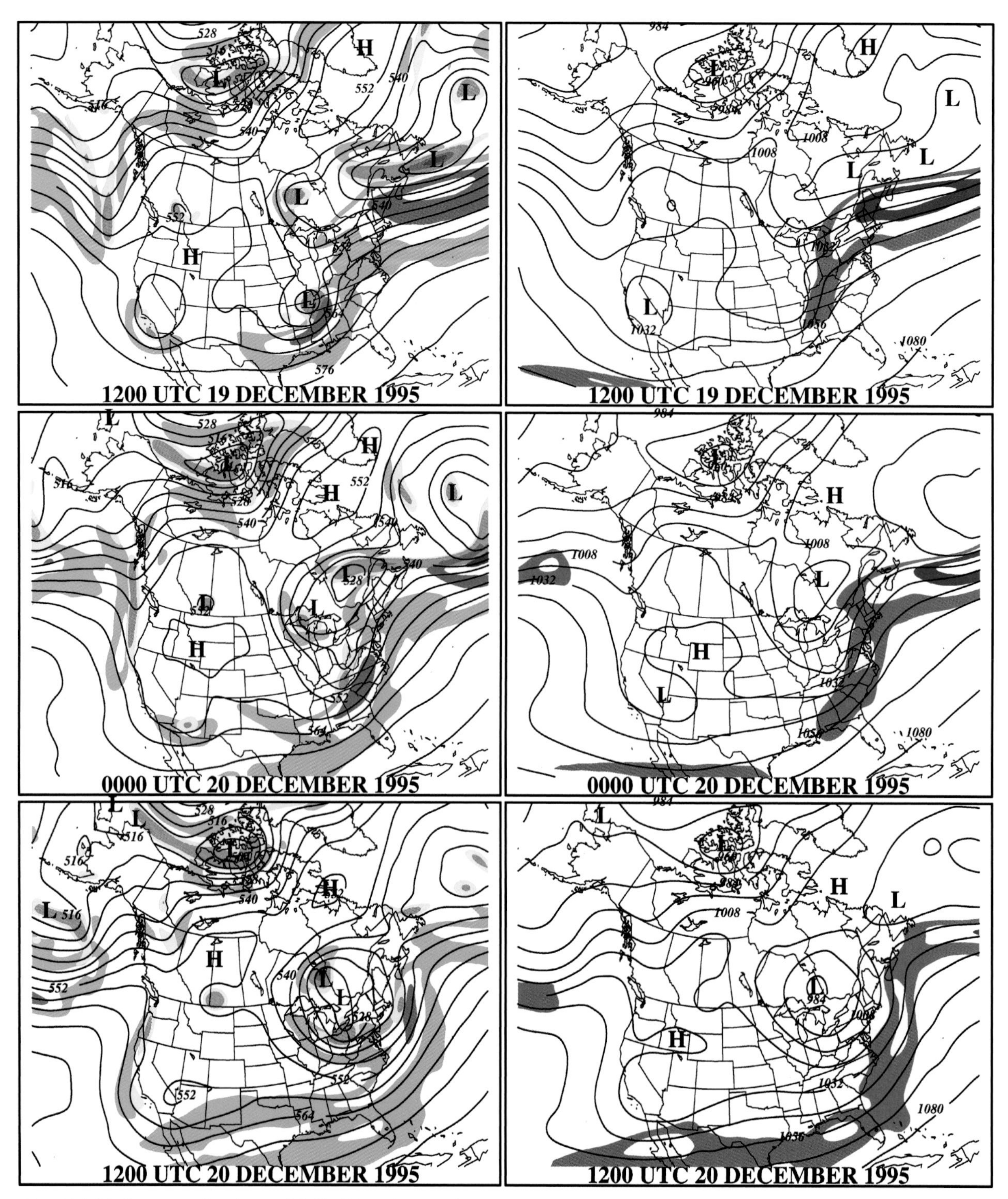

FIG. 11.2-36. Sequence of analyses of (left) 500- and (right) 250-hPa geopotential height and upper-level wind fields for the 24-h period between 1200 UTC 19 and 20 Dec 1995. See Fig. 11.1-3 for details.

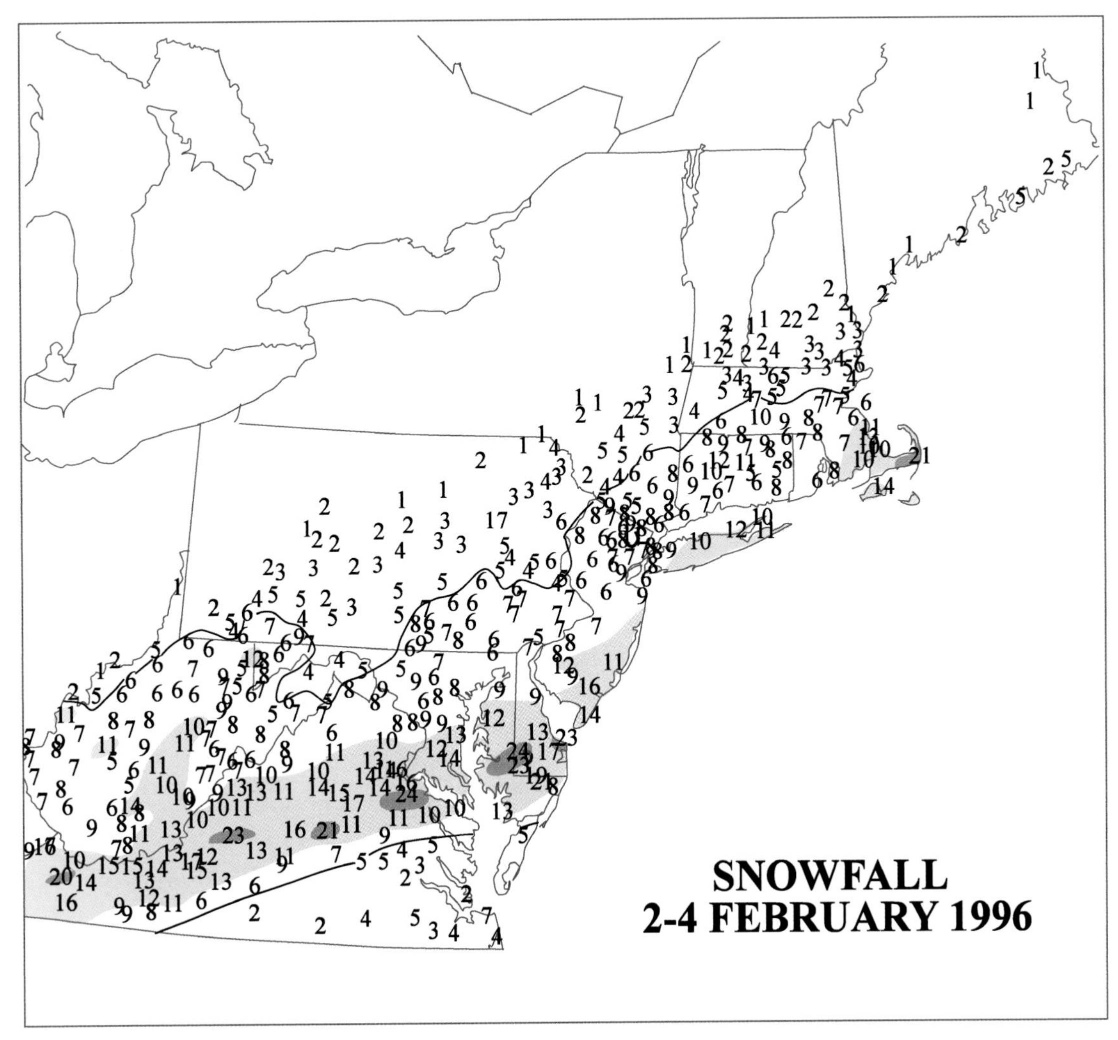

FIG. 11.2-37. Snowfall (in.) for 2–4 Feb 1996. See Fig. 11.1-1 for details.

35 and 36). A well-defined jet streak within the upper-level confluent flow is also evident over the Northeast, supporting the cold anticyclone and the expanding precipitation shield. A surface low over the Ohio Valley at 1200 UTC 19 December moved toward West Virginia as secondary cyclogenesis commenced east of Virginia by 0000–1200 UTC 20 December. As the upper-level trough neared the East Coast by 1200 UTC 20 December, the trough amplitude decreased as a broad closed upper low center formed north of New York. Although a secondary surface cyclone formed off Virginia and moved northeastward off the southern New England coast, the 850-hPa low center deepened only slowly and the advection patterns at the 850-hPa level did not amplify appreciably on 19–20 December. As the 850-hPa low extended eastward toward the East Coast, no easterly low-level jet greater than 20 m s^{-1} appeared to the north of the low. These combined factors, including the complex interaction of the upper troughs and unimpressive advection patterns at 850 hPa contributed to the moderate, rather than heavy, snowfall rates that accompanied this storm.

m. 2–4 February 1996: NESIS = 2.03

This moderate snowstorm during the snowy winter of 1995/96 evolved within a particularly complex flow field. The snowfall was associated with a series of surface lows tracking northeastward from the Gulf of Mexico along a frontal zone along the southeast coast, ushering in a very cold air mass associated with a strong high pressure system centered in the middle United States during the entire event. Most snowfall amounts within the Northeast urban corridor were generally between 6 and 10 in. (15 and 25 cm; Fig. 11.2-37), although a mesoscale snow band produced up to 15 in. (38 cm) of snow across portions of eastern Virginia, southern Maryland, and southern Delaware (see volume I, chapter 6).

This case is marked by some of the highest wind

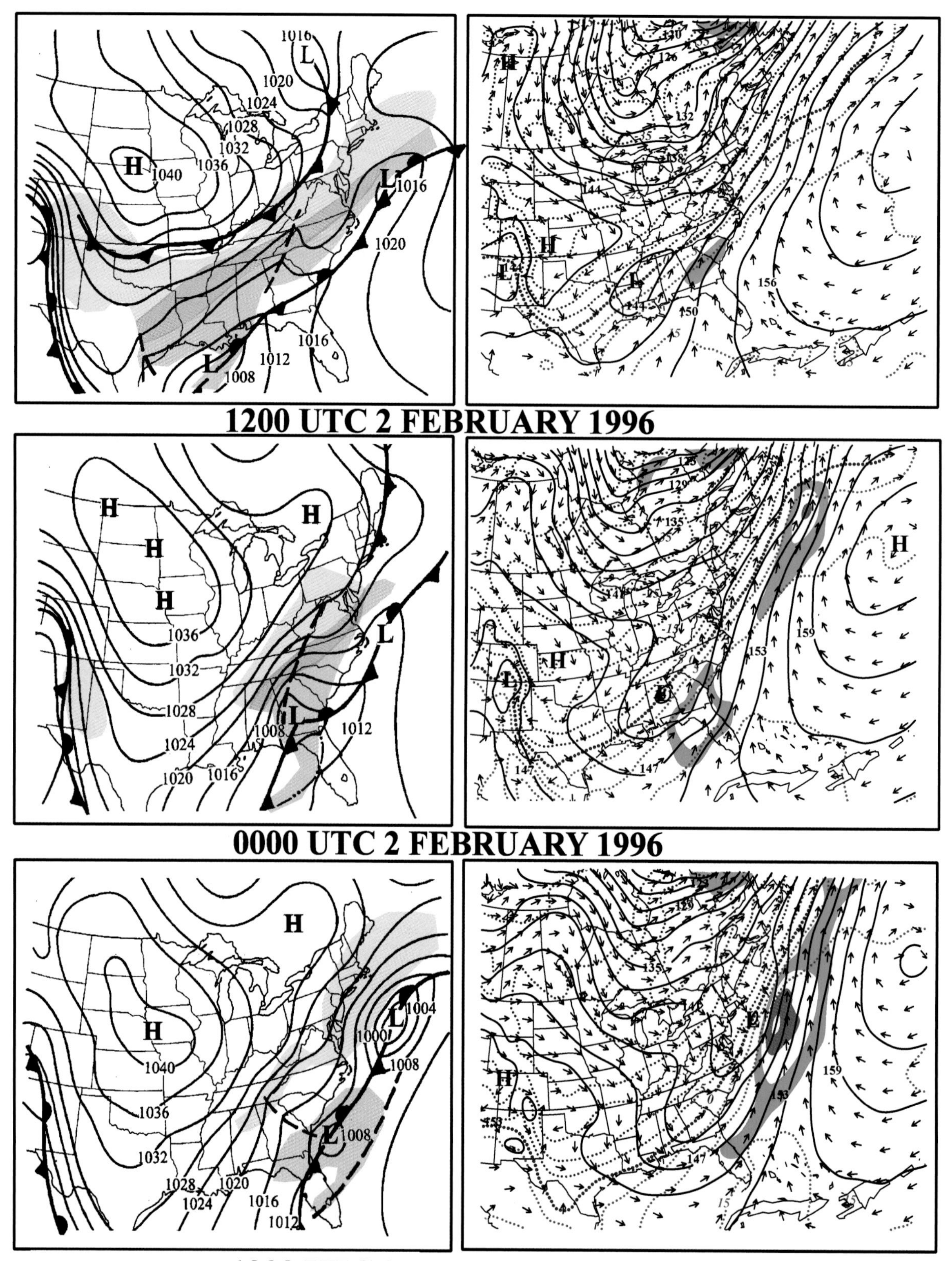

FIG. 11.2-38. Sequence of (left) surface and (right) 850-hPa charts for the 24-h period between 1200 UTC 2 and 3 Feb 1996. See Fig. 11.1-2 for details.

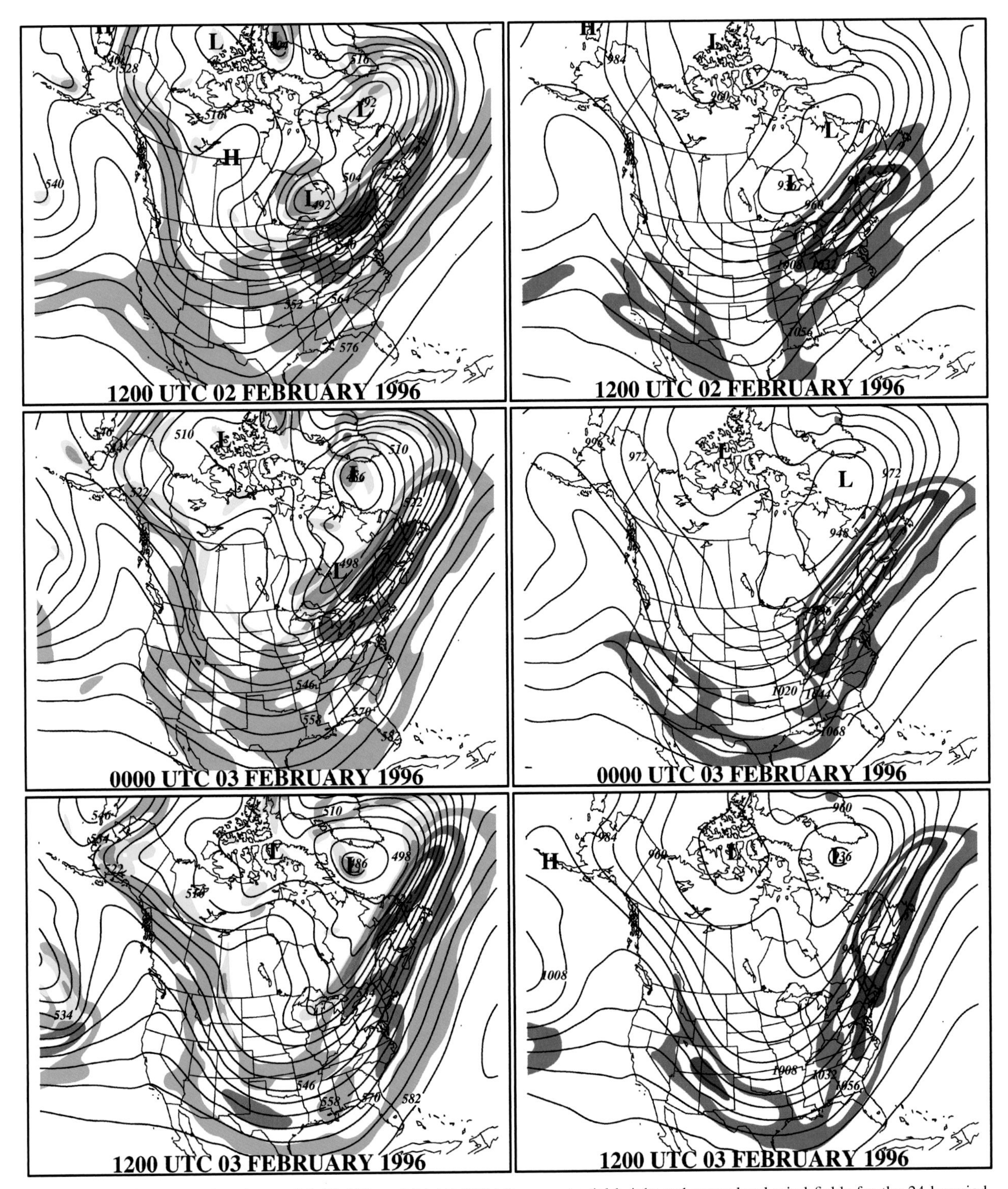

FIG. 11.2-39. Sequence of analyses of (left) 500- and (right) 250-hPa geopotential height and upper-level wind fields for the 24-h period between 1200 UTC 2 and 3 Feb 1996. See Fig. 11.1-3 for details.

speeds within the upper-level confluent jets observed in any of the cases examined in this monograph, with maximum wind speeds north of New England exceeding 110 m s^{-1} at the 250-hPa level at 0000 UTC 3 February (Fig. 11.2-39). Temperatures cold enough for snow were provided by the large, cold anticyclone located in the central United States and extending toward New York beneath the confluent entrance region of the upper-level

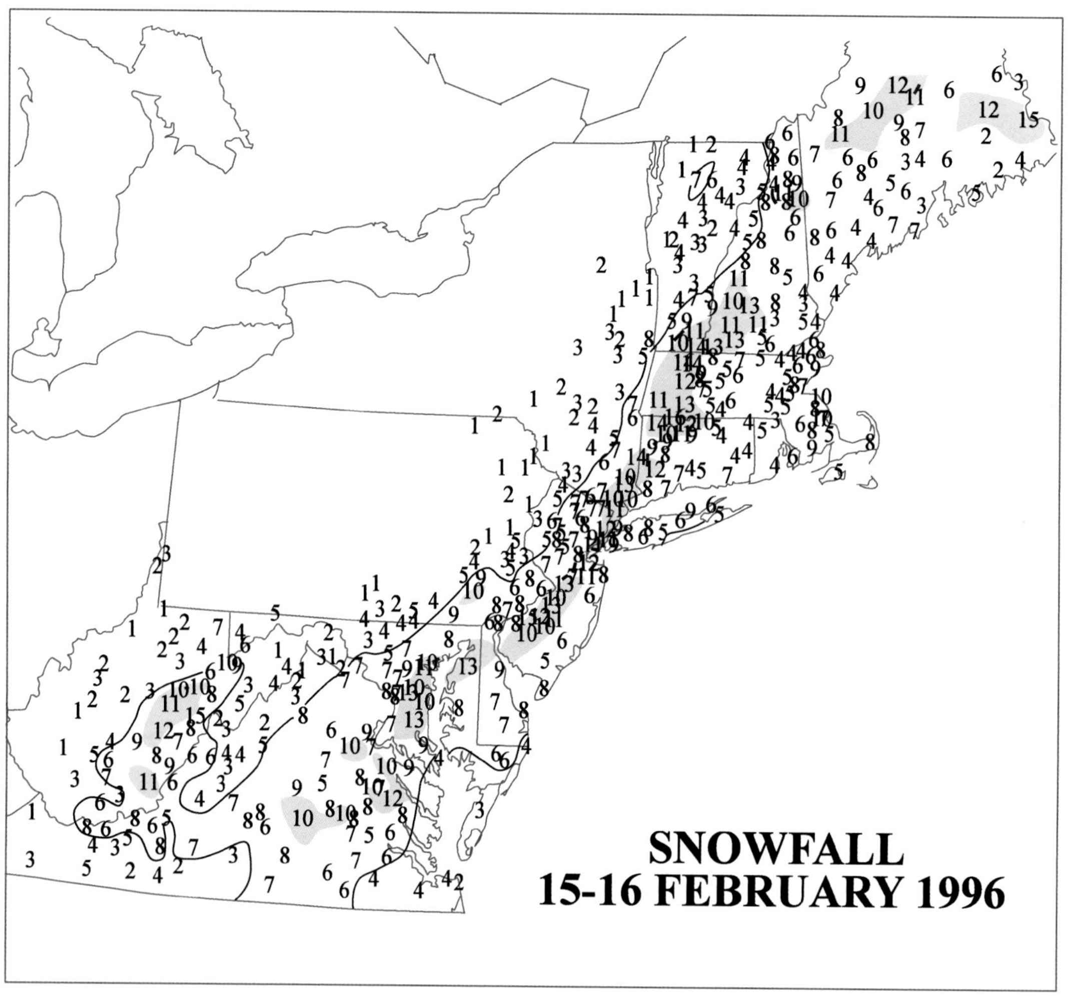

FIG. 11.2-40. Snowfall (in.) for 15–16 Feb 1996. See Fig. 11.1-1 for details.

jet. The cold air associated with this anticyclone broke numerous cold temperature records in the plains states and upper Midwest on 2 February 1996.

As with some of the other moderate cases, this event was marked by an ill-defined and weakening upper-level short-wave trough that appeared to become stretched out and absorbed within the confluent flow regime over the Northeast (Fig. 11.2-39), with only modest surface cyclogenesis (Fig. 11.2-38). A first short-wave trough on 2–3 February spawned a weak cyclone off the mid-Atlantic coast and generally light snows, except for the previously mentioned mesoscale snow band, which produced heavy snow [up to 15 in. (38 cm)] from central Virginia to Maryland's eastern shore and southern Delaware, accounting for the greater than 20-in. (50 cm) totals depicted in the snowfall chart (Fig. 11.2-37).

The more widespread snowfall event occurred over much of the Northeast urban corridor later on 3–4 February in association with the development of a modest cyclone along the Southeast coast. This cyclone, which moved from Georgia at 0000 UTC 3 February to well east of the Virginia coast by 1200 UTC 3 February (Fig. 11.2-38), was associated with a second, weak upper-level trough that also did not amplify as it moved into the strong confluent entrance region covering the entire eastern United States. This case could be summarized as one dominated by the vertical ascent pattern associated with the strong upper confluent jet streak in the Northeast. The weak and ill-defined upper-level short-wave features and related ill-defined jet streak exit regions moving northeast from the south-central United States appeared to limit the potential for this case. The stretched-out surface lows along the front and weakly defined circulation and advection patterns at 850 hPa are also consistent with the moderate nature of this snow event.

n. 16–17 February 1996: NESIS = 1.65

The second moderate snowstorm during February 1996 developed quite differently than the snowstorm earlier in the month (see previous case). This snowstorm produced a long, narrow band of 6–10-in. (15–25 cm) snows from the Carolinas to New England, with 10–

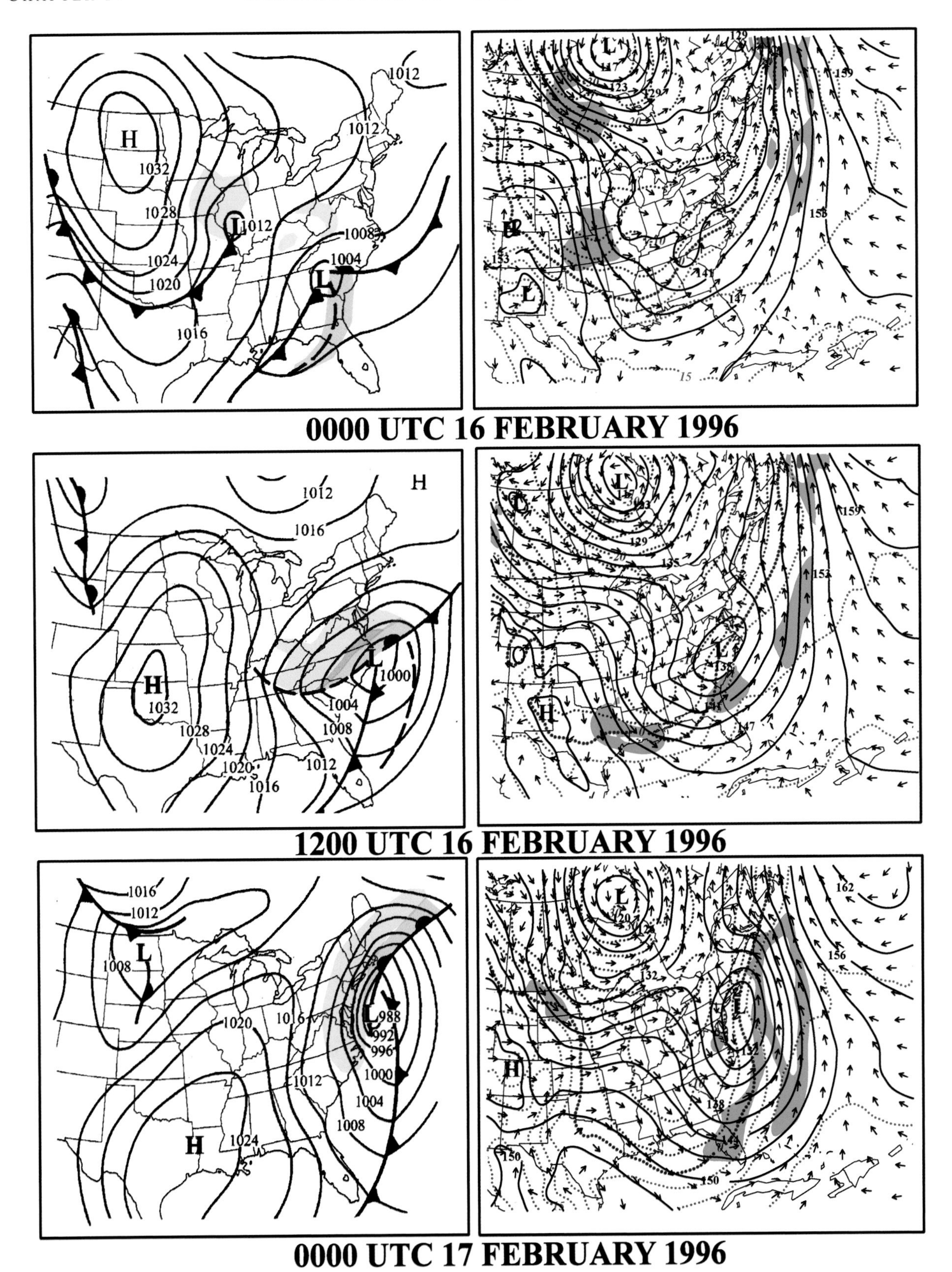

FIG. 11.2-41. Sequence of (left) surface and (right) 850-hPa charts for the 24-h period between 0000 UTC 16 and 17 Feb 1996. See Fig. 11.1-2 for details.

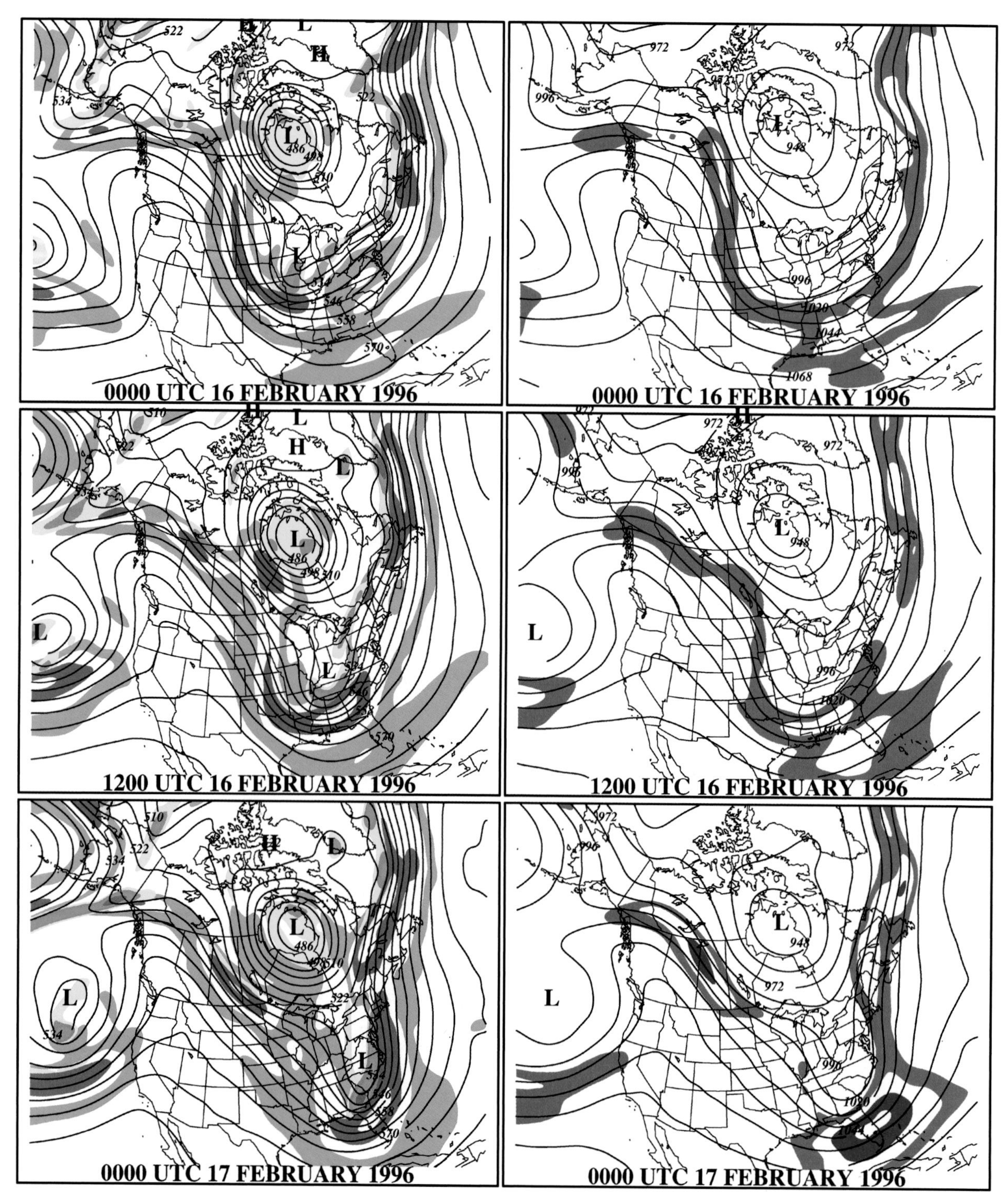

FIG. 11.2-42. Sequence of analyses of (left) 500- and (right) 250-hPa geopotential height and upper-level wind fields for the 24-h period between 0000 UTC 16 and 17 Feb 1996. See Fig. 11.1-3 for details.

15-in. totals (25–38 cm) within the Northeast urban corridor and across western New England (Fig. 11.2-40).

Unlike the 2–4 February 1996 snowstorm, this case was marked by the lack of a confluent jet pattern across the northeast United States during the precyclogenetic period at either 500 or 250 hPa, with a cold surface high moving southward across the central United States and little if any ridging of the cold air across the Great Lakes into the Northeast (Fig. 11.2-41). This case also featured a very strong, upper-level trough and associated

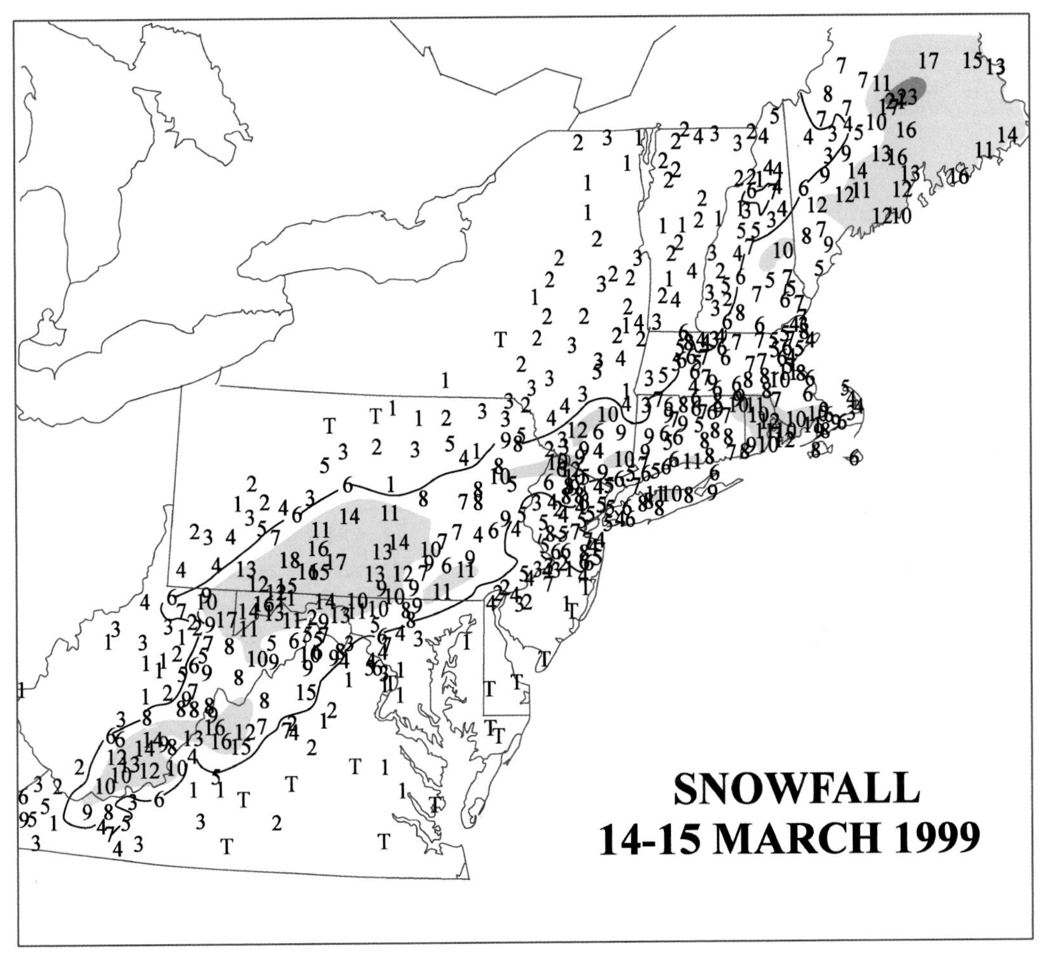

FIG. 11.2-43. Snowfall (in.) for 14–15 Mar 1999. See Fig. 11.1-1 for details.

jet streak amplifying southeastward into the central United States on 16 February (Fig. 11.2-42). Weak low pressure developed along a preexisting baroclinic zone over the southeast United States, responding to the vigorous upper trough diving southeastward across the Midwest at 0000–1200 UTC 16 February. The cyclone continued to develop along the Carolina coast within the diffluent exit region of the jet streak and moved northeastward along the coast producing moderate to heavy snows as the 850-hPa circulation intensified along the East Coast (Fig. 11.2-41). The surface low deepened about 12 hPa in 12 h and the storm structure resembled the "T-bone" scenario described by Shapiro and Keyser (1990), rather than the traditional occluded cyclone observed for many of the cases.

The heavy snow remained as a narrow band as the surface low pressure system became elongated along the coast on 16 February. This is perhaps related to the trough taking on a more southwest-to-northeast, or "positive," tilt between 1200 UTC 16 February and 0000 UTC 17 February, with little along-stream variation in the height and wind fields east of the trough axis (Fig. 11.2-42). Furthermore, there was a noticeable lack of ridging along the East Coast, precluding the shortening wavelengths between the upper trough and downstream ridge observed with many of the major snowstorms, thus contributing toward more moderate snowfall amounts.

o. 14–15 March 1999: NESIS = 2.20

This storm affected the Northeast urban corridor with a mix of rain, sleet, and snow and proved very difficult to forecast. It also occurred during a portion of the winter when daily values of the NAO were distinctly negative (Table 2-5, volume I) despite the winter's dominance by positive NAO values. Heaviest snow fell in central and western Pennsylvania, with a separate area of heavy snow occurring in Maine. Moderate to heavy snow fell within the metropolitan areas of Washington and Boston, with snowfall of up to 10 in. (25 cm) to the northwest of Washington, Baltimore, and Philadelphia. Snowfall totals in and around the suburbs of New York City and Boston (Fig. 11.2-43) were generally in the 6–10-in. (15–25 cm) range.

This case is marked by a cold surface high pressure

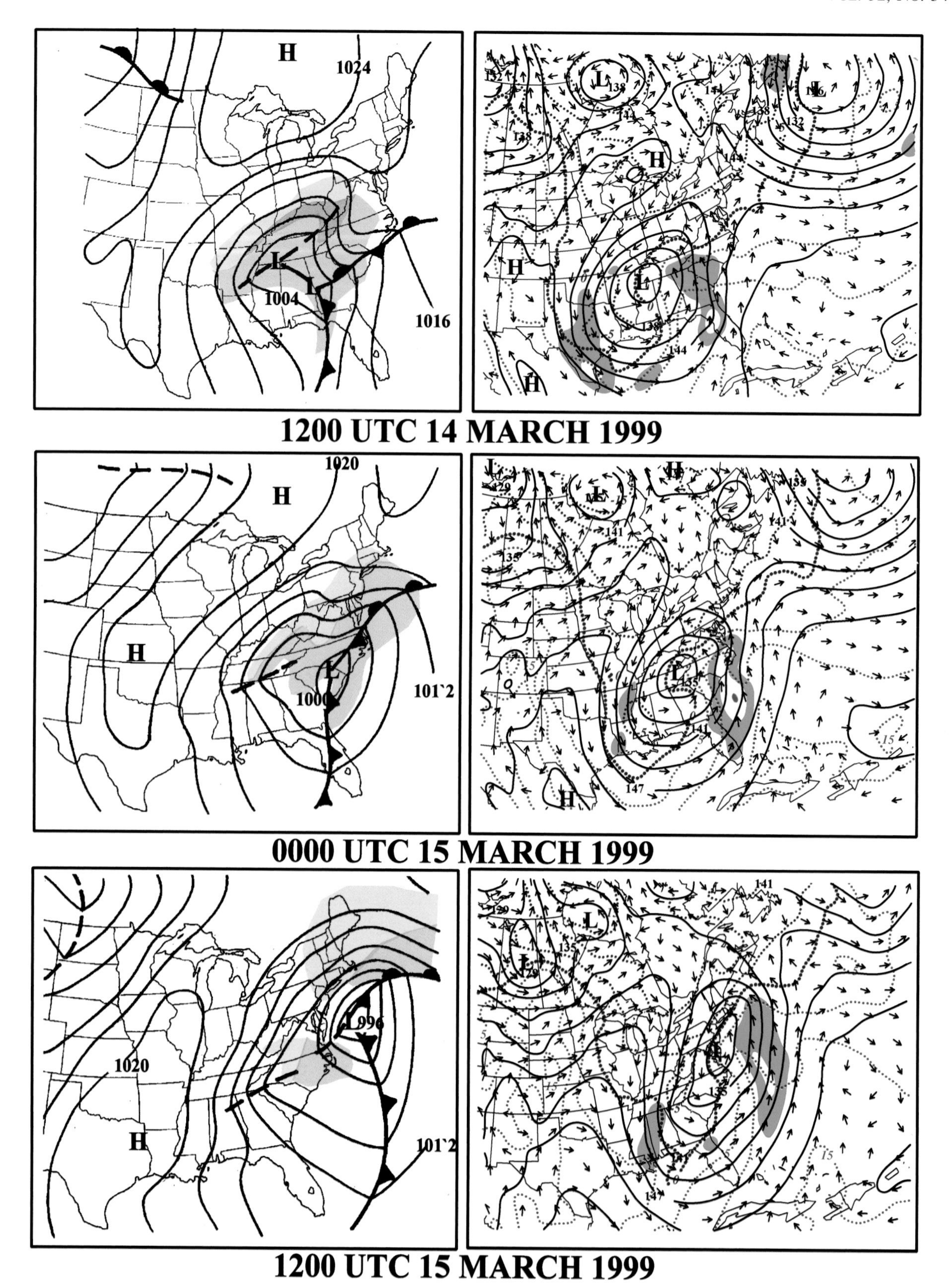

FIG. 11.2-44. Sequence of (left) surface and (right) 850-hPa charts for the 24-h period between 1200 UTC 14 and 15 Mar 1999. See Fig. 11.1-2 for details.

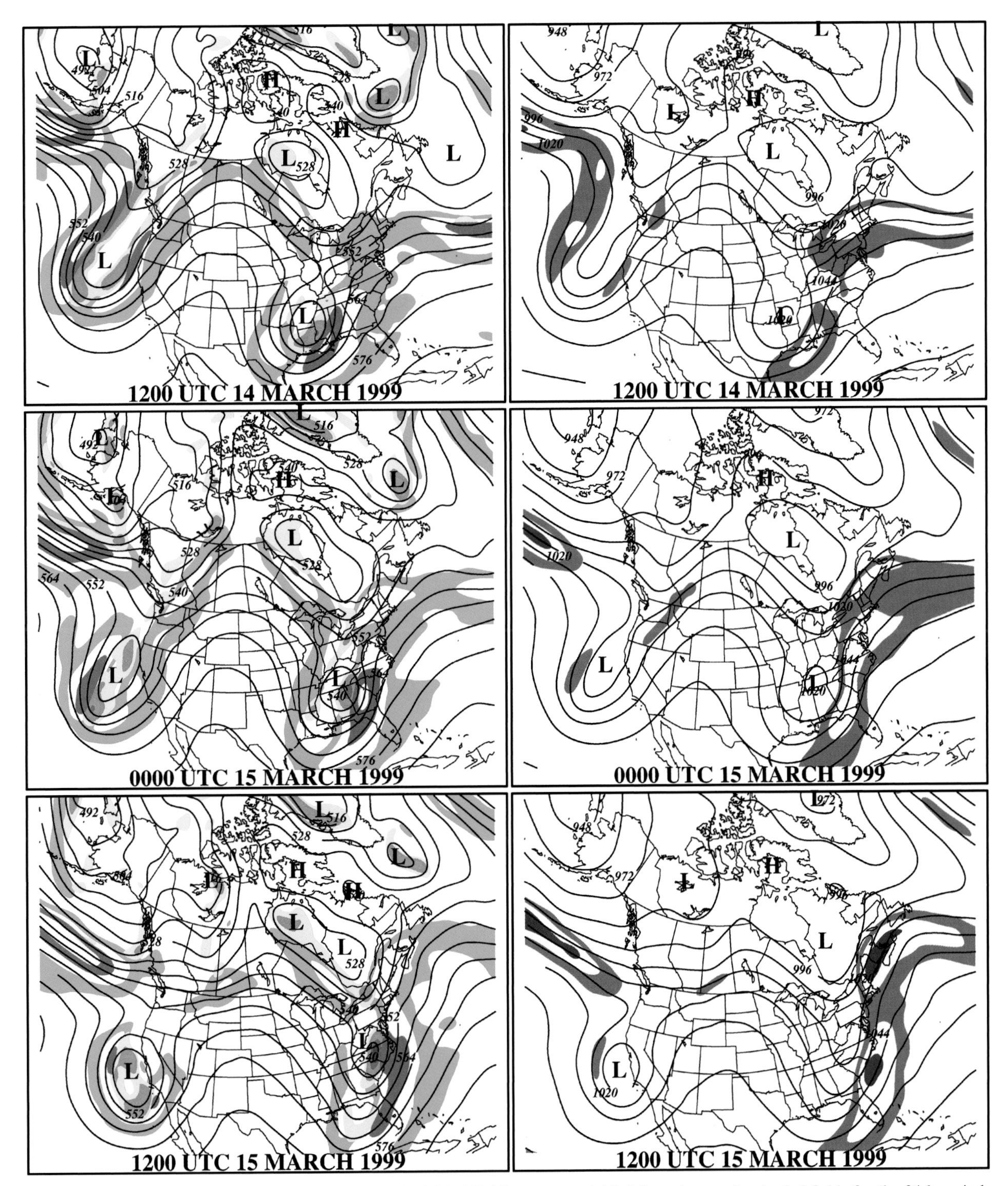

FIG. 11.2-45. Sequence of analyses of (left) 500- and (right) 250-hPa geopotential height and upper-level wind fields for the 24-h period between 1200 UTC 14 and 15 Mar 1999. See Fig. 11.1-3 for details.

system positioned over the Great Lakes region. A deep surface low developed in the southeastern United States by 1200 UTC 14 March marked also by a distinct 850-hPa circulation and developing 25–30 m s^{-1} low-level jet along the East Coast at 0000 UTC 15 March as the surface low moved toward the Carolina coast (Fig. 11.2-44). The moderate to heavy snows that occurred north and west of Washington and Philadelphia fell during this period.

The surface low developed as a distinct closed low

FIG. 11.3-1. Icing coverage (violet) for 7–9 Jan 1956.

aloft moved into the south-central United States and a confluent jet streak was sustained over New England during the initial cyclogenetic period (Fig. 11.2-45). Furthermore, at 1200 UTC 14 March, the dual-jet pattern that has been noted for more significant snow events was well established, especially at 250 hPa. By 1200 UTC 15 March, the upper-level system appears to lose its closed circulation and remains positively tilted along the East Coast. Concurrently, as the surface low moved northeastward from the Carolinas, the snowfall became less intense, accounting for the smaller accumulations that occurred between New York City and Boston than over southwestern Pennsylvania (Fig. 11.2-43). Later, as the upper-level trough moved northeastward toward New England, it became more negatively tilted (not shown), the surface low intensified, and heavy snowfall commenced over northern New England. The delay in the upper-level trough assuming a negative tilt spared New York City and southern New England from a major snowstorm for this case, with more moderate amounts generally occurring.

3. Ice storms

Seven ice storms are described in this section. These seven cases produced large areas of significant icing, some severely impacting transportation, while others resulted in widespread power losses as heavy accumulations of ice downed power lines. In some cases, freezing rain was the predominant mode of frozen precipitation (such as January 1956 and 27–28 January 1994) while in other cases, snow and freezing rain were both significant problems (e.g., December 1973, 13–14 January 1978, and 7–9 January 1994). The analyses emphasize the surface and upper-level weather patterns that support a vertical temperature profile that allows subfreezing air to remain at the earth's surface while a wedge of warm air aloft between 900 and 700 hPa allows snow to melt, in contrast to the heavy snow events shown in chapter 10.

While only seven cases are included in this section, it is apparent there are a number of scenarios that favor significant occurrences of freezing rain (see volume I, chapter 5). One element necessary is a source of cold air at the earth's surface. All cases are marked by the presence of large anticyclones whose locations vary greatly, especially when compared to the heavy snow cases. In some cases, an anticyclone is located to the north-northwest of the Northeast urban corridor, similar to the major snow events. Other cases are associated with anticyclones that continue to drift eastward across

FIG. 11.3-2. Sequence of (left) surface and (right) 850-hPa charts for the 24-h period between 1200 UTC 8 and 9 1956 (surface) and 1500 UTC 8 and 9 Jan 1956 (850 hPa). See Fig. 11.1-2 for details.

FIG. 11.3-3. Sequence of analyses of (left) 500- and (right) 250-hPa geopotential height and upper-level wind fields for the 24-h period between 1500 UTC 8 and 9 Jan 1956. See Fig. 11.1-3 for details.

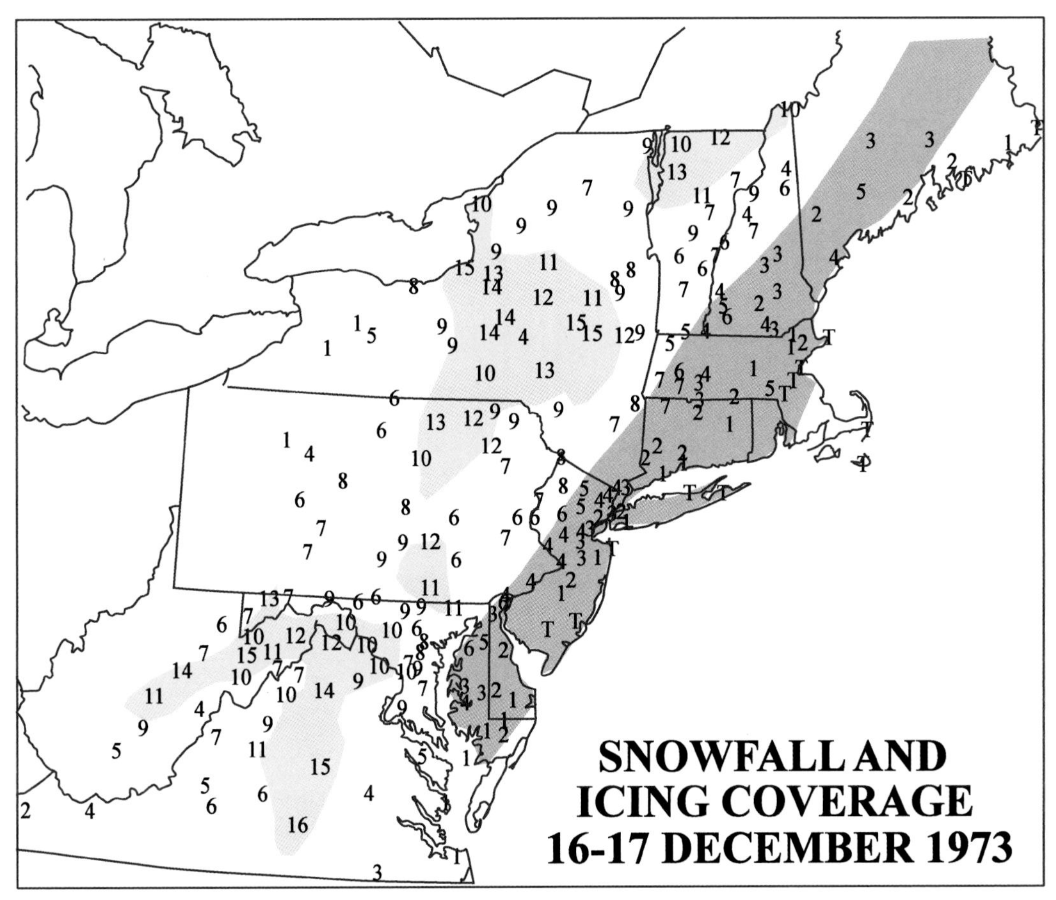

FIG. 11.3-4. Snowfall (in.) and icing coverage (violet) for 16–17 Dec 1973. See Fig. 11.1-1 for details.

eastern Canada, similar to the "interior snowstorms" described in chapter 11.1. While some anticyclones drift eastward completely off the North American continent (such as the storms on 17–18 January and 27–28 January 1994), cold low-level air still remains locked in place east of the Appalachian Mountains. With the exception of January 1956, upper-level confluent flow is common over the northeast or southeastern Canada. January 1956 is an extraordinary case in which a large blocking upper-level anticyclone is associated with a surface anticyclone that provides cold air at low levels and a large blocking cyclone provides the flow of relatively warm, moist air above the shallow, cold air. Hart and Grumm (2001) have described this case as the most anomalous departure from climatology in a 53-yr period.

Factors that allow warmer air to move above the shallow, cold surface air include a number of scenarios described in chapter 5 of volume I. One scenario is a deep upper-level trough associated with a strong meridional frontal zone west of a strong ridge over the Atlantic Ocean (e.g., December 1973 and January 1978). Warmer air from the south and east rises above the shallow, cold air. A second scenario, perhaps more common, is one where an anticyclone drifts eastward across eastern Canada while an upper trough moves to the northwest of the urban corridor (such as 17–18 January 1994, 27–28 January 1994, and 14–15 January 1999), allowing cold surface air to remain while southwesterly flow aloft brings in warmer air. This scenario shows the presence of an extensive region of southwesterly winds aloft west of the building upper-level ridge that allows rising 850-hPa temperatures to build northeastward. A third scenario is one in which the transition region between significant snow and rain is found with conditions similar to those found with major snowstorms. However, the cold air typically does not extend as far south as the heavy snow cases (7–9 January 1994).

a. 7–9 January 1956

This was a highly unusual weather event (see Mook 1956) that produced a large region of mostly light to moderate icing accumulations across nearly the entire Northeast (see distribution in Fig. 11.3-1). The area of icing was the largest of the seven cases (see Fig. 5-20, volume I) that spread first across northern and eastern New England and then southwestward across the rest

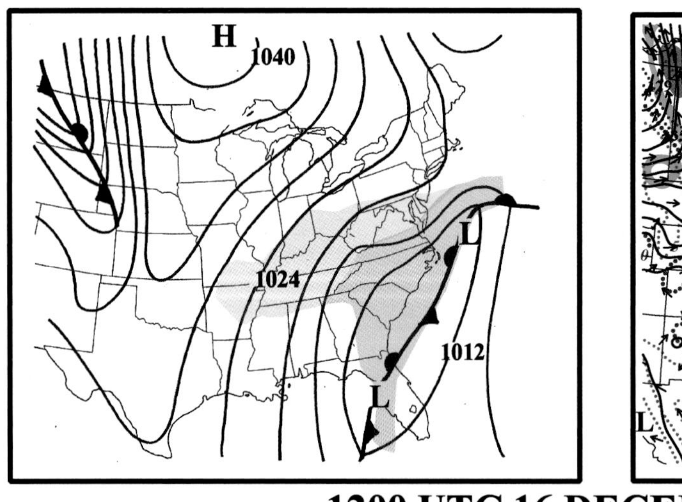

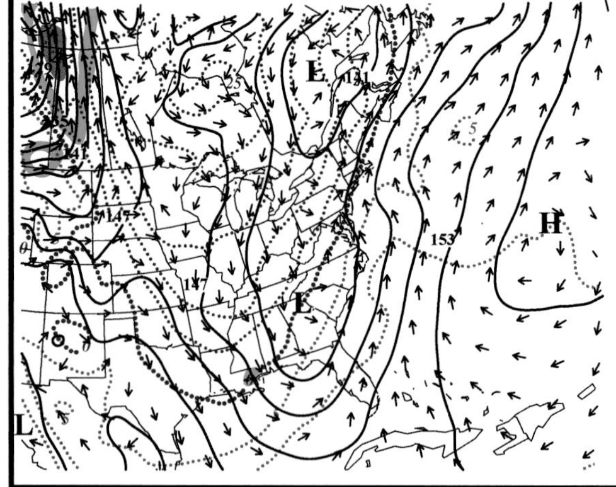

1200 UTC 16 DECEMBER 1973

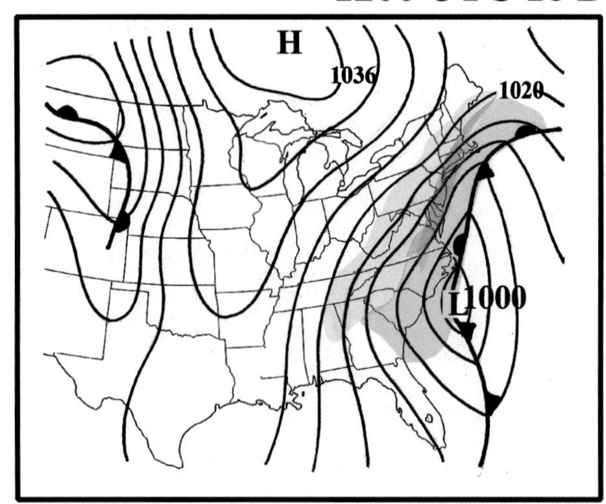

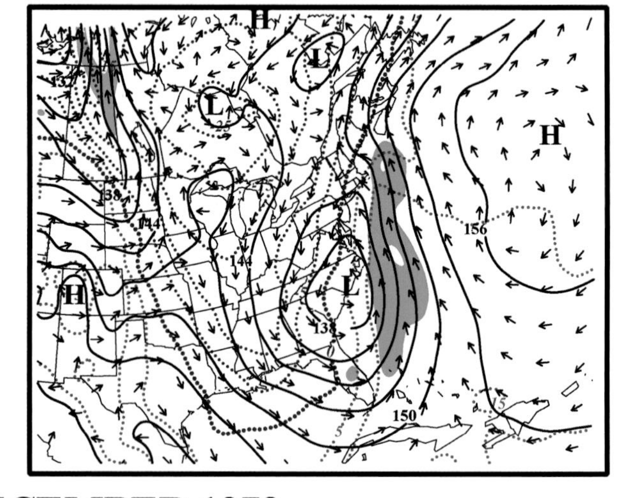

0000 UTC 17 DECEMBER 1973

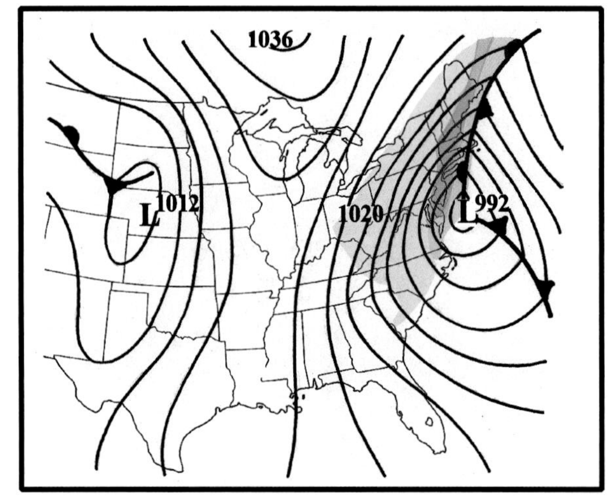

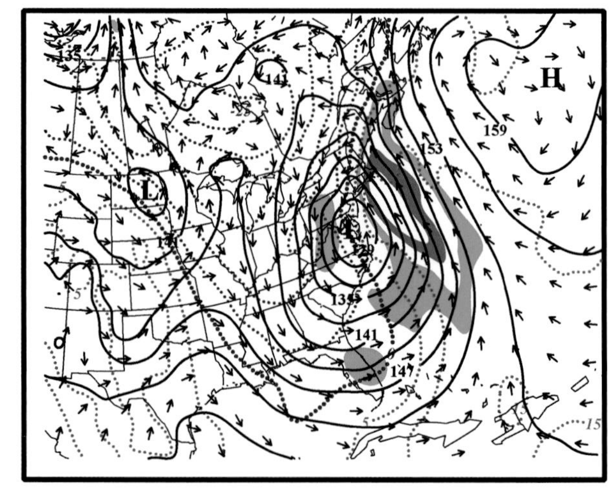

1200 UTC 17 DECEMBER 1973

FIG. 11.3-5. Sequence of (left) surface and (right) 850-hPa charts for the 24-h period between 1200 UTC 16 and 17 Dec 1973. See Fig. 11.1-2 for details.

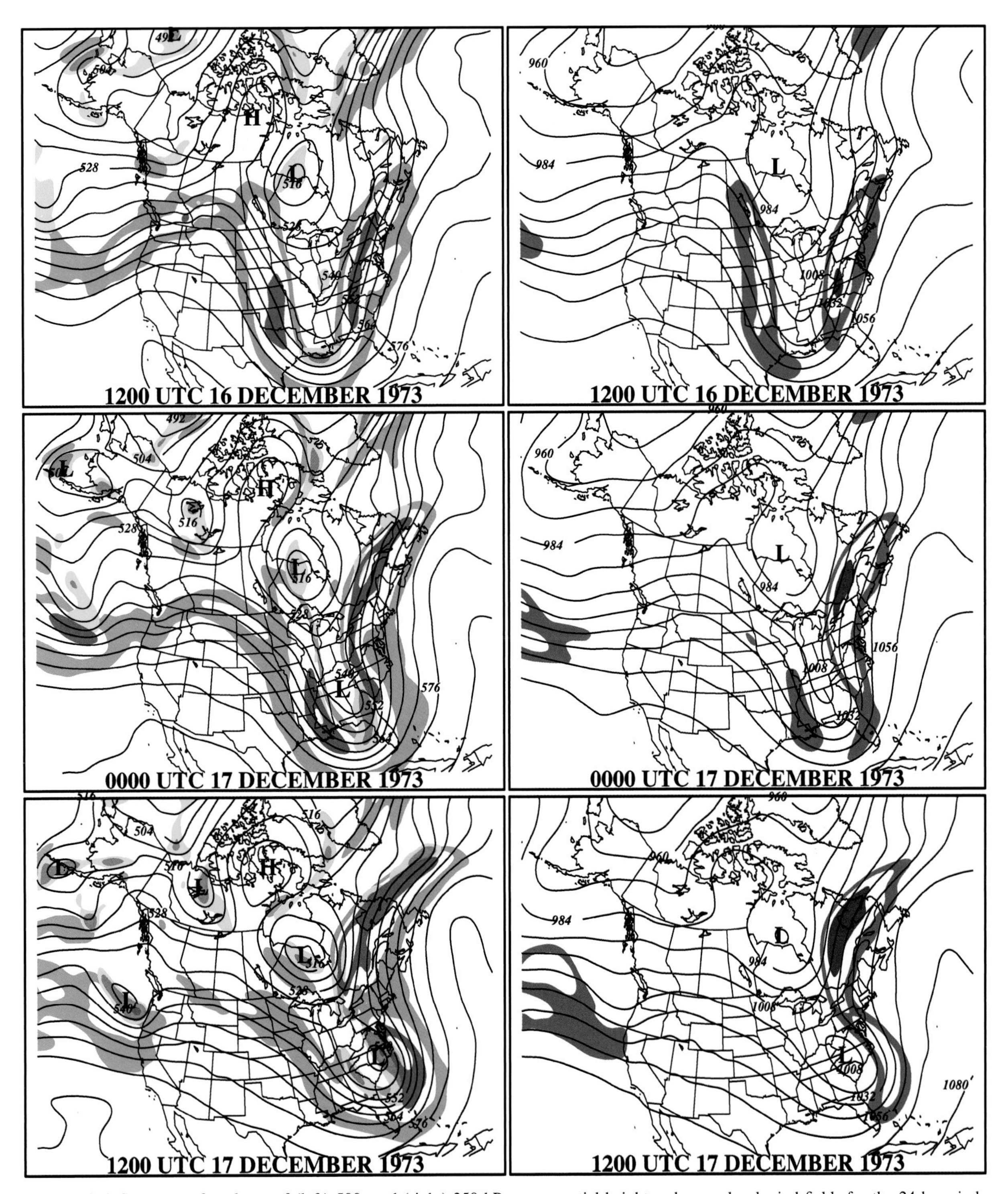

FIG. 11.3-6. Sequence of analyses of (left) 500- and (right) 250-hPa geopotential height and upper-level wind fields for the 24-h period between 1200 UTC 16 and 17 Dec 1973. See Fig. 11.1-3 for details.

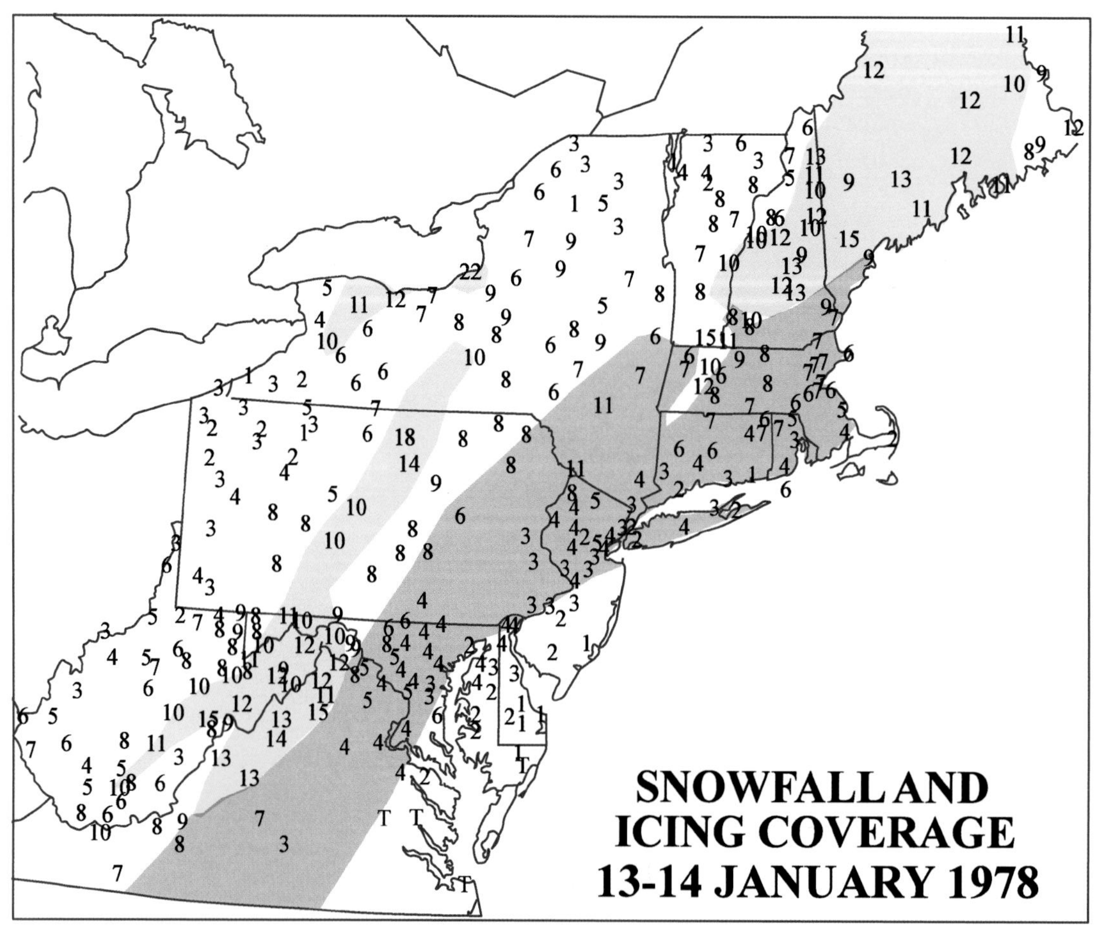

FIG. 11.3-7. Snowfall (in.) and icing coverage (violet) for 13–14 Jan 1978. See Fig. 11.1-1 for details.

of New England, New York, and much of the middle Atlantic states (rather than spreading northeastward in a typical winter storm). Unlike the other icing cases, there was little snow associated with this storm as much of the storm precipitation fell first as freezing rain and then as rain. While many icing events are associated with warm air that moves northeastward over the low-level dome of shallow, cold air, the presence of an anomalous upper-level cyclone and anticyclone (see Fig. 11.3-3) over the western Atlantic allowed warm air to move *west* and *southwestward* across New England and the middle Atlantic states. The upper-level pattern is so anomalous that Hart and Grumm (2001) found this case to be the highest-ranked storm of all storms during a 53-yr period from 1948 to 2001 on a scale that uses normalized departures from climatology of tropospheric values of height, temperature, winds, and moisture. Most ice accumulations were less than half an inch (1 cm), but the icing caused significant problems due to its widespread coverage.

The synoptic evolution of this event occurred during a significant atmospheric blocking episode. With a large upper-level anticyclone parked over northeastern Canada (many of the heavy snow cases in chapter 10 are associated with a closed upper-level low over eastern Canada), a short-wave trough amplified to its south and generated a very large cyclone over the central Atlantic (Fig. 11.3-2). To the north of the cyclone center, moisture moved westward off the Atlantic toward New England, and a very strong frontal boundary was established between cold air over the eastern United States and the maritime air offshore. With cold air in place at the lowest levels of the atmosphere, much of the initial precipitation fell as freezing rain. However, as the blocking pattern persisted, the warmer maritime air quickly spread westward, first across New England and then across much of the rest of the Northeast and middle Atlantic states, changing the freezing rain to rain over New England and as far south as Baltimore.

b. 16–17 December 1973

This ice storm was also a significant snowstorm that left more than 10 in. (25 cm) of snow across central New York, portions of east-central Pennsylvania, eastern West Virginia, and Washington, and portions of central Virginia (Fig. 11.3-4). It occurred during a period when the El Niño–Southern Oscillation (ENSO) was in

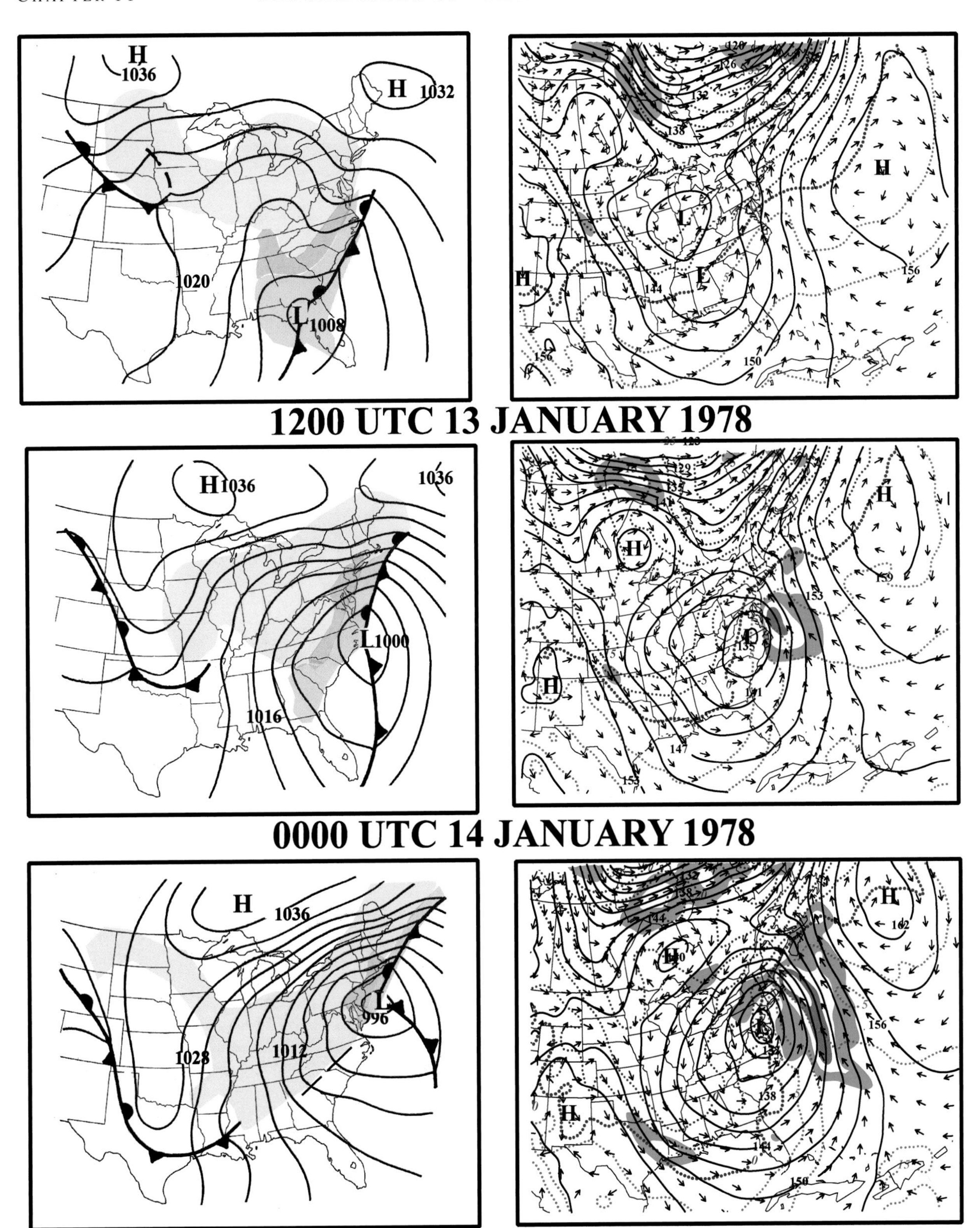

FIG. 11.3-8. Sequence of (left) surface and (right) 850-hPa charts for the 24-h period between 1200 UTC 13 and 14 Jan 1978. See Fig. 11.1-2 for details.

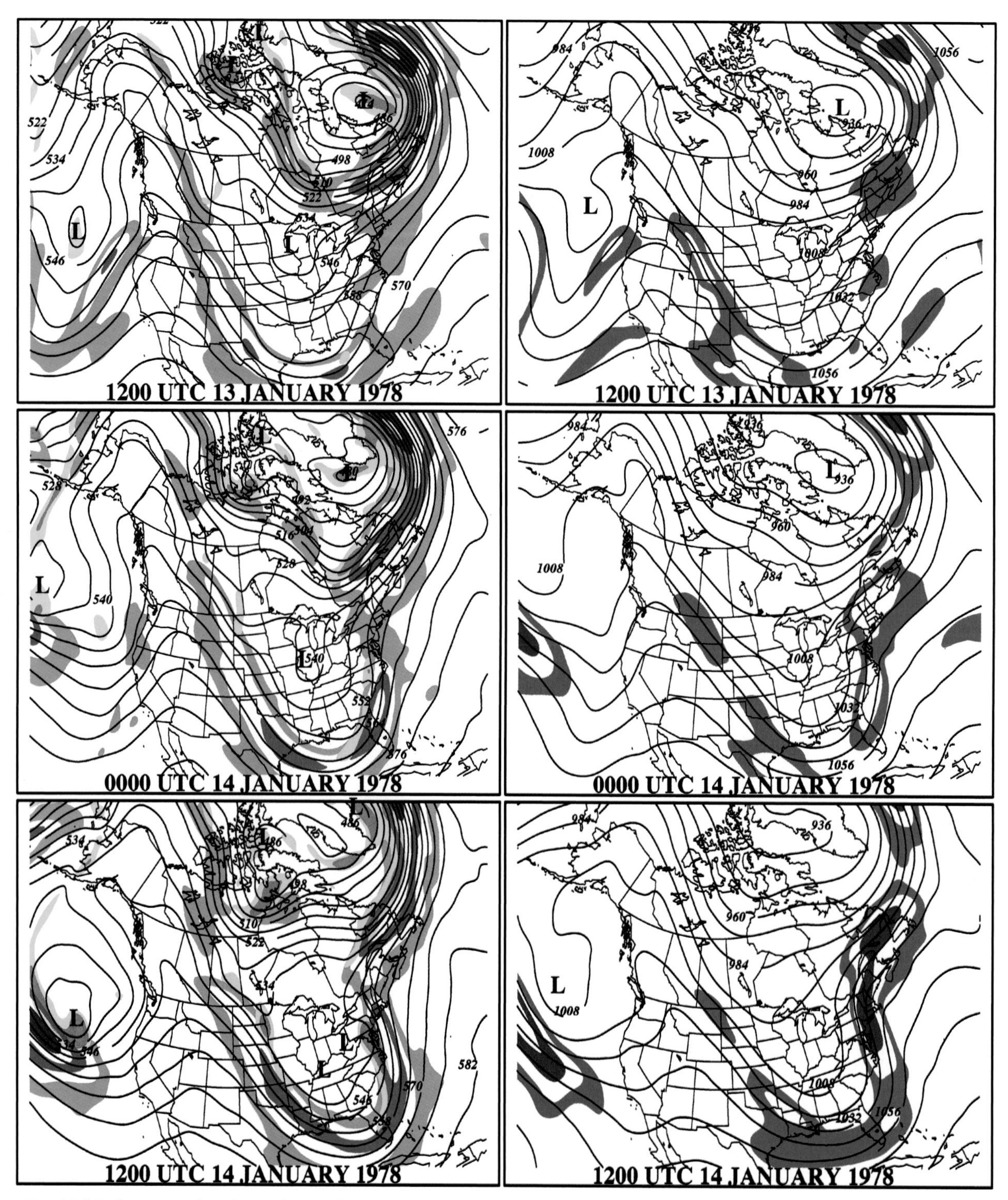

FIG. 11.3-9. Sequence of analyses of (left) 500- and (right) 250-hPa geopotential height and upper-level wind fields for the 24-h period between 1200 UTC 13 and 14 Jan 1978. See Fig. 11.1-3 for details.

a distinct negative phase (La Niña; Fig. 2-19, volume I) and daily values of the North Atlantic Oscillation (NAO) were positive (not shown), two factors not necessarily favorable to winter storm formation (see volume I, chapter 2). However, this storm is primarily remembered as a significant ice storm across portions of Delaware, New Jersey, New York City, Long Island, Connecticut, Rhode Island, and eastern Massachusetts. In

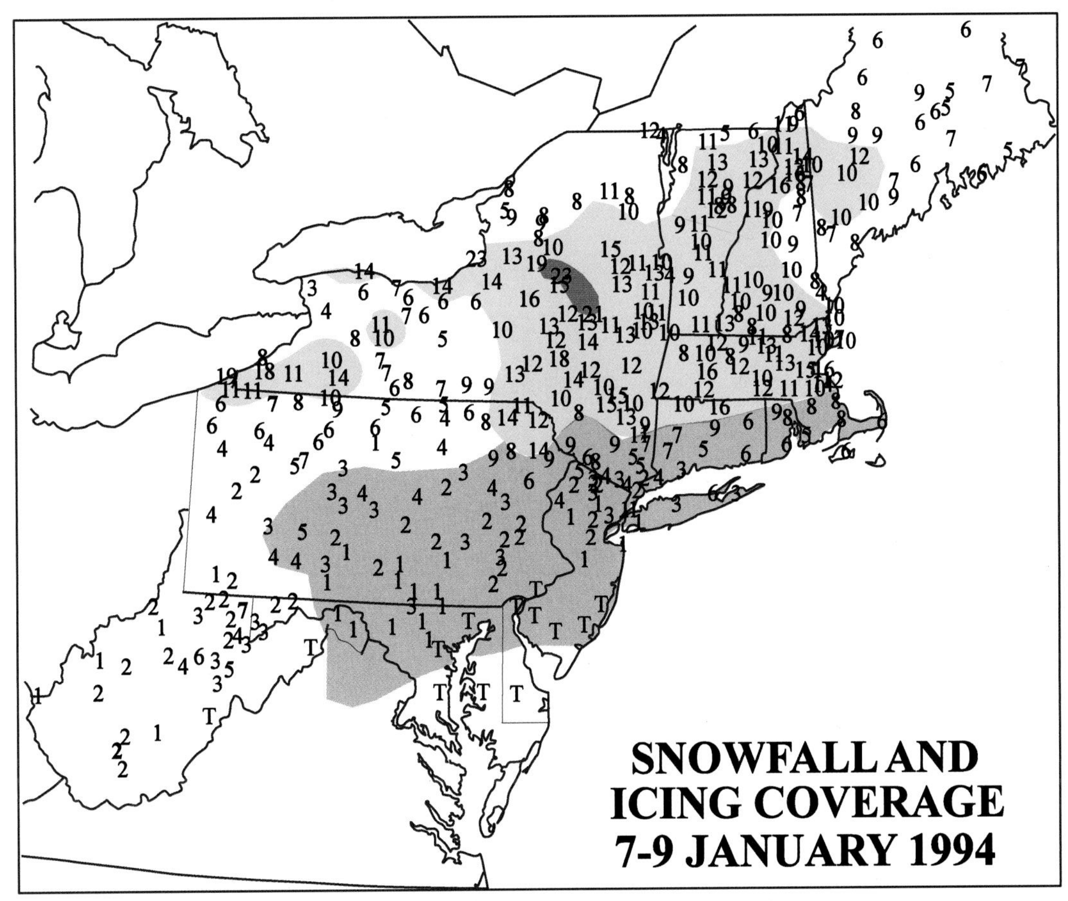

FIG. 11.3-10. Snowfall (in.) and icing coverage (violet) for 7–9 Jan 1994. See Fig. 11.1-1 for details.

Connecticut, the storm was described as the worst ice storm in history with damage to trees greater than that during the September 1938 hurricane. In New Jersey, New York City, and Long Island, this was one of the most damaging ice storms in history, resulting in power losses to millions.

This major storm developed in conjunction with a long-wave trough amplification across the eastern United States (Fig. 11.3-6). One of the major differences between this storm and many of the other Northeast snowstorms is that the orientation of the typical upper trough–confluence differs from that noted for many of the heavy snow cases. In this case, upper-level confluence is occurring along a more north–south axis, especially over the Great Lakes at 0000 UTC 17 December 1973 and over Quebec at 1200 UTC 17 December. Therefore, much of the Northeast is located with the entrance region of an amplifying upper-level jet streak over Quebec. The orientation of the confluence–upper-level jet supports a west-to-east temperature gradient with a source of very cold air poised over the Great Lakes and warmer air flooding northward over the western Atlantic Ocean. Therefore, the storm system was associated with a significant north–south frontal boundary and 850-hPa baroclinic zone (Fig. 11.3-5) that represented the contrast between the continental cold air and the northward stream of warmer air just east of the coastline. Low pressure moved slowly northward along the meridional surface front (Fig. 11.3-5) to a position east of the Maryland coast at 1200 UTC 17 December. While cold air remained at the surface from Delaware northward, the upper trough began to develop a closed low and became negatively tilted, allowing warmer air aloft to move westward, changing the snow to freezing rain as the low-level cold air remained dammed just east of the Appalachian Mountains. Warmer air did reach into extreme eastern New England as the surface low moved west of the coastline, with Boston's temperature rising to near 60°F (15°C), changing the precipitation to rain, while farther south, precipitation fell as freezing rain and snow during the entire event.

c. 13–14 January 1978

This ice storm was the first in a series of three major winter storms that occurred within a memorable week in the Northeast urban corridor. The other cases are classified as an interior snowstorm (16–17 January

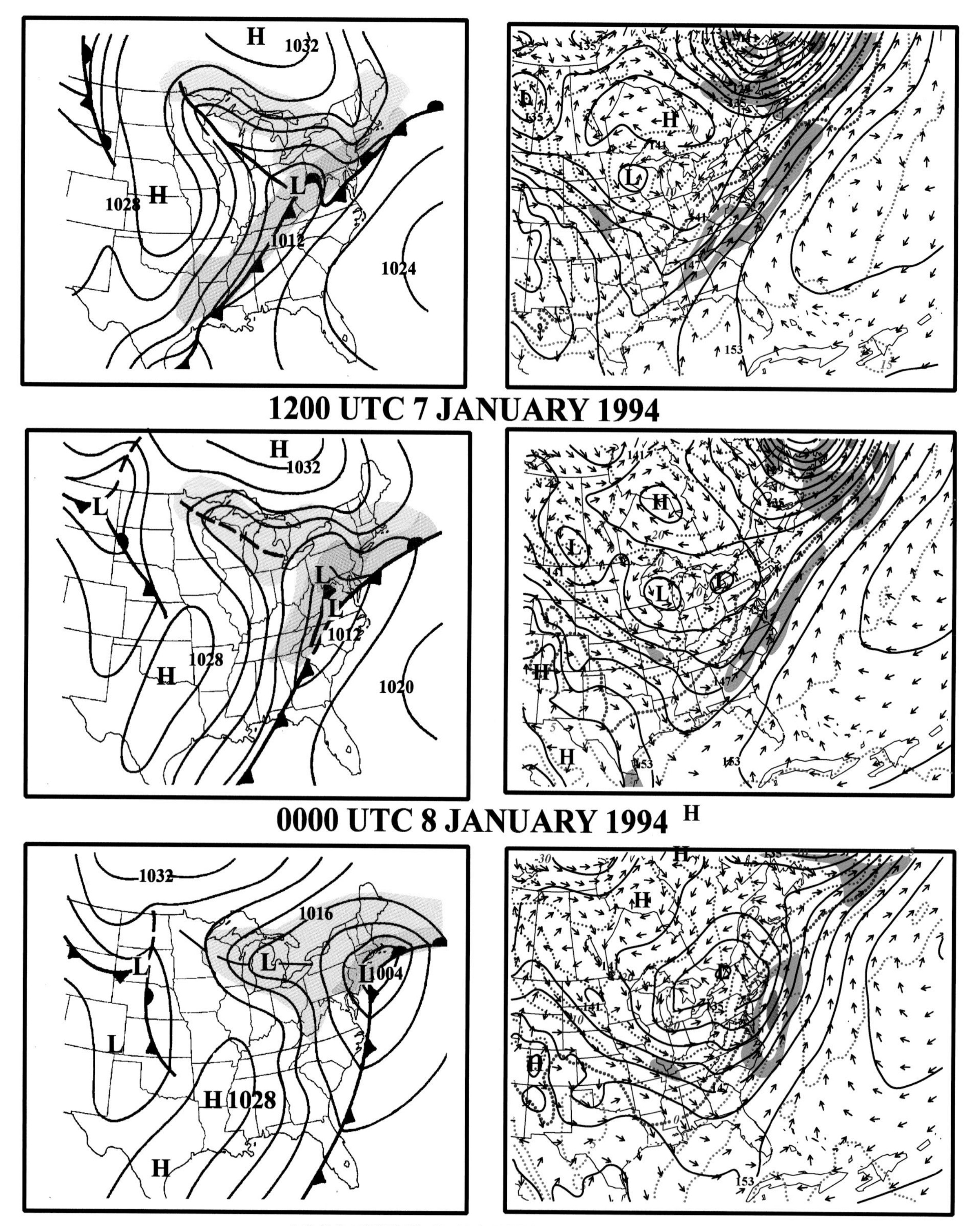

FIG. 11.3-11. Sequence of (left) surface and (right) 850-hPa charts for the 24-h period between 1200 UTC 7 and 8 Jan 1994. See Fig. 11.1-2 for details.

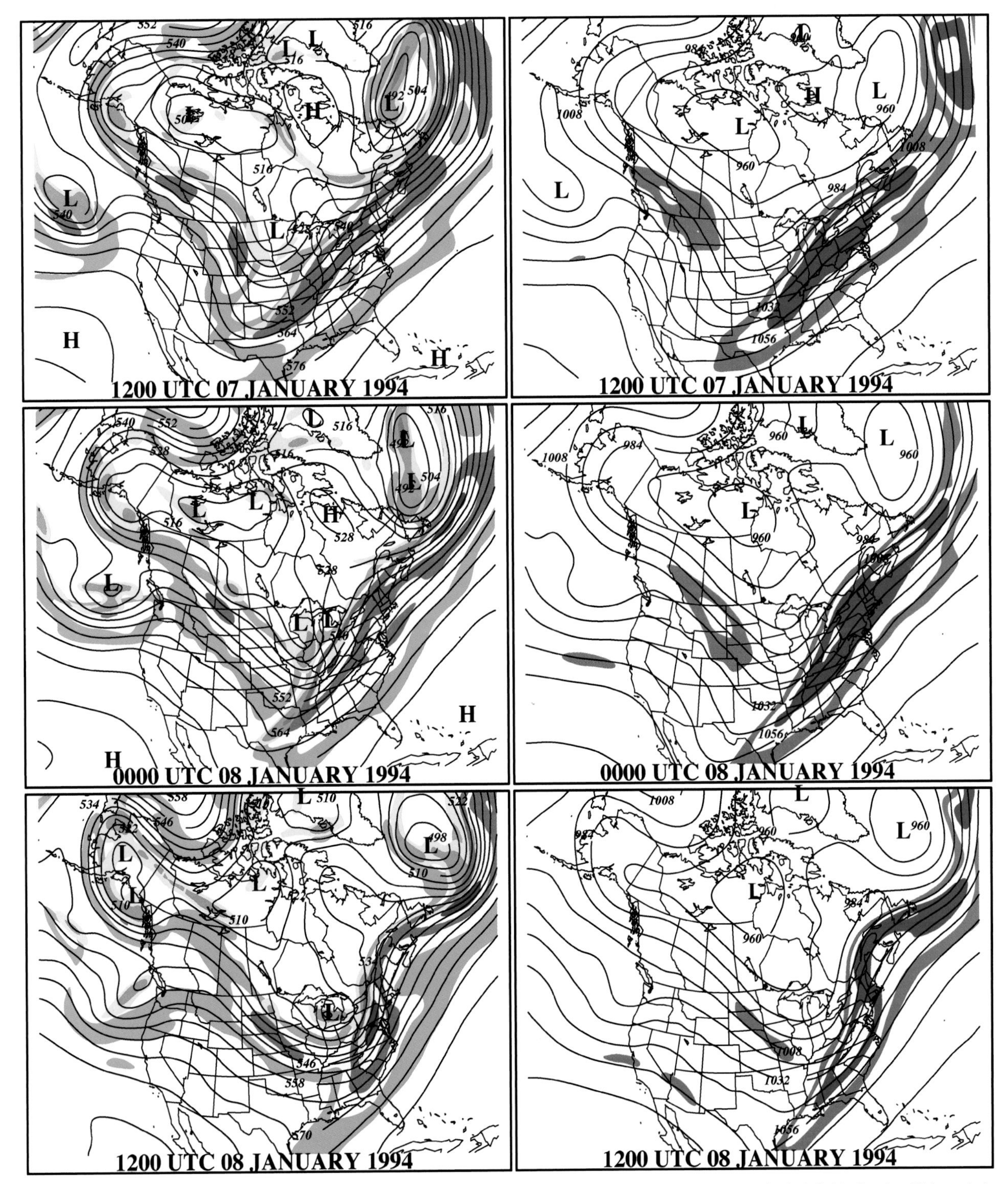

FIG. 11.3-12. Sequence of analyses of (left) 500- and (right) 250-hPa geopotential height and upper-level wind fields for the 24-h period between 1200 UTC 7 and 8 Jan 1994. See Fig. 11.1-3 for details.

1978; see chapter 11.1) and as a major Northeast snowstorm (19–20 January 1978; chapter 10.16). This case was a snowstorm across much of the Northeast (Fig. 11.3-7) that changed to freezing rain in Virginia, Maryland, Delaware, northern New Jersey, eastern Pennsylvania, southeastern New York, Connecticut, Rhode Island, and eastern Massachusetts. Eastern Pennsylvania (from the Philadelphia suburbs north to Allentown and

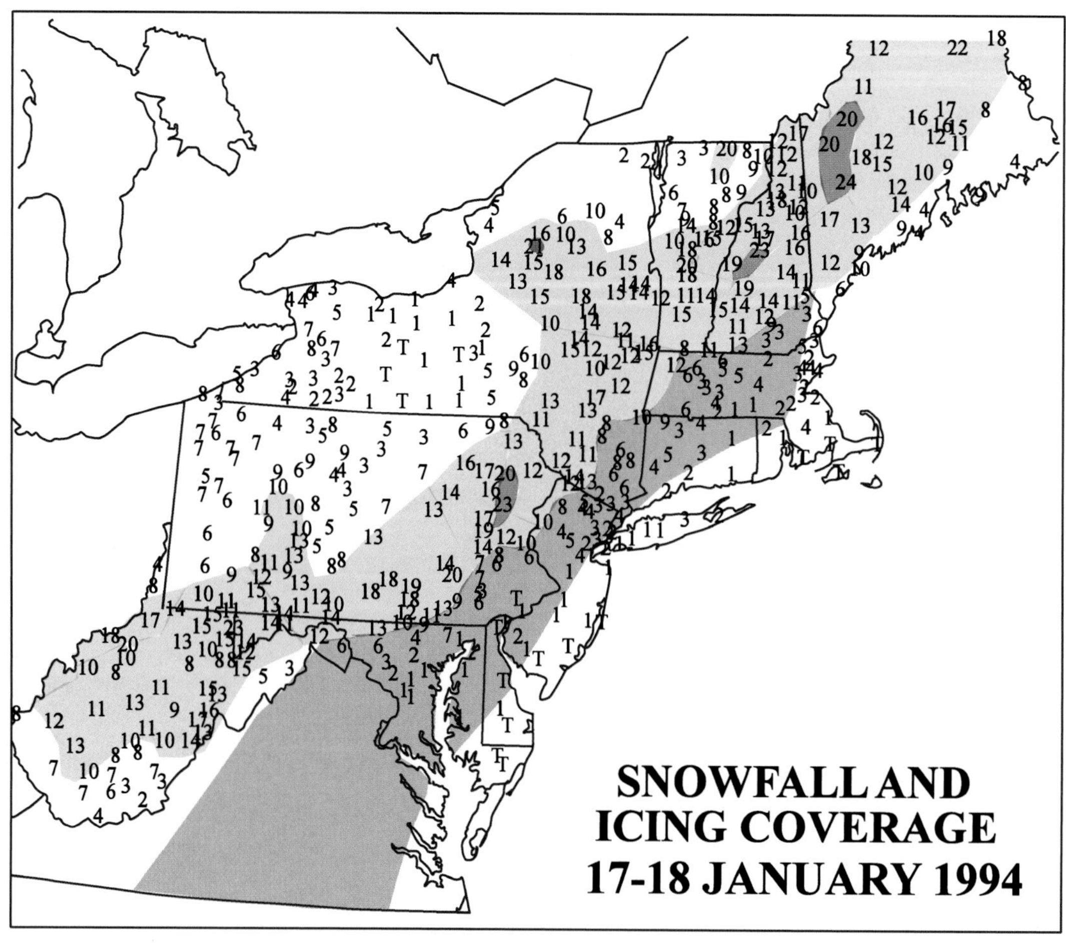

FIG. 11.3-13. Snowfall (in.) and icing coverage (violet) for 17–18 Jan 1994. See Fig. 11.1-1 for details.

west to Reading and Lancaster) was especially hard hit by freezing rain and ice pellets, where 0.5–1 in. (1–2 cm) of ice accumulated. Much of the area from northern New Jersey, New York City, and Long Island, northeastward into eastern Massachusetts, was also hard hit with one of the most severe ice storms on record, with many households losing electricity for periods of up to a week.

As with the other storms during January 1978, this storm developed during a period when ENSO was slightly positive (El Niño; Fig. 2-19, volume I) while daily values of the NAO were also positive (Table 2-5, volume I). The storm evolved as a number of upper-level troughs consolidated slowly into a broad major trough over the eastern United States (Fig. 11.3-9). A large area of upper confluence covered much of eastern Canada as a trough with a cutoff low lifted northeastward from Quebec to Greenland, allowing cold high pressure to cover much of southern Canada, providing a source of low-level cold air. A surface low formed in the Gulf of Mexico on 12 January and developed slowly as it moved northeastward along the East Coast through 14 January (Fig. 11.3-8). As the upper trough amplified over the eastern United States and the eastern Canadian trough lifted northeastward toward Greenland, 500-hPa heights built rapidly across the western Atlantic, allowing warm air to ascend west of a strong frontal boundary that extended along the coast, well north of the surface low. With cold air remaining at the surface east of the Appalachian Mountains as the warmer air moved in aloft from the south and east, a widespread area of freezing rain developed from the southeast United States northeastward along the entire coast. This case is similar to the December 1973 storm with the development of a strong north–south front, a large baroclinic zone at 850 hPa (Fig. 11.3-8), and the development of an upper-level negatively tilted trough that allowed cold air to remain at the surface while warmer air moved in aloft to produce freezing rain and ice pellets in the urban corridor, rather than snow.

d. 7–9 January 1994

The winter of 1993/94 will long be remembered, not only for the number of heavy snow events over much of the Northeast, but also for the large number of freez-

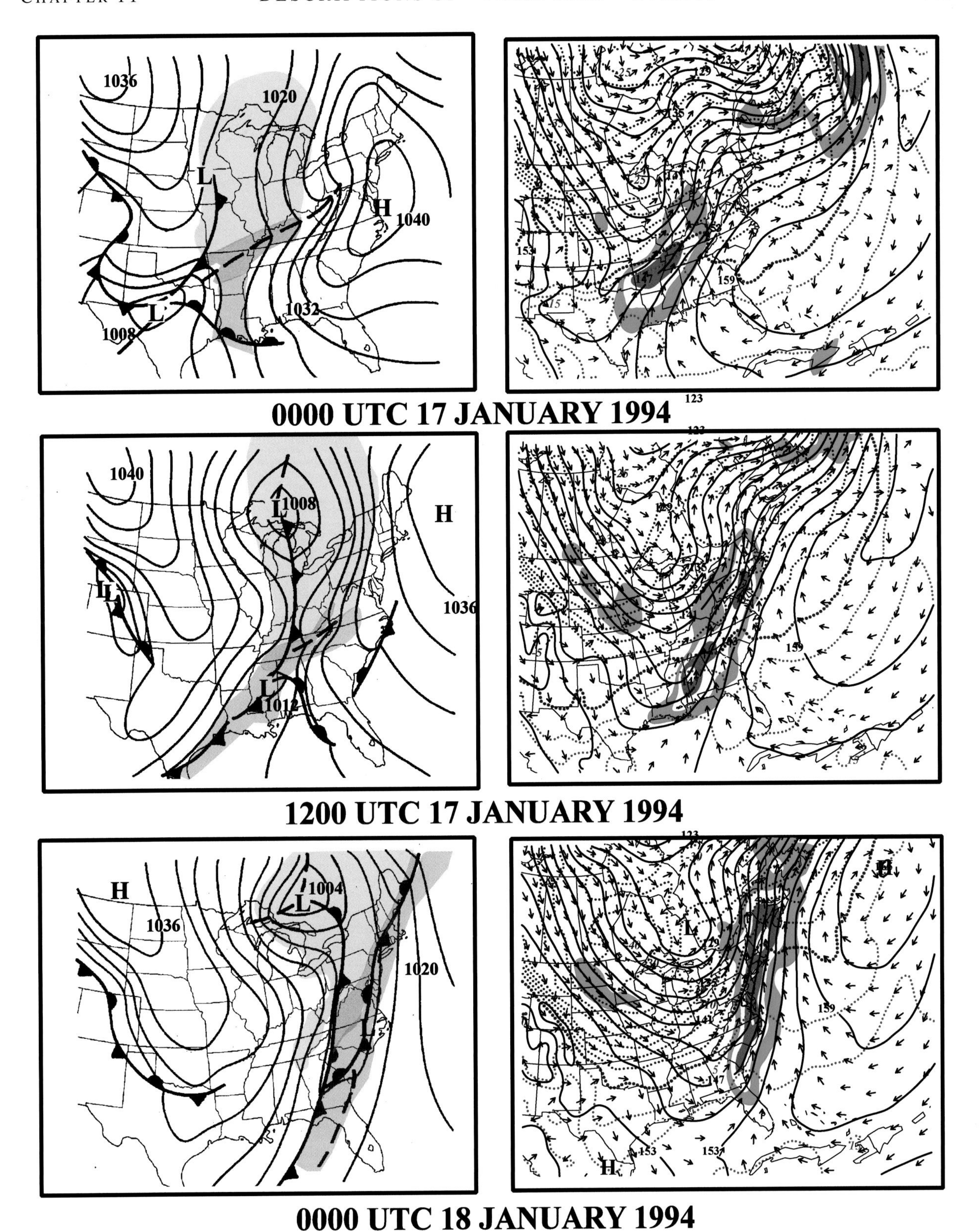

FIG. 11.3-14. Sequence of (left) surface and (right) 850-hPa charts for the 24-h period between 0000 UTC 17 and 18 Jan 1994. See Fig. 11.1-2 for details.

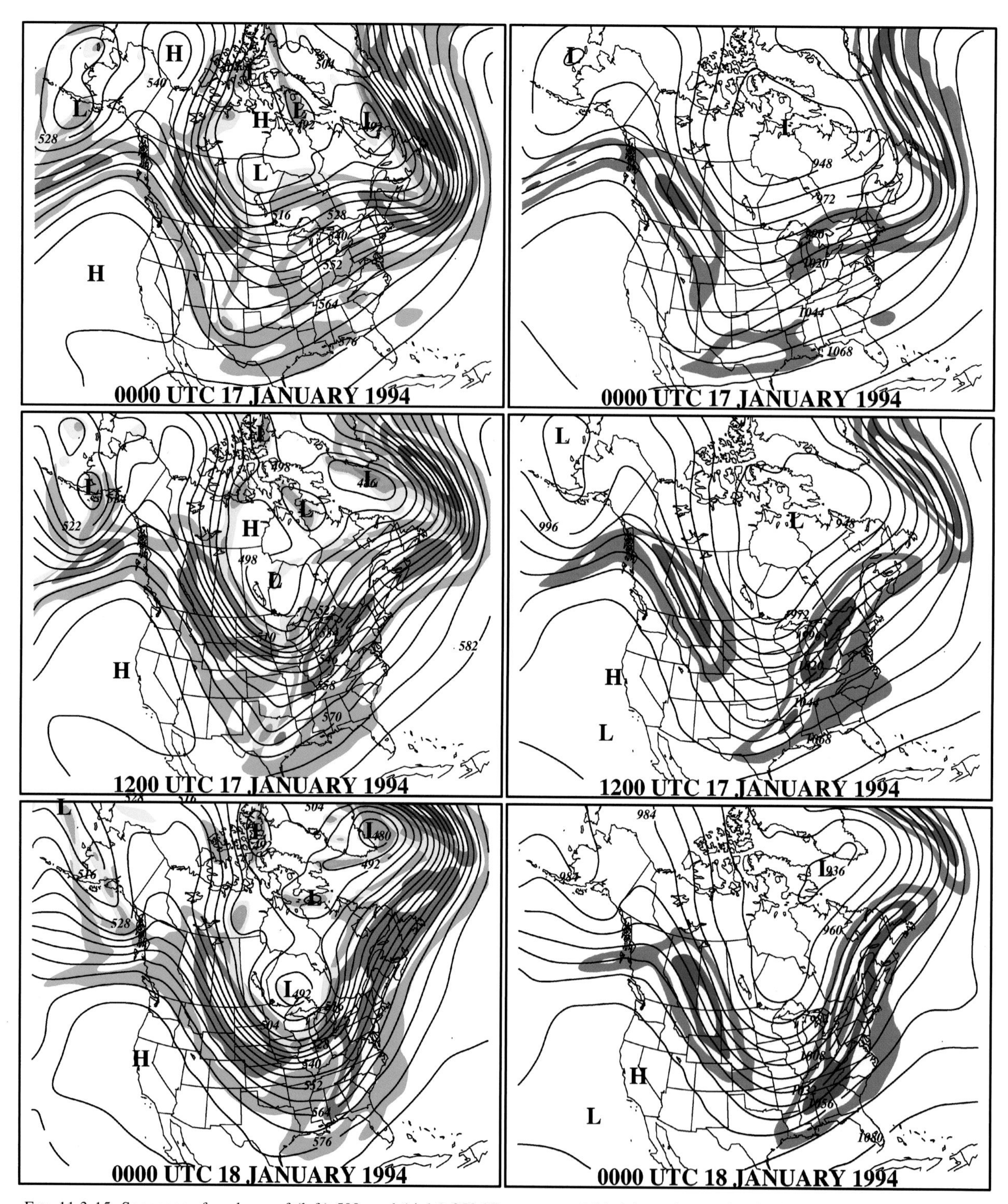

FIG. 11.3-15. Sequence of analyses of (left) 500- and (right) 250-hPa geopotential height and upper-level wind fields for the 24-h period between 0000 UTC 17 and 18 Jan 1994. See Fig. 11.1-3 for details.

ing rain events in the middle Atlantic States. This storm produced heavy amounts of snow across eastern New England (Fig. 11.3-10), including more than 15 in. (38 cm) in Boston, but also produced a very significant ice storm from Maryland northeastward toward Connecticut. In the Philadelphia metropolitan area, this may have been the most devastating ice storm in the region's history, with thousands of downed trees and power outages

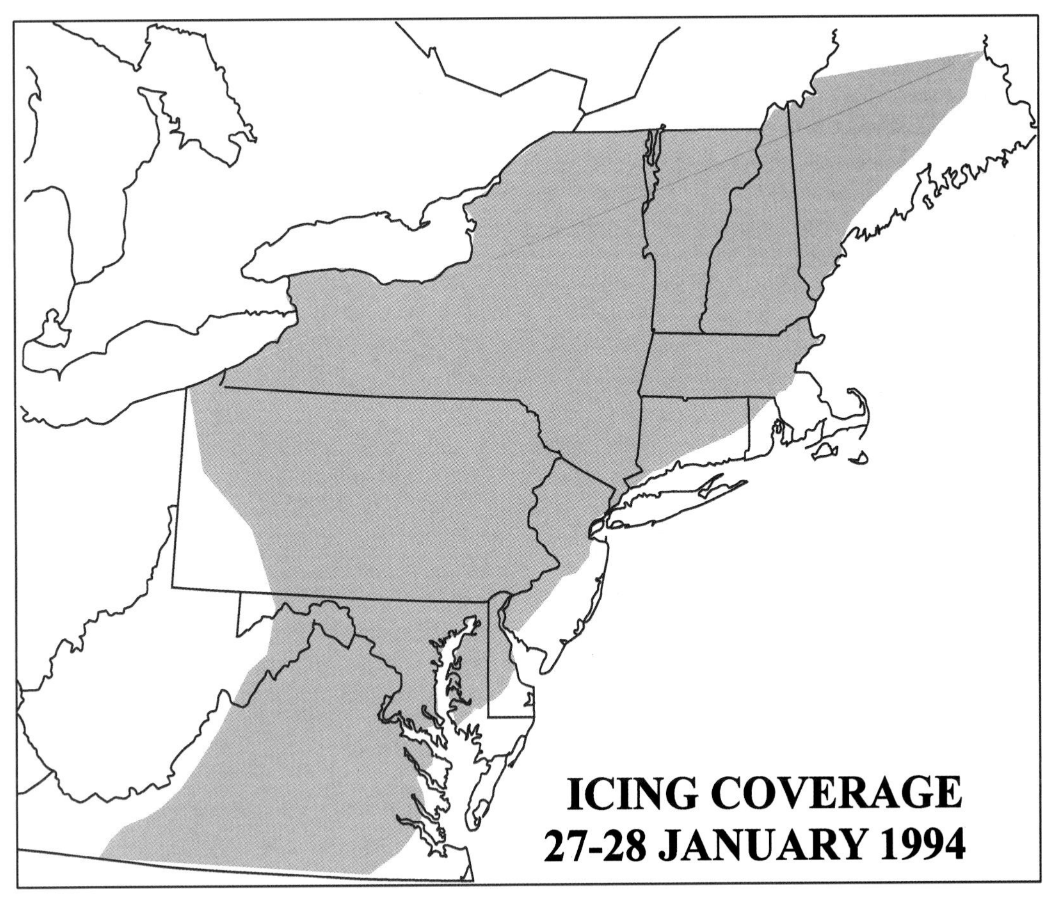

FIG. 11.3-16. Icing coverage (violet) for 14–15 Jan 1999

that affected hundreds of thousands of people for a period of up to a week. As with other significant winter storms during January 1994, this was a period in which ENSO was near zero or slightly positive, while the NAO was also near zero or slightly positive (not shown).

A major feature of this storm was the strong cell of high pressure anchored north of the Great Lakes that supplied very cold air to the northeastern United States during the event (Fig. 11.3-11). The high pressure system was located beneath a pronounced region of upper confluence over eastern Canada, observed between a slow-moving upper trough south of Greenland and a cutoff ridge over northern Quebec (Fig. 11.3-12). At the same time, a developing upper trough over the central United States moved northeastward toward the Ohio Valley as low pressure developed over the Ohio Valley and moved northeastward. The interaction of these features is also marked by a strong confluent jet flow over the eastern third of the United States, with the surface trough slowly evolving into a low pressure center east of New Jersey within the confluent entrance region of the jet. A strong frontal boundary east of the low separated a flow of very mild air over the southeastern United States from the cold, Canadian air located over the northeast. The slow northeastward movement of the upper trough to the west of the Northeast urban corridor allowed warmer air to flow over the cold low-level air from the south-southwest (see the 850-hPa analyses in Fig. 11.3-11), resulting in freezing rain rather than snow, from Washington to New York City, with below freezing air remaining locked in over much of the Northeast as the cold high pressure system remained north of New England.

e. 17 January 1994

This second major ice storm of January 1994 occurred between two exceptionally cold air masses that invaded much of the eastern United States during mid-January 1994. The cold-air outbreaks occurred during a period during which the NAO was distinctly positive, contrary to expectations (see volume I, chapter 2), and ENSO was nearly neutral but within a multiyear period in which it tended toward positive values (El Niño; Fig. 2-19, volume I).

This was a highly unusual winter storm in that heavy snow and significant icing occurred with a particularly unusual synoptic setting (Fig. 11.3-14). High pressure

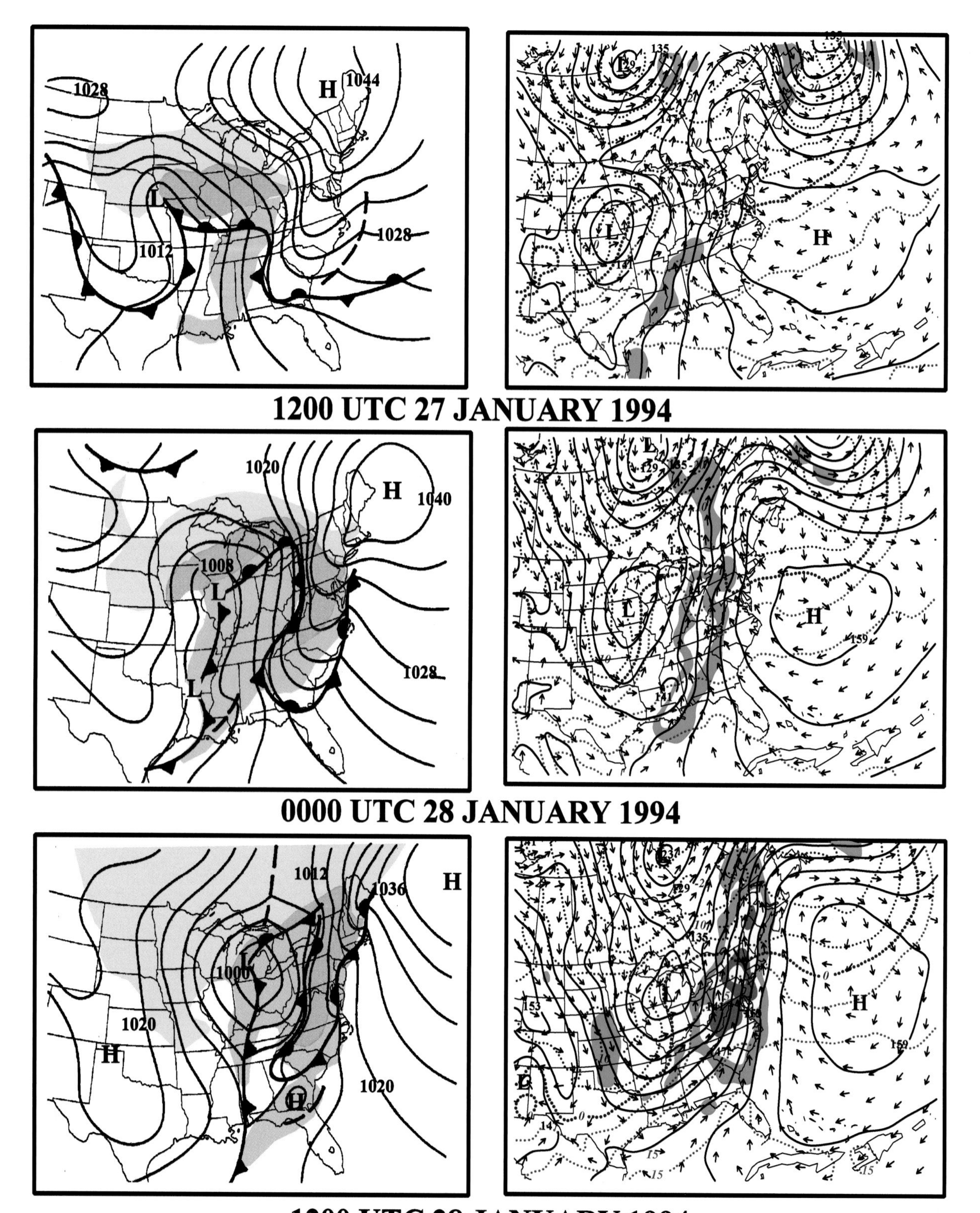

FIG. 11.3-17. Sequence of (left) surface and (right) 850-hPa charts for the 24-h period between 1200 UTC 27 and 28 Jan 1994. See Fig. 11.1-2 for details.

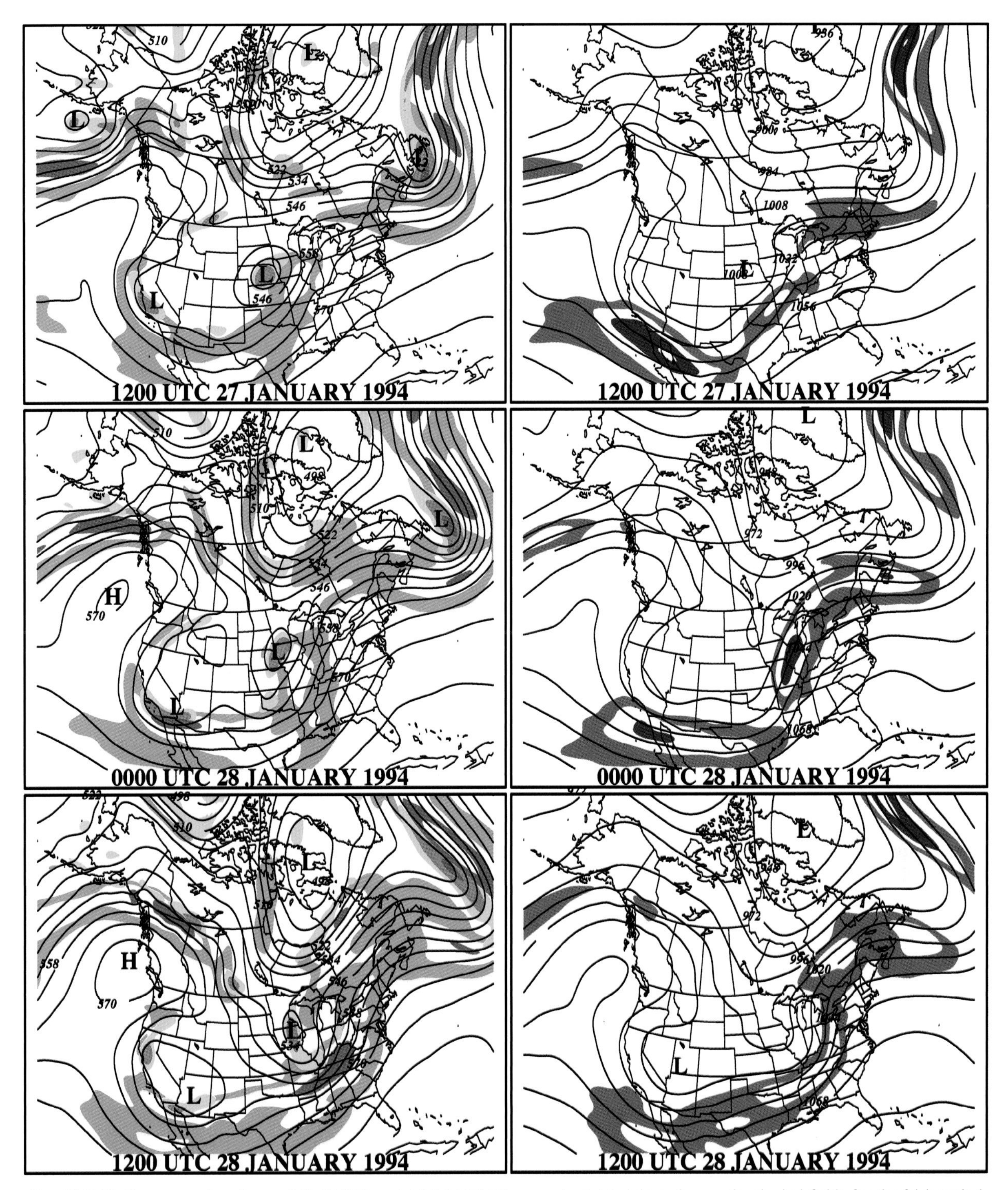

FIG. 11.3-18. Sequence of analyses of (left) 500- and (right) 250-hPa geopotential height and upper-level wind fields for the 24-h period between 1200 UTC 27 and 28 Jan 1994. See Fig. 11.1-3 for details.

exited eastward off the East Coast and the heavy precipitation developed along a complex frontal system, rather than a well-developed cyclone. Although high pressure moved off the east coast, extremely cold air associated with this system modified only slowly, and remained over much of the eastern United States. A band of snow exceeding 10 in. (25 cm) fell across West Virginia, northern Maryland, portions of Pennsylvania, and New York, with as much as 30 in. (75 cm) of snow over southern Ohio. Significant icing occurred across central

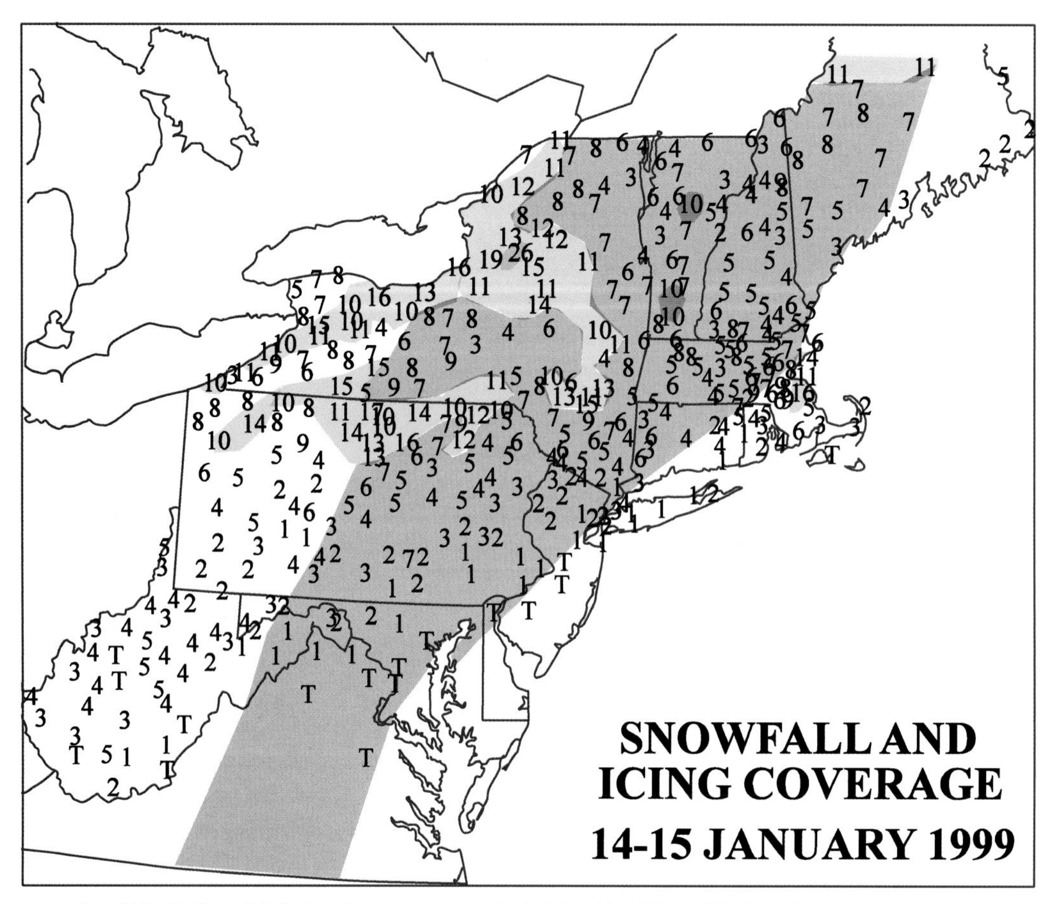

FIG. 11.3-19. Snowfall (in.) and icing coverage (violet) for 14–15 Jan 1999. See Fig. 11.1-1 for details.

and western Virginia, Maryland, and southeastern Pennsylvania (Fig. 11.3-13). In southern Maryland, this ice storm was considered the worst in 15 years with up to an inch (2.5 cm) of ice accumulating on exposed surfaces.

This storm was associated with an amplifying trough over the central United States composed of northern and southern short-wave systems (Fig. 11.3-15). This upper trough system was associated with a north–south-oriented frontal system with a small area of low pressure developing over the Tennessee Valley. As an upper trough over eastern Canada retreated eastward, 500-hPa heights rose over the western Atlantic, and the upper trough over the central United States moved eastward. High pressure moved off the East Coast, resulting in a broad southerly flow of air, but the air mass was initially so cold that it remained below freezing at low levels while above freezing air was advected over the area from the southwest at the 700-hPa level and above, resulting in widespread frozen precipitation, rather than rain or snow, from Virginia all the way to New England. As in other cases, the precipitation in this case was located within the entrance region of a well-defined, confluent jet streak extending northeastward along the east coast of the United States.

f. 27–28 January 1994

This was another significant ice storm in January 1994 that was associated with a classic example of "cold-air damming" (see volume I, chapter 3). A large, intense, cold anticyclone moved over the northeastern United States early on 27 January as low pressure was developing over the central United States. This anticyclone was accompanied by temperatures that were well below 0°F across much of New England and near 0°F (−17°C) in New York City and Boston. Cold air drained as far south as northern Florida and although the storm system was moving northeastward toward the Great Lakes, warmer air at the surface remained west of the Appalachians and an intense coastal front developed along the Southeast and mid-Atlantic coasts on 27 January. The area of icing during this event covered a large area from Virginia to Maine (Fig. 11.3-16).

The surface map at 1200 UTC 28 January (Fig. 11.3-17) illustrates that the surface high moved well east of Maine, but a wedge of cold air remained over interior New England, southwestward across eastern Pennsylvania, central Maryland, Virginia, and the western Carolinas into northern Georgia. It was here that severe icing conditions occurred. Interestingly, Washington re-

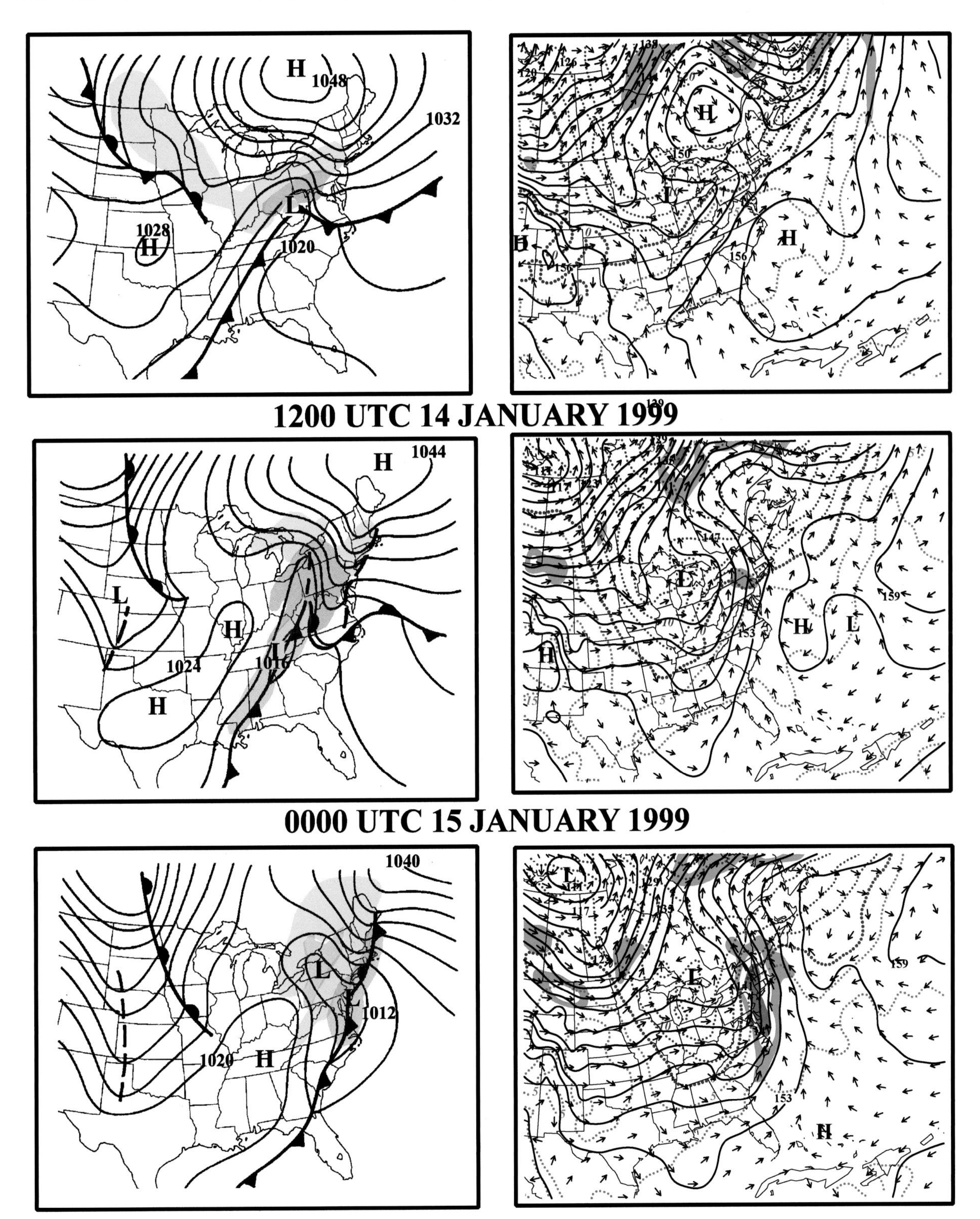

FIG. 11.3-20. Sequence of (left) surface and (right) 850-hPa charts for the 24-h period between 1200 UTC 14 and 15 Jan 1999. See Fig. 11.1-2 for details.

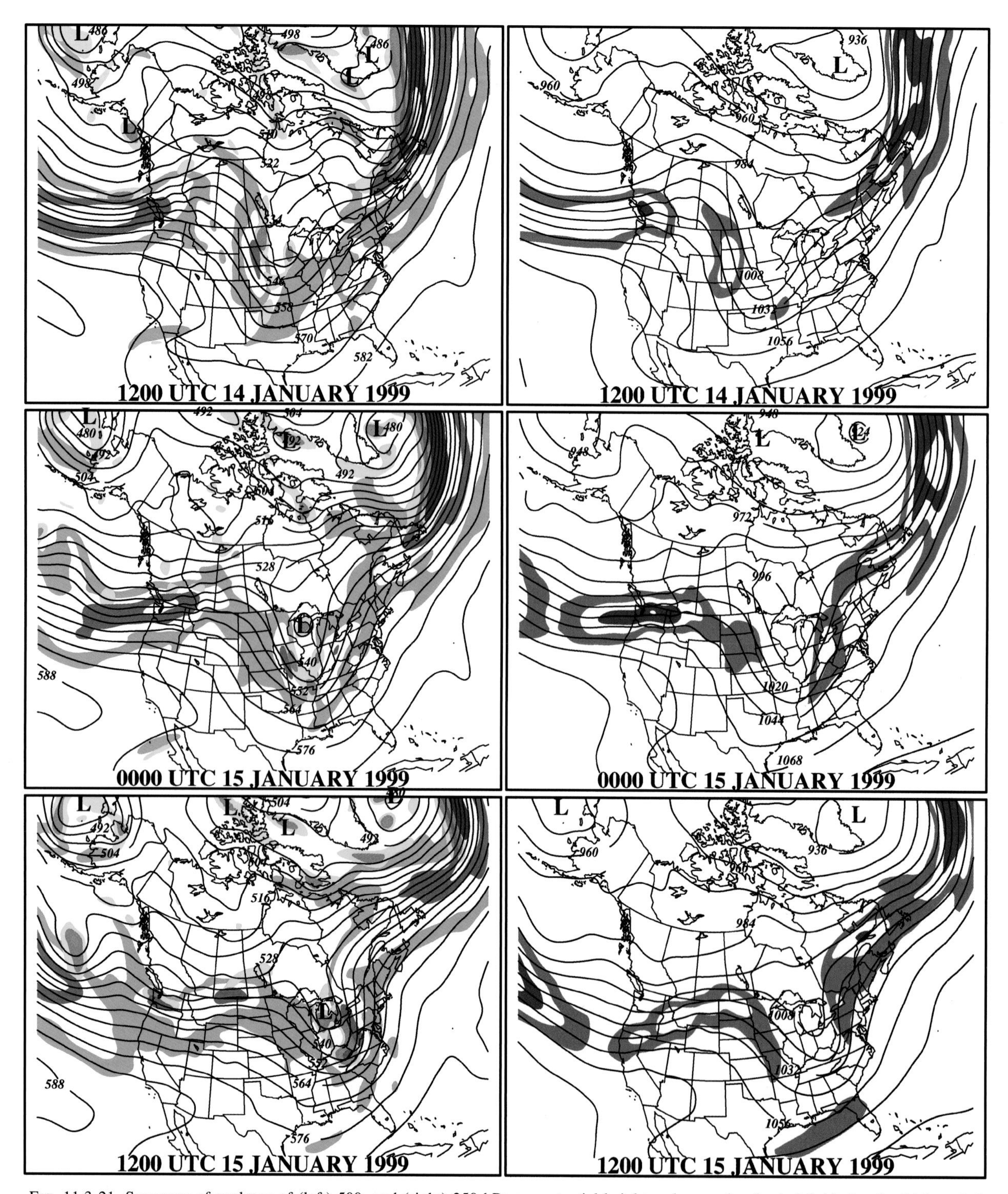

FIG. 11.3-21. Sequence of analyses of (left) 500- and (right) 250-hPa geopotential height and upper-level wind fields for the 24-h period between 1200 UTC 14 and 15 Jan 1999. See Fig. 11.1-3 for details.

ported severe icing and temperatures in the mid to upper 20s (°F) while temperatures in Philadelphia rose into the mid 40s (°F) with rain. To the west of the region, warmer air and temperatures in the 40s (°F) to near 50°F streamed northward into the Ohio Valley, western New York, and Pennsylvania and southern Ontario while surface temperatures remained below freezing just east of the Appalachian Mountains to Washington. To the east of the region, warm air streamed northward along the immediate coastline from Virginia to southern New

England. As the day progressed, the warmer air spread northward across much of eastern New England, raising temperatures as much as 90°F from the previous day!

The 500-hPa charts (Fig. 11.3-18) show that the strong high was associated with a region of confluence that moved rapidly eastward over the North Atlantic in association with an upper trough that moved from Newfoundland over the North Atlantic. As a result, the surface high also moved rapidly eastward over the Atlantic, as a storm system over the central United States moved eastward across the Ohio Valley and Great Lakes states with heights rising rapidly over the eastern United States. The strong ridging over the eastern United States and loss of upper confluence yielded a very unfavorable pattern for snow, but provided a flow regime that advected warmer air from the south and southwest over cold low-level air just east of the Appalachian Mountains, creating conditions favorable for a major icing event.

g. 14–15 January 1999

This ice storm is another case of strong cold-air damming preceded by a very strong cold anticyclone to the north of New England. The strong surface high pressure system and associated cold surface temperatures were difficult to dislodge east of the Appalachian Mountains as a large-amplitude upper trough moved eastward from the Midwest. This event produced heavy snow from northern Pennsylvania through much of interior New York (Fig. 11.3-19) and one of the most damaging ice storms on record in the Washington metropolitan area, as many homes lost power for several days.

This storm developed during a period in which ENSO was negative (La Niña) and daily values of the NAO were positive (not shown), conditions that appear to be unfavorable for a heavy Northeast snowstorm (see volume I, chapter 2). Unlike the heavy snow cases, the surface high north of New England responded to a very progressive upper trough moving from eastern Canada to the North Atlantic (Fig. 11.3-20). Upper-level confluent flow over eastern Canada gave way to rapidly building heights as a ridge developed along the East Coast and the surface high moved quickly eastward. While this occurred, cold low-level air drained southward into North Carolina but was not dislodged by a short-wave trough moving northeastward from the Ohio Valley and Great Lakes states, although milder air aloft was advected from the southwest over the shallow, cold air (see the 850-hPa analyses in Fig. 11.3-20). A coastal front developed along the mid-Atlantic and New England coastlines, separating the shallow cold air from rapidly modifying milder air over the western Atlantic. Low pressure over the Ohio Valley redeveloped along the coastal front as warm air moved northeastward along the coast from Virginia to Massachusetts, changing freezing rain to rain in those areas. However, shallow subfreezing air remained west of the frontal boundary, contributing to a serious ice storm.

As in other cases, the precipitation for this event was located in the right-rear entrance region of a confluent jet streak that remained west of the Northeast coast (Fig. 11.3-21). The combination of the ascent pattern associated with the jet entrance region, the warming aloft related to the southwesterly flow and building upper-level East Coast ridge, and the low-level cold-air damming produced this major ice storm.

From Amer. Tel. & Tel. Co.

THE HEAVY HAND OF WINTER ON TREES AND WIRES
An ice-storm in Rhode Island, December 1, 1921

Tremendous ice storm, Nov 1921. Source unknown.

The National Bank of Washington, D.C., 5 Dec 1957 (courtesy of Kevin Ambrose, *Blizzards and Snowstorms of Washington, D.C.*, Historical Enterprises, 1993).

Harvey Cedars, New Jersey, during the Great Atlantic Storm of Mar 1962. Photo by Dorothy Oldham (from *Great Storms of the Jersey Shore*, The Shore Publishing Corp. and The SandPaper, Inc., 1993).

Storm of 11–12 Dec 1992, New Jersey. Photo by Tom Moersh (from *Great Storms of the Jersey Shore*, The Shore Publishing Corp. and The SandPaper, Inc., 1993).

Crash of Air Florida Flight 90 into the Potomac River in Washington D.C.; tail recovered on 19 Jan 1982 (courtesy of Kevin Ambrose, *Washington Weather*, Historical Enterprises, 2002).

Mooers, New York, Jan 1998 (courtesy of John Ferguson, NWS Office, Burlington, Vermont).

Chapter 12

EARLY AND LATE SEASON SNOWS

In this chapter, snow events that occur during autumn or spring and define the margins of the snow season are described. The primary months for significant snowfall are December, January, February, and March (see volume I, chapter 2). However, major snowstorms can occur with much less regularity in November and April, and even in some rare instances in October and May. Most of the snow events occur inland, although there are cases where measurable snow occurred near the immediate coast. Examples of early and late season events are provided to illustrate several of the more unusual storms and to illustrate that, every now and then, wintry weather can visit the northeast United States early in the autumn and well into spring.

1. October and November events

Early season snows have occurred in the northern and hilly sections of the Northeast in September (the highest peaks in late August), but October is the first month that accumulating snow has occurred in the Northeast urban corridor. Ludlum (1966, 1968) provides a summary of many early snowstorms, with several of the more significant events summarized here. On 9 October 1804, an intense storm left greater than 20 in. (50 cm) of snow across western Massachusetts into Vermont and western New Hampshire, and more than 10 in. (25 cm) across the Catskills in New York. Another early season snowstorm occurred in an unusually snowy autumn on 5–6 October 1836, with amounts of 10–20 in. (25–50 cm) across portions of Pennsylvania and central New York and lesser amounts farther south into Virginia along the slopes of the Appalachians. Another storm on 3–4 October 1841 brought heavy snows to interior Connecticut and Massachusetts and possibly the earliest snow on record for New York City and Baltimore. Nearly 150 years later, a similar storm would occur in the interior Northeast on 3–4 October 1987, the greatest October snowstorm of the last half of the 20th century for the Northeast.

Closer to the coastline, one of the earliest measurable snows occurred on 8–9 October 1703, when snow was reported in Philadelphia and was measured at 3–4 in. (7.5–10 cm) in Boston. A late October storm in 1746 left 10–20 in. (25–50 cm) in the Boston area. New York City's heaviest October snowfall occurred on 26 October 1859, when a total of 4 in. (10 cm) was reported. Other notable October snows occurred in New York City in 1925 and 1972.

In November, interior snows are not a rarity, but major snowstorms are very rare events, especially in the heavily populated areas of the Northeast. Examples of major interior snowstorms include the great Appalachian Storm of 25 November 1950 (described in chapter 9), and the storms of 11–12 November 1968 and 24–25 November 1971 (see chapter 11). During the snowy winter of 1995/96, a major snowstorm affected interior portions of the Northeast on 14–15 November 1995. The two most notable early heavy snow events that affected at least some of the major urban areas include the storm of 6–7 November 1953 and the "Veterans' Day Storm" of 11–12 November 1987.

Following are brief summaries of several of these early season snowstorms.

a. 3–4 October 1987

This storm was the earliest significant snowfall of the 20th century for the Northeast and has been the subject of studies by Bosart and Sanders (1991) and Gedzelman and Lewis (1990). Measurable snow fell across interior eastern New York, western Connecticut, western Massachusetts, and Vermont (Fig. 12.1-1), accompanied at times by thunder and lightning. Snowfalls of greater than 10 in. (25 cm) occurred in New York's Catskill Mountains, and the highlands of western Massachusetts, extreme eastern New York, northwestern Connecticut, and southern Vermont. Slide Mountain, New York, reported 17 in. (43 cm); Grafton, New York, measured 22 in. (55 cm); and Pownal, Vermont, measured 18 in. (45 cm). Measurable snow fell as far south as coastal Connecticut with 0.5 in. (1 cm) at Bridgeport. Power outages were extensive since trees had a full canopy and tree-related damage to automobiles and houses was extensive. There were 20 storm-related deaths and more than 300 injuries (Bosart and Sanders 1991).

Surface charts (Fig. 12.1-2) show that the snowstorm was associated with a front oriented north–south along the East Coast, in which several areas of weak low pressure consolidated into a strengthening low pressure system that moved northward from east of Delaware at 0000 UTC 4 October 1987 to eastern New England at

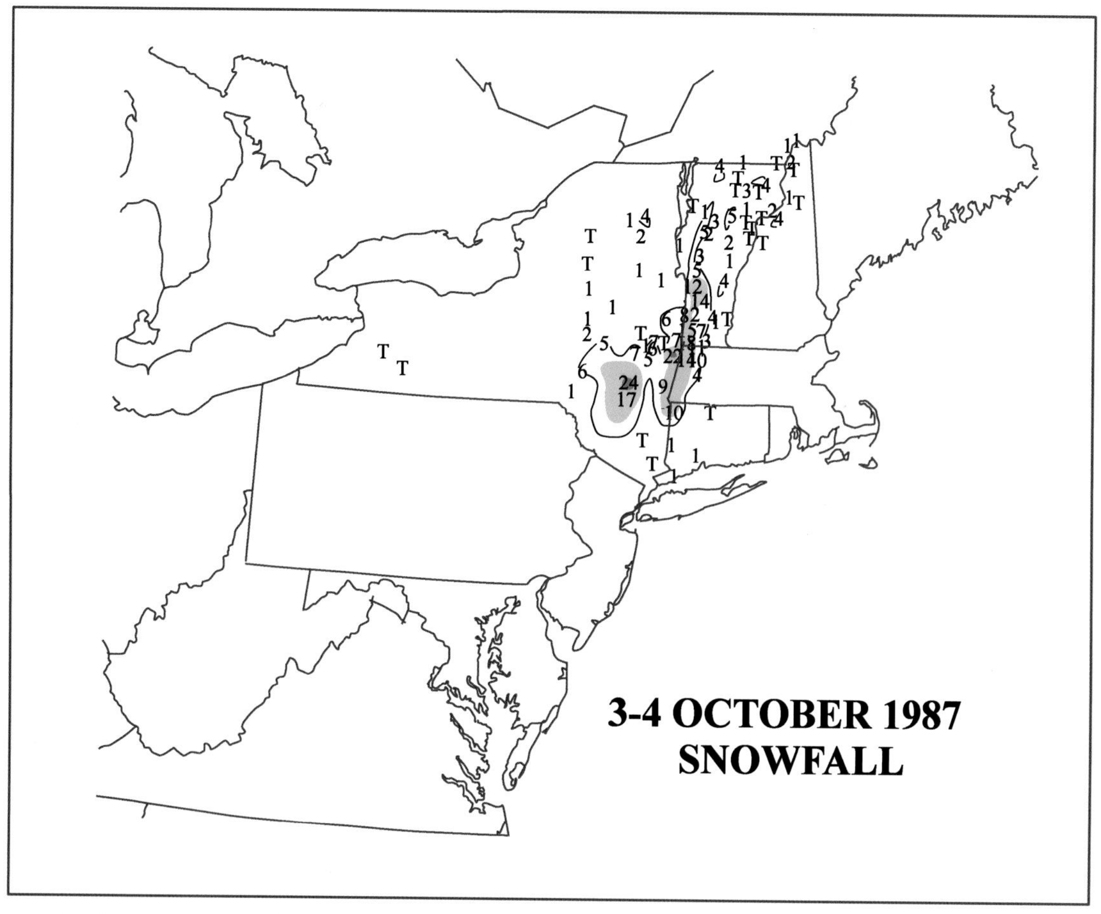

FIG. 12.1-1. Total snowfall (in.) for 3–4 Oct 1987. Shading contours are for 10(25) and 20 in. (50 cm). Solid line represents 4 in. (10 cm) contour. (Snowfall totals were kindly provided by Steve Maleski.)

1200 UTC 4 October. A central pressure of approximately 1002 hPa was analyzed at 0000 UTC 4 October and fell 12 hPa to 990 hPa across eastern New England by 1200 UTC 4 October 1987. As this surface low moved north into New England, an intense 850-hPa circulation developed along the 0°C isotherm (Fig. 12.1-2) with below freezing air well established in east-central New York by 0000 UTC 4 October, supporting the heavy snowfall in that region.

The 500-hPa charts (Fig. 12.1-3) show that the snowstorm was associated with a deepening, negatively tilted trough, within which a deep, closed circulation developed, while the low pressure–frontal system evolved into a deepening cyclone over New England. There is some evidence for a distinct polar jet near the base of the trough off the Massachusetts coast at 500 hPa at 1200 UTC 4 October and a distinct jet entrance region at 250 hPa downwind of the developing storm within a north–south-oriented confluent zone over eastern Canada at 0000 and 1200 UTC 4 October. The north–south confluent zone and jet entrance region over eastern Canada are collocated with a strengthening 850-hPa temperature gradient extending northward from the Northeast into Quebec (Fig. 12.1-2). But, these features developed as the cyclone intensified and did not serve as a precursor signal for the impending snow event. Bosart and Sanders (1991) note that prior to cyclogenesis, lower-tropospheric temperatures remained well above 0°C in the cold air west of the surface front. A cold pool developed over the snowfall area in New York largely as a result of cooling by melting snow, allowing heavy snow to reach low elevations (also see Wexler et al. 1954; Kain et al. 2000).

b. 10 October 1979

One of the earliest measurable snows on record occurred for portions of the Northeast urban corridor on 10 October 1979. For the most part, this was not a heavy snowstorm but it was the earliest measurable snow during the 20th century for a good portion of the middle Atlantic states and southern New England. The heaviest snows fell in the mountains of Virginia where 4–10-in. amounts (10–25 cm) were common (Fig. 12.1-4) with up to 17 in. (43 cm) of snow measured at Big Meadows, Virginia. More than 6 in. (15 cm) were reported in north-

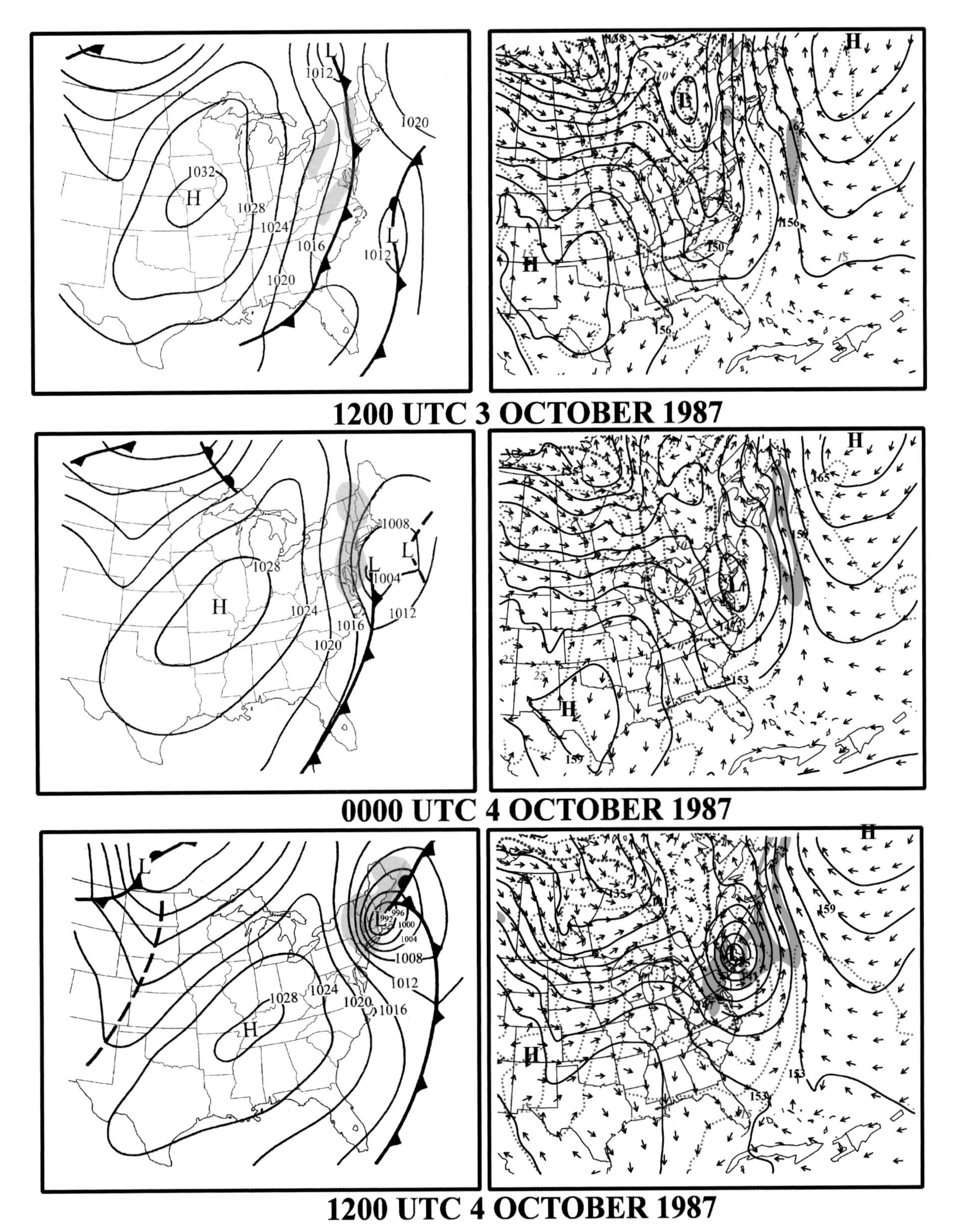

FIG. 12.1-2. Sequence of (left) surface and (right) 850-hPa charts for the 24-h period between 1200 UTC 3 and 4 Oct 1987. Surface maps include surface high and low pressure centers and fronts. Shading indicates the following: blue, snow; violet, mixed precipitation; and green, rain. Solid lines are isobars (4-hPa intervals), and dashed lines represent axes of surface troughs not considered to be fronts. The 850-hPa analyses include contours of geopotential height (solid, at 30-m intervals; 156 = 1560 m), isotherms (dotted, °C, at 5°C intervals; blue, 0°C and less; red, 5°C and greater), and intervals of wind speed greater than 20 m s^{-1} (at 5 m s^{-1} intervals; alternating blue/white shading; red shading, 40 m s^{-1}).

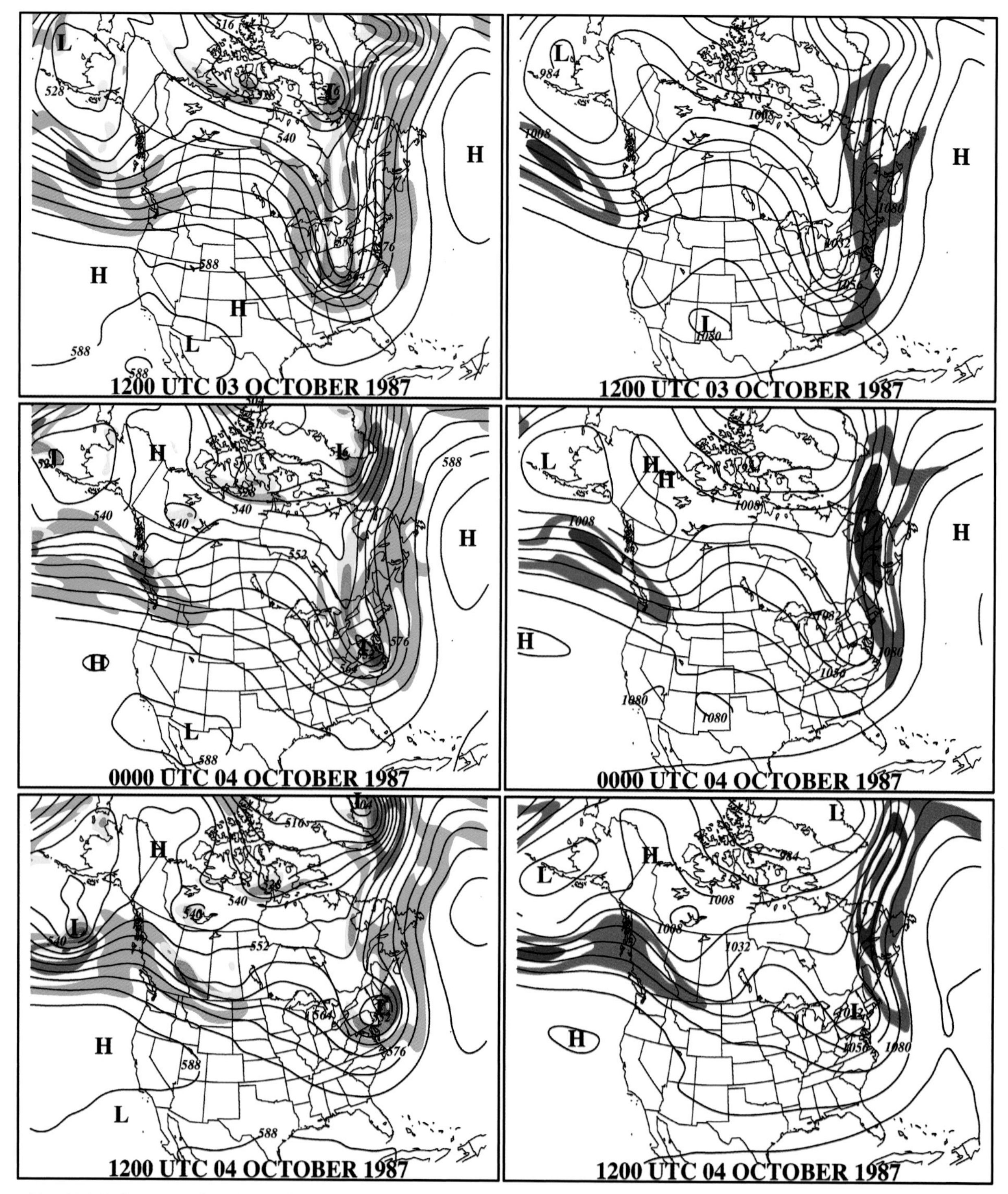

FIG. 12.1-3. Sequence of analyses of (left) 500- and (right) 250-hPa geopotential height and upper-level wind fields for the 24-h period between 1200 UTC 3 and 4 Oct 1987. Twelve-hourly analyses of 500-hPa geopotential height and upper-level wind include locations of geopotential height maxima (H) or minima (L), contours of geopotential height (solid, at 60-m intervals; 522 = 5220 m), locations of 500-hPa absolute vorticity maxima (yellow/orange/brown areas beginning at 16×10^{-5} s^{-1}; intervals of 4×10^{-5} s^{-1}), and 400-hPa wind speeds exceeding 30 m s^{-1} (at 10 m s^{-1} intervals; alternate blue/white shading). Analyses of 250-hPa geopotential height and winds (rhs) include heights (solid at 120-m intervals; 1032=10 320 m), and wind speeds exceeding 50 m s^{-1} (at 10 m s^{-1} intervals; alternate blue/white shading).

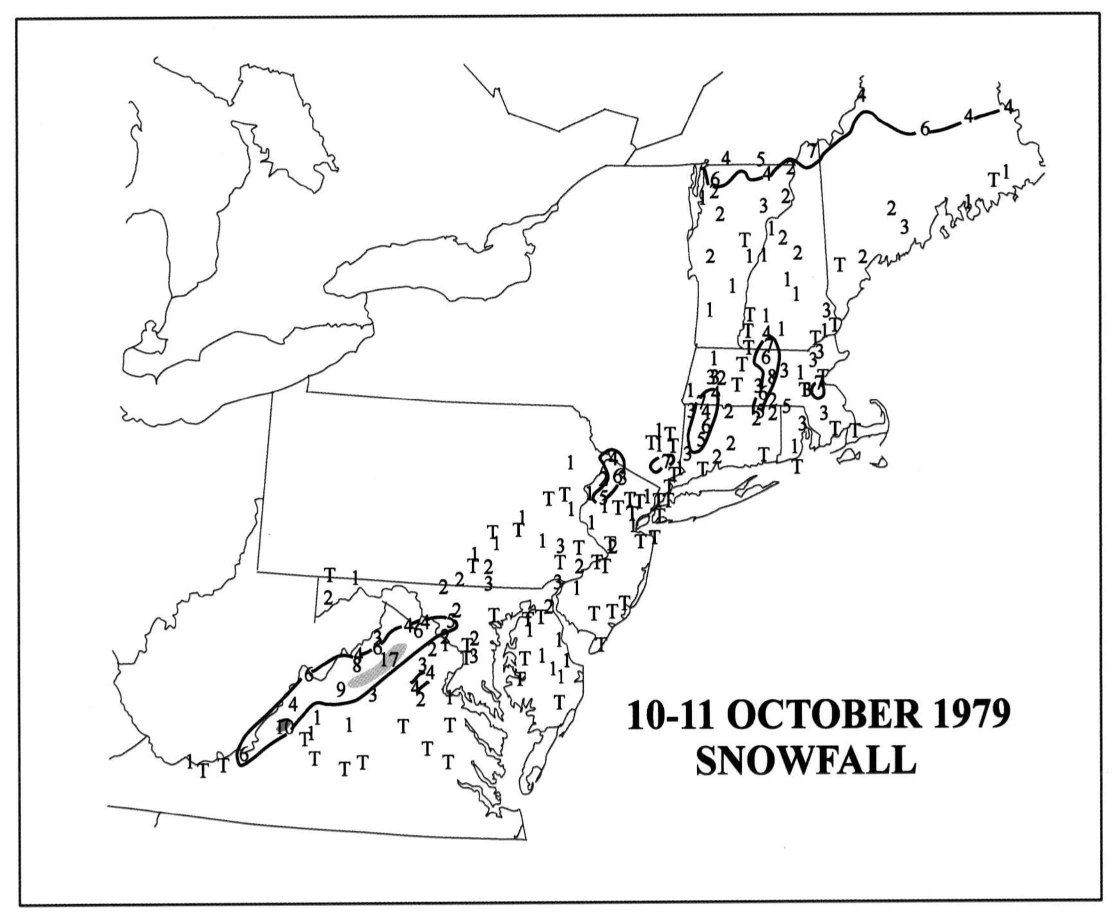

FIG. 12.1-4. Total snowfall (in.) for 10–11 Oct 1979. See Fig. 12.1-1 for details.

western Connecticut, central Massachusetts, and central Maine. Up to 5 in. (13 cm) of the snow fell in the Maryland suburbs of Washington, and 6.5 in. (16 cm) fell at Yorktown Heights, New York, just north of New York City. Measurable snows fell in the following areas: Worcester, Massachusetts [7.5 in. (18 cm)]; Providence, Rhode Island [2.5 in. (6 cm)]; Philadelphia, Pennsylvania [2.1 in. (5 cm)]; Wilmington, Delaware [2.5 in. (6 cm)]; Baltimore, Maryland [0.3 in. (<1 cm)]; Washington's National Airport [0.3 in. (< 1 cm)] and Dulles Airport [1.3 in. (3 cm)]; and Boston, Massachusetts [0.2 in. (< 1 cm)]. The World Series baseball game at Baltimore was cancelled due to snow and cold. Numerous power outages affected many residents and waterspouts were reported in New York Harbor near the Verrazano Narrows Bridge because of the extreme instability due to the cold air over relatively warm waters.

The surface charts (Fig. 12.1-5) show that the snow was associated with a relatively weak low pressure system that developed along a slow-moving cold front with nearly all the precipitation falling in the cold air behind the front. At 0000 UTC 10 October, high pressure was located over western Kansas while a slow-moving cold front cut across the eastern half of the United States. A broad band of rain fell to the north of the front as a weak 1009-hPa low developed over central North Carolina. At this time, rain was beginning to mix with and change to snow and sleet across portions of West Virginia, as colder air with temperatures below 0°C at 850 hPa slowly moved southwestward toward Virginia on 1200 UTC 10 October.

By 1200 UTC 10 October, the front continued to move slowly to the southeast with a weak surface low pressure system (1007 hPa) moving northeastward along the front. Rain had spread northeastward across the middle Atlantic states into New York and New England and quickly mixed with and changed to snow. Snow fell from the mountains of North Carolina through much of the central and western middle Atlantic states and southern New England. While 850-hPa analyses (Fig. 12.1-5) show the snow falling in a region characterized by a slow drop of 850-hPa temperatures slightly below 0°C, the occurrence of a large area of snowfall developing within the precipitation area may be indicative of cooling associated with melting snow (Kain et al. 2000). During the day, the surface low intensified and moved offshore and the rains and snows gradually ended from west to east. By 0000 UTC 11 October, the surface low

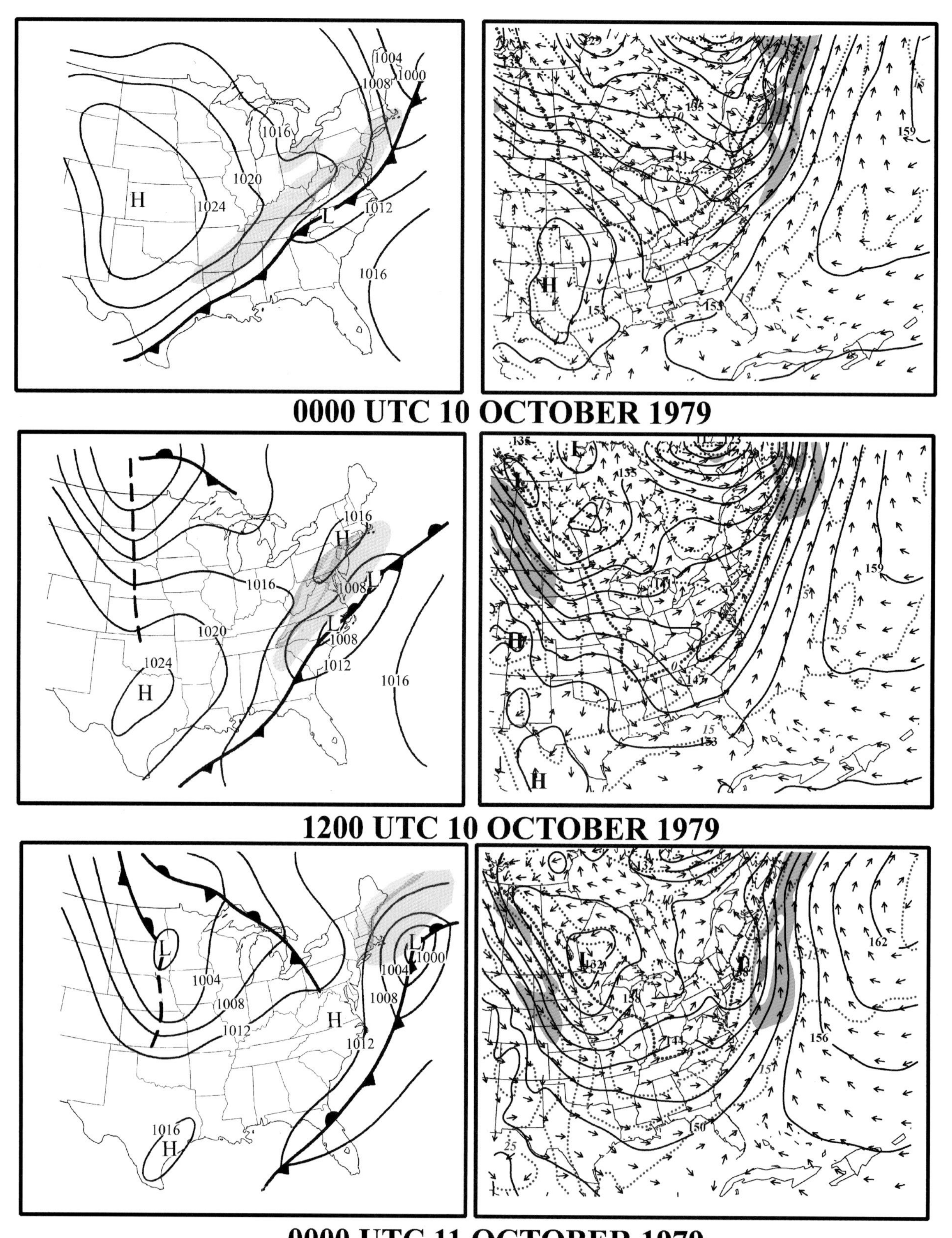

FIG. 12.1-5. Sequence of (left) surface and (right) 850-hPa charts for the 24-h period between 0000 UTC 10 and 11 Oct 1979. See Fig. 12.1-2 for details.

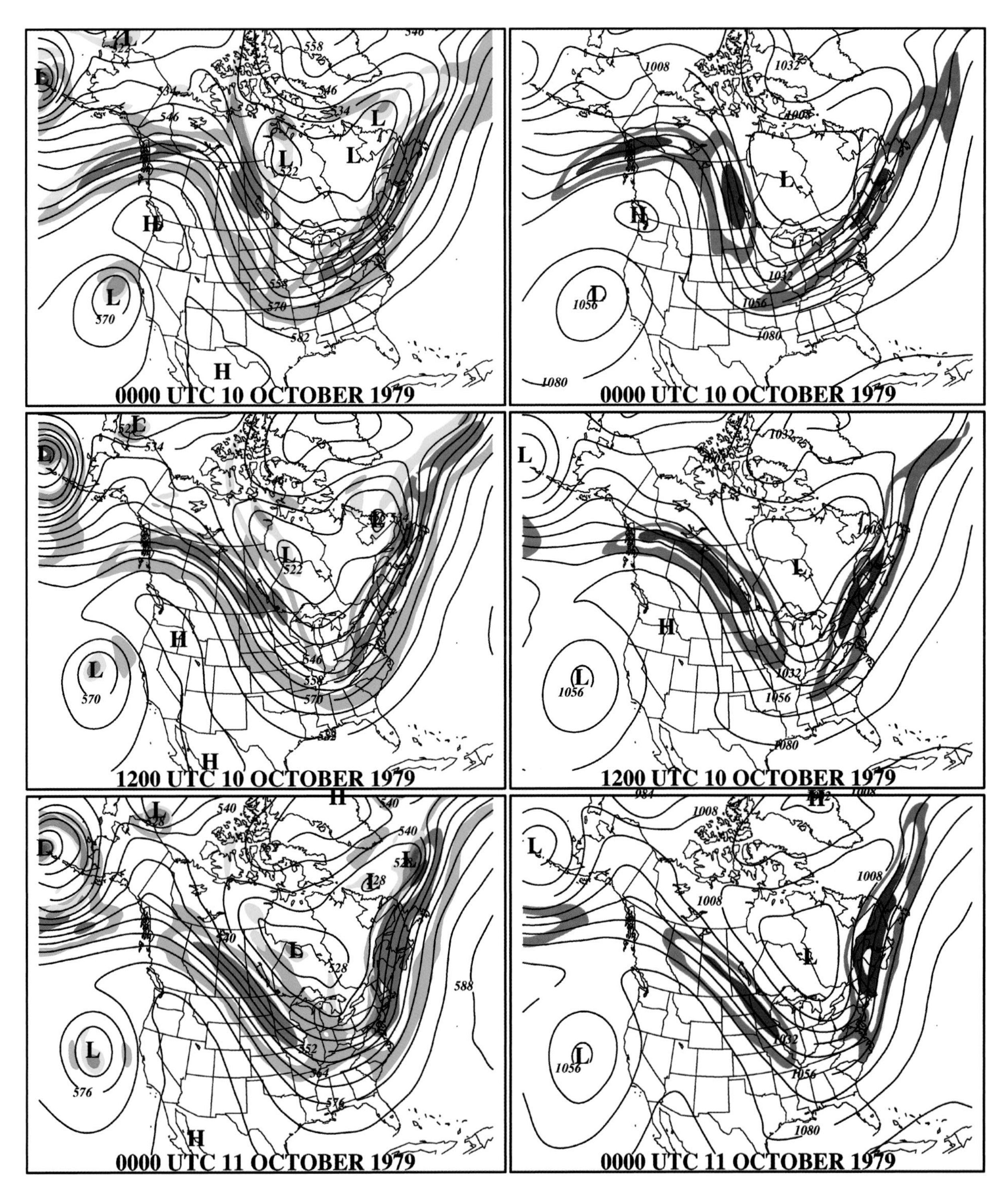

FIG. 12.1-6. Sequence of analyses of (left) 500- and (right) 250-hPa geopotential height and upper-level wind fields for the 24-h period between 0000 UTC 10 and 11 Oct 1979. See Fig. 12.1-3 for details.

was southeast of Nantucket, Massachusetts, and had deepened to 998 hPa and snows were affecting much of eastern New England.

The corresponding 500- and 250-hPa charts (Fig. 12.1-6) show a broad upper trough over the central and eastern United States that moved eastward over 24 h with little to no amplification during this period. The precipitation appears to develop within the right-entrance region of a jet streak at 400 hPa that extends from Tennessee to southeastern Canada at 1200 UTC 10 Oc-

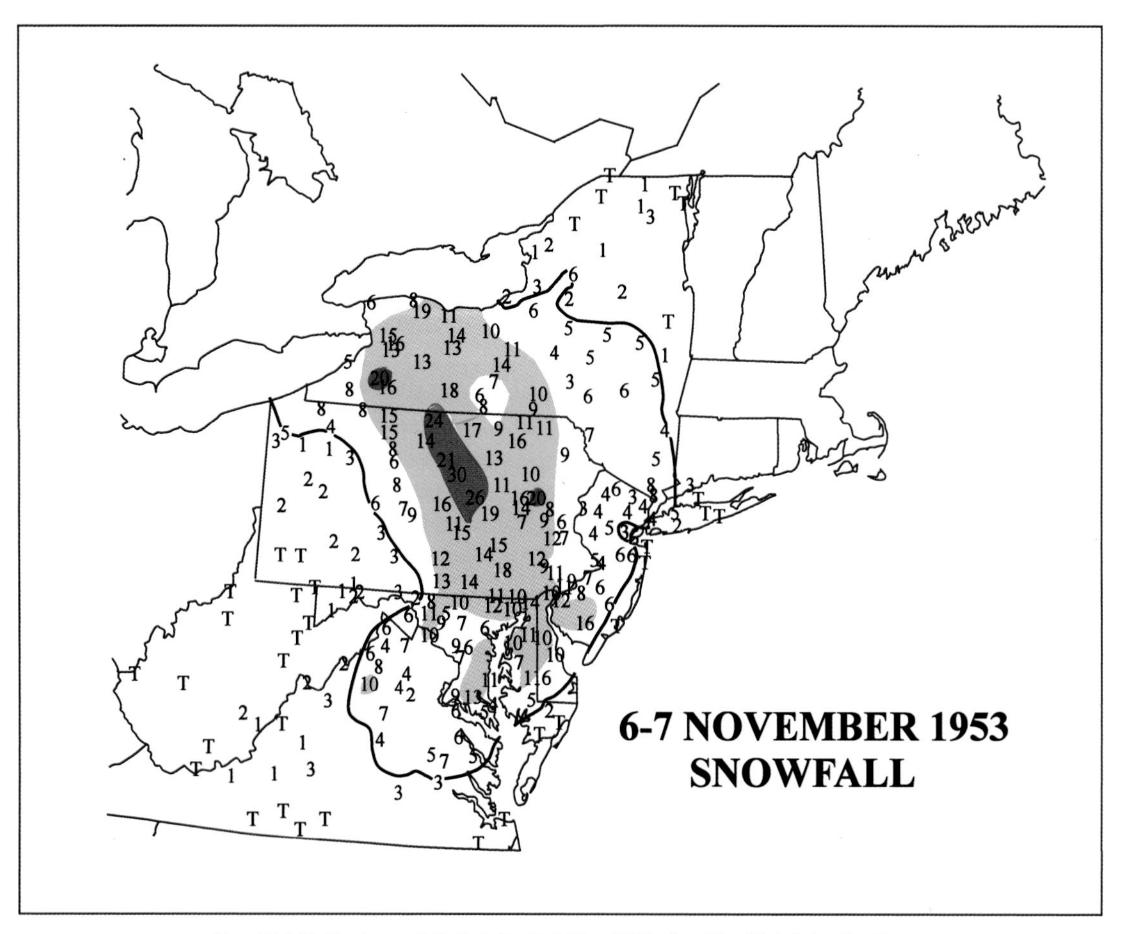

FIG. 12.1-7. Total snowfall (in.) for 6–7 Nov 1953. See Fig. 12.1-1 for details.

tober, with developing snow elongated northeastward along and just south of the axis of the jet.

c. 6–7 November 1953

This early season snowstorm was known for record-setting snowfall in Virginia, Maryland, Delaware, Pennsylvania, and western New York. Easterly and northeasterly gales coincided with high tides and produced severe coastal flooding, with some of the highest tides on record in the New York City area. The heaviest snows fell in a band from central Pennsylvania into western New York where reports of 15 in. (35 cm) of snow or more were common (Fig. 12.1-7). The greatest snow total occurred in central Pennsylvania, where 29.8 (74) and 27.5 in. (69 cm) fell at Lock Haven and Middleburg, Pennsylvania, respectively.

This storm produced the heaviest early season snowstorm in the following areas: Richmond, Virginia [7.3 in. (19 cm)]; Washington [6.6 in. (18 cm)]; Baltimore [5.9 in. (15 cm)]; Wilmington [11.9 in. (30 cm)]; Philadelphia [8.8 in. (22 cm)]; Harrisburg, Pennsylvania [15.4 in. (39 cm)]; and Williamsport, Pennsylvania [13.1 in. (33 cm)]. Record early season snowfall also affected southern New Jersey [Millville: 15.5 in. (39 cm)] and the Chesapeake Bay region [Charlotte Hall, Maryland: 13 in. (33 cm)].

The storm developed with a weak wave of low pressure east of Florida and south of an unusually strong and cold anticyclone that covered much of the eastern United States. Early on 6 November, the storm began to intensify and move north-northeastward off the Southeast coast. Late on 6 November and early on 7 November the surface low began moving in a more northerly direction toward Long Island as it developed rapidly and encountered the strong anticyclone centered near Maine (Fig. 12.1-8). The storm's rapid intensification and decreasing distance between the low pressure center and the anticyclone to the north created extensive northeasterly gales ahead of the storm and combined with high tides (some of the highest on record) to yield severe coastal flooding. The intensification of the storm is also reflected by the significant increase of 850-hPa circulation on 7 November, with an easterly low-level jet exceeding 35 m s^{-1} developing north of the storm center by 0300 UTC 7 November (Fig. 12.1-8). As the storm neared Long Island, it began to turn northwestward, and by the morning of 7 November, the center had passed near New York City before eventually mov-

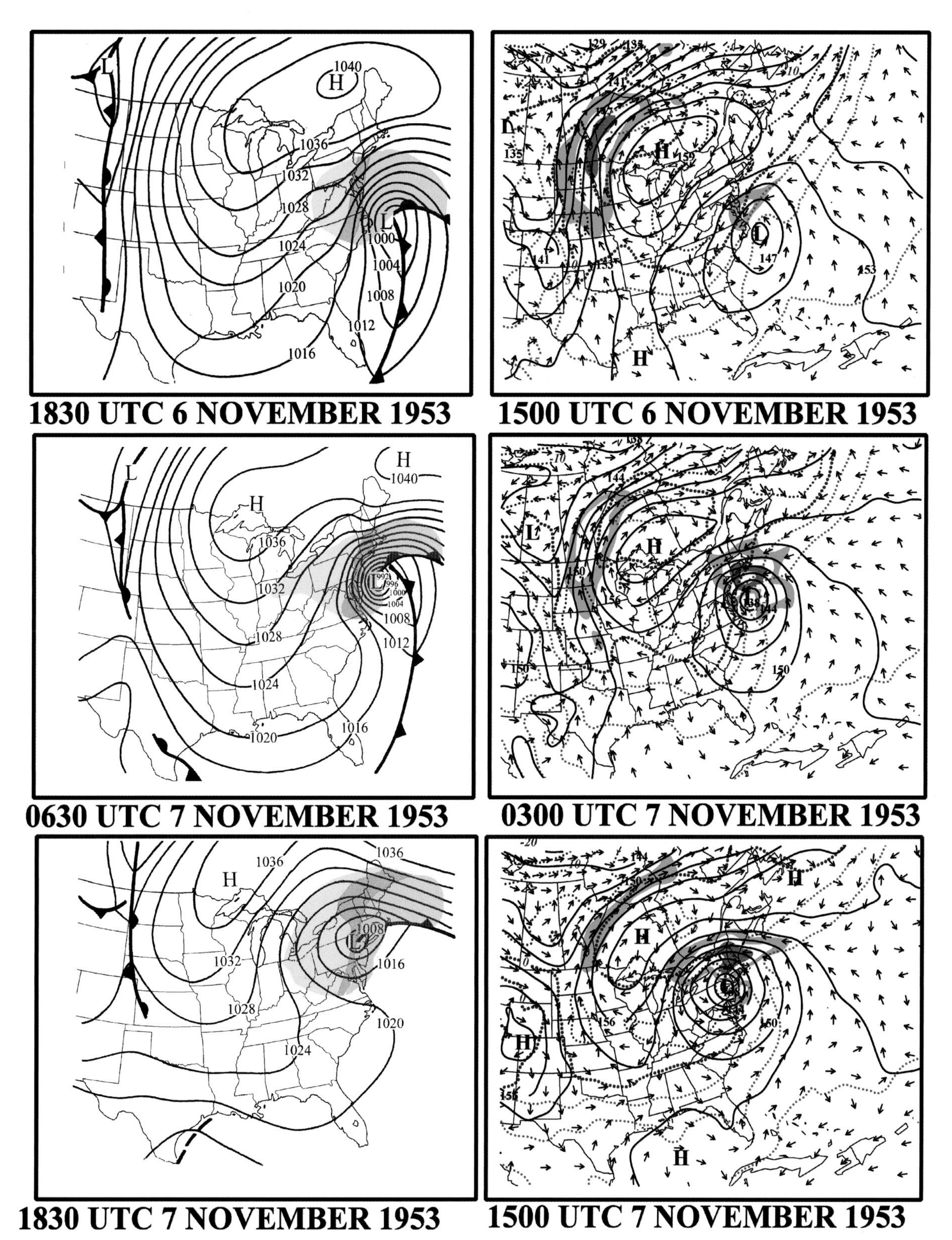

FIG. 12.1-8. Sequence of (left) surface and (right) 850-hPa charts for the 24-h period between 1500 UTC 6 and 7 Nov 1953. See Fig. 12.1-2 for details.

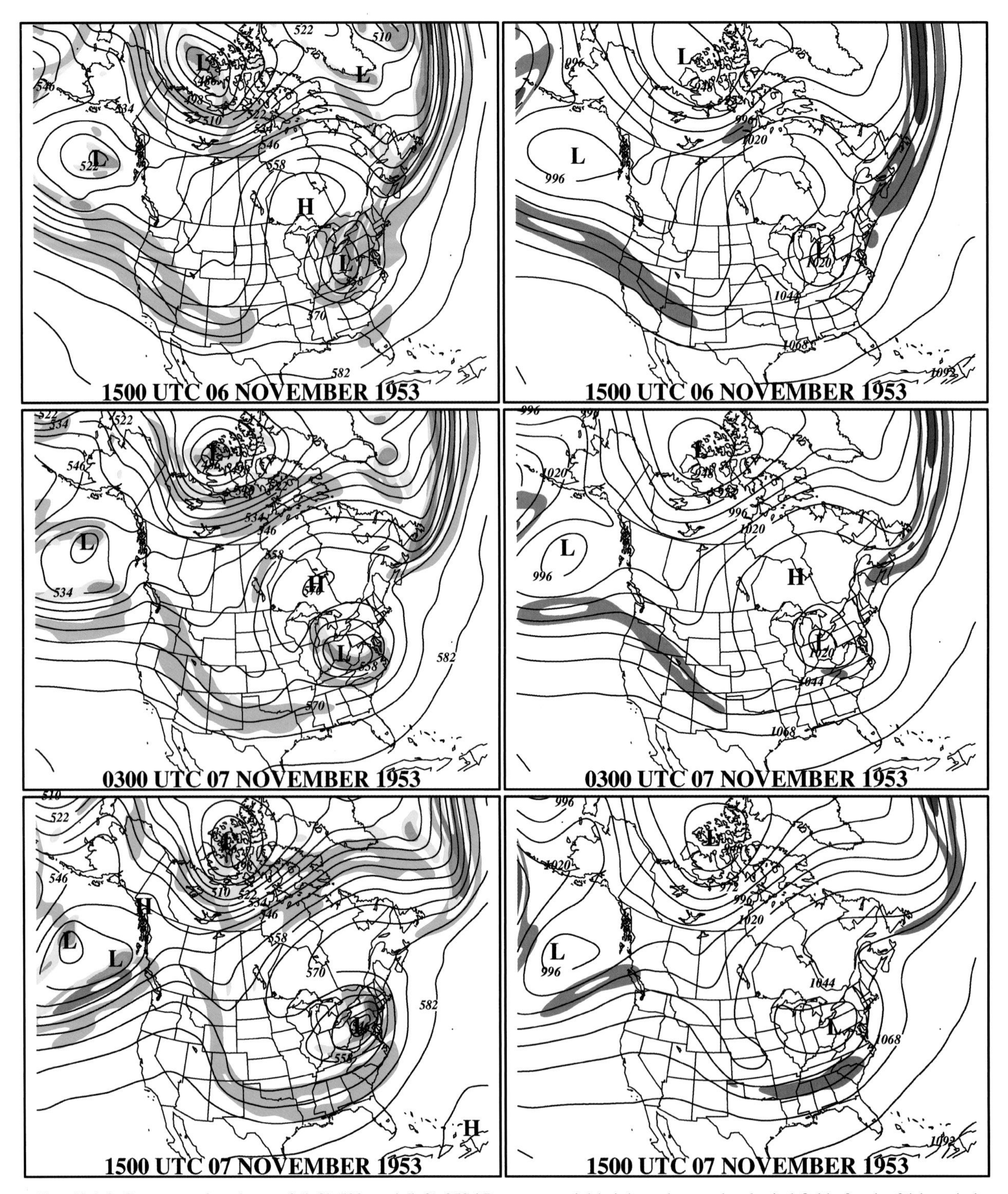

FIG. 12.1-9. Sequence of analyses of (left) 500- and (left) 250-hPa geopotential height and upper-level wind fields for the 24-h period between 1200 UTC 11 and 12 Nov 1987. See Fig. 12.1-3 for details.

ing over the Catskill Mountains early in the afternoon and weakening dramatically.

The storm initially began as a weak disturbance in the subtropical jet stream (not shown). At the same time, a developing cutoff low at the 500- and 250-hPa levels over the Ohio Valley and ridging north of the system over the Great Lakes provided the upper-air support for the low pressure system as it moved up the East Coast and intensified (Fig. 12.1-9). As the cutoff low deepened over Pennsylvania, the surface low continued to inten-

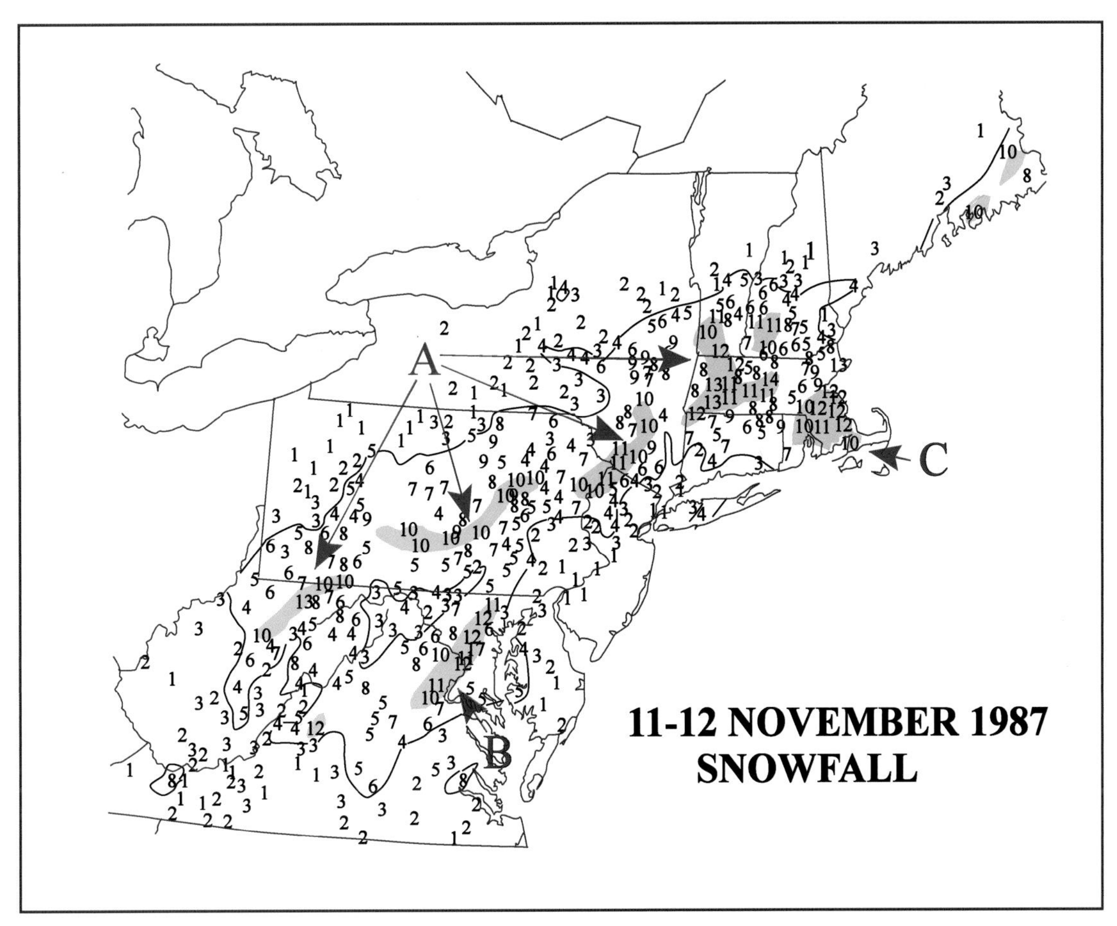

FIG. 12.1-10. Snowfall distribution for the Veterans' Day Snowstorm of 10–11 Nov 1987. Letters A, B, and C refer to separate bands of snowfall attributed to several distinct periods described in the text. See Fig. 12.1-1 for details.

sify and move slowly to the north and then northwest. The slow movement of the system and cold upper-level air associated with the cutoff low both contributed to the very heavy early season snowfall over central Pennsylvania. As the storm developed and slowly turned to the west toward New York on 7 November, the colder air penetrated farther south toward the Carolinas. Precipitation that started as rain across Virginia and North Carolina late on 5 November began to mix with and change to snow across interior North Carolina and along the Virginia–Maryland border by early in the morning hours on 6 November. By the morning of 6 November, much of the precipitation across central and eastern North Carolina and Virginia, Maryland, and Delaware had changed over to snow. It is also possible that the widespread changeover from rain to snow was associated with increasing vertical motions and cooling due to melting snow especially in the southern most areas of snowfall (Kain et al. 2000).

d. 10–12 November 1987: The Veterans' Day Storm

The Veterans' Day Snowstorm of November 1987 (see Maglaras et al. 1995) was an early season snowstorm that left 11.5 in. (29 cm) of snow in Washington, D.C., and up to 18 in. (45 cm) in the eastern and northern suburbs, the heaviest November snowfall on record and the earliest major snowstorm since 6–7 November 1953. Cars and trucks were stranded on roadways during the storm and some people spent the night in their vehicles. The 8 in. (20 cm) of snow at Boston was the heaviest so early in the season and was accompanied by winds that gusted to 30–50 mi h^{-1}, downing power lines across much of southeastern Massachusetts.

This storm was a remarkably complex event that occurred in three distinct and sometimes overlapping stages (labeled A, B, and C in Fig. 12.1-10). The first stage (A) was a moderate band of snowfall with maximum amounts of 6–12 in. (15–30 cm) that occurred on 10 November 1987 from northern West Virginia into central and northeastern Pennsylvania, eastern New York, and western New England. The second stage (B) was a mesoscale band of snow that developed over a 6–10-h period on 11 November 1987, centered on the Washington area (also see Fig. 6-6, volume I), leaving only 2 in. (5 cm) in the western suburbs, but as much as 17 in. (43 cm) in the eastern suburbs. The snowfall was accompanied by thunder and lightning related to mesoscale processes addressed in chapter 6 of volume I. The third and final stage (C) was the snowfall over eastern New England associated with a rapidly developing cyclone and 850-hPa circulation common to other snowstorms documented in chapter 10, which produced 8–12 in. (20–30 cm) in Rhode Island and eastern Massachusetts, one of the heaviest early season snows at those locations.

The heavy snow was associated with a series of upper-level short-wave troughs, a favorable jet streak pattern, and a slow-moving surface front (Figs. 12.1-11 and 12.1-12). The colder air required for the separate snow events was confined to the northeastern quarter of the country as the first area of low pressure developed east of the middle Atlantic states and southern New England late on 10 November and early on 11 November, and was perfectly situated from the Carolinas to New England at the 850-hPa level to support these three separate snow events that marked this case (Fig. 12.1-11).

The first band of heavy snow fell on 10 November as one short-wave trough spawned a weak cyclone that moved northeastward along the front from Louisiana to a position off the middle Atlantic and southern New England coastlines. This first area of heavy snow fell in the entrance region of an upper-level jet streak along the northeast U.S.–Canadian border (not shown). As the trough and related surface cold front moved eastward, the main short-wave trough (shown in the 500- and 250-hPa charts in Fig. 12.1-12) began to consolidate by 11 November. The mesoscale band of snow in the Washington area developed during the morning of 11 November within a distinct diffluent exit region of a 60–70 m s^{-1} jet streak extending to the northeast near the base of the trough at 250 hPa. This snowfall continued through the afternoon as a surface low was developing off the Virginia capes. However, this band developed and dissipated locally and the axis of heavy snow did not progress northeastward, as is often observed in other snowstorms.

As the upper-level trough and associated jet streak moved northeastward and became negatively tilted between 0000 and 1200 UTC 12 November, the surface low deepened rapidly off the southern New England coast late on 11 and early on 12 November (Fig. 12.1-11), heavy snows redeveloped in southeastern New England, resulting in the snow area (C) in Fig. 12.1-10. This snowfall episode continued as the main upper-level trough amplified and cut off as a closed low at 500 hPa by 1200 UTC 12 November while the surface low continued to deepen within the diffluent exit region of the upper-level jet.

e. 26–27 November 1898: The Portland *Gale*

This rapidly developing snowstorm in November 1898 is well known as one of the most destructive storms ever to affect the New England coast, as more than 100 ships (including the steamer *Portland*) were wrecked and sank with the loss of more than 200 lives. Heavy snows developed from New Jersey to New England (Fig. 12.1-13), with the heaviest snows falling along the immediate coastline accompanied by hurricane force winds. The heaviest snow fell in a band from New Jersey northeastward across Long Island, Connecticut, Rhode Island, most of Massachusetts, southern New Hampshire, and Maine (with Connecticut recording one of its heaviest snows on record). In the following locations, reports of 10–20 in. (25–50 cm) of snow were common: New Haven, Connecticut, 16.8 in. (43 cm); Atlantic City, New Jersey, 14 in. (36 cm); Portland, Maine, 16.4 in. (42 cm); and Boston, 12.5 in. (32 cm). Some amounts exceeded 20 in. (50 cm), including New London, Connecticut, with 27 in. (69 cm), and Bridgeport, Connecticut, with 24 in. (61 cm). Winds were reported to be the highest ever recorded at the time at Block Island, Rhode Island (90 mi h^{-1}, fastest mile); Woods Hole, Massachusetts (78 mi h^{-1}); and Nantucket (72 mi h^{-1}).

The only data available for this case are surface analyses, some of which are displayed in Fig. 12.1-13. Prior to the storm, cold air had spread across much of the eastern United States and was retreating off the east coast by early morning on 26 November. As a result of the cold air already in place, precipitation began as snow as far south as North Carolina and Virginia, indicating that the melting processes (Kain et al. 2000) and resultant cooling may not have been as significant of a contributor to the heavy snow for this case as it was for the October and early November snows discussed earlier.

A description of the storm in the November 1898 edition of *Monthly Weather Review* states: "A distinct feature of these storms is found in the fact that a development of destructive strength begins with a union at some point off the middle Atlantic or South New England coasts of two storms, one from the west or northwest and the other from the South Atlantic coast" (Garriot 1898). In the late 19th century, this case was described as an interesting case of secondary development (see volume I, chapter 3). The description of the storm includes the notion of a surface low "digging" southeastward across the Ohio Valley, as a separate surface low developed and moved northeastward along the South Atlantic coast and suddenly intensified at the expense of the weakening primary cyclone.

At 0800 local time (LT) 26 November 1898, two 1008-hPa surface lows over northwestern Ohio and northern Michigan were located at the leading edge of a significant cold-air outbreak. Meanwhile, a weak surface low just east of the South Atlantic coast was associated with light rain and snows over the Carolinas and Virginia. During 26 November, the Midwest lows continued south and eastward and weakened while the Southeast low deepened off the mid-Atlantic coast. By 2000 LT, the surface lows consolidated into one strengthening surface low east of the mid-Atlantic coast-

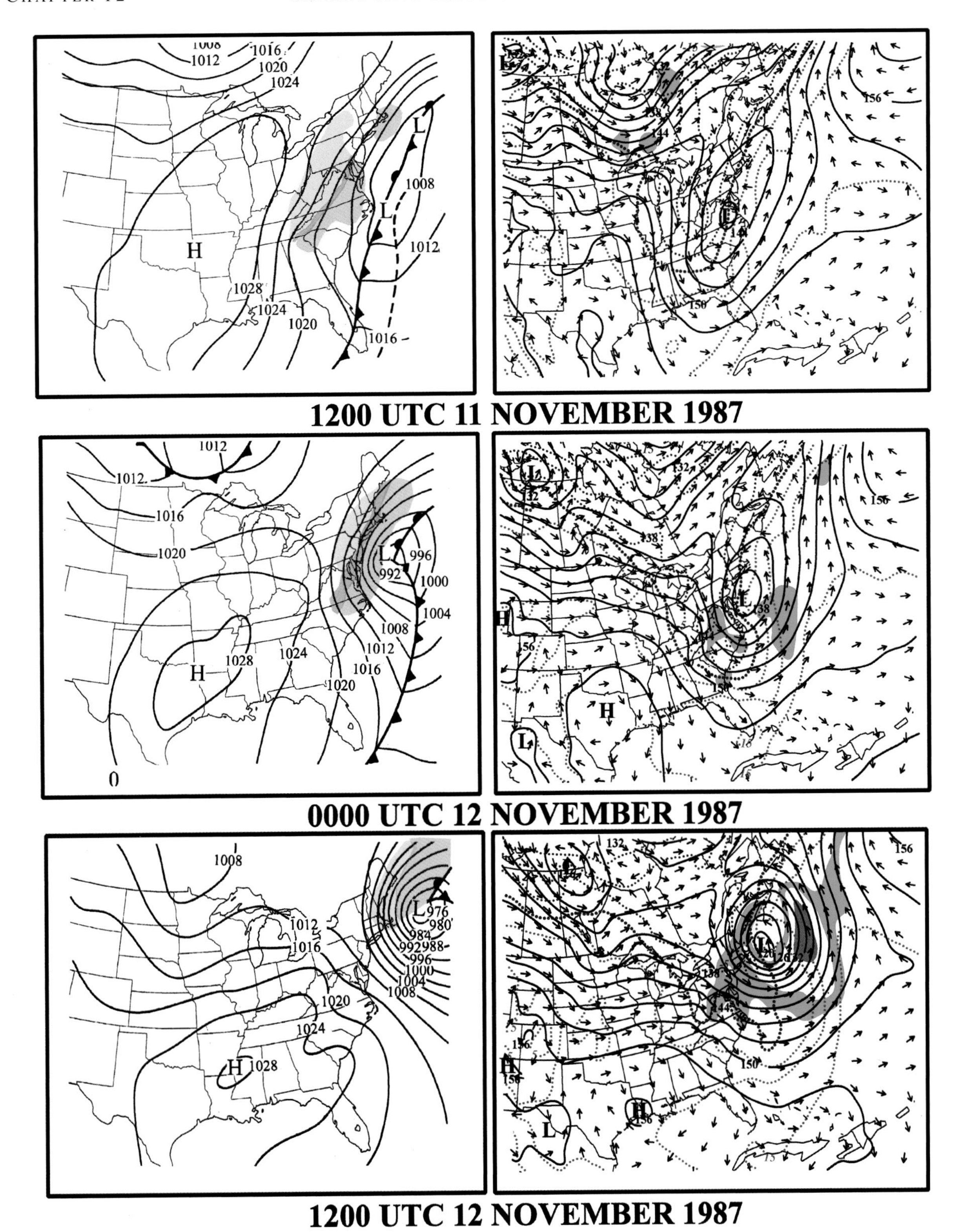

FIG. 12.1-11. Sequence of (left) surface and (right) 850-hPa charts for the 24-h period between 1200 UTC 11 and 12 Nov 1987. See Fig. 12.1-2 for details.

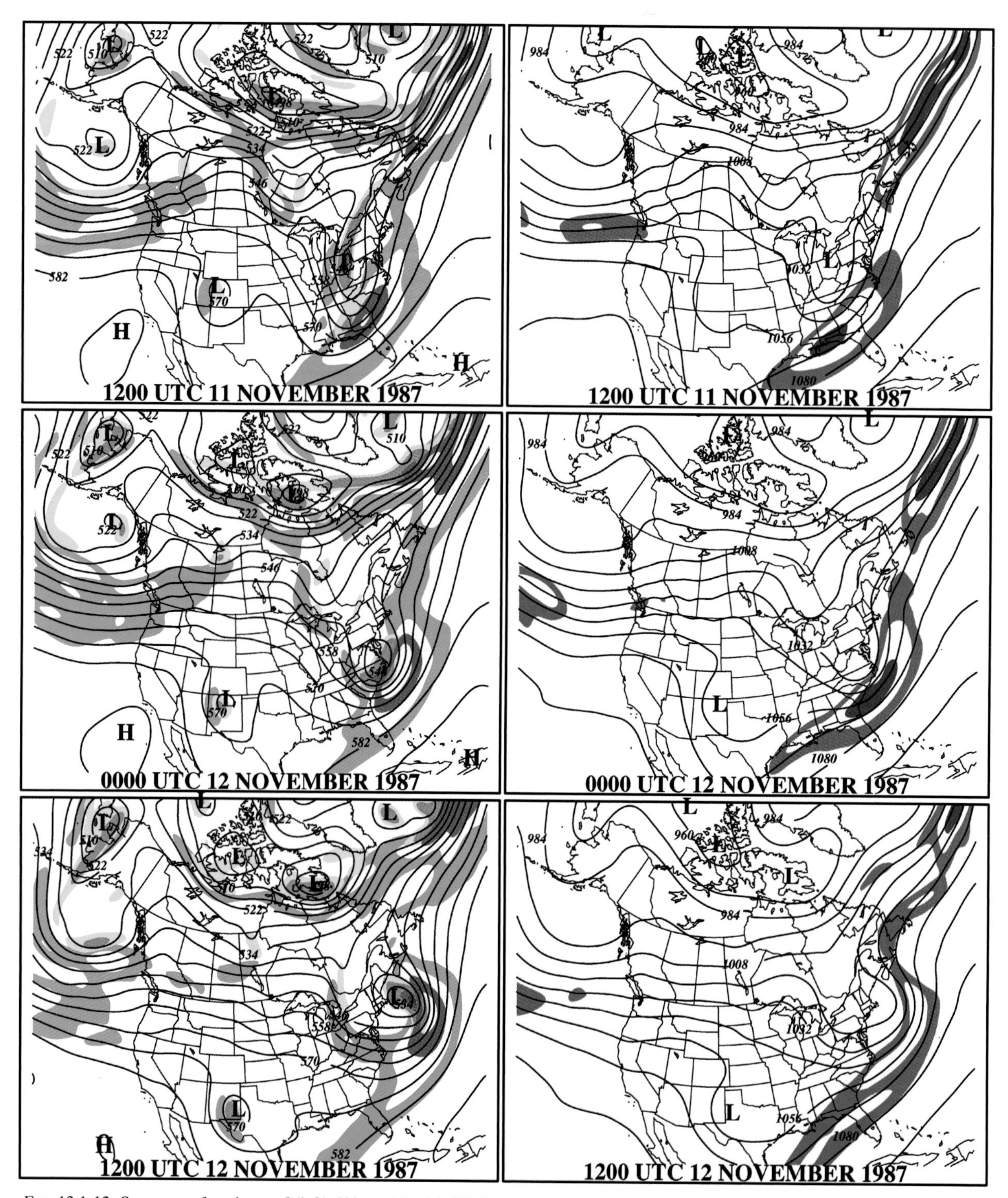

FIG. 12.1-12. Sequence of analyses of (left) 500- and (right) 250-hPa geopotential height and upper-level wind fields for the 24-h period between 1200 UTC 11 and 12 Nov 1987. See Fig. 12.1-3 for details.

line. According to newspaper reports, the snows spread northeastward across the northeast United States late on 26 November in conjunction with the developing coastal low. The tremendous snow and winds developed at night, moving into eastern New England by morning. At this time, the surface low had deepened explosively, reaching an estimated 972 hPa by 0800 LT 27 November. One interesting report notes clearing skies and calm winds over eastern Cape Cod between Chatham and Barnstable during the midmorning of 27 November, ap-

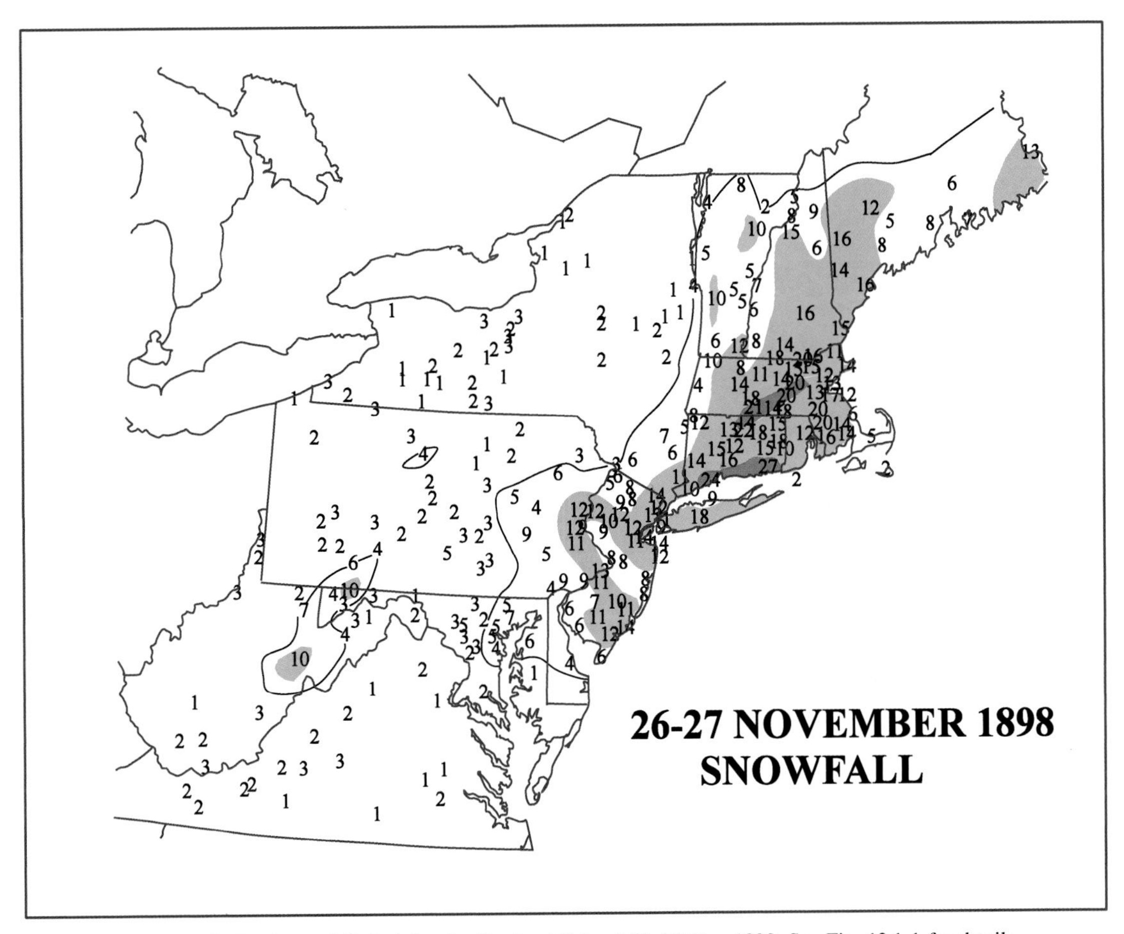

FIG. 12.1-13. Total snowfall (in.) for the *Portland* Gale of 26–27 Nov 1898. See Fig. 12.1-1 for details.

parently as the surface low passed over the region, followed by the backlash of the storm that brought great destruction to trees.

2. April and May events

Occasionally, winter hangs on into April and, on rare occasions, into May. Two April snow events since 1950 have been classified as major snow events and are included as part of the 32-case sample in chapter 10 (see chapters 11.19 and 11.28). Historically, the middle 19th century also produced several notable snowstorms during April. A major storm on 12 April 1841 yielded 10–12 in. (25–30 cm) of snow in Philadelphia, 18 in. (45 cm) in New York City, 16–18 in. (40–45 cm) at Providence, and 18–20 in. (45–50 cm) at New Bedford, Massachusetts. This storm was preceded and followed by significant snowfall only a few days before and after. On 14–18 April 1854, a storm that Ludlum describes as "one of the great coastal storms of all time" produced great shipping losses; snow from Washington to New England; 9 in. (22 cm) of snow in Newark, New Jersey; and over 20 in. (50 cm) of snow across northern and western New Jersey. Two major snowstorms in April 1857 affected interior New York and Pennsylvania with snows of 20–40 in. (50–100 cm) or greater. Significant snows affected the Northeast during April 1873, April 1875, and April 1886.

The most notable April snowstorm of the early 20th century affected the middle Atlantic states and southern New England on 3–4 April 1915, when more than 10 in. (25 cm) fell from Virginia to southern coastal New England. Recent notable April events include the snowstorm of 6–7 April 1982, which is considered one of the most anomalous springtime snowfall events since the intensity and record-breaking cold accompanying this storm were more typical of January than most of the wet snows that characterize springtime snowstorms. The late season storm of 18–19 April 1983 produced significant snows as far south as North Carolina and Virginia, and then produced heavy snow amounts across much of interior New York and Pennsylvania. On 9–10 April 1996, the final event of the snowiest winter season of the 20th century generated heavy snows across central and southern New Jersey, Long Island, Connecticut, Rhode Island, and Massachusetts. The snowstorm of 31 March–1 April 1997 produced some of the heaviest snows on record across portions of interior southeastern New York and Massachusetts, including 33 in. (84 cm) at Worcester and 25.4 in. (65 cm) at Boston, the third

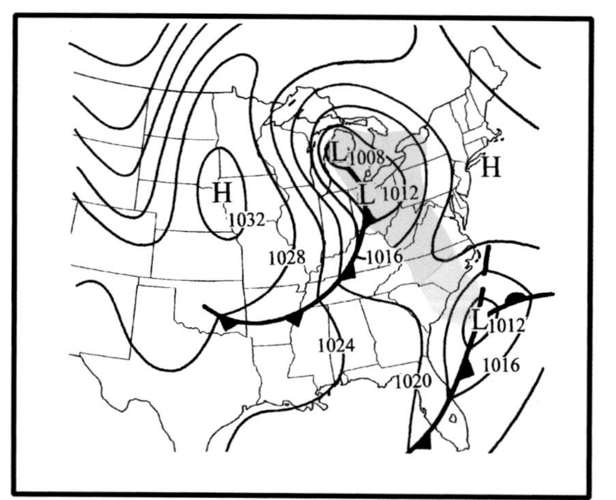

1300 UTC 26 NOVEMBER 1898

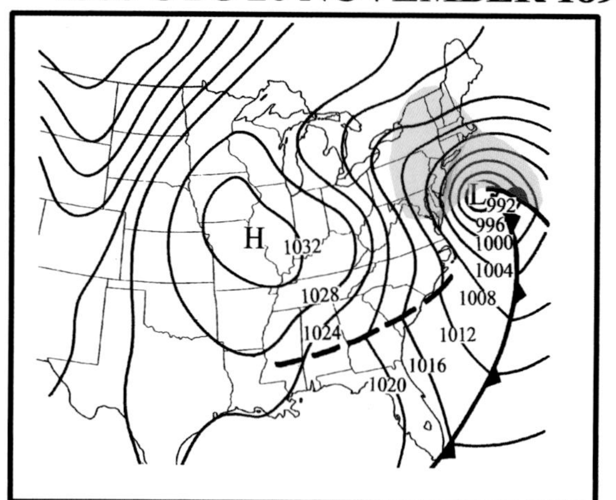

0100 UTC 27 NOVEMBER 1898

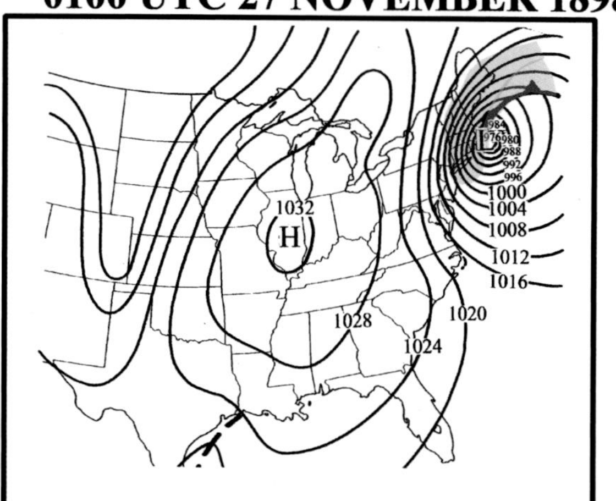

1300 UTC 27 NOVEMBER 1898

FIG. 12.1-14. Sequence of surface charts for the 24-h period between 1200 UTC 26 and 27 Nov 1898.

greatest snowstorm in its history and greatest 24-h snowfall. A late season snowstorm on 9–10 April 2000 affected the Washington metropolitan area through New York City and much of interior New York. In Washington, afternoon temperatures fell from 80°F (26°C) in the afternoon to the low 30s [°F (0°C)] overnight with heavy wet snow.

May has also had its share of rare snowfall events as well. On 3–4 May 1812, snow fell in Philadelphia and fell heavily in New York and New England. No mention is made of accumulation in New York City but 4 in. (10 cm) of snow were reported in Boston and amounts exceeding 10 in. (25 cm) were reported from interior New England. On 15–16 May 1834, a major storm left 6 in. (15 cm) at Erie, Pennsylvania; 4 in. (10 cm) at Rochester, New York; and 10 in. (25 cm) at Bradford, Pennsylvania, before intensifying over northern New England. Burlington, Vermont, received 12 in. (30 cm), while northeastern Vermont reported amounts exceeding 20 in. (50 cm). Snow showers extended as far southward as the Boston area and New York City with a hard freeze as far south as Washington. More recent events include 28–29 April 1987 over eastern New England, 17 May 1973 and 19 May 1976 over central New York, and 25 May 1967 across northern portions of New England. One of the most interesting late season snowfall events occurred over portions of the Northeast on 9 May 1977 when a coastal cyclone off New England yielded snowflakes along the mid-Atlantic coast and record-setting accumulating snow in Boston, Providence, and Worcester. Finally, the latest measurable snow on record visited Binghamton and Albany, New York, on 18 May 2002, and up to 8 in. (20 cm) fell in the Catskill Mountains.

During 1816, the infamous "year without a summer," snow fell in interior portions of New York and New England during June! Snow accumulated at higher elevations of northern New England on 6–8 June and snow showers extended to coastal Maine, the Berkshire Mountains of Connecticut and Massachusetts, New York's Catskill Mountains (where an inch was reported), and a few flakes were observed in the Boston suburbs. The latest general snow event occurred on 11 June 1842 when up to 12 in. (30 cm) fell in northern New England and snow was observed in the Boston area, the Pennsylvania and Maryland Appalachians, and at Harrisburg. In Philadelphia, snowflakes were reported as late as 1 June 1843. In New York City, the latest report is from 25 May 1845. Snowflakes have been reported on the highest peaks in the Northeast even into July. Brief case histories of several late season snowstorms are now provided below, with the April 1982 and 1997 cases discussed in detail in chapter 10.

a. 3–4 April 1915

The Easter storm of 3–4 April 1915 brought heavy snows to the coastal regions of the middle Atlantic states and southern New England, breaking April records from

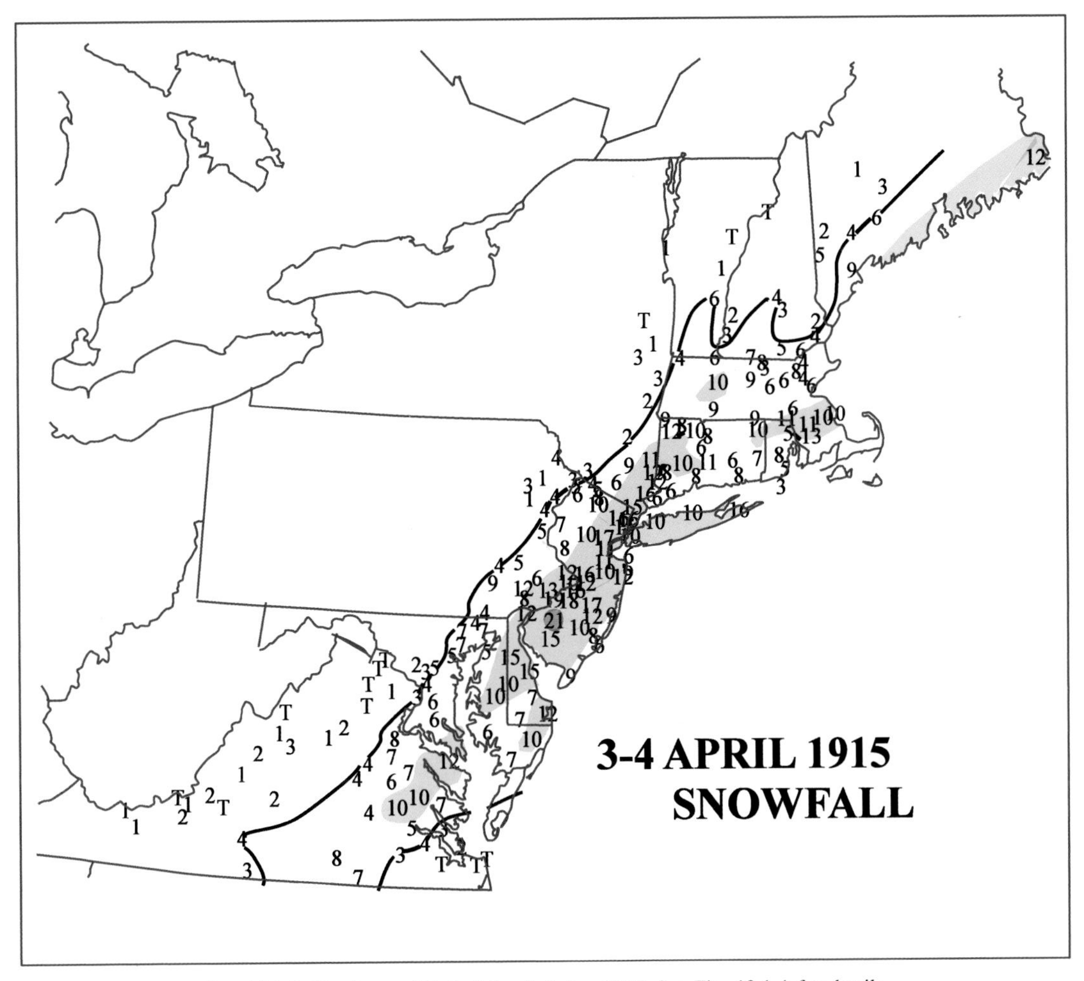

FIG. 12.2-1. Total snowfall (in.) for 3–4 Apr 1915. See Fig. 12.1-1 for details.

North Carolina to New York (Fig. 12.2-1). The greatest amount of snow reported from the storm was 21.2 in. (54 cm) at Clayton, New Jersey, and 19.4 in. (49 cm) at Philadelphia, its fourth greatest snowfall on record. As much as 10 in. (25 cm) of snow or greater was common from interior North Carolina, to eastern Virginia, eastern Maryland, Delaware, New Jersey, southeastern New York, and portions of Connecticut. The storm was especially intense off the middle Atlantic coast and maximum winds (fastest miles) were excessive: 62 mi h^{-1} at Norfolk, 62 mi h^{-1} at New York, 65 mi h^{-1} at Block Island, and 79 mi h^{-1} at Nantucket. This storm is easily the greatest April snowstorm on record for region stretching from the coast of Virginia to New Jersey. No other snowstorm occurred in the middle Atlantic region so late in the season during the rest of the 20th century.

The surface maps show that the storm developed in the eastern Gulf of Mexico as a large late season outbreak of relatively cold air covered all of the eastern United States in late March and early April (Fig. 12.2-2). A 1010-hPa surface low was located over Florida early on 2 April, which intensified rapidly and moved slowly northeastward to a position off Cape Hatteras early on 3 April where its central pressure is estimated at 975–980 hPa. In the following 24 h, the storm continued northeastward, apparently far enough offshore to keep snows away from much of the interior of the Northeast and appeared to weaken. Snowfall amounts in New England were less than those over the middle Atlantic states.

b. 9–10 April 1996

The last significant snowstorm of the long winter of 1995/96 produced snowfall in excess of 10 in. (25 cm) across Long Island, eastern Connecticut, Rhode Island, Massachusetts, New Hampshire, and Maine (Fig. 12.2-3). A few locations in Massachusetts, northern Rhode Island, and southern New Hampshire reported over 20 in. (50 cm) of snow from the storm. Where precipitation was only light to moderate, the snow either did not accumulate much or mixed with rain. However, in those areas where the precipitation fell heavily, it fell in the form of snow and accumulated rapidly. In New York City, the light to moderate snow accumulated slowly to

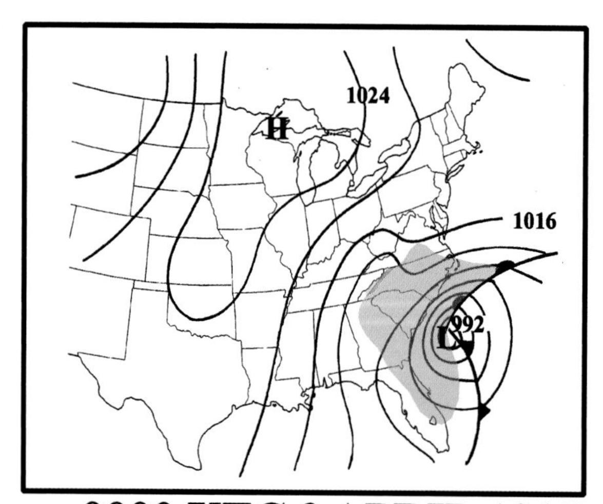

0000 UTC 3 APRIL 1915

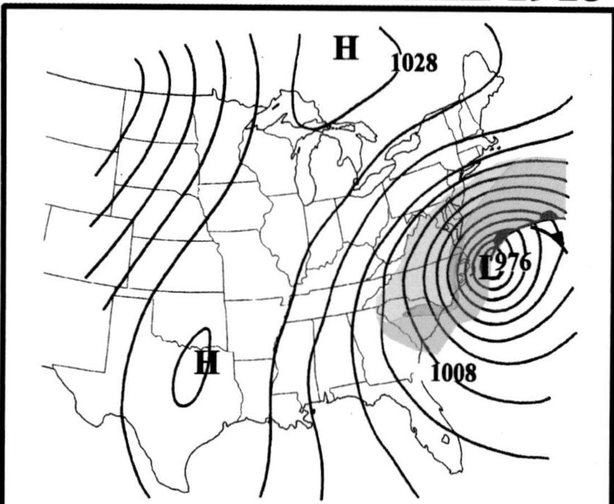

1200 UTC 3 APRIL 1915

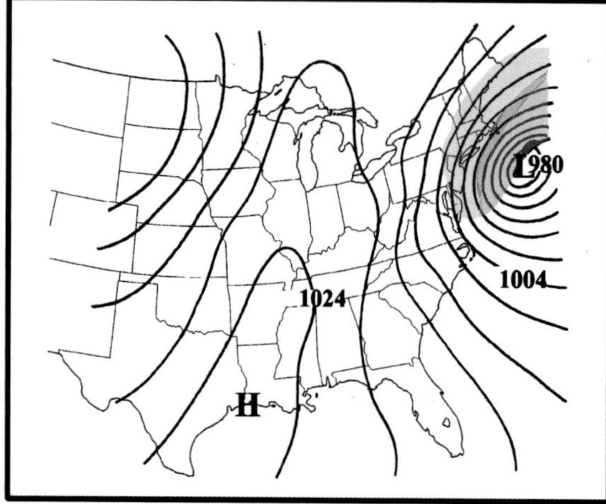

0000 UTC 4 APRIL 1915

FIG. 12.2-2. Sequence of surface charts for 3–4 Apr 1915.

only about an inch (2.5 cm) but in the eastern sections of the city, heavier snow fell with as much as 6 in. (15 cm) near the city's eastern limits. Accumulations include 12.6 in. (32 cm) at Brookhaven National Laboratory on Long Island, 16.0 in. (41 cm) at Worcester, and 15.0 in. (38 cm) at Blue Hill Observatory in Milton, Massachusetts.

The surface map (Fig. 12.2-4) shows that the storm system developed as a large cyclonic circulation appeared to consolidate into one major low pressure system near the North Carolina coastline by 1200 UTC 9 April. High pressure was located over the central United States but, unlike other cases of major snowstorms, the high pressure did not extend into New England as lower pressure dominated the eastern third of the country. Colder, subfreezing air can be found under the northeastern quarter of the United States at the 850-hPa level (Fig. 12.2-4).

By 0000 UTC 10 April, the surface low consolidated off the Carolina coastline and was developing rapidly into a significant cyclone east of the Delaware coast with a central pressure near 994 hPa, accompanied by an increasingly intense, but small circulation pattern at 850 hPa (Fig. 12.2-4). The surface low continued to deepen overnight and was located over southeastern New England with a central pressure near 984 hPa by 1200 UTC 10 April. Heavy snows developed across portions of eastern New Jersey and Long Island late on 9 April and then spread northeastward into much of eastern New England.

The storm developed in an upper-level regime dominated by a blocking ridge over Newfoundland (consistent with daily values of the NAO that were negative prior to and during the development of the storm; not shown; also see volume I, chapter 2) that appeared to suppress short-wave trough systems to its south (Fig. 12.2-5). An upper cutoff low developed over the Great Lakes region on 8–9 April, with a number of short-wave troughs spinning around the low. One major short-wave and associated jet streak reoriented the upper vortex and focused a diffluent exit region of the jet over the East Coast (evident at both the 400- and 250-hPa levels), appearing to support the intensification of the surface cyclone along the East Coast by 0000 UTC 10 April. The low deepened fairly rapidly by over 20 hPa from 1200 UTC 9 April to 984 hPa by 1200 UTC 10 April, while the heavy, wet snow was accompanied by 40–50 mi h^{-1} wind gusts in eastern New England as the entire system moved slowly up the coast.

c. 18–19 April 1983

The April 1983 snowstorm occurred only one year after one of the great April snowstorms of the century occurred (see chapter 10.19). Most of the heavy snow fell inland with this storm but it is an unusual case of a particularly late season event that yielded as much as 6 in. (15 cm) of snow across portions of North Carolina

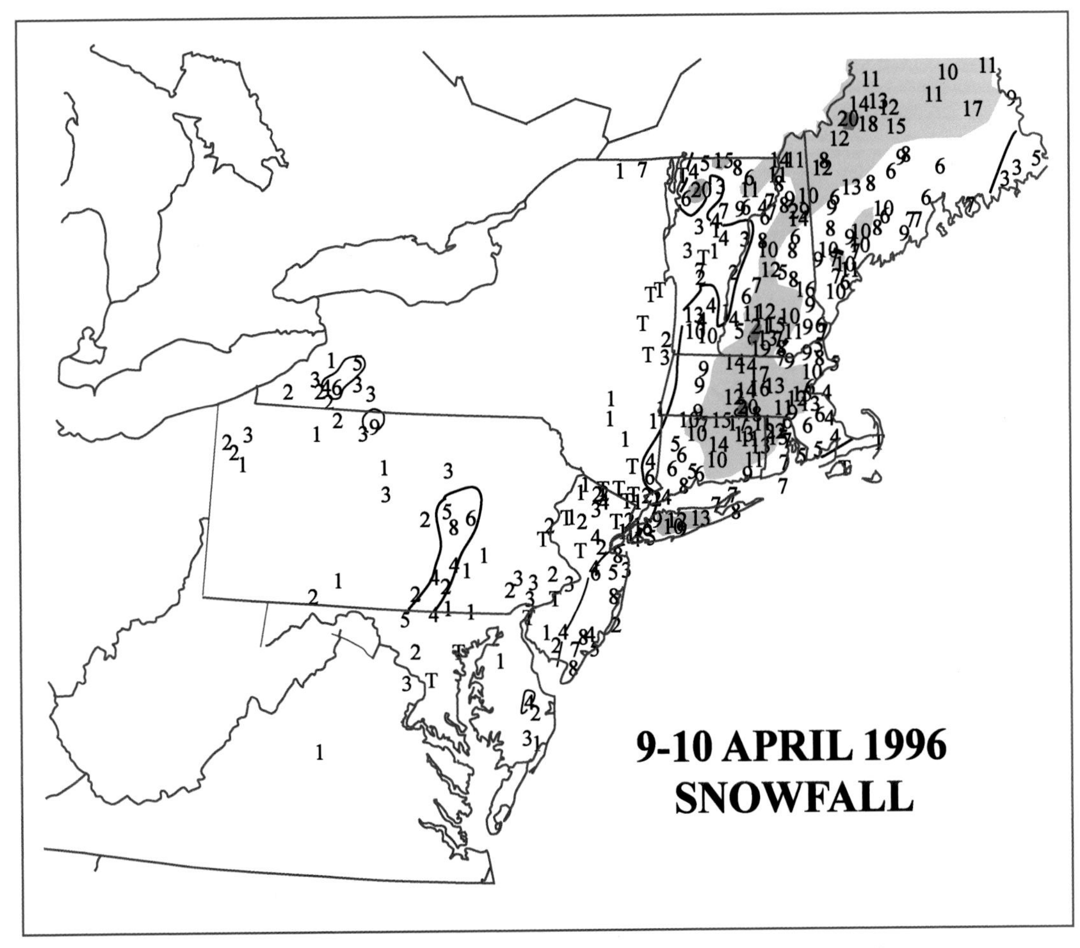

FIG. 12.2-3. Total snowfall (in.) for 9–10 Apr 1996. See Fig. 12.1-1 for details.

and southeastern Virginia before affecting the Northeast. Accumulating snow also fell along the New Jersey and New York coasts with up to 3 in. (7.5 cm) across Long Island. Accumulations of 10–20 in. (25–50 cm) were common in northeastern Pennsylvania and south-central New York (Fig. 12.2-6). Heaviest snows fell across interior Pennsylvania and New York with Scranton, Pennsylvania, reporting 19.6 in. (50 cm), its fifth heaviest snowfall on record (see the appendix in volume I). Many stations in eastern Pennsylvania reported that this snowstorm was the heaviest so late in the season. Stalled vehicles and hazardous driving conditions resulted in the closure of a number of major highways with travelers seeking refuge.

Surface charts (Fig. 12.2-7) show the eastern half of the United States under the domination of an unusually cold late season high pressure system centered over northern Canada, extending southward to the Gulf coast at 0000 UTC 19 April. A complex series of low pressure systems began to consolidate into one center near the North Carolina coastline by 0000 UTC. This low pressure system then intensified as it moved slowly northward off the New Jersey coast at 1200 UTC 19 April, deepening from 999 to 990 hPa. It then curved to the north-northwest near southwestern Connecticut by 0000 UTC 20 April with a central sea level pressure near 985 hPa. The intensifying surface low was accompanied by an expanding and intensifying 850-hPa circulation (Fig. 12.2-7) that became centered on the New Jersey coast by 0000 UTC 10 April with subzero temperatures (°C) at 850 hPa over most of the eastern half of the country.

Snow developed over southeastern Virginia and northeastern North Carolina late on 18 April. The snow then spread northeastward across New Jersey, eastern Pennsylvania, New York, and much of central and western New England by the morning of 19 April. Heaviest snows developed in Pennsylvania and New York on 19 April and continued into the night as the surface low continued to deepen and stalled off the New Jersey coast.

The development of this storm is similar to the April 1996 case described earlier in that a major cutoff low at 500 and 250 hPa dominated the northeastern United States during the precyclogenetic period and a downstream ridge was located just east of Labrador in the precyclogenetic phase (Fig. 12.2-8). The storm occurred during the start of a 10-day period in which the NAO had transitioned from positive to negative (not shown;

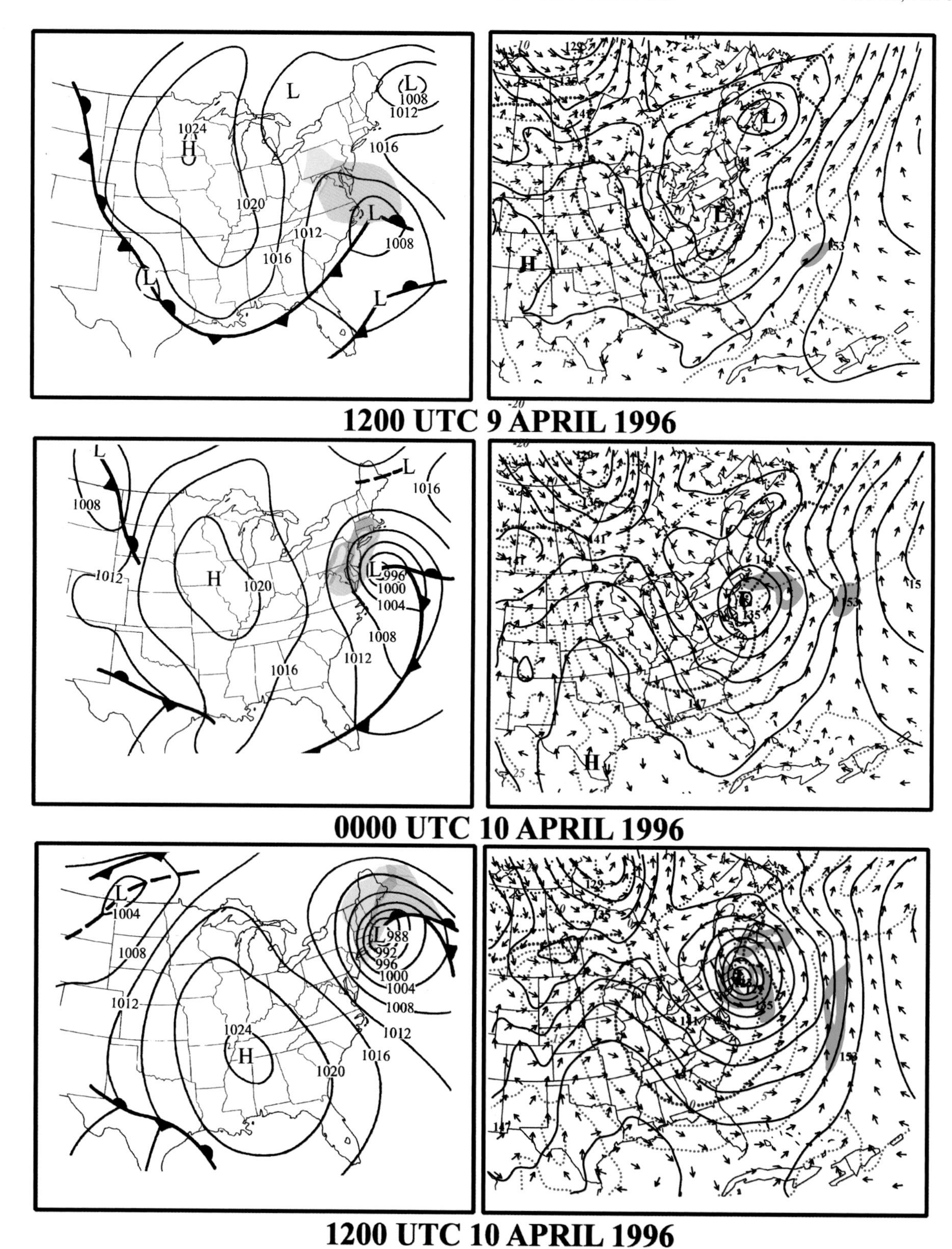

FIG. 12.2-4. Sequence of (left) surface and (right) 850-hPa charts for the 24-h period between 1200 UTC 9 and 10 Apr 1996. See Fig. 12.1-2 for details.

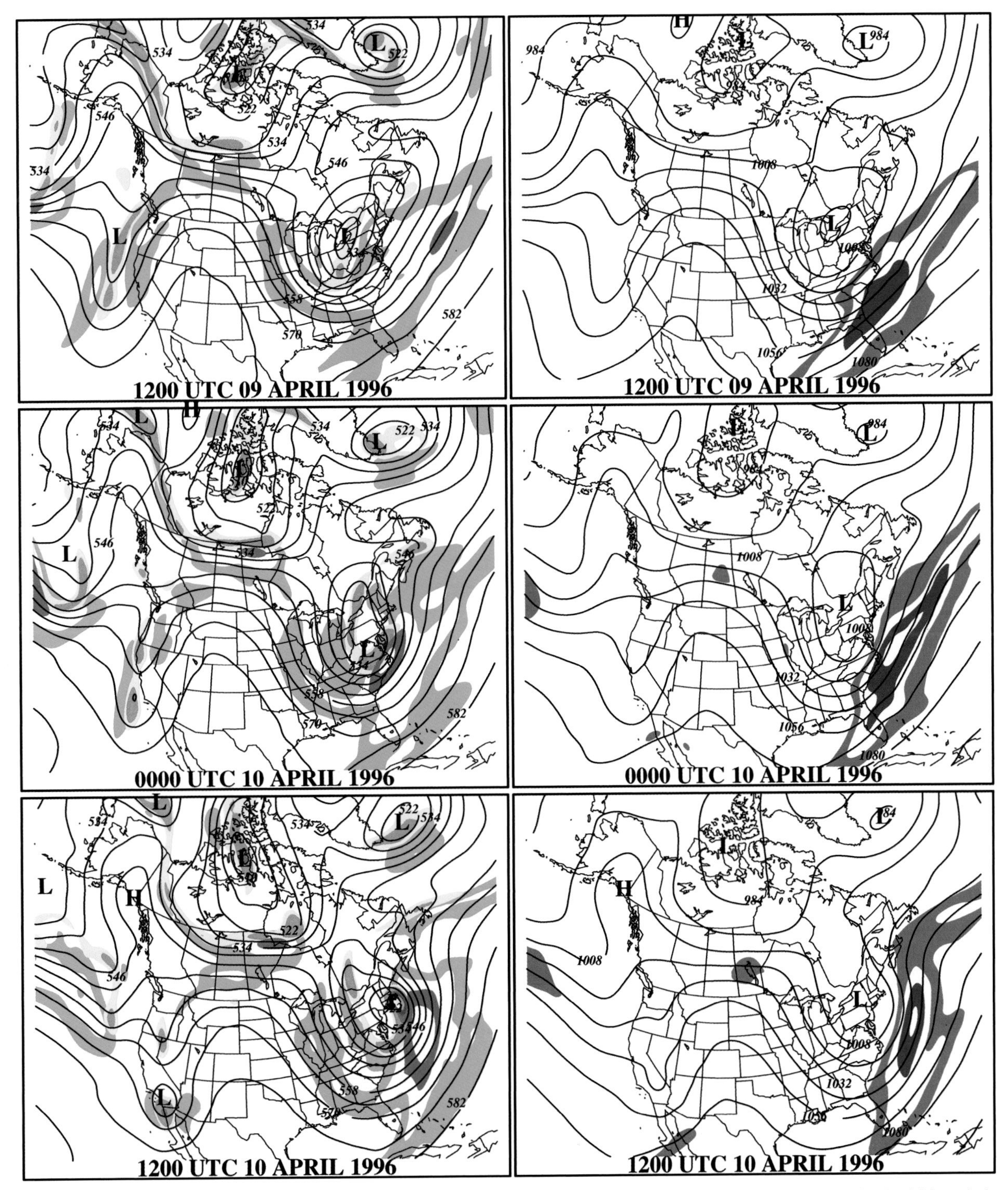

FIG. 12.2-5. Sequence of analyses of (left) 500- and (right) 250-hPa geopotential height and upper-level wind fields for the 24-h period between 1200 UTC 9 and 10 Apr 1996. See Fig. 12.1-3 for details.

also see volume I, chapter 2). As a short-wave feature rotated around the western and southern portions of the 500-hPa closed low, a strongly negatively tilted trough evolved over the middle Atlantic states by 1200 UTC 19 April (Fig. 12.2-8), with the surface low developing in the diffluent exit region of a jet streak over the mid-Atlantic region.

Storms such as these are not the classic heavy snow producers in that the upper-level confluent zone seems to play less of a role than the presence of a deep cutoff

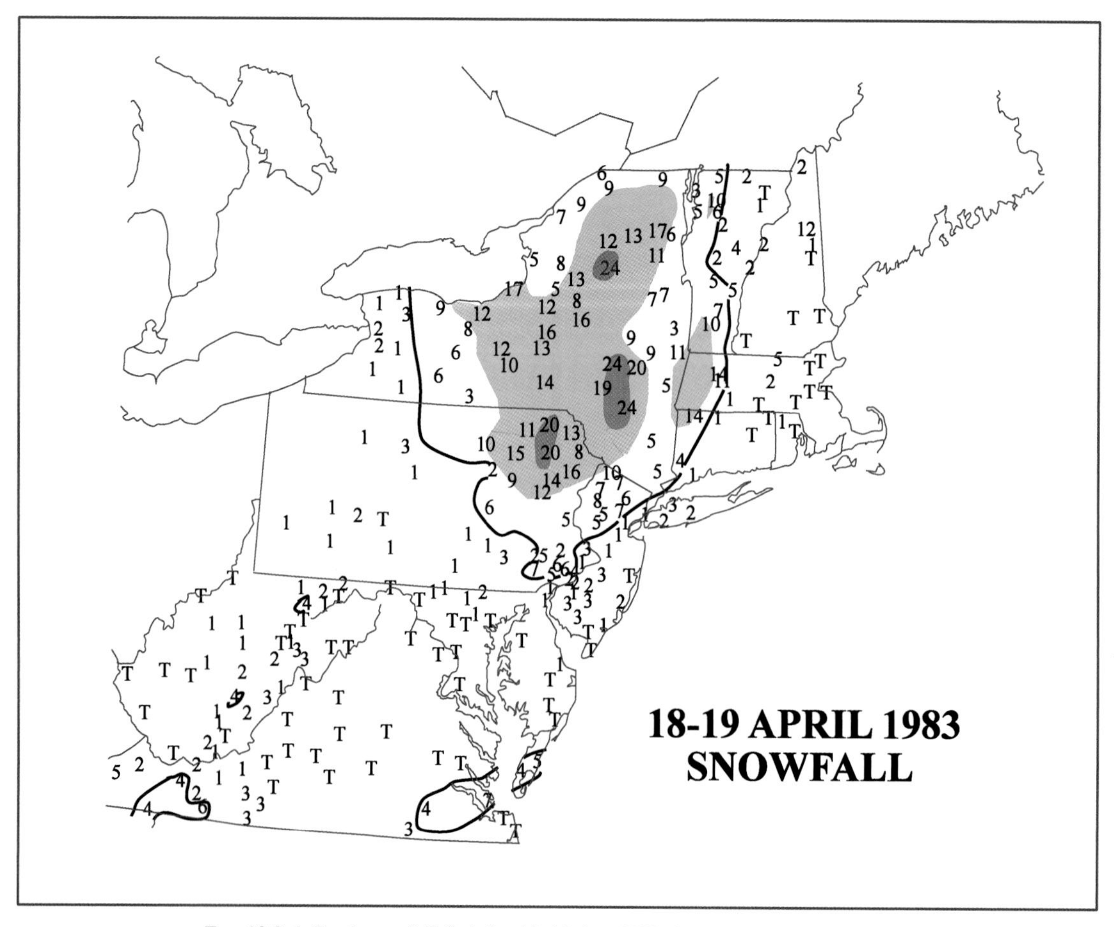

FIG. 12.2-6. Total snowfall (in.) for 18–19 Apr 1983. See Fig. 12.1-1 for details.

low, and there is no strong surface anticyclone located over the Northeast. These snowstorms are associated with pools of deep, cold tropospheric air that are cut off from the main synoptic-scale jet stream. The development of heavy snowfall occurs as a major cyclone develops along the coast and moves very slowly northeastward along the eastern edge of the cutoff low, upstream of the blocking ridge.

d. 9–10 May 1977

The 9–10 May 1977 snowstorm is a particularly unique situation since heavy snow fell so late into spring with few historical precedents (see the beginning of the chapter). Record late season snow fell in New York and New England with some astounding totals, including 12.7 in. (32 cm) at Worcester and 7.5 in. (19 cm) at Providence, the only measurable snow to fall in May during the 20th century. Throughout New York and southern New England, extensive tree and shrub damage was reported, with 100 000 and 500 000 people without power in Rhode Island and Massachusetts, respectively. In the higher elevations of the Catskills and Berkshires, snowfall amounts of 10–20 in. (25–50 cm) were common with a few greater amounts being reported (Fig. 12.2-9). Even along the coastline, Boston recorded its latest measurable snow with 0.5 in. (1.3 cm) and New York City reported the latest snowflakes recorded in modern times. Without the perspective provided from the historical accounts of cases in the 18th and 19th centuries, this case would look like a one-of-a-kind anomaly. As we have seen, however, the month of May (and during one year at least, the month of June) is still capable of providing a few wintry reminders, especially over northern and interior locations.

Surface charts (Fig. 12.2-10) show an impressive cold front accompanied by several low pressure centers dropping southward across the Midwest and Northeast late on 8 May. A strong surface anticyclone associated with temperatures near 0°C over a large portion of southeastern Canada followed the passage of the cold front. The cold front passage was accompanied by very strong, gusty winds and thunderstorms that developed from Pennsylvania southward. Wind gusts up to 40–50 mi h^{-1} downed power lines and trees throughout western Pennsylvania. In Maryland, 50–70 mi h^{-1} wind gusts also downed trees, disrupting power and blocking roads. As the cold front continued southward, low pressure remained stalled over southern New England early on 9 May while a new low pressure center developed farther

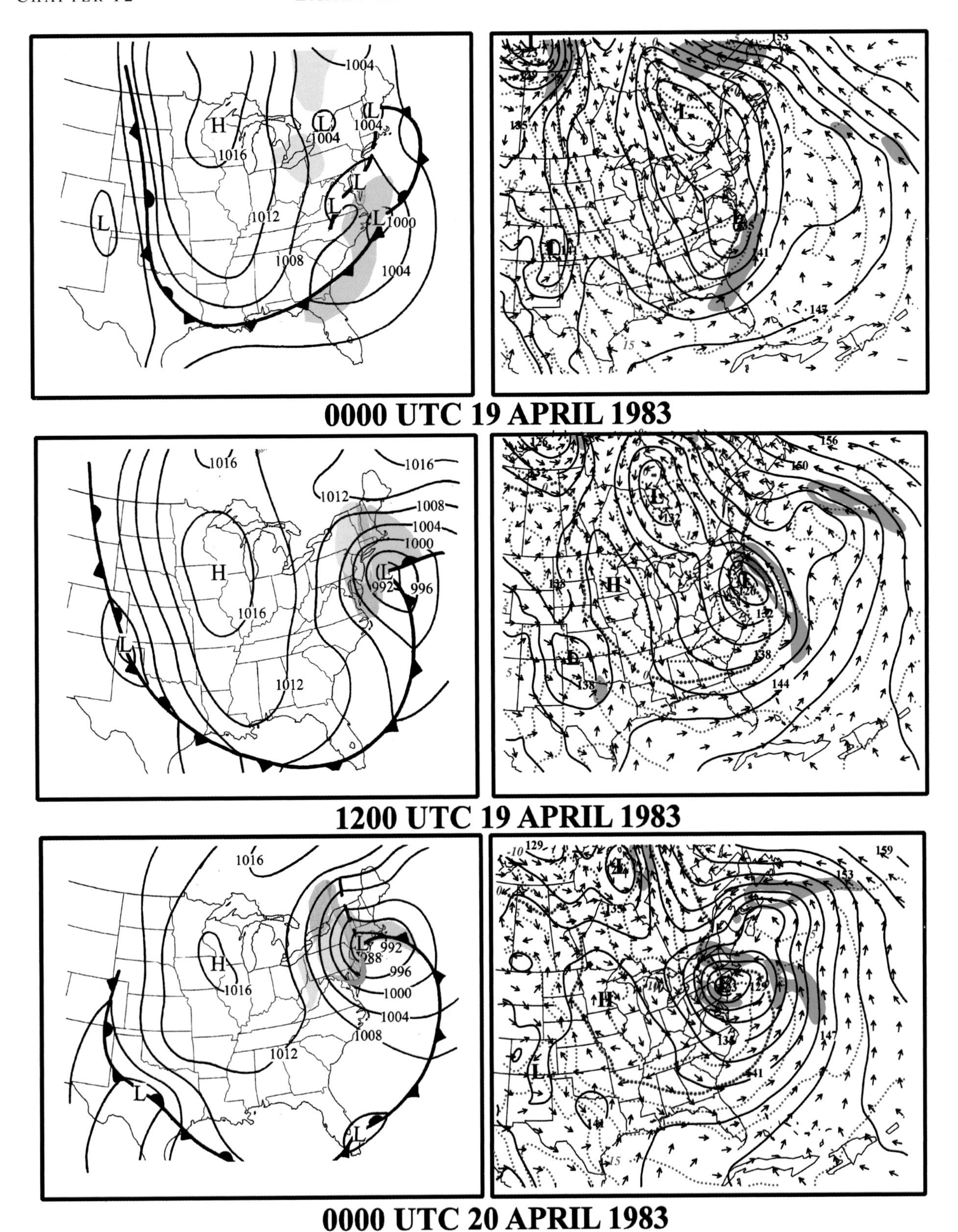

FIG. 12.2-7. Sequence of (left) surface and (right) 850-hPa charts for the 24-h period between 0000 UTC 19 and 20 Apr 1983. See Fig. 12.1-2 for details.

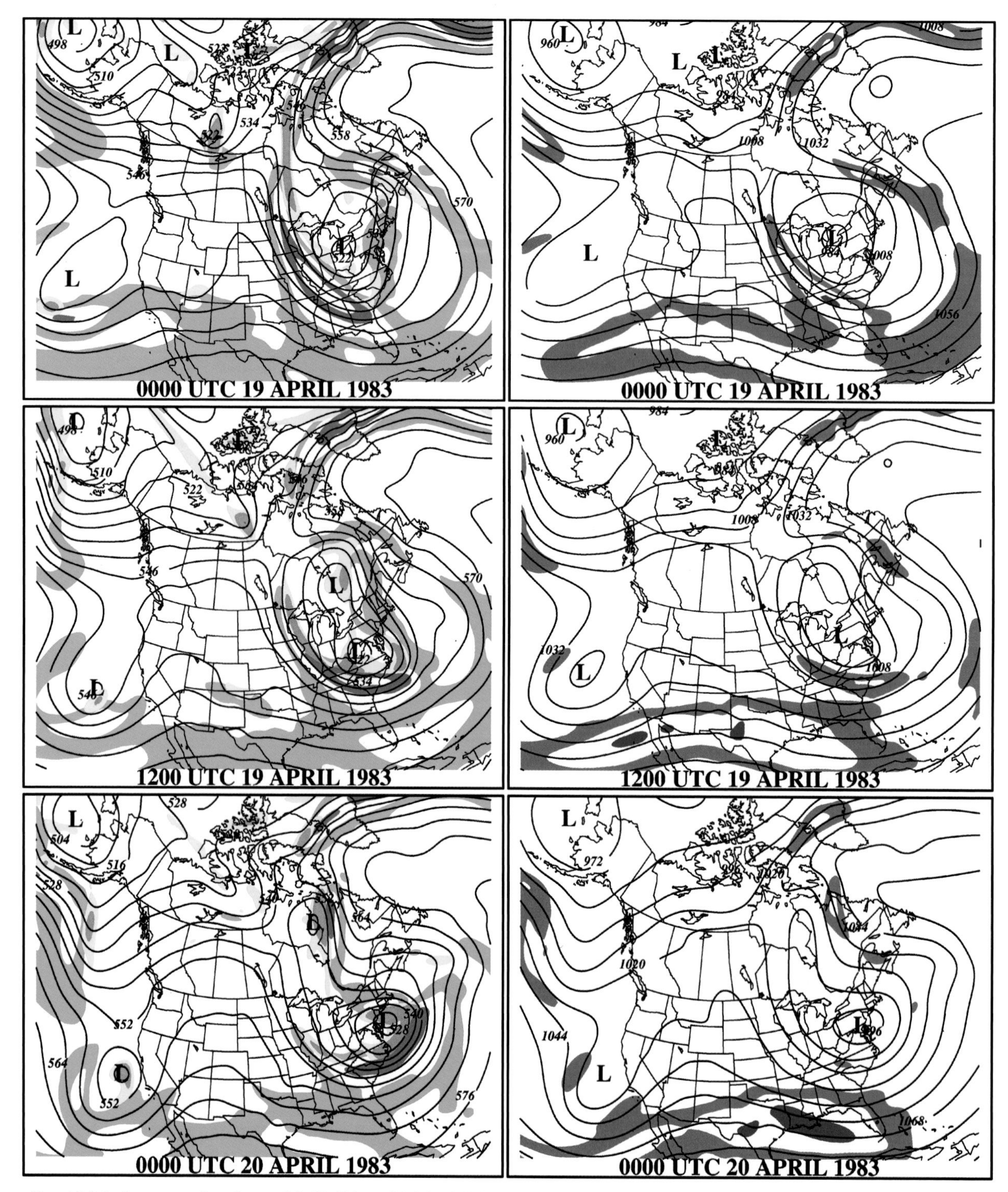

FIG. 12.2-8. Sequence of analyses of (left) 500- and (right) 250-hPa geopotential height and upper-level wind fields for the 24-h period between 0000 UTC 19 and 20 Apr 1983. See Fig. 12.1-3 for details.

southeastward over the western Atlantic Ocean. During this period a cold pool of air with subfreezing temperatures (−5°C) at 850 hPa (Fig. 12.2-10) dropped southward over the Northeast as a small circulation intensified along the New England coast. Rain changed to snow across interior New York and southern New England during the night and continued during the day, mixing with rain at lower elevations. By late on 9 May, this low pressure center became the primary storm system.

This was another example of the amplification and

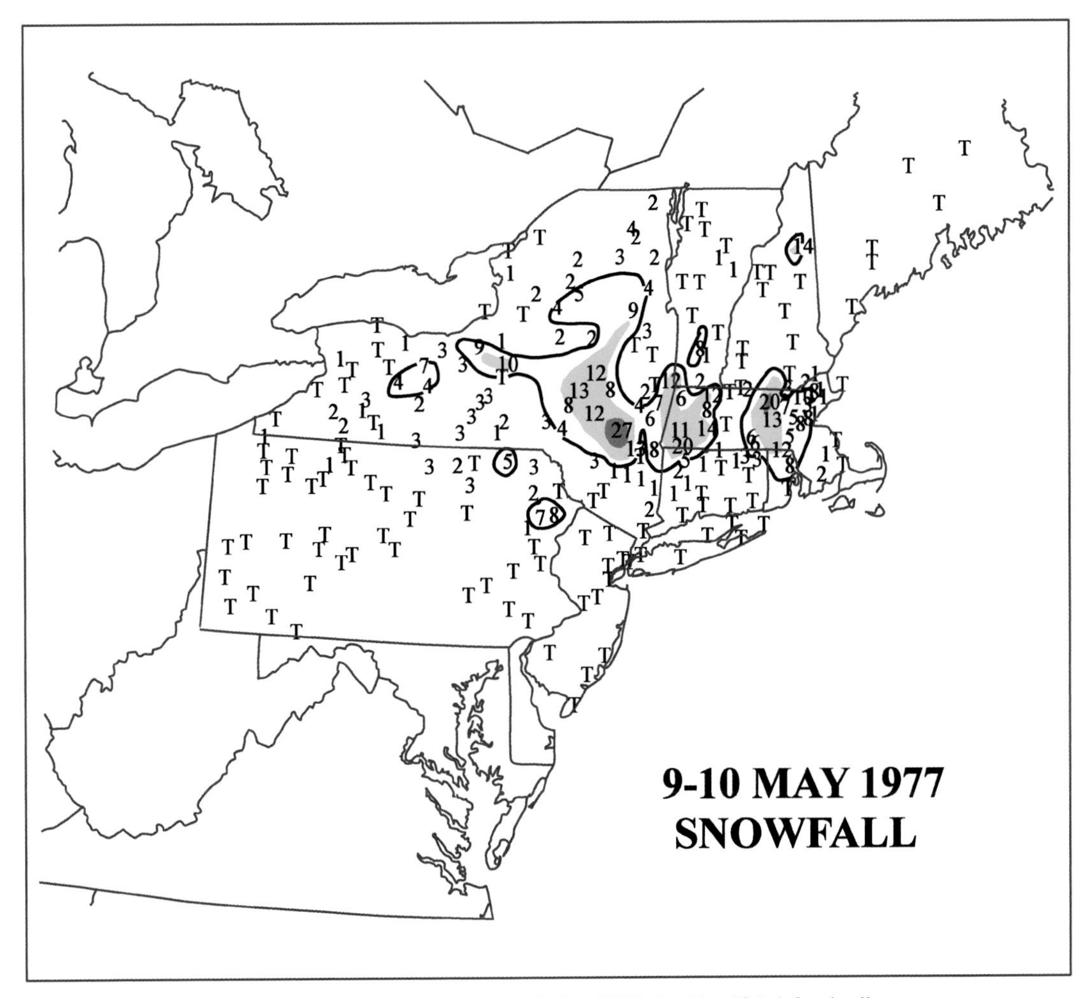

FIG. 12.2-9. Total snowfall (in.) for 9 May 1977. See Fig. 12.1-1 for details.

closing off of a short-wave trough that spawns a surface cyclone along the New England coast. The 500- and 250-hPa charts (Fig. 12.2-11) show the cold front was associated with a strongly amplifying upper trough plunging southward from eastern Canada at 0000 UTC 9 May and developing into a cutoff low near New York City by 1200 UTC 9 May and just south of southern New England by 0000 UTC 10 May. There is evidence from this reanalysis of a small-scale jet maximum near the base of the trough at 0000 UTC 9 May and this maximum is identifiable by 0000 UTC 10 May at both 400 and 250 hPa. The surface low deepened off the New England coast within the exit region of this jet as this trough amplified and eventually cut off, providing the support for the developing low and sustaining the cold air aloft that led to the late season snowfall.

3. Summary

An examination of a number of early and late season snow events points to some common characteristics with their winter season counterparts described in chapter 10. The similarities are basic in that these cases involve a deep surface low pressure system and a source of unseasonably cold air. However, the differences between these cases and snowstorms that occur during the more typical December–March period are significant. Several of these early and late season cases do not involve the presence of a strong high pressure system to the north of New York and New England to provide the source of the cold air needed for snow. In the cases of 3–4 October 1987, 6–7 November 1953, 11–12 November 1987, 31 March–1 April 1997 (see chapter 10.28), 9–10 April 1996, 18–19 April 1983, and 9–10 May 1977, the snow was associated with a deepening upper-level trough that evolved into a closed low at 500 and 250 hPa and with the development of a deep pool of cold air aloft. Only the case of November 1953 appears to involve a large anticyclone associated with an upper-level confluent zone and moving short-wave trough over extreme eastern Canada. The relatively minor snowstorm of 10–11 October 1979 was associated with a low-amplitude upper trough associated with a strong confluent jet stream at 500 and 400 hPa, and a weak

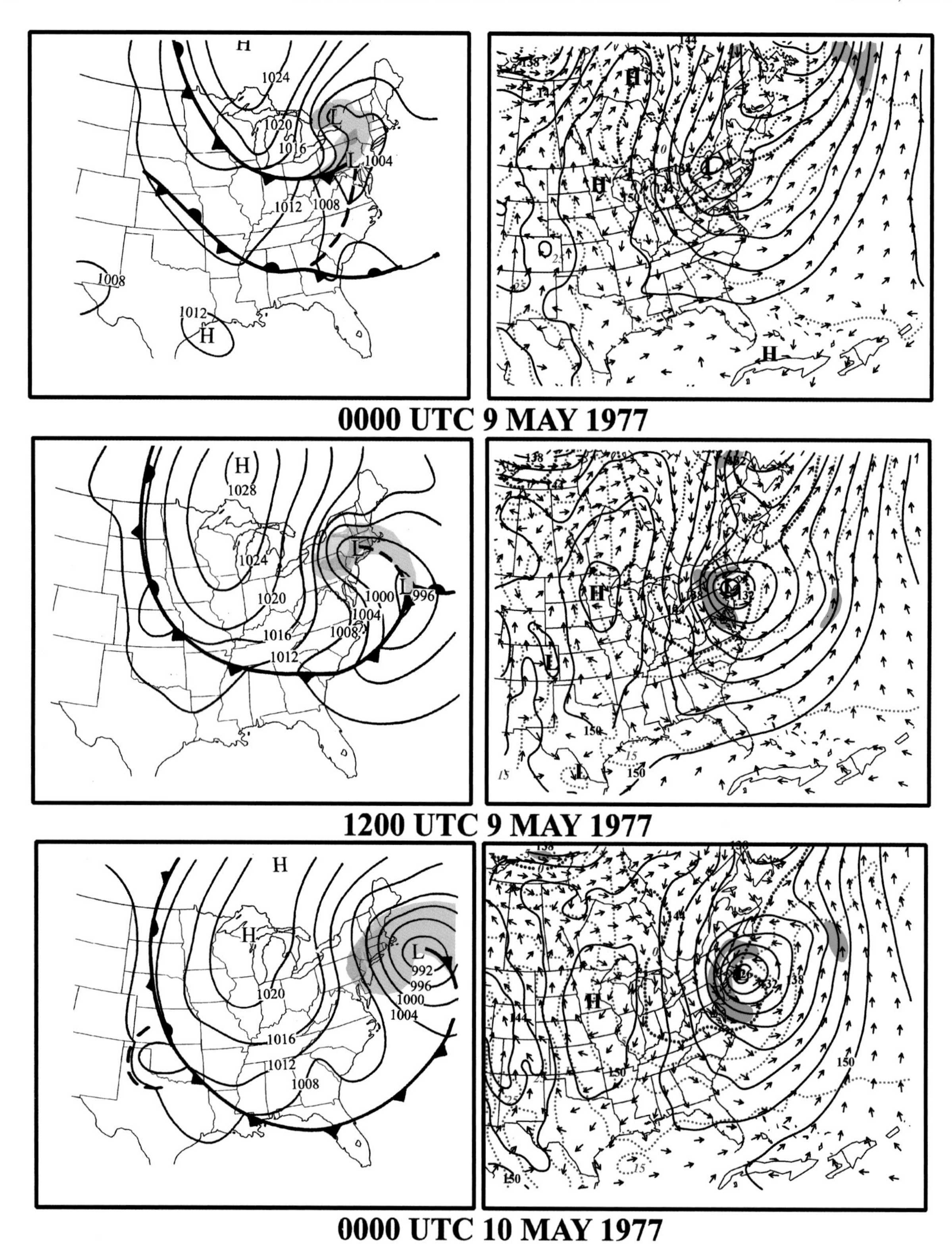

FIG. 12.2-10. Sequence of (left) surface and (right) 850-hPa charts for the 24-h period between 1200 UTC 9 and 10 May 1977. See Fig. 12.1-2 for details.

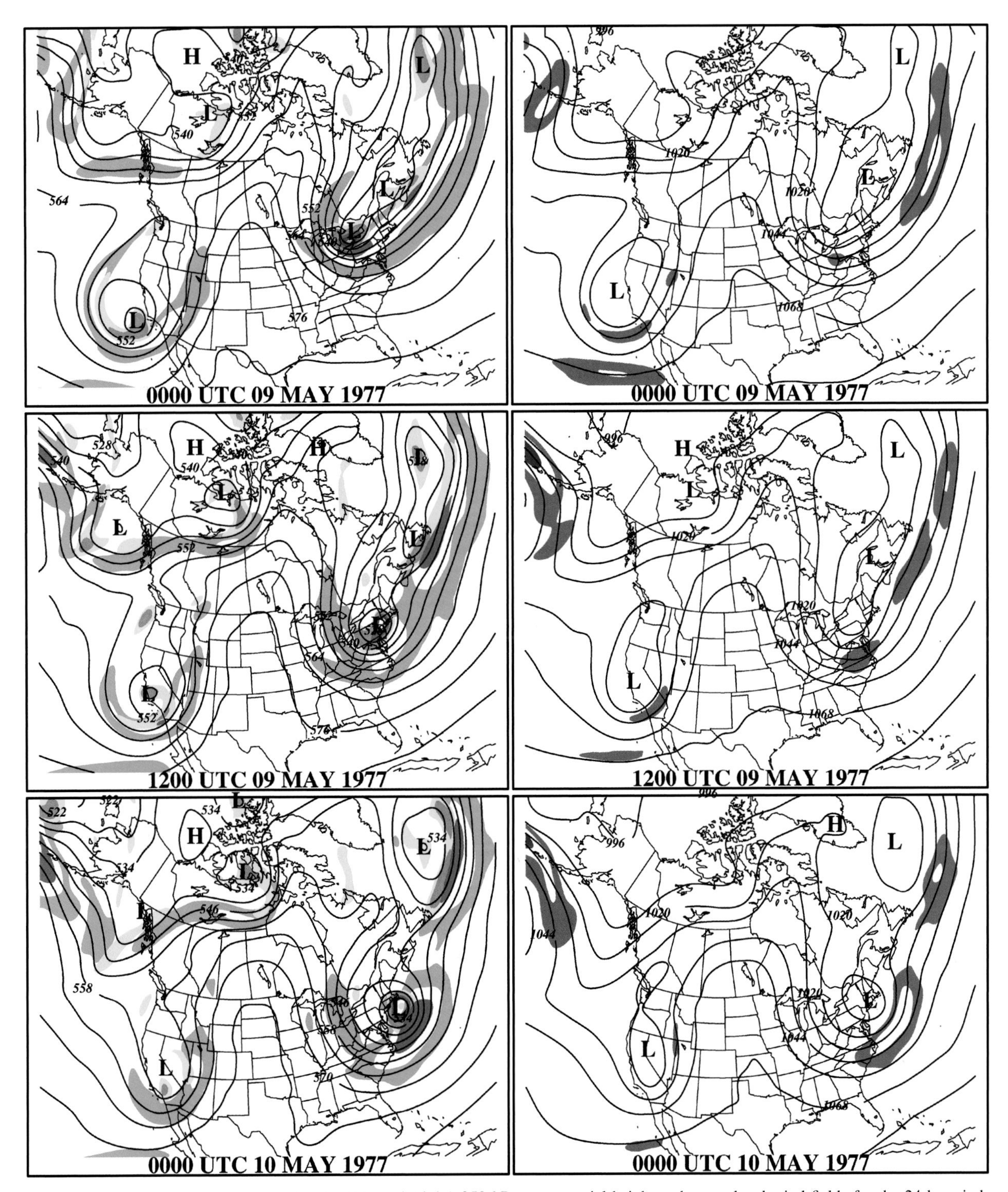

FIG. 12.2-11. Sequence of analyses of (left) 500- and (right) 250-hPa geopotential height and upper-level wind fields for the 24-h period between 0000 UTC 9 and 10 May 1977. See Fig. 12.1-3 for details.

surface high pressure center that extended eastward into the northeast United States during the snow event.

Another aspect of the early and late season snowfalls is an apparent role of melting snow to cool the atmosphere in the early stages of the storm and the subsequent changeover from rain to snow (Kain et al. 2000; Gedzelman and Lewis 1990). This factor appears to be important in the cases of 3–4 October 1987, 10–11 October 1979, 6–7 November 1953, and 9–10 May 1977, in which temperatures at 850 hPa were generally near or above 0°C at the onset of precipitation, beginning as rain, and then fell to near 0°C as the precipitation became heavier, and subsequently changed over to all snow. Several cases noted in this section are characterized by the changeover to snow as the precipitation intensity increased, with localized heavy snows generally found in higher elevations and inland locations.

Some of the unique aspects of several cases are summarized as follows. The snowstorm of 3–4 April 1915 is much like a classic snowstorm in that there is a rapidly deepening surface low pressure system off the mid-Atlantic states and the presence of an anticyclone to the northwest of the Northeast urban corridor. The snowstorm of November 1898 appears to be a classic example of trough "merging" or phasing (see volume I, chapter 4.3a) as a low pressure system over the Midwest dove southeastward and a weak surface low moved northeastward along the Southeast coast. The Midwest low pressure system quickly diminished as the Southeast low pressure system strengthened rapidly off the Northeast coast and developed into a very intense cyclone on 27 November. The case of 10–12 November 1987 was a complex combination of features involving a slowly amplifying upper-level trough across the eastern United States. In this case, one moderate snow event over the Northeast on 10 November was followed by the development of a mesoscale area of snowfall in the Washington area on 11 November, which was then followed by the development of heavy snow with a rapidly deepening cyclone off the New England coast early on 12 November.

In summary, early and late season snowstorms are rare events that develop in a manner similar to the major snowstorms described in chapter 10 in that they are directly related to intense cyclones that develop near and off the East Coast. However, these cases have characteristics that are distinctly different from their winter season counterparts in that the source of cold air is often linked directly to a deepening upper-level trough or a melting process in which falling snow cools the atmosphere sufficiently to change rainfall at the surface to snow. In these cases, the presence of a strong surface anticyclone does not appear to be a common characteristic. Yet, the results are the same with heavy snow falling in a relatively short period of time that can paralyze a portion of the Northeast, usually in a manner that can surprise a population ready for an early season snowfall, or tease the same population anticipating an early spring by a late season snowfall.

Josh Kocin, the coauthor's son, in Silver Spring, Maryland, 11 Nov 1987.

Matt Kocin, the coauthor's son, at NASA/Goddard Spaceflight Center, Greenbelt, Maryland, 12 Nov 1987.

A tree down in Albany, New York, on 4 Oct 1987 (from *Weatherwise Magazine*, The American Meteorological Society, 1990).

Snow on pumpkins. Berwyn, Pennsylvania, 10 Oct 1979 (photo by Sam Sandor, from *The Philadelphia Area Weather Book*, Temple University Press, 2002, p. 151).

Lilacs in the snow in Albany, New York, 18 May 2002 (photo by Brian King, courtesy of the Department of Computer Science, Albany, New York).

REFERENCES

Achtor, T. H., and L. H. Horn, 1986: Spring season Colorado cyclones. Part I: Use of composites to relate upper and lower tropospheric wind fields. *J. Climate Appl. Meteor.,* **25,** 732–743.

Alpert, P., M. Tsidulko, and U. Stein, 1995: Can sensitivity studies yield absolute comparisons for the effects of several processes? *J. Atmos. Sci.,* **52,** 597–601.

Ambrose, K., 1993: *Blizzards and Snowstorms of Washington, D.C.* Historical Enterprises, 115 pp.

——, 1994: *Great Blizzards of New York City.* Historical Enterprises, 123 pp.

Anthes, R. A., and D. Keyser, 1979: Tests of a fine-mesh model over Europe and the United States. *Mon. Wea. Rev.,* **107,** 963–984.

Atlas, R., 1987: The role of oceanic fluxes and initial data in the numerical prediction of an intense coastal storm. *Dyn. Atmos. Oceans,* **10,** 359–388.

Austin, J. M., 1941: Favorable conditions for cyclogenesis near the Atlantic coast. *Bull. Amer. Meteor. Soc.,* **22,** 270.

——, 1951: Mechanisms of pressure change. *Compendium of Meteorology,* T. F. Malone, Ed., Amer. Meteor. Soc., 630–638.

Bailey, R. E., 1960: Forecasting of heavy snowstorms associated with major cyclones. *Weather Forecasting for Aeronautics,* J. J. George, Ed., Academic Press, 468–475.

Baker, D. G., 1970: A study of high pressure ridges to the east of the Appalachian Mountains. Ph.D. dissertation, Massachusetts Institute of Technology, 127 pp.

Ballentine, R. J., 1980: A numerical investigation of New England coastal frontogenesis. *Mon. Wea. Rev.,* **108,** 1479–1497.

——, A. J. Stamm, E. E. Chermack, G. P. Byrd, and D. Schleede, 1998: Mesoscale model simulation of the 4–5 January 1995 lake-effect snowstorm. *Wea. Forecasting,* **13,** 893–920.

Barnes, S. L., 1964: A technique for maximizing details in numerical weather map analysis. *J. Appl. Meteor.,* **3,** 396–409.

——, and B. R. Colman, 1993: Quasigeostrophic diagnosis of cyclogenesis associated with a cutoff extratropical cyclone—The Christmas 1987 storm. *Mon. Wea. Rev.,* **121,** 1613–1634.

Barnston, A. G., and R. E. Livezey, 1987: Classification, seasonality and persistence of low-frequency atmospheric circulation patterns. *Mon. Wea. Rev.,* **115,** 1083–1126.

Beckman, S. K., 1987: Use of enhanced IR/visible satellite imagery to determine heavy snow areas. *Mon. Wea. Rev.,* **115,** 2060–2087.

Bell, G. D., and L. F. Bosart, 1988: Appalachian cold-air damming. *Mon. Wea. Rev.,* **116,** 137–161.

——, and ——, 1989: The large-scale atmospheric structures accompanying New England coastal frontogenesis and associated North American East Coast cyclogenesis. *Quart. J. Roy. Meteor. Soc.,* **115,** 1133–1146.

——, and M. S. Halpert, 1998: Climate assessment for 1997. *Bull. Amer. Meteor. Soc.,* **79** (5), S1–S49.

Bennetts, D. A., and B. J. Hoskins, 1979: Conditional symmetric instability—A possible explanation of frontal rainbands. *Quart. J. Roy. Meteor. Soc.,* **105,** 945–962.

Bjerknes, J., 1919: On the structure of moving cyclones. *Geofys. Publ., Norske Videnskaps-Akad. Oslo,* **1** (1), 1–8.

——, 1951: Extratropical cyclones. *Compendium of Meteorology,* T. F. Malone, Ed., Amer. Meteor. Soc., 577–598.

——, 1954: The diffluent upper trough. *Arch. Meteor. Geophys. Bioklimatol.,* **A7,** 41–46.

——, 1969: Atmospheric teleconnections from the equatorial Pacific. *Mon. Wea. Rev.,* **97,** 163–172.

——, and H. Solberg, 1922: Life cycle of cyclones and the polar front theory of atmospheric circulation. *Geofys. Publ., Norske Videnskaps-Akad. Oslo,* **3** (1), 1–18.

——, and J. Holmboe, 1944: On the theory of cyclones. *J. Meteor.,* **1,** 1–22.

Bleck, R., 1973: Numerical forecasting experiments based on the conservation of potential vorticity on isentropic surfaces. *J. Appl. Meteor.,* **12,** 737–752.

——, 1974: Short-range prediction in isentropic coordinates with filtered and unfiltered numerical models. *Mon. Wea. Rev.,* **102,** 813–829.

——, 1977: Numerical simulation of lee cyclogenesis in the Gulf of Genoa. *Mon. Wea. Rev.,* **116,** 137–161.

——, and C. Mattocks, 1984: A preliminary analysis of the role of potential vorticity in Alpine lee cyclogenesis. *Beitr. Phys. Atmos.,* **57,** 357–368.

Bonner, W., 1965: Statistical and kinematical properties of the low-level jet stream. Research Paper, 38, Satellite and Mesometeorology Research Project, University of Chicago, 54 pp.

——, 1989: NMC overview: Recent progress and future plans. *Wea. Forecasting,* **4,** 275–285.

Bosart, L. F., 1975: New England coastal frontogenesis. *Quart. J. Roy. Meteor. Soc.,* **101,** 957–978.

——, 1981: The Presidents' Day snowstorm of 18–19 February 1979: A subsynoptic-scale event. *Mon. Wea. Rev.,* **109,** 1542–1566.

——, 1999: Observed cyclone life cycles. *The Life Cycles of Extratropical Cyclones,* M. A. Shapiro and S. Grønås, Eds., Amer. Meteor. Soc., 187–213.

——, 2003a: Whither the weather analysis and forecast process? *Wea. Forecasting,* **18,** 520–529.

——, 2003b: Tropopause folding, upper-level frontogenesis, and beyond. *A Half Century of Progress in Meteorology: A Tribute to Richard J. Reed, Meteor. Monogr.,* Amer. Meteor. Soc., 13–47.

——, and J. P. Cussen, Jr. 1973: Gravity wave phenomena accompanying East Coast cyclogenesis. *Mon. Wea. Rev.,* **101,** 5445–5454.

——, and S. C. Lin, 1984: A diagnostic analysis of the Presidents' Day storm of February 1979. *Mon. Wea. Rev.,* **112,** 2148–2177.

——, and F. Sanders, 1986: Mesoscale structure in the Megalopolitan snowstorm of 11–12 February 1983. Part III: A large amplitude gravity wave. *J. Atmos. Sci.,* **43,** 924–939.

——, and A. Seimon, 1988: A case study of an unusually intense atmospheric gravity wave. *Mon. Wea. Rev.,* **116,** 1857–1886.

——, and J. A. Bartlo, 1991: Tropical storm formation in a baroclinic environment. *Mon. Wea. Rev.,* **119,** 1979–2013.

——, and F. Sanders, 1991: An early-season coastal storm: Conceptual success and model failure. *Mon. Wea. Rev.,* **119,** 2831–2851.

——, and W. E. Bracken, 1996: Coastal cyclonic development in the northwest periphery of a larger parent cyclone. Preprints, *14th Conf. on Weather Analysis and Forecasting,* Dallas, TX, Amer. Meteor. Soc., 551–554.

——, C. J. Vaudo, and J. H. Helsdon Jr., 1972: Coastal frontogenesis. *J. Appl. Meteor.,* **11,** 1236–1258.

——, G. J. Hakim, K. R. Tyle, M. A. Bedrick, W. E. Bracken, M. J. Dickinson, and D. M. Schultz, 1996: Large-scale antecedent conditions associated with the 12–14 March 1993 cyclone ("Superstorm '93") over eastern North America. *Mon. Wea. Rev.,* **124,** 1865–1891.

——, W. E. Bracken, and A. Seimon, 1998: A study of cyclone mesoscale structure with emphasis on a large-amplitude inertia–gravity wave. *Mon. Wea. Rev.,* **126,** 1497–1527.

Boucher, R. J., and R. J. Newcomb, 1962: Synoptic interpretation of some TIROS vortex patterns: A preliminary cyclone model. *J. Appl. Meteor.,* **1,** 127–136.

Boyle, J. S., and L. F. Bosart, 1983: A cyclone/anticyclone couplet over North America: An example of anticyclone evolution. *Mon. Wea. Rev.,* **111,** 1025–1045.

——, and ——, 1986: Cyclone–anticyclone couplets over North America. Part II: Analysis of a major cyclone event over the eastern United States. *Mon. Wea. Rev.,* **114,** 2432–2465.

Brandes, E. A., and J. Spar, 1971: A search for necessary conditions for heavy snow on the East Coast. *J. Appl. Meteor.,* **11,** 397–409.

Branick, M. L., 1997: A climatology of significant winter-type weather events in the contiguous United States 1982–94. *Wea. Forecasting,* **12,** 193–207.

Brill, K. F., L. W. Uccellini, R. P. Burkhart, T. T. Warner, and R. A. Anthes, 1985: Numerical simulations of a transverse indirect circulation and low-level jet in the exit region of an upper-level jet. *J. Atmos. Sci.,* **42,** 1306–1320.

——, ——, J. Manobianco, P. J. Kocin, and J. H. Homan, 1991: The use of successive dynamic initialization by nudging to simulate cyclogenesis during GALE IOP1. *Meteor. Atmos. Phys.,* **45,** 15–40.

Bristor, C. L., 1951: The Great Storm of November 1950. *Weatherwise,* **4,** 10–16.

Brooks, C. F., 1914: The distribution of snowfall in cyclones of the eastern United States. *Mon. Wea. Rev.,* **42,** 318–329.

Brown, H. E., and D. A. Olson, 1978: Performance of NMC in forecasting a record-breaking winter storm, 6–7 February 1978. *Bull. Amer. Meteor. Soc.,* **59,** 562–575.

Browne, R. F., and R. J. Younkin, 1970: Some relationships between 850 millibar lows and heavy snow occurrences over the central and eastern United States. *Mon. Wea. Rev.,* **98,** 399–401.

Browning, K. A., 1971: Radar measurements of air motion near fronts. *Weather,* **26,** 320–340.

——, 1986: Conceptual models of precipitation systems. *Wea. Forecasting,* **1,** 23–41.

——, 1990: Organization of clouds and precipitation in extratropical cyclones. *Extratropical Cyclones: The Erik Palmén Memorial Volume,* C. W. Newton and E. O. Holopainen, Eds., Amer. Meteor. Soc., 129–153.

——, and T. W. Harrold, 1969: Air motion and precipitation growth in a wave depression. *Quart. J. Roy. Meteor. Soc.,* **95,** 288–309.

Brunk, I., 1949: The pressure pulsation of 11 April 1944. *J. Meteor.,* **6,** 181–187.

Buizza, R., and P. Chessa, 2002: Prediction of the U.S. storm of 24–26 January 2000 with the ECMWF Ensemble Prediction System. *Mon. Wea. Rev.,* **130,** 1531–1551.

Burnham, G. H., 1922: Economic effects of New England's unprecedented ice storm of November 25–29, 1921. *J. Geogr.,* **21,** 161–168.

Buzzi, A., and S. Tibaldi, 1978: Cyclogenesis in the lee of the Alps: A case study. *Quart. J. Roy. Meteor. Soc.,* **104,** 271–287.

Byrd, G. P., R. A. Anstett, J. E. Heim, and D. M. Usinski, 1991: Mobile sounding observations of lake-effect snowbands in western and central New York. *Mon. Wea. Rev.,* **119,** 2323–2332.

Cahir, J. J., 1971: Implications of circulations in the vicinity of jet streaks at subsynoptic scales. Ph.D. thesis, The Pennsylvania State University, 170 pp.

Caplan, P., 1995: The 12–14 March 1993 Superstorm: Performance of the global model. *Bull. Amer. Meteor. Soc.,* **76,** 201–212.

Caplovich, J., 1987: *Blizzard! The Great Storm of '88.* Vero Publishing, 242 pp.

Carlson, T. N., 1961: Lee-side frontogenesis in the Rocky Mountains. *Mon. Wea. Rev.,* **89,** 163–172.

——, 1980: Airflow through midlatitude cyclones and the comma cloud pattern. *Mon. Wea. Rev.,* **108,** 1498–1509.

——, 1991: *Mid-latitude Weather Systems.* HarperCollins Academic, 507 pp.

Carpenter, D. M., 1993: The lake effect of the Great Salt Lake: Overview and forecast problems. *Wea. Forecasting,* **8,** 181–193.

Chang, C. B., D. J. Perkey, and C. W. Kreitzberg, 1982: A numerical case study of the effects of latent heating on a developing wave cyclone. *J. Atmos. Sci.,* **39,** 1555–1570.

Charney, J. G., 1947: The dynamics of long waves in a baroclinic westerly current. *J. Meteor.,* **4,** 135–162.

——, and N. A. Phillips, 1953: Numerical integration of the quasi-geostrophic equations for barotropic and simple baroclinic flows. *J. Meteor.,* **10,** 71–99.

Colucci, S. J., 1976: Winter cyclone frequencies over the eastern United States and adjacent western Atlantic, 1963–1973. *Bull. Amer. Meteor. Soc.,* **57,** 548–553.

——, 1985: Explosive cyclogenesis and large-scale circulation changes: Implications for atmospheric blocking. *J. Atmos. Sci.,* **42,** 2701–2717.

Cressman, G., 1970: Public forecasting: Present and future. *A Century of Weather Progress,* J. E. Caskey Jr., Ed., Amer. Meteor. Soc., 71–77.

Crum, T. D., and R. L. Alberty, 1993: The WSR-88D and the WSR-88D Operational Support Facility. *Bull. Amer. Meteor. Soc.,* **74,** 1669–1687.

Cunningham, R. M., and F. Sanders, 1987: Into the teeth of the Gale: The remarkable advance of a cold front at Grand Manan. *Mon. Wea. Rev.,* **115,** 2450–2462.

Danard, M. B., 1964: On the influence of released latent heat on cyclone development. *J. Appl. Meteor.,* **3,** 27–37.

——, and G. E. Ellenton, 1980: Physical influences on East Coast cyclogenesis. *Atmos.–Ocean,* **18,** 65–82.

Danielsen, E. F., 1966: Research in four-dimensional diagnosis of cyclone storm cloud systems. Air Force Cambridge Research Lab. Rep. 66–30, Bedford, MA, 53 pp. [NTIS-AD-632668.]

——, 1968: Stratospheric–tropospheric exchange based upon radioactivity, ozone and potential vorticity, *J. Atmos. Sci.,* **25,** 502–518.

Day, P. C., and S. P. Fergusson, 1922: The great snowstorm of January 27–29, 1922 over the Atlantic Coast states. *Mon. Wea. Rev.,* **50,** 21–24.

Dean, D. B., and L. F. Bosart, 1996: Northern Hemisphere 500-hPa trough merger and fracture: A climatology and case study. *Mon. Wea. Rev.,* **124,** 2644–2671.

DeGaetano, A. T., 2000: Climatic perspective and impacts of the 1998 northern New York and New England ice storm. *Bull. Amer. Meteor. Soc.,* **81,** 237–254.

Dickinson, M. J., L. F. Bosart, W. E. Bracken, G. J. Hakim, D. M. Schultz, M. A. Bedrick, and K. R. Tyle, 1997: The March 1993 Superstorm cyclogenesis: Incipient phase synoptic- and convective-scale flow interaction and model performance. *Mon. Wea. Rev.,* **125,** 3041–3072.

Dines, W. H., 1925: The correlation between pressure and temperature in the upper air with a suggested explanation. *Quart. J. Roy. Meteor. Soc.,* **51,** 31–38.

Dirks, R. A., J. P. Kuettner, and J. A. Moore, 1988: Genesis of Atlantic Lows Experiment (GALE): An overview. *Bull. Amer. Meteor. Soc.,* **69,** 148–160.

Doesken, N. J., and A. Judson, 1996: *The Snow Booklet: A Guide to the Science, Climatology and Measurement of Snow in the United States.* Colorado State University, 85 pp.

Dole, R., 1986: Persistent anomalies of the extratropical Northern Hemisphere wintertime circulation: Structure. *Mon. Wea. Rev.,* **114,** 178–207.

——, 1989: Life cycles of persistent anomalies. Part I: Evolution of 500 mb height fields. *Mon. Wea. Rev.,* **117,** 177–211.

——, and N. D. Gordon, 1983: Persistent anomalies of the extratropical Northern Hemisphere wintertime circulation: Geographical distribution and regional persistence characteristics. *Mon. Wea. Rev.,* **111,** 1567–1586.

Doyle, J. D., and T. T. Warner, 1993a: A three-dimensional numerical investigation of a Carolina coastal low-level jet during GALE IOP2. *Mon. Wea. Rev.,* **121,** 1030–1047.

——, and ——, 1993b: A numerical investigation of coastal frontogenesis and mesoscale cyclogenesis during GALE IOP2. *Mon. Wea. Rev.,* **121,** 1048–1077.

——, and ——, 1993c: Nonhydrostatic simulations of coastal mesobeta-scale vortices and frontogenesis. *Mon. Wea. Rev.,* **121,** 3371–3392.

Dunn, L. B., 1987: Cold-air damming by the Front Range of the Colorado Rockies and its relationship to locally heavy snows. *Wea. Forecasting,* **2,** 177–189.

——, 1988: Vertical motion evaluation of a Colorado snowstorm from a synoptician's perspective. *Wea. Forecasting,* **3,** 261–272.

——, 1992: Evidence of ascent in a sloped barrier jet and an associated heavy-snow band. *Mon. Wea. Rev.,* **120,** 914–924.

Eady, E. T., 1949: Long waves and cyclone waves. *Tellus,* **1,** 33–52.
Egger, J., 1974: Numerical experiments on lee cyclogenesis. *Mon. Wea. Rev.,* **102,** 847–860.
Eliassen, A., and E. Kleinschmidt, 1957: Dynamic meteorology. *Handbuch der Physik,* S. Flugge, Ed., Vol. 48, Springer-Verlag, 1–154.
Emanuel, K., 1979: Inertial instability and mesoscale convective systems. Part I: Linear theory of inertial instability in rotating viscous fields. *J. Atmos. Sci.,* **36,** 2425–2449.
——, 1983: The Lagrangian parcel dynamics of moist symmetric instability. *J. Atmos. Sci.,* **40,** 2368–2376.
——, 1985: Frontogenesis in the presence of low moist symmetric stability. *J. Atmos. Sci.,* **42,** 1062–1071.
Evans, M. S., D. Keyser, L. F. Bosart, and G. M. Lackmann, 1994: A satellite-derived classification scheme for rapid maritime cyclogenesis. *Mon. Wea. Rev.,* **122,** 1381–1416.
Farrell, B., 1984: Modal and nonmodal baroclinic waves. *J. Atmos. Sci.,* **41,** 668–673.
——, 1985: Transient growth of damped baroclinic waves. *J. Atmos. Sci.,* **42,** 2718–2727.
Ferber, G. K., C. F., Mass, G. M. Lackmann, and M. W. Patnoe, 1993: Snowstorms over the Puget Sound lowlands. *Wea. Forecasting,* **8,** 481–504.
Ferretti, R., F. Einaudi, and L. W. Uccellini, 1988: Wave disturbances associated with the Red River Valley severe weather outbreak of 10–11 April 1979. *Meteor. Atmos. Phys,* **39,** 132–168.
Forbes, G. S., R. A. Anthes, and D. W. Thomson, 1987: Synoptic and mesoscale aspects of an Appalachian ice storm associated with cold air damming. *Mon. Wea. Rev.,* **115,** 564–591.
Foster, J. L., and R. J. Leffler, 1979: The extreme weather of February 1979 in the Baltimore–Washington area. *Natl. Wea. Dig.,* **4,** 16–21.
Friday, E. W., 1994: The modernization and associated restructuring of the National Weather Service: An overview. *Bull. Amer. Meteor. Soc.,* **75,** 43–52.
Fujita, T. T., 1971: Proposed characterization of tornadoes and hurricanes by area and intensity. Satellite and Meteorology Research Paper 91, The University of Chicago, Chicago, IL, 42 pp.
Gall, R., 1976: Structural changes of growing baroclinic waves. *J. Atmos. Sci.,* **33,** 374–390.
Garriot, E. B., 1898: The North Atlantic Coast Storm of November 26–27, 1898. *Mon. Wea. Rev.,* **26,** 493–495.
——, 1899: Forecasts and warnings. *Mon. Wea. Rev.,* **27,** 41–44.
Gaza, R. S., and L. F. Bosart, 1990: Trough merger characteristics over North America. *Wea. Forecasting,* **5,** 314–331.
Gedzelman, S. D., and E. Lewis, 1990: Warm snowstorms: A forecaster's dilemma. *Weatherwise,* **43,** 265–270.
Gilhousen, D. B., 1994: The Value of NDBC observations during March 1993's "Storm of the Century." *Wea. Forecasting,* **9,** 255–264.
Godev, N., 1971a: The cyclogenetic properties of the Pacific coast: Possible source of errors in numerical prediction. *J. Atmos. Sci.,* **28,** 968–972.
——, 1971b: Anticyclonic activity over south Europe and its relation to orography. *J. Appl. Meteor.,* **10,** 1097–1102.
Goree, P. A., and R. J. Younkin, 1966: Synoptic climatology of heavy snowfall over the central and eastern United States. *Mon Wea. Rev.,* **94,** 663–668.
Graves, C. E., J. T. Moore, M. J. Singer, and S. Ng, 2003: Band on the run: Chasing the physical processes associated with heavy snowfall. *Bull. Amer. Meteor. Soc.,* **84,** 990–995.
Green, J. S. A., F. H. Ludlam, and J. F. R. McIlveen, 1966: Isentropic relative-flow analysis and the parcel theory. *Quart. J. Roy. Meteor. Soc.,* **92,** 210–219.
Grumm, R. H., 1993: Characteristics of surface cyclone forecasts in the aviation run of the global spectral model. *Wea. Forecasting,* **8,** 87–112.
——, R. J. Oravec, and A. L. Siebers, 1992: Systematic model forecast errors of surface cyclones in NMC's Nested-Grid Model, December 1988 through November 1990. *Wea. Forecasting,* **7,** 65–87.
Gurka, J. J., E. P. Auciello, A. F. Gigi, J. S. Waldstreicher, K. K. Keeter, S. Businger, and L. G. Lee, 1995: Winter weather forecasting throughout the eastern United States. Part II. An operational perspective of cyclogenesis. *Wea. Forecasting,* **10,** 21–41.
Gyakum, J. R., 1983a: On the evolution of the *QE II* storm. I: Synoptic aspects. *Mon. Wea. Rev.,* **111,** 1137–1155.
——, 1983b: On the evolution of the *QE II* storm. II: Dynamic and thermodynamic structure. *Mon. Wea. Rev.,* **111,** 1156–1173.
——, 1987: On the evolution of a surprise snowfall in the United States Midwest. *Mon. Wea. Rev.,* **115,** 2322–2345.
——, and E. S. Barker, 1988: A case study of explosive subsynoptic-scale cyclogenesis. *Mon. Wea. Rev.,* **116,** 2225–2253.
——, and P. J. Roebber, 2001: The 1998 ice storm—Analysis of a planetary-scale event. *Mon. Wea. Rev.,* **129,** 2983–2997.
——, J. R. Anderson, R. H. Grumm, and E. L Gruner, 1989: North Pacific cold-season surface cyclone activity. *Mon. Wea. Rev.,* **117,** 1141–1155.
Hadlock, R., and C. W. Kreitzberg, 1988: The Experiment on Rapidly Intensifying Cyclones over the Atlantic ERICA) field study: Objectives and plans. *Bull. Amer. Meteor. Soc.,* **69,** 1309–1320.
Hakim, G. J., and L. W. Uccellini, 1992: Diagnosing coupled jet-streak circulations for a northern plains snowband from the operational Nested Grid Model. *Wea. Forecasting,* **7,** 26–48.
——, L. F. Bosart, and D. Keyser, 1995: The Ohio Valley wave-merger cyclogenesis event of 25–26 January 1978. Part I: Multiscale case study. *Mon. Wea. Rev.,* **123,** 2663–2692.
——, D. Keyser, and L. F. Bosart, 1996: The Ohio Valley wave-merger cyclogenesis event of 25–26 January 1978: Part II: Diagnosis using quasigeostrophic potential vorticity inversion. *Mon. Wea. Rev.,* **124,** 2176–2205.
Halpert, M. S., and C. F. Ropelewski, 1992: Surface temperature patterns associated with the Southern Oscillation. *J. Climate,* **5,** 577–593.
——, and G. D. Bell, 1997: Climate assessment for 1996. *Bull. Amer. Meteor. Soc.,* **78** (5), S1–S49.
Harlin, B. W., 1952: The great southern glaze storm of 1951. *Weatherwise,* **5,** 10–13.
Harrold, T. W., 1973: Mechanisms influencing the distribution of precipitation within baroclinic disturbances. *Quart. J. Roy. Meteor. Soc.,* **99,** 232–251.
Hart, R. E., and R. H. Grumm, 2001: Using normalized climatological anomalies to rank synoptic-scale events objectively. *Mon. Wea. Rev.,* **129,** 2426–2442.
Hayden, B. P., 1981: Secular variations in Atlantic coast extratropical cyclones. *Mon. Wea. Rev.,* **109,** 159–167.
Hayden, E., 1888: *The Great Storm off the Atlantic Coast of the United States March 11–14, 1888. Nautical Monogr.,* No. 5, U.S. Government Printing Office, 65 pp.
Hibbard, W., D. Santek, L. Uccellini, and K. Brill, 1989: Application of the 4-D McIDAS to a model diagnostic study of the Presidents' Day cyclone. *Bull. Amer. Meteor. Soc.,* **70,** 1394–1403.
Hjelmfelt, M. R., 1990: Numerical study of the influence of environmental conditions on lake effect snowstorms over Lake Michigan. *Mon. Wea. Rev.,* **118,** 138–150.
——, 1992: Orographic effects in simulated lake-effect snowstorms over Lake Michigan. *Mon. Wea. Rev.,* **120,** 373–377.
Holton, J. R., 1979: *An Introduction to Dynamic Meteorology.* 2d ed. Academic Press, 391 pp.
Homan, J., and L. W. Uccellini, 1987: Winter forecast problems associated with light to moderate snow events in the mid-Atlantic states on 14 and 22 February 1986. *Wea. Forecasting,* **2,** 206–228.
Hooke, W. H., 1986: Gravity waves. *Mesoscale Meteorology and Forecasting,* P. Ray. Ed., Amer. Meteor. Soc., 272–288.
Hoskins, B. J., M. E. McIntyre, and A. W. Robertson, 1985: On the use and significance of isentropic potential vorticity maps. *Quart. J. Roy. Meteor. Soc.,* **111,** 877–946.

Houghton, J. T., Y. Ding, D. J. Griggs, M. Noguer, P. J. van der Linden, and D. Xiaosu, Eds., 2001: Climate Change 2001: The Scientific Basis. Contribution of Working Group I to the Third Assessment Report of the Intergovernmental Panel on Climate Change. Cambridge University Press, 881 pp.
Houze, R. A., Jr., 1993: *Cloud Dynamics.* Academic Press, 573 pp.
Hovanec, R. D., and L. H. Horn, 1975: Static stability and the 300 mb isotach field in the Colorado cyclogenetic area. *Mon. Wea. Rev.,* **103,** 628–638.
Howard, K. W., and E. I. Tollerud, 1988: The structure and evolution of heavy-snow-producing Colorado cyclones. Preprints, *Palmén Symp. on Extratropical Cyclones and Their Role in the General Circulation,* Helsinki, Finland, Amer. Meteor. Soc., 168–171.
Huffman, G. J., and G. A. Norman Jr., 1988: The supercooled warm rain process and the specification of freezing precipitation. *Mon. Wea. Rev.,* **116,** 2172–2182.
Hughes, P., 1976: *American Weather Stories.* NOAA, 116 pp.
——, 1981: The blizzard of '88. *Weatherwise,* **34,** 250–256.
Huo, Z., D.-L. Zhang, and J. R. Gyakum, 1995: A diagnostic analysis of the Superstorm of March 1993. *Mon. Wea. Rev.,* **123,** 1740–1761.
——, ——, and ——, 1998: An application of potential vorticity inversion to improving the numerical prediction of the March 1993 Superstorm. *Mon. Wea. Rev.,* **126,** 424–436.
——, ——, and ——, 1999a: Interaction of portential vorticity anomalies in extratropical cyclogenesis. Part I: Static piecewise inversion. *Mon. Wea. Rev.,* **127,** 2546–2561.
——, ——, and ——, 1999b: Interaction of potential vorticity anomalies in extratropical cyclogenesis. Part II: Sensitivity to initial perturbations. *Mon. Wea. Rev.,* **127,** 2563–2575.
Hurell, J., 1995: Decadal trends in the North Atlantic Oscillation: Regional temperatures and precipitation. *Science,* **269,** 676–679.
Johnson, D. R., and W. K. Downey, 1976: The absolute angular momentum budget of an extratropical cyclone: Quasi-Lagrangian diagnostics. *Mon. Wea. Rev.,* **104,** 3–14.
Junker, N. W., J. E. Hoke, and R. H. Grumm, 1989: Performance of NMC's regional models. *Wea. Forecasting,* **4,** 368–390.
Kain, J. S., S. M. Goss, and M. E. Baldwin, 2000: The melting effect as a factor in precipitation-type forecasting. *Wea. Forecasting,* **15,** 700–714.
Kalnay, D. E., and Coauthors, 1996: The NCEP/NCAR 40-Year Reanalysis Project. *Bull. Amer. Meteor. Soc.,* **77,** 437–471.
Kaplan, M. L., J. W. Zack, V. C. Wong, and J. J. Tuccillo, 1982: A sixth-order mesoscale atmospheric simulation system applicable to research and real-time forecasting problems. *Proc. CIMMS 1982 Symp.,* Norman, OK, 38–84.
Karl, T., Ed.,1996: *Long-Term Climate Monitoring by the Global Climate Observing System.* Kluwer Academic, 648 pp.
Keeter, K. K., and J. W. Cline, 1991: The objective use of observed and forecast thickness values to predict precipitation type in North Carolina. *Wea. Forecasting,* **6,** 456–469.
——, S. Businger, L. G. Lee, and J. S. Waldstreicher, 1995: Winter weather forecasting throughout the eastern United States. Part III. The effects of topography and the variability of winter weather in the Carolinas and Virginia. *Wea. Forecasting,* **10,** 42–60.
Keshishian, L. G., and L. F. Bosart, 1987: A case study of extended East Coast frontogenesis. *Mon. Wea. Rev.,* **115,** 100–117.
——, ——, and W. E. Bracken, 1994: Inverted troughs and cyclogenesis over interior North America: A limited regional climatology and case studies. *Mon. Wea. Rev.,* **122,** 565–607.
Keyser, D., and D. R. Johnson, 1984: Effects of diabatic heating on the ageostrophic circulation of an upper tropospheric jet streak. *Mon. Wea. Rev.,* **112,** 1709–1724.
——, and M. Shapiro, 1986: A review of the structure and dynamics of upper-level frontal zones. *Mon. Wea. Rev.,* **114,** 452–499.
——, B. D. Schmidt, and D. G. Duffy, 1989: A technique for representing three-dimensional vertical circulations in baroclinic disturbances. *Mon. Wea. Rev.,* **117,** 2463–2494.
Klein, W. J., and J. S. Winston, 1958: Geographical frequency of troughs and ridges on mean 700 mb charts. *Mon. Wea. Rev.,* **86,** 344–358.
Kleinschmidt, E., 1950: On the structure and origin of cyclones (Part 1). *Meteor. Rundsch.,* **3,** 1–6.
——, 1957: Cyclones and anticyclones. *Handbuch der Physik,* S. Flugge, Ed., Vol. 48, Springer-Verlag, 1–154.
Koch, S. E., and P. B. Dorian, 1988: A mesoscale gravity wave event observed during CCOPE. Part III: Wave environment and probable source mechanisms. *Mon. Wea. Rev.,* **116,** 2570–2592.
——, and ——, 1988: A mesoscale gravity wave event observed during CCOPE. Part I: Multiscale statistical analysis of wave characteristics. *Mon. Wea. Rev.,* **116,** 2527–2544.
——, and C. O'Handley, 1997: Operational forecasting and detection of mesoscale gravity waves. *Wea. Forecasting,* **12,** 253–281.
——, M. L. des Jardins, and P. J. Kocin, 1983: An interactive Barnes objective map analysis scheme for use with satellite and conventional data. *J. Climate Appl. Meteor.,* **22,** 1487–1503.
Kocin, P. J., 1983: An analysis of the "Blizzard of '88." *Bull. Amer. Meteor. Soc.,* **64,** 1258–1272.
——, and L. W. Uccellini, 1990: *Snowstorms along the Northeastern Coast of the United States: 1955 to 1985. Meteor. Monogr.,* No. 44, Amer. Meteor. Soc., 280 pp.
——, and ——, 2004a: *Overview.* Vol. 1. *Northeast Snowstorms.* No. 54, Amer. Meteor. Soc., in press.
——, and ——, 2004b: *The Cases.* Vol. 2. *Northeast Snowstorms of the 20th Century. Meteor. Monogr.,* No. 54, Amer. Meteor. Soc., in press.
——, and ——, 2004c: A northeast snowfall impact scale. *Bull. Amer. Meteor. Soc.,* **85,** 177–194.
——, L. W. Uccellini, J. W. Zack, and M. L. Kaplan, 1985: A mesoscale numerical forecast of an intense convective snowburst along the East Coast. *Bull. Amer. Meteor. Soc.,* **66,** 1412–1424.
——, ——, and R. A. Petersen, 1986: Rapid evolution of a jet stream circulation in a pre-convective environment. *Meteor. Atmos. Phys.,* **35,** 103–138.
——, A. D. Weiss, and J. J. Wagner, 1988: The great Arctic outbreak and East Coast blizzard of February 1899. *Wea. Forecasting,* **3,** 305–318.
——, P. N. Schumacher, R. F. Morales Jr., and L. W. Uccellini, 1995: Overview of the 12–14 March 1993 Superstorm. *Bull. Amer. Meteor. Soc.,* **76,** 165–182.
——, L. W. Uccellini, K. F. Brill, and M. Zika, 1998: Northeast snowstorms: An update. Preprints, *16th Conf. on Weather Analysis and Forecasting,* Phoenix, AZ, Amer. Meteor. Soc., 421–423.
Koppel, L. L., L. F. Bosart, and D. Keyser, 2000: A 25-yr climatology of large-amplitude hourly surface pressure changes over the conterminous United States. *Mon. Wea. Rev.,* **128,** 51–68.
Krishnamurti, T. N., 1968: A study of a developing wave cyclone. *Mon. Wea. Rev.,* **96,** 208–217.
Kristovich, D. A. R., and Coauthors, 2000: The Lake-Induced Convection Experiments and the Snowband Dynamics Project. *Bull. Amer. Meteor. Soc.,* **81,** 519–542.
Kuo, Y.-H., and S. Low-Nam, 1990: Prediction of nine explosive cyclones over the western Atlantic Ocean with a regional model. *Mon. Wea. Rev.,* **118,** 3–25.
Lackmann, G. M., L. F. Bosart, and D. Keyser, 1996: Planetary- and synoptic-scale characteristics of explosive wintertime cyclogenesis over the western North Atlantic Ocean. *Mon. Wea. Rev.,* **124,** 2672–2702.
——, ——, and ——, 1997: A characteristic life cycle of upper-tropospheric cyclogenetic precursors during the Experiment on Rapidly Intensifying Cyclones over the Atlantic (ERICA). *Mon. Wea. Rev.,* **125,** 2729–2758.
——, ——, and ——, 1999: Energetics of an intensifying jet streak during the Experiment on Rapidly Intensifying Cyclones over the Atlantic (ERICA). *Mon. Wea. Rev.,* **127,** 2777–2795.
Lai, C.-C., and L. F. Bosart, 1988: A case study of trough mergers in split westerly flow. *Mon. Wea. Rev.,* **116,** 1838–1856.

Langland, R. J., M. A. Shapiro, and R. Gelaro, 2002: Initial condition sensitivity and error growth in forecasts of the 25 January 2000 East Coast snowstorm. *Mon. Wea. Rev.,* **130,** 957–974.

LaPenta, W. M., and N. L. Seaman, 1990: A numerical investigation of East Coast cyclogenesis during the cold-air damming event of 27–28 February 1982. Part I: Dynamic and thermodynamic structure. *Mon. Wea. Rev.,* **118,** 2668–2695.

——, and ——, 1992: A numerical investigation of East Coast cyclogenesis during the cold-air damming event of 27–28 February 1982. Part II: Importance of physical mechanisms. *Mon. Wea. Rev.,* **120,** 52–76.

Lau, K.-M., and H. Weng, 1999: Interannual, decadal-interdecadal, and global warming signals in sea surface temperature during 1955–97. *J. Climate,* **12,** 1257–1267.

Leathers, D. J., D. R. Kluck, and S. Kroczynski, 1998: The severe flooding event of January 1996 across north-central Pennsylvania. *Bull. Amer. Meteor. Soc.,* **79,** 785–797.

Leese, J. A., 1962: The role of advection in the formation of vortex cloud patterns. *Tellus,* **14,** 409–421.

Lindzen, R. S., and K.-K. Tung, 1976: Banded convective activity and ducted gravity waves. *Mon. Wea. Rev.,* **104,** 1602–1617.

Livezey, R. E., and T. M. Smith, 1999: Covariability of aspects of North American climate with global sea surface temperatures on interannual to interdecadal timescales. *J. Climate,* **12,** 289–302.

Loughe, A., C.-C. Lai, and D. Keyser, 1995: A technique for diagnosing three-dimensional circulations in baroclinic disturbances on limited-area domains. *Mon. Wea. Rev.,* **123,** 1476–1504.

Ludlum, D. M., 1956: The great Atlantic low. *Weatherwise,* **9,** 64–65.

——, 1966: *Early American Winters I 1604–1820.* Amer. Meteor. Soc., 198 pp.

——, 1968: *Early American Winters II 1821–1870.* Amer. Meteor. Soc., 257 pp.

——, 1976: *The Country Journal New England Weather Book.* Houghton Mifflin, 148 pp.

——, 1982: *The American Weather Book.* Houghton Mifflin, 296 pp.

——, 1983: *The New Jersey Weather Book.* Rutgers University Press, 256 pp.

Maddox, R. A., D. J. Perkey, and J. M. Fritsch, 1981: Evolution of upper tropospheric features during the development of a mesoscale convective complex. *J. Atmos. Sci.,* **38,** 1664–1674.

Maglaras, G., J. F. Waldstreicher, P. J. Kocin, A. F. Gigi, and R. A. Marine, 1995: Winter weather forecasting throughout the eastern United States. Part 1: An overview. *Wea. Forecasting,* **10,** 5–20.

Mahoney, J. L., J. M. Brown, and E. I. Tollerud, 1995: Contrasting meteorological conditions associated with winter storms at Denver and Colorado Springs. *Wea. Forecasting,* **10,** 245–260.

Mailhot, J., and C. Chouinard, 1989: Numerical forecasts of explosive winter storms: Sensitivity experiments with a meso- scale model. *Mon. Wea. Rev.,* **117,** 1311–1343.

Manney, G. L., J. D. Farrara, and C. R. Mechoso, 1994: Simulations of the February 1979 stratospheric sudden warming: Model comparisons and three-dimensional evolution. *Mon. Wea. Rev.,* **122,** 1115–1140.

Manobianco, J., L. W. Uccellini, K. F. Brill, and Y.-H Kuo, 1992: The impact of dynamic data assimilation on the numerical simulations of the *QE II* cyclone and an analysis of the jet streak influencing the precyclogenetic environment. *Mon. Wea. Rev.,* **120,** 1973–1996.

Marks, F. D., Jr., and P. M. Austin, 1979: Effects of the New England coastal front on the distribution of precipitation. *Mon. Wea. Rev.,* **107,** 53–67.

Martin, J. E., 1998: The structure and evolution of a continental winter cyclone. Part II: Frontal forcing of an extreme snow event. *Mon. Wea. Rev.,* **126,** 329–348.

——, 1999: Quasigeostrophic forcing of ascent in the occluded sector of cyclones and the Trowal airstream. *Mon. Wea. Rev.,* **127,** 70–88.

Marwitz, J. D., and J. Toth, 1993: A case study of heavy snowfall in Oklahoma. *Mon. Wea. Rev.,* **121,** 648–660.

Mass, C. F., and D. M. Schultz, 1993: The structure and evolution of a simulated midlatitude cyclone over land. *Mon. Wea. Rev.,* **121,** 889–917.

Mather, J. R., H. Adams, and G. A. Yoshioka, 1964: Coastal storms of the eastern United States. *J. Appl. Meteor.,* **3,** 693–706.

Mattocks, C., and R. Bleck, 1986: Jet streak dynamics and geostrophic adjustment processes during the initial stages of lee cyclogenesis. *Mon. Wea. Rev.,* **114,** 2033–2056.

McKelvey, B., 1995: *Snow in the Cities. A History of America's Urban Response.* University of Rochester Press, 202 pp.

McQueen, H. R., and H. C. Keith, 1956: The ice storm of January 7–10, 1956 over the northeastern United States. *Mon. Wea. Rev.,* **84,** 35–45.

Miller, J. E., 1946: Cyclogenesis in the Atlantic coastal region of the United States. *J. Meteor.,* **3,** 31–44.

Mook, C. P., 1956: The "Knickerbocker" snowstorm of January 1922 at Washington, D.C. *Weatherwise,* **9,** 188–191.

——, and K. S. Norquest, 1956: The heavy snowstorm of 18–19 March 1956. *Mon. Wea. Rev.,* **84,** 116–125.

Moore, J. T., and P. D. Blakely, 1988: The role of frontogenetical forcing and conditional symmetrical instability in the Midwest snowstorm of 30–31 January 1982. *Mon. Wea. Rev.,* **116,** 2153–2176.

——, and V. Vanknowe, 1992: The effect of jet-streak curvature on kinematic fields. *Mon. Wea. Rev.,* **120,** 2429–2441.

Mote, T. L., D. W. Gamble, S. J. Underwood, and M. L. Bentley, 1997: Synoptic-scale features common to heavy snowstorms in the southeast United States. *Wea. Forecasting,* **12,** 5–23.

Mullen, S. L., and B. B. Smith, 1990: An analysis of sea-level cyclone errors in NMC's Nested Grid Model (NGM) during the 1987–1988 winter season. *Wea. Forecasting,* **5,** 433–447.

Murray, R., and S. M. Daniels, 1953: Transverse flow at entrance and exit to jet streams. *Quart. J. Roy. Meteor. Soc.,* **79,** 236–241.

Naistat, R. J., and J. A. Young, 1973: A linear model of boundary layer flow applied to the St. Patrick's Day storm of 1965. *J. Appl. Meteor.,* **12,** 1151–1162.

Namias, J., and P. F. Clapp, 1949: Confluence theory of the high tropospheric jet stream. *J. Meteor.,* **6,** 330–336.

Newton, C. W., 1954: Frontogenesis and frontolysis as a three-dimensional process. *J. Meteor.,* **11,** 449–461.

——, 1956: Mechanisms of circulation change during a lee cyclogenesis. *J. Meteor.,* **13,** 528–539.

——, and A. V. Persson, 1962: Structural characteristics of the subtropical jet stream and certain lower-stratospheric wind systems. *Tellus,* **14,** 221–241.

——, and A. Trevisan, 1984: Clinogenesis and frontogenesis in jet stream waves. Part II: Channel model numerical experiments. *J. Atmos. Sci.,* **41,** 2735–2755.

Nicosia, D., and R. H. Grumm, 1999: Mesoscale band formation in three major northeastern United States snowstorms. *Wea. Forecasting,* **14,** 346–368.

Nielsen, J. W., 1989: The formation of New England coastal fronts. *Mon. Wea. Rev.,* **117,** 1380–1401.

——, and P. P. Neilley, 1990: The vertical structure of New England coastal fronts. *Mon. Wea. Rev.,* **118,** 1793–1807.

Niziol, T. A., 1987: Operational forecasting of lake effect snow in western and central New York. *Wea. Forecasting,* **1,** 311–321.

——, W. R. Snyder, and J. S. Waldstreicher, 1995: Winter weather forecasting throughout the eastern United States. Part IV: Lake effect snow. *Wea. Forecasting,* **10,** 61–77.

O'Handley, C., and L. F. Bosart, 1996: The impact of the Appalachian Mountains on cyclonic weather systems. Part I: A climatology. *Mon. Wea. Rev.,* **124,** 1353–1373.

Oravec, R. J., and R. H. Grumm, 1993: The prediction of rapidly deepening cyclones by NMC's Nested-Grid Model in winter 1989 through autumn 1991. *Wea. Forecasting,* **8,** 248–270.

Orlanski, I., 1975: A rational subdivision of scales for atmospheric processes. *Bull. Amer. Meteor. Soc.,* **56,** 527–530.

——, and K. M. Chang, 1993: Ageostrophic geopotential fluxes in downstream and upstream development of baroclinic waves. *J. Atmos. Sci.,* **50,** 212–225.

Palmén, E., 1951: The aerology of extratropical disturbances. *Compendium of Meteorology,* T. F. Malone, Ed., Amer. Meteor. Soc., 599–620.

——, and C. W. Newton, 1969: *Atmospheric Circulation Systems.* Academic Press, 603 pp.

Petterssen, S., 1955: A general survey of factors influencing development at sea-level. *J. Meteor.,* **12,** 36–42.

——, 1956: *Weather Analysis and Forecasting.* Vol. 1. McGraw-Hill, 428 pp.

——, D. L. Bradbury, and K. Pedersen, 1962: The Norwegian cyclone models in relation to heat and cold sources. *Geofys. Publ., Norske Viderskaps-Akad. Oslo,* **24,** 243–280.

Phillips, N. A., 1951: A simple three-dimensional model for the study of large-scale extratropical flow patterns. *J. Meteor.,* **8,** 381–394.

——, 2000: A review of theoretical question in the early days of NWP. *50th Anniversary of Numerical Weather Prediction Commemorative Symposium,* A. Spekat, Ed., Deutsche Meteorologische Gesellschaft, 13–28.

——, 2001: The start of numerical weather prediction in the United States. Manuscript in preparation.

Platzman, G. W., 1952: Some remarks on high-speed computers and their use in meteorology. *Tellus,* **4,** 168–178.

Platzman, G. W., 1979: The ENIAC computations of 1950—Gateway to numerical weather prediction. *Bull. Amer. Meteor. Soc.,* **60,** 302–312.

Pokrandt, P. J., G. J. Tripoli, and D. D. Houghton, 1996: Processes leading to the formation of mesoscale waves in the Midwest cyclone of 15 December 1987. *Mon. Wea. Rev.,* **124,** 2726–2752.

Powers, J. G., and R. J. Reed, 1993: Numerical simulation of the large-amplitude mesoscale gravity-wave event of 15 December 1987 in the central United States. *Mon. Wea. Rev.,* **121,** 2285–2308.

Ramamurthy, M. K., R. M. Rauber, B. P. Collins, M. T. Shields, P. C. Kennedy, and W. L. Clark, 1991: UNIWIPP: A University of Illinois field experiment to investigate the structure of mesoscale precipitation in winter storms. *Bull. Amer. Meteor. Soc.,* **72,** 764–776.

Rasmussen, R., and Coauthors, 1992: Winter Icing and Storms Project (WISP). *Bull. Amer. Meteor. Soc.,* **73,** 951–974.

Rauber, R. M., M. K. Ramamurthy, and A. Tokay, 1994: Synoptic and mesoscale structure of severe freezing rain event: The St. Valentine's Day ice storm. *Wea. Forecasting,* **9,** 183–208.

Reed, R. J., 1955: A study of a characteristic type of upper-level frontogenesis. *J. Atmos. Sci.,* **12,** 226–237.

——, and F. Sanders, 1953: An investigation of the development of a mid-tropospheric frontal zone and its associated vorticity field. *J. Atmos. Sci.,* **10,** 338–349.

——, and E. F. Danielsen, 1959: Fronts in the vicinity of the tropopause. *Arch. Meteor. Geophys. Bioklimatol.,* **A11,** 1–17.

——, M. T. Stoelinga, and Y.-H. Kuo, 1992: A model-aided study of the origin and evolution of the anomalously high potential vorticity in the inner region of a rapidly deepening marine cyclone. *Mon. Wea. Rev.,* **120,** 893–913.

——, Y.-H. Kuo, and S. Low-Nam, 1994: An adiabatic simulation of the ERICA IOP 4 storm: An example of quasi-ideal frontal cyclone development. *Mon. Wea. Rev.,* **122,** 2688–2708.

Reitan, C. H., 1974: Frequencies of cyclones and cyclogenesis for North America, 1950–1970. *Mon. Wea. Rev.,* **102,** 861–868.

Reiter, E. R., 1963: *Jet Stream Meteorology.* The University of Chicago Press, 515 pp.

——, 1969: Tropospheric circulations and jet streams. *Climate of the Free Atmosphere,* D. F. Rex, Ed., Vol. 4, *World Survey of Climatology,* Elsevier Science, 85–203.

Richwein, B. A., 1980: The damming effect of the southern Appalachians. *Natl. Wea. Dig.,* **5,** 2–12.

Riehl, H., and Coauthers, 1952: *Forecasting in the Middle Latitudes. Meteor. Monogr.,* No. 5, Amer. Meteor. Soc., 80 pp.

Riordan, A. J., 1990: Examination of the mesoscale features of the GALE coastal front of 24–25 January, 1986. *Mon. Wea. Rev.,* **118,** 258–282.

Roebber, P. J., 1984: Statistical analysis and updated climatology of explosive cyclones. *Mon. Wea. Rev.,* **112,** 1577–1589.

——, 1993: A diagnostic case study of self-development as an antecedent conditioning process in explosive cyclogenesis. *Mon. Wea. Rev.,* **121,** 976–1006.

——, J. R. Gyakum, and D. N. Trat, 1994: Coastal frontogenesis and precipitation during ERICA IOP 2. *Wea. Forecasting,* **9,** 21–44.

Ropelewski, C. F., and M. S. Halpert, 1986: North American precipitation and temperature patterns associated with the El Niño/Southern Oscillation (ENSO). *Mon. Wea. Rev.,* **114,** 2352–2362.

——, and ——, 1987: Global and regional scale precipitation patterns associated with the El Niño/Southern Oscillation (ENSO). *Mon. Wea. Rev.,* **115,** 1606–1626.

——, and ——, 1989: Precipitation patterns associated with the high index phase of the Southern Oscillation. *J. Climate,* **2,** 268–284.

——, and ——, 1996: Quantifying Southern Oscillation–precipitation relationships. *J. Climate,* **9,** 1043–1059.

Rosenblum, H. S., and F. Sanders, 1974: Meso-analysis of a coastal snowstorm in New England. *Mon. Wea. Rev.,* **102,** 433–442.

Saffir, H. S., 1977: Design and construction requirements for hurricane resistant construction. American Society of Civil Engineers Preprint 2830, 20 pp.

Salmon, E., and P. J. Smith, 1980: A synoptic analysis of the 25–26 January 1978 blizzard cyclone in the central United States. *Bull. Amer. Meteor. Soc.,* **61,** 453–460.

Sanders, F., 1986a: Explosive cyclogenesis in the west-central North Atlantic Ocean, 1981–84. Part I: Composite structure and mean behavior. *Mon. Wea. Rev.,* **114,** 1781–1794.

——, 1986b: Frontogenesis and symmetric stability in a major New England snowstorm. *Mon. Wea. Rev.,* **114,** 1847–1862.

——, 1987: Skill of NMC operational dynamical models in prediction of explosive cyclogenesis. *Wea. Forecasting,* **2,** 322–336.

——, 1990: Surface analysis over the oceans—Searching for sea truth. *Wea. Forecasting,* **5,** 596–612.

——, 1992: Skill of operational dynamical models in cyclone prediction out to five-days range during ERICA. *Wea. Forecasting,* **7,** 3–25.

——, and J. R. Gyakum, 1980: Synoptic dynamic climatology of the "bomb." *Mon. Wea. Rev.,* **108,** 1589–1606.

——, and ——, 1983a: On the evolution of the *QE II* storm. Part I: Synoptic aspects. *Mon. Wea. Rev.,* **111,** 1137–1155.

——, and L. F. Bosart, 1985a: Mesoscale structure in the Megalopolitan snowstorm of 11–12 February 1983. Part I: Frontogenetical forcing and symmetric instability. *J. Atmos. Sci.,* **42,** 1050–1061.

——, and ——, 1985b: Mesoscale structure in the Megalopolitan snowstorm, 11–12 February 1983. Part II: Doppler radar study of the New England snowband. *J. Atmos. Sci.,* **42,** 1398–1407.

——, and E. Auciello, 1989: Skill in prediction of explosive cyclogenesis over the western North Atlantic Ocean, 1987/1988: A forecast checklist and NMC dynamical models. *Wea. Forecasting,* **4,** 157–172.

Sanderson, A. N., and R. B. Mason Jr., 1958: Behavior of two East Coast storms, 13–24 March 1958. *Mon. Wea. Rev.,* **86,** 109–115.

Santer, B. D., K. E. Taylor, J. E. Penner, T. M. L. Wigley, U. Cubasch, and P. D. Jones, 1996: Towards the detection and attribution of an anthropogenic effect on climate. *Climate Dyn.,* **12,** 77–100.

Scherhag, R., 1937: Bermerkurgen über die bedeutung der konvergenzen und divergenzen du geschwindigkeitsfeldes fur die Druckänderungen. *Beitr. Phys. Atmos.,* **24,** 122–129.

Schneider, R. S., 1990: Large-amplitude mesoscale wave disturbances within the intense Midwest extratropical cyclone of 15 December 1987. *Wea. Forecasting,* **5,** 533–558.

Schultz, D. M., 1999: Lake-effect snowstorms in northern Utah and western New York with and without lightning. *Wea. Forecasting,* **14,** 1023–1031.

——, 2001: Reexamining the cold conveyor belt. *Mon. Wea. Rev.,* **129,** 2205–2225.

——, and C. Mass, 1993: The occlusion process in a midlatiutde cyclone over land. *Mon. Wea. Rev.,* **121,** 918–940.

——, and P. N. Schumacher, 1999: The use and misuse of conditional symmetric instability. *Mon. Wea. Rev.,* **127,** 2709–2732.

Seimon, A., L. F. Bosart, W. E. Bracken, and W. R. Snyder, 1996: Large-amplitude inertia–gravity waves. Part II: Structure of an extreme gravity wave event over New England on 4 January 1994 revealed by WSR-88D radar and mesoanalysis. Preprints, *14th Conf. on Weather Analysis Forecasting,* Dallas, TX, Amer. Meteor. Soc., 434–441.

Shapiro, M. A., and P. J. Kennedy, 1981: Research aircraft measurements of jet stream geostrophic and ageostrophic winds. *J. Atmos. Sci.,* **38,** 2642–2652.

——, and D. Keyser, 1990: Fronts, jet streams and the tropopause. *Extratropical Cyclones: The Erik Palmén Memorial Volume,* C. W. Newton and E. O. Holopainen, Eds., Amer. Meteor. Soc., 167–193.

——, H. Wernli, N. A. Bond, and R. Langland, 2000: The influence of the 1997–1999 ENSO on extratropical baroclinic life cycles over the eastern North Pacific. *Quart. J. Roy. Meteor. Soc.,* **126,** 1–20.

Shields, M. T., R. M. Rauber, and M. K. Ramamurthy, 1991: Dynamical forcing and mesoscale organization of precipitation bands in a Midwest winter cyclonic storm. *Mon. Wea. Rev.,* **119,** 936–964.

Shuman, F. G., 1989: History of numerical weather prediction at the National Meteorological Center. *Wea. Forecasting,* **4,** 286–296.

Simmons, A. J., and B. J. Hoskins, 1979: The downstream and upstream development of unstable baroclinic waves. *J. Atmos. Sci.,* **36,** 1239–1260.

Sinclair, M. R., and R. L. Ellsberry, 1986: A diagnostic study of baroclinic disturbances in polar air streams. *Mon. Wea. Rev.,* **114,** 1957–1983.

Smith, B. B., and S. L. Mullen, 1993: An evaluation of sea-level cyclone forecasts produced by NMC's Nested-Grid Model and Global Spectral Model. *Wea. Forecasting,* **8,** 37–56.

Smith, C. D., Jr., 1950: The destructive storm of November 25–27, 1950. *Mon. Wea. Rev.,* **78,** 204–209.

Smith, R. B., 1979: The influence of the mountains on the atmosphere. *Advances in Geophysics,* Vol. 21, Academic Press, 87–230.

——, 1984: A theory of lee cyclogenesis. *J. Atmos. Sci.,* **41,** 1159–1168.

Smith, S. R., and J. J. O'Brien, 2001: Regional snowfall distributions associated with ENSO: Implicatons for seasonal forecasting. *Bull. Amer. Meteor. Soc.,* **82,** 1179–1191.

Snook, J. S., and R. A. Pielke, 1995: Diagnosing a Colorado heavy snow event with a nonhydrostatic mesoscale numerical model structured for operational use. *Wea. Forecasting,* **10,** 261–285.

Spiegler, D. B., and G. E. Fisher, 1971: A snowfall prediction method for the Atlantic seaboard. *Mon. Wea. Rev.,* **99,** 311–325.

Staley, D. O., 1960: Evaluation of potential-vorticity changes near the tropopause and the related vertical motions, vertical advection of vorticity, and transfer of radioactive debris from stratosphere to troposphere. *J. Atmos. Sci.,* **17,** 591–620.

Stauffer, D. R., and T. T. Warner, 1987: A numerical study of Appalachian cold-air damming and coastal frontogenesis. *Mon. Wea. Rev.,* **115,** 799–821.

Stein, U., and P. Alpert, 1993: Factor separation in numerical simulations. *J. Atmos. Sci.,* **50,** 2107–2115.

Stewart, G. R., 1941: *Storm.* Random House, 349 pp. [Reprinted 2003, Heyday Books, 352 pp.]

Stewart, R. E., 1985: Precipitation types in winter storms. *Pure Appl. Geophys.,* **123,** 597–609.

——, 1992: Precipitation types in the transition region of winter storms. *Bull. Amer. Meteor. Soc.,* **73,** 287–296.

——, and P. King, 1987a: Freezing precipitation in winter storms. *Mon. Wea. Rev.,* **115,** 1270–1279.

——, and ——, 1987b: Rain–snow boundaries over southern Ontario. *Mon. Wea. Rev.,* **115,** 1894–1907.

——, J. D. Marwitz, and R. E. Carbone, 1984: Characteristics through the melting layer of stratiform clouds. *J. Atmos. Sci.,* **41,** 3227–3237.

——, R. W. Shaw, and G. A. Isaac, 1987: Canadian Atlantic Storms Program: The meteorological field project. *Bull. Amer. Meteor. Soc.,* **68,** 338–345.

Stokols, P. M., J. P. Gerrity, and P. J. Kocin, 1991: Improvements at NMC in numerical weather prediction and their effect on winter storm forecasts. Preprints, *First Int. Symp. on Winter Storms,* New Orleans, LA, Amer. Meteor. Soc., 15–19.

Suckling, P. W., 1991: Spatial and temporal climatology of snowstorms in the Deep South. *Phys. Geogr.,* **12,** 124–139.

Sutcliffe, R. C., 1939: Cyclonic and anticyclonic development. *Quart. J. Roy. Meteor. Soc.,* **65,** 518–524.

——, 1947: A contribution to the problem of development. *Quart. J. Roy. Meteor. Soc.,* **73,** 370–383.

——, and A. G. Forsdyke, 1950: The theory and use of upper air thickness patterns in forecasting. *Quart. J. Roy. Meteor. Soc.,* **76,** 189–217.

Tepper, M., 1954: Pressure jump lines in midwestern United States, January–August 1951. U.S. Weather Bureau Research Paper 376, 70 pp.

Thorncroft, C. D., B. J. Hoskins, and M. E. McIntyre, 1993: Two paradigms of baroclinic-wave life-cycle behavior. *Quart. J. Roy. Meteor. Soc.,* **119,** 17–56.

Tracton, M. S., 1993: On the skill and utility of NMCs medium-range central guidance. *Wea. Forecasting,* **8,** 147–153.

——, and E. Kalnay, 1993: Operational ensemble prediction at the National Meteorological Center: Practical aspects. *Wea. Forecasting,* **8,** 379–398.

Trenberth, K. E., 1976: Spatial and temporal variations of the Southern Oscillation. *Quart. J. Roy. Meteor. Soc.,* **102,** 639–653.

——, 1997: The definition of El Niño. *Bull. Amer. Meteor. Soc.,* **78,** 2771–2778.

Uccellini, L. W., 1975: A case study of apparent gravity wave initiation of severe convective storms. *Mon. Wea. Rev.,* **103,** 497–513.

——, 1984: Comments on "Comparative diagnostic case study of East Coast secondary cyclogenesis under weak versus strong synoptic forcing." *Mon. Wea. Rev.,* **112,** 2540–2541.

——, 1990: Processes contributing to the rapid development of extratropical cyclones. *Extratropical Cyclones: The Eric Palmén Memorial Volume,* C. W. Newton and E. O. Holopainen, Eds., Amer. Meteor. Soc., 81–105.

——, and D. R. Johnson, 1979: The coupling of upper- and lower-tropospheric jet streaks and implications for the development of severe convective storms. *Mon. Wea. Rev.,* **107,** 682–703.

——, and S. E. Koch, 1987: The synoptic setting and possible energy sources for mesoscale wave disturbances. *Mon. Wea. Rev.,* **115,** 763–786.

——, and P. J. Kocin, 1987: An examination of vertical circulations associated with heavy snow events along the east coast of the United States. *Wea. Forecasting,* **2,** 289–308.

——, ——, R. A. Petersen, C. H. Wash, and K. F. Brill, 1984: The Presidents' Day cyclone of 18–19 February 1979: Synoptic overview and analysis of the subtropical jet streak influencing the precyclogenetic period. *Mon. Wea. Rev.,* **112,** 31–55.

——, D. Keyser, K. F. Brill, and C. H. Wash, 1985: The Presidents' Day cyclone of 18–19 February 1979: Influence of upstream trough amplification and associated tropopause folding on rapid cyclogenesis. *Mon. Wea. Rev.,* **115,** 2227–2261.

——, R. A. Petersen, K. F. Brill, P. J. Kocin, and J. J. Tuccillo, 1987: Synergistic interactions between an upper-level jet streak and diabatic processes that influence the development of a low-level jet and a secondary coastal cyclone. *Mon. Wea. Rev.*, **115,** 2227–2261.

——, P. J. Kocin, R. S. Schneider, P. M. Stokols, and R. A. Dorr, 1995: Forecasting the 12–14 March 1993 Superstorm. *Bull. Amer. Meteor. Soc.*, **76,** 183–199.

——, J. M. Sienkiewicz, and P. J. Kocin, 1999: Advances in forecasting extratropical cyclogenesis at the National Meteorological Center. *The Life Cycles of Extratropical Cyclones,* M. A. Shapiro and S. Grønås, Eds., Amer. Meteor. Soc., 317–336.

Upton, W., 1888: The storm of March 11–14, 1888. *Amer. Meteor. J.*, **5,** 19–3733.

van Loon, H., and J. C. Rogers, 1978: The seesaw in winter temperatures between Greenland and northern Europe. Part I: General description. *Mon. Wea. Rev.*, **106,** 296–310.

Wagner, A. J., 1957: Mean temperature from 1000 mb to 500 mb as a predictor of precipitation type. *Bull. Amer. Meteor. Soc.*, **38,** 584–590.

Walker, G. T., and E. W. Bliss, 1932: World weather V. *Mem. Roy. Meteor. Soc.*, **4,** 53–84.

Wallace, J. M., and D. S. Gutzler, 1981: Teleconnections in the geopotential height field during the Northern Hemisphere winter. *Mon. Wea. Rev.*, **109,** 784–812.

Wash, C. H., J. E. Peak, W. E. Calland, and W. A. Cook, 1988: Diagnostic study of explosive cyclogenesis during FGGE. *Mon. Wea. Rev.*, **116,** 431–451.

Weisman, R. A., 1996: The Fargo snowstorm of 6–8 January 1989. *Wea. Forecasting,* **11,** 198–215.

Weismuller, J. L., and S. M. Zubrick, 1998: Evaluation application of conditional symmetric instability, equivalent potential vorticity, and frontogenetic forcing in an operational environment. *Wea. Forecasting,* **13,** 84–101.

Weldon, R. B., 1979: Satellite training course notes. Part IV. Cloud patterns and upper air wind field. AWZ/TR-79/0003, United States Air Force.

Werstein, I., 1960: *The Blizzard of '88.* Thomas Y. Crowell Company. 157 pp.

Wesley, D. A., R. M. Rasmussen, and B. J. Bernstein, 1995: Snowfall associated with a terrain-generated convergence zone during the Winter Icing and Storm Project. *Mon. Wea. Rev.*, **123,** 2957–2977.

Wexler, R., R. J. Reed, and J. Honig, 1954: Atmospheric cooling by melting snow. *Bull. Amer. Meteor. Soc.*, **35,** 48–51.

Whitaker, J. S., L. W. Uccellini, and K. F. Brill, 1988: A model-based diagnostic study of the explosive development phase of the Presidents' Day cyclone. *Mon. Wea. Rev.*, **116,** 2337–2365.

Widger, W. K., Jr., 1964: A synthesis of interpretations of extratropical vortex patterns as seen by TIROS. *Mon. Wea. Rev.*, **92,** 263–282.

Wolfsberg, D. G., K. A. Emanuel, and R. E. Passarelli, 1986: Band formation in a New England snowstorm. *Mon. Wea. Rev.*, **114,** 1552–1569.

Younkin, R. J., 1968: Circulation patterns associated with heavy snowfall over the western United States. *Mon. Wea. Rev.*, **96,** 851–853.

Zhang, F., C. Snyder, and R. Rotunno, 2002: Mesoscale predictability of the "surprise" snowstorm of 24–25 January 2000. *Mon. Wea. Rev.*, **130,** 1617–1632.

Zielinski, G., 2002: A classification scheme for winter storms in the eastern and central United States with an emphasis on Nor'easters. *Bull. Amer. Meteor. Soc.*, **83,** 37–51.

DVD INSTRUCTIONS

The purposes of the enclosed DVD are to give the user access to all data for the East Coast snowstorms discussed in the monograph, to provide access to the PC version of the Gridded Analysis and Display System (GrADS) graphical display software, and to allow the user to view some standard meteorological fields for each storm event.

***DISCLAIMER*:** We provide no technical support for the information or software included on the DVD, although every effort was made to ensure that GrADS and the demo viewing programs work successfully. The validity of information and online links for various documentation on GrADS and Grib mentioned herein are the responsibility of those who manage them. Therefore, we make no claim as to the validity or accuracy of such information or links.

***NOTE*:** For convenience in this document, the DVD ROM drive letter is referred to as "D". Please change accordingly to the appropriate DVD drive letter on your system. References to "Microsoft" and "Windows" are registered trademarks of the Microsoft Corporation. GrADS is trademarked by Brian Doty, Center for Ocean-Land-Atmosphere Studies (COLA). For GrADS copyright and no-warranty information, read "License.txt" in D:\PCGrads\win32e.

There is one DVD enclosed, which contains Grib data, a PC version of GrADS, and a viewing demo program. The different storm data types are as follows:

- Thirty Selected Storms (divided into two 15-storm parts, for simplicity)
  - Part A, 1956–1972
  - Part B, 1978–2000
- Moderate Storms
- Interior Storms
- Ice Storms
- Early-Late Storms
- Historical Storms
- Other Recent Storms

The disk has two main directories: (1) "Grib_Data" which contains all the Grib data for the storms, and (2)"PCGrads" which contains all the files and executables to run the GrADS graphical software. There are also four files: (1)"INSTRUCTIONS_README.txt" which contains the same information and instructions given here, (2)"Data_Specifications.txt" which gives a brief description of what data are contained in the Grib files, and (3)"Display_Demo_1024x768.bat" and (4)"Display_Demo_800x600.bat" which run the display demo program at either 1024x768 or 800x600 screen resolution.

***DATA*:** The high-resolution analyses for each event were obtained by employing the National Centers for Environmental Prediction's (NCEP) Global Forecast System (GFS) model data assimilation. Data from the NCEP/NCAR Reanalysis (Kalnay et al. 1996; Kistler et al. 2001) as well as other observed and satellite data were incorporated into the GFS's assimilation system to create the resulting analyses.

All data are in the gridded binary (Grib) format, which is a standard for the exchange of meteorological data. For more information on Grib, see the office note on Grib at http://www.nco.ncep.noaa.gov/pmb/docs/on388. Further information on Grib files and their relation to GrADS can also be found on the GrADS Web site at http://grads.iges.org/grads. If desired, one is free to download or copy the Grib files from the DVD onto their own system.

The data are stored on the disk in D:\Grib_Data. Within that directory, there are subdirectories for each storm type (e.g., "Thirty_Selected_B" or "Moderate"), and within each of those subdirectories are further subdirectories for each storm event of that type. The event subdirectory names indicate the beginning year, month, day, and hour of the storm, and have the format YYYYMMDDHH (YYYY=year, MM=month, DD=day, HH=hour, such as 1993031200 for the 12 March 1993 storm). These contain all the Grib files for that particular event at 6-hourly increments as well as the GrADS control (.ctl) and index (.idx) files required for graphical display by GrADS. Grib file names within the event subdirectories have a similar YYYYMMDDHH format with a "pgb" prefix (e.g., pgb1993031200 for the Grib data at 00 GMT on 12 March 1993). Note that the hour (HH) is given in GMT, Greenwich Mean Standard (or "Z") Time.

As an example, the 12 March 1993 snowstorm Grib data would be in D:\Grib_Data\Thirty_Selected_B\1993031200. The Grib files within that directory would be pgb1993031200, pgb1993031206, . . . , etc., and the GrADS control and index files would be pgb1993031200.ctl and pgb1993031200.idx, respectively (note that there is only one each of the GrADS .ctl and .idx files to cover an entire event and these files are denoted by the beginning YYYYMMDDHH).

Data in all the Grib files cover a broad geographical area in the horizontal, extending from 0–180 degrees west longitude and 10–80 degrees north latitude, at 1×1 degree longitude/latitude resolution. In the vertical, data are stored on pressure surfaces. Variables are in general defined vertically from 1000 hPa up to either 100 or 10 hPa (26 or 21 levels, respectively); however, there are also several defined at other specific levels (e.g., at the surface, 10 m above the surface, at the 2-PV unit level, etc.). See the "Data_Specifications.txt" file for a listing of all variables that are in the Grib files.

Because the GEMPAK graphical software is also widely used, particularly in educational institutions, it is possible to easily convert Grib files into the corresponding GEMPAK grid file format via the use of the "nagrib" (or similar) routine in GEMPAK. Those familiar with GEMPAK and who have it installed should refer to any documentation they may have on this software. As noted above, one can copy/download the Grib files from the DVD onto their own system.

GrADS DISPLAY SOFTWARE (requires Microsoft Windows 95 or higher):

GrADS is a convenient interactive desktop tool for displaying and manipulating meteorological data. Because it can directly work with Grib files, there is no need to convert from Grib to another gridded format. For further information and documentation on GrADS, see the GrADS Web site at http://grads.iges.org/grads. There is also information contained on the DVD, such as "Getting_Started.html" in D:\PCGrads\win32e, and "Users_Guide.html" or "Tutorial.html" in D:\PCGrads\doc\gadoc. For those familiar with GrADS or who are more adventurous, it is possible to run GrADS on your own directly from the DVD by double-clicking on "Grads.exe" in D:\PCGrads\win32e. Please note that the program might be a bit slow in loading after the initial choice is made for landscape/portrait mode, and thus take a couple of minutes for the GrADS prompt to appear.

PC GrADS is included on the disk, and is self-contained in the sense that there is no need to download it onto one's system in order to run it. This version of GrADS is intended to be used only on a system running Microsoft Windows 95 or higher, and does not require an X-Server. See http://grads.iges.org/grads/Getting_win32e_Started.html for more information on this PC version of GrADS. If you want to install the software on your own computer, visit the GrADS software download site at http://grads.iges.org/grads/downloads.html. Here, you'll have download options for many different platform types. Note that the GrADS scripts which run the display demo (discussed below) are only contained on the DVD, and are not a part of the standard GrADS software download.

SNOWSTORM DISPLAY DEMO (requires Microsoft Windows 95 or higher):

The display demo supplied on the DVD enables the user to loop through several fields for each Kocin-Uccellini event. To run the display program, access the DVD ROM drive in which you placed this DVD (e.g., via "My Computer"). In the root directory of the drive (D:\), there are two files called "Display_Demo_1024x768.bat" and "Display_Demo_800x600.bat" for running the demo at either 1024x768 or 800x600 screen resolution. Double-click on either file, and the display demo program will start at the chosen resolution. You may have to click on the GrADS graphics window (identified by "GrADS" on the window title bar) to bring it to the foreground if it is "covered up" by other windows once the program begins.

You are first presented with a choice of storm type(s) contained on the disk. Single-click on the desired type (for all menus in the display demo, you'll only need to single-click). After making a choice, a listing of events is shown in DDMMMYYYY format (DD=day, MMM=3-letter abbreviated month, YYYY=year, such as 12MAR 1993); the event dates reflect the beginning dates of the storms. Click on an event, and you are then presented with a choice of fields to display and loop through. These fields include:

- **200 hPa heights and isotachs:** The plots display the evolution of the heights (solid black contours every 12 decameters) and wind isotachs (color shaded every 10 m/s from 50 to 90+ m/s).
- **300 hPa heights and isotachs:** The plots display the evolution of the heights (solid black contours every 12 decameters) and wind isotachs (color shaded every 10 m/s from 50 to 90+ m/s).
- **500 hPa heights and vorticity:** The plots display the evolution of the heights (solid black contours every 6 decameters) and vorticity (color shaded every 4 units from 20 to 40+ units, where a "unit" here is 1×10^{-5} s^{-1}).
- **700 hPa heights, relative humidity, vertical velocity:** The plots display the evolution of the heights (solid black contours every 3 decameters), relative humidity (color shaded at the 70% and 90% levels), and vertical velocity (solid orange contours every 5 μbar/s, negative values only, up to zero).
- **850 hPa heights and temperature:** The plots display the evolution of the heights (solid black contours every 3 decameters) and temperature (dashed contours every 5°C, red >= 5°C, blue <= 0°C).
- **Sea-level pressure and 1000–500 hPa thickness:** The plots display the evolution of the sea-level pressure (solid black contours every 4 hPa) and 1000–500 hPa thickness (dashed contours every 6 decameters, red >=546 dm, blue <= 540 dm).

After selecting one of the fields, the screen will clear and a plot of the first time period of that field will be displayed. There are 3 buttons on the upper left labeled "Fwd", "Rev", and "Save" (note that the first time period only has the "Fwd" and "Save" buttons, while the last time period only has "Rev" and "Save"). Click on "Fwd" or "Rev" to loop the field forward or backward in time, respectively, at 6-h intervals. On the upper right, there are 2 buttons labeled "New Field" and "Exit". These enable you to return to the "choose field" menu or to exit the demo display completely. A title at the top of each plot indicates the field being displayed as well as the date/time.

On any given plot, click the "Save" button to store the image in GIF file format (.gif) on your hard drive. This causes a dialog box to appear verifying whether you wish to save the image file to C:\. Click "Yes" to proceed with the save or "No" to not save. Choosing "Yes" stores the plot in the root directory of your C drive, i.e., in C:\ (*note that if you do not have a hard drive called "C", this saving mechanism will not work correctly*). Filenames have the format HHZDDMMMYYYY_PPP_grads.gif (HH=hour in GMT, or "Z", time, DD=day, MMM=3-letter abbreviated month, YYYY=year, and PPP=level; e.g., 12Z12MAR1993_500_grads.gif is the saved plot for 12Z 12 March 1993 at 500 hPa). You can then re-name and move any saved files as desired. This file naming convention should be unique enough to avoid overwriting any existing files in your C:\ directory, *but do be cautious and make sure before saving any of the images.*

Note that it is possible to "back out" through the menus, i.e., from the plot display you can click on "New Field" to return to the field choice menu for that event; from the field menu you can click on "New Event" to get the event listing of storms for that type; and finally from the event listing you can click "Storm Type Menu" to return to the original main menu listing the storm types on that disk. On all menus, there is an "Exit" button to exit the display demo program.

REFERENCES

Kalnay, E., M. Kanamitsu, R. Kistler, W. Collins, D. Deaven, L. Gandin, M. Iredell, S. Saha, G. White, J. Woolen, Y. Zhu, M. Chelliah, W. Ebisuzaki, W. Higgins, J. Janowiak, K. C. Mo, C. Ropelewski, J. Wang, A. Leetmaa, R. Reynolds, Roy Jenne, and D. Joseph, 1996: The NCEP/NCAR 40-year reanalysis project. *Bull. Amer. Meteor. Soc.*, **77,** 437–471.

Kistler, R., E. Kalnay, W. Collins, S. Saha, G. White, J. Woolen, M. Chelliah, W. Ebisuzaki, M. Kanamitsu, V. Kousky, H. van den Dool, R. Jenne, and M. Fiorino, 2001: The NCEP-NCAR 50-year reanalysis: Monthly means CD-ROM and documentation. *Bull. Amer. Meteor. Soc.*, **82,** 247–267.